Ecology of Marine Protozoa

Ecology of Marine Protozoa

Edited by

Gerard M. Capriulo

State University of New York at Purchase

New York Oxford
OXFORD UNIVERSITY PRESS 1990

Oxford University Press

Oxford New York Toronto
Delhi Bombay Calcutta Madras Karachi
Petaling Jaya Singapore Hong Kong Tokyo
Nairobi Dar es Salaam Cape Town
Melbourne Auckland

and associated companies in
Berlin Ibadan

Published by Oxford University Press, Inc.,
200 Madison Avenue, New York, New York 10016

Library of Congress Cataloging-in-Publication Data
Ecology of marine protozoa / edited by Gerard M. Capriulo.
p. cm.
Bibliography: p. Includes index.
ISBN 0-19-504316-2
1. Protozoa Ecology. 2. Marine invertebrates—Ecology.
I. Capriulo, Gerard M.
QL362.E29 1990 593.1'0452636—dc20 88-9356 CIP

9 8 7 6 5 4 3 2 1

Printed in the United States of America
on acid-free paper

To my wife Amelia
for her eternal support

Preface

Recent years have seen a surge of interest in the role of protozoa in the ecology of the world's oceans. They represent both an important sink for and source of carbon, translating microbial biomass into secondary production important to the growth of numerous invertebrate and vertebrate marine organisms, many of which represent economically important species. The metabolic activities of protozoa result in the production of large amounts of recycled nutrients, which are a significant component of global nutrient cycles. Marine protozoa also exhibit some of the most interesting symbiotic associations presently known, spanning the gamut from parasitism to mutualism, which in the case of certain foraminifera, for example, has allowed the evolution of extremely large size. Study of such symbiotic associations has done much to advance our knowledge of the evolution of eukaryotes. Additionally, stratigraphic examination of extinct forms has provided, and continues to provide, important records on past world climate and ocean conditions, and the study of the zoogeographic distribution of extant forms gives us major insights into present-day ocean conditions and circulation patterns.

Despite the obvious importance of protozoa to the ecology of marine ecosystems, few organized sources of information on the ecology of these forms can be found. Those that are found are primarily organized along taxonomic lines or are dedicated solely to specific groups. With this in mind I have organized this book in a manner that I believe most marine ecologists, biological oceanographers, and marine biologists would like to see, that is, based on ecological function. I hope the result succeeds in filling what I perceive to be a void in the marine ecological literature.

The text provides a general overview of the ecology of marine protozoa with an abbreviated look at the taxonomy and ecology of some free-living and parasitic forms and also considers the paleoecology and oceanography, zoogeography, feeding activities, nutrition and growth, nutrient regeneration, respiration and metabolism, and symbioses of free-living marine protozoa. Because of the extremely specialized nature of the field, its overlap with various fields of medicine, the voluminous published material on the subject, and space limitations that constrained the present work, the parasitic protozoa are excluded from consideration in this text, except for a brief consideration in the introductory overview chapter. I feel that the parasitic marine protozoa would be best treated as the subject matter of a separate volume.

This text is intended for use by research scientists and graduate students primarily in the fields of marine ecology, biological oceanography, and marine biology, as well as by microbiologists, cell biologists, mathematical ecologists, paleontologists, marine geologists, and marine chemists wishing to update themselves on what is known about and is being done with marine protozoa, and/or who are considering crossing over to work in this area. This book is also designed for beginning graduate students and advanced undergraduates taking specialty courses. It is my hope that this text will serve the reader well as a reference text, even though the included references, although extensive, are by no means exhaustive.

I would like to give special thanks to Ken Gold, who helped with the initial planning phase of this endeavor; John Lee, for his expert advice concerning parts of this book and generous support, including use of his laboratory and darkroom facilities; and my sister Clare, for sacrificing large blocks of her time to type major portions of this text. I would also like to thank Jere Lipps, Martin Buzas, O. R. Anderson, Barry Sherr, John Lee, Art Repak, and Tom Fenchel, as well as several anonymous reviewers for critically reviewing various parts of this work. I am grateful to Gertrude Fisher for her help with emergency drafting requirements and to Vicki Careccia, Betti Bennett, and Mary Giuseffi for their assistance in manuscript typing.

This work was supported in part by grants from the Hudson River Foundation for Science and Environmental Research and Sea Grant (NOAA), and by a State University of New York at Purchase President's Junior Faculty Development Award to the editor, as well as by various other organizations that are acknowledged at the end of individual chapters.

Purchase, N.Y. G.M.C.
September 1989

Contents

Contributors

Gerard M. Capriulo
Division of Natural Sciences, State University of New York at Purchase, New York, 10577, USA

David A. Caron
Biology Department, Woods Hole Oceanographic Institution, Woods Hole, Massachusetts 02543, USA

Richard E. Casey
Marine Studies Program, University of San Diego, San Diego, California 92110, USA

Tom Fenchel
Marine Biological Laboratory, University of Copenhagen, DK-3000 Helsingor, Denmark

Joel C. Goldman
Biology Department, Woods Hole Oceanographic Institution, Woods Hole, Massachusetts 02543, USA

John J. Lee
Department of Biology, City College of City University of New York, Convent Ave. and 138th Street, New York 10031, USA

Arthur J. Repak
Biology Department, Quinnipiac College, Hamden, Connecticut 06518, USA

Howard A. Rubin
Department of Mathematics, Hunter College of City University of New York, Park Ave. at 68th Street, New York 10021, USA

P. Lewis Steineck
Division of Natural Sciences, State University of New York at Purchase, New York, 10577, USA

F.J.R. Taylor
Departments of Oceanography and Botany, University of British Columbia, Vancouver, British Columbia V6T 1W5, Canada

Bernt Zeitzschel
Institute of Oceanography, University of Kiel, Düsternbrooker WEG 20, 2300 Kiel, West Germany

Ecology of Marine Protozoa

We shall not cease from exploration
And the end of all our exploring will be
To arrive where we started
And know the place for the first time.

T.S. ELIOT

1

The Ecology of Marine Protozoa: An Overview

JOHN J. LEE
GERARD M. CAPRIULO

WHAT ARE THE BOUNDARIES OF THE GROUP?

Protozoa is the collective term used by zoologists to encompass all the unicellular and primitive colonial animals. During the course of evolution these "first" animals have become so diverse that contemporary experts have elevated seven groups of protozoa to phylum status. The boundaries of the subkingdom or superphylum are not clear, particularly in the Sarcomastigophora. Almost every order placed in the protozoan class Phytomastigophorea is considered a phylum of phycologists. The organisms in each order range from those with typical algal photosynthetic and storage organelles to colorless holozoic or saprozoic forms. The transition between the extremes is gradual and many holozoic or saprozoic forms also have some photosynthetic capacity. The classes Acarpomyxea, Acrasea, Eumycetozoea, and Plasmodiophorea within the Sarcodina and the phylum Labyrinthomorpha are considered to be various forms of slime molds by mycologists. On the other hand, certain algal groups closely related to those considered to be protozoa are artificially excluded from the classification of protozoa. Although there are many scientists with strong feelings about the categories in which specific organisms should be placed, it must be remembered that classification schemes are merely means devised to sort out and organize information on the living world, and that allowance has to be made for differing points of view.

Some of the recent major general references on protozoa or major groups of protozoa are the books written or edited by Sleigh (1973), Grell (1975), MacKinnon and Hawes (1961), Kudo (1954), Corliss (1961, 1979), Hall (1953), Grassè (1952), Reichenow (1953), Jepps (1956), Chen (1967, 1972), Jahn and Jahn (1949), Jahn, Bovee, and Jahn (1979), Hutner (1964), Sieburth (1979), Lee et al. (1985), Fenchel (1987), Anderson (1983), Anderson (1988), and Nisbet (1984).

CLASSIFICATION

Classification schemes of protozoa rely most heavily on morphological and cytological features. For the most part nutritional, biochemical, physiological, parasitological, and genetic information is not used in classification above the generic (genus) level. The following abbreviated scheme of classification, which follows that of Lee et al. (1985), met with broad general acceptance at the 1977 International Congress of Protozoology, and is based on Levine et al. (1980). In this abbreviated outline only groups particularly abundant in the marine environment are included. An asterisk (*) placed next to a taxon indicates that some lower subgroups of the taxon with little or no marine distribution have been deleted from this list.

Phylum I Sarcomastigophora: With a single nuclear type (except in some foraminifera); sexuality essentially syngamic; locomotion by means of flagella or pseudopodia.
- Subphylum 1. Mastigophora: Locomotion by means of flagella; asexual reproduction by longitudinal binary fission.
  - Class 1. Phytomastigophorea*: Typically with chloroplasts or closely related to forms with chloroplasts.
    - Orders: Cryptomonadida
      Dinoflagellida
      Euglenida
      Chrysomonadida
      Heterochlorida

Chloromonadida
Prymnesiida
Volvocida
Prasinomonadida
Silicoflagellida

Class 2. Zoomastigophorea*: No chromatophores.
Orders: Choanoflagellida
Kinetoplastida

Subphylum II. Opalinata*: Not marine.

Subphylum III. Sarcodina: Locomotion by pseudopodia or protoplasmic flow; flagella may be present at certain life cycle stages.

Superclass 1. Rhizopodea

Class 1. Lobosea*: Broad pseudopods, but not anastomosing, usually uninucleate; no sorocarps, sporangia, or similar fruiting bodies.

Subclass 1. Gymnamoebia: Without tests.
Orders: Amoebida
Schizopyrenida
Pelobiontida

Subclass 2. Testacealobosia: Body enclosed by a test, tectum, or other complex membrane external to the plasma membrane and glycocalyx.
Orders: Arcellinida
Trichosida

Class 2. Acarpomyxea: Small plasmodia; no reversal of streaming, tests, spores, or fruiting bodies.
Orders: Stereomyxida

Class 3. Acrasea*: Amoebae with eruptive lobose pseudopodia to form pseudoplasmodia that give rise to a fruiting body without a stalk tube.

Class 4. Eumycetozoea*: Amoebae with filose pseudopodia producing plasmodia, some groups with aerial fruiting bodies held up by stalk tubes.

Class 5. Plasmodiophorea*: Obligate intracellular parasites with minute plasmodia; zoospores with unequal flagella produced in sporangia.

Class 6. Filosea*: Hyaline thin pseudopods often branching, sometimes forming webs.
Orders: Gromiida

Class 7. Granuloreticulosea: Fine pseudopodial network with fine granules.
Orders: Athalamida
Monothalamida
Foraminiferida

Class 8. Xenophophorea: Multinucleate plasmodium enclosed by branched tubular system containing cemented external particles.
Orders: Psamminida
Stannomida

Superclass 2. Actinopodea: Often spherical, usually planktonic, with axopodia, skeletons, when present, organic, containing silica or strontium sulfate.

Class 1. Acantharea: Skeleton of strontium sulfate with 10, 16, 20, or 32 radial spines usually more or less joined in the center of the cell; capsular membrane lining central cell mass.
Orders: Holacanthida
Symphyacanthida
Chaunacanthida
Arthracanthida

Class 2. Polycystinea: Most forms with solid siliceous latticed shells; central capsular membrane with many pores built up from polygonal plates.
Orders: Spumellarida
Nassellarida
Crypotoaxnoplastida
Centroaxoplastida
Periaxoplastida

Class 3. Phaeodarea: Skeleton built of organic and hollow silaceous elements; central capsule with three principal openings.

Orders: Phaeocystida
Phaeogymnocellida
Phaeosphaerida
Phaeocalpida
Phaeogromida
Phaeoconchida
Phaeodendrida

Class 4. Heliozoea*: Without central capsule; mostly freshwater species; a few brackish water and coastal forms.

Orders: Desmothoracida
Actinophyrida
Taxopodida
Centrohelida
Ciliophryida
Rotosphaerida

Phylum II. Labyrinthomorpha

Class: Labyrinthulea: Trophic stage with a net plasmodium of spindle-shaped gliding cells. Heterokont zoospores. Saprobic and parasitic on algae in marine and estuarine habitats.

Orders: Labyrinthulida

Phylum III. Apicomplexa: All parasites; apical complex consisting of polar ring(s), rhoptries, micronemes, conoid and subpellicular microtubules present at some stage; cilia and flagella absent except for some flagellated microgametes; syngamy.

Class 1. Perkinsasida: Incomplete conoid; flagellated zoospores with anterior vacuole; no sexuality.

Class 2. Sporozoasida: Well-developed apical complex; both sexual and asexual reproduction with life cycle that characteristically has stages of merogony, gametogony, and sporogony; some groups with flagellated gametes.

Subclass 1. Gregarinasina*: Large, mature extracellular gamonts; conoid modified into mucron or epimerite parasites of digestive tract or body of lower invertebrates, including annelids, sipunculids, hemichordates, and ascidians.

Orders: Archigregarinorida
Eugregarinorida

Sublcass 2. Coccidiasina: Small, mature gamonts, typically intracellular; conoid not modified into mucron or epimerite; anisogamous; life cycle consists of stages with merogony, gametogony, and sporogony.

Orders: Protococcidiorida
Eucoccidiorida

Subclass 3. Piroplasmasina*: Small puriform, rod-shaped, or amoeboid, reduced apical complex; no spores or cysts; reproduce asexually by binary fission or schizogony; parasitic in erythrocytes.

Phylum IV. Microspora*: Spores formed from single sporoplasm discharged from spore via tubular polar filament; all parasites of invertebrates, especially arthropods.

Phylum V. Ascetospora: Multicellular spore with one or more sporoplasms; without polar capsules or polar filaments.

Class 1. Stellatosporea: Haplosporosomes present; spore with one or more sporoplasms.

Orders: Occlusosporida
Balanosporida

Class 2. Paramyxea: Bicellular spores consisting of a parietal cell and one sporoplasm.

Order: Paramyxida

Phylum VI. Myxozoa: Spores of multicellular origin; one or more polar capsules and sporoplasms with one, two, or three valves; species parasitic in cold-blooded vertebrates and invertebrates.

Class 1 Myxosporea*

Orders: Bivalvulida
Multivalvulida

Class 2. Actinosporea: Spores with three polar capsules each enclosing a coiled polar filament; membrane with three valves; several to many sporoplasms; trophozoite stage reduced; parasitic in invertebrates, particularly annelids.

Order: Actinomyxida

Phylum VII. Ciliophora: Cilia present in at least one stage of the life cycle; reproduction typically by binary transverse fission, but budding and multiple fission also occur. Sexuality involving conjugation, autogamy, and cytogamy. Most species free-living. Includes eight classes.

Class 1. Karyorelictea* Long, vermiform, flattened; many genera with one denuded surface; contractile, body kinetids have overlapped postciliary ribbons or postciliary fibrils similar to those of spirotrichian dikinetids; two to many macronculei.

Orders: Protostomatida (e.g., *Tracheloraphis*)
Loxodida (e.g., *Remanella*)

Class 2. Spirotrichea*: Have conspicuous right and left oral and/or preoral ciliature, with serial oral polykinetids leading usually clockwise, into the oral cavity, either around a broad anterior end or along anterior and left margins of the body. The cytostome is deep in some groups and shallow in others. Some are loricate, others are not; the lorica may be attached to substrate or not.

Orders: Heterotrichida (heterotrichs, including, e.g., *Fabrea, Condylostoma, Folliculina*)
Oligotrichida (oligotrichs, including, e.g., *Laboea, Strombidium*)
Choreotrichida (strobilids, e.g., *Strobilidium*; tintinnids, e.g., *Tintinnopsis,* etc.)

Class 3. Prostomatea*: Cytostome apical to subapical; oral dikinetids approximately tangential to perimeter of oral area in some species, less than so in others; shallow precytostomal cavity; thin tela corticalis usually present.

Orders: Prostomatida (e.g., *Metacystis*)
Prorodontida (e.g., *Balanion, Urotricha*)

Class 4. Litostomatea*: Simple oral cilia, usually not as polykinetids.

Orders: Haptorida (e.g., *Didinium, Mesodinium*)
Pleurostomatida

Class 5. Phyllopharyngea*: Oral region includes radially arranged microtubular ribbons.

Orders: Cyrtophorida (e.g., *Dysteria*)
Rhynchodida

Subclass Suctoria

Orders: Exogenida
Endogenida

Class 6. Nassophorea*: Polykinetids are cirral in several groups (e.g., hypotrichs); body alveoli well developed; cyrtos well developed in some groups; left oral polykinetids may form prominent adoral zone with alveoli between kinetosomal rows; extrusomes if present are fibrous trichocysts.

Orders: Peniculida (e.g., *Frontonia*)

Subclass Hypotrichia

Order: Euplotida (e.g., *Euplotes, Aspidisca*)

Class 7. Oligohymenophorea*: Oral apparatus, when present, distinct from body cilia; oral structures in ventral oral cavity or deeper infundibulum may extend onto and surround peristome; often microphagous, some macrophagous carnivores, some histophagous.

Orders: Hymenostomatida (e.g., *Colpidium*)
Scuticociliatida (= scuticociliates)

Subclass: Peritrichia (= peritrichs)

Orders: Sessilida
Mobilida

Subclass: Astomatia

Order: Astomatida

Subclass: Apostomatia

Order: Apostomatida (e.g., *Collinia beringensis*)

Class 8. Colpodea*: Resting cysts common; freshwater and edaphic, few marine; some habitate fecal pellets of marine crustaceans.

Order: Colpodida

THE ROLES OF PROTOZOA IN THE ECONOMY OF THE SEA

Protozoa are widely distributed in every marine and brackish-water habitat and play important roles at the lower ends of grazing marine food webs and in biogeochemical mineral cycling. With the exception of the giant foraminifera, most protozoa fall into the micro- and meiofauna assemblages of communities. It is now generally believed that the microzooplankton (operationally defined as all phagotrophic organisms that pass through 202-μm-mesh plankton net; Dussart 1965) are an important component of the food webs of both oceanic and near-shore environments. Current conceptualizations view pelagic protozoa as important intermediary packagers of energy. They graze actively on the bacteria and small algae in the pico-, nano-, and microplankton fractions and are in turn consumed by larger animals. Some estimates suggest that an average of 20–70 percent of the daily organic carbon production of small algae passes through the microzooplankton fraction. Choreotrich (e.g., tintinnids) and oligotrich ciliates, radiolarians, acantharians, and foraminiferans constitute the most abundant groups of microzooplankters in coastal zone and open ocean waters. However, problems of differential quality in fixation and preservation of the forms still leave the question of the population structure of microzooplankton assemblages in the controversial realm.

Many of the protozoa, particularly the ciliates, have very complex adaptations for bacterivory. Membranes and membranelles are used to create currents, filter, sort, and trap suspended bacteria and small algae or scrape them from the surfaces of particles. Certain species of ciliate may filter as much as 20,000 times their own volume per hour. During logarithmic growth, a medium-sized ciliate such as *Tetrahymena pyriformis,* consumes 500–600 bacteria per hour. Some planktonic ciliates have recently been found to ingest bacteria at extremely high rates, exhibiting specific clearance rates 10 to 100 times higher than benthic ciliates (Sherr and Sherr, 1987).

In addition to "packaging" food for higher steps in the food web, protozoan grazing regenerates nutrients and affects the species composition of microfloral assemblages. For example, several investigators have found that PO_3 and PO_4 are cycled much more rapidly in grazed systems than in ungrazed ones. In this respect benthic protozoa, along with meiofauna and slightly larger metazoa such as amphipods, seem to play key roles in detritial food webs. Various investigators have found that the rate of decomposition of marsh grasses and similar cellulose sources is greatly accelerated in the presence of these small animals (two to three times faster) than in their absence. It is reasoned that by selective and heavy grazing pressures they become an important factor in the regulation of rates of detrital decomposition. Their activities also mix and aerate their habitats and release and cycle nutrients. All are factors that should enhance microbial growth (Lee, 1980b).

Protozoan populations are quite high in marine sediments and the epibenthos where detrital decomposition takes place. Prominent among the protozoa of such communities are small zooflagellates (e.g., *Bodo, Monas, Oikomonas,* choanoflagellates, colorless euglenoids) amoebae, foraminifera, and ciliates. Densities of protozoa are estimated to be as high as 5×10^8/g dry weight detritus in well-sorted sands; ciliates seem to dominate the protozoan assemblage. Densities of ~2,000 animals/cm^2 surface have been reported. Foraminifera, a conspicuous group which is more heterogeneously distributed, have mean standing crops in the order of 10–50/cm^2 surface, but patches, or "blooms," can reach densities two or three orders of magnitude higher (e.g., Matera and Lee, 1972). In very-fine-grained (0.1 mm) sediments zooflagellates dominate the protozoan assemblage.

There is a great deal of evidence that small marine herbivores and bacterivores often grow at different rates on diets of similar-sized food organisms. This really is not surprising since it is well known that there is a wide spectrum in the chemical composition of the cell walls, storage products, and general cell structure of different groups of microflora. What is surprising is the great range in ecological efficiency associated with different diets. The range of ecological efficiencies was as large as 2–21 percent in two species of marine ciliates, *Uronema marinum* and *Euplotes vannus,* when they were fed a variety of different food organisms (Rubin and Lee, 1976). Work with other ciliates has extended this range to greater than 70 percent. It is reasoned that at least some protozoa have metabolic and genetic capabilities for maximizing the materials and energy gains from their food. Thus, food quality and differential grazing may be an important aspect of the energy transformations by protozoa in marine ecosystems (Lee, 1980b and Chapter 5).

Many foraminifera, radiolaria, acantharia, and some ciliates harbor endozoic algae, an apparent adaptation for growing in nutrient-impoverished seas where phytoplankton and bacterioplankton densities are relatively low. The association of endozoic algae and protozoa has apparently been a long and continuous one. The calcareous tests of foraminifera have left an impressive record tending to suggest that endosymbiosis with algae was apparently a driving force in the evolution of larger foraminifera. Some of these protozoan giants, such as *Parafusulina kingorum* and *Nummulites gizehensis,* reached sizes of 6–12 cm near the close of the Cretaceous and the beginning of the Tertiary. Though the dinoflagellate *Symbiodinium microadviaticum* is by far the most common endosymbiont in corals, anemones, giant clams, and most invertebrate groups, the endosymbionts found in protozoa are more varied.

Quite a variety of endosymbiotic algae, including chlorophytes, dinoflagellates, red algae, and diatoms, have been described from larger Foraminifera. The interesting red-water ciliate, *Mesodinium rubrum,* has cryptomonad endosymbionts. Because of the wide use of the transmission electron microscope, it has become apparent that many marine protozoa also harbor endosymbiotic bacteria and algae. In particular, planktonic ciliates are now being reported to often contain algal symbionts or chloroplasts derived from algae. Another aspect of the ecology of protozoa in the sea is their association as endo- or ectocommensals or parasites in a great variety of marine invertebrates. Quite a number of specialized ciliates have been described in lengthy papers on the fauna of such invertebrates as lamellibranchs (mantle cavities), polychaetes (intestine), gastropods, nemerteans, crustaceans, and echinoderms (various habitats on and in them).

Apart from the role that protozoa play in the carbon cycle of the sea, certain groups, the foraminifera, the coccolithophorids, and the radiolaria, play significant role in the cycling of calcium and silicon. The contributions of the foraminifera to the formation of limestones are very well known because of the attention the Darwinists drew to the fact that the great pyramids of Egypt are built from limestones made largely from the shells of these protozoa. The rate of production of foraminiferal shells varies in different marine habitats but various estimates place the standing crop of living foraminifera on the shelf and continental margins to be ~20,000–50,000/10cm^2 (observed range 0–50,000) and annual production to contribute in the neighborhood of 8 to 10 times that number in the sediments. The major fossil-forming groups (foraminifera, radiolaria, coccolithophorids, dinoflagellates) have been very useful to paleoecologists. Particular fossil assemblages have been compared with the zoogeography of present living assemblages in order to provide clues to particular paleoenvironments and habitats (see Chap. 2). Stable isotope ratios and the coiling directions of various species of planktonic foraminifera have been used to make paleoclimatic estimates.

COLLECTION AND CULTIVATION OF MARINE PROTOZOA

The reader is directed to the UNESCO Working Group publications on Phytoplankton (1978), Zooplankton Sampling (1968), and Zooplankton Preservation (1976), Sieburth (1979, Chap. 4–7) and Lee et al. (1985) for reviews on some protist sampling and cultivation methodologies, as well as to primary literature sources (e.g., Uhlig, 1968; Spoon and Burbanck, 1967; Gifford, 1985; Soldo and Brickson, 1980; Soldo and Merlin, 1972; Gold, 1968, 1973; Gold and Morales, 1975, 1976). Such detail is beyond the scope of this chapter and this text. We do, however, here present a simplified overview of various methodologies after considering general microhabitat characterization.

A. *Microhabitat Characterization* (based on unpublished notes of E.B. Small)
   1. *Estuarine and coastal*
      a. *Overlying water*: Loaded with dissolved organics, microscopic-sized detritus particles, as well as clay particles, often supporting high numbers of autotrophic and heterotrophic microalgae, bacteria as well as heterotrophic microzooplankton (2–200 μm).
      b. *Emergent and submergent living and decaying vegetation*: Site of attachment of the "aufwuchs" community (e.g., diatoms, photoauxotrophic protists); *Zoothamnium* and *Vorticella* peritrich genera with many species, as well as *Stentor* and folliculinid filter-feeding ciliates; stalked suctorian ciliate carnivores; various naked amoebae; and myriad other organisms that may also be attached directly to the substrate or at least feed on attached organisms.
      c. *Detritus*: Plant particles macroscopic to microscopic in size range, that settle to the bottom of the sediment–water interface. Populated by a mixture of some planktonic organisms, aufwuchs organisms attached to particulates and/or planktonic organisms.

In quiet backwaters of salt marshes where pockets of water are entrained with minimal disturbance from tidal elevation or current, this detritus site is frequently luxuriant with protists, often several species in the same genus or family occupying what at first may appear to be the same microhabitat or niche.

d. *Sand, silt and clay sediments* (see Sieburth, 1979, and Fenchel, 1987, for a general review): Divisible into strata based on the abundance or absence of dissolved O_2 and/or H_2S, as well as on the dominant bacterial microbiota responsible for the chemistry of the major inorganic compounds that undergo changes in redox state. The stasis of such sediment layers can be altered by water movement and turbulence. The separation into distinct layers with protists specially adapted to endemic chemistry as well as microbial food conditions is obvious. Phototrophic and heterotrophic protists are found in the top few centimeters of the aerobic sediments. The redox discontinuity layer frequently contains those organisms physiologically adapted to prevalent conditions (= microaerophiles). The anaerobic black sulfide–laden sediments contain their own group of protists that (1) can incur an oxygen debt and temporarily feed in the this zone, (2) are true anaerobes, or (3) some mix of the two (see Fenchel, 1987, for a review of this).

2. *Open ocean*: The open ocean environment contains many heterogeneous habitats produced by various unique combinations of hydrographic conditions and anomalies. Included here as "distinct" habitats are regions of oceanic fronts (Bowman and Esaias, 1978), water column density discontinuity layers, marine snow aggregates, oceanic gyres, warm core and cold core Gulf Stream rings, the Sargasso Sea, floating ice, and ice-edge and sea-ice undersurface zones. Protozoan populations of these habitats have been only rarely considered (see Chap. 4).
3. *Hydrothermal vents*: Protozoan populations of these regions have only recently been considered by Small and Gross (1985) and Sniezek (1987) (see Chap. 4).

B. *Collection Methods*

1. *Open water organisms*
    a. Plankton nets—Nitex nylon mesh sizes to 10 μm.
    b. Whole water sample preservation (e.g., Bouin's, Lugols, gluteraldehyde, formaldehyde; see Lee et al., 1985, and UNESCO publications).
    c. Sterile syringe–slurp gun collection with possible additional Millipore® filter filtration–concentration.
    d. Whole water sample with further filtration concentration, including standard or reverse filtration and/or enrichment.
    e. Whole water sample with subsequent concentration by centrifugation.
    f. Bait trap with gnotobiotic food supply (e.g., Small et al.; 1986, Lee et al., 1985).
    g. Small- or large-volume water pump vacuum sampling.
    h. Differential chemical signal, Nitex covered tube traps (Levandowsky, unpublished DCS method).
    i. Sediment traps (e.g., Berger and Soutar, 1967; Gowing and Silver, 1983; Honjo, 1978; Wiebe et al., 1976).
    j. SCUBA (direct collection).
    k. Submersibles (direct collection).
    l. For all of the preceding methods, further concentration of organisms can be achieved by micropipetting.
2. *Organisms associated with natural substrates including detritus*
    a. Clipping of plants, algae, blue greens (cyanobacteria) coupled with direct microscopic observation and/or enrichment.
    b. Slide or cover slip traps (Bamforth, 1982; Cairns, 1982; Hentschel, 1925; Holm, 1928; Nusch, 1970, Bissonnette, 1930; Hammann, 1952; Sladekova, 1962).
    c. Polyurethane foam float sampling (Cairns et al., 1969, 1973; Cairns', 1982 "sponge" infiltration method).
    d. Micropetri dish float sampling (Spoon and Burbanck, 1967).
    e. Whole sample direct microscopic observation.
    f. Extraction by filtration.
    g. Extraction by the sediment tube method (Spoon, unpublished).
    h. Numerous enrichment techniques.

3. *Sediment-associated organisms*
   a. Core plus Uhlig extraction (Uhlig, 1968).
   b. Core plus 3 d sediment tube method.
   c. Bait traps method (Small et al., 1986).
   d. Levandowsky (1 h) DCS method.
   e. Slide, petri dish, and "sponge" sampling as in 2.

C. *Cultivation*: Marine protozoa from coastal zones, estuaries, salt marshes, and so on, are easily isolated in agnotobiotic culture or grown in circulating marine aquaria. Continuous cultivation is more of an art than a science. There are several "rules of thumb" that are helpful to beginners:

1. If there are many small molluscs, microcrustacea, annelids, and other small invertebrates in the samples, it is best to decant the sample through a 1- or 0.5-mm sieve to remove them.
2. Care must be taken to avoid placing too much sample in a container. If the sample is high in organics (e.g., epiphytic, epibenthic, or plankton haul), it should not occupy more than 10–20 percent of the sample container. Fresh filtered seawater should be added to bring the volume up to 70 percent of the container volume.
3. Samples should be placed in an ice chest or submerged under water at the time of collection to avoid killing many organisms by the combined effects of elevated temperature, rapid bacterial growth, and anoxia.
4. In the laboratory the samples should be placed in containers with large surface-to-volume ratios (e.g., tissue culture flasks, Erlenmeyer flasks, stacking aquaria, etc.).
5. As a general rule it is also good practice to remove small micrometazoa (meiofauna) from raw cultures since some of them consume protozoa or compete with them for food.
6. Crude cultures generally do well if incubated in moderate light (250 W/cm^2) at approximately the water temperature at the time of collection or a few degrees lower.
7. In the first few weeks of crude cultures it is advisable to decant the seawater and replace it with fresh filtered seawater frequently. One system is to change the water daily for the first week and every third day for the second week. Later media changes can be extended to weekly or monthly intervals.
8. Because agnotobiotic (raw, crude) cultures are notoriously finicky, it is generally good practice to separate the organisms in which one has particular interest from most others. Crude cultures of protozoa with few species of interest tend to have greater stability.

Certain judgmental aspects quickly enter the culture picture. To put it simply, it is desirable to try to achieve a balance between the growth of food organisms and protozoa; although this goal is reasonable, it is not always attainable. One of the most popular media used to grow protozoa with their food is Erdschreiber. It is an enriched seawater medium composed of seawater 95 ml; $Na_2HPO_4{\cdot}H_2O$, 5 mg; soil extract, 5 ml; and $NaNO_3$, 10 mg. The medium is sterilized by filtration or carefully autoclaved to prevent precipitation. Although it is easy to make Erdschreiber, the preparation of soil extract is tricky. The growth-promoting qualities of the extracts vary greatly with the properties of the soil. Some soils are even toxic for protozoa. A black forest topsoil is usually very good. Directions for making soil extract are detailed in an article by Lee and co-workers (Lee et al., 1970). A variety of benthic foraminifera, ciliates, and amoebae have been cultured in Erdschreiber. Some workers prefer to grow their protozoa and food separately. Algal overgrowth is a real problem in some foraminiferal cultures. Workers who have this problem have overcome it by plunging the tubes of separately grown algae into a 60°C water bath to kill the algae prior to feeding.

At the other extreme is the problem in maintaining the amazing marine amoeba *Pontifex maximus*. This creature literally eats its way across a culture dish. To maintain it one must grow dishes of algae in advance. The progressive reproduction of the amoebae from a small drop of inoculum in the center of a culture dish can be followed by eye, as day by day, the area of the dish totally cleared of algae is progressively enlarged.

Very few holozoic marine protozoa have been grown in axenic or even monaxenic culture. The media used for the successful rearing of the ciliates are similar to those used for freshwater forms (Hanna and Lilly, 1970, Hanna, 1974; Soldo and Merlin, 1972). Axenic marine ciliates have been grown in crude natural media containing 100 ml seawater, 1 percent proteose peptone, 0.5 percent dextrose, and 0.5 percent brewer's yeast. Many synthetic marine media for protozoa are based on those first developed by Provasoli and his co-workers (Provasoli et al., 1957). If cultures are not axenic, it is important for a beginner to realize that nutrient levels in marine environments are usually quite a bit lower than those encountered in freshwater habitats. Care must be taken to keep the levels of enrichments fairly low if synxenic cultures of most marine protozoa are to be successful (e.g., Lee and Pierce, 1963).

BIOLOGY OF PARTICULAR GROUPS

A. *Zooflagellates*

Many nonpigmented flagellates are the colorless counterparts of pigmented forms and as such are treated along with them. Although a large number of the zooflagellates are small and simple in structure, many groups of parasitic zooflagellates are as complex in structure as any other protozoa. Zooflagellates are found in almost every marine habitat, including the hyponeuston. Among the most common in many shallow marine waters are the small kinetoplastid flagellates *Bodo* (Fig. 1.1e) and *Rhynchomonas*. Characteristically, they have one flagellum directed anteriorly and a second one trailing to its posterior. *Trimastix*, another ubiquitous zooflagellate, has three flagella, two of which trail the organism. Some species of *Tetramitus* (Fig. 1.1p), an interesting flagellate with four flagella that has a cystic and amoeboid stage in its life cycle, are marine. Both freshwater and marine fish are hosts of different species of trypanosomes.

About 50 species of *Trypanosoma* (Fig. 1.1k) have been described from marine fishes from all seas. They are transmitted by leeches. They reproduce in the digestive tract of the leeches and form ineffective stages that are passed on to the next host. Skates in general are infected by very large trypanosomes; the largest (115 pm long) is *T. gargantua* in *Rajanasuta*, from New Zealand waters. The fingerlings of Pacific anadromous salmonids are sometimes massively invaded by *T. salmositica*.

Choanoflagellates, which are simple flagellates with a collar around their single anterior flagellum, are now known to be common in the nanoplankton and are widely distributed (Fig. 1.1h,j,m,o). Their phylogenetic relationships are still not clear. They have been considered a class or subclass of algae, Craspedophyceae, linked to the Chrysophyta by many phycologists. Leadbeater (1972, 1973, 1978) and Hibberd (1975, 1986) are convinced that these organisms are better placed in the zooflagellates. Many choanoflagellates are stalked and make very beautiful complex, basketlike loricae from siliceous rods (costae). The funnel-shaped part of the lorica, the lorica chamber, is woven from spiraling costae. Costae converge at the hind end to produce the stalk. Some loricae have free longitudinal costae, or spines, at their anterior ends. During the reproduction of the loricate forms, one of the daughter cells remains attached to the lorica while the other, the juvenile, swims away and forms a new lorica.

In general, zooflagellates tend to be either saprozoic or holozoic (some are both). Holozoic species tend to be largely bacterivorous. Zooflagellates are found in many marine aquaria and in many crude (agnotobiotic) cultures of natural collections. They are easily maintained by adding small quantities of natural materials (e.g., 0.01–0.1 g/L yeast extract, proteose peptone, trypticase, etc.) to sterile seawater. In the absence of predators (e.g., ciliates), they usually achieve some sort of balance between themselves and bacteria in enrichment cultures.

B. *Amoebae*

Marine amoebae are often overlooked, but they are very common in most shallow marine habitats (Fig. 1.2a–i). Many amoebae may be common in deeper benthic communities or in the pelagic zone, but few investigators have looked for them there. They have also been found on the surfaces and body cavities of marine animals, where they graze upon associated bacteria. Some of them, for example, *Paramoeba perniciosa*, which causes gray crab disease of blue crabs along the eastern U.S. coast, are pathogens. In heavy infections most of the blood cells of the animal are replaced by the amoeba.

Sewer sludge dumping and sewer outfalls in the marine environment have introduced many terrestrial forms into the sea. Species of *Acanthamoeba*, some of which are capable of causing diseases or death of warm-blooded animals, have been identified recently in sewage sludge–impacted sediments. On the other hand, marine amoebae appear to be active consumers of *Escherichia coli* in impacted sediments.

Most marine amoebae are so small and translucent that they often are overlooked by all but the most experienced observer. The presence of most species is detected only by examination under a compound microscope equipped with dark field, phase, or interference optics. An inverted microscope is ideal for searching for amoeboid movements in collections at relatively high magnifications (100–400 x). The dark-field illuminating systems of some of the better-quality dissection microscopes are also adequate at high magnifications. It is best to examine samples as soon as possible after collection or to establish temporary cultures to which bacteria are added as a food source.

Some amoebae from the littoral and near-shore sublittoral zones are easily collected and cultured; other species survive for only a few hours after col-

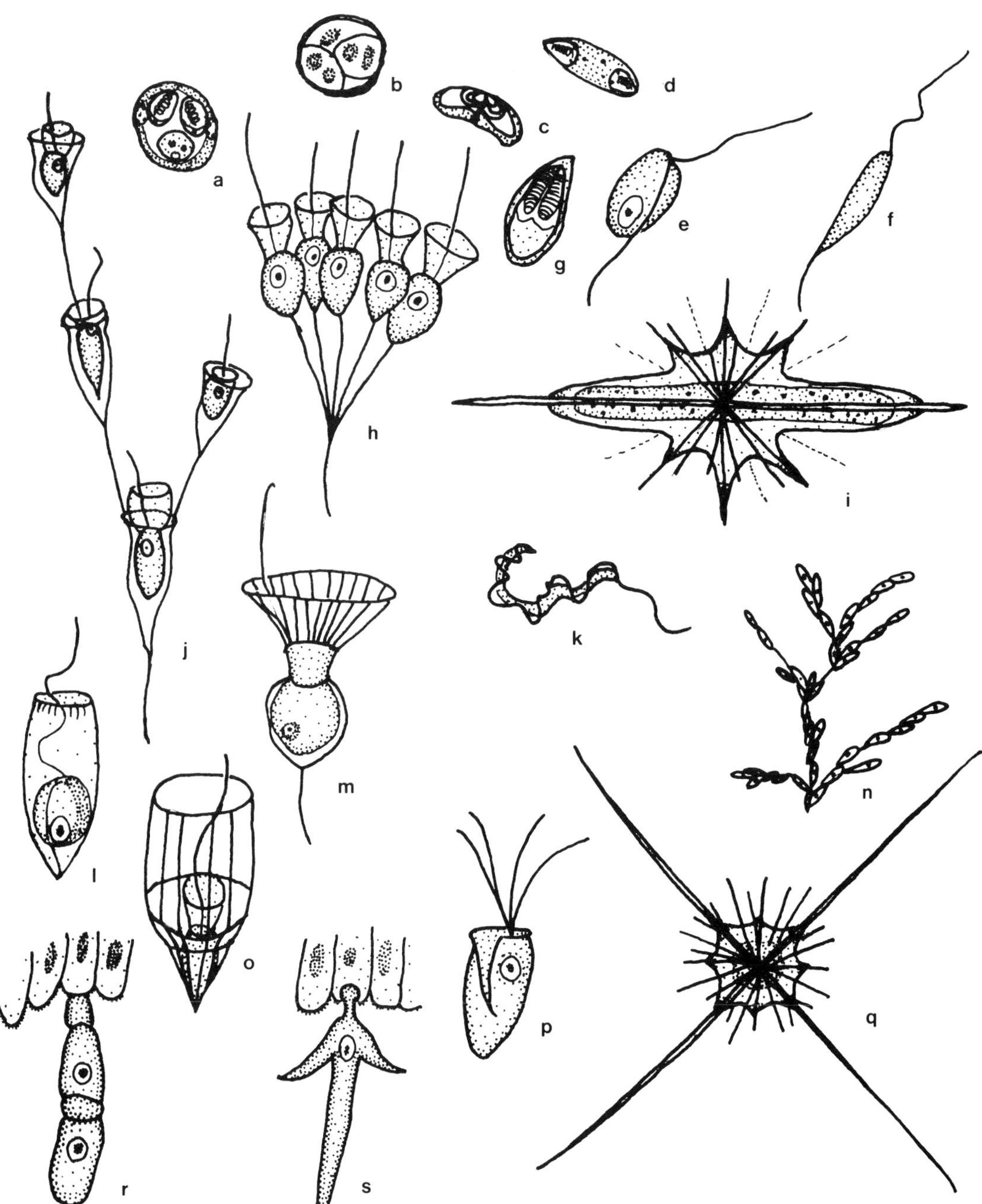

FIGURE 1.1 Various protozoa belonging to the phyla Sarcomastigophora, Labyrinthomorpha, Apicomplexa, and Myxozoa. (a) Bivalvulida, *Myxobolus* (after Lom, 1970); (b) Occlusporida, *Minchina refrigens* (after Perkins, 1970); (c) Bivalvulida, *Ceratomyxa* (after Lom, 1970); (d) Bivalvulida, *Mixidium* (after Lom, 1970); (e) Kinetoplastida, *Bodo* (after Ruinen, 1938); (f) Kinetoplastida, *Cryptobia* (after Lom, 1970); (g) Bivalvulida, *Myxosoma* (after Lom, 1970); (h) Choanoflagellida, *Codoncladium* (after Hollande, 1953); (i) Arthracanthida, *Amphilonche elongata* (after Schewiakoff, 1926); (j) Choanoflagellida, *Polyoeca dumosa* (after Dunkerly, 1910); (k) Kinetoplastida, *Trypanosoma* (after Becker, 1970); (l) Choanoflagellida, *Bicoeca* (after Prowazek, 1903); (m) Choanoflagellida, *Salpingoeca* (after Boucaud-Camou, 1966); (n) Labyrinthulea, *Labyrinthula* (after Pokorny, 1967); (o) Choanoflagellida, *Pleurosiga* (after Thomsen, 1976); (p) Kinetoplastida, *Tetramitus* (after Hollande, 1953); (q) Chaunacanthida, *Stauracon pallidus* (after Schewiakoff, 1926); (r) Eugregarinida, pair of *Nematopsis legeri* (after Grasse, 1953); (s) Eugregarinida, *Ancora sagittata* (after Grasse, 1953).

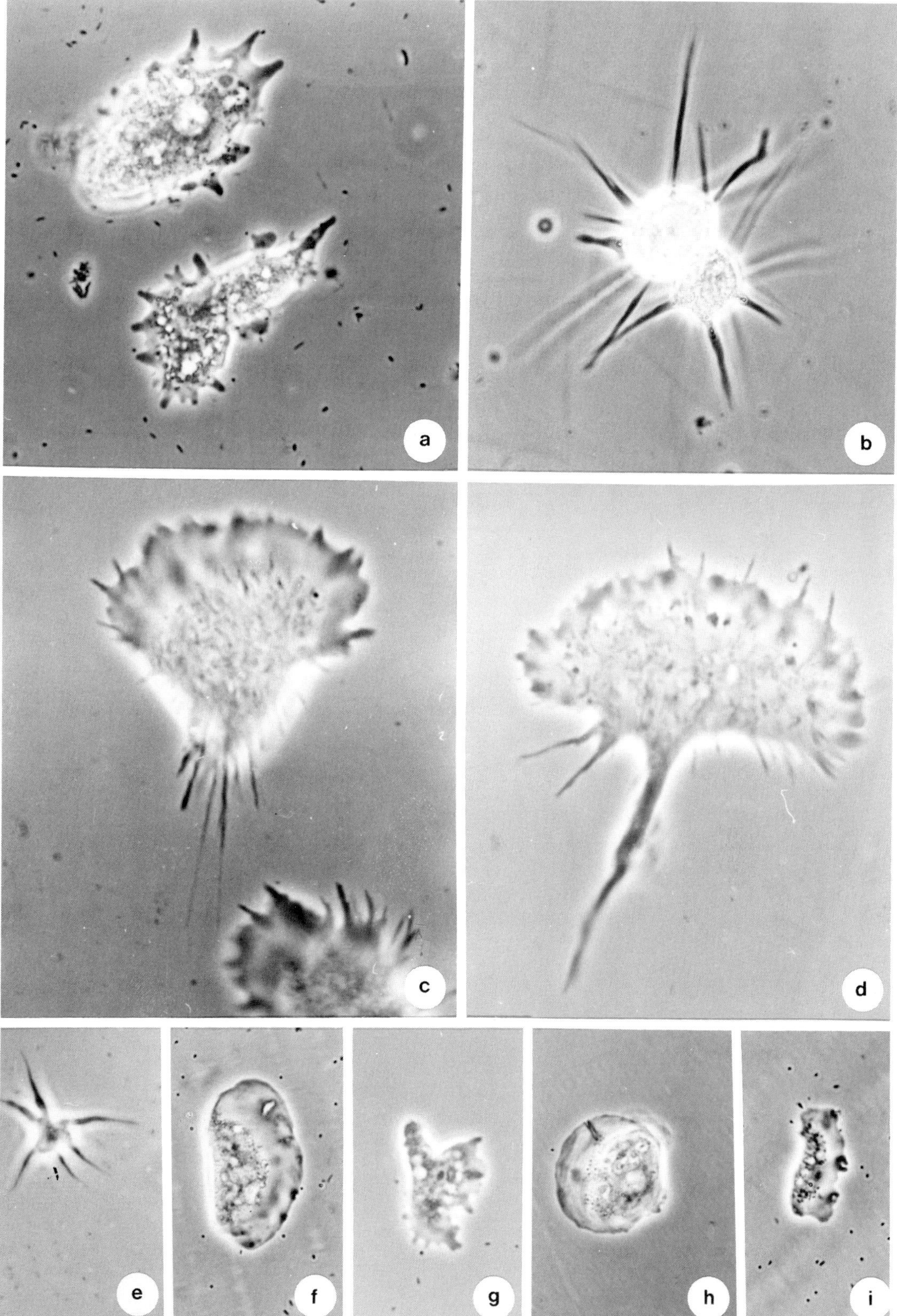

FIGURE 1.2 Amoebida. (a) *Mayorella corlissi,* motile stage, ~ ×640; (b) same species, rayed stage, ~ ×640; (c–d) *Flabellula hoguae,* motile stage showing uroid, ~ ×1,600; (e) same species, rayed stage, ~ ×640; (f) *Unda maris* showing food vacuole after pinocytosis, ~ ×640; (g) *Paramoeba pemaquidensis* showing nucleus and nebenkorper, ~ ×640; (h) *Hyalodiscus angelovica* showing food channel, ~ ×640; (i) *Gibbodiscus newmani* showing exophagopods, ~ ×640. (All figures are phase micrographs by Dr. Thomas Sawyer.)

lection. A small sample of a seaweed with an epiphytic community usually is a good source of inoculum for an amoeba culture. A few milliliters pipetted from the surface of most undisturbed sediments is also a good inoculum. Amoebae in the water column can be cultured by gently (<5 lb/in.^{2}) filtering (Millipore RA; 1.2-μm porosity) a liter or more of seawater and then transferring the filter to a culture dish containing seawater.

Petri plates are the most convenient culture vessels since the amoebae can be located with a dissecting microscope and picked up with a pipette for study under a compound microscope. A thin layer of seawater over nutrient agar, Erdschreiber medium, or Sawyer's yeast extract and malt extract medium, or ground cord grass (*Spartina*) media solidified with 1.5 percent agar with a thin overlay of seawater, are good media. Very few species of marine amoebae are capable of adapting to growth on media prepared with distilled water. Transfers from actively growing cultures can be made with pipettes or by excising with a scalpel a small chunk of agar containing amoebae and transferring the subsample to freshly bacterized agar plates. Many amoebae can be grown together with their algal or bacterial food (e.g., *Vexillifera* spp., *Mayorella* spp. and *Vannella*, sp.) but other species (e.g., *Pontifex maximus*) are such voracious eaters that they must be fed regularly or transferred to previously prepared algal lawns. *Enterobacter aerogenes* is often inoculated as a food organism, since it supports the growth of many well-known species. To guard against evaporation, petri dish cultures of amoebae are often placed in sealable plastic bags or in moist chambers.

Although most amoebae have plastic or elastic shapes, the characteristics of their pseudopods, locomotion, uroids, and nuclei are distinct enough to enable even a beginner to recognize and distinguish members of different families from each other (see Fig. 1.2). Amoebae can develop one or more pseudopods, which can be broad, lobose, tubulate, cylindrical, ridged, short, slender, radiate, conical, serrate, blunt, bulged, stubby, digitate, branched, and ramose. Pseudopodial formation and locomotion in many amoebae is continuous, but in other forms it is eruptive, or wavelike. Although many amoebae do not have uroids or have only temporary uroids, many other species have characteristic permanent or semipermanent uroids that may be bulb-shaped, grouped into small lobes, bristlelike, papulate, or filamentous. Nuclei, especially those undergoing mitosis, must be stained with protargol or hematoxylin to distinguish their characteristics. Different species are either uninucleate or multinucleate. The nuclei usually have a distinct central endosome (which can sometimes be indistinct), or they may be ring-shaped between polar caps. The distribution and characteristics of the ecto- and endoplasm are other diagnostic features. Some marine amoebae also have water expulsion vesicles (formerly called pulsating or contractile vacuoles), which aid in their identification.

Many marine amoebae have two distinct forms and easily slip from one to another (e.g., Fig. 1.2a,b). The radiate or pelagic forms are found in the water column. Disturbances caused by waves, currents, and large-animal activities may dislodge extended actively feeding forms, which then change into radiate or floating forms. Some species may change from the extended locomotive form to a radiate one that floats away from the substrate without any obvious external motivation. When radiate forms contact the bottom of a culture dish, they may move about or rotate for a while on the tips of their pseudopods. Eventually the pseudopods are retracted and the protozoa revert to the extended form. The ratio of the radiating pseudopods to the central body mass often is used to separate closely related genera. Marine amoebae reproduce asexually by binary fission. Some species form characteristic cysts. Knowledge of the nutrition of marine amoebae has come largely from microscopic observations. Many species are bacterivorous but many other species eat small algae and dinoflagellates. A few species (e.g., *Pelomyxa ostendensis*) have bacterial endosymbionts. Some important monographs and papers on marine and freshwater amoebae are those of Bovee (1970, 1972); Bovee and Sawyer (1979); Page (1970, 1971); Sawyer (1975a,b); Schaeffer (1926), and Sawyer (1980).

C. *Radiolaria*

The general similarity of the organization of the actinopod protozoa (Figs. 1.3, 1.4, 1.5)—the Radiolaria, Acantharia, and Heliozoa—justifies their placement in a separate superclass. Recent fine structural studies of problematic organisms have helped clarify the boundaries and relationships of the various groups. *Sticholonche zanclea* is a good example (Fig. 1.4f). This animal was originally placed in the heliozoan order Taxopodida but transferred nearly 70 years ago to the Radiolaria (Phaeodaria). Fine structural studies of *Sticholonche* show that the packing of its microtubules in the axonema is closer to that of some of the centrohelid Heliozoa and some Acantharia than to that of any of the Radiolaria that have been studied in this respect. The insertion of the axonemes on the nuclear capsule is dissimilar to any described in the radiolarian central capsule. The Acantharia are sepa-

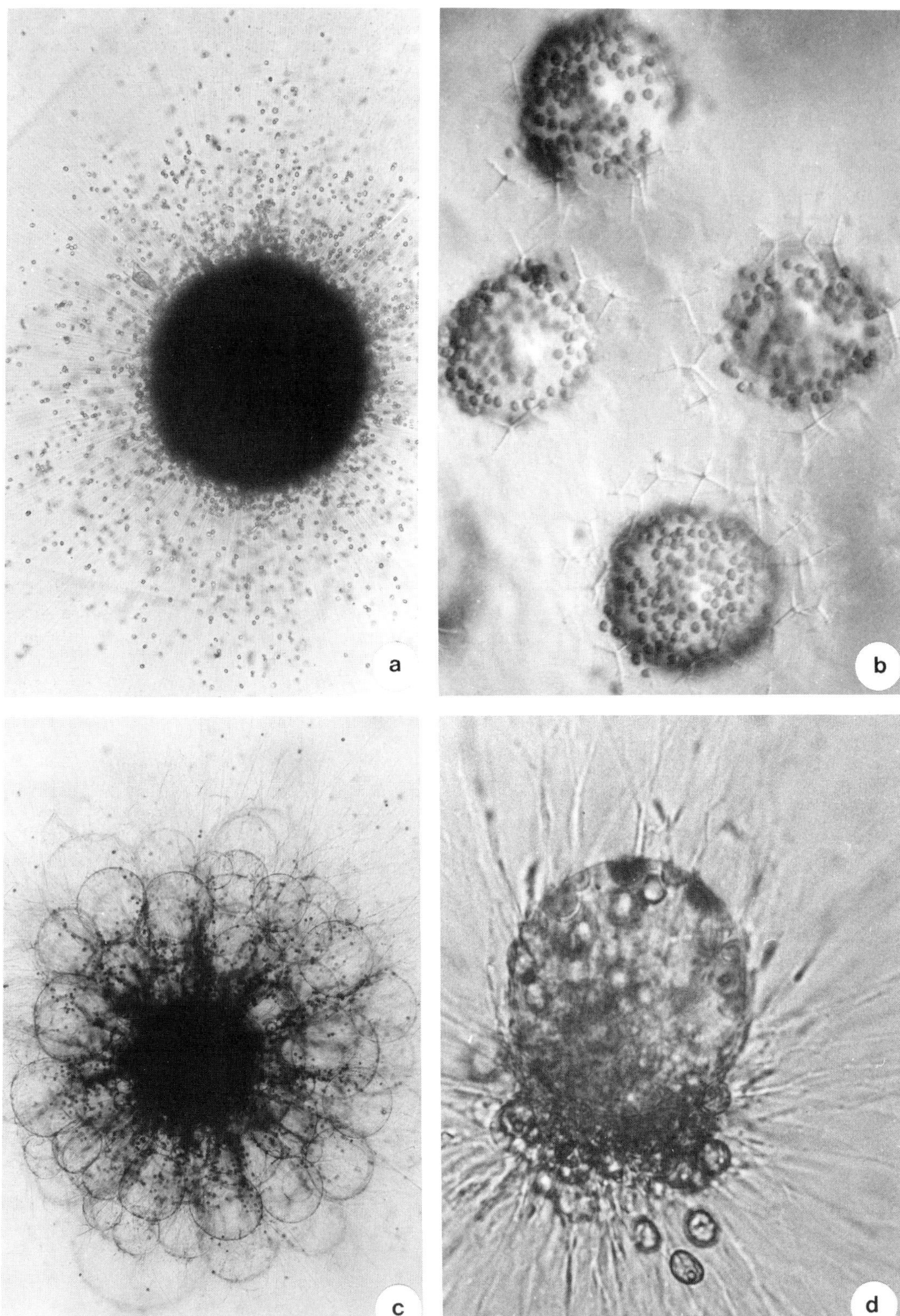

FIGURE 1.3 Spumellarida and Nassellarida. (a) A living radiolarian is surrounded by a corona of radial axopodia containing numerous symbionts. ×60. (b) Cells in *Sphaerozoum punctatum*, a colonial radiolarian, contain siliceous spicules surrounding the central cell body. Numerous symbionts, appearing as dense spheroids on the surface of each cell, are enclosed in ectoplasmic vacuoles surrounding the cell. ×130. (c) *Thalassicola nucleata*, a skeletonless radiolarian, contains numerous alveoli (bubblelike cytoplasmic compartments) surrounding the central cell body called the central capsule. ×20. (d) A skeletonless nassellarian radiolarian possesses algal symbionts near the base of the central capsule where the halo of rhizopodia emerges. ×400. (All figures are phase photomicrographs by Dr. O. Roger Anderson.)

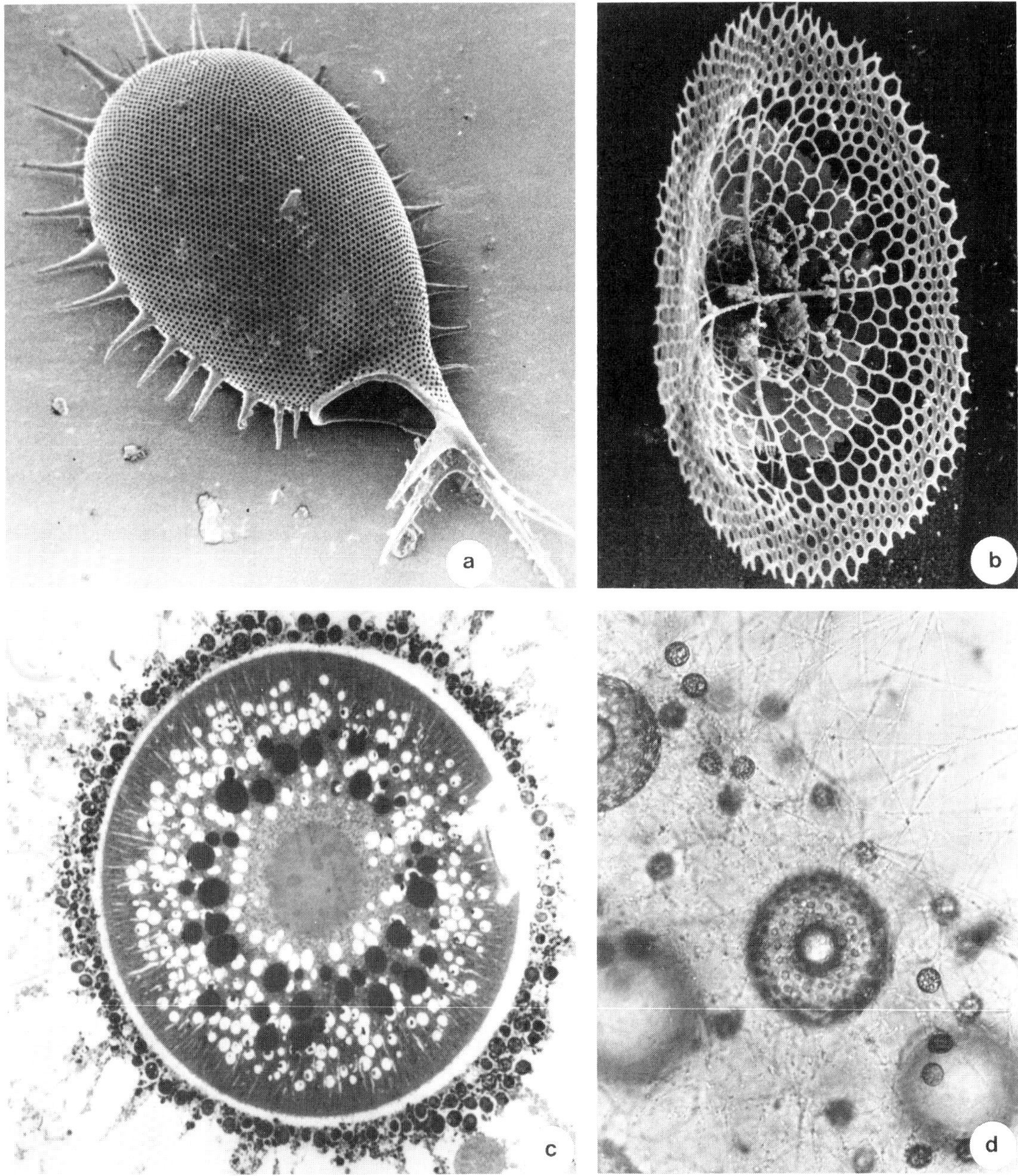

FIGURE 1.4 Phaeodarea, Polycystinea, and Heliozoea. (a) *Challengeron* sp., a phaeogromid phaeodarian, ~ ×200. (b) *Sethophormis*, a nassellarian. (c) A light micrograph of a thin-section through the central capsule of *Thalassicola nucleata* (Fig. 1.3c) shows the nucleus, dense granules (possibly food reserves), hyaline vacuoles, and a dense layer of cytoplasm surrounding the central capsular wall; ×200. (d) Cells of *Collosphaera globularis*, a colonial radiolarian (Spumellarida, suborder Sphaerocollina) are surrounded by a spheroidal, porous shell penetrated by rhizopodia that interconnect the cells and enclose the algal symbionts; ×250.

rated from the other groups by their possession of a skeleton of radial spines, composed of strontium sulfate, which are always joined in the center and by their general lack of a capsular membrane. Heliozoa also lack a central capsule and their skeletal elements are siliceous or chitinoid. The extracapsular cytoplasm (ectoplasm) of polycystines is differentiated more or less into a frothy vacuolated layer, directly under the shell, the calymma, which serves for flotation, and an inner layer, the sarcomatrix, which contains food vacuoles and other organelles. Dinoflagellate zooxanthellae (Fig. 1.3b,d; Fig. 1.4d) have been observed in the ectoplasm of many species but not in the Phaeodarea. Electron

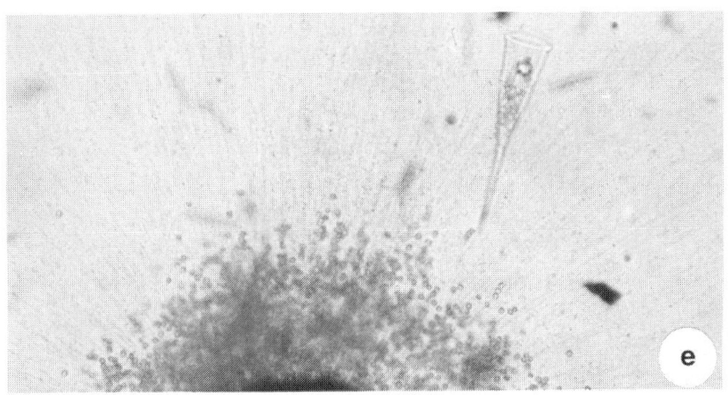

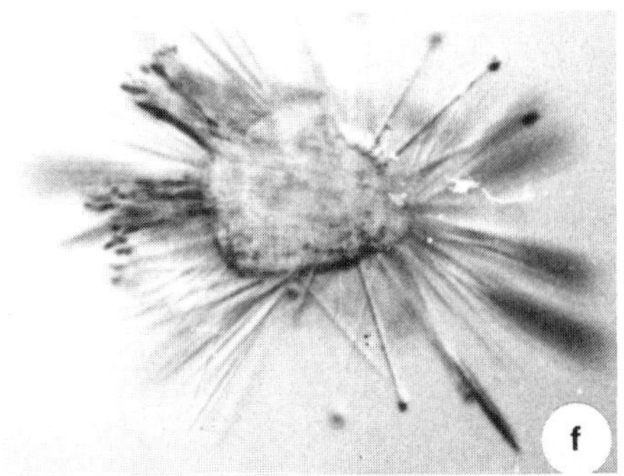

FIGURE 1.4 *(continued)* (e) A tintinnid prey is snared within the rhizopodial system of a solitary radiolarian possessing numerous symbionts surrounding the opaque central capsule. A nucleus, mitochondria, Golgi bodies, and membranous organelles are suspended within the dense cytoplasm in the central capsule; ×50. (f) *Sticholonche zanclea*, a taxopodid heliozoan; ~ ×60. (a, b, and f by Drs. Jean and Monique Cachon; c–e by Dr. O. Roger Anderson. a and b, SEM; c, light transmission; d–f, phase micrographs.)

microscopic observations leave no doubt as to their dinoflagellate identity. Prasinophyte algal symbionts have also been found in some Radiolaria (Anderson, 1976). The capsular membrane has been described as tough or horny. It is pierced by a number of pores (fusules). In the polycystine Radiolaria the capsule is built up from polygonal plates that in the Spumellarida (Peripylea of some authors) are pierced by a great number of pores uniformly distributed on the whole surface of the capsule. The Nassellarida (Monopylea of some authors) have a single large pore (several small fusules are gathered at a single pore). The phaeodarian radiolaria (Tripylea of some authors) have a central capsule with only three openings, a tubular main opening (the astropyle), and opposite it on the other side of the capsule, two smaller openings (the parapyles).

The nuclei of those few Radiolaria that have been carefully examined are quite unusual and one would like to know much more about their biology. Nuclei are generally quite large, contain many nucleoli, and increase in size during the vegetative life of the cell. The polyploid nuclei of Phaeodaria, which undergo successive endomitoses, can contain an estimated 2–3 × 10^3 chromosomes. There are differences in opinion in the interpretation of the observations various researchers have made during the steps of sporogenesis. The picture is complicated because the endosymbiotic dinoflagellates of those species that have them also may be released into the sea at the same time as radiolarian isospores.

Because of the difficulty in raising or maintaining most Radiolaria in the laboratory or obtaining adequate material for study, their life cycles are still not well known. Some of the Spumellarida (e.g., *Thalassophysa*) have colonial stages in their life cycles. The larger Radiolaria capture and eat copepods and other small crustaceans, algae, and other protozoans that come in contact with their actinopodia (Anderson, 1983).

Radiolaria are widely distributed in the plankton of the world's oceans and have been present since the lower Silurian and possibly the Cambrian (see Chap. 2). More than 700 genera and thousands of species of fossil and recent organisms have been described. Species tend to be larger in the tropics and subtropics than in the temperature zone. The majority of species live within the upper 500 m of the water column, but some species have been captured as deep as 5,000 m. In general, the Spumellarida live in the upper 50–60 m. Cachon-Enjumet (1961) found that both the Spumellarida and the Nassellarida occurred in the surface waters of the Mediterranean in the colder months (October to May) and then descended as the water temperature rose to 20°C. The Phaeodarina, in general, are deeper-water species, many being captured at 3,000–5,000 m. Certain taxonomic groups are apparently captured at different depths.

Differences between modern radiolarian assemblages at various latitudes and climatic boundaries have been used by paleobiologists for interpretations of paleotemperatures and climatic fluctuations (e.g., Hays, 1965; Hays and Opdyke, 1967).

Much of our knowledge of the comparative morphology of the Radiolaria goes back to the last quarter of the last century and the first quarter of this century, and the classical studies of Haeckel, Hertwig, Brandt, Haecker, and Schröder. Although some of the species they described later turned out to be stages in the life cycles of the same species, their work underlies much of our contemporary taxonomic thinking. In addition to the basic structure of the central capsule, radiolaria are further subdivided into suborders and families on the basis of the structure of their skeletons. The simplest, and perhaps the most primitive, Spumellarida, belong to the suborder Collodaria (Fig. 1.3b,

Fig. 1.4d). Some have only disjointed, needlelike spines; other, slightly more advanced species have siliceous spines consisting of centrally jointed elements, which look superficially like some sponge spicules and which have 3, 4, 6, 8, and 10 equidistant radiating arms. Still others belonging to the suborder have more-developed siliceous latticelike shells bearing secondary spines. Members of the suborder Sphaerellaria have well-developed spherical, oval, cylindrical, lens-shaped, and lentelliptical siliceous skeletons that are ornamented with spines, branched spines, or small secondary spines (Fig. 1.4b; Fig. 1.5b,c).

The central capsule of the Nassellarida is encircled by a thick membrane. A flattened part of it, the pore field, is pierced by a circlet of fusules (Plate 1.6). The endoplasm contains a spherical nucleus, red-pigmented fat droplets, and an axoplast. The axoplast lies next to the nucleus and is the origin of many axial roots. The cone formed by the axial roots as they pass through the fusules is the podoconus. Nasselaridan skeletons have three main elements: a basal tripod, the cephalis or helmet, and the sagittal ring. The simplest basal tripod is composed of three simple divergent rods united at a common center. More complicated tripods have four, five, six or more branches. The other two elements, the cephalis and sagittal ring, are believed to have been derived from fusion of tripod elements. Three suborders of Nassellarida are recognized: the Nassoidea, a group that has no skeleton or isolated spine; the Plectellaria, a group having a rudimentary skeleton that in the most advanced member of the group looks like a wicker work (Fig. 1.4b); and the Cyrtellaria, a group having complete latticed shells composed of a single bilocular cephalis or one divided by transverse furrows into two or three segments (thorax, abdomen, and postabdomen).

The Phaeodrea are easily distinguished from other Radiolaria in plankton hauls by their central capsules and the dark-greenish globules that constitute the phaeodium. Axopodial roots pass through the parapyles, which are homologous to the fusules of Sphaerellarida. The edges of the astropyle may be folded out to form a tube, the proboscis, which represents an oral apparatus. The nucleus occupies almost all of the central capsule. There are 17 families of Phaeodarea. Some contain species without skeletons and others have radial spines and needles, or latticelike, meshlike spherical, polyhedral, spindle-shaped, hemispherical, or lenticular skeletons or shells with various types of ornamentation.

Some good general references on the Radiolaria or particular groups are the chapter by Tregouboff, Deflandre, and Grasse in Tregouboff (1953); the monographs of Campbell (1954); Haeckel (1887); Hollande and Enjumet (1960); Popofsky (1909, 1913, 1917); and Shroder (1913) and the text by Anderson (1983). Some important papers on various aspects of their biology are those of Grell (1953); Cachon-Enjumet (1961); Hollande et al. (1970); and Cachon and Cachon (1971 and 1985); and Cachon (1972, 1976a,b, 1977, 1978). Recent fine structural studies have been published by Anderson (1976, 1977, 1978, 1983).

D. *Acantharia*

Acantharia are all pelagic. They are characterized by possession of a skeleton made of 10 diametrical or 20 joined radial spines composed of strontium sulfate. The spines are arranged in a characteristic pattern (Müllerian law). Acantharia can have a single large nucleus or can be multinucleate. Contractile strands (myonemes) run from the spines to a gelatinous vacuolated layer just below the cell envelope.

The life cycles of Acantharia generally include sporogenesis and often an encystment stage (Fig. 1.5a). Binary fission has been observed in one species. Encystment is characterized by complete loss of skeletal structure and by the elaboration of a gelatinous theca on which small plates of $SrSO_4$ are deposited. Cyst walls have three types of openings: those for the passage of spicules, those for cytoplasmic passage, and larger oscules through which the spores are liberated.

The taxonomy of Acantharia is based on the way in which the spines are joined in the center of the animal. See Cachon and Cachon (1985) for a recent review of the group. Members of the order Holacanthida have 10 diametrical spines that simply cross, or are temporarily joined, in the center of the animal. Those belonging to the Symphyacantha have 20 radial spines fused into a solid star (or central body). Members of the Arthracantha, the most highly evolved group, have radial spines with pyramidal faces. Those belonging to the Chaunacantha have saw-toothed, loosely articulated spines. One suborder has species with 30–500 irregularly distributed radial spines, and the other suborder has species with 18–32 regularly distributed radial spines.

Two types of zoozanthellae have been found in Acantharia, dinoflagellates and a prymnesiophycean (Febvre and Febvre-Chevalier, 1979). In the primitive acantharia the zooxanthellae are located singly in the ectoplasm. In some more advanced Acantharia species, 15–30 individuals have been found enclosed as a group in a common envelope.

Some general references on the biology and tax-

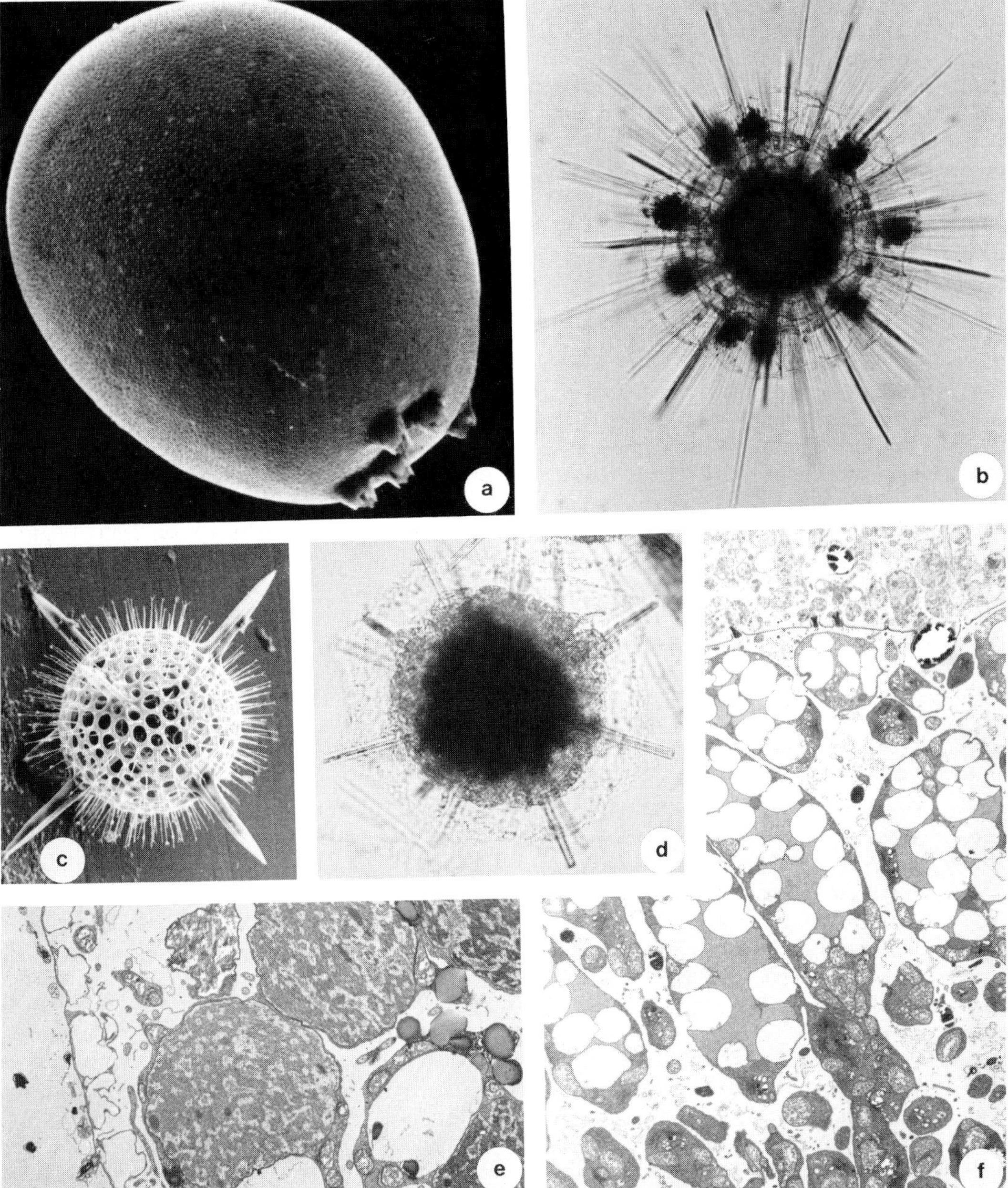

FIGURE 1.5 Acantharea and Polysystinea. (a) A cyst of *Gigartacon,* a chaunacanthid acantharian. (b) A skeleton-bearing sphaerellarinid spumellarian radiolarian with two concentric lattice shells produces numerous radial spines. ×200. (c) *Hexacontium,* a sphaerellarinid spumellarian radiolarian with six large spines. (d) An unidentified acantharian from a plankton haul in the northwest Atlantic, about 200 miles southeast of New York. (e) An ultrathin section through the central capsule of *Sphaerozoum punctatum* exhibits several nuclei containing cordlike masses of chromatin. A layer of vacuoles lies adjacent to the thin capsular wall. ×5,000. (f) An electron micrograph shows cytoplasmic detail of the intracapsular lobes. The inner segment of each lobe contains dense cytoplasm possessing mitochondria, Golgi bodies, and endoplasmic reticulum. The outer lobes are filled with a granular substance and numerous vacuoles. A nonliving organic capsular wall surrounds the cytoplasmic lobes. ×2,070. (a and c by Drs. Jean and Monique Cachon; b, e, and f by Dr. O. Roger Anderson. d by Dr. John J. Lee; a and c, SEM; b and d, transmission light; e and f, TEM.)

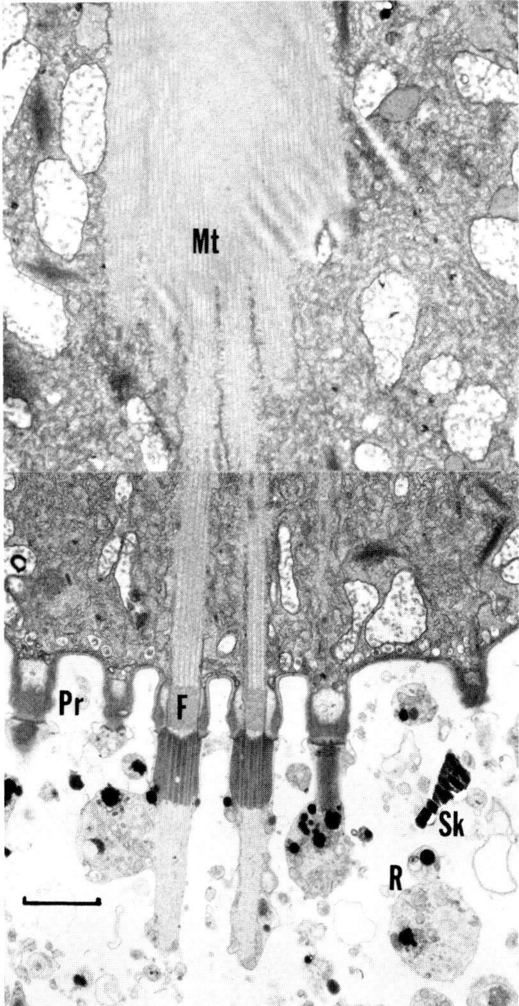

FIGURE 1.6 Radiolaria. A composite electron micrograph of the intracapsular cytoplasm of a nassellarian radiolarian shows the pores (Pr) at the base of the capsule. Each pore is penetrated by a thin rhizopodial strand called a fusule (F) connecting intracapsular cytoplasm with the extracapsular rhizopodia (R). A cone-like array of microtubules called a podoconus supplies the microtubules passing through the fusules into the ectoplasm which surrounds a segment of skeleton (Sk). ×8,000. (Electron micrograph by Dr. O. Roger Anderson.)

onomy of the Acantharia are the section by Tregouboff in the *Traité de Zoologie* (1953) and the monograph by Schewiakoff (1926). Papers by Hollande and Enjumet (1957) and Hollande and co-workers (1970) are good sources of information on encystment, reproduction, and symbiosis in Acantharia. Febvre (1974, 1977) has published on the fine structure of the cortex and myonemes. The papers of Massera-Bottazzi and Vanucci (1964, 1965a,b) give some information on the depth and spatial distribution of many species.

E. *Heliozoa*

Heliozoa are far more common in freshwater than they are in marine environments. They are characterized by their axopodia, which are fairly slender pseudopods, each with a central fibrous axis extending from the tip of the axopod through the body to the nuclear membrane or a structure known as an axoplast. In recent years several workers have obtained beautiful electron micrographs of the central microtublar fibers. In transverse section the fibrous axis is seen as a double spiral of cross-linked microtubular fibrils or a prismatic bundle of hexagonally arranged microtublar fibrils. Axopodia, which may be more than 100 μm long, extend and shorten relatively slowly. When marine Heliozoa are studied in relatively nonconfining preparations (e.g., deep slides, flasks, etc.) they can be seen to float, roll, and glide on the tips of their axopods. Bacteria, small algae, flagellates, and ciliates are trapped by contact with the adhesive pseudopods of heliozoans. These pseudopods retract to bring the food to the body surface for enclosure in a food vacuole. Three different kinds of extrusions involved in feeding and mucous secretions are seen in the ectoplasm and axopodial cytoplasm of heliozoa.

In common with marine amoebae, many Heliozoa are so small and translucent that they are easily overlooked in superficial examinations of natural collections. They are best observed in phase-contrast, interference-contrast, and dark-field microscopy. Some centrohelid heliozoans have siliceous skeletal spines or scales. Some species such as *Actinolaphus pedunculatus* are stalked. Heliozoans multiply by binary fission or budding. Algal symbionts have been reported in *Heterophrys myripoda*. Nothing is known about the biology of the relationship.

Marine Heliozoa are not a well-studied group. Some additional information about their taxonomy and structure can be found in the books of Kudo (1954); Hall (1953); Hepps (1956); Grell (1968); and the section by Tregouboff in Grasse's *Traité de Zoologie* (1953) and by Febvre-Chevalier in *The Illustrated Guide to the Protozoa* (1985). The latter has a taxonomic key based on the ultrastructure of the microtubular organizing center. The papers by Bardele (1969), Cachon and Cachon (1978), and Watters (1968) have good illustrations of the central microtubular fibers of the axopods of several heliozoans.

F. *Labyrinthulids*

Labyrinthulid amoebae belong to a separate phylum. They are also considered by mycologists to be members of a separate order Labyrinthulales. They do not have any sort of orthodox pseudopodia, nor are they phagotrophic. Individual cells are spindle shaped or round and move through channels of slime secreted by the cell (Fig. 1.1n). The individual slime filaments are thin and form branching and anastomosing networks. They are found in sea grasses (e.g., *Zostera marina, Ruppia maritima,* and *Zanichellia palustis*) and sea weeds (e.g., *Cladophora refracta, Laminaria iberica,* and *Ectocarpus confervoides*) and have even been reported as parasites of diatoms and chlorophytes (e.g., *Nitzschia* sp., *Coscinodiscus* sp., and *Chaetoceros* sp.). Some species form sori after congregation of spine cells or pseudosori from single cells. Their nutrition is saprozoic. It has been suggested that labyrinthulids are responsible for the wasting disease of eel grass. They are apparently easy to culture and require sterols for their growth. Some important papers on their biology are by Pokorny (1967, 1985), Schmoller (1971), Vishniac (1955a,b), and Watson (1957); in addition, see the book by Johnson and Sparrow (1961).

G. *Foraminifera*

Foraminifera are widely distributed in marine habitats and have been for most of recorded animal history. Their durable calcareous tests (shells) are easily identified in cores of marine sediments and sedimentary rocks. More than 34,000 species of Foraminifera have been described, all but 4,000 of which are found in the rich and highly diversified fossil record they have left since the Cambrian. Aside from their characteristic tests, Foraminifera are also easily recognized by their distinctive granuloreticulose pseudopodial nets, which extend considerable distances (2–10 × the test diameter) from the tests (Fig. 1.7a). This is one of the most fascinating microbial phenomena that can be observed with a compound microscope. In the same optical plane small granules and mitochondria move swiftly past each other in an endless, often anastomosing stream. Biophysicists have had lively disagreements on the mechanisms by which this two-way streaming is accomplished (e.g., Allen 1961a,b; Jahn and Bovee, 1965; Travis and Bowser, 1986; Travis and Allen, 1981; Travis et al., 1983; Bowser et al.. 1985).

In common with ciliates, some groups of rotaliinid and miliolinid Foraminifera have another distinctive biological feature: nuclear dimorphism (heterokaryosis). These heterokaryotic animals, at one stage of the life cycle, the agamont, have two kinds of nuclei, one or more large somatic nuclei that apparently have control of metabolic functions in cells containing them, and a large number of smaller-sized generative nuclei that divide to give rise to the nuclei of the next generation. The somatic nuclei degenerate during the process leading to asexual reproduction.

The numbers of nuclei involved in different heterokaryotic species are as few as four in the tiny *Rotaliella heterocaryotica* (three generative nuclei, one somatic nucleus) to hundreds in the giant foraminiferan *Sorites marginalis* (several hundred generative nuclei and two- or three-score somatic nuclei).

Their tests can be very simple (unilocular, single chamber), membranous or tectinous spherical or plastic sacs (e.g., *Allogromiina* Fig. 1.7a); a variety of different types of multichambered tests constructed of quartz, grains, other minerals or materials of biogenic origin held in an organic and partially mineralized matrix bound to an organic inner shell (e.g., Textulariina) (Fig. 1.8b); or three major varieties of mostly multioculate calcareous test types (Fig. 1.8). One extinct calcareous group, the Fusulinina, is characterized by microgranular calcite. Another major calcareous group, the Miliolina, is characterized by small calcite crystals arranged so that they give a porcelaneous appearance in reflected light (Fig. 1.8c). The third major group of calcareous foraminifera, the Rotaliina, has hyaline lamellar walls with characteristic perforations or pores (Fig. 1.8f; Fig. 1.9b,d). Pseudopods may extend through the pores or the pores may contain plugs or sieve plates.

The initial chamber of a multilocular foraminiferan is called a proloculum. One can usually determine the stage in the life cycle of a foram from the relative size of the proloculum; it is usually large (megalospheric) in the generation that reproduces sexually, and small (microspheric) in the asexually reproducing generation. Multilocular species follow a dozen or so patterns (Fig. 1.8). Regardless of whether growth is in a linear, coiled, fan-shaped, or branched pattern, the protoplasm within the test is free to flow from chamber to chamber and to the pseudopodial net on the exterior. The greatest variety of tests occur among those that are secreted and calcareous. During the course of evolution many changes have taken place in the structure of the test walls, their ornamentation, and the principal opening(s) for the pseudopods (aperture[s]). Classification of Foraminifera is based on test composition (including microstructure) and gross structure. The fine structural details of pores, apertural tooth plates, lips, flanges, keels, ribs, plugs, spines, and ornamentation of various kinds are also important characteristics for distinguishing various intermediate and lower taxonomic groups. The identification of Foraminifera and their classification is a very specialized science. The *Catalog of Foraminifera,* which lists almost every

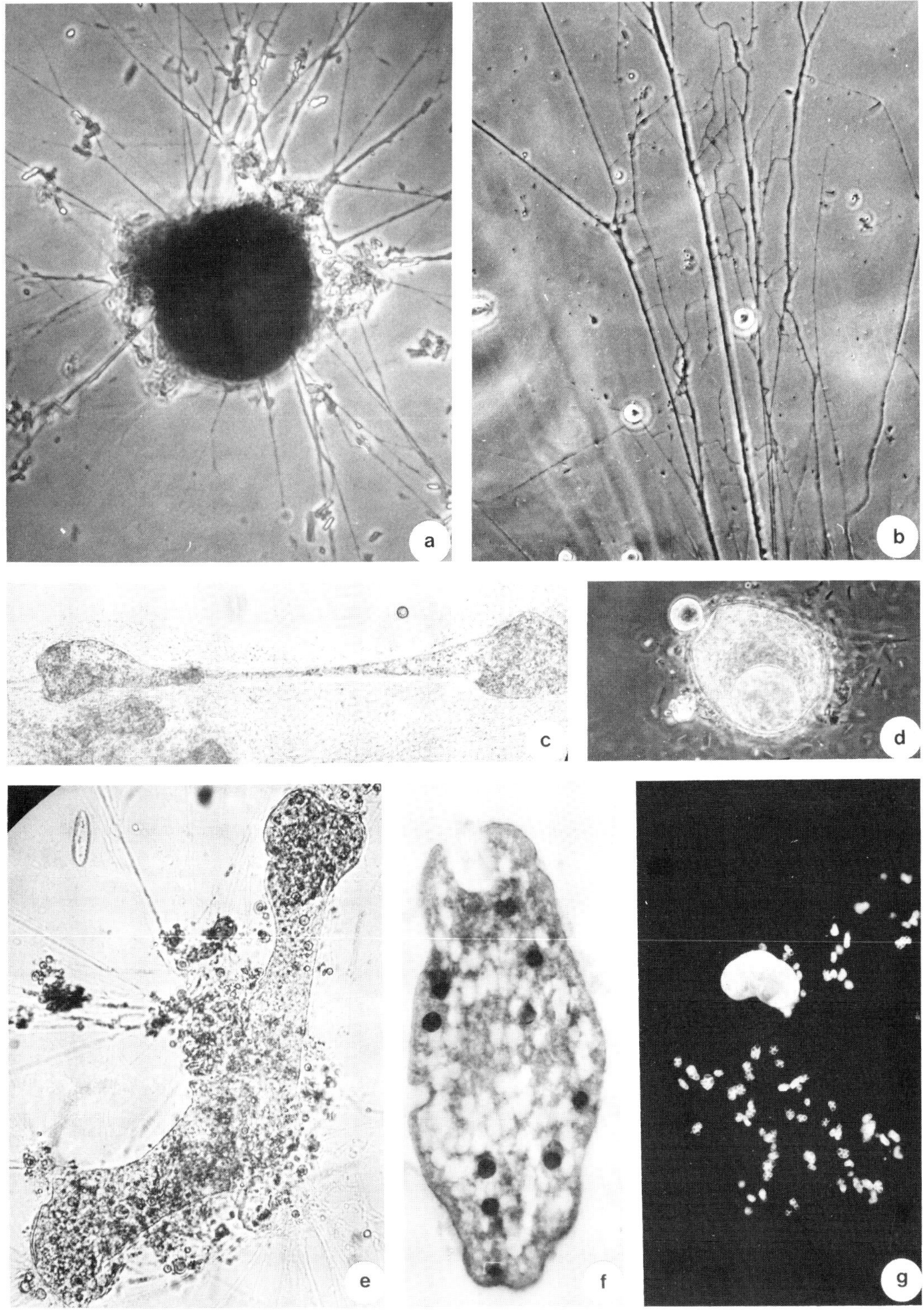

FIGURE 1.7 Foraminiferida. (a) The granuloreticulopod network surrounding the foraminifer *Allogromia laticollaris*. ~ ×160. (b) A magnified view of the same network showing the fine branches of the feeding net and captured algae. ~ ×1,200. (c) *Allogromia* sp. in the process of binary fission. ~ ×100. (d) A specimen of *Allogromia* sp. in the process of forming both an endogenous and an exogenous bud. ~ ×200. (e) An *Allogromia* sp. in the process of dividing by cytotomy. ~ ×200. (f) A stained histological section of *Allogromia* sp. showing its nuclei. ~ ×250. (g) Schizozoites just released from a parent schizont of *Rosalina leei*. ~ ×50. (All figures by Dr. John J. Lee. a–e, phase; f, transmission light; g, dark field photomicrographs.)

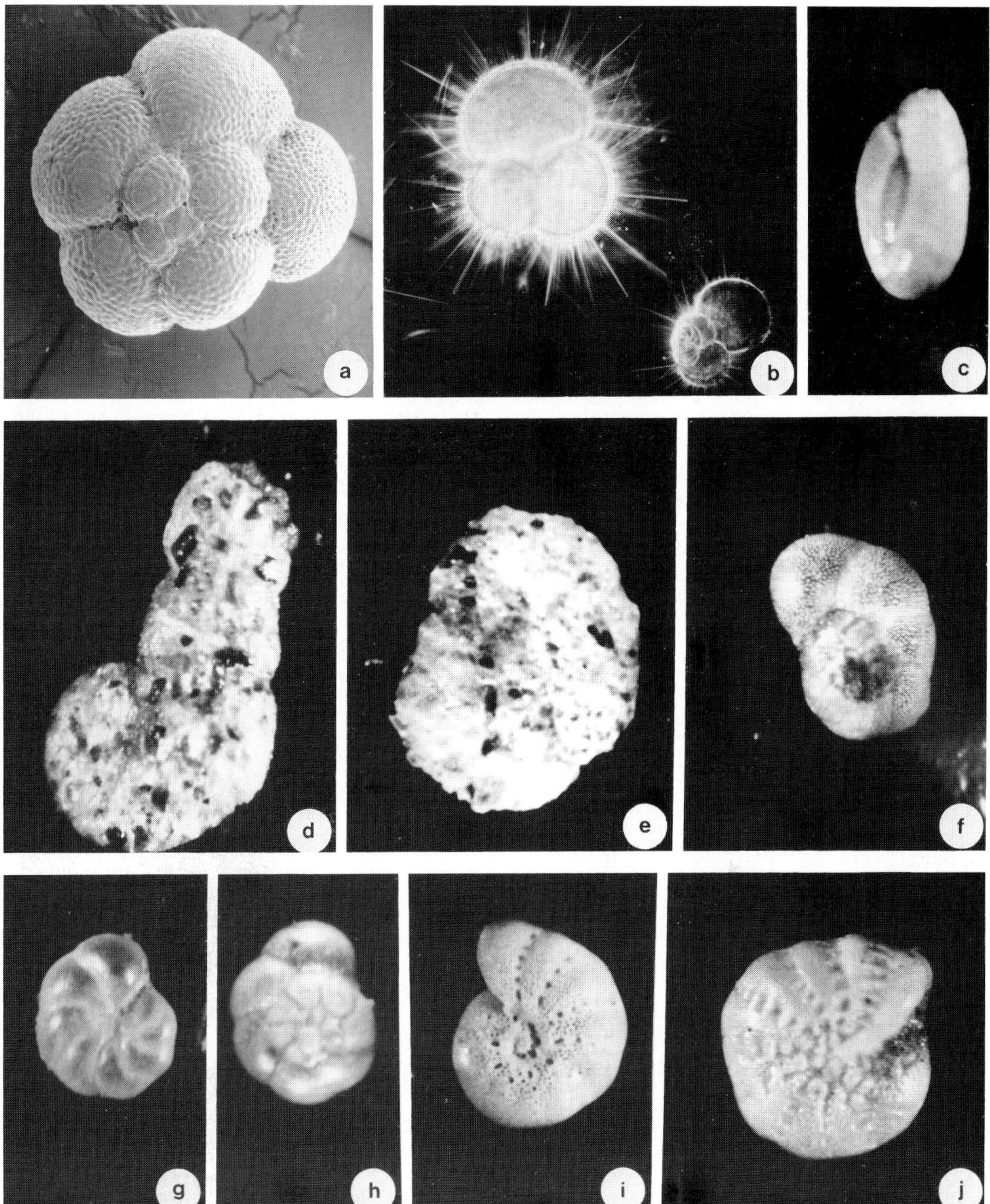

FIGURE 1.8 Foraminiferida. (a) *Globoquadrina dutertrei*, ~ ×60; (b) *Globigerina bulloides*, ~ ×30; (c) *Quinqueloculina seminula*, ~ ×60; (d) *Ammobaculites dilatatus* showing the variety of grains agglutinated to the test of this arenaceous foraminifer, ~ ×60; (e) *Rosalina leei*, ~ ×60; (f) *Elphidium incertum*, ~ ×60; (g) *Ammonia beccarii*, ~ ×60; (h) *Elphidium translucens*, ~ ×60; (i) *Elphidium gunteri*, ~ ×60. (a by Drs. Wayne Bock and William Hay. All other figures by Dr. John J. Lee. a, SEM; b, dark field, all other figures epi-illumination photomicrographs.)

species of Foraminifera that has ever been described, has more than 70 volumes (each 5 cm thick) (Ellis and Messina, 1940 and following). Briefer, but comprehensive treatments of foram structure and taxonomy are found in the books of Loeblich and Tappan (1964; classification updated in 1984); Murray (1971); Cushman (1959); and Gallowav (1933).

The life cycles of very few species of Foraminifera, perhaps fewer than two dozen, have been studied in detail. Because so much variation has been found among those cycles that have been

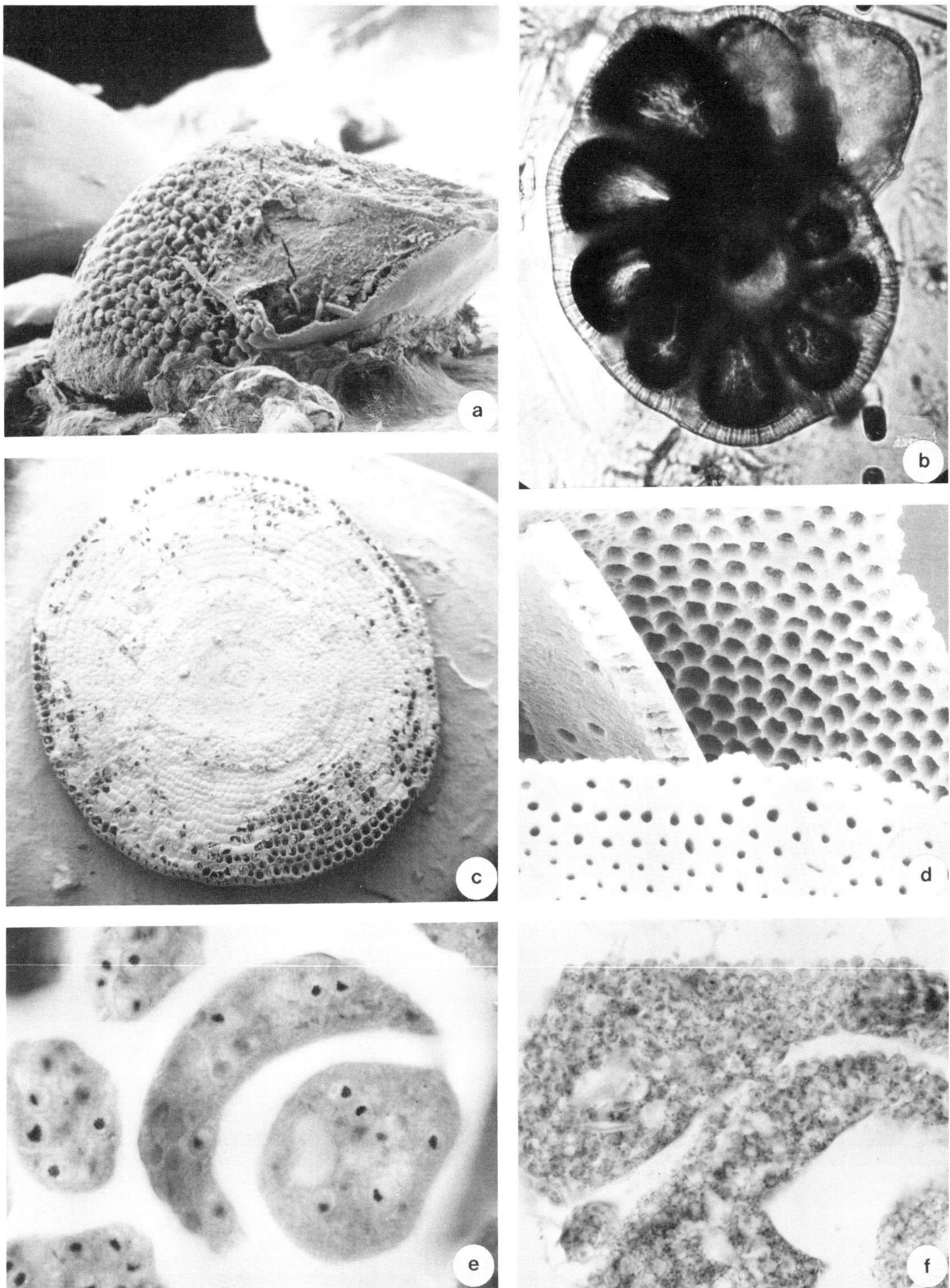

FIGURE 1.9 Foraminiferida. (a) Lateral view of a slightly damaged rubber cast of the inner shell surface of *Amphistegina lessonii*. ~ ×40. (b) Optical section of *Protelphidium tisburyensis* showing perforate nature of the test and internal chamber arrangements. ~ ×120. (c) *Amphisorus hemprichii*, some of the surface plates removed during processing to reveal the internal chamberlet walls. ~ ×80. (d) Portion of the outer test wall of *Amphistegina lessonii* showing the pores and the cuplike depressions formed by the pore rims on the internal surface of the test wall. ~ ×2,000. (e) Histological section of *Amphisorus hemprichii* showing the dinoflagellate zooxanthellae. ~ ×1,000. (f) Histological section of *Amphistegina lessonii* showing diatom zooxanthelzae. The zooxanthellae fit into the cuplike pore–rim depressions in the test wall. ~ ×450. (All figures by Drs. John J. Lee and Marie E. McEnery. a, c, and d, SEM; d, e, and f, transmission light micrographs.)

studied, generalizations are given with caution. The "classical" life cycle of Foraminifera was partially worked out at the turn of the twentieth century by Schaudinn and Lister on a species, *Elphidium crispum,* which used to bloom in great numbers in the English Channel near Plymouth. In species with classical life cycles (Fig. 1.10) there is a regular alternation of generations between an asexual and a sexual phase. The sexual-phase individuals (gamonts) are haploid and uninucleate. Meiosis takes place in mature asexual individuals (agamonts), which are multinucleate and diploid. Although a metagenic life cycle is unusual for animal species, it is a common pattern in plants. In many foram species the two generations are distinct enough morphologically to be easily recognized. Gamonts are smaller in overall size but have a larger proloculum (megalospheric). Agamonts characteristically have small prolocula (microspheric) but are much larger in overall size. Later work by Jepps and Myers has shown that aspects of the *Elphidium crispum* life cycle are seasonally driven. In the winter in the English Channel the proportion of megalospheric forms to microspheric forms is approximately 30:1. The proportion changes to 1:1 or less in the spring as schizogony (asexual multiple fission) reaches a peak. Gamonts grow all summer and reach maturity in one year in the warm East Indian seas, or two years in the cold English Channel. The gamonts of *E. crispum* are small (2–5 μm) and biflagellated; they are released into the open ocean in large numbers where they meet and fuse. The gamonts of one interesting aufwuchs community species, *Rosalina bulloides,* encyst during the formation of gametes. Just before the release of gametes, gamonts form a very large flotation chamber, are released from their cysts, and float to the surface. The pelagic phase of the life cycle ends with the release of biflagellate gametes, which then fuse to form zygotes and sink to begin the cycle anew. Free-swimming gametes (with one, two, or three flagella) have been described for almost a dozen species belonging to monothalamic and many different polythalamic groups. It has been shown experimentally that at least one species, *Myxotheca arenilega,* is autogamous. (Gametes from the same parent fuse and produce viable offspring.)

Several other species of forams produce autogamous gametes that are never released from the parent. In *Rotaliella hetercaryotica* the walls between successive chambers dissolve during gametogenesis. After the cytoplasm is divided up among the amoeboid gametes, they fuse to form zygotes within the parental test (Grell, 1954). A similar picture was found in the sexual phase of the monothalamic species *Allogromia laticollaris* (Arnold, 1954; McEnery and Lee, 1976).

Union of amoeboid gamonts takes place in other species of Foraminifera (gamontogamy). The gamonts creep toward each other and attach by their ventral sides. They then surround themselves by a fairly tough organic membrane. At the beginning of gametogenesis the walls separating the chambers and most of the ventral walls of the gamonts are dissolved. In *Patellina corrugata, Spirillina vivipara, Rubratella intermedia,* and *Metarotaliella parva* amoeboid gametes are formed that subsequently fuse in the common space between the two shells. In gamontogamous species of *Discorbis* flagellated gametes are formed. Although the gamonts of most gamontogamous species are generally similar in size, pairing of unequal-sized gamonts has also been observed in *Rubratella intermedia.* The gametes of such pairing also appear anisogamous. The smaller parent produces gametes with smaller, more-condensed nuclei. Sexual differentiation among gamonts have been studied experimentally in *Metarotaliella parva* (Weber, 1965). Young gamonts cannot mate with each other but

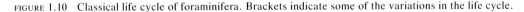

FIGURE 1.10 Classical life cycle of foraminifera. Brackets indicate some of the variations in the life cycle.

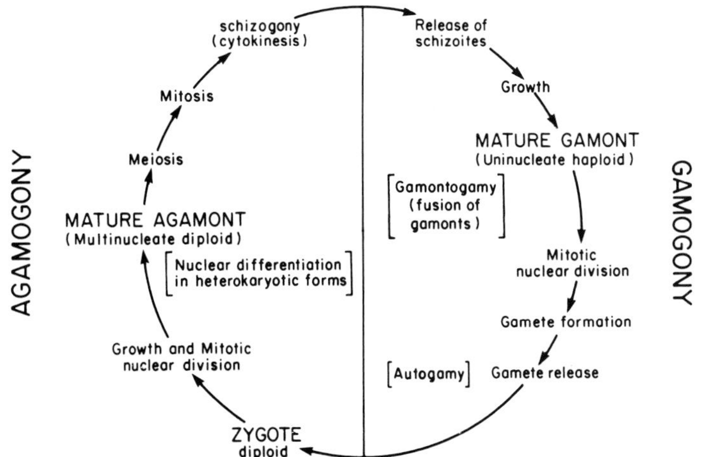

only with "mature" and "older" ones. "Mature" gamonts can mate with each other and "older" gamonts as well. Older gamonts are decreasingly reactive to each other but are stimulated to mate with young gamonts.

A number of diverse species, including *Cornuspira lavolaensis*, *Entosolenia marginata*, *Discorinopsis aguagui*, *Discorbis orbicularis*, *Spiroloculina hyalina*, *Calcituba polymorpha*, and *Allogromia laticollaris*, have life cycle options that include successive asexual generations (apogamic or partially apogamic cycles). The most unusual of these species is *Allogromia laticollaris*. Each of the four strains that have been studied have slightly different emphases on different phases of the life cycle (Figs. 1.11, 1.12). In one strain, CSH, there is a regular metagenic alternation between a diploid uninucleate agamont (agamont I) and a diploid multinucleate agamont (agamont II). Another strain, SIP, produces only successive asexual generations of agamont II. Still a third strain has been found that produces successive generations of agamont II about 75 percent of the time and agamont I in the remaining 25 percent of the divisions. The organism can also reproduce by binary fission, budding, and cytotomy (a form of multiple budding). The sexual phase of the cycle is extremely rare in the preceding three strains, occurring once in 1×10^3 generations. Some of the large agglutinated Foraminifera reproduce by budding (*Halyphysema*) or fragmentation (e.g., *Astrorhiza*, *Bathysiphon*).

Characteristic assemblages of all types of Foraminifera occur in all marine environments above the lysocline (the deep zone where their shells dissolve). Below the lysocline only the Textulariina have been found. A very few species, *Protelphidium tisburyensis*, *Miliammina fusca*, *Psammosphaera* sp., *Ammonia beccarii*, *Ovammina*, and some allogromiids have been found in fresh waters or will survive in fresh waters for extended periods of time. Although the planktonic species are best known from their frequent capture in the world's oceans and for the extensive *Globigerina* oozes on the sea floor, most Foraminiferan species are benthic. Different assemblages of species are characteristic of different water depths, habitats, and latitudes (e.g., Lee 1982). The distribution of the planktonic Foraminifera is largely temperature controlled, maximum densities being found at 6–30 m. This knowledge has been usefully applied to paleoecological interpretations of fossil assemblages. An interesting group of large Foraminifera with tests (Textulariina, Komokiacea) has been found in abyssal oligotrophic areas and in hadal trenches (Tendal and Hessler, 1977). They are fairly large (1–5 mm) and look either like balls or scouring pads woven out of interconnective tubules or like branches or shrubs. Although the group has been found at depths between 400 and 9,600 m, the group is most abundant in samples at depths between 1,000 and 7,000 m. Although our knowledge of the abundance of the group is understandably sketchy, it has been reported that there were about 2,000 specimens in a 500-cm² box core.

Foraminifera are very patchy in their distribution. In various samples they have been reported to

FIGURE 1.11 The nonclassical life cycle of *Allogromia laticollaris*.

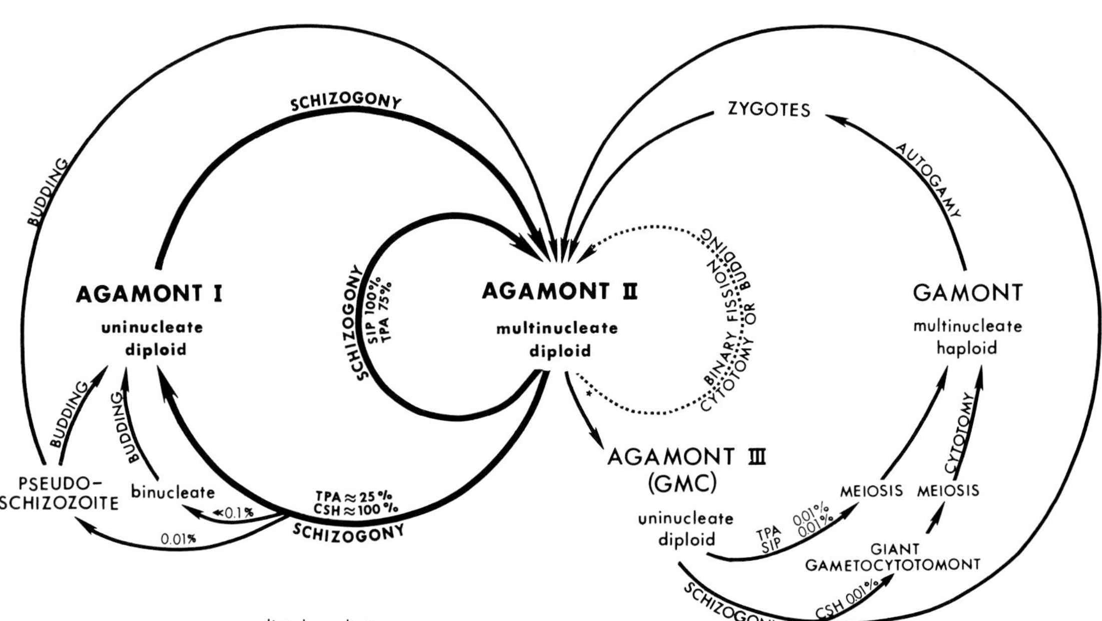

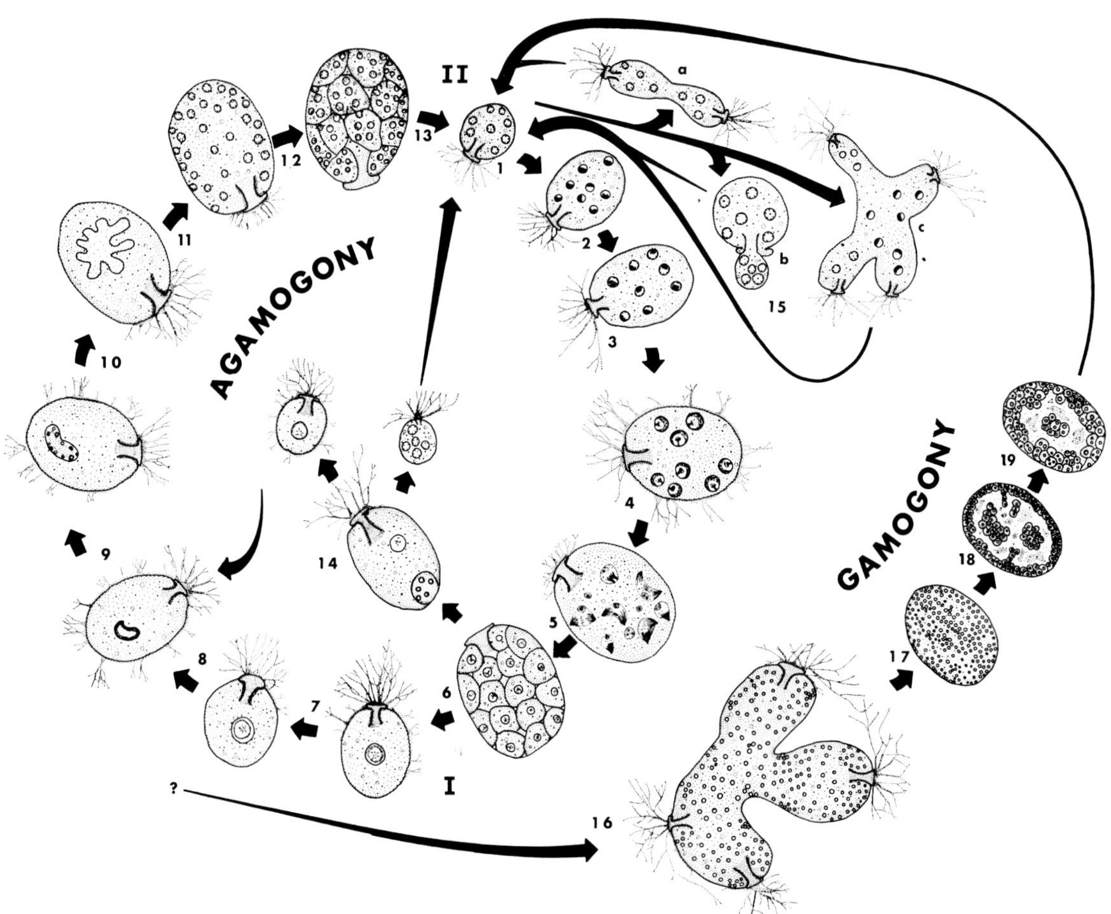

FIGURE 1.12 Life cycle of *A. laticollaris*. *Agamont phase*: (1) juvenile agamont II, early G_1; (2) young agamont II, mid-G_2 (RNA synthesis also occurs in this phase); (3) growing agamont II, late G_1; (4) mature agatom II—chromosomes in "mushroomlike configuration," RNA accumulated at the periphery; (5) karyokinesis; (6) cytokinesis (schizogony); (7) young agamont I, S phase; (8) maturing agamont I, G_2 phase; (9) mature agamont I, nucleus differentiating, RNA granules at the periphery of the nucleus; (10) mature agamont I, early "amebaiform" nucleus; (11) agamont I, amebiform nucleus; (12) agamont I, post *Zerfall*; (13) agamont I (schizogony); (14) agamont I, relatively uncommon life cycle alternate pathway in which budding gives rise to an agamont II; (15) agamonts II, relatively uncommon alternate life cycle pathways including (a) binary fission, (b) budding, (c) cytotomy. *Gamont phase*: (16) giant gametocytotomont; (17) multinucleate gamont prior to the formation of gametes; (18) gamont filled with gametes; (19) gamont with some gametes and zygotes.

be as abundant as 4,500/10 cm^2 or as few as none. One informed estimate suggests that average foram benthic values are in the range of 50–200/10 cm^2 on the shelf and marginal seas. In the water column the abundances of planktonic Foraminifera have been reported to vary as much as $1\text{–}1 \times 10^8$/1,000 m^3 sample. Highest average concentrations in some regions (North Pacific) are about $10^3\text{–}10^4$ greater than in low-density regions (e.g., the Sargasso Sea).

The books of Phleger (1960) and Murray (1973), the chapter by Bé (1977), and the paper by Tendal and Hessler (1977) are excellent summaries of the extensive literature on the ecology and distribution of recent Foraminifera. Foraminifera are easily maintained in the laboratory. One of the easiest cultivation methods is to place them in circulating marine aquaria in which the water is gently flowing. They are also maintained in ordinary laboratory glassware with large surface to volume ratios (tissue culture flasks, finger bowls, deep petri dishes). Sterile filtered seawater or Erdschreiber are good media (see Lee and Muller, 1975). Most cultures of benthic Foraminifera do better if micrometazoa and ciliates are excluded. Selected species of small diatoms and chlorophytes may be added as food. Artificial light is generally used so that overgrowth by algae is prevented. Depending on the species, incubation temperature is generally controlled so that it is within the ranges found in the collection sites during the summer.

Careful laboratory studies suggest that many species of Foraminifera are quite selective in their feeding habits (reviewed in Lee, 1974, 1980). In general, they eat small species of diatoms, chlorophytes, and bacteria. Experimental studies suggest that many littoral foram species are finely tuned to the annual bloom cycles of particular algal species and feed voraciously at those times. Benthic Foraminifera are also food gatherers, surrounding themselves and sometimes burying themselves within balls of food which they gather with their pseudopods. The food is digested later, often by the offspring of the gatherers. The feeding rates of juveniles of several species have been shown to greatly exceed (>100–200 percent) the rates of more mature individuals. The feeding rates of even the juveniles of many species slow down as cultures become crowded. Although this has not been rigorously studied, it appears that individuals set up smaller feeding nets or feeding territories in crowded cultures. Laboratory studies show that the reproductive and growth rates of all species so far examined are tied into the particular qualities of certain species or mixtures of algae.

In general, Foraminifera grow and are more fecund on specific mixtures of algae than on a diet of any single algal species in the mixture. All the species of Foraminifera that have been studied in gnotobiotic culture seem to require some bacteria in their diet. The bacteria apparently supply some growth factor(s) that are not found in sufficient quantities in the algae. There are enough differences in feeding preferences of the various benthic species that have been examined to suggest that niche separation in the group is largely through resource partitioning (see the paper by Muller, 1975).

In the last few years there has been great interest in the larger Foraminifera found in the tropical and semitropical seas. These animals are easily seen with the naked eye and often reach small-coin size! Although not every species of larger Foraminifera from the photic zones of tropical marine habitats has been examined cytologically, those that have are filled with endosymbiotic algae. Unlike the majority of marine algal–invertebrate symbioses, which usually involve a dinoflagellate partner, different species of larger Foraminifera are hosts to endosymbiotic chlorophytes, unicellular rhodophytes, and diatoms as well as to dinoflagellates. Many species with brown endosymbionts also harbor small numbers of green symbionts. Experimental studies suggest that algal endosymbionts in different species contribute from 10 to > 90 percent to the carbon budgets of their hosts. The calcification rates of symbiont-bearing Foraminifera are enhanced in the light suggesting that symbionts play some role in this process as well.

H. *Apicomplexa* ("Sporozoa"), *Microspora, Myxospora,* and *Ascetospora*

The protozoa *Apicomplexa* ("Sporozoa"), *Microspora, Myxospora,* and *Ascetospora* are all parasitic forms and widely distributed in or on marine fishes and invertebrates. The majority invade tissues and body cavities and can cause extensive cellular and tissue damage.

The gregarines are typically extracellular parasites of the digestive tract and body cavities of invertebrates. They have life cycles with both sexual and asexual phases. Though the early development of some gregarines occurs in tissue cells, the trophozoites usually emerge to complete the cycle in the host's body cavity or intestine. The range of size of the spindle-shaped or tubular mature trophozoites of gregarines is astounding—some species are as small as 10 μm long, other species are as long as 3–4 mm! One major group of gregarines, the Septasorina, has a differentiated anterior portion (protomerite), equipped with a hold-fast organelle (epimerite), which is embedded in the tissues of the host. The trophozoites of a few gregarines undergo multiple fission (merogony), to form other trophozoites. Trophozoites give rise to gamonts that pair or form larger groups (syzygy). The associated gamonts secrete an enclosing membrane to become a gametocyst. Gametes are formed in the gametocysts, which fuse to form zygotes. In most gregarines the zygote encysts soon after syngamy. Within the cyst (oocyst) the zygote divides to produce sporozoites. Although the life cycles of many gregarines living in marine species are incompletely known, the life cycles of others seem timed to coincide with the life cycles of their hosts. In *Gonospora,* for example, the production of spores coincides with spawning in its polychaete host. In one family, Porosporidae, both crustacean and molluscan hosts are involved in the life cycle. Trophozoites mature in the intestine of a crab or lobster. Individual trophozoites encyst and undergo rapid nuclear divisions to form spherical gymnospores, which contain "merozoites." The gymnospores are released into the sea where they encounter and enter molluskan hosts. Within the host tissues the "merozoites" become differentiated into gametes, which fuse to form zygotes, which produce sporozoites.

The cycle is completed when the mollusk is eaten by a crab or lobster. Gregarines have been described from the intestines, gills, or coelom of many types of crabs, barnacles, copepods, lobsters, penaeids, shrimps, polychaetes, holothurians, and mollusks.

The coccidia are predominantly intracellular parasites of epithelial tissues and are widely distributed in fish and some marine invertebrates (Annelida, Arthropoda, and Mollusca). Life cycles can include both asexual and sexual reproductive phases (Fig. 1.13). Hosts ingest oocysts or spores. The sporozoites enter epithelial cells and develop into multinucleate schizonts. The schizonts, as their name implies, undergo multiple fission (merogony). The offspring, merozoites, enter other cells and repeat the cycle. After two or more successive generations of schizonts, the merozoites develop into two types: microgametocytes or macrogamonts. The microgametocytes produce microgametes. Syzygy occurs in the more primitive groups of coccidia. The macrogametes and microgametes are usually morphologically differentiated, with the smaller microgamets often being flagellated. After syngamy the zygote with oocyst usually divides into sporoblasts, which secrete sporocyst membranes around themselves. Sporozoites are formed within the sporocysts. Sporogony may or may not be completed in the host. In some coccidia, mites, ticks, and leeches obtain gametonts, and further development takes place in these vectors. In one unusual life cycle, merogony of *Aggregata eberthi* takes place in the intestinal connective tissue of a crab that has ingested sporocysts. If an infected crab (*Portunus*) is eaten by a squid (*Sepia*), some of the merozoites develop into gametonts and syngamy and sporogony follow. The coccidia are divided into two orders on the basis of whether gametonts are associated in syzygy during differentiation and whether a few or many microgametes are produced.

Perhaps the best-known marine coccidium is *Selenococcidium intermedium,* which occurs in the gut of the European lobster (*Homarus*). Species of *Aggregata* are found in different kinds of crabs and shrimp around the world. *Hemogregarina bigemiha* is widely distributed in the red blood corpuscles of marine teleosts and elasmobranchs along the European, North American, South African, South Pacific, and Red Sea coasts. Almost 40 species of coccidia have been found in marine fishes. Most belong to the genus *Eimeria,* characterized by an oocyst with four sporocysts each containing two sporozoites. *E. sardinae* attacks the seminiferous tubules of the tests of many European coastal fishes. The swim bladder of the cod is commonly infected with another species.

Microsporidia are also small (2–20 μm) intracellular parasites that are largely found as parasites of arthropods and few fishes (Fig. 1.14). The infective stage is an oval spore that has an extrusible polar filament. After a spore is ingested by a new host, the sporoplasm emerges as an amoeboid trophozoite. It passes through gut walls into tissue space or the blood to become intracellular in target tissue cells. The trophozoites grow and reproduce by binary fission (schizogony). The offspring may reproduce again or become sporonts which produce spores. The invaded cells in, most cases, become enormously hypertrophied. Spores can mature in the same host and start the cycle all over again. Microsporidoses occur regularly in marine fishes. *Nosema lophii* attacks the ganglion cells of the central nervous systems of the species *Lophius* from Eu-

FIGURE 1.13 General life cycle of Telosporea.

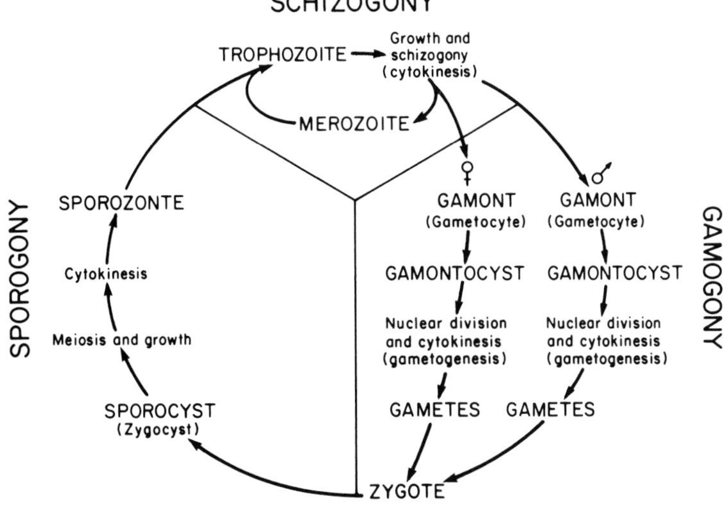

FIGURE 1.14 Microspora. (a) Yearling smelt, *Osmerus mordax*, infected with the microsporidian *Glugea hertwigi*. This severe example involves the entire digestive tract with several large parasite cysts protruding from the anal vent. Length of specimen 11 cm. (b and c) *Glugea hertwigi* spore before discharge. Ultrastructure of the spore reveals an outer wall (W), a coiled polar tube (PT), tightly packed membrane pleats of the polarplast (arrows), and the infective sporoplasm nucleus (N). After ingestion by a smelt, the spore discharges the polar tube with a velocity suitable for penetrating the intestinal epithelium and inoculates the sporoplasm through the tube into a host cell. Bar represents 0.5 μm.

rope and the western Atlantic. *N. stephani* is commonly found in species of *Pleuronectes, Pseudopleuronectes* and related forms in Europe and the United States. Many microsporidian species (e.g., *N. hertwigi, Plistophora* spp.) attack muscle tissue, producing tumorlike masses. *Nosema branchialis* is a frequent parasite of the gills of gadid fishes. *Ameson michaelis* can cause heavy chalky-grayish infections of the appendages and abdominal muscle tissue of blue crabs. *A. nelsoni* attacks the abdominal muscles of brown, white, and pink shrimp. *Pleistophora lintoni* and *Indosporus spraguei* infect grass shrimp, crabs, and shallow-water fish. In the grass shrimp the parasitism makes the tail opaque. *Agmasoma penaei* can replace much of the abdominal muscle tissues in heavily parasitized penaeid shrimp. It is also found in their blood vessels, gonads, fore gut, and hind gut.

Myxosporida are mostly parasites of fishes and other lower vertebrates. They are multinucleate. The characteristic spores of Myxosporida have one to six polar capsules with extrusible polar filaments grouped at one end of the cell. A sporoplasm occupies the rest. The capsules and sporoplasm are surrounded by a chitinous spore membrane with two to six valves. After spores containing zygotes are ingested by specific hosts, amoebulae emerge from the spores and migrate through the intestinal wall to target tissues (Fig. 1.15). The amoebulae become trophozoites and grow at the expense of the host tissues. Nuclear division accompanies growth. In some species asexual reproduction (plasmotomy

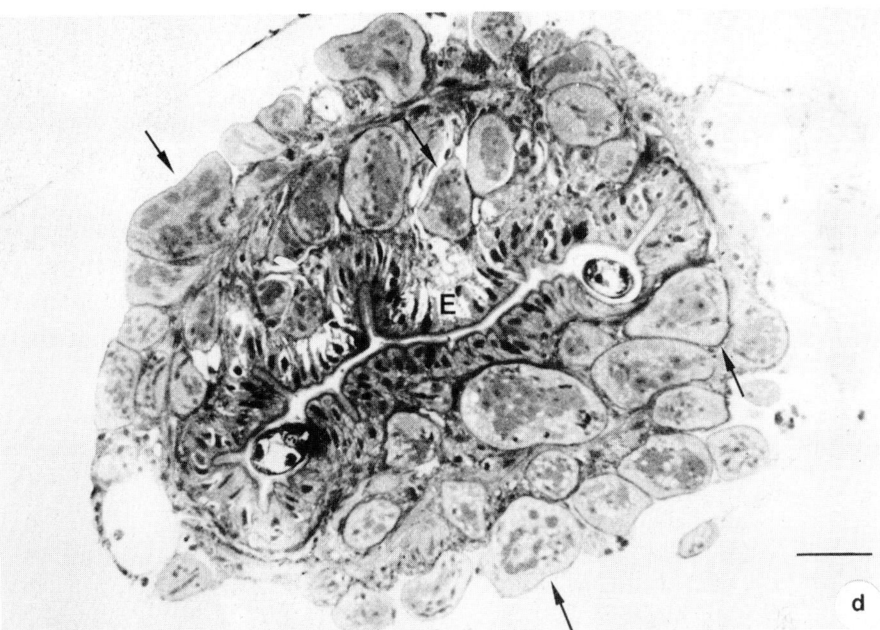

FIGURE 1.14 *(continued)* (d) Gross section of intestine from experimentally infected smelt larvae, one week after exposure to *G. hertwigi* spores. At least 40 xenomas (arrows) 20–50 μm in diameter develop with the submucosa. "Xenoma" refers to the unique association between a viable hypertrophied host cell and developing intracellular parasite. Infected host cells contain early schizont stages of the parasite. E, epithelium. Bar represents 50 μm. (All figures by Dr. Ann Scarborough.)

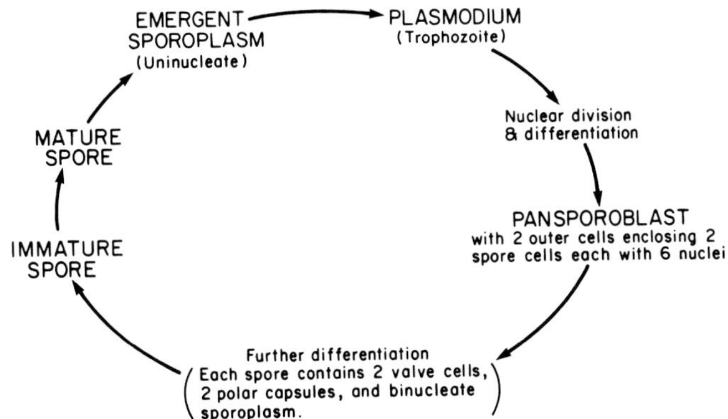

FIGURE 1.15 General life cycle of Myxosporea.

or budding) occurs. Parts of the cell become differentiated around groups of nuclei that become sporonts (sporoblasts). The sporonts grow and, depending on the species, their nuclei divide several times. The surrounding host tissue becomes necrotic and inflamed during trophozoite growth. The fibrous tissue forming an envelope around the trophozoite, and the trophozoite, itself, is known as a myxosporidan cyst. It has been observed that in some species the sporont nuclei divide. In other species it has been suggested that only the zygote is diploid. Nuclei fuse to form zygotes and the sporonts (monosporoblast if one spore is formed, disporoblast if two are formed, or pansporoblasts if more are formed) differentiate into a spore(s) with polar capsules and sporoplasm. Spores are transmitted directly. Most marine myxosporida are found in the gallbladder and urinary bladder. The liver, head cartilage, skin, skeletal musculature, and other organs are also targets for these parasites. The most widely distributed myxosporidan genera are *Myxidium, Ceratomyxa, Sphaeromyxa, Leptotheca, Chloromyxum,* and *Sinuolinea.* Various species of these genera have been identified in bladders from fishes caught in almost every sea. In severe infections the gallbladder can be partially degenerated or even functionally eliminated. More seriously pathogenic are species of *Unicapsula*, *Kudoa*, and *Hexacapsula*, which invade the skeletal musculature of different teleosts. The infections can become so heavy as to make them unmarketable. Infections with *Unicapsula muscularis* cause the "wormy halibut" disease. *Kudoa thrysites* cause the "milky barracuda" disease. The parasites turn large parts of muscle bundles into mucus-filled cavities containing parasites. *Myxobus cerebralis* invades and erodes cartilage supporting the central nervous system of salmonid fish, causing "circling disease."

Ascetosporida produce cysts without polar filaments. The spore coat sometimes has spines and some species have an operculum. Species are found in fishes, tunicates, insects, mollusks, annelids, nemertines, trematodes, and rotifers. They parasitize the coelom or other cavities. Some species are found in tissues or cells. After spores are ingested by the host, an amoebula emerges that invades tissue cells or makes its way into a body cavity. Growth of the trophozoite is accompanied by nuclear divisions. The multinucleate plasmodium that is formed divides into uninucleate sporoblasts, which then develop directly into spores. *Minchinia nelson* (MSX) has apparently been responsible for mass mortality of the oyster (*Crassostrea virginica*) on the mid-Atlantic seaboard of the United States. Wood-boring shipworms (*Teredo navalis*, *T. furcifera*, and *T. bartschi*) are apparently alternate hosts for *Minchinia nelsoni* in at least part of the mid-Atlantic seaboard. *Marteilia refringens* and *Minchinia armoricana* are pathogenic in European flat oysters (*Ostrea edulis*). Mortalities from this disease seem higher in France and Spain than they are in Dutch estuaries. *M. refringens* has also been found in the Japanese oyster, *Crassostrea gigas*, a commercially significant species in the coastal water of Brittany. *Haplosporidium tumefacientis* is a parasite of the digestive gland and kidney of the California sea mussel, *Mytilus californianus*. Another ascetosporidan, *Marteilia* sp., is a parasite of the amphipod *Orchestia gammarellus*, where its growth in testes, ovaries, sperm ducts, and androgenic glands is the apparent cause of intersexuality of males and thelygeny in females. Some *Haplosporidia* hyperparasitize worms parasitic in marine animals. *Urosporidium crescens* parasitizes the metacercariae of *Carneophallus* sp., a parasite of the southern and eastern U.S. blue crab, *Callinectes sapidus*. The metacercariae become black when the protozoan parasite sporulates causing a speckling resembling pepper; hence, the disease is known as the pepper crab disease. A similar speckling occurs in the grass shrimp, *Paleomonetes pugio*, when the metacercariae of *Microphallus* become infected by *Urosporidium*. In the Rhone delta, the sporulation of *Urosporidium liroveci* causes similar speckling in the trematode, *Gymnophallus nereicola*, which itself is a parasite of the clam *Abra ovata*.

The best single source of review articles and references on all aspects of microsporidia is the two volumes edited by Bulla and Cheng (1976, 1977). Articles by Becker (1970a), Lom (1970), Putz and McLaughlin (1970), Sanders et al. (1970), and Sprague (1970a,b) in the American Fisheries Society Symposium on diseases of fishes and shellfishes are also excellent sources of information on piscine flagellates. Sawyer et al. (1975) provide information on leeches as vectors for protozoan diseases. Sprague and Couch (1971) have published an annotated list of protozoan parasites, hyperparasites, and commensals of decapod crustaceans. Standard reference works edited by Sindermann (1970, 1977) and Ribelin and Migaki (1975) are excellent sources of information on the diagnoses, pathology, and control of the principal protozoan diseases of marine fishes and shellfishes. The proceedings of an international Symposium on Ascetosporian (haplosporidan and haplosporidanlike) disease of shellfish, which fills a single issue of *Marine Fisheries Review* (January–February 1979), is an excellent single source of information on these protists.

I. *Ciliates*

Ciliates are perhaps the most widely distributed and diverse group of marine protozoa (Fig. 1.16). Representatives are found in every conceivable marine habitat. Their exceptional beauty and morphological complexity have attracted many students to the study of their comparative anatomy, which, in itself, has become a very important subdiscipline. In view of the publications of Corliss' (1979) monograph on this phylum and the work of Small and Lynn (1985), both of which lucidly outline the general morphology of ciliates, the characteristics separating the various classes, and the rationale for classification, we shall not attempt to duplicate or abridge these excellent sources of information on the group.

One of the most widely distributed and best-known groups of marine ciliates is the suborder Tintinnina (Fig. 1.17, Fig. 1.18a). These zooplankters, which vary in size from less than 10 μm to 3 mm, have a fossil record that stretches back to the Silurian. In the Jurassic and Lower Cretaceous, they were extremely abundant in the deeper-water facies of the Tethyan seas and are useful as strati-

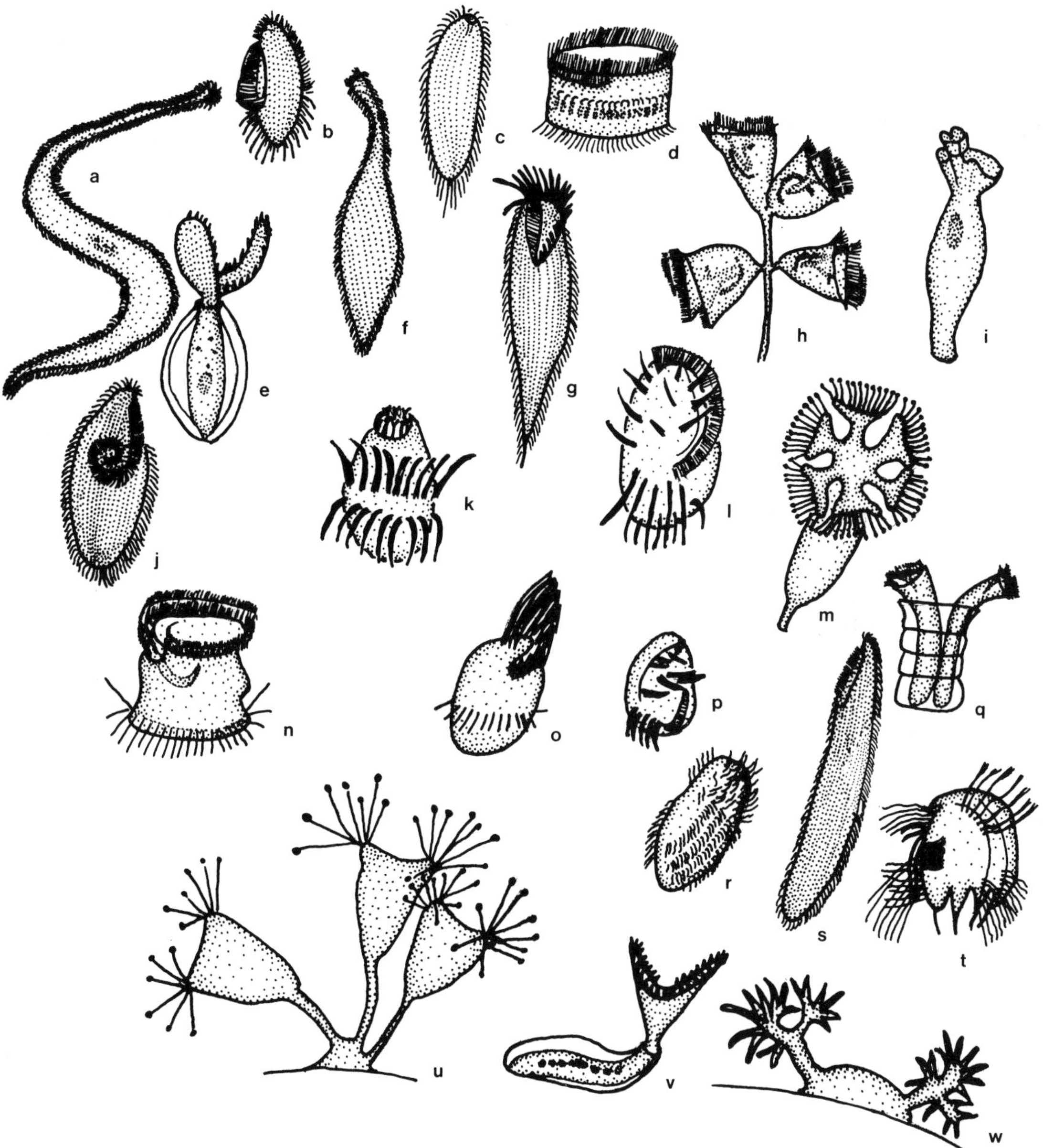

FIGURE 1.16 Ciliophora. (a) Karyorelictea, *Tracleloraphis* (after Dragesco, 1958); (b) Scuticociliatida, *Pleuronema marina* (after Kahl, 1928); (c) Prostomatida, *Prorodon elegans* (after Kahl, 1928); (d) Peritrichia, *Trichodina* (after Corliss, 1979); (e) Heterotrichida, *Folliculina simplex* (after Dass, 1947); (f) Haptorida, *Lacrymaria color* (after Kahl, 1928); (g) Heterotrichida, *Condylostoma longissimum-caudatum* (after Kahl, 1928); (h) Peritrichia, *Zoothamnium* (after Corliss, 1979); (i) Chonotrichia, *Spirochronia gemmipara* (after Guilcher, 1971); (j) Heterotrichida, *Fabrea salina* (after Kirby, 1934); (k) Haptorida, *Mesodinium pulex* (after Borror, 1972); (l) Hypotrichida, *Euplotes vannus* (after Heckmann, 1976); (m) Suctoria, *Asterifer faurei* (after Guilcher, 1971); (n) Peritrichia, *Urcelolaria spinicola* (after Beers, 1964); (o) Oligotrichida, *Strombidium sulcatum* (after Borror, 1965); (p) Hypotrichia, *Aspidisca steini* (after Kahl, 1928); (q) Peritrichia, *Cothurnia* (after Kahl, 1928); (r) Primociliatida, *Stephanopogon apogon* (after Borror, 1965); (s) Karyorelictea, *Geleia simplex* (after Fauré Fremiet, 1951); (t) Odontostomatida, *Saprodinium integrum* (after Kahl, 1928); (u) Suctoria, *Acineta tuberosa* (after Grell, 1978); (v) Heterotrichida, *Metafolliculina andrewsi* (after Grell, 1973); (w) Suctoria, *Dendrocometes paradoxus* (after Grell, 1973).

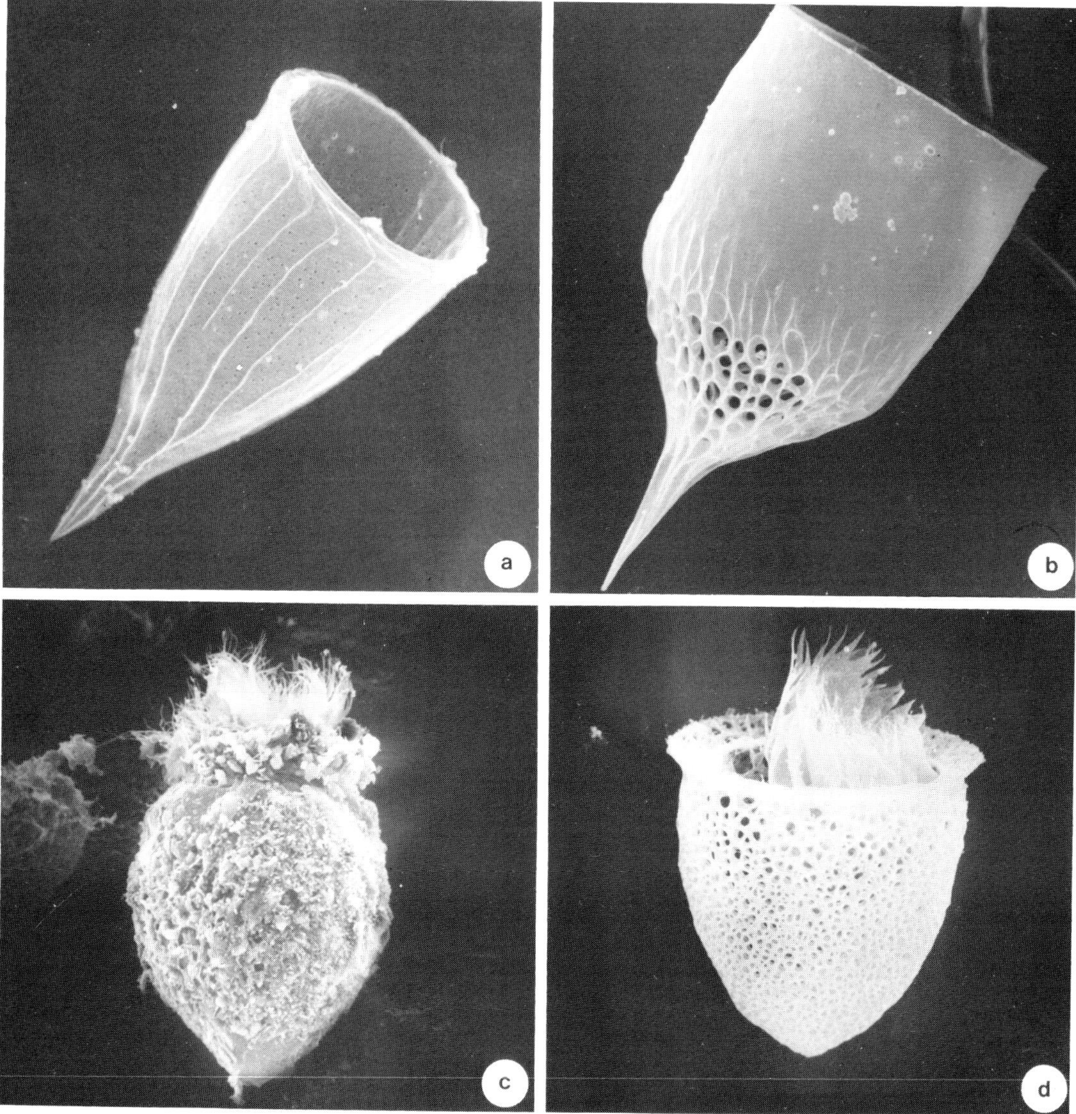

FIGURE 1.17 Ciliophora, class Spirotrichea, order Choreotrichida, suborder Tintinnina. (a) Hyaline lorica of *Rhabdonella armor*; ~ ×1,125. (b) Reticulate lorica of *Epiplocylis pacifica*; ~ ×875. (c) A tintinnid with agglutinated materials on its lorica; ~ ×1,000. (d) Cilia protruding from the reticulate lorica of *Dictyocysta* sp.; ~ ×1,000. (a and b by Dr. Kenneth Gold; c and d by Dr. Eugene Small. All figures, SEM micrographs.)

graphic markers. Tintinnines have reduced body ciliature and extensive and conspicuous apically located buccal paramembranelles. Their lorica are tubular and cup- or goblet-shaped. The loricae of several species have been studied under both TEM and SEM microscopes. They are built of three series of fibers: (1) a set running longitudinally, (2) a set of helicoidally wound fibers, and (3) a layer of transverse bundles. The exterior of the lorica of many species includes agglutinated foreign materials, commonly quartz grains, coccoliths from coccolithophorids, or diatom frustules. The loricae of species without agglutinated materials can be delicate and membranous or relatively firm. The exterior is smooth in some species and ornamented in others. There is quite a bit of morphological variation in the structure of the lorica collar and in the form of the aboral end. At the present time the characteristics of the lorica are used in the taxonomy of the group. Tintinnines are only partially sampled by the continuous plankton recorder and some other plankton nets so that we have only a fragmentary picture of their distribution patterns in the oceans (e.g., Lindley, 1975; Chap. 3 of this book). Data suggest that more species are present in the neritic and warm regions of the North Atlantic than in the colder

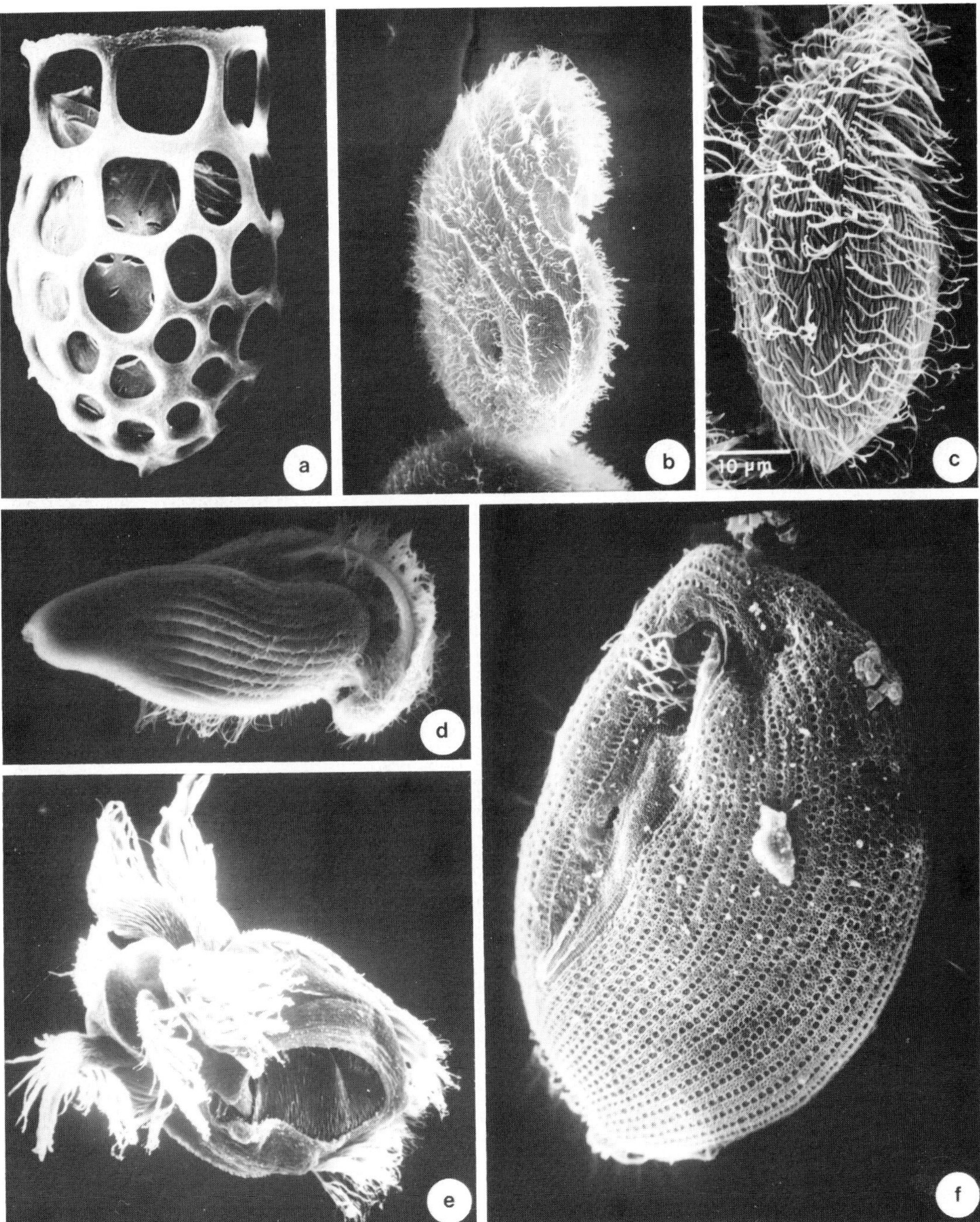

FIGURE 1.18 Ciliates belonging to various classes. (a) Lorica of the tintinnid ciliate *Dictyocysta* sp.; ~ ×1,800. (b) *Lechriopyla mastax* from the gut of the sea urchin, *Strongylocentrotus purpuratus*; ~ ×600. (c) *Cyclidium stercolis*; ~ ×1,660. (d) *Metopus circumlabens* from the gut of the sea urchin *Strongylocentrotus purpuratus*; ~ ×600. (e) *Uronychia* sp.; ~ ×800; (f) The ventral surface of a deciliated *Pleurocoptes furgasoni*. ~ ×2,000. (All figures SEM micrographs by Dr. Eugene Small.)

northwestern areas of the Atlantic. Bimodal seasonal cycles have been found in warmer waters. Tintinnines often reach amazingly high densities in the range of 10^4/L in the water column. Some tintinnines have been successfully cultured in the laboratory, giving hope that much more will be learned of the biology of these important ciliates. In culture, generation times for *Tintinnopsis beroidea*, *T. acuminata*, *Helicostomella subulata*, and *Eutintinnus pectinis* were less than a day. The latter two species had doubling times of only 12 hours under ideal conditions. Growth rates appear to be slower under more natural food conditions and in the field. Feeding rates at high densities of food are as much as 10–20 percent of the ciliates' body weight per hour. With the growth efficiencies attributed to them in one series of laboratory experiments exceeding 50 percent, these animals could at times exercise control over nanoplanktonic primary producers and perhaps bacterioplankton in the sea.

The scuticociliates (Scuticociliatida, an order of Oligohymenophorea) are particularly abundant in marine habitats (e.g., Fig. 1.16b). Their body ciliature is generally uniform and they may have one or more caudal cilia. Some species have thigmotactic cilia. The buccal ciliature consists of a dominating paroral membrane and several (usually three) membranelles. In one very common and well-known family, the Pleuronematidae, the paroral membrane is developed into a saillike velum stretching from the anterior end of the body along the right side of the buccal cavity and curling around the cytostome in a large semicircle. These forms spiral as they swim through the water column, creating a funnel-shaped vortex that drives bacteria and nanoplankton into their buccal cavity. One marine scuticociliate species, *Uronema marinum*, has been cultured in axenic culture in a chemically defined medium consisting of 17 amino acids and nucleotides, five fatty acids, stigmasterol, thiamine, riboflavin, pyridoxal, nicotinamide, folic acid, and panthothenate (Hanna and Lilly, 1974). These complex nutritional requirements are apparently translated into differences in growth rates and ecological efficiencies when the organism forages in nature. This point was tested in the laboratory. In gnotobiotic cultures logarithmically growing *Uronema marinum* had generation times as short as 2.7 hours on diets of some chlorophytes, and as long as 7.9 hours on other test diets. Ecological growth efficiency was as high as 21 percent on the former diet and as low as 6 percent on the latter (Rubin and Lee, 1976).

Besides being very common in the water column, scuticociliates abound in intertidal sands and are frequently found as endosymbionts in many marine invertebrates including bivalve mollusks, echinoids, crinoids, echiuroids, polychaetes, crustaceans, and sipunculids. Many different feeding types are found among the group. In addition to bacterivorous and algivorous types, carnivorous, histophagous, and omnivorous types have been described.

Littoral benthic marine habitats have been the favorite hunting grounds of many very distinguished marine protozoologists (e.g., Kahl, Fauré-Fremiet, Tuffrau, Dragesco, Bock, Fenchel, Deroux, Borror, and Small). Because of their interest we know much about the ciliates from these habitats. Ciliates are always abundant in sand ($\sim 5 \times 10^6$–$10^7/m^2$) or other types of benthic sediments if there are abundant bacteria, flagellates, or algae. More than 400 species have been described in sandy interstitial habitats alone. Many of the interstitial ciliates are elongated and ribbonlike (e.g., *Geleia gigas, Remanella caudata, Kentrophoros longissimus, Tracheloraphis* spp. (e.g., Fig. 1.16a,s).

Benthic microfaunal communities are dominated by herbivorous ciliates. In the summer and in shallow waters the most numerous species are those that graze on diatoms, phytoflagellates, chlorophytes, other small algae, and cyanobacteria. In colder months and in deeper water bacterivorous species are more common. Carnivorous ciliates, those that mainly prey on other ciliates, constitute about 10 percent of the microfauna.

The work of Fenchel (e.g. 1967, 1968a,b, 1987) have given us much insight into the structure and function of the benthic ecosystem and the role of ciliates in it. Ciliates are the dominant animal group in fine sands (125–250 μm), exceeding even meiofauna in biomass. Fenchel's work suggests that the ciliate fauna of sublittoral sands shows a characteristic vertical zonation. One group of diatomivorous and bacterivorous species (*Strombidium sauerbreyi, S. latum, Chlamydodon triquestrus, Chilodontopsis elongata, Lynchella gradata, Discotricha papillifera, Frontonia arenaria, Pleuronema coronatum, Conchostoma longissimum*) is more or less restricted to the oxidized surface layers. A very numerous second group of characteristic ciliates (e.g., *Stomella verminforme, Aspidisca* spp.) (Fig. 1.16p) lives in the vicinity of the redox discontinuity layer several centimeters below the surface. A third group (*Trachelorophis* spp., *Cardiostomella vermiforme, Kentrophoros* spp., *Ramanella* spp., *Geleia* spp., *Homalozoon caudatum*, and *Coleps* spp.) lives in reduced, anaerobic, and sulfide-containing layers. Strictly confined to H_2S-containing sediments are *Caenomorpha levandi, Saprodiniuim halophila, Myelostoma biartitum*, and *Parablepharisma clamydorpherum*. In an estuarine sand microbiocenose (sands rich in organic detritus found in shallow bays and lagoons) the redox discontinuity layer is situated close to the surface or is at the surface. The species that characterize estuarine microbiocenoses (*Tracheloraphis kahli, Chlamydodon* spp., *Strombidium* spp., (Fig. 1.16o), *Uronema* spp., *Sonderia* spp., *Peritromus faurei, Frontonia* spp., *Aspidisca* spp.) feed on certain bacterial and cyanobacterial species rarely found in cleaner sands.

A sulfuretum microbiocenose is found in association with large accumulations of drifted seaweeds, cord grasses, and so on, decaying in lentic places. In sulfureta, reducing properties are found at all sediment levels and in the overlying seawater. The ciliates found here (e.g., *Prorodon* spp., *Plagiopogon* spp., *Caenomorpha* spp., *Metopus* spp., *Parablepharisma* spp., *Gruberia sp., Mesodinium* spp., (Fig. 1.16k), *Litonotus lamella*) are species that eat the various types of chemolithotrophic, photolithotrophic, and photoorganotrophic bacteria likely to be found in the sulfureta.

Large numbers of sedentary ciliates occur in various marine habitats (e.g., Fig. 1.16 i,m,u). These animals belong mainly to the Chonotrichida, Suctoria, and Peritrichia. Chonotrichida are vase-shaped ectocommensals on marine crustaceans. Common genera of chonotrichs include *Chilodochona, Stylochona,* and *Trichochona*. Peritrichous ciliates, particularly vorticellids such as *Vorticella* and *Zoothamnium,* are found in a variety of marine habitats, including plant and animal surfaces wherever gentle currents carry potential food. In some cases they are the beneficiaries of "fallout" from the feeding or respiratory currents of the animals to which they are attached. Colonies of *Zoothamnium* spp. and *Epistylis* are commonly associated with the carapace, uropods, and pleopods of decapods and copepod crustaceans. Some stalkless peritrichs (e.g., *Lagenophrys*) have low-profile limpet shapes, which produce less turbulence and drag in the water passing through the gills and appendages to which they are attached. Suctoria are also epizoic on a variety of marine invertebrates and also attach to seaweeds. Various species of *Ephelota,* a genus with stout striated stalks, for instance, are common on hydroids and bryozoans. Many species capture and feed on other ciliates; others feed on bacteria, algal zoospores, and small flagellates. One of the best-known species is *Acineta tuberosa*. German workers have studied laboratory cultures of this suctorian feeding on the ciliate *Strombidium* in order to learn more about the mechanisms of suctorian feeding and their life cycles. Many genera (e.g., *Trichophyra, Lernaeophyra,* and *Dendrosomides*) form endogenous buds that lead to the formation of small colonies on the surfaces to which they are attached. Some genera (e.g., *Acineta, Thecacineta* and *Acinetopsis*) form loricae from which only part of the body or the tentacles extend.

Certain genera of ciliates are well known as fish parasites. *Trichodina* (Fig. 1.17d) and *Trichodinella* are ectoparasites of the gills of marine fishes. They either glide along the gill epithelium or attach to epithelial cells by means of an adhesive disc equipped with complex denticles having hooks and/or spikes. There is at least one report that the gill epithelium can be completely destroyed by a heavy invasion of *Trichodina*. Although freshwater species invade invertebrates, marine trichodinids are transmitted directly.

Another ciliate, *Cryptocaryon irritants,* ubiquitous in all seas, causes white spot disease in marine fishes. Its structure and life cycle are quite similar to those of *Ichthyophthirius,* the common and better-known ciliate parasite that causes "Ich," a disease that plagues freshwater aquarists. The surfaces of all infected fish are covered with whitish or gray pustules which are vesicles or pockets of ciliates burrowing under the epidermis. They feed on the host's cells and undermine the epithelium. Lesions accompanied by the host's mucus production often become secondarily infected. Epizootics caused by *Cryptocaryon* can affect practically all marine teleosts under aquarium conditions.

One of the most interesting parasitic groups are the apostomatid (Apostomatida) ciliates. As their name suggests, they have no cytostomes in certain stages. Typically they have a rosette in the oral area and complex polymorphic life cycles. Most commonly they are found in hermit crabs, shrimps, and copepods, with sea anemones as alternating hosts. Some are found in polychaete annelids, cephalopods, ctenophores, isopods, amphipods, and decapods. A new species of apostome (*Collinia beringensis*) has recently been found in the haemocoel cavity of the euphausiid *Thysanoessa inermis* (Capriulo and Small, 1986). The classical life cycle of the group consists of (1) a trophont (vegetative phase), (2) a protomont (phase during which the nucleus and body ciliature reorganize), (3) a tomont (encysted stage in which the animals divide into a great many small ciliates), (4) a protomite (stage in which the ciliature undergoes torsion), (5) a tomite (a free-swimming but nonfeeding infective stage), and (6) a phoront (an encysted stage formed after the tomite becomes attached to the host and which undergoes morphogeneisis to form the trophont). Some of the best-known apostomes are *Gymnodinioides calkinsi* from the gills and moult of the grass shrimp *Palaeomonete, Foettingeria actiniarum* which has phoronts on various copepods, ostracods, amphipods, isopods and decapods, and trophonts in anemones such as *Anemonia, Metridium, Sagartia, Astrangia* and *Chromidina,* and *Opalinopsis,* which are found in the liver, kidneys and gonads of species of squid and octopus.

The ciliate *Mesodinium rubrum,* a planktonic gymnostome associated with "red water," has attracted a great deal of attention in recent years. Blooms of it, extending as large as 100 mi^2, have been reported in neritic locations such as bays and fjords and in upwelling areas. The ciliate, which is relatively small (10–70 μm; ~ 30–50 μm in blooms), has a wide distribution in the seas. It has been found in the polar waters of the David strait and the Kara-Barents sea area, temperate localities off New Zealand; Peru; Vancouver, B.C.; California;

Maine; Long Island; North Sea; South Africa; and the equatorial waters off Equador. Densities of ciliates in strongly colored red waters reach as high as 2,000 cells/ml.

Mesodinium has an atypical cytopharynx reduced body ciliature and bifurcated oral tentacles. The cilia, which are united into cirri, arise in a broad equatorial ciliary belt around the cell. The unusual feature of the ciliate is that each one contains cryptomonad endosymbiont remnants with functioning mitochondria and chloroplasts. Details of the fine structure of populations of different blooms seem to indicate considerable variability in symbiont and host structure (Hibberd, 1977; Lindholm et al., 1987; Oakley and Taylor, 1978). Perhaps a species complex is involved or several species have been given one name. The outer envelope of the alga is quite reduced since only a single-unit membrane, or two in one population studied (Lindholm et al., 1987), separate the cryptomonad endosymbiont from the cytoplasm of the host. Experimental studies suggest that the ciliate-endosymbiont combination behave autotrophically. All aspects of this unique photosynthetic ciliate and its symbiont, including morphology, ultrastructure, physiology, ecology, and evolution, have recently been reviewed by Lindholm (1985).

The best single source of information on the ciliates is Corliss' (1979) book. It contains descriptions and illustrations of the characteristics of various ciliates, their classification, and a guide to the literature. The monographs of Campbell (1942, 1954); Kofoid and Campbell (1929, 1939); Jorgensen (1924); Tappan and Leoblich (1968); and the papers of Gold (1968, 1973, 1975, 1976), Capriulo (1982), Capriulo and Carpenter (1983), Capriulo et al. (1982, 1986), Capriulo and Ninivaggi (1982), Verity (1985, 1987), Verity and Langdon (1984), Verity and Villareal (1986), Stoecker et al. (1981, 1983), Spittler (1973), Rassoulzadegan (1978), Rassoulzadegan and Etienne (1981), and Heinbokel (1978a,b), are good sources of information on the tintinnids. The monographic papers of Kahl (1928–1935); Fauré-Fremiet (1924); Fenchel (1967; 1968a,b), Dragesco (1960; 1963a,b; 1965) and the book by Fenchel (1987) contain a wealth of information for those beginning studies on benthic and interstitial ciliates. The ecology of ciliates in a northeastern U.S. salt marsh is discussed by Borror (1972). He has also written an illustrated key, annotated systematic list, and bibliography (Borror 1973) which covers the ciliates of coastal and estuarine waters of New England. Lom and Laird (1969) and Lom (1970) are excellent sources of information on parasitic ciliates of marine fishes. Much information about the ciliates living on or attached to the surfaces, gills, or body cavities of various marine invertebrates can be found in the papers of Fenchel (1965); Chatton and Lwoff (1949, 1950); Kahl (1934a,b); Kozloff (1946, 1966); Raabe and Raabe (1959); Beers (1961, 1966); Lynn and Berger (1973); Powers (1935); Sprague and Couch (1971); Stein (1967). The classical paper of Chatton and Lwoff (1935) and the monograph by Lwoff (1950) are still excellent starting points for more information on the Apostomatida. Bradbury (1966a,b, 1973, 1974) and Bradbury and Trager (1967a,b) provide some up-to-date fine structural and morphogenetic information. A review of the distribution and fine structure of *Mesodinium rubrum* by Taylor and co-workers (1971), updated by Hibberd (1977), Lindholm (1985), and Lindholm et al. (1987), provides information on this very interesting organism. The chonotrichs are the subject of excellent monographs by Jankowski (1973) and Kahl (1935). Grell's (1973) textbook has an excellent account of the ultrastructural aspects of suctorian feeding. Kudo's (1954) text, Kahl's (1934b) monograph, and Borror's (1973) key are good places to start looking for information on suctoria. Many other interesting papers on individual species or groups are cited in Corliss (1979). Many papers have been written on marine peritrichs. Earlier observations are reviewed in Kahl (1935); newer papers are cited in Corliss (1979).

ACKNOWLEDGMENTS

The authors are grateful to the following colleagues for their critical review of this chapter: Jean and Monique Cachon, John O. Corliss, Norman E. Levine, Thomas K. Sawyer, and Earl Weidner. James Atz, O. Roger Anderson, the late Allen Bé, Kenneth Gold, Ann Scarborough and Arthur Repak, Thomas Sawyer, and Eugene Small generously supplied various photographs.

REFERENCES

Allen, R.D. (1961a) In Brachet J. and Mirsky A. (eds.). *The Cell*, Vol. 2, New York: Academic Press, pp. 135–216.

Allen, R.D. (1961b) A new theory of amoeboid movement and protoplasmic streaming. *Exp. Cell Res.* S8:17–31.

Anderson, O.R. (1976) A cytoplasmic fine-structure study of two spumellarian radiolaria and their symbionts. *Mar. Micropaleontol.* 1:81–89.

Anderson, O.R. (1977) Cytoplasmic fine structure of nassellarian radiolaria. *Mar. Micropaleontol.* 2:251–264.

Anderson, O.R. (1978) light and electron microscopic observations of feeding behavior, nutrition, and respiration in laboratory cultures of *Thalassicolla nucleata*. *Tiss. Cell. 10:401–412.*

Anderson, O.R. (1983) *Radiolaria*. Springer-Verlag, 355 pp.

Anderson, O.R. (1988) *Comparative Protozoology, Ecology, Physiology, Life History*. New York: Springer-Verlag, 482 pp.

Arnold, Z. (1954) *Discorinopsis aguayoi* Bermudez and *Discorinopsis vadescens* Cushman and Bronniman: a study of variation in cultures of living Foraminifers. *Cushman Found. Foram. Res. Contr.* 5:4–13.

Bamforth, S.S. (1982) The variety of artificial substrates used for microfauna. In Cairns, J., Jr. (ed.). *Artificial Substrates*. Ann Arbor, Mich.: Ann Arbor Science Publishers, pp. 115–130.

Batisse, A. (1969) Acinétiens nouveaux ou mal connus des côtes méditerranéennes françaises. I. *Ophryodendron hollandei* n. sp. (Suctorida, Ophryodendridae). *Vie Milieu* 20:251–277.

Bé, A.W.H. (1977) An ecological, zoogeographic and taxonomic review of recent planktonic foraminifera. In Ramsay, A.T.S. (ed.). *Oceanic Micropaleontology*. London: Academic Press, pp. 1–88.

Becker, C.D. (1970a) In *A Symposium on Diseases of Fishes and Shellfishes*, Amer. Fish. Soc. Spec. Publ. No. 5, pp. 82–100.

Becker, C.D. (1970b) In *Parasitic Protozoa* Vol. 1. Kreier, J.P., ed. New York: Academic Press, pp. 357–416.

Beers, C.D. (1961) The obligate commensal ciliates of *Strongylocentrotus drobachiensis*: occurrence and division in urchins of diverse ages; survival in seawater in relation to infectivity. *Biol. Bull.* 121:69–81.

Beers, C.D. (1964) *Urceoleria spinicola* n. sp., an Epizoic ciliate (Peritrichida, Mobilina) of sea-urchin spines and pedicellariae. *J. Protozool.* 11:430–435.

Beers, C.D. (1966) Distribution of *Urceolaria spinicola* (Ciliata, Peritrichida) on the spines of the sea urchin *Strongylocentrotus droebachiensis*. *Biol. Bull.* 131:219–229.

Berger, W.H., and Soutar, A. (1967) Planktonic Foraminifera: field experiment on production rate. *Science* 1561:1495–1497.

Bissonnette, T.H. (1930) A method of securing marine invertebrates. *Science* 71:464–465.

Borror, A.C. (1965) Morphology and ecology of some uncommon ciliates from Alligator Harbor, Florida. *Trans. Amer. Microsc. Soc.* 84:550–565.

Borror, A.C. (1972) Tidal marsh ciliates (Protozoa): morphology, ecology, systematics. *Acta Protozool.* 10:29–71.

Borror, A.C. (1973) Marine flora and fauna of the Northeastern United States. Protozoa: Ciliophora NOAA Tech. Rep. NMFS Circ. No. 378, 62 pp.

Bovee, E. (1972) The lobose amebas. 4. A key to the order Granulopodida Bovee and Jahn, 1966, and descriptions of some new and little known species in the order. *Arch. Protistenk* 114:371–403.

Bovee, E., and Sawyer, T.K. (1979) Marine flora and fauna of the northeastern United States. Protozoa: Sarcodina:Amoebae. *Natl. Mar. Fish. Serv. Circ.* 419:1–57.

Boucaud-Camou, E. (1966) Les choanoflagelles des côtes de la Manche: I. Systématique. *Bull. Soc. Linn. Normandie*, Ser. 10, 7:191–209.

Bowman, M.J., and Esaias, W.E. (1978) *Oceanic Fronts and Coastal Processes*. Springer-Verlag, Berlin.

Bowser, S.S., McGee-Russell, S.M., and Rieder, C.L. (1985) Digestion of prey in foraminifera is not anomalous: a correlation of light microscopic, cytochemical, and HVEM technics to study phagotrophy in two allogromids. *Tiss. Cell.* 17:823–839.

Bradbury, P.C. (1966a) The life cycle and morphology of the Apostomatous ciliate, *Hyalophysa chattoni*. n.g., n. sp. *J. Protozool.* 13:209–225.

Bradbury, P.C. (1966b) The fine structure of the mature tomite of *Hyalophysa chattoni*. *J. Protozool.* 13:591–607.

Bradbury, P.C. (1973) The fine structure of the cytostome of the apostomatous ciliate *Hyalophysa chattoni*. *J. Protozool.* 20:405–414.

Bradbury, P.C. (1974) The fine structure of the apostomatous ciliate, *Hyalophysa chattoni*. *J. Protozool.* 21:112–120.

Bradbury, P.C., and Trager, W. (1967a) Excystation of apostome ciliates in relation to molting of the crustacean hosts. II. Effect of glycogen. *Biol. Bull.* 133:310–316.

Bradbury, P.C. (1967b) Metamorphosis from the phoront to the trophont in *Hyalophysa*. *J. Protozool.* 14:307–312.

Bulla, L.A., Jr., and Cheng, T.C., eds. (1976) *Biology of the Microsporidia*. New York: Plenum.

Bulla, L.A., Jr., and Cheng, T.C., eds. (1977) *Systematics of the Microsporidia*. New York: Plenum.

Cachon-Enjumet, M. (1961) Contribution à l 'étude des Radiolaires Phaeodaries. *Arch. Zool. Exp. Gen.* 100(3):151–237.

Cachon, J., and Cachon, M. (1971) Origin, organisation 115 relation with the other cell organelles, general considerations on the macromolecular organisation of the stereoplasm in actinopods. *Arch. Protistenk.* 113:80–97.

Cachon, J. (1972) The axopodial system of Radiolaria Sphaeroidae I. Centroaxoplastididae. *Arch. Protistenk.* 114:51–64.

Cachon, J. (1976a) The axopodial system of *Collodoria* (Polycystin Radiolaria) I. The exo-axoplastidiata. *Arch. Protistenk.* 118:227–234.

Cachon, J. (1976b) The axopods of Radiolaria in their free and ectoplasmic part: structure and function. *Arch. Protistenk.* 118:310–320.

Cachon, J. (1977) The axopodial system of Collodariae (Polycystin Radiolaria) 2. *Thalassolampe margarodes* Haeckel. *Arch. Protistenk.* 119:401–406.

Cachon, J. (1978) Infrastructure constitution of the miorotuali of the axopodial system of Radiolaria. *Arch. Protistenk.* 120:229–231.

Cachon, J., and Cachon, M. (1985) I. Class Acantharea Haeckel, 1881. In Lee, J.J., Hutner, S.H. and Bovee, E.C. (eds.) *An Illustrated Guide to the Protozoa*. Soc. of Protozool. and Allen Press, Lawrence, Kansas.

Cairns, J., Jr., ed. (1982) *Artificial Substrates*. Ann Arbor, Mich. Ann Arbor Science Publishers. 279 pp.

Cairns, J. Jr., Dahlberg, M.L., Dickson, K.L., Smith, N., and Waller, W.T. (1969) The relationship of freshwater protozoan communities to the Macarthur–Wilson equilibrium model. *Amer. Nat.* 103:439–454.

Cairns, J., Jr., Yongue, W.H., and Boatin, Jr., H. (1973) The protozoan colinization of polyurethane foam units anchored in the benthic area of Douglas Lake, Michigan. *Trans. Amer. Microsc. Soc.* 92:648–656.

Campbell, A.S. (1942) Scientific results of cruise VII of the Carnegie during 1928–1929 under command of Captain J.P. Ault. Biology. II. The oceanic tintinnoina of plankton gathered during the last cruise of the Carnegie. Carnegie Inst. Wash. Pub. 537, pp. 1–163.

Campbell, A.S. (1954) Tintinnina. In Moore, R.C. (ed.), *Treatise of Invertebrate Paleontology,* Part D. Protista 3, Lawrence: University of Kansas Press, pp. 166–180.

Capriulo, G.M. (1982) Feeding of field collected tintinnid microzooplankton on natural food. *Mar. Biol.* 71:73–86.

Capriulo, G.M., and Carpenter, E.J. (1983) Abundance, species composition and feeding impact of tintinnid microzooplankton in Central Long Island Sound. *Mar. Ecol. Prog. Ser.* 10:277–288.

Capriulo, G.M., Gold, K., and Okubo, A. (1982). Evolution of the lorica in tintinnids: a possible selective advantage. *Ann Inst. oceanoqr.* Paris, 58(s):319–324.

Capriulo, G.M., and Ninivaggi, D.V. (1982). A comparison of the feeding activities of field collected tintinnids and copepods fed identical natural particle assemblages. *Ann. Inst. oceanoqur.* Paris, 58(s):325–334.

Capriulo, G.M., and Small, E.B. (1986). Discovery of an apostome ciliate (*Collinia beringensis* n. sp.) endoparasitic in the Bering Sea euphausiid *Thysanoessa inermis. Dis. Aquat. Org.* 1:141–146.

Capriulo, G.M., Taveras, J., and Gold, K. (1986) Ciliate feeding: effect of food presence or absence on occurrence of striae in tintinnids. *Mar. Ecol. Prog. Ser.* 30:145–158.

Chatton, E., and Lwoff, A. (1935) Les Ciliés apostomes. Morphologie Cytologie, éthologie, évolution, systématique. Première partie. Aperçu historique et général. Etude monographique des genres et des espèces. *Arch. Zool. Exp. Gen.* 77:1–453.

Chatton, E., and Lwoff, A. (1949) Recherches sur les ciliés thigmotriches. I. *Arch. Zool. Exp. Gen.* 86:169–253.

Chatton, E., and Lwoff, A. (1950) Recherches sur les ciliés thigmotriches. II. *Arch. Zool. Exp. Gen.* 86:393–485.

Chen, T., ed. (1967) *Research in Protozoology,* Vol. 1. London: Pergamon Press.

Chen, T. (1972) *Research in Protozoology,* Vol. 4. London: Pergamon Press.

Corliss, J. (1961) *The Ciliated Protozoa.* Oxford: Pergamon Press.

Corliss, J. (1979) *The Ciliated Protozoa,* 2nd ed. Elmsford, N.Y.: Pergamon Press.

Couch, J.A. (1978) Diseases, parasites, and toxic responses of commercial panaeid shrimps of the Gulf of Mexico and South Atlantic coasts of North America. *Fishery Bull.* 76:1–44.

Cushman, J.A. (1959) Foramifera: Their Classification and Economic Use, 5th ed. Cambridge, Mass.: Harvard University Press.

Das, S.M. (1947) The biology of two species of Folliculinidae (Ciliata, Heterotrichia) found at Cullercoats, with a note on the British species of the family. *Proc. Zool. Soc.* 117:441–456.

DeFlandre, G., and Grassé, P. (1953) Sous-embranchement des Actinopodes. Generalités. In Grasse, P. (ed.), *Traité de Zoologie,* Tôme 1, Fascicule 2, Paris: Masson et Cie. pp. 267–268.

Dragesco, J. (1960) Ciliés mesopsammiques littoraux. Systématique morphologie, écologie. *Trav. Stat. Biol. Roscoff.* (N.S.) 12:49–58.

Dragesco, J. (1962) L'orientation actuelle de la systématique des ciliés et la technique d'impregnation au protéinate d'argent. *Bull. Microsc. Appl.* (ser. 2).

Dragesco, J. (1963a) Compléments à la Connaissance des Ciliés Mesopsammiques des Roscoff. *Cah. Biol. Mar.* 4:91–119.

Dragesco, J. (1963b) Compléments à la Connaissance des Ciliés Mesopsammiques des Roscoff. *Cah. Biol. Mar.* 4:251–275.

Dragesco, J. (1965) Ciliés Mesopsammiques d'Afrique Noire. *Cah. Biol. Mar.* 6:357–399.

Dussart, B.M. (1965) Les différentes catégories de planction. *Hydrobiologia* 26:72–74.

Ellis, B., and Messina, A. (1940) *Catalogue of Foraminifera.* New York: American Museum of Natural History.

Farley, C.A. (1975) Epizootic and enzootic aspects of *Minchinio nelson:* (Haplosporida) disease in Maryland oysters. *J. Protozool.* 22:418–427.

Fauré-Fremiet, E. (1924) Contribution à la Connaissance des infusoires planktoniques. *Bull. Biol. Fr. Belg.* S6:1–171.

Fauré-Fremiet, E. (1950) Écologie des Ciliés Psammophiles Littoraux. *Bull. Biol. Fr. Belg.* 84:35–75.

Fauré-Fremiet, E. (1951) The marine sand-dwelling ciliates of Cape Cod. *Biol. Bull.* 100:59–70.

Febvre-Chevalier, C. (1985) IV. Class Heliozoea Haeckel 1866. In Lee, J.J., Hutner, S.H., and Bovee, E.C. An illustrated Guide to the Protozoa. Soc. of Protozool. and Allen Press, Lawrence Kansas pp. 302–317.

Febvre, J. (1974) Relations morphologiques entre les constituants de l'envelope, les myonèmes, le squelette et le plasmalemme chez les arthracantha shew (Acantharia). *Protistologica* 10:141–158.

Febvre, J. (1977) Etude ultrastructurale et position systématique d'une Zooxanthelle (Haptophycee) symrionte d'Acanthaire. *Biol. Cell.* 29:33a.

Febvre, J., and Febvre-Chevalier, C. (1979) Ultrastructural study of zooxanthellae of three species of Acantharia (Protozoa: Actinopoda), with details of their taxonomic position in the Prymnesiales (Prymnasiophyceae, Hibberd, 1976). *J. Mar. Biol. Ass. U.K.* 59:215–226.

Fenchel, T. (1965a) Ciliates from Scandinavian mulluses. *Ophelia* 2:71–174.

Fenchel, T. (1965b) On the ciliate fauna associated with the marine species of the amphipod genus *Gammarus* J.G. Fabricius. *Ophelia* 2:281–303.

Fenchel, T. (1967) The ecology of marine microbenthos. I. The quantitative importance of ciliates as compared to metazoans in various types of sediments. *Ophelia* 4:121–137.

Fenchel, T. (1968a) The ecology of marine microbenthos. II. The food of marine benthic ciliates. *Ophelia* 5:73–121.

Fenchel, T. (1968b) The ecology of marine microbenthos. III. The reproductive potential of ciliates.*Ophelia* 5:123–136.

Fenchel, T. (1987) *Ecology of Protozoa: The Biology of Free-living Phagotrophic Protists.* New York: Science Tech/Springer-Verlag, 197 pp.

Galloway, J. (1933) *A Manual of Foraminifera.* Bloomington, Ind.: University of Indiana Press.

Gifford, D.J. (1985) Laboratory culture of marine planktonic oligotrichs (Ciliophora, Oligotrichida). *Mar. Ecol. Prog. Ser.* 23:257–267.

Gold, K. (1968) Some observations on the biology of *Tintinnopsis* sp. *J. Protozool.* 15:193–194.

Gold, K. (1973) Methods for growing Tintinnida in continuous culture. *Amer. Zool.* 13:203–208.

Gold, K. (1979) Scanning electron microscopy of *Tintinnopsis parva* sp. *J. Protozool.* 26:415–419.

Gold, K., and Morales, E.A. (1975) Seasonal changes in the lorica sizes and the species of Tintinnida in the New York Bight. *J. Protozool.* 22:520–528.

Gold, K., and Morales, E.A. (1976) Studies on the sizes, shapes, and the development of the lorica of agglutinated tintinnida. *Biol. Bull.* 150:377–392.

Gowing, M.M., and Silver, M.W. (1983) *Mar. Biol.* 73:7–16.

Grassé, P. (1952) *Traité de Zoologie. I. Premier Fascicule Phylogénie Protozaires.* Paris: Masson et Cie.

Grassé, P. (1952) *Traité de Zoologie I. Fascicule II. Protozaires.* Paris: Masson et Cie.

Grell, K.G. (1953) Der Stand unserer Kenntnisse uber den Bau der Protistenkerne. *Verh. Deutsch. Zool. Ges.* 89(1952):212–251.

Grell, K.G. (1954) Der Generationswechsel der polythalamen Foraminifere *Rotaliella heterocaryotica. Arch. Protistenk.* 100:268–286.

Grell, K.G. (1967) Sexual reproduction in protozoa. In Chen, T.- T. (ed.), *Research in Protozoology,* Vol. 2. Oxford: Pergamon Press, pp. 147–213.

Grell, K. (1968) *Protozoology.* Berlin: Springer-Verlag.

Grell, K. (1973) *Protozoology.* New York: Springer-Verlag.

Guilcher, Y. (1950) Morphogenèse et morphologie comparée chez les ciliés gemmipares: chonotriches et tentaculiferes. *Ann. Biol.* 26:465–478.

Guilcher, Y. (1951) Sur quelques Acinethens nouveaux ectoparasites des Copepods Harpacticoides. *Arch. Zool. Exp. Gen.* 87:24–30.

Haeckel, E. (1887) Report on radiolaria collected by H.M.S. Challenger during the years 1873–1876. In Thompson, C.W., and Murray, J. (eds.), *The Voyage of H.M.S. Challenger,* Vol. 18. pp. 1–1760, London: Her Majesty's Stationery Office.

Hall, R.P. (1953)*Protozoology.* Englewood Cliffs, N.J.: Prentice-Hall.

Hammann, I. (1952) Okologische und biologische Untersuchwungen an Susswasserperitrichen. *Arch. Hydrobiol.* 47:177–228.

Hanna, B.A., and Lilly, D.M. (1974) Axenic culture of *Uronema maslnum. Amer. Zool.* 10:539–540.

Hanna, B.A. (1974) Growth of *Uronema marinum* in chemically defined medium. *Mar. Biol.* 26:153–160.

Hays, J.D. (1965) Radiolaria and late Tertiary and Quaternary history of Antarctic Seas. In Llano, G.A. (ed.), *Biology of Antarctic Seas II.* Amer. Geophys. Union, Antarct. Res. Ser. 5:125–184.

Hays, J.D., and Opdyke, N. (1967) Antarctic radiolaria, magnetic reversals, and climatic change. *Science* 158:1001–1011.

Heckmann, K. (1963) Paarungssystem und genabhangige Paarungstypdifferenzierung bei dem hypotrichen ciliaten *Euplotes vannus.* O.F. Muller. *Arch. Protistenk.* 106:393–421.

Heinbokel, J.F. (1978a) Studies on the functional role of tintinnids in the Southern California Bight. I. Grazing and growth rates in laboratory cultures. *Mar. Biol.* 47:177–189.

Heinbokel, J.F. (1978b) Studies on the functional role of tintinnids in the Southern California Bight. II. Grazing rates of field populations *Mar. Biol.* 47:191–197.

Hentschel, E. (1925) Alwasserbiologie in *Abderhalden: Hard. Deutsch. Biol. Arbeitsmeth IX*(2):233–280.

Hibberd, D.J. (1975) Observations on the ultrastructure of the choanoflagellate *Codosiga botrytis* (Ehr) Saville-Kent with special reference to the flagellar apparatus. *J. Cell Sci.* 17:191–219.

Hibberd, D.T. (1977) Observations on the ultrastructure of the cryptomonad endosymphont of the red-water ciliate *Mesodinium rubrum. J. Mar. Biol. Ass. U.K.* 57:45–61.

Hibberd, D.T. (1986) Ultrastructure of the Chrysophyceae–phylogenetic implications and taxonomy In Kristiansen, J., and Anderson, R.A. (eds.), *Chrysophytes: Aspects and Problems.* New York: Cambridge University Press.

Hollande, A., and Enjumet, M. (1957) Enkystement et reproduction isosporogénétigue chez les Acanthaires. *C.R. Acad. Sci.* 244:508–510.

Hollande, A., and Enjumet, M. (1960) Cytologie, évolution et systématique des Sphaeroides (Radiolaires). *Arch. Mus. Nat. Hist. Natur.* (7e Serie) 7:1–134.

Hollande, A., Cachon, M., and Valentin, J. (1967) Infrastructure des axopodes et organisation générale de *Sticholonche zanclea.* Hertwig (Radiolaire Sticholonchidea). *Protistologica* 3:155–166.

Hollande, A., Cachon, J., and Cachon-Enjumet, M. (1970) La signification de la membrane capsulaire des Radiolaires et ses rapports avec le plasmalemme et les membranes du reticulum endoplasmique. Affinités entre Radiolaires, Heliozoaires et Peridiniens. *Protistologica* 6:311–318.

Holm, E. (1928) Uber die Suctorien in der Elbe bei Hambrug und ihre Lebensbedingungen. *Arch. Hydrobiol.* Supp. 4:389–440.

Honjo, S. (1978) Sedimentation of material in the Sargasso Sea at a 5,367 meter deep station. *J. Mar. Res.* 36:469–492.

Hunter, S., ed. (1964) *Biochemistry and Physiology of Protozoa.* Vol. 1–3. New York: Academic Press.

Jahn, T.L., and Bovee, E.C. (1965) Mechanisms in movement in taxonomy of sarcodina. I. As a basis for a new major dichotomy in two classes, Autotractea and Hydraulea. *Amer. Mid. Nat.* 73:30–40.

Jahn, T., and Jahn, F. (1949) *How to Know the Protozoa.* Dubuque, Iowa: W.C. Brown and Co.

Jahn, T., and Jahn, F. (1972) *The Protozoa.* Dubuque, Iowa: W.C. Brown and Co.

Jahn, T.L., Bovee, E.C., and Jahn, F.F. (1979) *How to Know the Protozoa*. Dubuque, Iowa: W.C. Brown and Co.

Jankowski, A.W. (1973) *Fauna of the USSR: Infusoria Subclass Chonotricha*. Vol. 2, No. 1, Acad. Nauka, Leningrad: Nauk. SSSR.

Jepps, M.W. (1956) *The Protozoa: Sarcodina*. Edinburgh: Oliver and Boyd.

Johnson, P.T. (1976) Paramoebiases in the blue crab, *Callinectes sapidus*. *J. Invert. Pathol.* 29:308–320.

Johnson, T.W., and Sparrow, F.K. (1961) *Fungi in Oceans and Estuaries*. Weinheim, Germany: J. Cramer.

Jørgensen, E. (1924) Mediterranean Tintinnidae. Rep. Danish Oceanogr. Exped., 1908–1910, *Mediterranean* 2(Biol.)J. 3:1–110.

Kahl, A. (1928a) Die Infusorien (Ciliata) der Obdesloer Salzwasserstellen. *Arch. Hydrobiol.* 19:50–123.

Kahl, A. (1928b) Die Infusorien (Ciliata) der Obdesloer Salzwasserstellen. *Arch. Hydrobiol.* 19:189–246.

Kahl, A. (1934a) Ciliata endocommensalia et parasitica. In Grimpe, G., and Wagler, E. (eds.), *Die Tierwelt der Nord- und Ostsee*, Lief 26 (Teil 11, C4) Leipzig, pp. 147–183.

Kahl, A. (1934b) Suctoria. In Grimpe, G., and Wagler, E. (eds.), *Die Tierwelt der Nord- und Ostsee, Lief 26 (Teil 11, C5), Leipzig, pp. 184–226.*

Kahl, A. (1935) In Dehl, F. (ed.), *Die Tiefwelf Deutschlands*. Jena: G. Fischer, pp. 651–805.

Kahn, R.A. (1977) Susceptibility of marine fish to Trypanasomes. *Can. J. Zool.* 55:1235–1241.

Kirby, H. (1934) Some ciliates from salt marshes in California. *Arch. Protist.* 82:114–133.

Kofoid, C.A., and Campbell, A.S. (1929). A conspectus of the marine and freshwater Ciliata belonging to the suborder Tintinnoines, with descriptions of new species principally from the Agassiz Expedition to the Eastern Tropical Pacific. *U. Calif. Pub. Zool.* 34:1–403.

Kofoid, C.A., and Campbell, A.S. (1939) The Ciliata: the Tintinnoinea. Reports on the scientific result of the expedition to the eastern tropical Pacific, 1904–1905. *Bull. Mus. Comp. Zool. Harvard* 84: 1–473.

Kozloff, E.N. (1946) Studies on ciliates of the family Anclstrocomoidae Chatton and Lwoff (order Holotricha, suborder Thigmotricha). IV. *Heterocineta janickii* Jorock: *Heteocineta goniobasidas* sp. nov., *Hetesocineta fluminicolae* sp. nov., and *Enerthecoma properans* Jorocki. *Biol. Bull.* 91:200–209.

Kozloff, E.N. (1966) *Phalacrocleptes verriciformic* gen. nov., sp. nov., an unciliated ciliate from the sabellid polychaete *Schizobranchia insignis*. *Biol. Bull.* 130:202–210.

Kudo, R. (1954) Protozoology. Springfield, Ill.: Charles C Thomas.

Leadbeater, B. (1972) Fine-structural observations on some marine choanoflagellates from the coast of Norway. *J. Mar. Biol. Ass. U.K.* 52:67–69

Leadbeater, B. (1973) External morphology of some marine choanoflagellates from the coast of Yugoslavia. *Arch. Protistenk.* 115:234–252.

Leadbeater, B.S.C. (1978) Developmental and ultrastructural observations on two stalked marine choanoflagellates, *Aconthoecopsis spiculifera* Ellis. *Proc. R. Soc. Lond.* 204:57–66.

Lee, J.J., and Pierce, S. (1963) Growth and physiology of Foraminifera in the laboratory; Part 4—monozenic culture of an allogromiid with notes on its morphology. *J. Protozool.* 10:404–411.

Lee, J.J., Tietjen, J.H., Stone, R.J., Muller, W.A., Rullman, J., and McEnery, M. (1970). The cultivation and physiological cultivation ecology of members of salt marsh epiphytic community. *Helgolander Wis. Meeres.* 20:136–156.

Lee, J.J. (1974) Towards understanding the niche of foraminifera. In Hedley, R.H., and Adams, C.G. (eds.), *Foraminifera*, Vol. 1. London: Academic Press, pp. 207–260.

Lee, J.J., and Muller, W.A. (1974) Culture of salt marsh microorganisms and micrometazoa. In Smith, W.L., and Chanley, M.H., (eds.), *Culture of Marine Invertebrate Animals*. New York: Plenum, pp. 84–107.

Lee, J.J. (1980a) Nutrition and physiology of the foraminifera. In Levandowsky, M., and Hunter, S.H. (eds.), *Biochemistry and Physiology of the Protozoa*, Vol. 3, 2nd ed. New York: Academic Press, pp. 43–66.

Lee, J.J. (1980b) A conceptual model of marine detrital decomposition and the organisms associated with the process. In Droop, M.R. and Jannasch, H.W., eds. *Advances in Aquatic Microbiology* V. 2. New York: Academic Press, pp. 257–291.

Lee, J.J., Hutner, S.H., and Bovee, E.C., eds. (1985) *An Illustrated Guide to the Protozoa*. Society of Protozoologists Lawrence, Kansas: Allen Press.

Levine, N.D., Corliss, J.D., Cox, F.E.G., Deroux, G., Grain, J., Honigberg, B.M., Leedale, G.F., Loeblich, A.R., III, Lomi, J., Lynn, D., Merinfeld, E.G., Page, F.C., Poljansky, G., Sprague, V., Vavra, J., Wallace, F.G., and Weiser, J. (1980). A newly revised classification of the Protozoa. *J. Protozool.* 27:37–58.

Lindholm, T. (1985) *Mesodinium rubrum*—a unique photosynthetic ciliate. In Jannasch, H.W., and Williams, P.J. LeB. (eds.), *Advances in Aquatic Microbiology*, Vol. 3. New York: Academic Press, pp. 1–48.

Lindholm, T., Lindross, P., and Mork, A.-C. (1987) Ultrastructure of the photosynthetic ciliate *Mesodinium rubrum*. *BioSystems* 21:141–149.

Lindley, J.A. (1975) Continuous plankton records: a plankton atlas of the North Atlantic and North Sea: Supplement 3—Tintinnide (Protozoa, Ciliophora) in 1965. *Bull. Mar. Ecol.* 8:201–213.

Loeblich, A.R., and Tappan, H. (1964) Sarcodina chiefly "Thecamoebians" and Foraminiferida: treatise on invertebrate paleontology, Part C. In Moore, R.C. (ed.) *Protista*, Vols. 1–2. Geological Society of America and University of Kansas Press, 900 pp.

Loeblich, A.R., and Tappan, H. (1968) Annotated index to genera, subgenera and suprageneric taxa of the ciliate order tintinnida. *J. Protozool.* 15:185–192.

Loeblich, A.R., and Tappan, H. (1984) Suprageneric classification of the foraminiferida (Protozoa). *Micropaleontology* 30:1–70.

Lom, J., and Larid, M. (1969) Parasitic protozoa from marine and eurhaline fish of Newfoundland and New Brunswick. I. Peritrichous ciliates. *Can. J. Zool.* 47:1367–1380.

Lom, J. (1970) Protozoa causing disease in marine fishes. In Snieszko, S.F. (ed.), *A Symposium on diseases of fishes and shellfishes*. Spec. Pub. 5, Amer. Fish. Soc. Washington, D.C., pp. 101–123.

Lwoff, A. (1950) *Problems of Morphogenesis in Ciliates*. New York: Wiley.

Lynn, D.H., and Berger, J. (1973) Morphology, systematics, and demic variation of *Plagiopyliella pacifica,* 1951 (Ciliata: Philasterina), an entocommensal of stronglocentrotid echinoids. *Trans. Amer. Microsc. Soc.*, 91:310–336.

Mackinnon, D.L., and Hawes, R.S. (1961) *An Introduction to the Study of Protozoa*. Oxford: Oxford University Press.

Massera-Bottazzi, E., and Vanucci, A. (1964) Acantharia in the Atlantic Ocean. *Estratto Arch. Oceanogr. Limnol.* 13:315–385.

Massera-Bottazzi, E., and Vanucci, A. (1965a) Acantharia in the Atlantic Ocean. *Estratto Arch. Oceanogr. Limnol.* 14:1–68.

Massera-Bottazzi, E., and Vanucci, A. (1965b) Acantharia in the Atlantic Ocean. *Estratto Arch. Oceanogr. Limnol.* 14:69–255.

Matera, N.J., and Lee, J.J. (1972) Environmental factors affecting the standing crop of foraminifera in sublittoral and psammolittoral communities of a Long Island salt marsh. *Mar. Biol.* 14:89–103.

McEnery, M.E., and Lee, J.J. (1976) *Allogromia laticollaris:* a foraminiferan with an unusual apogamic metagenic life cycle. *J. Protozool.* 23:94–108.

Myers, E.H. (1938) The present state of our knowledge concerning the life cycle of the Foraminifera. *Proc. Natl. Acad. Sci., Washington* 24:10–17.

Myers, E.H. (1943) Life activities of foraminifera in relation to marine ecology. *Proc. Amer. Phil. Soc.* 86:439–459.

Muller, W.A. (1975) Competition for food and other niche-related studies of three species of salt marsh foraminifera. *Mar. Biol.* 31:339–351.

Murray, J.W. (1971) *An Atlas of British Recent Foraminiferids*. London: Heinemann.

Murray, J. (1973) *Distribution and Ecology of Living Benthic Foraminiferids*. London: Heinemann.

Nigrini, C., and Moore, T.C. (1979) *A Guide to Modern Radiolaria*. Cushman Foundation Foram. Res., Spec. Publ. 16.

Nisbet, B. (1984) *Nutrition and Feeding Strategies in Protozoa*. CROOM HELM, 280 pp.

Noble, E.R., and Noble, G.A. (1971) *Parasitology: The Biology of Animal Parasites,* 3rd ed. Philadelphia: Lea & Febiger.

Nusch, E.A. (1970) Okologische und systematische Untersuchungen der Peritricha (Protozoa, Ciliata) in Aufwuchs von Talspernen und Flusstauen mit verschiedenen Saprobitotograd (mit model versucken) *Arch. Hydrobiol. Suppl.* 37:243–386.

Oakley, B.R., and Taylor, F.J.R. (1978) Evidence for a new type of endosymbiotic organization in a population of the ciliate *Mesodinium rubrum* from British Columbia. BioSystems 10:361–369.

Overstreet, R.M. (1978) Marine Maladies? Worms, Germs and Other Symbionts from the Northern Gulf of Mexico. Mississippi–Alabama Sea Grant Consortium MASGP-78-021.

Page, F.C. (1970) Two new species of Paramoeba from Maine. *Trans. Amer. Microsc. Soc.* 90:157–173.

Page, F.C. (1971) A comparative study of five freshwater and marine species of Thecamoebidae *Trans. Amer. Microsc. Soc.* 90:152–183.

Phleger, F. (1960) *Ecology and Distribution of Recent Foraminifera*. Baltimore: Johns Hopkins University Press.

Pokorny, K.S. (1967) *Labyrinthula. J. Protozool.* 14:697–708.

Popofsky, A. (1909) Die nordisches plankton. *Fassicules* 3,6.

Popofsky, A. (1913) Die Sphaerellarien des Warmwassergebietes. *Deutsche Sudpolar-Exp 1901-1903, Berlin, Zool. Bd. 8,* H, 3:235–278.

Popofsky, A. (1917) Die Collosphaereden der Deutschen Sudpolar-Expedition 1901-1903. Mit Nachtrag zu den Spumellarien und Nasselarien. *Deutsche Sudpolar Exped. 1901-1903, Berlin Zool. Bd. 8,* H, 3:235–278.

Powers, P.B.A. (1935) Studies on the ciliates of sea urchins. A general survey of the infestation occurring in Tortugas echinoids. *Rap. Tortugas Lab.* 29:293–326.

Provasoli, L., McLaughlin, J.J.A., and Droop, M.R. (1957) The development of artificial media for marine algae. *Arch. Mikrobiol.* 25:392–428.

Prowazek, S. (1903) *Arch. Protistenk.* 2:195–212.

Putz, R.E., and McLaughlin, J.J.A. (1970) In Snieszka, S.F. (ed.), *A Symposium on Disease of Fishes and Shellfishes,* Am. Fish. Soc. Spec. Publ. 5, pp. 124–132.

Raabe, J., and Raabe, Z. (1959) Urceolarudae of molluscs of the Baltic Sea. *Acta Parasitol. Paleontol.* 7:453–465.

Rassoulzadegan, F. (1978) Dimensions et taux d'ingestion des particles consummées par un tintinnide *Favella ehrenbergii* (Clap. et Lachm.) Jørg. Cilié pelagique marin. *Ann. Inst. Oceanogr. Paris* 54: 17–24.

Rassoulzadegan, F., and Etienne, M. (1981) Grazing of the tintinnid *Stenosemella ventricosa* (Clap. and Lachm.) Jørg. on the spectrum of the naturally occurring particulate matter from the Mediterranean neritic area. *Limnol. Oceanogr.* 26:258–270.

Reichenow, E. (1953) *Lehrbuch der Protozoenkunde*. Jena: G. Fisher.

Ribelin, W.E., and Migaki, G. (eds.) (1975) *The Pathology of Fishes*. Madison: University of Wisconsin Press.

Riedel, W.R. (1971) Systematic classification of polycystine radiolaria. In Funnell, B.M. and Riedel, W.R. (eds.), *The Micropaleontology of Oceans*. London: Cambridge University Press, pp. 649–661.

Rogers, W.A., and Gaines, J.J., Jr. (1975) In Ribelin, W.E., and Migaki, G. (eds.), *The Pathology of Fishes,* Madison: University of Wisconsin Press.

Rubin, H.A., and Lee, J.J. (1976) Informational energy flow as an aspect of the ecological efficiency of marine ciliates. *J. Theor. Biol.* 62:69–91.

Ruinen, J. (1938) Notizen uber Salyflagellaten 2. Uber die verbreitung der Salzflagellaten. *Arch. Protistenk* 90:210–258.

Sanders, J.E., Fryer, J.L., and Gould, R.W. (1970) In Snieszko, S.F. (ed.), *A Symposium on Diseases of*

Fishes and Shellfishes, Amer. Fish Soc. Spec. Publ. 5, pp. 133–141.

Sawyer, T. (1975a) Marine amoeba from surface waters of Chincoteague Bay, Virginia: one new genus and eleven new species within the families Thecamoeridae and Hyalodiscidae. *Trans. Amer. Microsc. Soc.* 94:305–323.

Sawyer, T. (1975b) Marine amoeba from surface waters of Chincoteague Bay Virginia: two new genera and nine new species within the families Mayorellidae, Flabellolidae, and Stereomyxidae. *Trans. Amer. Microsc. Soc.* 94:71–92.

Sawyer, T. (1975c) Clydonella n.g. (Amoebida: Thecamoebidae), proposed to provide an appropriate generic home for Schaeffer's marine species of *Rugipes, C. vivex* (Schaeffer, 1926) n. comb. *Trans. Amer. Microsc. Soc.* 94:395–400.

Sawyer, T. (1980) Marine amebae from clean and stressed bottom sediments of the Atlantic Ocean and the Gulf of Mexico. *J. Protozool.* 27:13–33.

Sawyer, T., Hnath, J., and Conrad, J. (1975) *Thecamoeba hoffmani* sp. n. (Amoebida: Thecamoebidae) from gill of fingerling salmonid fish. *J. Parasitol.* 60:677–682.

Schaeffer, A. (1926) *Taxonomy of the Ameras with Descriptions of Thirty-nine New Marine and Freshwater Species*, Washington, D.C.: Carnegie Institution of Washington.

Schewiakoff, W. (1926) Die Acantharia des Golfes von Neapel. Fauna und Flora des Golfes von Neapel 37:xxiv–755.

Schmoller, H. (1971) Die Labrinthulem und ihre Beziehung zu den Amoben. *Naturwiss.* 58:142–146.

Schröder, O. (1914) *Zool. Anz.* 43:320.

Sherr, E.B., and Sherr, B.F. (1987) High rates of consumption of bacteria by pelagic ciliates. *Nature* 325:710–711.

Sieburth, J. McN. (1979) *Sea Microbes*. New York: Oxford University Press, 491 pp.

Sindermann, C.J. (1970) *Principal Disease of Marine Fish and Shellfish*. New York: Academic Press.

Sindermann, C.J. (ed.) (1977) *Disease Diagnosis and Control in North American Marine Aquaculture*. Amsterdam: Elsevier.

Sladekova, A. (1962) Limnological investigation methods for the periphyton (Aufwuchs) Community. *Bot. Rev.* 28:286–350.

Sleigh, M. (1973) *The Biology of Protozoa*. London: Edward Arnold, 315 pp.

Small, E.B., and Gross, M.E. (1985) Preliminary observations of protistan organisms, especially ciliates, from 21°N hydrothermal vents of the eastern Pacific: an overview. *Bull Biol. Soc. Wash.* 6:453–464.

Small, E.B., Heisler, J., Sniezek, J., and Iliffe, T. (1986) *Glauconena bermudense* n.sp. "Scuticociliatida oligohymenophorea," a troglobitic Ciliophoran from Bermudian marine caves. *Stygologia* 2:167–179.

Small, E.B., and Lynn D. (1985) Phylum Ciliophora. In Lee, J.J., Hutner, S.H., and Bovee, E.C., (eds.), *An Illustrated Guide to the Protozoa*. Lawrence, Kansas: Allen Press.

Sniezek, J. Jr. (1987) An examination of the ciliate fauna at the black smoker hydrothermal vents of 10° 57′ N, 103° 46′ W. M.S. Thesis, University of Maryland at College Park, 161 pp.

Soldo,, A.T., and Brickson, S.A. (1980) A simple method for plating and cloning ciliates and other protozoa. *J. Protozool.* 27:328–331.

Soldo, A., and Merlin, E. (1972) A cultivation of symbiote-free marine ciliates in axenic medium. *J. Protozool.* 19:519–524.

Sprague, V. (1970a) in *A Symposium on Diseases of Fishes and Shellfishes*. Amer. Fish. Soc. Spec. Publ. 5, pp. 416–430.

Sprague, V. (1970b) In *A Symposium on Diseases of Fishes and Shellfishes. Amer. Fish. Soc. Spec. Publ.* 5, pp. 511–526.

Sprague, V., and Couch, J. (1971) An annotated list of protozoan parasites, hyperparasites and commensals of decapod crustacea. *J. Protozool.* 18:526–537.

Sprague, V. (1971) Disease of oysters. *Ann. Rev. Microbiol.* 25:211–230.

Spittler, P. (1973) Feeding experiments with tintinnids. *Oikos* (s) 15:128–132.

Spoon, D.M., and Burbanck, W.D. (1967) A new method for collecting sessile ciliates in plastic petri dishes with tight fitting lids. *J. Protozool.* 14:735–739.

Stein, G.A. (1967) Parasitic ciliates (Peritricha, Urceolariidae) of some fishes of the Kamchatka. *Acta Protozool.* 4:291–306.

Stoecker, D., Davis, L.H., and Provan, A. (1983) Growth of *Favella* sp. (Ciliata: Tintinnina) and other microzooplankters in cages incubated *in situ* and comparison to growth *in vitro*. *Mar. Biol.* 75:293.

Stoecker, D., Guillard, R.R.L., and Kavee, R.M. (1981) Selective predation by *Favella ehrenbergii* (Ciliata, Tintinnia) on and among dinoflagellates. *Biol. Bull.* 160:136–145.

Tappan, H., and Loeblich, A.R., Jr. (1968) Lorica composition of modern and fossil Tintinnida (Ciliate Protozoa), systematics, geologic distribution and some new Tertiary taxa. *J. Paleontol.* 42:1378–1394.

Taylor, F.J.R., Blackbourn, D.J., and Blackbourn, J. (1971) The red-water ciliate *Mesodinium rubrum* and its incomplete symbionts: a review including new ultrastructural observations. *J. Fish. Res. Bd. Can.* 28:391–407.

Tendal, O.S., and Hessler, R.R. (1977) In *Galathea Report*, Vol. 14. Copenhagen: Scandinavian Sci. Press.

Thomsen, H.A. (1976) Studies on marine dinoflagellates: II. Fine structure observations on some silicified choanoflagellates from the Isefjord (Denmark), including the description of two new species. *Norw. J. Bot.* 23:33–51.

Travis, J.L., and Allen, R.D. (1981) Studies on the motility of the foraminifera. I. Ultrastructure of the reticulopodial network of *Allogromia laticollaris* (Arnold). *J. Cell Biol.* 90:211–221.

Travis, J.L., Kenealy, J.F.X., and Allen, R.D. (1983) Studies on the motility of the foraminifera. 2. The dynamic microtubular cytoskeleton of the reticulopodial network of *Allogromia laticollaris. J. Cell. Biol.* 97:1668–1676.

Travis, J.L., and Bowser, S.S. (1986) Microtubule-dependent reticulopodial motility. Is there a role for actin? *Cell Motility and the Cytoskeleton* 6:146–152.

Tregouboff, G. (1953) Classe des Radiolaires. In Grasse, P.P. (ed.), *Traité de Zoologie Anatomie Systématique, Biologie*, Vol. 1. Paris: Masson, pp. 322–436.

Uhlig, G. (1968) Quantitative methods in the study of interstitial fauna. *Trans. Amer. Microsc. Soc.* 87:226–232.

UNESCO monograph on Zooplankton fixation and preservation (1976) H.P. Steedman (ed.). Paris: The UNESCO Press, 350 pp.

UNESCO monograph Phytoplankton manual (1978) Sournia, A. (ed.). Paris: The UNESCO Press, 337 pp.

Verity, P.G. (1985) Grazing, respiration, excretion and growth rates of tintinnids. *Limnol. Oceanogr.* 30:1268–1282.

Verity, P.G. (1987) Abundance, community composition, size distribution and production rates of tintinnids in Narragansett Bay, Rhode Island. *Est. Coast Shelf Sci.* 24:671–690.

Verity, P.G., and Langdon, C. (1984) Relationships between lorica volume, carbon, nitrogen and ATP content of tintinnids in Narragansett Bay *J. Plankton Res.* 6:859–868.

Verity, P.G., and Villareal, T.A. (1986) The relative food value of diatoms, dinoflagellates, flagellates and cyanobacteria for tintinnid ciliates. *Arch. Protistenk.* 131:71–84.

Vishniac, H.S. (1955a) The nutritional requirements of isolates of *Labyrinthula* spp. *J. Gen. Microbiol.* 12:455–463.

Vishniac, H.S. (1955b) The activity of steroids of growth factors for a *Labyrinthula* spp.

Watson, S.W. (1957) Cultural and cytological studies on species of *Labyrinthula*. Ph.D. Thesis, University of Wisconsin, Madison.

Watters, C. (1968) Studies on the motility of the Heliozoa. I. The locomotion of *Actionosphaerium eichhorn* and *Actinophrys* sp.

Weber, H. (1965) *Arch. Protistenk.* 108:217–270.

Wiebe, P.S., Boyd, C., and Winget, C. (1976) Particulate matter sinking to the deep sea floor at 2,000 m in the Tongue of the Ocean, Bahamas, with description of a new sedimentation trap. *J. Mar. Res.* 34:341–354.

2

Ecology and Paleobiology of Foraminifera and Radiolaria

P. LEWIS STEINECK
RICHARD E. CASEY

PART A: INTRODUCTION

Purpose

With few exceptions, paleontological investigations of the phagotrophic protistans are limited to the polycystine Radiolaria and to those groups of Foraminifera that possess an agglutinated or biomineralized test. These organisms, which preserve readily in most depositional and diagenetic settings, are represented by an extensive fossil record in sedimentary rocks of varied age, lithology, and origin (Fig. 2.1). Investigators in the early and middle periods of the twentieth century focused on traditional paleontological pursuits, such as biostratigraphic correlation between distant strata, taxonomic description, and the interpretation of paleoenvironments. The interest and encouragement of the fossil fuels industries provided a major impetus to such studies inasmuch as microfossils are often the only ones found in well cuttings and cores. Since the 1960s, development of a new theoretical framework for the earth sciences (plate tectonics) and the advent of new technologies and analytical approaches (deep-sea drilling, multivariate statistics, electron microscopy) have contributed to major advances in areas of existing concern and to the emergence of specialized subdisciplines such as paleoceanography and paleoclimatology. Most important, the continuous fossil record of Radiolaria and Foraminifera present in deep-sea sediments has permitted micropaleontologists to contribute significant and unique perspectives to the current debate about the process and tempo of evolution.

This chapter has three objectives: (1) to survey briefly the basic morphology, taxonomy, ecology, and stratigraphic distribution of the Forminifera and Radiolaria; (2) to provide examples of how these fossils are used in geohistorical reconstruction; and (3) to discuss interdisciplinary topics of potential interest to workers in protozoology, marine biology, and related fields. Of necessity, our review is both limited in scope and selective in nature. Lipps (1981) has calculated that an average of 1,200 papers was published annually on fossil protistans between 1972 and 1978, a rate of production that continues unabated. Consequently, much important work could not be incorporated into this essay or could be touched on only briefly. Different authors might well have made other choices and established other priorities. Nevertheless, it is hoped that our efforts will illustrate the breadth, significance, and intellectual vigor of contemporary micropaleontological research.

Introduction to the Foraminifera

The Foraminifera are phagotrophic, amoeboid protistans belonging to the class Granuloreticulosa of the subphylum Sarcodina (Fig. 2.2). They are characterized by an alternation of generations, one or more nuclei, and a test that encloses the nucleate endoplasm. The test is surrounded by an elongate, branching to weblike net of pseudopodia (granuloreticulopodia) that are the sites of two-way fibrillar streaming (Lipps, 1982). In phylogenetically primitive forms, the test is composed of a flexible tectinous substance thought be an acid mucopolysaccharide. Advanced taxa build their tests from adventitious particles obtained from the environment (agglutinated) or from mineral matter secreted by the organism. The test may contain only a single undivided chamber (unilocular) or may consist of linear or coiled chains of chambers (multilocular) constructed in discrete episodes. Typically, an opening in the wall of the last chamber (aperture) provides communication between the endoplasm and the pseudopodial net and the environment. An internal passageway between the compartments of the test is maintained during ontogeny in most groups by stolons, canals, or the apertures (foramina) of successive chambers.

Foraminifera are among the most widely distributed and ecologically varied groups of marine organisms. The greater majority of species are mo-

Era (Erathem)	Period (System)	Epoch (Series)	Ma
Cenozoic	Quaternary	Recent (Holocene)	~11,000 yr BP
		Pleistocene	1.2
	Neogene	Pliocene	5.2
		Miocene	23.8
	Paleogene	Oligocene	36.5
		Eocene	58
		Paleocene	66.4
Mesozoic	Cretaceous		144
	Jurassic		208
	Triassic		248
Paleozoic	Permian		286
	Pennsylvanian ⎤ Carboniferous		320
	Mississippian ⎦ Carboniferous		360
	Devonian		408
	Silurian		438
	Ordovician		505
	Cambrian		590
"Precambrian"	Sinian		

FIGURE 2.1 The Geologic Time Scale is based on standardized hierarchy of arbitrarily defined units originally proposed by geologists in the nineteenth century based on the superposition and fossil content of rock sequences in Europe. Geochronologic calibration is accomplished by radiometric dating and extrapolation from the geomagnetic (polarity reversal) time scale. The notation Ma represents millions of years before present; for the Quaternary, yr BP signifies years before present. Ages are taken from Harland et al. (1982), Berggren et al. (1985), and Kent and Gradstein (1985). Time-stratigraphic units (erathem, system, series) refer to discrete rock bodies deposited during intervals of geologic time (era, period, epoch). See Harland et al. (1982) for a discussion of the historical development of the geologic time scale.

bile, benthic, and epifaunal, but infaunal, epiphytic, adherent, epizoic–commensal, parasitic, and planktonic forms are common. These organisms utilize a wide variety of trophic strategies to obtain the nutrients necessary for growth and reproduction (Lipps, 1982); at least two species are capable of the direct intake of dissolved organic matter (DeLaca et al., 1981). Foraminifera occupy an important position in marine food chains by virtue of their ability to assimilate energy from minute autotrophs and partially degraded organic material and convert it into a form that can be used by secondary consumers (Lipps and Valentine, 1970).

The holoplanktonic Globigerinina are an important component of marine euphotic communities. They first appeared in the early Cretaceous and occupied new geographic and depth habitats by the middle Cretaceous. Today, as in the past, vast areas of the seafloor (e.g., beneath fertile surface waters and above calcium compensation depth [CCD]) are carpeted with calcareous oozes in which the tests of globigerinines are the dominant sand-sized component.

Numerous and diverse foraminiferal faunas are associated with euhaline environments from the inner-neritic to the hadal zones. Hyposaline marshes,

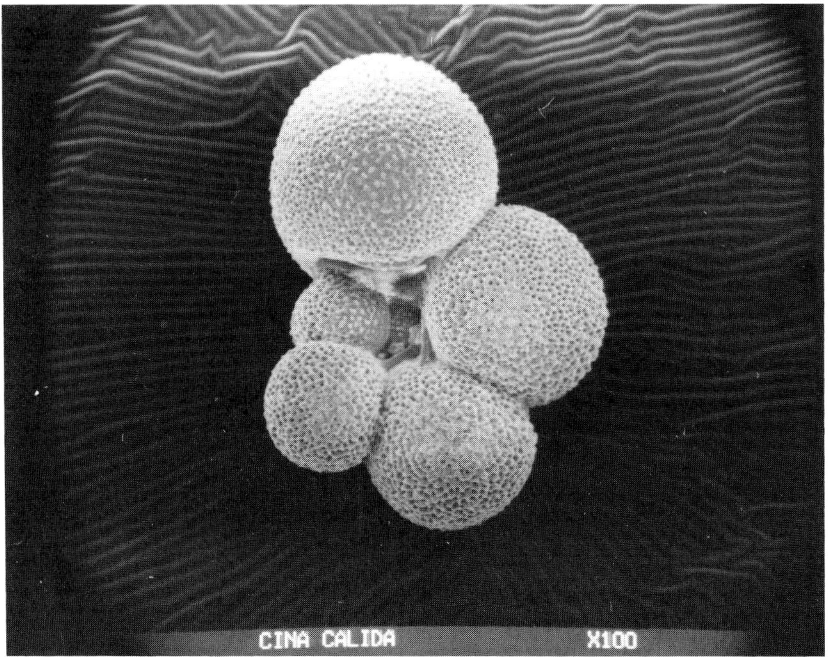

FIGURE 2.2 *Globigerina calida*. A typical multilocular-septate species of Foraminifera. This species is planktonic and possesses a calcareous hyaline-perforate wall. The aperture is visible in the umbilicus (axis of coiling) and extends into the base of the last-formed (largest) chamber. Recent of the Atlantic Ocean. × 100. (SEM micrograph.). (See also Figs. 2.4, 2.9, and 2.10, and Plate 1).

bays, and estuaries are inhabited by low-diversity but often abundant assemblages. Foraminifera have been reported from a few freshwater locales (Boltovskoy and Lena, 1971), including land-locked lakes (Resig, 1974), but such occurrences are rare. Taxonomically distinctive and diverse faunas of large agglutinated species comprise the greatest fraction of biomass present in abyssal soft-bottom communities below the CCD.

The time of origin of the Foraminifera has not yet been established with certainty. It is possible that the group had a long history prior to the evolution of preservable tests sometime in the latest Precambrian or early Paleozoic. Simple vase-shaped cups, hemispheres, and straight and coiled tubes of finely agglutinated material from the late Precambrian of Canada and Europe (Conway-Morris and Fritz, 1980; Brasier, 1982a) can be assigned only tentatively to the Foraminifera. Similar fossils (e.g., *Platysolenites*) from the Early Cambrian of northern Europe are generally considered to represent the oldest well-documented examples of Foraminifera (Haynes, 1981, p. 87). Reports of calcareous-microgranular forms inhabiting Lower Cambrian archaeocyathid reefs (Riding and Brasier 1975; Kobluk and James, 1979) are intriguing but difficult to evaluate. The fossils in question may represent other microorganisims, such as calcareous algae. Undoubted calcareous foraminifers have been recovered in Early Ordovician strata (Schallreuter, 1983). Other taxa present in rocks of this age are simple, agglutinated tubes with a distinct proloculus (Conkin and Conkin, 1982). By the Upper Ordovician agglutinated faunas were abundant, were taxonomically diverse, and contained genera with a variety of coiling shapes and chamber arrangements. All major subgroups of the Foraminifera had evolved by the Carboniferous. The sequential appearance of progressively more complex and functionally advanced forms of chamber arrangements and shapes during the Paleozoic supports an adaptive interpretation for the evolution of the remarkable architectural diversity displayed by the Foraminifera (Brasier, 1982a,b).

Introduction to the Radiolaria

Radiolarians have not been as rigorously or completely studied as have the foraminifera. Since radiolarians are all pelagic and mainly accumulate in deep-sea sediments as fossil silicious skeletons, it was not until the initiation of the deep-sea drilling project (DSDP) that they became appreciated for their worth in dating and interpreting sediments. In 1983 an excellent review of radiolaria was published in book form (Anderson, 1983). In order not to repeat some of the excellent and extensive documentation in that book, the radiolarian section herein will only briefly review aspects dealt with in detail in there and will expand on radiolarian research, findings, and speculations since the publication of that book. These studies mentioned are selected studies, not a comprehensive survey, and rely heavily on the work of the junior author and his associates. Others would undoubtedly choose different examples; however, this selection is believed to be representative of recent advances.

The Radiolaria are marine holoplanktonic, phagotrophic (in part), amoeboid protistans belonging to the class Actinopoda of the subphylum Sarcodina. They are characterized by division of the cell contents into an inner zone separated by a membrane (central capsule) from an outer zone. They are further divided within the Radiolaria on the nature of the openings on this central capsule. The subgroups preserved in the fossil record are the Polycystina (or polycystins), which possess solid opaline skeletal structures, and the Phaeodaria (or phaeodarians), which possess hollow skeletal structures of an admixture of silica and organic matter.

The polycystin radiolarians are chronologically the longest ranging (Cambrian to Holocene), geographically the greatest ranging (pole to pole, surface to abyss), and taxonomically the most diverse of the well-preserved microzooplankton. Although there have been accounts of pre-Cambrian radiolarians, the earliest well-documented radiolarians to date appear to be early Cambrian (Lipps, 1985). (See Fig. 30 of Table 12 for a depiction of one of the earliest radiolarians.)

PART B: THE FORAMINIFERA

Test Morphology

Basic Test Characteristics

Test growth in the Foraminifera follows three basic patterns (Brasier, 1982a). Nonseptate, contained growth is associated with primitive groups or degenerate, parasitic taxa (Fig. 2.3c). Nonseptate, continuous growth allows for an increase in size through addition to a straight or coiled tube (Fig. 2.3a & b). Foraminifera that expand in this manner possess a large, inflexible nucleus. The supposition that they are ancestral to other groups of foraminifers is at least partly supported by the fossil record. Multilocular-septate growth evolved independently in five or more separate lineages during the late Paleozoic, perhaps in conjunction with the development of a smaller and more mobile nucleus (Brasier, 1982a; Fig. 2.4). Septate growth allows a more rapid rate of size increase, more efficient interior–exterior lines of communication, and the morphologic innovation and experimentation that led to the occupation of new niches and the use of new feeding strategies (Brasier, 1982a).

The growth plan for multilocular tests (Fig. 2.5) can be generated by a simple logarithmic equation that includes (1) shape of the generating curve, (2) rate of translation, (3) angular increment between succeeding chambers, (4) rate of expansion of the curve, (5) chamber shape, and (6) position of the curve relative to the coiling axis (Brasier, 1980; Arnold, 1982a). Similar patterns of chamber shape and arrangement occur in each of the major taxonomic subdivisions of the Foraminifera. Individuals of a species produced by the fusion of gametes have

FIGURE 2.3 Agglutinated deep-sea Foraminifera. (A & B) *Involutina anguillae*: an example of nonseptate, continuous growth. (From Brady, 1884, Plate 38, Figs. 1 and 6.) Note the terminal, unrestricted aperture in both species. All illustrations ×15. (C) *Saccammina sphaerica*: an example of nonseptate, contained growth. (From Brady, 1884, Plate 18, figs. 11 and 13.)

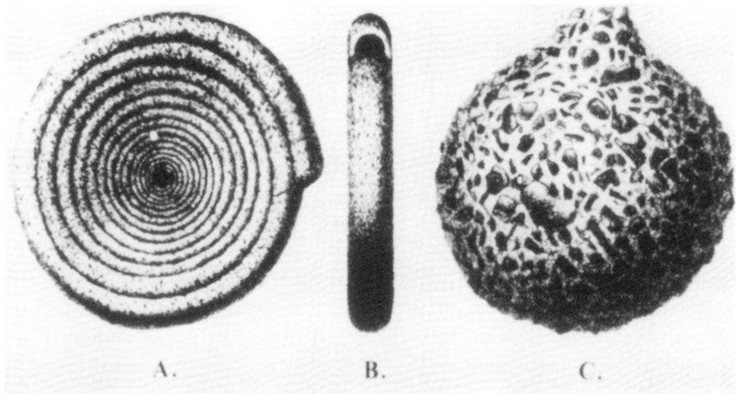

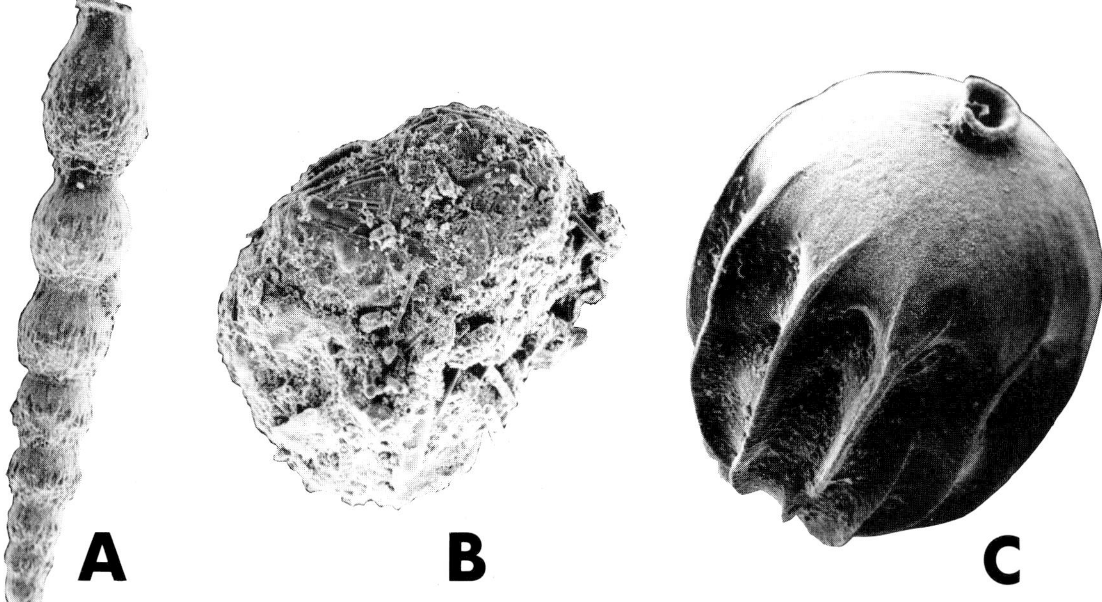

FIGURE 2.4 Diversity of multilocular–septate test form in Foraminifera. (A) *Stilostomella*, calcareous hyaline perforate wall with rectilinear (uniserial) chamber arrangement. (B) *Haplophragmoides*, agglutinated wall with planispirally coiled chambers. Wall composed of sponge spicules, foraminiferal test fragments, and cement. (C) *Siphogenerima*, calcareous hyaline-perforate wall with triserial chamber arrangement (partially obscured by ornamental ribbing). Note that the aperture is placed on a neck and contains a "tooth." All specimens × 105. (Courtesy of David Yozzo, SUNY at Purchase, N.Y.)

a smaller initial chamber (proloculus) and a larger test compared with those resulting from asexual reproduction. More extreme polymorphism is associated with successive stages of the reproductive cycle in some genera (Haynes, 1981, Figs. 2.3, 2.4; Banner et al., 1985). Multiform chamber arrangements (e.g., planispiral→biserial) are common and may recapitulate the recent phylogenetic history of the taxa involved.

Almost every species of Foraminifera possesses one or more apertures. Primitive unilocular or tubular forms may have an unconstricted opening at one end of the test (Fig. 2.3). More typical is a constricted and consistently placed aperture that varies in form and position in different groups (Fig. 2.2, 2.4). Multiple apertures are common in some families. The aperture may be partially closed by a lip or flange (Fig. 2.2). A tooth plate, a structure attached to the penultimate septal face and projecting through the lumen of the last chamber into the apertural opening, is present in some taxa (Fig. 2.4c).

Wall Composition and Structure

In the Foraminifera, the test wall consists of a tectinous envelope or a combination of one or more organic layers with agglutinated particles or biomineralized grains (Haynes, 1981).

Tectinous Wall Type

The tectinous wall is composed of a thin, flexible membrane containing layers of protein, carbohydrate, and other organic compounds (Schwab and Plapp, 1983). Because this structure often disintegrates quickly following burial, the fossil record of tectinous groups is restricted to a few and widely scattered occurrences.

The tectinous wall may have originally developed as a byproduct of the phagocytosis-excretion process in naked-celled foraminifers. Banner and others (1973) have speculated that otherwise indigestible cellulose was broken down by intracellular bacteria into mucropolysaccharides, which were then transported to and retained on the cell surface.

Agglutinated Wall Type

The agglutinated or "brick and mortar" wall is composed of particles procured from the environment, attached to an inner organic membrane and bound together by a calcareous or ferruginous cement (Fig. 2.3b). Some species use any available grain; others select grains of a specific size, shape, or composition. The degree of cementation (and hence preservability) varies. Advanced forms develop a tight, polished matrix with a large component of cement that may be difficult to distinguish at first glance from the biomineralized wall dis-

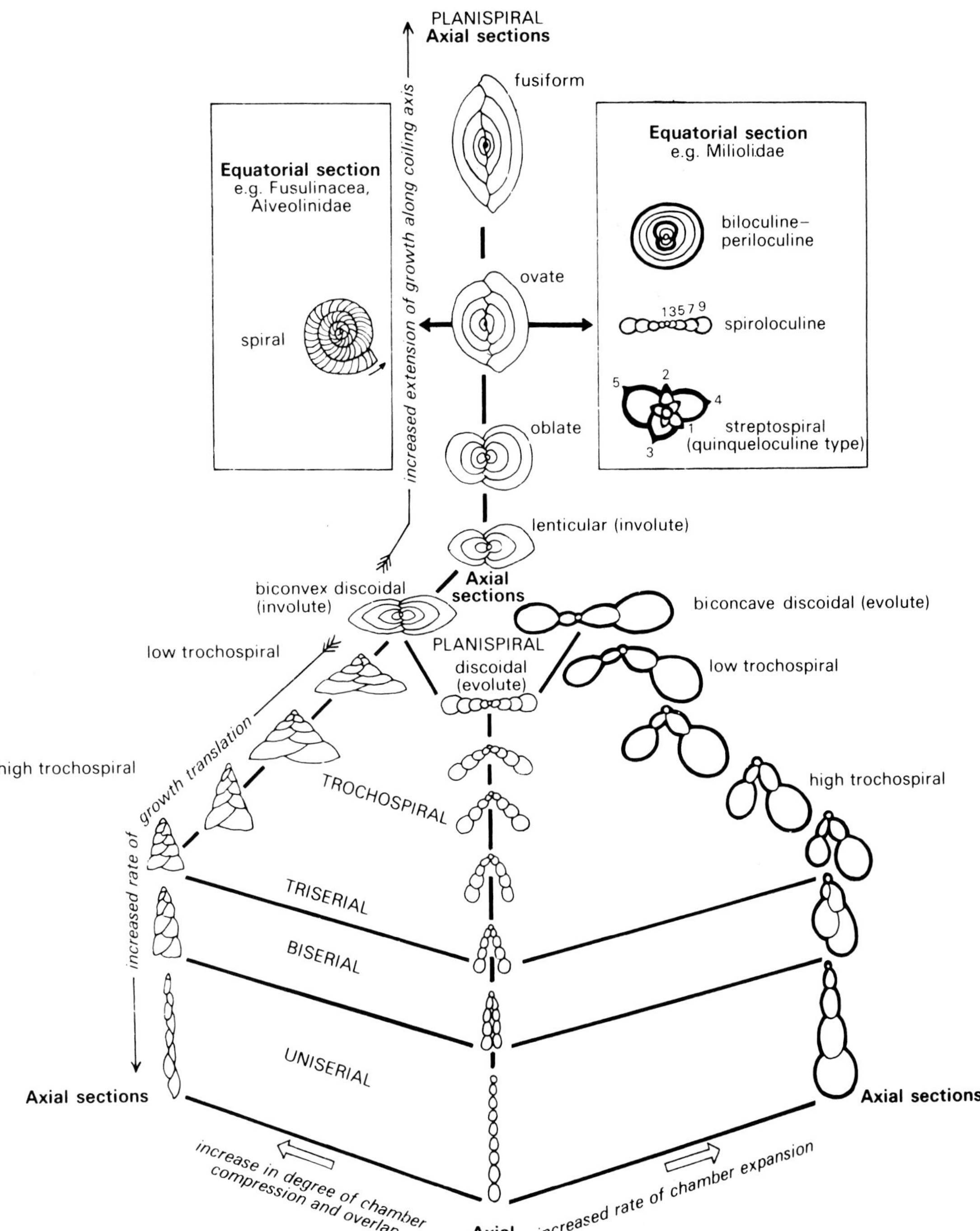

FIGURE 2.5 Derivation of multilocular test form by changes in the parameters of the logarithmic growth equation. (From Brasier, 1980, Fig. 13.6; reprinted by permission of Allen & Unwin Publishers, copyright 1980.)

cussed in subsequent paragraphs. Some eurytopic species with calcareous cement produce less intergranular matrix in response to corrosive bottom waters (e.g., below the CCD) (Weston, 1984).

The wall may be simple, laminated, pierced by convoluted passageways (labyrinthine) or more rarely, composed of structurally and compositionally differentiated layers (Hansen, 1979; Haynes, 1981; De Laca et al., 1980). Most agglutinated taxa are imperforate. In perforate forms, the pores may be sealed by a proximal organic membrane (Banner and Pereira, 1981) or may open into the chamber (Coleman, 1980; Banner and Desai, 1985).

Towe (1967) has argued that the agglutinated wall

developed as a consequence of the phagocytosis of mineral particles coated with organic matter. The coating provided nourishment while the grains were shifted to the cell exterior and held against the tectinous membrane. Iron chelated to the organic surface of the mineral grains was metabolized into an insoluble, colloidal, hydrated ferric-oxide that was then used to secure the grains more firmly to the cell surface.

Porcelaneous Wall Type

The porcelaneous wall is composed of randomly arranged microcrystallites of calcite encased in an organic meshwork (Fig. 2.6b). Some forms add an oriented inner and/or outer layer composed of elongate calcite laths (Towe and Ciffelli, 1967; Hansen, 1979). A superficial layer of coarse agglutinated grains is present in a few species. Although the porcelaneous wall has been traditionally thought of as imperforate, Anderson (1984) demonstrated that many porcelaneous forms occurring in the Gulf Coast Paleogene possess large pits that in some cases are sealed with organic plugs, membranes, or sieve plates. It is not clear whether such structures are functionally analogous to the pores of the hyaline-perforate wall discussed later.

Formation of the porcelaneous wall has been discussed by Angell (1980). Encystment initiates the calcification sequence. A mass of cytoplasm moves outward, its vacuoles releasing crystals that are placed into position by pseudopodia. The secretory organelle passes along the margin of the preformed chamber outline in a continuous process of calcification until the new chamber is completed. Fine pseudopodia may add a surface veneer of euhedral rods. Calcite deposition may also occur on preexisting exterior walls to produce a pseudolamellar structure. The internal surface is then lined with a thick organic membrane.

Crystallization of the porcelaneous wall appears to conform to Lowenstam's (1981) category of "biologically induced" mineralization characterized by bulk cytoplasmic secretion. The absence of a preexisting organic template results in much less effective control of shape, mineralogy, and crystallographic orientation. This "primitive" method of biomineralization, coupled with Arnold's (1979) studies of the origin of the distinctive miliolid

FIGURE 2.6 Schematic representation of the major types of secreted, calcareous walls in the Foraminifera. (A) Suborder Fusulinia. (B) Suborder Miliolina. (C–F) Suborder Rotaliina; black arrows denote the crystallographic orientation of individual grains. For a discussion of the ultrastructure of the rotaline wall see Bellemo (1974a,b, 1976, 1979) and Conger et al. (1977). After Haynes (1981, Fig. 14.4).

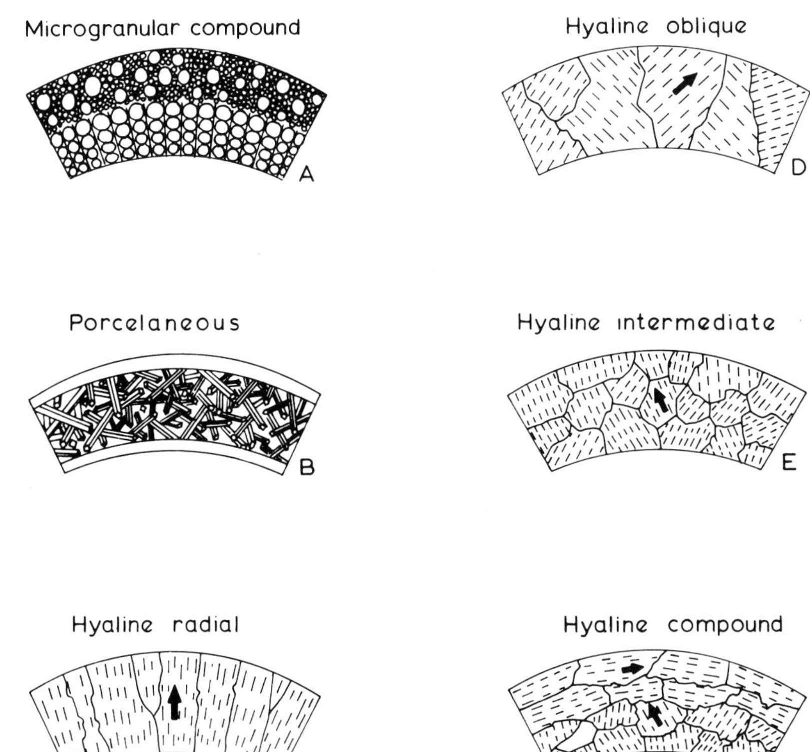

chamber arrangement, suggests that porcelaneous-walled foraminifers (Miliolina) were derived independently from allogromidlike ancestors and may require co-equal taxonomic rank (order) with the Foraminifera.

Micrograular Wall Type

The micrograular wall type is associated with an extinct group of middle to late Paleozoic shallow-water foraminifers, the Fusulinina (Fig. 2.6a). The coarse layering and structure of the fusulinine wall are now well known and have been used in the phylogenetic reconstruction of lineages. On an ultrastructural level, the wall is composed of tiny (1–5 μm) inequigranular, blocky crystals of calcite that contain and are surrounded by voids that were probably, at the time of test formation, filled with organic matter (Hansen, 1979; Green et al., 1980). The numerous reentrant angles and facets of the microcrystals in these fossils are consistent with growth in an isotropic fluid or fluidlike substance. A close similarity to the process of test secretion observed in living porcelaneous Miliolina is suggested (Green et al., 1980).

Miscellaneous Biomineralized Wall Types

One species, *Silicoloculina profunda* (suborder Silicoloculinina), secretes a trilaminate test composed of closely spaced, hollow tubules of amorphous opalina silica (Resig et al., 1980). In the suborder Spirillinina, the wall consists of an optically continuous single crystal of calcite ($CaCO_3$) encased by inner and outer organic membranes (Towe et al., 1977). The C-axis of the crystal is oriented parallel to the base of the test, whereas the A-axis corresponds to the axis of trochospiral coiling (Towe et al., 1977). Representatives of the suborder Robertinina secrete small pinacoidal crystals of the orthorhombic mineral aragonite ($CaCO_3$) (Hansen, 1979; Loeblich and Tappan, 1984). Aragonite precipitation appears to be controlled by the amino acid composition of the inner organic lining, which differs considerably from that of calcitic foraminifers (King, 1977).

Hyaline–Perforate Wall Type

The hyaline–perforate wall is composed of pore-bearing layers of calcite crystals enclosed in organic envelopes and deposited in close association with one or more organic membranes (Towe and Cifelli, 1967; Figs. 2.1, 2.4a–b, 2.6c–f). The pores are tubular openings to the exterior that are sealed internally by one or more organic membranes. They appear to be formed by the selective resorption of calcite at predetermined sites (Hemleben et al., 1977). Lamellar wall growth, where the formation of a new chamber is accompanied by secretion of a layer that covers all or part of the exterior of the existing test, is present in most groups (Hansen, 1979). In the rotalines, the inner layer of the new chamber covers the previous apertural face to create a bilamellar septum. The complex hyaline–perforate wall is considered to be the most advanced mode of test construction present in the Foraminifera (Hansen, 1979).

Crystal growth may proceed either on the basal pinacoidal or rhombohedral face and assumes one of a number of possible crystallographic configurations (see Fig. 2.6c–f). The significance of these differences in wall structure and crystallographic orientations for suprageneric classification is a matter of dispute.

In the planktonic foraminifers, a bilamellar wall structure results from simultaneous crystallization on both surfaces of a primary organic membrane (Hemleben et al., 1977). An additional shell layer is deposited over the entire test prior to gametogenesis in spinose species such as *Globigerinoides sacculifer* and *G. ruber* (Bé, 1982). Many species add a veneer of large euhedral crystals to the primary wall as individuals sink into colder subsurface waters during ontogeny (Hemleben et al., 1985).

Models of Calcification

The morphological and crystallographic organization of the hyaline–perforate wall is mediated by an organic precursor whose macromolecular configuration acts as a template for the ordered nucleation and growth of the mineral phase (Towe and Cifelli, 1967). The organic membrane provides a capsule or mold into which appropriate ions are introduced and then induced to crystallize into microarchitectural units of genetically determined size, shape, and crystallographic orientation (Lowenstam, 1981; Lowenstam and Weiner, 1983). In *Heterostegina*, high-performance liquid chromatography and infrared spectroscopy revealed that the organic matrix is composed of a heterogeneous mixture of polysaccharides and proteins (Weiner and Erez, 1984). An EDTA-soluble fraction contains oversulphated glycosaminogol; the insoluble fraction includes at least two classes of proteins. Because of their highly acidic nature, both fractions may have a joint role in the precipitation of calcite by providing Ca^{+} binding sites at precise locations along their amino acid sequences (Weiner and Erez, 1984; Weiner, Traub, and Lowenstam, 1983).

Towe (1972) has critically reviewed the two widely held theories that attempt to explain the function of the organic layer in biomineralization. In the compartment or "passive" theory, the shape and spacing of organic envelopes provide the environment for crystal growth, whereas in the epitaxial theory, the stereochemical properties of a reactive matrix dictate the pattern of subsequent calcification. As Towe (1972) has emphasized, neither viewpoint fully explains the three-dimensional shape and varied crystallographic orientations of biogenic calcite. Some investigators attempt to remedy their defects by combining them into an active–passive model of precipitation. A more satisfactory theory of biomineralization will require consideration of the potential roles of gels and biochemical processes and a more detailed understanding of the composition and configuration of the organic layer (e.g., Weiner, Traub, and Lowenstam, 1983; Weiner and Erez, 1984) and the nature of its contact with newly secreted crystallites (Towe, 1972; Lipps, 1973).

Function of the Test

Many workers have speculated about the functional significance of the foraminiferal test (Marszalek et al., 1969; Banner et al., 1973; Braiser, 1975a, 1982a,b; Haynes, 1981; Hottinger, 1983; Lipps, 1982). On a basic level, it acts as a protective barrier against unfavorable physical, chemical, or biological conditions. In numerous instances, the ectoplasm has been observed to withdraw within the test when an individual was disturbed or exposed to harmful chemicals. The different shapes and internal structures of the test facilitate a variety of specialized functions among various taxa (e.g., feeding in specialized habitats or modes of nutrition; De Laca et al., 1980), the culturing of endosymbiotic algae, providing ballast for stability or maintaining a preferred flotational level in the water column. Lipps (1975, 1982, 1983) and Price (1980) related various test forms to different feeding strategies and to the influence of biotic interactions such as competition.

On theoretical grounds, Brasier (1982a,b) argued that new coiling modes evolved as attempts to shorten internal lines of communication. This hypothesis is supported by the fossil record and by the present dominance of planispiral and multiform (e.g., minimum lines of communication) tests in the ecologically favorable but intensely competitive neritic environment.

The osmiophilic inner organic lining of the perforate Foraminifera is thought to have significant biochemical and functional diversity within a species and from species to species (Banner et al., 1973; Leutenegger, 1977; Hemleben et al., 1977). It insulates the endoplasm from harmful conditions, serves as a filter for sunlight in species with endosymbiotic algae and via the pores, plays a role in many metabolic activities (Banner et al., 1973). In littoral and marsh environments, where adverse water chemistry (low pH, low salinities) may preclude calcification, the organic membrane permits foraminifers to survive and reproduce.

The organic linings of pores appear to function in such processes as ionic transport, osmoregulation, gas exchange, respiration, and the transfer of dissolved nutrients and excreta (Berthold, 1976; Leutenegger, 1977; Leutenegger and Hansen, 1979). To facilitate gas exchange, dense concentrations of mitochondria are located beneath the pores of benthic species living in anoxic conditions (Leutenegger and Hansen, 1979). More evenly dispersed mitochondria are associated with species inhabiting aerated environments. The pores of symbiont-bearing larger Foramnifera are used for the rapid absorption of CO_2 (Leutenegger and Hansen, 1979).

Considerable intra- and interspecific variation in the size and density of pores occurs in the planktonic Foraminifera. The relative surface area of pores may help maintain test buoyancy (by adding to or subtracting from overall test density) and an optimal rate of ionic exchange. In symbiont-bearing epipelagic species, the microperforated pore membrane is the site of the intake and excretion of soluble and colloidal material (Bé et al., 1980). Mesopelagic species have sealed pore membranes, possibly signifying a lower metabolic rate, caused by the lack of endosymbionts and the occupation of colder water masses.

In many agglutinated species, the pores penetrate the full thickness of the chamber wall. Coleman (1980) speculated that pseudopodia emerge from these openings to collect and properly position grains during the process of chamber formation.

The aperture allows the conveyance of food particles, vacuoles, and vegetative nuclei between the endoplasm and the pseudopodial net and is the place where offspring are discharged to the exterior (Braiser, 1980). Supplementary sutural and umbilical apertures, where present, permit the rapid withdrawal of ectoplasm into the test when stressful environmental conditions are encountered. In one genus, *Haynesina,* they also channel nutritive materials to the core of the test without passage through the aperture and outer whorls of chambers (Alexander and Banner, 1984).

Classification

The classification of the order Foraminifera is now under comprehensive review by Loeblich and Tappan (1981, 1982, 1984, 1985; Table 2.1). The heterophasic (haploid gamont) life cycle and polynucleate schizont of the Foraminifera suggest that this group is a monophyletic taxon (Grell, 1979). However, important differences in karyology, cytology, and life cycles that are now being discovered (e.g.,

TABLE 2.1 Classification of the Order Foraminiferida (Eichwald 1830) According to Loeblich and Tappan (1984)

Suborder Allogromiina (Loeblich and Tappan, 1961)

Test unilocular or may tend to become multilocular; wall membranaceous or proteinaceous, may have ferruginous encrustations or small quantity of agglutinated particles. Upper Cambrian to Holocene.

Suborder Textulariina (Delage and Herouard, 1896)

Test agglutinated, foreign particles held in organic or mineralized ground mass. Cambrian to Holocene.

Suborder Fusulinina (Wedekind, 1937)

Test wall of homogeneous microgranular calcite, of tightly packed equidimensional subangular crystals, a few micrometers in diameter. Advanced forms with wall differentiated into two or more layers. Ordovician to Triassic.

Suborder Involuntinina (Hohenegger and Piller, 1977)

Proloculus followed by enrolled tubular second chamber; wall calcareous, perforate, radiate, probably originally aragonitic, but may be recrystallized to homogeneous microgranular structure, thickened or with pillarlike structures in umbilical region of one or both sides. Triassic to Cretaceous.

Suborder Miliolina (Delage and Herouard, 1896)

Test of porcelaneous calcite, commonly with organic lining and may have added adventitious material; generally imperforate in postembryonic stage; may have spiral passage between proloculus and later chambers. Carboniferous to Holocene.

Suborder Silicoloculinina (Resig, Lowenstam, Echols, and Weiner, 1980)

Wall imperforate, of secreted opaline silica. Middle Miocene to Holocene.

Suborder Spirillinina (Hohenegger and Piller, 1975)

Proloculus followed by enrolled tubular undivided chamber, or by a few chambers per whorl, coiling planispiral to high trochospiral; wall of calcite, optically a single crystal or few to a mosaic of crystals; *a*-axis preferred orientation along axis of coiling and *c*-axis parallel to umbilical surface (*Patellina*); wall formed by accretion at edge, not by calcification of an organic template produced by pseudopodia. ? Triassic, Jurassic to Holocene.

Suborder Lagenina (Delage and Herouard, 1896)

Wall of monolamellar, optically and ultrastructurally radiate calcite, with crystal *c*-axes perpendicular to surface; crystal units enveloped by organic membranes; primitive taxa without secondary lamination, later ones secondarily lamellar. Middle Carboniferous to Holocene.

Suborder Robertinina (Loeblich and Tappan, 1984)

Test planispirally to trochospirally enrolled; wall of hyaline perforate, ultrastructurally and optically radiate aragonite (orthorhombic crystal form of calcium carbonate), the hexagonal prisms with *c*-axis normal to the wall surface and basal pinacoid parallel to the surface. middle Triassic to Holocene.

Suborder Globigerinina (Delage and Herouard, 1896)

Planktonic habit; test wall of perforate hyaline calcite, optically radiate, preferred crystal orientation with *c*-axis normal to surface; primary lamination bilamellar, with secondary lamination due to addition of material at formation of new chamber. Lower Jurassic to Holocene.

Suborder Rotaliina (Delage and Herouard, 1896)

Test multilocular, typically enrolled, but may be reduced to biserial or uniserial, or chambers may proliferate with an encrusting habit; chambers simple or subdivided by secondary partitions; wall calcareous, of perforate hyaline, lamellar calcite (hexagonal crystal form of calcium carbonate), formed by calcification at each side of an organic membrane; aperture may be simple or have an internal toothplate, entosolenian tube, or hemicylindrical structure; internal canal systems or stolon systems present in some. Triassic to Holocene.

Arnold, 1984) raise some uncertainty on this question. Most investigators (e.g., Tappan, 1976) derive the biomineralized suborders from agglutinated ancestors (Fig. 2.7). Equally likely are separate derivations of these taxa directly from different stocks within the tectinous lagynacids (Allogromiina) (Arnold, 1978).

Loeblich and Tappan (1982, 1984; Tappan, 1976) consider suprageneric taxonomic categories as subjective constructs, in contrast to the reality of species. Their approach to classification involves a simultaneous assessment of key aspects of test morphology, biology, and data on stratigraphic distributions. Taxonomic criteria used in the classification are weighted a priori as to their significance: wall composition defines suborders; wall structure, internal features, and septal patterns distinguish superfamilies and families. The stated goal is to re-

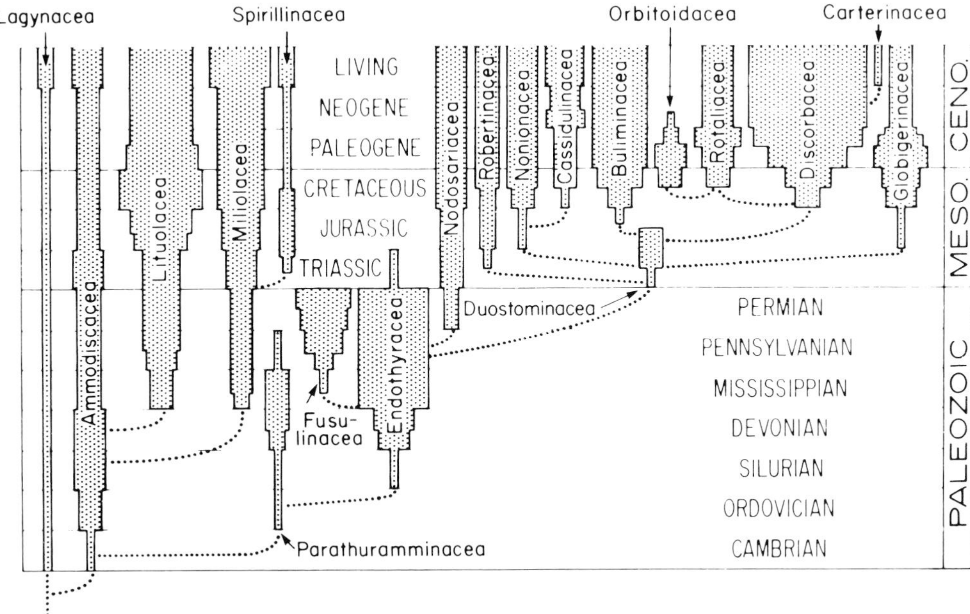

FIGURE 2.7 Suggested evolutionary relationships of the superfamilies of Foraminifera. In this reconstruction biomineralized taxa are derived from the agglutinated Ammodiscacea, which is turn evolved from the tectinous Lagynacea. (From Tappan, 1976, Fig. 4.)

flect the phylogenetic history of the Foraminifera in a classification composed of "natural" groups based on total morphological affinities (Tappan, 1976).

Their study, when completed and published in the Paleontological Society's *Treatise on Invertebrate Paleontology,* is not likely to be universally accepted. Many contentious problems remain in identifying the origin of major groups from the available fossil record and in assessing the evolutionary polarity (primitive–advanced) and possible monophyletic origin of important morphologic features. Loeblich and Tappan view their forthcoming volume as a provisional synthesis that will be modified as new data become available. In the future, a better understanding of the biochemistry of the organic matrix (King and Hare, 1972; King, 1977; Weiner and Erez, 1984), the structure and function of organelles (Grell, 1979; Arnold, 1984), and cytoplasmic fine structure (Anderson and Tuntivate-Choy, 1984) of Foraminifera will permit refinement of a classification that remains to a large degree based on the preservable morphology of the test.

Ecophenotypic Variation

The Biological Species Concept

The accurate discrimination of foraminiferal species is essential in paleoenvironmental, biostratigraphic, and evolutionary studies. Ideally, the description of a species should be accompanied by biometrical analysis of large populations collected from different locations and stratigraphic horizons. In practice, however, many new species have entered the literature supported only by brief qualitative characterizations of a small assemblage drawn from a single sample that is unlikely to contain the full expression of genotypic variability. Several investigations of intraspecific polymorphism (Hermelin and Malmgren, 1980; Buzas et al., 1985) have suggested that a significant fraction of the over 40,000 described species of Foraminifera (Tappan, 1971) are synonyms (Fig. 2.8). The recognition and elimination of redundant taxa are important steps toward a more realistic appraisal of biogeographic and stratigraphic distributions and within-habitat diversity.

Modern systematic theory considers the species as a discrete "individual" on which evolutionary processes act (Eldredge and Novacek, 1985). The species is seen as a morphologically and ecologically cohesive entity composed of individuals connected in time and space by a shared genetic heritage. Some workers question whether this "biological species" concept can be applied to heterophasic protistans (Hull, 1976; Gould and Eldredge, 1977; Lazarus, 1983). However, the presence of genetic recombination and gametic fusion during part of the life cycle of almost all foraminifers indicates that this view of the species provides the only meaningful basis for foraminiferal taxonomy.

Successful recognition of biologically valid fora-

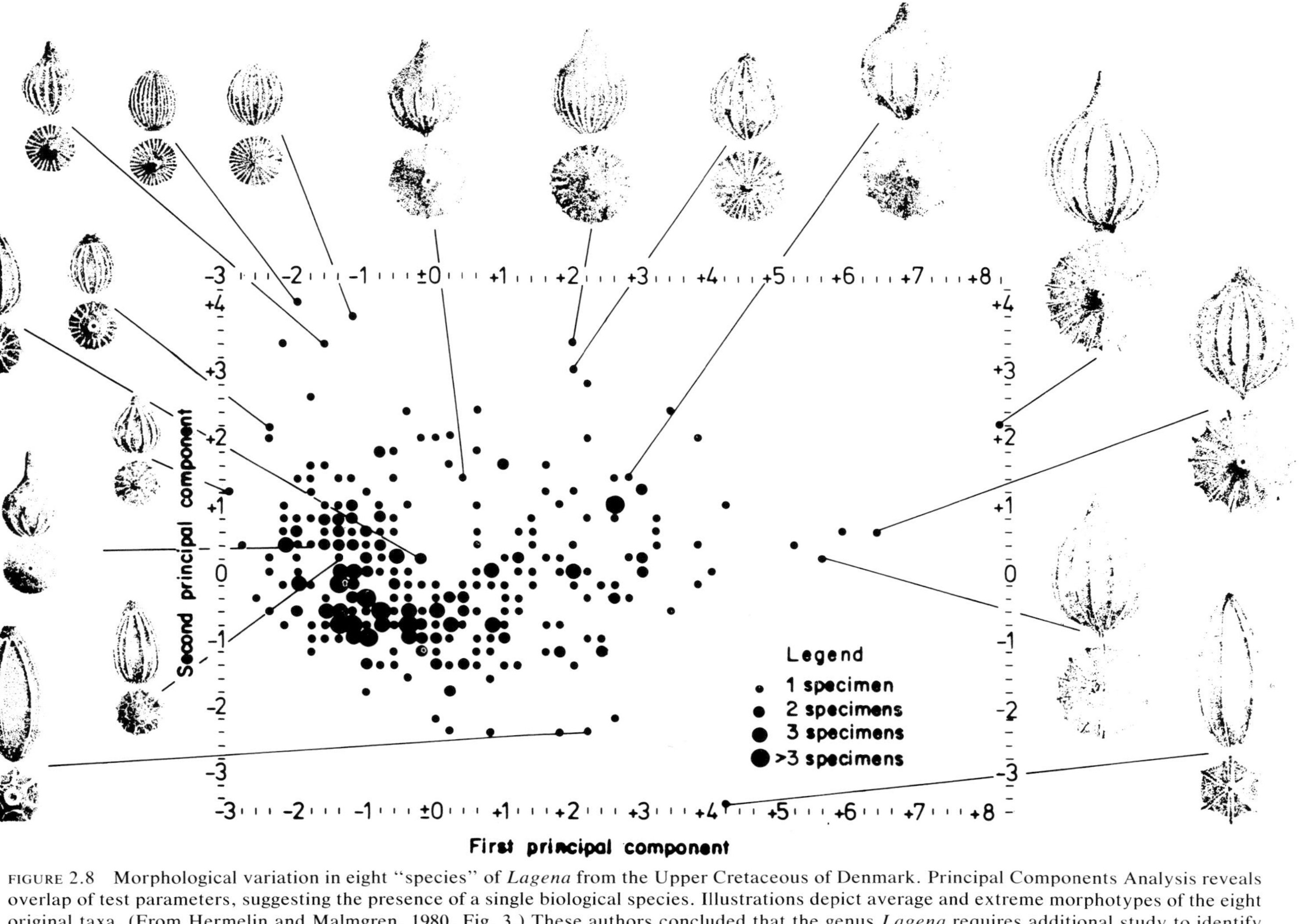

FIGURE 2.8 Morphological variation in eight "species" of *Lagena* from the Upper Cretaceous of Denmark. Principal Components Analysis reveals overlap of test parameters, suggesting the presence of a single biological species. Illustrations depict average and extreme morphotypes of the eight original taxa. (From Hermelin and Malmgren, 1980, Fig. 3.) These authors concluded that the genus *Lagena* requires additional study to identify possible synonyms among over 1,000 described species.

miniferal species for use in applied research requires an appreciation of the potential ecophenotypic and genetic polymorphism that can occur within and between geographically isolated populations. Fortunately, powerful multivariate statistical methods and automated shape-digitizing procedures are now available for the quantification and analysis of morphological variation (Scott, 1974; Arnold, 1982a; Healy-Williams and Williams, 1981; Werdelin and Hermelin, 1983; Reyment, 1983; Lohman, 1983; Buzas et al., 1985): These techniques permit the objective recognition of morphologically overlapping populations, which can be inferred to be reproductively compatible, and the discontinuities that set them apart from other species.

Benthic Species

In many fossil and recent species, a wide array of morphological traits change significantly in response to ecological gradients present within their geographic and statigraphic range (*larger forms:* Fermont, 1977; Fermont et al., 1983; Hallock, 1979; Hallock and Hansen, 1979; Drooger, 1984; *smaller taxa:* Feyling-Hansen, 1972; Poag, 1978; Corliss, 1979; Ivert, 1980; Miller et al., 1982; Reyment, 1982; Hermelin, 1983; Lamb and Miller, 1984; Painter and Spencer, 1984). Reyment (1983) has proposed that intraspecific variation in benthic foraminifers often results from the combined effects of weak, divergent selection and deme-specific ecophenotypic polymorphism in semiisolated populations of a species occurring in a heterogeneous (in time or space) environment.

The hyaline–perforate species *Elphidium excavatum* is common and widely distributed in marsh, estuarine, and shelf environments on both sides of the North Atlantic from Arctic to warm-temperate latitudes. Its phenotypes differ in ornamentation, test shape, and sutural morphology (Fig. 2.9). European assemblages exhibit a clinal polymorphism that correlates with temperature (Feyling-Hansen, 1972; Wilkinson, 1979). In the western North Atlantic, this species is represented by five sympatric "formae" that can be found in an intergradational morphological series at any given location. Each forma, however, is most abundant in a restricted environmental setting within the overall range of the species. The extent of morphological divergence present at a given site is determined by the degree of climatic and environmental fluctuations. An isolated and morphologically distinct upper-bathyal population is, predictably, more uniform than those existing in littoral and shelf locales (Schafer and Cole, 1982; Buzas et al., 1985).

Significant morphological variation occurs in all the dominant constituents of the foraminiferal fauna of San Antonio Bay (Poag, 1978; Fig. 2.10). Each species changes clinally along temperature–salinity gradients between two end members that had been previously considered separate taxa. Within calcareous species, populations with thin-walled, small tests and fewer total chambers live in optimum environments and reach reproductive maturity early. Large, thickly calcified tests are associated with marginal conditions and the delayed onset of reproduction. This pattern of morphological variation can be seen in environmentally analogous faunas in the Gulf Coast dating back to the Miocene.

FIGURE 2.9 Morphotypes of *Elphidium excavatum* from the northeast Atlantic. (A) forma *selseyensis*—temperate to polar estuaries. (B) f. *magna*—nearshore, turbulent environments. (C) f. *clavatum*—normal to slightly hyposaline environments. Approximate width of each specimen = 200 μm. (Courtesy of Dr. Anne B. Miller, Dalhousie University.)

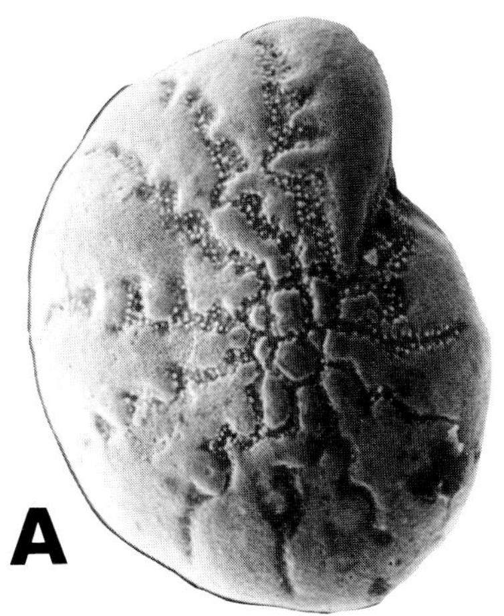

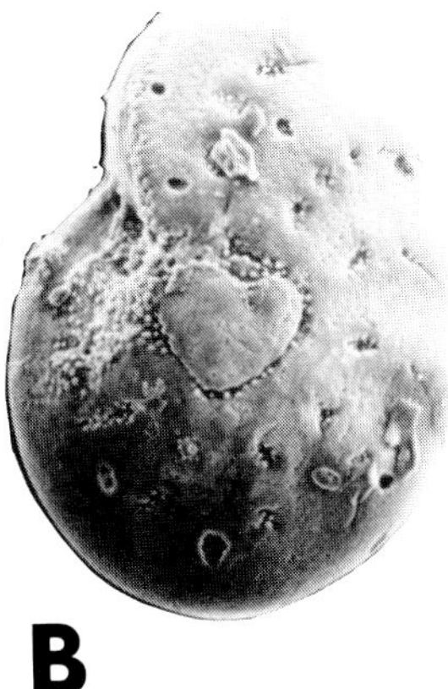

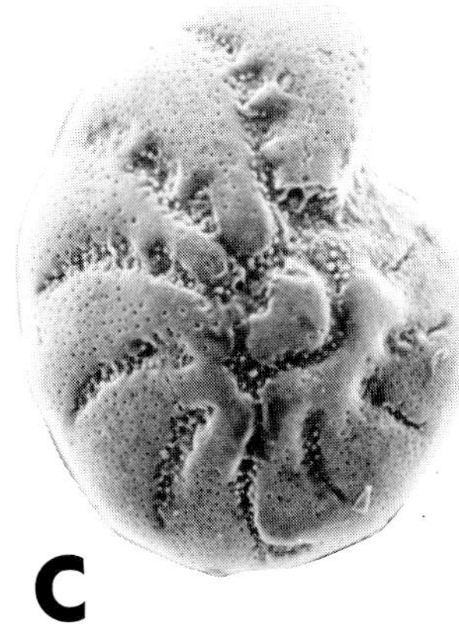

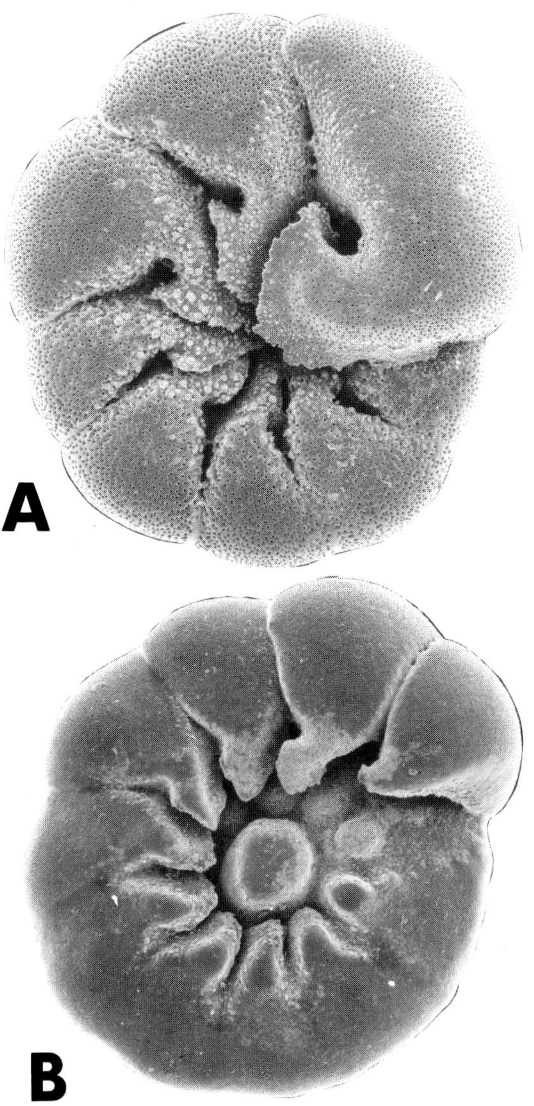

FIGURE 2.10 Paired ecophenotypes of *Ammonia parkinsonia*, San Antonio Bay, Texas. (A) forma *tepida*—optimal conditions (approximate width of specimen - 600 μm). (B) f. *typica*—marginal conditions (approximate width 1,200 μm). (Courtesy of Dr. C. Wyllie Poag, U.S. Geological Survey.)

Planktonic Species

Ecophenotypic responses to changes in the hydrographic properties of water masses in different latitudes and depths are known to occur in almost all living species of planktonic Foraminifera (Table 2.2, Kennet, 1976; Vincent and Berger, 1981). This morphological variation may be expressed in a gradational series (e.g., test porosity) or as discrete states of a threshold character (e.g., coiling direction; Fig. 2.11). In many forms the shift in phenotypic parameters takes the form of a simple cline with respect to latitude and water temperature (Kennett and Srinivasan, 1980). In others it results from fluctuations in the relative proportions of distinct phenotypes with contrasting ecological requirements (Healy-Williams et al., 1985). Many aspects of test polymorphism in planktonic species have been interpreted as biomechanical solutions to the problem of passive sinking in waters of dissimilar densities (Scott, 1973a,b; Lipps, 1979; Healy-Williams et al., 1985).

Opinions vary on the ecological significance of some ecophenotypic characters. Malmgren and Kennett (1977) argued that a high percentage of individuals with dwarfed (kummerform, see Fig. 2.11a) final chambers would result when an assemblage was subjected to unfavorable conditions. Alternatively, Kahn (1981) concluded that this feature represented a genetically controlled reduction in the rate of chamber expansion as a maximum test size was approached. In his view, optimal environments, encouraging rapid, early growth, would lead to a preponderance of kummerforms within a population.

"Larger" Foraminifera

An important aspect of the historical development of the Foraminifera since the late Paleozoic has been the iterative and heterochronous evolution of "giant" taxa (to 15 cm) from normal-sized microgranular, porcelaneous, and hyaline–perforate ancestors (for taxonomic and morphological review see Haynes, 1982, Chaps. 7, 8, 13; Fig. 2.12). Living examples (e.g., Fermont, 1977; Hallock, 1979, 1981a, 1984; Hottinger, 1982; Leutenegger, 1984) dwell exclusively in well-lit tropical shallows, cultivating endosymbiotic algae in an advanced form of mixotrophic nutrition. By extrapolation, morphologically similar forms in the fossil record are thought to have utilized the same trophic strategy. (Ross, 1972; Hottinger, 1982). The obligate and host-specific symbiotic relationship, found in Recent species, probably developed early in the history of each group (e.g., Leutenegger, 1984). Striking parallels in the evolutionary development of text shape and internal structures between and within the different groups of larger foraminifers provide some of the best-documented cases of convergence (Ross, 1972; Hottinger, 1983). Haynes (1981) considered that, in their physiology and ecology, these forms resemble acellular colonial syncytia more than heterotrophic smaller Foraminifera.

TABLE 2.2 Selected Examples of Phenotypic Variation in Recent Planktonic Foraminifera

Example	Species	Comments	References
Coiling ratios	*Neogloboquadrina pachyderma*	Correlates with sea-surface temperature	1
Test shape	*Globorotalia truncatulinoides*	Conicalness of test correlates with temperature	2
Shape of final chamber	*Globorotalia cultrata*	Flexuous forms associated with upwelling	3
Number of chamber in final whorl	*Globigerina bulloides*	Decreases with decreasing water temperature	4
Size	*Orbulina universa*	Small test size develops in optimum conditions	1
Pigmentation	*Globigerinoides ruber*	Intensity varies with seasonal temperature variation and latitude	3
Development of imperforate keel	*Globorotalia crassaformis*	Varies with water temperature	1
Apertural size and shape	*G. bulloides*	Increases with decreasing temperature	4
Secondary calcification	*Globorotalia* spp.	Deposition of calcite crust correlated with decreasing water temperatures encountered during ontogenetic sinking	5
Porosity	*O. universa*	Highest porosity in warmer waters	1
Umbilical plates	*Neogloboquadrina dutertrei*	Tropical assemblages develop a series of umbilical plates that partially close the aperture	6

Sources: (1) Kennett (1976); (2) Healy-Williams (1983), Lohmann and Malmgren (1983), Healy-Williams et al. (1985); (3) Vincent and Berger (1981); (4) Malmgren and Kennett (1976, 1977, 1978); (5) Hemleben et al. (1985); (6) Kennett and Srinivasan (1980).

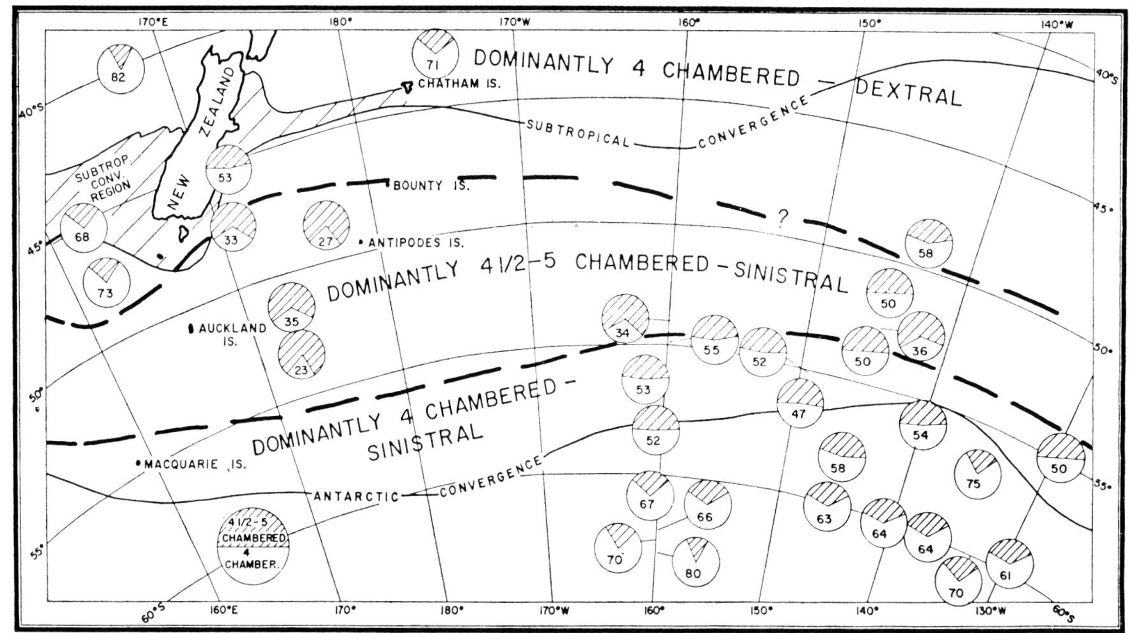

FIGURE 2.11 Ecophenotypic variation in test morphology and surface wall texture in the planktonic species *Neogloboguadrina pachyderma* from bottom sediments of the south Pacific Ocean. Line drawing = clinal changes with latitude in overall test parameters. (From Kennett, 1976, Fig. 9.)

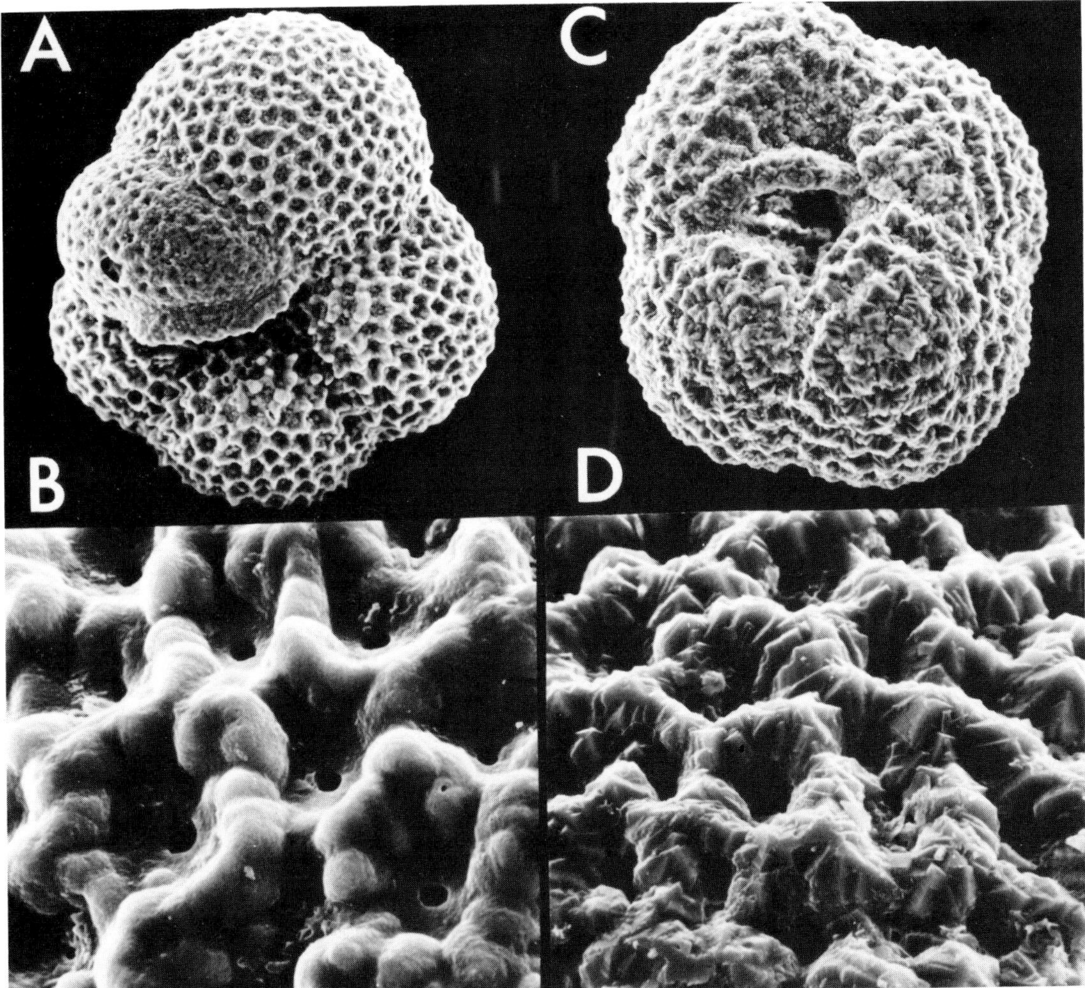

FIGURE 2.11 *(continued)* Micrographs A–D indicate differences in wall texture in specimens from contrasting water masses. (From Srinivasan and Kennett, 1974, fig. 2; see also Kennett and Srinivasan, 1980.) (A and B) Reticulate-walled form with four and a half chambers in the final whorl. Note the small kummer-form final chamber in (A). (C and D) Specimen with crystalline wall texture consisting of a veneer of blocky, euhedral calcite crystals. Among populations of this species, reticulate wall texture is predominant in arctic and subarctic regions, whereas crystalline wall texture is most frequently observed in warm-temperate water masses. (SEM reproduced courtesy of Dr. James P. Kennett, University of California, Santa Barbara.)

Advantages of Mixotrophic Nutrition

Hallock (1981b) has analyzed the energetic benefits of obligate endosymbiosis in larger Foraminifera (Fig. 2.13). With plentiful radiant energy and the efficient recycling of nutrients between the host and the symbiont, both the production of organic matter (two to three orders of magnitude) and the availability of energy for growth (one to two orders of magnitude) increases in comparison to heterotrophic species. Maximal advantages accrue to mixotrophs in stable, oligotrophic (low quantities of dissolved nutrients) environments where heterotrophs must expend large quantities of energy to capture organic matter concentrated in particulates (Hallock, 1981b, Table 2.2; 1982, 1985). Under such conditions, a mixotroph is capable of gradually accumulating the nutrients and calcium ions necessary for large test size by slow but sustained growth and the delay of reproductive maturation. The species profits from this strategy because the number and size of young eventually produced in semelperous (suicidal) reproduction is increased (see Hottinger, 1982). The larger the daughter cells, the greater their chance of survival (Hallock, 1982, 1985). However, the potential for a rapid increase in population size is correspondingly reduced. Sexual reproduction is suppressed because asexual propa-

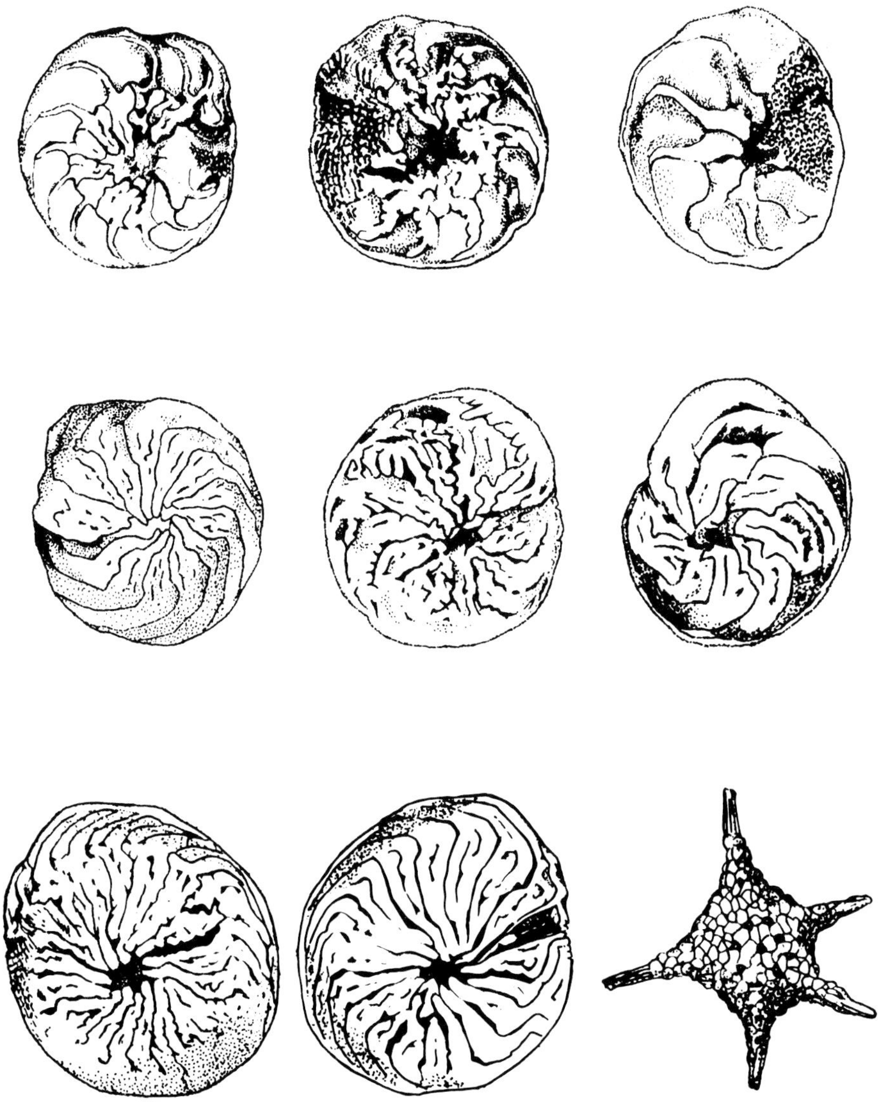

FIGURE 2.12 Examples of larger symbiont-bearing Foraminifera from Pacific reefs. All specimens *Amphistegina* spp. except lower right, which is *Baculogypsina*. (From Hallock, 1984, part of Fig. 1.)

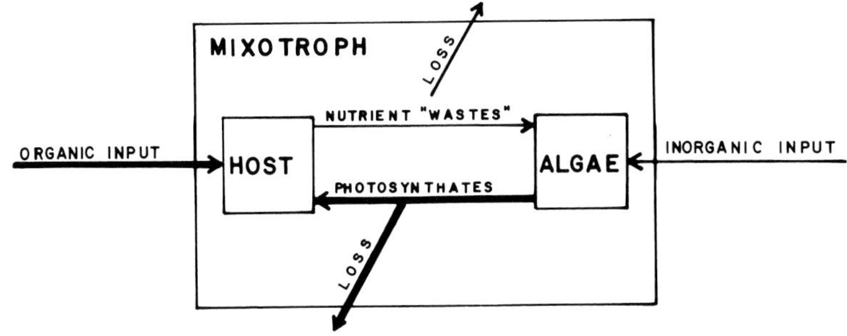

FIGURE 2.13 Simplied model of recycling in a foraminiferal–algal symbiotic system. Wide arrows represent transfer of nutrients in organic compounds. Narrow arrows depict movement of dissolved inorganic compounds. (From Hallock, 1981b, Fig. 1; reproduced with permission from *Marine Biology*, copyright 1981 by Springer-Verlag.)

gation can effectively reproduce clonal populations well adapted to predictable and enduring environmental conditions.

Morphological Adaptations

Many aspects of the morphology and life habits of larger symbiont-bearing foraminifers are correlates of mixotrophic nutrition in shallow, high-energy environments (see summaries in Haynes, 1981, and Hottinger, 1982, 1983).

Discoidal forms are, typically, sessile inhabitants of low-energy environments (Hottinger, 1982; Brasier, 1984). Their exterior walls and high surface-to-volume ratio maximize the capture of sunlight. Such species provide for the internal transfer of nutrients, vacuoles, and organelles by secreting annular or cyclic chambers (Braiser, 1984). In areas where turbulent conditions prevail, more robust fusiform (spindle-shaped) species are common (Brasier, 1984). Pseudopodia, emerging from numerous openings along the elongate apertural face or from the axial regions, firmly attach the test to the substrate and provide the capacity for locomotion when needed (Hottinger, 1982).

The symbionts of extant species are concentrated near the external walls to increase their exposure to sunlight (Leutenegger, 1984). Some forms incline their test upward at an angle of 45° so as to facilitate the photosynthetic activities of endosymbionts located on both the ventral and dorsal sides. The depth of occurrence of many species coincides with the limits of maximum photosynthetic efficiency of their symbionts (Leutenegger, 1984). Test sphericity increases with depth in many species; this trend may be a mechanism to increase surface area in low-light conditions (Fermont, 1977, 1982; Hallock, 1979; Hallock and Hansen, 1979; Hottinger, 1983; but see Rottger and Hallock, 1982). Within the test of some forms, a single alga is housed in a special cup located near a pore that provides for the efficient exchange of gases and nutrients (Leutenegger and Hansen, 1979). Complex canal and stolon systems (Hottinger, 1982) provide effective channels for internal cystoplasmic movement without disturbing the preferred position of the symbionts.

In porcelaneous species, thin, transparent segments of the exterior wall are situated adjacent to reentrants. These structures collectively reduce the thickness of the wall to a few millimicrons (Hansen and Dallberg, 1979) and may serve the same function as the pores in hyaline–perforate taxa.

Geologic History

Larger Foraminifera thrived and diversified during periods of high sea levels and equable climates (late Paleozoic, Jurassic–Cretaceous, Paleocene–Middle Eocene) (Hallock, 1982, 1985). The extensive epicontinental seas of these intervals were nutrient-poor as a consequence of diminished rates of oceanic upwelling and buried sources of clastic sediments. Thus, mixotrophs enjoyed a critical competitive advantage and expanded in numbers and types (Hallock, 1982). Each radiation was terminated by orogeny and regression, which augmented nutrient supply to the now restricted shelf habitats. These regions were quickly dominated by faster-growing, more rapidly reproducing, small heterotrophs.

During the Cenozoic, larger Foraminifera generally were restricted to a belt 35 degrees north and south of the equator. Their geographic distributions can be grouped into three persisting faunal provinces: Neotropical, Mediterranean and Indo-West Pacific (Adams, 1983). The Paleocene to Middle Eocene was a time of globally warm climates and high sea level. Recovering from their near-total demise at the end of the Cretaceous, larger Foraminifera diversified rapidly and achieved cosmopolitan distributions at this time (Adams, 1983). Many species existed as epiphytes on the newly evolved sea grass *Thalassia,* which actively colonized shallow, sublittoral level bottoms throughout the tropics and subtropics (Brasier, 1975b; Eva, 1980). Dispersal of rafts of sea grass by winds and currents may have assisted the transoceanic passage of some ecologically associated Foraminifera (Brasier, 1975b). During an Early Eocene climatic optimum, some genera of larger Foraminifera reached 60 degrees north and 45 degrees south (Adams, 1983).

The Late Eocene and Oligocene were characterized by regression and steepening of pole-to-equator climatic gradients. The diversity of larger Foraminifera decreased and endemism became increasingly pronounced. In the middle Miocene, deteriorating global climate, sea-level lowering, orogeny, and intensified oceanic circulation coincided with the extinction of many genera (Adams, 1983). Indigenous taxa in the Mediterranean were extirpated by the Late Miocene "salinity crisis" (Adams, 1976). Having experienced the severe climatic fluctuations of the Pleistocene, the faunas that now inhabit today's eutrophic world ocean are depauperate and isolated survivors of past assemblages (Hallock, 1985).

Fusulinina

This extinct order of late Paleozoic microgranular-walled Foraminifera was the first to evolve large test size, fusiform test shape, and internal modifications as adaptions to a mixotrophic mode of nutrition and life on agitated shallow-water bottoms (Ross, 1982a,b; Fig. 2.14). Early forms were small and lenticular, but advanced spindle-shaped taxa reached lengths of 14 cm. The test is planispiral and lacks primary apertures. Fusulinid taxonomy is based on the structure of the wall and modifications of the internal chamber space (Ross, 1982a,b; see Fig. 2.14). These organisms were restricted to well-aerated, near-shore, shelf, and upper-basin-slope settings, thriving in close association with corals and algae. Distinct habitats, defined by salinity, temperature, depth, and turbidity, were occupied by morphologically specialized and taxonomically unique faunas that maintained their own evolutionary identity during Carboniferous to Middle Permian time (Ross, 1967; Table 2.3). Fusulinids are rare to absent in macrofossiliferous limestone beds, except for cross-bedded, coarse-grained channel deposits where the mixing of faunas from different environments was likely. Species occupying high-energy environments may have formed a dense, cohesive mat whose intertwined pseudopodia provided resistance to involuntary dislodgement (Ross, 1972).

Many of the internal structures of the Fusulinina have been interpreted as functional adaptions to their mode of life. Alveoli in the keriotheca are large enough to harbor endosymbionts such as the dinoflagellates that are present in morphologically analogous, extant, tropical Foraminifera such as *Marginopora* (Ross, 1972; 1982a,b). The tunnel permitted the rapid movement of vegetative nuclei and nutrients within the test. Increasing amounts of secondary calcite deposition during ontogeny helped to eliminate excess metabolic calcite and added ballast for improved stability. The nepiont (proloculus plus the first few chambers) was retained in the adult until sufficient size and weight had been achieved to lessen its chances of being swept away from favorable environments by currents and storms.

Fusulinids were limited to tropical and warm-temperate seaways located within 35 degrees north and south of the equator (Ross, 1967). Early Carboniferous faunas dispersed freely across unrestricted circumequatorial corridors (Fig. 2.15a). Warm surface currents flowing parallel to the western margin of North America and northward through the Ural Seaway allowed fusulinids to reach the Franklin Geosyncline (arctic Canada), the northern coasts of Eurasia, and the midlatitudes of western South America. Subsequently, the coalescence of continental blocks during the formation

FIGURE 2.14 Cross sections of seven families of fusulinids. The inserts show enlargement of walls to illustrate differences in wall structure and thickness. (From Fig. 2 of Ross, 1982a)

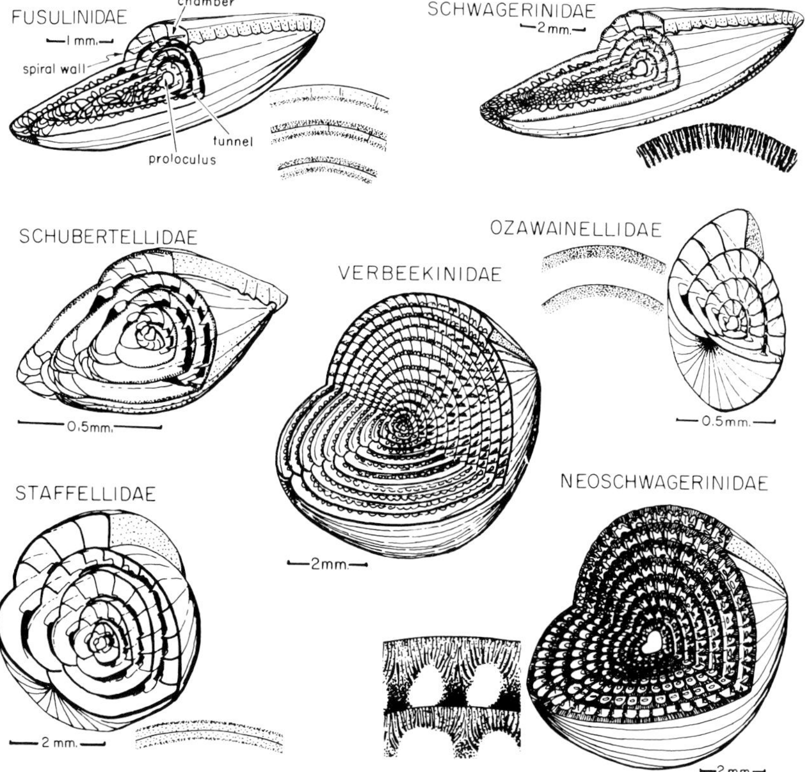

TABLE 2.3 Paleoenvironments of Pennsylvanian and Permian Fusulinid Faunas of Western North America (after Ross, 1971)

Fauna	Distinctive morphology	Depositional environment
T. carpaxoides, *Dunbariella* *Schwagerina*	Subcylindrical tests with dense axial fillings	Near shore
T. spp., *Schubertella* *Ozawainella*, *Stafella*	Thick fusiform tests with gently folded septa, symmetric chomata, and thick, coarsely alveolar walls	Shallow shelf with algal meadows
T. tenuis, *Verbeekina*		Shelf edge mounds
T. acutiuloides		Deep water (basin slope mounds)
T. jackboroensis lineage	Thick fusiform tests	Very shallow algal-rich platforms
T. ohioensis, *T. burgessae*	Elongate tests with irregular outline	Shallow, turbid waters associated with regressive and initial transgressive phases
Neoschwagerina, *Stafella*, *Verbeekina*	Inflated subspherical tests	Meroplanktonic (epipelagic)

Generic abbreviation: *T.* = *Triticites*.

of Pangaea and the concomitant orogenic uplift disrupted marine connections and increased environmental gradients between low and midlatitudes. Progressively more endemic faunas developed during the Middle Carboniferous and Permian (Fig. 2.15,b.c). At least five more or less isolated faunal realms can be distinguished in rocks of this age: Eurasian–Artic, midcontinent–Andean, middle Cordilleran, and an eastern and western Tethys (Ross, 1967; Ross and Ross, 1985a; Fig. 2.15b,c). The temperate faunas of the Franklin Geosyncline died out in the middle Permian as this region drifted further to the north. These extinctions were followed in rapid succession by the disappearance of fusulinids in the Andean and midcontinent regions. Tethyan assemblages underwent a brief diversification in the Late Permian, characterized by the appearance of the highly specialized Schubertellidae and Stafellidae, but became extinct by the end of that period (Ross, 1982a,b).

The complex internal morphology, rapid evolution, and relatively wide geographic distribution of the fusulinids make them excellent intra- and interregional index fossils for the Carboniferous to Permian interval (Douglass, 1977; Ross, 1984; Douglass and Nestell, 1984; Okimura et al., 1985). However, increasing provincialism during the middle and late Permian limits the degree of correlation possible between faunal realms. Fusulinid biostratigraphic zones in the Carboniferous to Middle Permian are contained within unconformity-bounded sequences that reflect globally synchronous transgressive–regressive cycles (Ross and Ross, 1985b). Thus, it is likely that the abrupt appearance and disappearance of fusulinid faunas resulted from episodic deposition and not from punctuated evolution (Ross and Ross, 1985b). Late Paleozoic sea-level cycles, now recognized in the United States, northwestern Europe, and the Soviet Union, may be related to sea-floor spreading events or to glacioeustatic fluctuations caused by the growth of continental ice sheets in the southern part of Gondwanaland (Ross and Ross, 1985b).

Fusulinid provinciality has helped to decipher the complex history and composite nature of the western Cordillera of North America (Douglass, 1967; Ross and Ross, 1983; Monger and Ross, 1984). During the late Paleozoic and Mesozoic, fragments of island arcs, back-arc basins, and oceanic plateaus, originally located in the central and western parts of Panthallassia, were swept eastward and collided with the western margin of North America (Fig. 2.15). At least eight fault-bounded "accreted terranes" are now juxtaposed in subparallel blocks, each with its own depositional and tectonic history. Some of these terranes contain far-traveled Tethyan fusulinid faunas with close affinities to those of Japan and the eastern margin of Eurasia.

Ecology and Paleoecology of Smaller Benthic Foraminifera

Ecology and Distribution

Salt Marshes and Estuaries

Foraminiferal faunas in salt marshes are characterized by low species diversity, widely variable densities over short periods and in small areas, the presence of cosmopolitan species, and differences in species composition and relative abun-

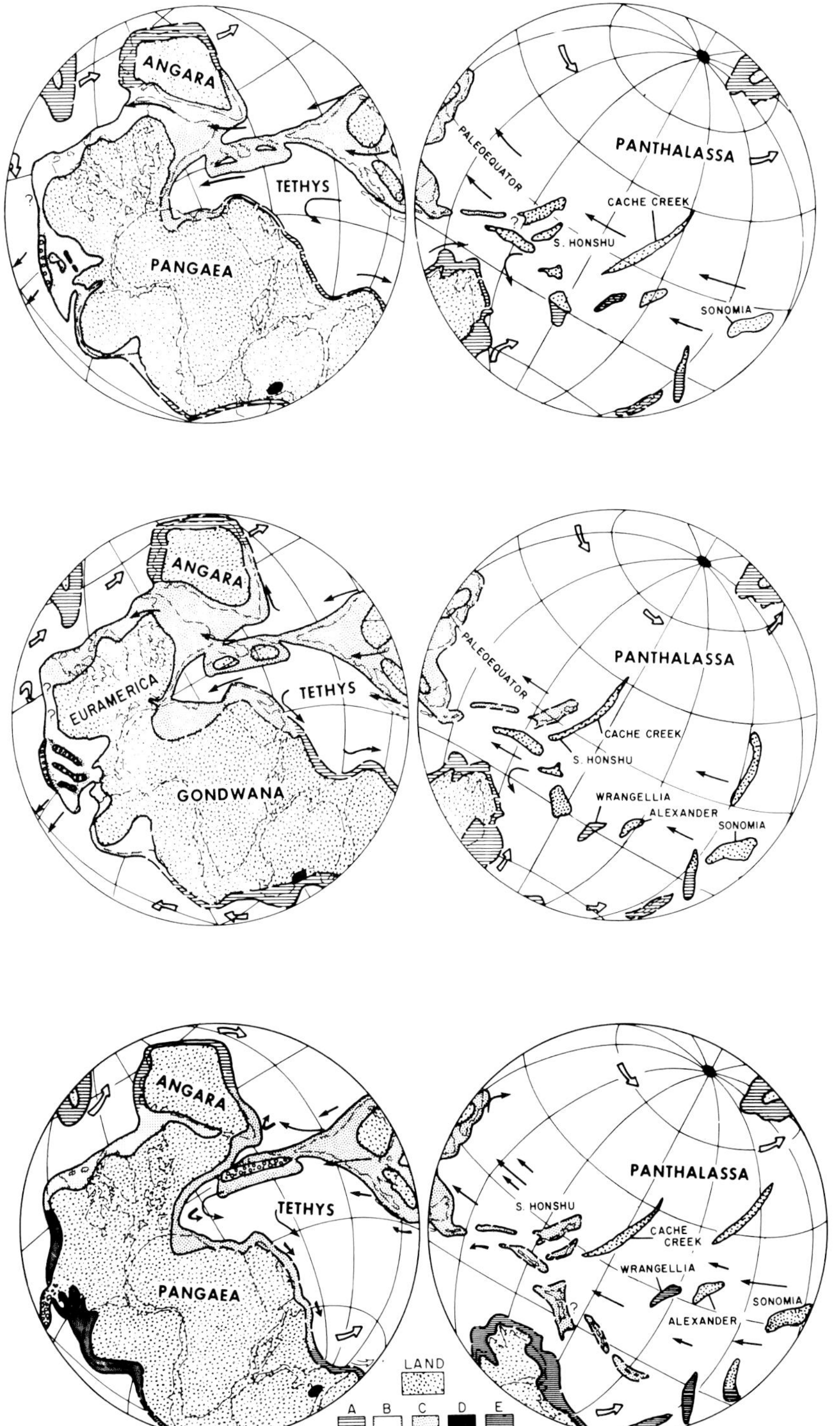

FIGURE 2.15 Paleogeographic reconstruction of late Paleozoic continental blocks and their fusulinid provinces. Top: early Carboniferous. Middle: late Carboniferous. (Note the coalescence of continental fragments into the supercontinent Pangaea.) Bottom: early Permian. Solid arrows show the path of warm surface currents; open arrows indicate cool surface currents. Fusulinid provinces: A, E = cool faunal regions; B = Franklinian—Ural province; C = cosmopolitan, tropical (Tethyan) province; D = southwestern North American—Andean province. Small land masses and oceanic plateaus in western Panthallasa were driven westward (S. Honshu) and eastward (Sonomia, Wrangellia, Cache Creek) into collision with the main body of Pangaea to form accreted terranes. (From Ross and Ross, 1985a, Figs. 1, 3, and 4. Reprinted from *Geology* by permission of the Geological Society of America.)

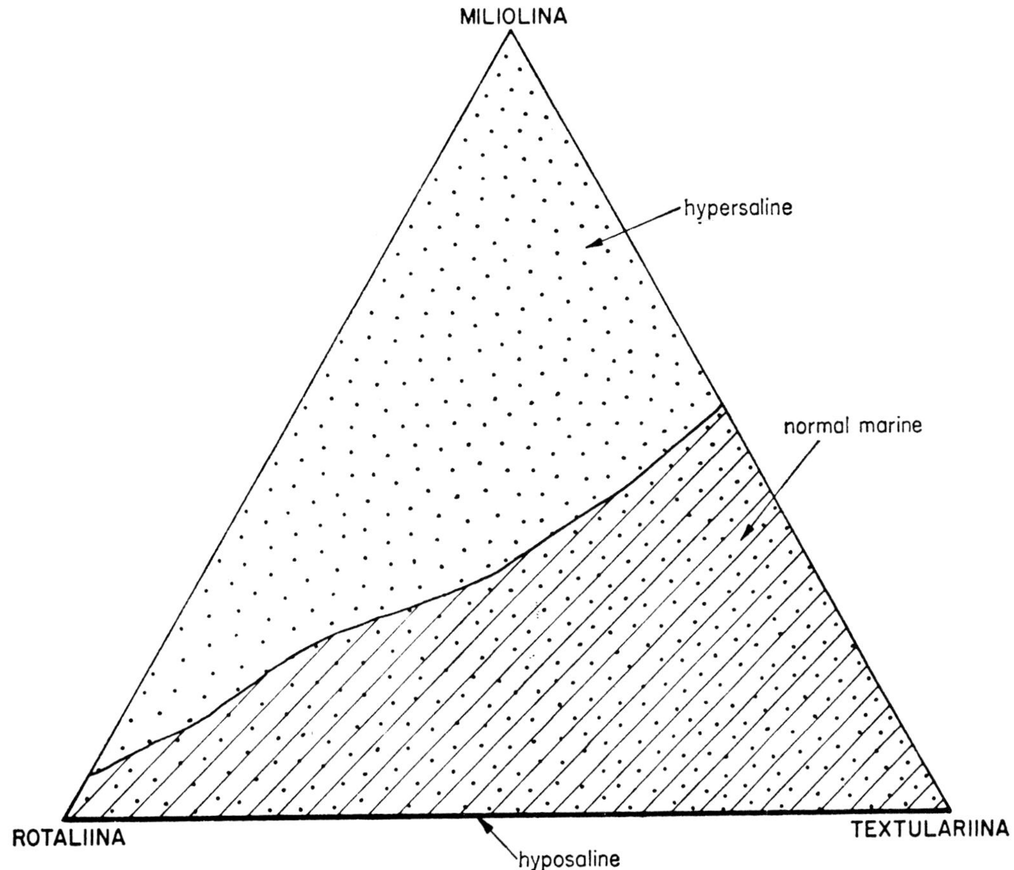

FIGURE 2.16 Triangular plot showing relative abundances of agglutinated, porcelaneous, and hyaline-perforate Foraminifera in salt marshes of different salinities. (From Murray, 1973, Fig. 15.)

dance between nearby sites (Murray, 1973; Scott, 1976; Scott and Medioli, 1980). Assemblages from hyposaline, normal-marine, and hypersaline salt marshes are readily discriminated on a triangular diagram of the relative abundance of calcareous, arenaceous, and porcelaneous species (Murray, 1973; Fig. 2.16). The distribution of species within a salt marsh is determined by elevation above sea level, which influences the time of subaerial exposure during a tidal cycle and the nature of the macrophytic cover (Scott and Medioli, 1980). The local abundance of a given species within its range is controlled by a complex nexus of abiotic and biotic factors (Buzas, 1969; Buzas et al., 1977; Buzas and Severin, 1972; Haman, 1983). Many species are able to increase their population size rapidly when conditions are favorable. The growth of these "blooms" may be terminated by predation and by other interspecific interactions (Buzas, 1982; Buzas and Severin, 1982).

Salt marsh and estuarine forms are capable of living and reproducing to depths of 14 cm beneath the surface in substrates with suitable organic content, porosity, and interstitial pore-waters (Frankel, 1975; Buzas, 1977; Collinson, 1980). In one such occurrence, the most frequent species at 10 cm was not present in living epifaunal assemblages at the time of sampling (Steineck and Bergstein, 1979). The infaunal standing crop may equal or exceed that of epifaunal populations (Buzas, 1977). The diversity and density of foraminiferal faunas in shallow-water environments cannot be expected to decrease uniformly with depth below the sediment surface (Buzas, 1974, 1977; Steineck and Bergstein, 1979).

Estuarine faunas are transitional in character (Scott et al., 1980; Buzas and Severin, 1982). Marsh species are prevalent in upper, low-salinity reaches, whereas open-shelf, stenohaline forms penetrate landward into the lower, tidally dominated segments. In estuaries where tidal and lotic forces are equal, high concentrations of suspended particulate matter within the transitional zone support dense populations of *Ammotium cassis* (Scott et al., 1980). Diversity tends to increase toward the estuary mouth as environmental predictability increases (Scott et al., 1980; Buzas and Severin, 1982).

Shelf and Slope Habitats

The Foraminifera of these environments are segregated into regional assemblages whose geographic and depth distributions are congruent with the prevailing water mass and current regime (Ingle and Keller, 1980; Ingle et al., 1980; Williamson, 1982, 1985; Williamson et al., 1984; Buzas and Culver, 1980; Culver and Buzas, 1981, 1982, 1983a; Lutze and Coulbourn, 1984; Arnold, 1983a; Streeter and Lavery, 1982; Miller and Lohmann, 1982; Murray, 1985; Weston, 1985; Fig. 2.17). In general, species diversity tends to increase across the shelf to a shelf-edge high and then decreases slowly downslope, only to increase again in some parts of the abyssal zone (Gibson and Buzas, 1973; Arnold and Sen Gupta, 1981; Douglas and Woodruff, 1981). Within any one region, however, indices of diver-

FIGURE 2.17 Distribution of benthic foraminiferal provinces along the eastern continental margin of North America. The 100- and 1,000-fathom submarine contours are shown. (From Culver and Buzas, 1981, Fig. 3.)

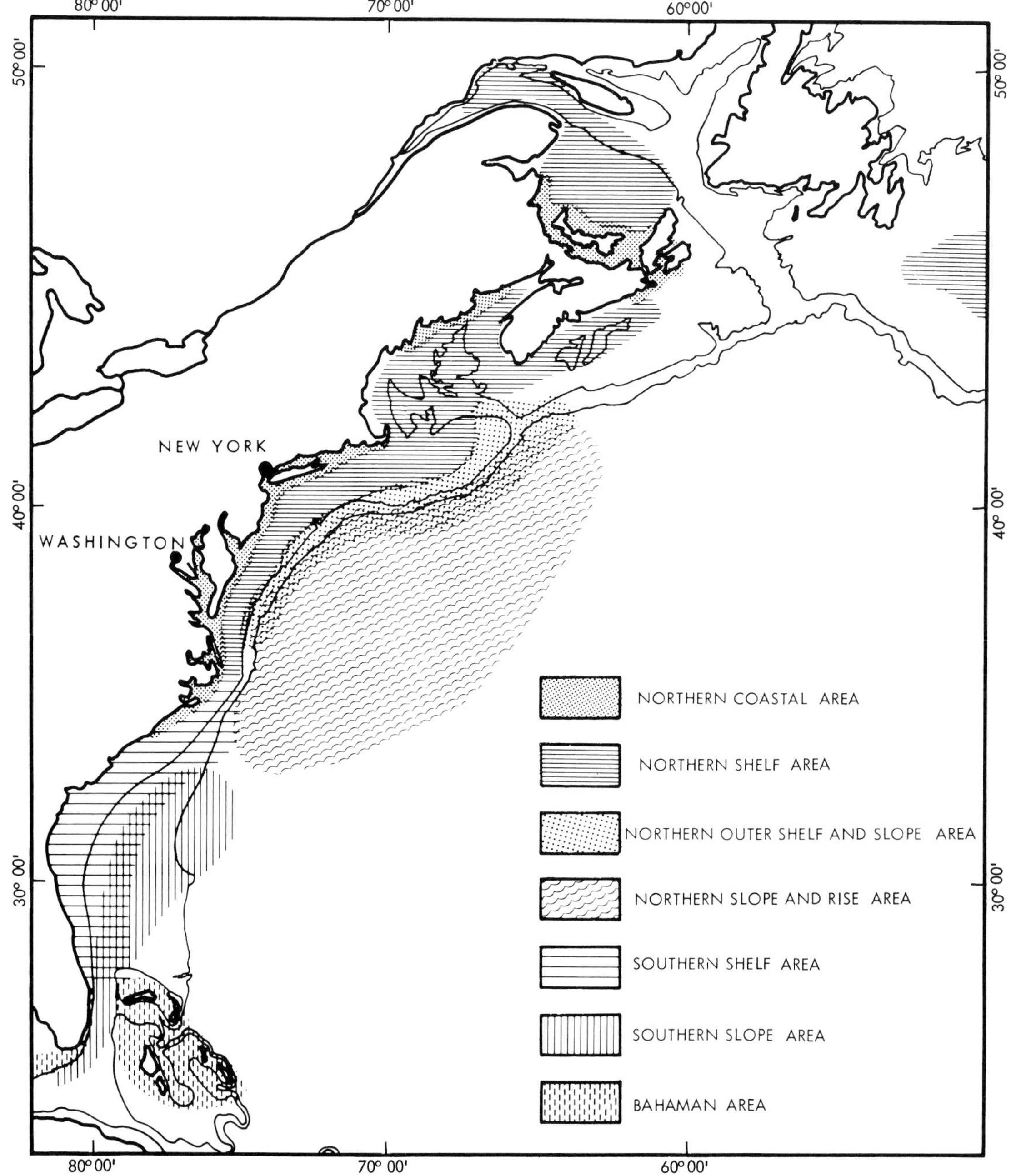

sity and equitability may depart from this trend because of local water-mass and sediment conditions (Williamson, 1985). Faunas are largely calcareous except in the Arctic and Antarctic, where corrosive water masses occur at relatively shallow depths (Ostermann and Kellogg, 1979; Milan and Andersen, 1981).

Inner and outer shelf assemblages can usually be distinguished and correlated with the prevailing hydrographic regime (Buzas and Culver, 1980; Williamson, 1982; Culver and Buzas, 1983a; Williamson et al., 1984), but the boundary between them is often gradational because of the seasonal fluctuation in water-mass distribution. Although high-latitude shelves in the North Atlantic have only recently been submerged, their faunas are in equilibrium with the modern-day environment (Williamson et al., 1984). Along the relatively well-studied Atlantic continental margin of North America, a major discontinuity in species distributions in southern Florida coincides with a change from clastic to carbonate substrates (Buzas and Culver, 1980). Modification of current strength and direction by topographic features creates areas of unique sediments that are occupied by disjunct, substrate-specific assemblages (Scott et al., 1983; Williamson et al., 1984). Local influences, such as turbid river discharges and the upwelling of deep, cold water onto the shelf, alter the regional depth-related pattern of species occurrences (Sen Gupta and Strickert, 1982; Culver and Buzas, 1983a; Lutze and Coulbourn, 1984).

A prominent faunal change is present at the shelf-edge along most Atlantic continental margins (Buzas and Culver, 1980; Culver and Buzas, 1982, 1983b; Arnold, 1983; Lutze and Coulbourn, 1984). The lower limit of shelf assemblages correlates with topography, sediment character, and depth of seasonally fluctuating water masses (Culver and Buzas, 1982; Arnold, 1983).

On the contintenal slope and rise, the adjacent water-mass stratification is accompanied by a corresponding sequence of depth-restricted, areally extensive faunas (Ingle and Keller, 1980; Ingle et al., 1980; Buzas and Culver, 1980; Culver and Buzas, 1982; Streeter and Lavery, 1982; Miller and Lohmann, 1982; Lutze and Coleman, 1984). Although downwardly displaced faunas and sediments are common on some slopes (e.g., Ingle et al., 1980), they are rare to absent on others (Schafer et al., 1981; Schafer and Cole, 1982). Rapid recolonization of disturbed areas on both the shelf and slope takes place by the accelerated reproduction of normally dominant species (e.g., Ellison and Peck, 1983).

Within the regional water-mass stratification and depth gradient, specific environmental factors that influence the distribution and abundance of slope species include topography (e.g., submarine canyons), light intensity, upwelling, sediment type, nutrient supply, current energy, and the intensity of the oxygen-minimum zone (Ingle and Keller, 1980; Ingle et al., 1980; Schafer et al., 1981; Schafer and Cole, 1982; Arnold, 1983a; Streeter and Lavery, 1982; Lutze and Coleman, 1984; Weston, 1985). Sedimentary environments where macrobenthic predators are rare or absent, such as oxygen-deficient or winnowed, coarse-grained substrates, often support abundant populations of benthic foraminifers (Ingle and Keller, 1980; Douglas, 1981; Schafer and Cole, 1982; Lutze and Coleman, 1984). A few stenotopic taxa appear to be confined in their distribution by a small number of environmental variables. For example, along the eastern continental margin of North America, *Globobulimina-Bulimina* faunas are restricted to the regional oxygen-minimum zone (200–1,000 m) (Streeter and Lavery, 1982; Miller and Lohman, 1982). Below these depths, peak frequencies of the infaunal species *Uvigerina peregrina* occur in organic-rich, fine-grained sediments where oxygen may be absent, even though it is present in the overlying water mass (Miller and Lohmann, 1982). Most species, however, are eurytopic on a global scale and respond to a multiplicity of biotic and abiotic factors.

The basin and swell topography of the southern California borderland presents a faunistically more complex pattern than the topographically uniform slopes of the Atlantic margin (Table 2.4). Because the hydrography of each basin is controlled by effective sill depth, many environmental parameters remain unchanged with depth down to the basin floor. Eight recurrent assemblages have been identified whose distribution is controlled by water-mass characteristics, dissolved oxygen and the nutrient content and grain size of sediments (Douglas and Heitman, 1979; Douglas, 1981).

In summary, the bathymetric zonation of continental margin Foraminifera in the modern world ocean is related to climatic gradients (shelf) and water-mass distributions (shelf and slope). These factors, together with other marine processes, determine the physical, chemical, and substrate properties of the environments in which benthic foraminifers reside. Bathymetry by itself (i.e., hydrostatic pressure) appears to play an inconsequential role in limiting the occurrence of assemblages; similar faunas occur at different depths from place to place because of regional differences in circulation patterns (Schafer and Cole, 1982).

The Abyssal Zone

In the deep sea, many ecologically important aspects of the benthic milieu inhabited by foraminifers are determined by the deep thermohaline circulation of the world ocean. Each unit, or water

TABLE 2.4 Recurrent-Species Associations, Southern California Borderland (after Douglas, 1981, Table 1)

Assemblage	Habitat	Depth range (m)	Water mass*
Nonionella spissa-Bulimina denudata	Outer continental shelf	50–200	CC
Cassidulina limbata-Hanzawaia limbata	Offshore bank and terrace	20–400	CC,T
Buccella angulata-Angulogerina angulosa	Offshore ridge and deep bank	100–400	CC,T
Suggrunda eckisi-Epistominella sandiegoensis	Upper continental slope	85–450	T
Bolivina argentea-Epistominella smithi	Lower slope, nearshore basins	400–950	EPI
Bolivina spissa-Hoeglundina elegans	Lower slope, offshore basins	300–2,000 +	EPI
Fursenkoina-Cassidulinoides cornuta	Central basin floor, nearshore basins	550–900	EPI
†	Basin floor, offshore basins	1,200–1,900	EPI

*Water-mass abbreviations: CC = California Current; T = Transitional Water; EPI = East Pacific Intermediate Water Mass.

†No unique recurrent species assemblage. Most of the dominant species are present in preceding assemblage.

mass, in this circulation is identified (and its vertical position in the density stratification determined) by small differences in conservative properties (temperature, salinity) acquired in a surface source region (e.g., Norwegian–Greenland Sea, Weddell Ice Sheet). The broad spatial extent of these water masses and the slight contrasts between them have produced a relatively uniform abyssal environment within which cosmopolitan faunas exist over a wide depth range (Douglas and Woodruff, 1981). However, statistically defined population maxima of one or more species are circumscribed by their optimal conditions existing in one of the present-day deep and bottom water masses (Schnitker, 1980; Lohmann, 1978; Peterson, 1984; Douglas and Woodruff, 1981; Woodruff and Douglas, 1981; Table 2.5). The association of a species with a water mass is often maintained in many basins of the world ocean. On the other hand, interregional differences in the dominant taxa of a given water mass have been documented. These appear to be related to progressive changes in nonconservative properties that vary as a function of the length of time since a water parcel was isolated from the surface (Peterson, 1984).

Because most environmental variables in the deep sea are covariant, it is difficult to identify those that influence species distributions (Lohmann, 1978). Some possibilities that have been suggested include temperature, amount of labile organic matter in sediments, salinity, oxygen content, and degree of calcite undersaturation (Lohmann, 1978; Streeter and Schackleton, 1979; Schnitker, 1980; Belanger and Streeter, 1980; Burke, 1981; Corliss, 1983; Peterson, 1984).

Depth-restricted species distributions occur in silled basins despite the absence of a well-defined water-mass stratification (Belanger and Streeter, 1980). These abyssal settings, where only subtle changes in hydrographic and sedimentary parameters take place between sill depth and the basin floor, offer important opportunities to relate faunal trends to variation in one or two ecological factors. In the topographically enclosed basins of the eastern Atlantic, the dominance of *Epistominella umbonifera* below 4,500 m statistically correlates with an increase in the calcite undersaturation of seawater (Bremer and Lohmann, 1982). The association of abundant numbers of this species with relatively corrosive (with respect to biogenic carbonate) bottom waters has also been noted in the Indian and Pacific Oceans (Burke, 1981; Corliss, 1983; Peterson, 1984). It may result from the more rapid dissolution of other forms or alternatively, may reflect a positive biological response to a preferred environmental setting. The latter hypothesis appears to be the most likely, because Corliss and Honjo (1981) have shown that *E. umbonifera* is as susceptible to solution as other calcareous deep-sea taxa.

In the eastern North Atlantic, *Globocassidulina subglobosa* is the most abundant component in a sparse, low-diversity fauna living beneath a hydrographically unstable mixture of North Atlantic Deep Water (NADW) and Mediterranean Sea Outflow (Weston and Murray, 1984). This species is usually restricted to typical NADW elsewhere in the Atlantic. Such anomalous distributions in the deep sea may be explained by the existence of localized subspecies or sibling species with differing environmental requirements or by the potential, on the part of some taxa, to opportun-

TABLE 2.5 Oceanic Water Masses and Their Benthic Foraminiferal Faunas

Water mass	Site	Dominant taxa	Reference
Antarctic Bottom Water	North Atlantic	*Epistominella umbonifera* *Pullenia bulloides*	1
	South Atlantic	*E. umbonifera* *P. bulloides*	2
	Indian Ocean	*E. umbonifera* *P. bulloides* ± *E. exiguus*	3,4
	Pacific	*E. umbonifera* *P. bulloides* ± *E. exiguus*	3,6
North Atlantic Deep Water (NADW)	North Atlantic		
	Lower	*Oridosalis umbonatus*	1,7
	Middle	*E. exiguus*	7
	Upper	*Pullenia spp., Nonionella*	7
	South Atlantic	*Globocassidulina subglobosa,* *Uvigerina peregrina* *Cibicides wuellerstofi*	2
Mediterranean Sea Outflow	North Atlantic	*G. subglobosa, U. peregrina* *Osangularia rugosa,* *Hoeglundina elegans*	7,8
Indian Ocean Deep Water		*G. subglobosa, E. bradyi*	3,4
Pacific Intermediate Water (deep oxygen-minimum zone)		*U. peregrina, H. elegans,* *Gyroidina* spp.	5,6
Pacific Deep Water		*E. exiguus, P. quinqueloba*	5,6

Sources: (1) Schnitker (1980); (2) Lohmann (1978); (3) Peterson (1984); (4) Corliss (1983); (5) Woodruff and Douglas (1981); (6) Burke (1981); (7) Weston and Murray (1984); (8) Streeter and Shackleton (1979).

istically exploit a benthic environment that excludes others.

Taphonomy

Taphonomy is the study of the processes that affect organic remains after death and prior to final burial (Behrensmeyer and Kidwell, 1985). The assemblage finally entombed in the rock record is referred to as a "death assemblage" because its composition is often more a reflection of postmortem history (Table 2.6) than ecological distributions. Under the best of circumstances, taphonomic changes create a "time-averaged" fauna representative of major long-term environmental conditions (Scott and Medioli, 1980; Behrensmeyer and Kidwell, 1985). Thus, in paleoecologic studies, fossil collections are most realistically compared with the total species composition (live and dead) of a modern environment (Scott and Medioli, 1980). Taphonomic analysis of ancient faunas is a prerequisite for their subsequent use in evolutionary, stratigraphic, and environmental studies (Douglas et al., 1980; Murray, 1982, 1984; Curry, 1982).

The preservation potential of shallow-water faunas is governed by topography, current regime, intensity of bioturbation, predation, sedimentation, interstitial geochemistry, and relative fragility of the tests of individual species (Murray, 1982, 1984a; Douglas et al., 1980; Leventer et al., 1982; Fig.

TABLE 2.6 Important Postmortem Effects on Living Foraminiferal Assemblages

1. Increase in diversity due to accumulation of rare species
2. Change in species proportions due to short-term productivity–fecundity patterns
3. Mechanical disintegration of fragile species (e.g., thin-wall agglutinated taxa)
4. Chemical dissolution of calcareous species (e.g., salt marsh and other acidic environments)
5. Transport by downslope mass-wasting processes
6. Transport by level-bottom traction currents (large species)
7. Transport in suspension (juveniles and small species)
8. Reworking of fossil or subfossil species
9. Mixing of faunas by bioturbation
10. Selective predation
11. Condensation due to low sedimentation rate
12. Weakening and destruction of tests by feeding mechanisms and digestive processes of infaunal deposit feeders
13. Transport from the living site while in the gut of mobile predators

Sources: Douglas et al. (1980); Murray (1982, 1984a); Murray et al. (1982); Thomas and Schafer (1982); Lipps (1983).

	DISSOLUTION OF CALCAREOUS TESTS				
TRANSPORT	NONE TO SLIGHT (< 30% loss) Shiny to dull	MODERATE (30–50% loss) Dull, opaque, broken	SEVERE (50–90% loss) Nearly all broken	VERY SEVERE (> 90% loss) Rare, corroded	LITHOLOGY
NONE OR SLIGHT	Juveniles & adults of same species I	Few small calcareous tests IA/IIA	No small calcareous tests IB/IIB	Mainly Textulariina, juveniles & adults IC/IIC	Mud, silt, muddy sand
DEPOSITION OF SMALL INDIVIDUALS (< 200μm) FROM SUSPENSION	Abundant juveniles & small tests II				
WINNOWED LAG DEPOSIT -LOSS OF SMALL TESTS (< 200μm) -RESIDUE OF MEDIUM-LARGE TESTS	Abraded III	Abraded & broken IIIA	Nearly all tests broken IIIB	Medium to large Textulariina IIIC	Fine to medium sand
TRANSPORTED-ABRASION AND DESTRUCTION IN BED LOAD -RESIDUE OF LARGE ABRADED TESTS	Calcareous tests present IV	Calcareous tests rare IVA	IVB	Textulariina only IVC	Medium to coarse sand; gravel
	Decrease in calcareous component →				

FIGURE 2.18 Progressive alteration of a living foraminiferal assemblage by dissolution and transport. The original assemblage is postulated to have contained a small proportion of agglutinated (Textulariina) species. (From Murray, 1984a, Fig. 4.)

2.18). In regions of complex bathymetry and strong currents, such as the southern California Borderland, the similarity between ecologically related associations of foraminifera and the sediment assemblage may be low; only a few environmental settings (e.g., central basin plains) incorporate reasonably intact faunas into the fossil record (Douglas et al., 1980).

Within ocean basins, the large quantities of $CaCO_3$ precipitated by plankton (pteropods, coccolithophoids, planktonic Foraminifera) in the photic zone creates seawater that is undersaturated with respect to calcite (Berger, 1981; Kennett, 1982). Chemical processes in the ocean redress this deficiency by dissolving biogenic calcite. Solution of calcareous tests by corrosive bottom water often produces significant differences between the "death assemblage" present in oceanic sediments and the ecological communities living on the seabed and in the water column (Adelseck and Berger, 1975; Thunnell and Honjo, 1981; Kellogg, 1984b; Cullen and Prell, 1984).

Carbonate dissolution intensifies with increasing depth, hydrostatic pressure, dissolved carbon dioxide, residence times of water masses, quantities of organic matter in sediments, and decreasing temperature and sedimentation rate (Berger, 1981; Belyaeva and Burmistrova, 1985). The lowest limit of the accumulation of sedimentary carbonate is referred to as the Calcium Compensation Depth (CCD). This horizon occurs at 5 km in the North Atlantic but rises to about 4 km in the North Pacific. The shallower Carbonate Lysocline (CL) represents the level at which the percentage of carbonate of sediments decreases abruptly (Belyaeva and Burmistrova, 1985). Higher still in the water column, the Foraminiferal Lysocline (FL) separates well-preserved from poorly preserved planktonic foraminiferal assemblages. The FL may be as shallow as 2 km in the northern Indian Ocean (Cullen and Prell, 1984).

Significant changes in the position of the CCD and FL have occurred during the Cenozoic (Berger, 1981; Vincent and Berger, 1981). On a global scale, long-term variations in the level of the CCD are influenced by the overall fertility of the world ocean and by change in the pattern and intensity of thermohaline flow (Berger, 1981; Kennett, 1982). Rapid, high-amplitude fluctuations in the depth at which foraminiferal assemblages were affected by solution occurred during the Pleistocene (Thompson, 1976). Recent studies have shown that Pleistocene dissolution episodes in the Atlantic and Pacific were not synchronous because of differences in deep-water circulation patterns and the residence times of water masses (Peterson and Prell, 1985; Crowley, 1985).

The chemical destruction of foraminiferal tests in the deep sea is a function of their morphology, of their residence time on the seabed surface, and of the relative undersaturation of the local bottom water mass. Only very small planktonic tests are etched in transit through the water column (Adelseck and Berger, 1975). Susceptibility to corrosion varies from species to species in both planktonic and benthic Foraminifera (Corliss and Honjo, 1981; Vincent and Berger, 1981). Spinose, thin-walled, planktonic forms dissolve more easily than thick-walled nonspinose taxa. The relative survivability of a test is determined by the ratio of its volume to the weighted average surface area, including the inner and outer pore surfaces (Adelseck, 1977). Several authors have compiled scales of relative susceptibility for recent and fossil planktonic species (Vincent and Berger, 1981; Thunnell and Honjo, 1981; Malmgren, 1983; Keir and Hurd, 1983).

Paleoecology of Smaller Benthic Foraminifera

Case Study 1: Cenozoic Foraminifera of the Southern California Borderland

Geohistory analysis combines the use of benthic foraminifers as indicators of depositional environments with quantitative stratigraphic techniques to calculate the rate and vector (subsidence, uplift) of the tectonic history of a basin (Van Hinte, 1978). Ingle's (1980) study of Cenozoic sediments in the southern California Borderland is an exemplar of this method. The area investigated has undergone a varied and complex geologic development in response to changes in the sea-floor-spreading regime of the eastern Pacific Ocean. Since the Miocene, translational and rotational plate movements and transform faulting have resulted in the uplift of anticlinal ridges and the subsidence and differential filling of isolated basins (Ingle, 1980; Douglas, 1981).

Investigations of this kind rely on two independently calibrated axes of interpretation. First, Ingle (1980) established a chronostratigraphic framework for the dating and correlation of the subaerial sections, which he studied using planktonic microfossil biostratigraphy, radiometric dating, and magnetostratigraphy. Second, he developed a paleobathymetric–paleoenvironmental model, applying multiple depth criteria for the discrimination of depositional environments (Fig. 2.19). Information used in this model included (1) upper depth limits of benthic foraminiferal species now living in the eastern Pacific (Ingle and Keller, 1980; Ingle et al., 1980) (the lower depth limits of many species in some regions are affected by downward displacement of dead tests); (2) comparisons with fossil faunas recovered at deep-ocean drilling sites where paleodepths can be determined geophysically; (3) dominance, diversity, and abundance trends; (4) relative proportions of various microfossil groups; and (5) lithology and sedimentary structures. Data of types 1 and 2 presuppose that the depth distribution of benthic foraminifers has been relatively static over time (for eloquent dissents, see Blake and Douglas, 1980; Douglas, 1982). In the "geohistory model," temporal faunal trends are explained by tectonic–geographic factors (e.g., changing paleobathymetry). The final element in the analysis is the calculation of residual estimates of basinal subsidence and uplift from inferred paleobathymetric fluctuations and rates of sedimentation per unit time (Fig. 2.20). Ingle (1980) has presented a detailed and thoroughly documented account of the changing paleogeography and structural evolution of a geologically active region over an interval of ~55 million years (my). The extent to which his conclusions are confounded by the migration of benthic faunas in response to changes in the distribution and character of water masses is an open question.

Case Study 2: Paleogene Deep-Sea Agglutinated Foraminifera

A cosmopolitan and taxonomically distinctive, agglutinated fauna, composed of coarse-grained, morphologically simple (unilocular, biserial, and uniserial) species, is restricted in its distribution to the fine-grained intervals of flysch-type sedimentary sequences of Paleogene age (Brouwer, 1965; Gradstein and Bergrren, 1981; Miller et al., 1982; Winkler and Van Stuijvenberg, 1982; Verdenius and Van Hinte, 1983). Recent studies have stressed the influence of a limited and interrelated set of depth-independent environmental parameters as prerequisites for their occurrence. These include rapid deposition of fine-grained sediments, presence of large quantities of anaerobically decomposing organic matter, and restricted circulation at the water–sediment interface (Moorkens, 1976; Gradstein and Berggren, 1981). Agglutinated faunas in the Paleogene of the central North Sea and the northeastern Canadian margin developed in silled basins (0.7–1.5 km paleodepths) on a subsiding and fractured slope located to the east of actively prograding deltas (Gradstein and Berggren, 1981).

In the central Labrador Sea (DSDP site 112, 3 km paleodepth) Eocene agglutinated faunas, present in calcareous silts, were replaced by a dominantly calcareous Oligocene assemblage without a concomitant shift in sediment type (Miller et al., 1982). This change coincided with the flushing of the waters of the Labrador Sea by a recently formed, well-oxygenated water mass of North Atlantic origin (Miller et al., 1982; Murray, 1984a). Miller et al. (1982) concluded that the development of either calcareous or arenaceous faunas in the upper abyssal zone is controlled by the hydrographic properties of bottom waters rather than by substrate character per se.

Paleoceanography

Introduction

The configuration, hydrographic properties, and intensity of mass transport systems in the modern world ocean are the results of complex interactions between the geography of continents and oceans (a result of plate tectonic processes) and patterns of global and regional climate (Hay, 1983). These sys-

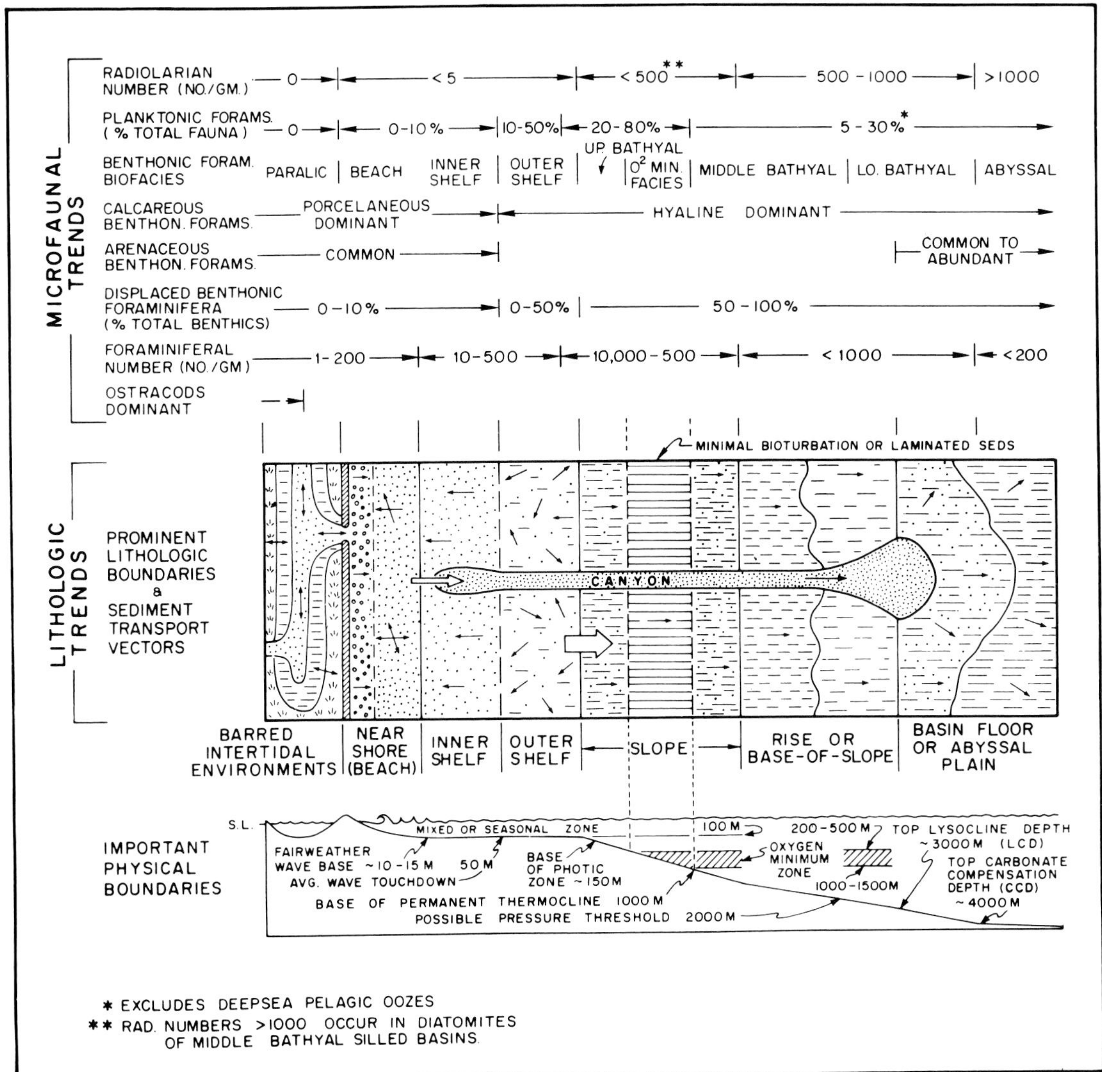

FIGURE 2.19 Faunal and sedimentologic characteristic of major depositional environments represented in Cenozoic sediments of the southern California borderland. A model based on the application of multiple depth criteria to the problem of recognizing the paleodepth of marine sediments. (From Ingle, 1980, Fig. 7.)

tems, such as drift, thermohaline and contour currents, and upwelling zones, redistribute latent heat of fusion, renew the fertility of surface and bottom environments, and determine the distribution of marine organisms. Present-day conditions have developed over a period of 100 my in a series of short-term, steplike transitions (threshold events) that were related to the fragmentation and dispersal of Pangaea, the severance and isolation of the Tethyan Seaway, the deterioration of global climate into a glacial mode, and the relocation of sources of bottom-water from low-latitude to high-latitude enclosed seas and ice shelves (Berggren and Hollister, 1977; Kennett, 1978, 1980; Berger, 1981; Vincent and Berger, 1981). The new multidisciplinary field of paleoceanography seeks to reconstruct this history and its effects on the paleobiogeography of microfossils. Although a recent addition to the geosciences, the literature of paleoceanography is already vast. Those individuals who wish to pursue this topic in greater depth should consult Berggren and Hollister (1977), Schopf (1980), Kennett (1978, 1980, 1982, 1983), Vincent and Berger, (1981), Berger (1981), Miller (1982), and Berggren and Schnitker (1983).

FIGURE 2.20 Geohistory analysis of Cenozoic sediments, central Santa Ynez Mountains, southern California. Paleobathymetry inferred from model shown in Figure 2.19. Tectonic trends shown in the right-hand column. (From Ingle, 1980, Fig. 9.)

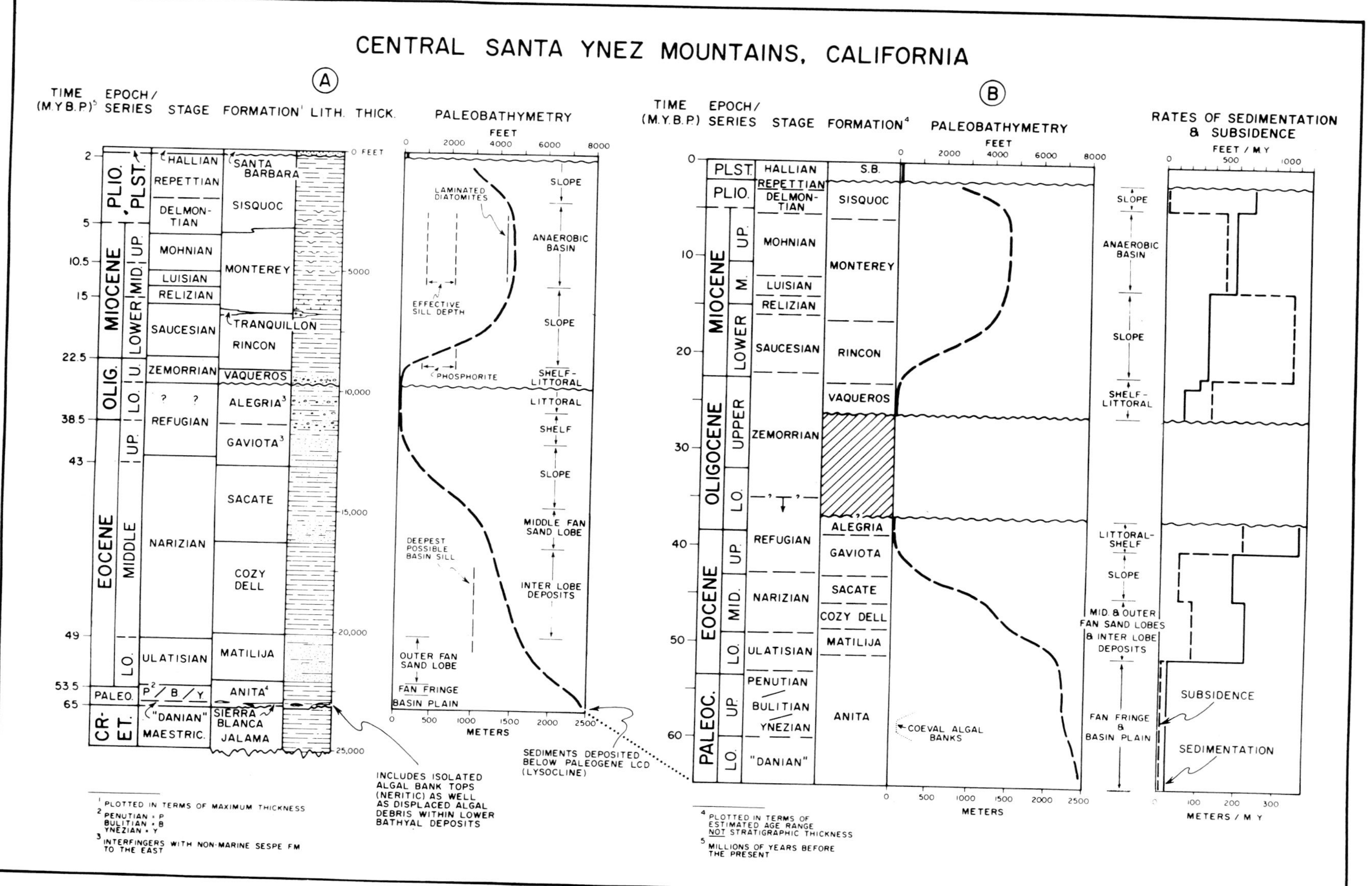

CENTRAL SANTA YNEZ MOUNTAINS, CALIFORNIA
A
TIME (M.Y.B.P.)[5]
EPOCH/ SERIES
STAGE
FORMATION[1]
LITH.
THICK.
PALEOBATHYMETRY
FEET
METERS
PLIO.
PLST.
MIOCENE
OLIG.
EOCENE
PALEO.
CR-ET.
UP.
MID.
LOWER
U.
LO.
MIDDLE
HALLIAN
REPETTIAN
DELMON-TIAN
MOHNIAN
LUISIAN
RELIZIAN
SAUCESIAN
ZEMORRIAN
REFUGIAN
NARIZIAN
ULATISIAN
P[2]/B/Y
"DANIAN"
MAESTRIC.
SANTA BARBARA
SISQUOC
MONTEREY
TRANQUILLON
RINCON
VAQUEROS
ALEGRIA[3]
GAVIOTA[3]
SACATE
COZY DELL
MATILIJA
ANITA[4]
SIERRA BLANCA
JALAMA
0 FEET
5000
10,000
15,000
20,000
25,000
LAMINATED DIATOMITES
EFFECTIVE SILL DEPTH
PHOSPHORITE
DEEPEST POSSIBLE BASIN SILL
SLOPE
ANAEROBIC BASIN
SHELF-LITTORAL
LITTORAL
SHELF
MIDDLE FAN SAND LOBE
INTER LOBE DEPOSITS
OUTER FAN SAND LOBE
FAN FRINGE
BASIN PLAIN
INCLUDES ISOLATED ALGAL BANK TOPS (NERITIC) AS WELL AS DISPLACED ALGAL DEBRIS WITHIN LOWER BATHYAL DEPOSITS
SEDIMENTS DEPOSITED BELOW PALEOGENE LCD (LYSOCLINE)
[1] PLOTTED IN TERMS OF MAXIMUM THICKNESS
[2] PENUTIAN = P BULITIAN = B YNEZIAN = Y
[3] INTERFINGERS WITH NON-MARINE SESPE FM TO THE EAST
B
TIME (M.Y.B.P.)
EPOCH/ SERIES
STAGE
FORMATION[4]
PALEOBATHYMETRY
FEET
METERS
PLST.
PLIO.
MIOCENE
OLIGOCENE
EOCENE
PALEOC.
UP.
M.
LOWER
UPPER
LO.
MID.
S.B.
SISQUOC
MONTEREY
RINCON
VAQUEROS
ALEGRIA
GAVIOTA
SACATE
COZY DELL
MATILIJA
ANITA
PENUTIAN
BULITIAN
YNEZIAN
COEVAL ALGAL BANKS
SLOPE
ANAEROBIC BASIN
SHELF-LITTORAL
LITTORAL-SHELF
MID. & OUTER FAN SAND LOBES & INTER LOBE DEPOSITS
FAN FRINGE & BASIN PLAIN
RATES OF SEDIMENTATION & SUBSIDENCE
FEET / M.Y.
METERS / M.Y.
SUBSIDENCE
SEDIMENTATION
[4] PLOTTED IN TERMS OF ESTIMATED AGE RANGE NOT STRATIGRAPHIC THICKNESS
[5] MILLIONS OF YEARS BEFORE THE PRESENT

Methods

The data-gathering techniques and analytical procedures of paleoceanography are as varied as the backgrounds of its practitioners (for a full review see Berger, 1981; Vincent and Berger, 1981; Kennett, 1982). For many workers the primary source of information is the micropaleontological content of deep-sea cores.

Paleoecological Transfer Functions

The geographic distribution of Recent planktonic foraminifers can be statistically related to present-day trends in the surface environment of the oceans by factor-analytical and multiple-regression techniques. When applied to down-core data, these equations, or "transfer functions," are capable of providing quantitative estimates of the salinity and temperature of Pleistocene oceans (Hutson, 1977; Prell et al., 1979, 1980; Crowley and Matthews, 1983). The full protocol of the transfer function procedure has been summarized by Vincent and Berger (1981, pp. 1067–1071) and Kennett (1982, pp. 642–644).

Routine use of transfer functions rely on three assumptions: species have not changed their ecological requirements over time; past oceanic conditions have fallen within the range of variation existing today; and modification of the sample by winnowing and/or dissolution has been negligible (Vincent and Berger, 1981; Kennett, 1982). Their validity in a given study area can be tested by comparing results derived from different microplanktonic groups (Molfino et al., 1983). Temperature and salinity maps for selected intervals in the late Pleistocene illustrate the resolving power of transfer functions (McIntyre et al., 1976; Kellogg, 1980; Moore et al., 1980; Ruddiman and McIntyre, 1984; Fig. 2.21). The interpretation of "no-analog" results (e.g., where one or more assumptions are not met) is dis-

FIGURE 2.21 Inferred sea-surface temperatures of the Pacific Ocean during August, 18,000 yr BP. Temperature (°C) values derived by paleoecological transfer functions applied to planktonic Foraminifera. (From Moore et al., 1980, Fig. 5.)

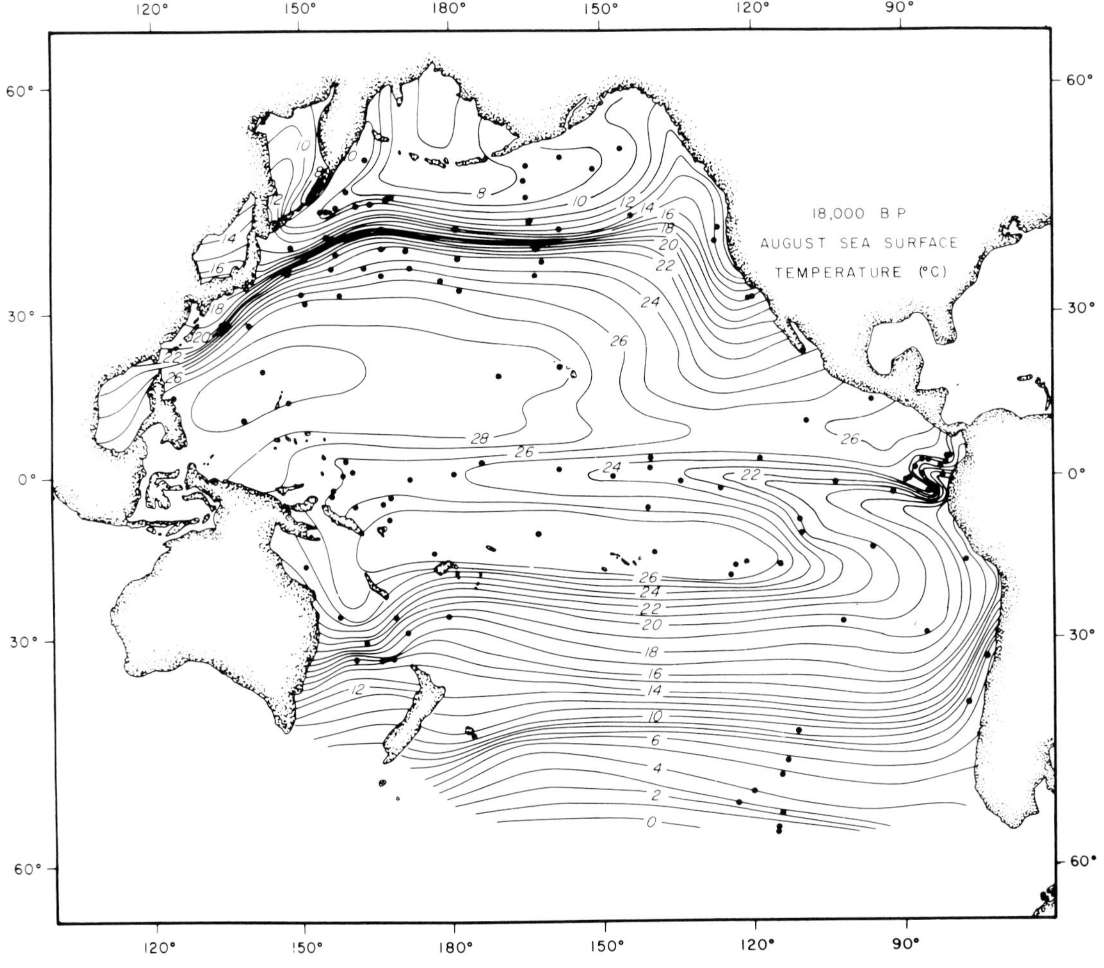

cussed by Hutson (1977), Hutson and Prell (1980), Kellogg (1984a), and Loubere (1982).

Stable Isotope Methods (Oxygen)

If calcite is secreted by an organism in thermodynamic equilibrium with seawater, the $^{18}O:{}^{16}O$ ratio of the solid phase differs from that of the liquid solely as a function of temperature. By this principle, the proportion of ^{18}O in biogenic carbonate increases with decreasing temperature (Kennett, 1982; Vincent and Berger, 1981). Isotopic ratios derived from calcareous microfossils are expressed in a $\delta\,^{18}O$ notation that measures deviation from a universal fossil standard. A paleotemperature estimate is calculated by the equation given in Kennett (1982, pp. 82–83) and Vincent and Berger (1981, p. 1075).

Problems in the interpretation of $\delta\,^{18}O$ data result from the variation in isotopic composition of seawater over time and the occurrence of disequilibrium precipitation in many species (Table 2.7). The isotopic ratio of seawater has varied over time on a global scale with the amount of isotopically light (0-16 enriched) ice stored in glaciers. More local and short-term changes are related to the precipitation–evaporation balance. Post–middle Miocene trends in $\delta\,^{18}O$ are assumed to include both ice volume and water temperature effects (Woodruff et al., 1981; Savin et al., 1981, 1985) (Fig. 2.22). Debate continues about the significance of Eocene to Lower Miocene data because of uncertainty over when the Antarctic ice cap first developed (compare Woodruff et al., 1981, Shackleton et al., 1985; Savin et al., 1985; Crowley and Matthews, 1983; Poore and Matthews, 1984a,b, and Miller and Fairbanks, 1985).

Oxygen–isotopic ratios have been used to estimate surface and bottom-water temperatures (Woodruff et al., 1981; Prell and Curry, 1981; Thomas, 1985; Savin et al., 1985), the periodicity of the sun–earth orbital relationship and its influence on global ice volume (Keigwin and Boyle, 1985; Imbrie, 1985; Schackleton and Pisias, 1985), glacial meltwater effects (Leventer et al., 1982), the depth habitats of extinct planktonic foraminifers (Douglas and Savin, 1978; Boersma and Premoli-Silva, 1983; Poore and Matthews, 1984b; Savin et al., 1985; Shackleton et al., 1985), the intensity of upwelling (Dunbar, 1983; Wefer et al., 1983; Ganssen and Sarntheim, 1983), vertical temperature gradients in past oceans (Keigwen et al., 1979; Loutit et al., 1983), and the annual range of surface temperatures (Prell and Curry, 1981; Ganssen and Sarntheim, 1983).

To accommodate the many sources of disequilibrium effects, paleotemperature studies of both planktonic and benthonic species require the analysis of large numbers of taxa whose departure from equilibrium is small and consistent (e.g., *Uvigerina*, some buliminids, *Globigerinoides ruber, Globigerina bulloides, Neogloboguardrina dutertrei*) (Wefer et al., 1983; Ganssen and Sarnthein, 1983; Grossman, 1984; Dunbar and Wefer, 1984). For planktonic species, favorable preservational conditions allow measurement of an annual average surface paleotemperature with a potential accuracy of $\pm 0.7°C$ (Graham et al., 1981). However, such results must be treated cautiously because of the increasing evidence that the greater proportion of calcite in tests found in bottom sediments was precipitated below 100 m (Bouvier-Soumagnac and Duplessy, 1985).

TABLE 2.7 Potential Sources of Variation in the Carbon and Oxygen Isotopic Ratios of Foraminferal Tests

Diagenetic	*Mineralogical*
Preferential leaching of ^{16}O by corrosive seawater (P,B)	Trace element chemistry of test (B,P,O)
Origin and decomposition state of organic matter in sediments (C,B)	Differential fractionation during precipitation of aragonitic test (A,C,O)
Ecological	*Metabolic*
Microhabitat preference (B,C,O)	Activity of algal symbionts (P, LB, O, C)*
Calcification at different depths during ontogeny (P,O,C)	Efficiency of gas exchange with the environment (B,P,O,C)*
Rapid growth of juveniles in fertile cold surface waters in upwelling regions (P,O)	High growth rate of smaller individuals (P,B,C)
Calcification across temperature and salinity gradients along a depth-varying isopycnal (P,O)	
Annual temperature cycle (C,B,O)	
Intraspecific variation (B,P,O)	
Trophic differences within sympatric populations (LB,C)	

P = planktonic species; B = smaller benthic species; LB = larger benthic species; A = aragonitic species; C = carbon isotopes; O = oxygen isotopes.

*Disequilibrium related to incorporation of metabolic CO_2 into test.

Sources: Kahn (1979); Killingsley et al. (1981); Curry and Matthews (1981); Belanger et al. (1981); Fairbanks et al. (1982), Zimmerman et al. (1983), Fermont et al. (1983), Dunbar (1983), Wefer et al. (1983), Ganssen and Sarntheim (1983), Grossman (1984), Dunbar and Wefer (1984).

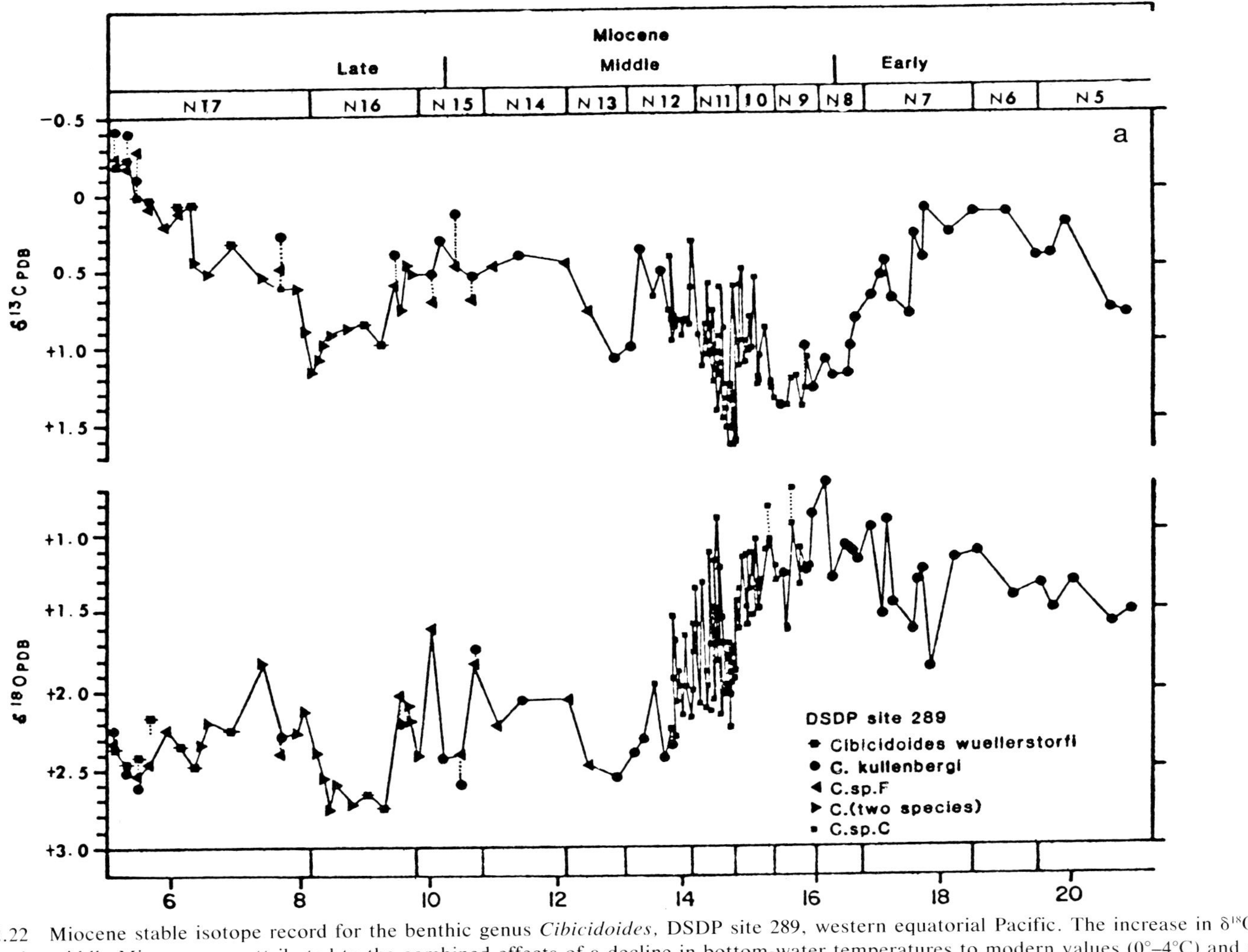

FIGURE 2.22 Miocene stable isotope record for the benthic genus *Cibicidoides*, DSDP site 289, western equatorial Pacific. The increase in $\delta^{18}O$ values in the middle Miocene was attributed to the combined effects of a decline in bottom-water temperatures to modern values (0°–4°C) and a change in the isotopic composition of seawater related to rapid growth of the Antarctic Ice Cap. (From Woodruff et al., 1981, Fig. 1; copyright 1981 by the AAAS.)

Stable Isotope Methods (Carbon)

The carbon isotopic ratio ($^{12}C/^{13}C$) of calcareous microfossils is a function of the overall isotopic composition of the seawater pool of total dissolved inorganic carbon (ΣCO_2), the redistribution of CO_2 by water masses and the fixation or decomposition of marine organic carbon (Curry and Lohmann, 1982; Vergnaud-Grazzini, 1983; Woodruff and Savin, 1985).

Major influences on the $\delta\ ^{13}C$ composition in the open ocean are summarized in Table 2.8. All hyaline–perforate species studied so far depart from equilibrium because of a combination of metabolic and environmental factors (Kahn, 1979; Curry and Matthews, 1981; Berger and Vincent, 1981; Luz and Reiss, 1983; Ganssen and Sarntheim, 1983; Dunbar, 1983; Wefer et al., 1983; Dunbar and Wefer, 1984; Grossman, 1984; Table 2.7). In populations of a few species (e.g., *Planulina wuellerstorfi*, *Cibicidoides* spp.) the degree of disequilibrium is consistent with interoceanic and bathymetric trends in the $\delta\ ^{13}C$ of ΣCO_2 (Graham et al., 1981; Belanger et al., 1981). Study of these forms in deep-sea cores can be used to unravel the distribution, flow rate, nutrient content, and preformed chemistry of past water masses (Curry and Lohmann, 1982, 1984; Vergnaud-Grazzini, 1983; Mix and Fairbanks, 1984; Vincent and Berger, 1985). Detailed $\delta\ ^{13}C$ data on composite benthonic assemblages are now available for Miocene sediments in the Pacific (Loutit et al., 1983; Woodruff and Savin, 1985). Basinwide fluctuations correlate with sea-level changes; between-site variations at a given point in time are thought to reflect local differences in surface fertility and the rate of incorporation of organic matter into the sedimentary microenvironments occupied by benthic foraminifers.

In the future, global $\delta\ ^{13}C$ events may provide a detailed and precise means of correlation between marine sequences of diverse origin and paleodepth (Kennett, 1983). One such datum, a -0.7 ‰ shift dated at 6.2 Ma (late Miocene), has now been identified in both deep and shallow marine sediments (Loutit and Kennett, 1979; Vincent et al., 1980; Loutit and Kiegwin, 1982; Vincent et al., 1980).

Benthic Foraminifera and Paleoceanography

The following section summarizes the responses of deep-sea benthic faunas to changes in abyssal hydrography during three intervals in the Cenozoic. The papers cited herein were chosen for their value as examples of methodology and interpretation. Inclusion in the following account does not imply acceptance of their conclusions as opposed to other studies on the same topic.

TABLE 2.8 Sources of Variation in the $\delta\ ^{13}C$ Content of Deep and Bottom Water Masses

Control	Effect	Paleoceanographic application
Global		
Rate of transfer between carbon reservoirs (marine organic carbon, atmospheric carbon dioxide, terrestrial biomass)	Regression releases terrestrial organic carbon to the oceans, lowering the $\delta\ ^{13}C$ of all oceanic waters by the same amount at the same time	Correlation and sea-level studies (Woodruff and Savin, 1985; Kennett, 1983)
Local		
Differences within a water mass over its flow path	Related to rate of oxidation of marine carbon and to residence time; both govern the release of ^{12}C-enriched decomposition products	Changes in the distribution, hydrographic properties, and flux of water masses (Curry and Lohmann, 1982, 1984; Vergnaud-Grazzini, 1983; Miller and Fairbanks, 1985)
Differences between water mass	Related to chemistry and preformed nutrient content of surface waters prior to downward convection and isolation from the surface	as above
Local surface productivity	Addition of ^{12}C-enriched decomposition products to sedimentary microenvironments where benthic foraminifers reside	Paleofertility patterns (Woodruff and Savin, 1985)

Case Study 1: Paleogene (Atlantic)

Principal components analysis and varimax-rotation techniques have resolved the composite fauna into biofacies whose spatial and temporal distribution can be charted at DSDP stations with different paleodepths (Tjalsma and Lohmann, 1983) and in subaerially exposed deep-marine sequences (Wood et al., 1984; Fig. 2.23). However, in relating Paleogene deep-sea faunal patterns to water-mass changes, one must be aware that the thermohaline stratification at this time was generated in a very different fashion from that of today (Hay, 1983). Negatively buoyant plumes of warm, saline, water, originating in evaporitic, Tethyan epicontinental seas, supplied sluggish, oxygen-poor, latitudinally flowing bottom currents that characterized much of the oceans until the Oligocene (Hay, 1983; Poore and Matthews, 1984b).

Deep-sea benthic Foraminifera experienced a significant but short-lived period of extinction near the end of the Cretaceous (Dailey, 1983). Surviving taxa were the most frequent components of a bathymetrically and geographically widespread Paleocene deep fauna in the North and Central Atlantic (Tjalsma and Lohmann, 1983). The lack of depth zonation at this time suggests a relatively homogeneous water column composed of poorly differentiated Tethyan outflows (Tjalsma and Lohmann, 1983; Hay, 1983; Fig. 2.23).

A Late Paleocene pulse of species extinctions eliminated Cretaceous relicts from the deep sea and restricted them to shallow-water environments (Tjalsma and Lohmann, 1983; Berggren and Schnitker, 1983). Following this crisis, a low-diversity, *Nuttalides*-dominant fauna was established at abyssal depths (>2 km) (Tjalsma and Johmann, 1983; Miller et al., 1984). No oceanographic mechanism for this event has yet been proposed. In the Early Eocene, the downslope migration of bathyal stocks restored deep-sea diversity to more normal levels. The presence of depth-restricted faunas during the Early and Middle Eocene indicates that a water-mass stratification had developed in the Atlantic. Varying proportions of *Nuttalides* in abyssal samples during this interval may be a function of the penetration of low-oxygen, corrosive bottom waters of Tethyan origin into middle and high latitudes (Miller, 1983; Miller et al., 1984).

The *Nuttalides* fauna was replaced suddenly at the Middle–Late Eocene boundary by a *Cibicidoides-Bulimina* fauna that persisted into the Oligocene (Tjalsma and Lohmann, 1983; Berggren and Schnitker, 1983; Snyder et al., 1984; Wood et al., 1985; Fig. 2.23). An extended but gradual faunal turnover (originations, extinctions, depth migrations) took place during the Late Eocene to earliest Oligocene interval (Berggren and Schnitker, 1983; Miller, 1983; Corliss et al., 1984; Snyder et al., 1984), as deep-sea faunas adjusted to the growing prevalence of cold, well-oxygenated bottom waters of North Atlantic origin (Miller et al., 1982; Miller, 1983). The absence of a basinwide mass extinction at the Eocene–Oligocence boundary argues against the occurrence of a major bolide impact at this time as advocated by some geophysicists (see Keller et al., 1983, and Van Valen, 1984a, for discussion).

Case Study 2: Miocene (Pacific)

In the Middle and Late Miocene, major modifications in the latitudinal and vertical thermal gradients of the Pacific resulted from the increasing production of frigid bottom water beneath newly formed ice shelves along the coastline of the Antarctic continent (Kennett, 1978, 1980; Woodruff and Douglas, 1981; Loutit et al., 1983; Savin et al., 1985). Woodruff (1986) studied the response of benthic foraminifers to these events in over 200 DSDP sites (1.5–4.5 km paleodepth) located between 45°N and S latitudes. Biofacies were identified using R-mode principal components analysis.

A warm, Early Miocene Pacific Ocean was characterized by the continued presence of Paleogene assemblages and the absence of a clearly defined depth zonation. Significant changes in depth distributions (including development of depth-stratified faunas) and a peak in extinction and origination rates between 16 and 13 Ma were attributed to the cooling and intensified flow of bottom waters of Antarctic origin. A second interval of basinwide faunal reorganization at 10–8 Ma correlated with further growth in the Antarctic ice cap and a concomitant increase in the supply of Antarctic bottom water to the Pacific.

Case Study 3: Quaternary (North Atlantic)

Within the past 700,000 years, global climatic cycles periodically altered the location and generative mechanisms of bottom-water formation in the North Atlantic. Under modern interglacial conditions, large volumes of cold, well-oxygenated, nutrient-depleted North Atlantic Deep Water (NADW) advect southward from the Norwegian–Greenland Seas. This water mass determines the distribution of benthic Foraminifera over much of the Atlantic, and through deep-circulation linkages, influences faunal patterns in the Indian and Pacific Oceans (Schnitker, 1980, 1982; Corliss, 1983). During periods of glacial climates, convective and refluxive processes ceased in the Norwegian–Greenland Sea because of perennial sea ice (Kellogg, 1977, 1980; Schnitker, 1980). The primary locus of bottom-water formation shifted to the southernmost portions of the North Atlantic ice sheet (Shackleton et al., 1983). Cooling of intermediate waters created a low-volume water mass whose hydrographic properties (low oxygen, high quantities

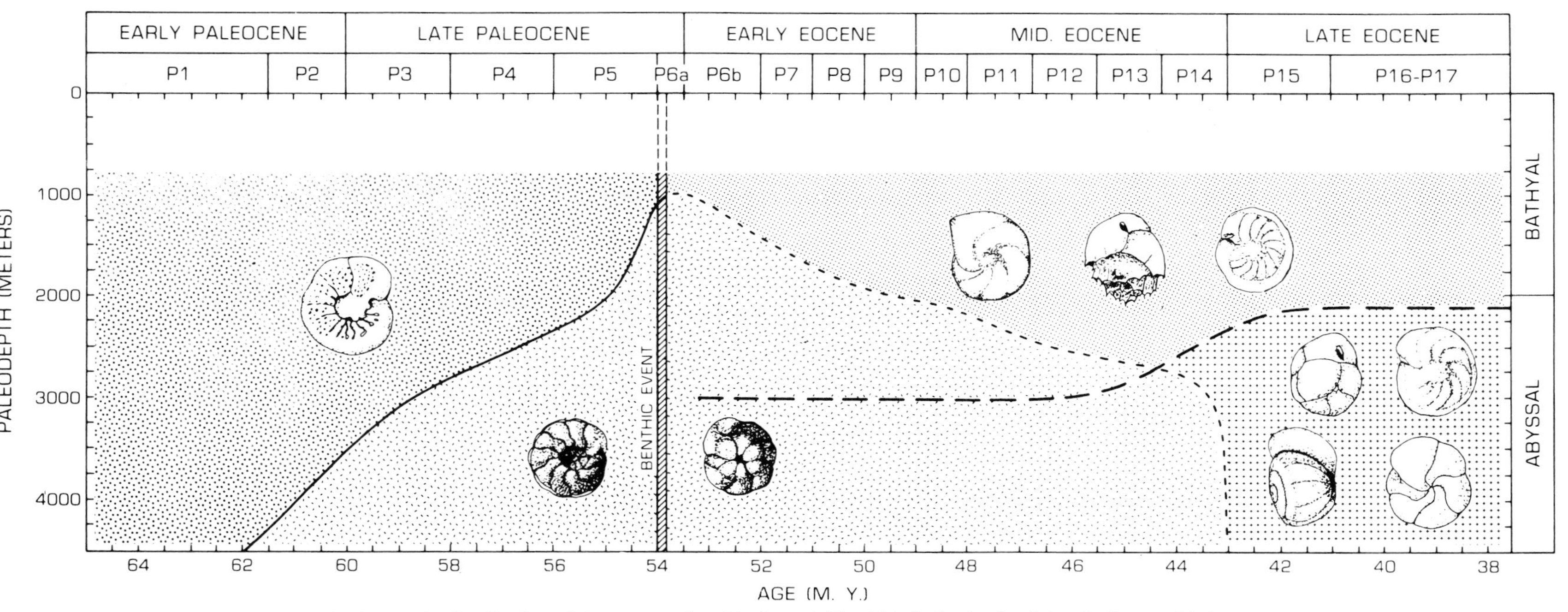

FIGURE 2.23 Paleocene and Eocene bathymetric distribution of deep-water benthic foraminiferal biofacies in the Atlantic Ocean. Biofacies represented include (left to right) *Gavellinella beccariformis*, *Nuttalides*, *Lenticulina*, and *Globocassidulina-Cibicidoides*. (From Fig. 55 of Tjalsma and Lohmann, 1983.)

of dissolved nutrients) differed significantly from that of interglacial NADW (Mix and Fairbanks, 1985). The reduced flux and increased residence time of the glacial water mass contributed to a relative stagnation of the Atlantic basins during periods of cold climate (Mix and Fairbanks 1985; Curry and Lohmann, 1982, 1984).

Large-scale modifications in the areal and depth distribution of benthonic Foraminifera occurred as global circulation patterns reacted to the absence of interglacial NADW (Schnitker, 1980; Streeter and Shackleton, 1979; Curry and Lohmann, 1982; Schnitker, 1984; Fig. 2.24). In the North Atlantic, the presence of relatively oxygen-poor, nutrient-rich bottom water enabled a *Uvigerina* fauna to expand its distribution to include most of the basin. Species characteristic of hydrographically similar water masses in the southern oceans penetrated northward into the basin at great depths. In the southeastern Indian Ocean, the synchronous appearance of a *Melonis–Uvigerina* assemblage may have resulted from changes in the chemistry of the Circumpolar Deep Water resulting from the elimination of interglacial NADW from the South Atlantic watermass stratification (Corliss, 1983).

During the approximately 11,000-year interglacials (e.g., 120,000 years BP), faunal distributions in the North Atlantic were generally similar to those existing today (Schnitker, 1980). The peak abundance of a few species coincided with times of climatic transition (Schnitker, 1984).

Following the termination of the last glacial age, the reestablishment of present-day oceanic circulation patterns was diachronous with respect to depth in the northwestern North Atlantic (Schnitker, 1979; Streeter and Lavery, 1982). Modern faunas were in place at 3 km water depth by 12,000 years BP, whereas their appearance at 4 km was delayed until 8,000 years BP. These findings are consistent with a model in which the renewed flow of the NADW progressively replaced glacial-age water masses from intermediate depths downward over time (Streeter and Lavery, 1982).

FIGURE 2.24 Areal distribution of three associations of benthic foraminifers in the North Atlantic. Left: glacial (18,000 yr BP). Right: glacial (120,000 yr BP). Interglacial faunal patterns are generally similar to present-day distributions. (From Schnitker, 1980, Figs. 3 and 4. Reprinted with permission from the *Annual Review of Earth and Planetary Science*, Vol. 8, copyright 1980 by Annual Reviews, Inc.)

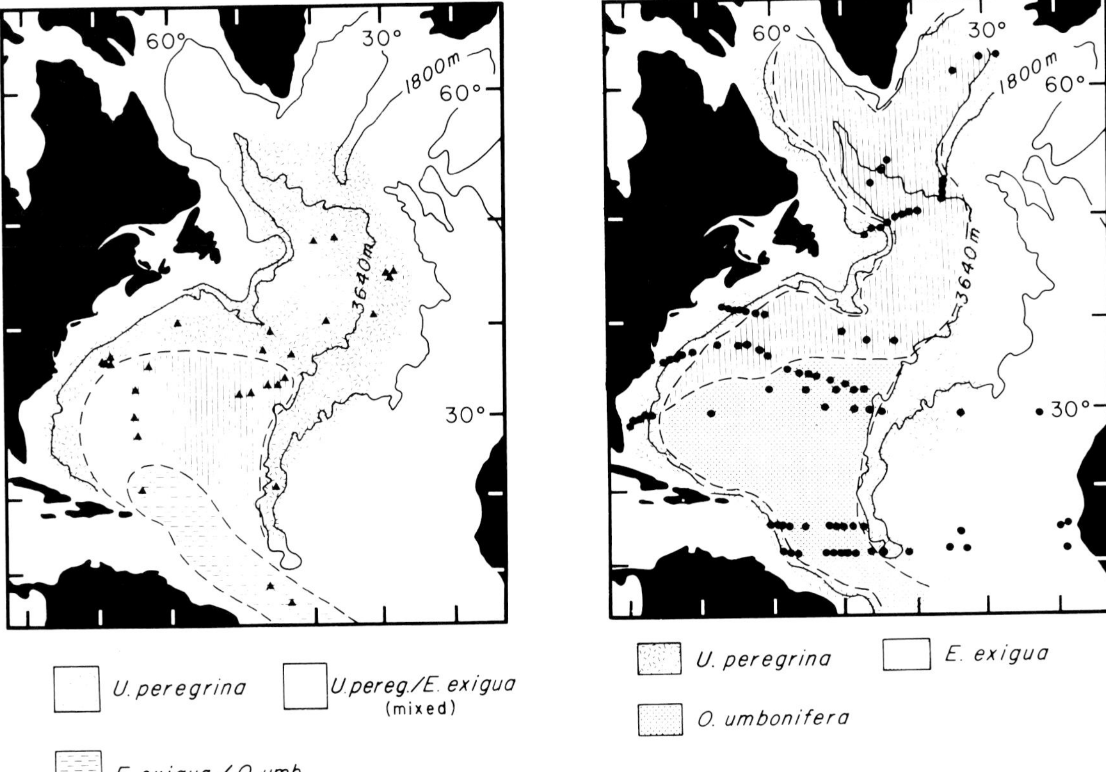

The Globigerinina

Introduction

The biogeographic distribution of the 40 or so extant species of this holoplanktonic suborder is controlled by the temperature, salinity, and productivity of surface waters (Vincent and Berger, 1981; Loubere, 1981; Bé, 1982). Most forms are restricted to one or two latitudinally segregated biofacies whose boundaries are modified near continents by boundary currents and zones of upwelling (Vincent and Berger, 1981; Thiede, 1983; Wefer et al., 1983; Ganssen and Sarntheim, 1983).

Low-latitude faunas have by far the greatest number of species (Vincent and Berger, 1981; Bé, 1982). Species diversity in the tropical pelagic realm is maintained by a well-defined depth stratification (in which some taxa are suspended at density interfaces or within oxygen-minimum zones), specific preferences for small-scale salinity variations within the near-surface mixed layer, and seasonal blooming in phase with the annual cycle of availability of preferred phytoplankton food sources (Shackleton and Vincent, 1978; Loubere, 1981; Curry and Matthews, 1981; Healy-Williams et al., 1985; Reynolds and Thunnell, 1985).

Peak standing crops of planktonic foraminifers (containing spinose, symbiont-bearing species and juveniles of other taxa) occur within the upper 65 m of the water column (Hemleben and Spindler, 1983). Adult tests of species without spines or symbionts are most frequent below 125 m (Hemleben and Spindler, 1983; Hemleben et al., 1985). Assisted by deposition of a thick layer of secondary calcite, these forms (e.g., *Globorotalia hirsuta*) sink during ontogeny into meso- and bathypelagic depths, possibly residing as geronts in close proximity to the bottom (Hemleben et al., 1985).

The relationship between test form and function is governed by the need to counterbalance the excess density of calcite relative to seawater (Lipps, 1966, 1979; Scott, 1973a,b). Various aspects of globigerinine morphology (porosity, test, and chamber shape) and the activity of specialized organelles may play a role in maintaining the buoyancy of the test (Scott, 1979; Malmgren and Kennett, 1976, 1977; Healy-Williams et al., 1985).

Fossil assemblages of planktonic foraminifera, by comparison with existing diversity, morphological and ecophenotypic trends, can be used as indicators of past surface circulation patterns. The widespread distribution and rapid evolution of most globigerinine lineages make them useful for stratigraphic subdivision and correlation. Although important, a full discussion of this topic falls outside the scope of the present review. An extensive literature on the principles and methods of modern high-resolution biostratigraphy and biochronology is available to the curious reader (Berggren et al., 1980, 1983, 1985; Srinivasan and Kennett, 1981a; Keller et al., 1982; Gradstein and Agterburg, 1982; Malmgren and Kennett, 1982; Kennett and Srinivasan, 1984; Berggren, 1984a,b; Miller et al., 1985).

Origins

The first representatives of the Globigerinina were Early Jurassic meroplanktonic species of discorbine ancestry (Grigelis and Gorbatchich, 1980; Banner, 1982). From the mid-Jurassic to the Early Cretaceous, small, morphologically simple forms produced local blooms in the shallow seaways overlying Tethyan shelves and platforms but were excluded from the open ocean by the dominance of calpionellid algae (Caron and Homewood, 1983). The cosmopolitan distributions of mid-Cretaceous species of *Globuligerina* and *Hedbergella* suggest that a holoplanktonic life mode had facilitated their invasion of the entire pelagic realm. Subsequently, several lineages of globigerinines were able to successfully occupy progressively deeper layers of the photic zone (Hart, 1980; Caron and Homewood, 1983).

Geologic History

Since the mid-Cretaceous, planktonic Foraminifera have undergone a cyclic pattern of evolutionary radiation and contraction, punctuated by a catastrophic extinction at the Cretaceous–Tertiary boundary (Olsson, 1982; Caron and Homewood, 1983; Table 2.9, Fig. 2.25). Each radiation was characterized by the appearance of trophically specialized species with longer life cycles (including ontogenetic depth migration), establishment of high-diversity tropical faunas, and iterative development of advanced test morphologies from generalized globigeriniform stocks. No single factor appears to explain this history of radiation by relay adequately (for discussions see Douglas and Savin, 1978; Vincent and Berger, 1981; Hart, 1980; Thunnell, 1981; Caron and Homewood, 1983). Episodes of diversification have coincided with some or all of the following environmental conditions: high sea level, nutrient-enriched surface waters, strong upwelling, moderate to extreme latitudinal temperature gradients, and a stable density stratification in the epi- and mesopelagic zones. Deep-dwelling taxa may have evolved at these times to take advantage of concentrations of labile organic matter suspended at density interfaces.

The intervening periods of contraction have generally correlated with times of surface-water cool-

TABLE 2.9 Summary of Generalized Characteristics of Periods of Radiation and Diversity Reduction in the History of Planktonic Foraminifera

Radiation*	Contraction†
High global diversity	Low global diversity
Strong latitudinal gradient in diversity	Weak latitudinal gradient in diversity
Appearance of complex and innovative morphologies	Absence of complex morphologies
High rates of species originations	Low rates of species originations
Keeled species common in tropics and subtropics	Keeled species absent
Depth stratification well developed	Depth-stratification weak or absent
Dominance of K-selected species (high efficiency in exploiting resources)	Dominance of R-selected species (high reproductive potential)

Sources: Thunnell (1978), Keller (1981b, 1982, 1983), Boersma and Premoli-Silva (1983); Caron and Homewood (1983), Berggren (1984b), Shackleton et al. (1985).

*Middle Cretaceous (Albian and Cenomanian stages), Late Cretaceous (Campanian and Maestrichtian stages), Late Paleocene–Middle Eocene, Early Miocene, Middle Pliocene.

†Late Cretaceous (basal Turonian and basal Campanian stages), Late Eocene to Oligocene, late Pliocene.

ing and a destabilized or destratified water column above the permanent thermocline (Thunnell, 1981; Keller, 1982, 1983; Stanley 1984a,b).

The collision of the earth with a large meteorite may have been the proximate cause for a sudden and near-total extinction of all planktonic groups at the end of the Cretaceous (Alvarez 1984; Alvarez et al., 1982, 1984; see Silver and Schultz, 1982, and Berggren and Van Couvering, 1984, for recent compendia of differing viewpoints). Smit (1982) believed that only one planktonic foraminiferal species survived the postulated impact and associated atmospheric and oceanographic phenomena to become the root for the subsequent expansion of early Paleocene faunas. Although the geochemical evidence of a bolide collision is compelling,the paleontological data can be as convincingly explained by alternative ecological theories (Perch-Nielsen et al., 1982; Dailey, 1983; Eckdale and Bromley, 1984). For example, several workers have suggested that a major interruption in the supply of nutrients to the photic zone occurred at this time because of the rapid spread of deciduous angiosperm floras (Tappan, 1982; see also Williams et al., 1983; Boersma, 1984). This event expanded continental soil cover, which produced a pronounced increase in nutrient retention on land (Tappan, 1982).

Classification

Modern systematics has as its primary objective the elucidation of patterns of descent by evolution and the encapsulation of these data into a classification. Regrettably, many influential classifications of the planktonic Foraminifera have ignored phylogenetic relationships and applied a rigid, monothetic hierarchy of arbitrarily selected morphological criteria in the assignment of species to higher taxa (e.g., Loeblich and Tappan, 1964; Blow, 1969, 1979). Evolutionary studies have now shown that all such form groupings are polyphyletic. As a result of pervasive convergent and iterative evolution, important character states in a typological taxonomy of globigerinines (e.g., coiling mode, umbilical position, chamber shape) have developed in isomorphous species belonging to unrelated lineages (Steineck and Fleisher, 1978, Table 2.2).

The repeated evolution of most key features of test morphology suggests that they confer adaptive advantages in a planktonic–protistan life strategy (Lipps, 1966). For example, a peripheral, imperforate keel is present in 10 distinct lineages of Cenozoic planktonic Foraminifera (Steineck and Fleisher, 1978), where it may serve a common purpose by reinforcing the test margin (Scott 1973a,b) and providing a site for the attachment of streams of ectoplasm (Hemleben, 1975). Neogene lineages often contain closely related unkeeled, semikeeled, and keeled species that would be segregated into separate genera in existing typological classifications (Kennett and Srinivasan, 1983).

Since the seminal contributions of Lipps (1966), Parker (1962, 1967), Berggren (1968), and McGowran (1968), several authors have advocated an evolutionary reclassification of the globigerinines based on the principles of classical evolutionary methods initially espoused in the works of Ernst Mayr and George Gaylord Simpson (Steineck, 1971; Bandy, 1972; Fleisher, 1974, 1975; Steineck and Fleisher, 1978; Vincent and Berger, 1981; Kennett and Srinivasan, 1983). In this approach, phylogenetic reconstruction precedes the formal task of erecting and delimiting supraspecific taxa. Evolutionary relationships can be established by the direct observation of ancestor–descendant pathways in continuous deep-marine sequences (Kennett and Srinivasan, 1983) or by the use of surficial wall texture (Steineck and Fleisher, 1978, p. 627) as a heuristic device for the placement of species with unknown origins.

The physiography of the globigerine wall is a stable and conservative feature within otherwise polymorphic lineages (Parker, 1967; Saito et al., 1976; Huang, 1981; Cifelli, 1982). It develops early in the history of a group of related species persisting with

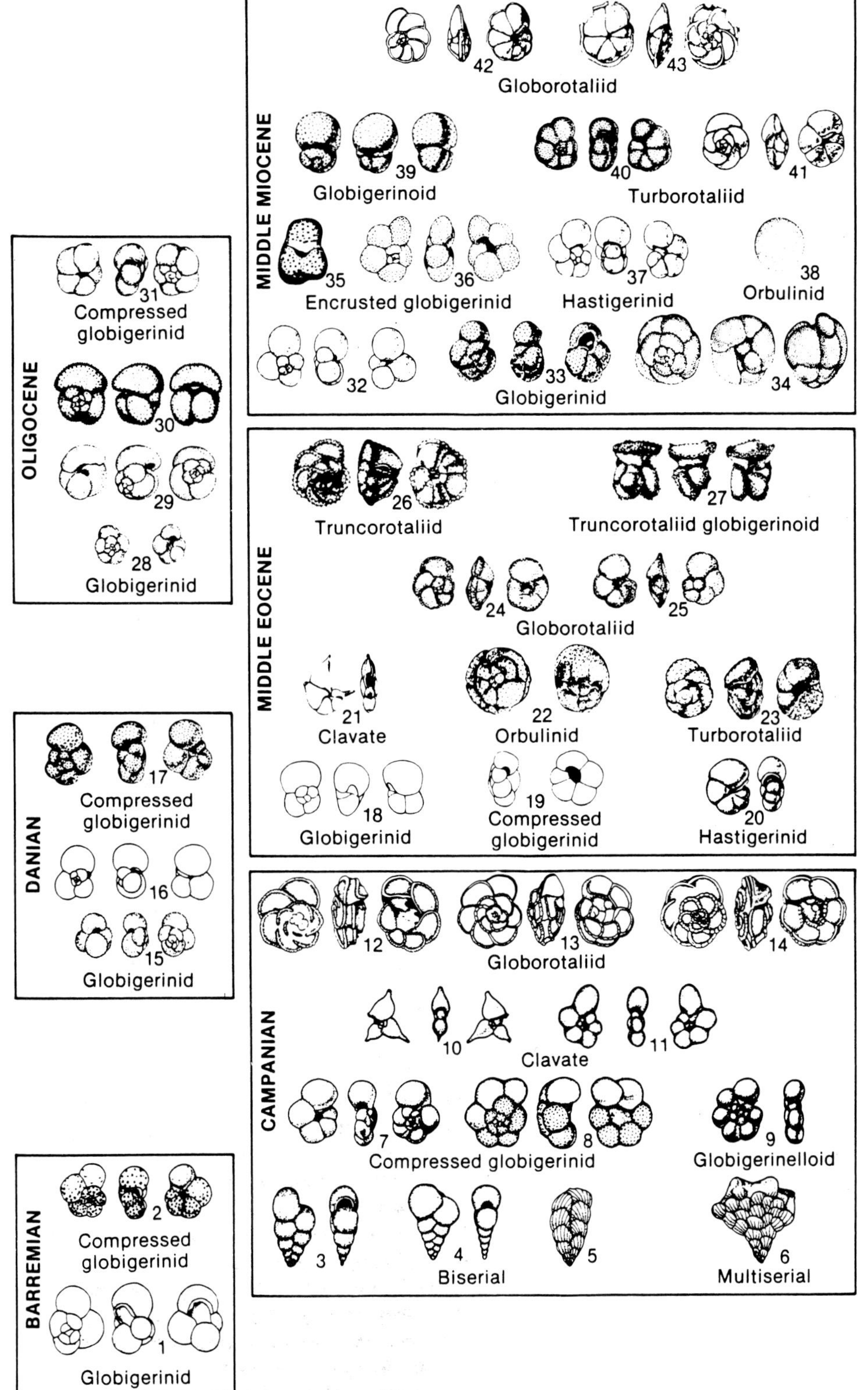

FIGURE 2.25 Iterative evolution of basic globigerinine test form from the early Cretacceous (Barremian) to the middle Miocene. Time intervals shown on the right represent periods of evolutionary radiation; those on the left represent periods of evolutionary contraction. Middle Eocene species with complex morphologies were not ancestral to similar taxa present in the middle Miocene. (From Vincent and Berger, 1981, Fig. 6.)

little or no modification during subsequent diversification and morphologic divergence. As a composite of many organic and mineralized elements, surficial wall texture is a functionally integrated adaptive complex that reflects the fundamental physiological, trophic, and depth requirements of a lineage. The phylogenetic significance of this feature has been confirmed by study of the amino acid sequences of recent species (King and Hare, 1972).

Classification per se follows upon the recognition of evolutionary affinities. Steineck and Fleisher (1978) suggested that three criteria are essential in the definition of genera: monophyly, diversity, and morphological innovation and divergence (Fig. 2.26). Numerous taxonomic revisions emerging from bioseries and wall texture studies have now been published (Bandy, 1972; Fleisher, 1974, 1975; Srinivasan and Kennett, 1975, 1981a,b; Kennett and Srinivasan, 1980; Huang, 1981; Keller, 1981a,b; Cifelli and Scott, 1983; Leckie and Webb, 1985). Saito et al. (1976) and Huang (1981) have proposed family-level classifications based on wall texture, but their application of this feature is at least partly typological. To date, a comprehensive classification of Cenozoic planktonic Foraminifera based explicitly on classical evolutionary methods has not been attempted.

Some systematists, following the cladistic philosophy of Willi Hennig, stress the difficulties inherent in attempting to identify actual ancestor–descendant relationships and denigrate the value of the stratigraphic record of fossil occurrences in the estimation of phylogeny. Cladistics emphasizes the presence or absence of shared, derived character states (synapomorphies) as the sole index of evolutionary relationship (Cracraft and Eldredge, 1979; Eldredge and Cracraft, 1980; Eldredge and Novacek, 1985). The resulting classification is depicted in a dichotomously branching diagram (cladogram) that pairs most-closely related taxa derived from an unspecified common ancestor. However, as pointed out by Scott (1976, 1983), comparative morphology by itself is not adequate to determine phylogeny in a group such as the globigerinines, where morphological characters are few and convergence is common.

FIGURE 2.26 Application of diversity and divergence as criteria in classification. The moderately divergent branch (right) is a phyletic dead end and is retained in the ancestral Taxon 1. The branch to the left, although less divergent, has served as a phyletic springboard and is given taxonomic status. New Taxon 2 can be delimited by the curved solid line at the point of initial divergence (A) or by the dotted line where the morphological innovations are consolidated (B). Although three adaptive zones (morphologies) are occupied by the radiation, only two taxa are recognized. (From Figure 3 of Steineck and Fleisher, 1978.)

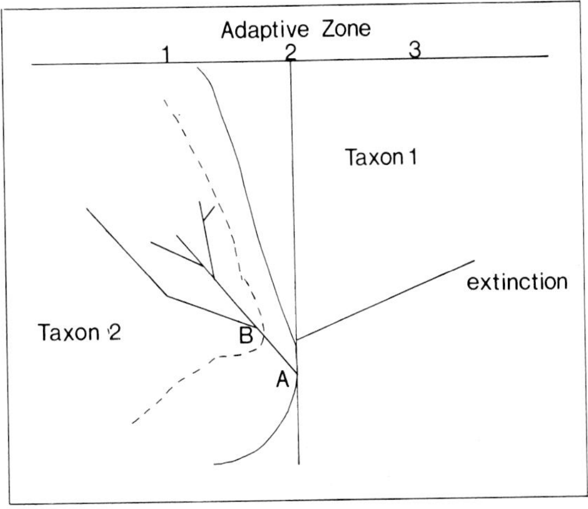

Evolution and Extinction

Speciation

The formulation and testing of evolutionary theories based on the fossil record is a relatively new phenomenon (Gould, 1980a). Some theorists assert that selection operates on a hierarchical organization of nested genomic individuals (including the species as a fundamental unit) and over a multiplicity of time scales (Vrba and Eldredge, 1984; Gould, 1985; Stanley, 1982, 1985). If this is the case, then the study of fossil groups with high preservation potential (e.g., the Foraminifera) will allow recognition of large-scale processes, operating over geologic time, that must be considered in any new and unified explanation for the history of life (Gould, 1980a,b; Vrba and Eldredge, 1984).

The prediction of Gould and Eldredge (1977, pp. 141–142) that speciation by punctuated equilibrium (rapid, divergent, macromutational change in peripheral isolates), however prevalent in metazoans, will be difficult to identify in protistans with polyphasic reproduction, has been borne out by recent investigations of foraminiferal evolution (Table 2.10). Nevertheless, some of the species-level transitions in benthic foraminifers cited in Table 2.10 have a strong quantum component (Dodson and Reyment, 1980; Drooger, 1984). In contrast to these findings, many planktonic lineages display gradualistic trends during speciation. Punctuated gradualism may be the dominant process by which species arise in the globigerinines (Malmgren et al., 1983; Fig. 2.27). A morphologically stable lineage undergoes a short, but not geologically instantaneous, interval of phyletic evolution and the resumption of stasis in the descendant taxon. In this

TABLE 2.10 Selected Examples of Species-Level Evolution in Foraminifera

Example	Geologic range and geographic coverage	Mode of evolution
GLOBIGERININA		
Globorotalia conoidea → *G. inflata*	Latest Miocene–late Pliocene (one core, Challenger Plateau, southern Pacific Ocean)	Gradualism (Malmgren and Kennett, 1981, 1983; Malmgren et al., 1982)
Globorotalia cibaoensis → *G. crassaformis*	5–4.2 Ma (one core, Rio Grande Rise, south Atlantic Ocean)	Gradualism (Arnold, 1983b)
Globorotalia sphericomiozea → *G. puncticulata*	5.21–2.6 Ma (one subaerial section, New Zealand)	Punctuated gradualism with lineage splitting (Scott, 1982)
Globorotalia plesiotumida → *G. tumida*	Late Miocene to Recent (one core, Ninety East Ridge, southern Indian Ocean)	Punctuated gradualism (Malmgren et al., 1983, 1984)
Globorotalia truncatulinoides	700,000 yr BP–Present (one core, Rio Grande Rise, south Atlantic Ocean)	Gradualism or shift in latitudinal position of clinally varying populations (Lohmann and Malmgren, 1983)
FUSULININA		
Lepidiolina multiseptata subspp.	Middle to Upper Permian (subaerial sections, southeast Asia)	Gradualism (Ozawa, 1975)
ROTALIINA (smaller)		
fusion of *Nodosaria liratella* and *N. striatoclavulata*	Middle Triassic (one subaerial section, Austria)	Hybridization (Hohenegger, 1978)
Afrobolivina afra → *A. africana*	Late Cretaceous–Paleocene (western Nigeria boreholes)	"Quantum speciation" (Dodson and Reyment, 1980; Reyment, 1983)
ROTALIINA (larger)		
Orbitoides spp.	Late Cretaceous–Early Paleocene (several subaerial sections, France)	Long-term gradual trends in encapsulation of early chambers superimposed on more abrupt changes in other nepionic parameters (Drooger, 1984)

Ma = million years before present.

type of evolution, panmixis limits divergence in semi-isolated peripheral demes and then allows the rapid propagation of new genetic traits when the main population is exposed to greater selective pressures induced by changing oceanographic conditions.

Evaluation of the results summarized in Table 2.10 must take into account the biological and taphonomic context of the original data (Schindel, 1980; Dingus and Sadler, 1982). Evolutionary patterns in shallow-water benthic taxa may assume a punctuational aspect because of the rapid and episodic nature of epicratonal sedimentation (Schindel, 1980; McKinney, 1985). A single, thick horizon will contain large numbers of individuals drawn from a population coexisting at a "moment" (a day to a few years) of geologic time, creating the impression of temporally extended stasis (McKinney, 1985; Behrensmyer and Kidwell, 1985). In such a variable depositional regime, however, "average" rates of sedimentation have little or no meaning (Behrensmyer and Kidwell, 1985). Phenotypic transformation over time appears abrupt because some bedding planes are unrecognized hiatuses caused by tectonic, glacioeustatic, or autocyclic fluctuations in sea level (Ross and Ross, 1985b; Anderson et al., 1984). On the other hand, in thinner but more continuous pelagic sequences, fewer fossils, with more intermediate morphotypes, accumulate per unit time (Schindel, 1980; McKinney, 1985). Analysis of such condensates will have a built-in bias toward the perception of evolution as gradualistic (McKinney, 1985).

It is also necessary to consider the differences between the population structure and dispersal capabilities of planktonic and benthonic organisms (Lazarus, 1983). Marine planktonic species are typified by a few large demes, weak homeostatic

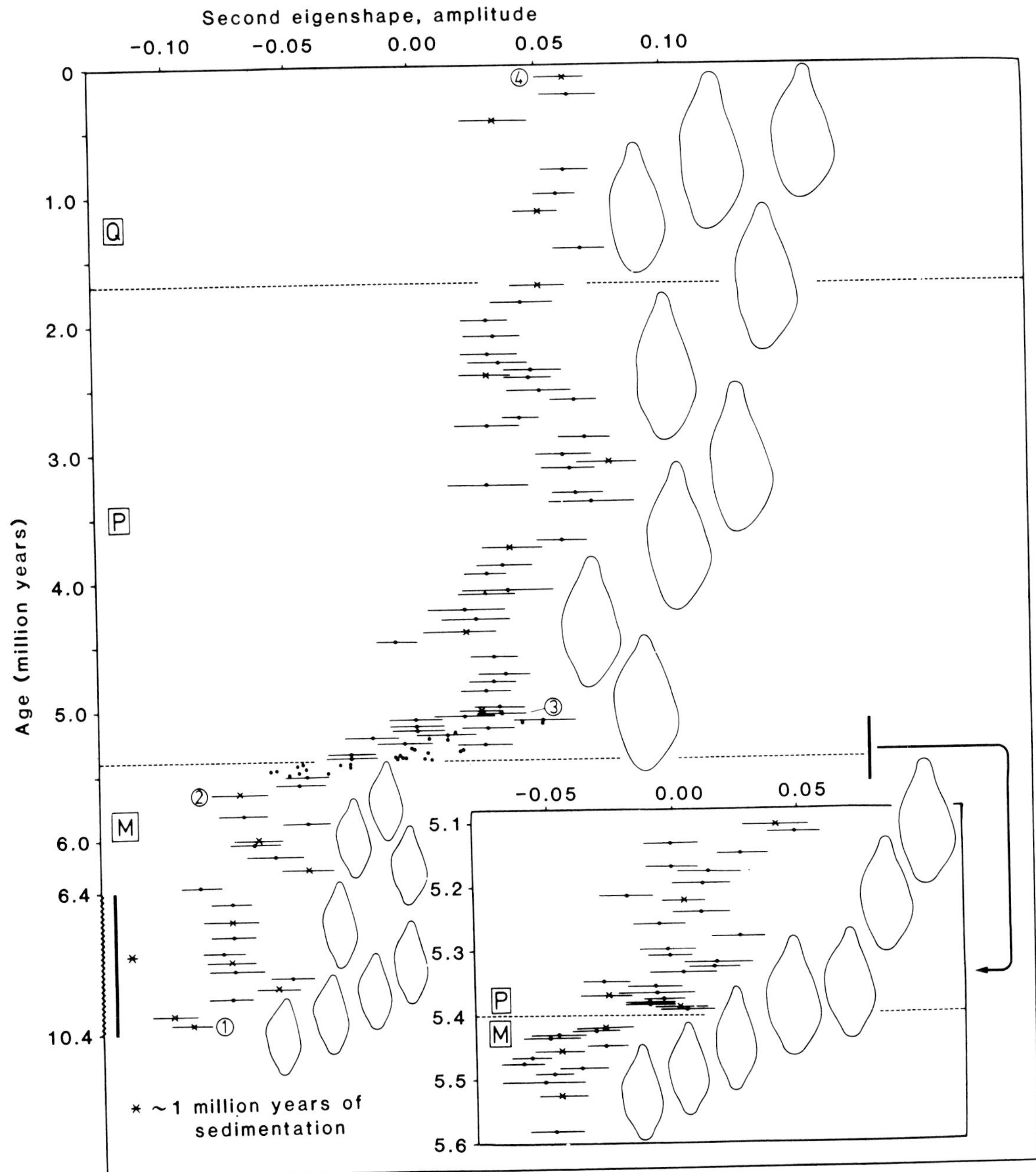

FIGURE 2.27 Punctuated gradualism in the *Globorotalia tumida* lineage, DSDP site 214, southern Indian Ocean. The transition from *G. pleisiotumida* to *G. tumida* occurred near the Miocene (M)–Pliocene boundary within an interval of 0.6 my (~12 percent of total duration of lineage). Anagenetic speciation is attributed to selective forces generated by density modifications in the water column. (After Malmgren et al. 1983, Fig. 1; copyright, 1983, by AAAS.)

mechanisms, long temporal ranges, and inherently high capacity for effective genetic interchange between clinally varying populations living in adjacent water masses (Scott, 1982; Lazarus, 1983). Adaptive, morphogenetic constraints may play a dominant role in stabilizing pelagic species (Stanley, 1982, 1985, p. 22). Allopatry (resulting from bipolarity or the emergence of land barriers) is rare. Speciation must take place by hybridization, sympatric phyletic change, or parapatric (clinal) differentiation (Lazarus, 1983). Gould and Eldredge (1977) have emphasized the role of small and genetically isolated populations as a prerequisite for punctuational evolution. In their absence, saltational speciation is unlikely to occur in planktonic lineages.

Evolutionary Rates

Buzas and Culver (1984) have shown that Cenozoic shallow-water (<200 m) taxa living on the eastern margin of North America evolved at a faster rate (based on mean species duration in millions of years) than those occurring in adjacent deep-water environments. Inasmuch as the most functionally advanced foraminiferal morphologies are prevalent on continental shelves (Brasier, 1982,a,b), overall faunal change in this example is positively correlated with architectural improvement and innovative speciation (sensu Kitchell, 1985, p. 100). Jablonski and Bottjer (1983) have described late Cretaceous macroinvertebrate mudbottom communities where the reverse is true.

The results of Buzas and Culver (1984) imply that in the inherently patchy neritic environment, the probability of speciation in benthic foraminifers is increased by the irregular distribution of semi-isolated and polymorphic demes (see Jablonski et al., 1985). This contrasts with the more homogeneous bathyal zone where gene flow between large and essentially continguous populations is sufficient to dampen tendencies toward divergent evolution.

Community Evolution

Patterns of multispecies evolution (e.g., at the guild or community level) may arise from two contrasting processes (Hoffman and Kitchell, 1984). The "Red Queen" hypothesis proposes that in a constant environment, evolution is driven by biotic interactions between species within the constraints of equilibrium diversity and constant community fitness (Van Valen, 1984b, 1985a,b). New adaptations in one species affect the success of others by disturbing the competitive balance within a community. The probability of extinction remains equal over time; individual species survival is influenced solely by random changes in interspecific relationships within an ecosystem. The "stationary" model of Stenseth and Maynard-Smith (1984; see also Hoffman and Kitchell, 1984, and Vrba, 1985) also assumes finite community diversity but invokes abiotic forces as the major impetus to evolution, which ceases in their absence.

Recent tests of these ideas, using the stratigraphic ranges of planktonic foraminifers, have yielded conflicting results. Arnold (1982b), Levinton and Ginsburg (1984), and Hoffman and Kitchell (1984) have demonstrated that log-linear survivorship curves can be calculated from various subsets of Cenozoic species. These conclusions, which are in accord with predictions of the "Red Queen," raise the possibility that the dynamics of globigerinine radiations are elicited by an intrinsic biological causality (Levinton and Ginsburg, 1984). Alternatively, Wei and Kennett (1983) found that per capita extinction rates have fluctuated over time since the Oligocene because of environmentally induced diversity-dependent and paleoceanographic factors.

The use of differing analytical methods, data bases, and time calibration procedures may be responsible for these inconclusive findings. Theoretically, however, in a nonconstant environment, faunal patterns predicted by "Red Queen" and stationary models converge to the point where they are difficult to separate on the basis of existing empirical data (Hoffman and Kitchell, 1984). In any case, the assessment of environmental stability in a pelagic ecosystem is elusive and judgmental. To the extent that abiotic changes are inferred from the fossil record, such assessments are based at least in part on circular reasoning. Thus, eventual clarification of the role of the "Red Queen" in regulating the species richness of globigerinine faunas may be hard to achieve.

Extinction

The cause of major biotic crises has engendered considerable debate in recent years (Fischer and Arthur, 1977; Nitecki, 1984). Some workers have suggested that a single global mechanism (atmospheric chemistry, Fischer, 1984; temperature, Stanley, 1984a,b) had modulated Phanerozoic diversity cycles. Whatever the merits of their arguments, both Stanley and Fischer have incorrectly concluded that during the latest part of Eocene time, the oceanic realm crossed a major environmental threshold linked to "severe marine extinctions" (e.g., Stanley, 1984a, p. 205). Recent studies of Middle Eocene to Oligocene foraminiferal faunas have revealed a series of climatic oscillations, accompanied by selective extinctions, none of which stand out from the others as a major crisis (Keller, 1982, 1983; Keller et al., 1983; Corliss et al., 1984; Snyder et al., 1984). Herman (1981) has pointed out that the rate, rather than the magnitude or duration, of oceanographic events may be primarily responsible for the observed history of planktonic and deep-sea benthic foraminifers that have shown varying degrees of resiliency to major perturbations in their environments.

PART C: THE RADIOLARIA

Morphology

Polycystin radiolarians are divided into Spumellaria, with skeletons based on a spherical symmetry

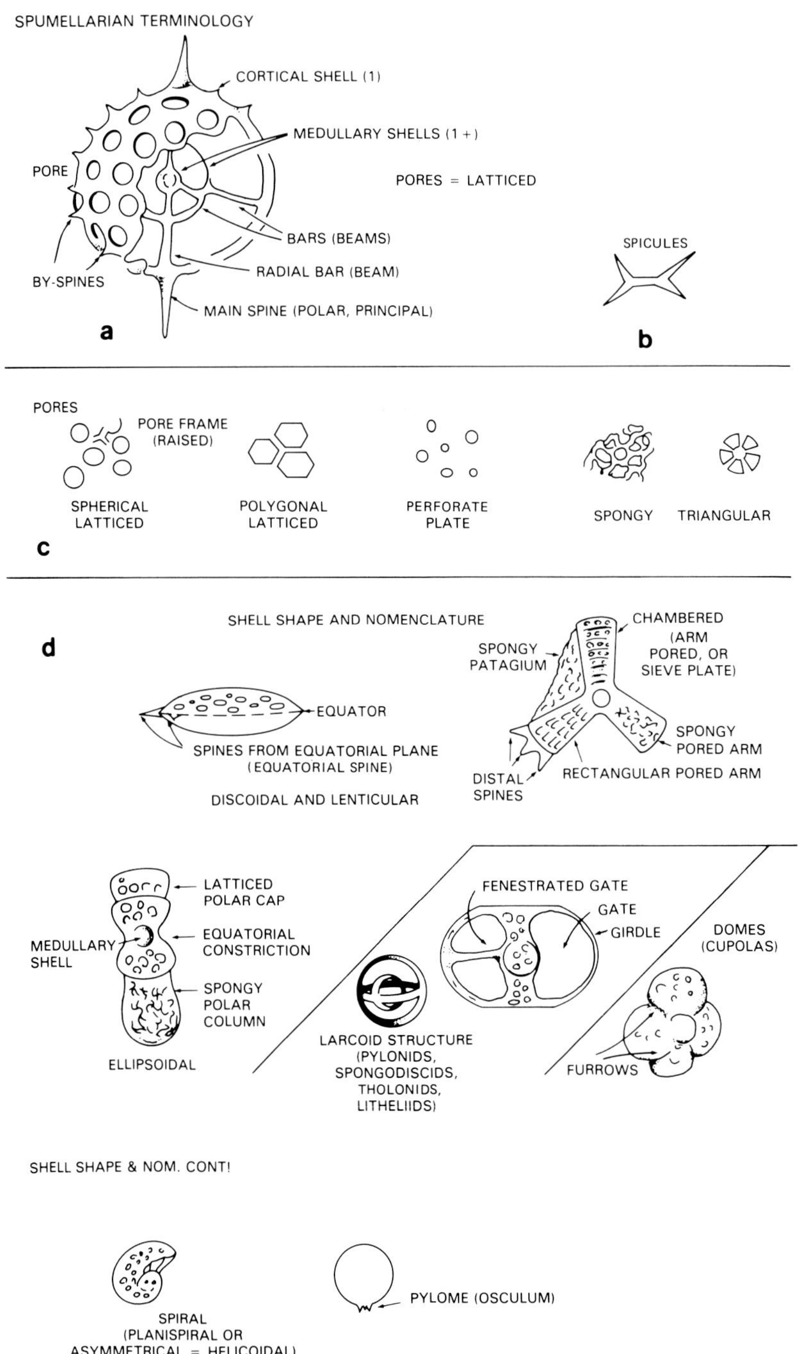

FIGURE 2.28 Basic spumellarian radiolarian skeletal terminology.

(Table 2.12), and Nassellaria, with a plan other than spherical—usually bilateral (Table 2.12). Nassellarians in turn are divided into Cyrtida, with generally conical skeletons (Table 2.12), and Spyroida, with skeletons based on a usually distinct, D-shaped sagittal ring (Table 2.12).

Basic spumellarian skeletal terminology is illustrated in Fig. 2.28. The skeletal terminology reviewed, aside from familiarizing the reader with radiolarians in general, is also very useful when reading Table 2.12. Sometimes the skeleton of a spumellarian is just a spicule (Fig. 2.28b); these spumellarian spicules can usually be distinguished from nassellarian spicules (Fig. 2.29a) in that the spumellarian spicules are usually round in cross section, whereas the nassellarian spicules are usually faceted. Spicules differ from spines in that the spicule is the entire skeleton and a spine is a pro-

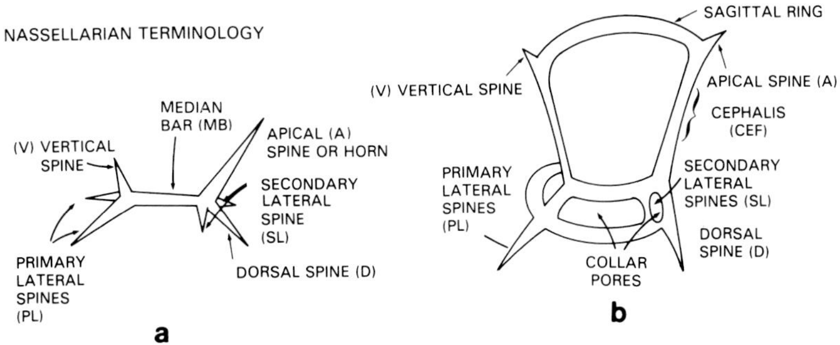

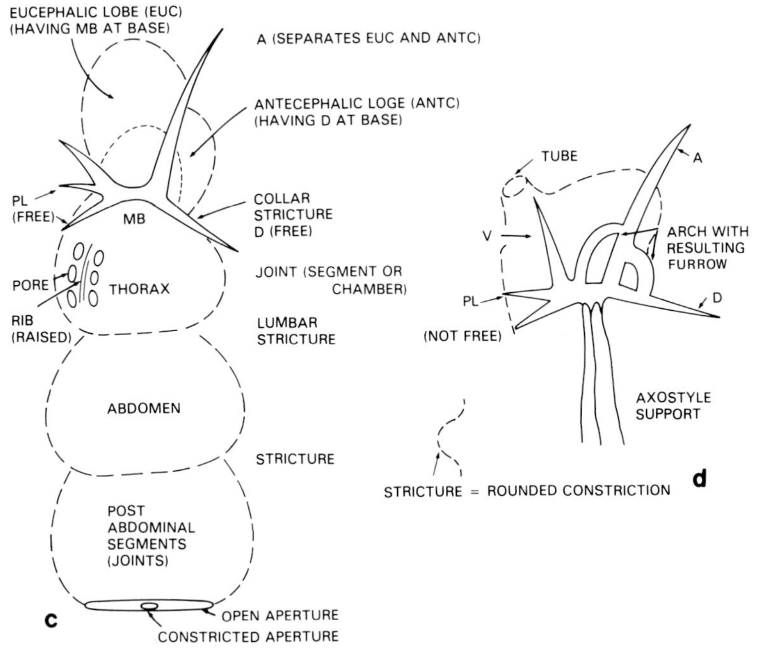

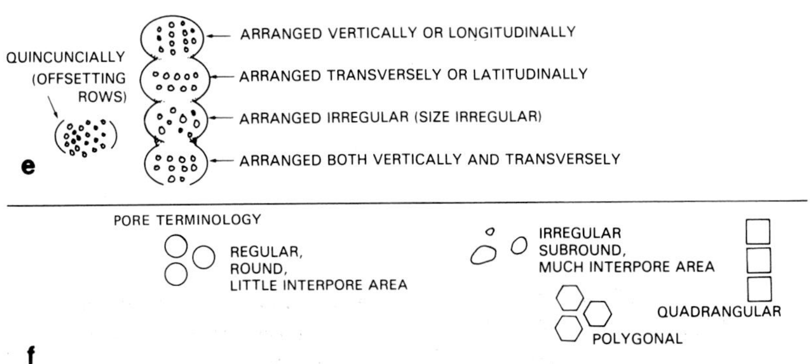

FIGURE 2.29 Basic nassellarian radiolarian skeletal terminology.

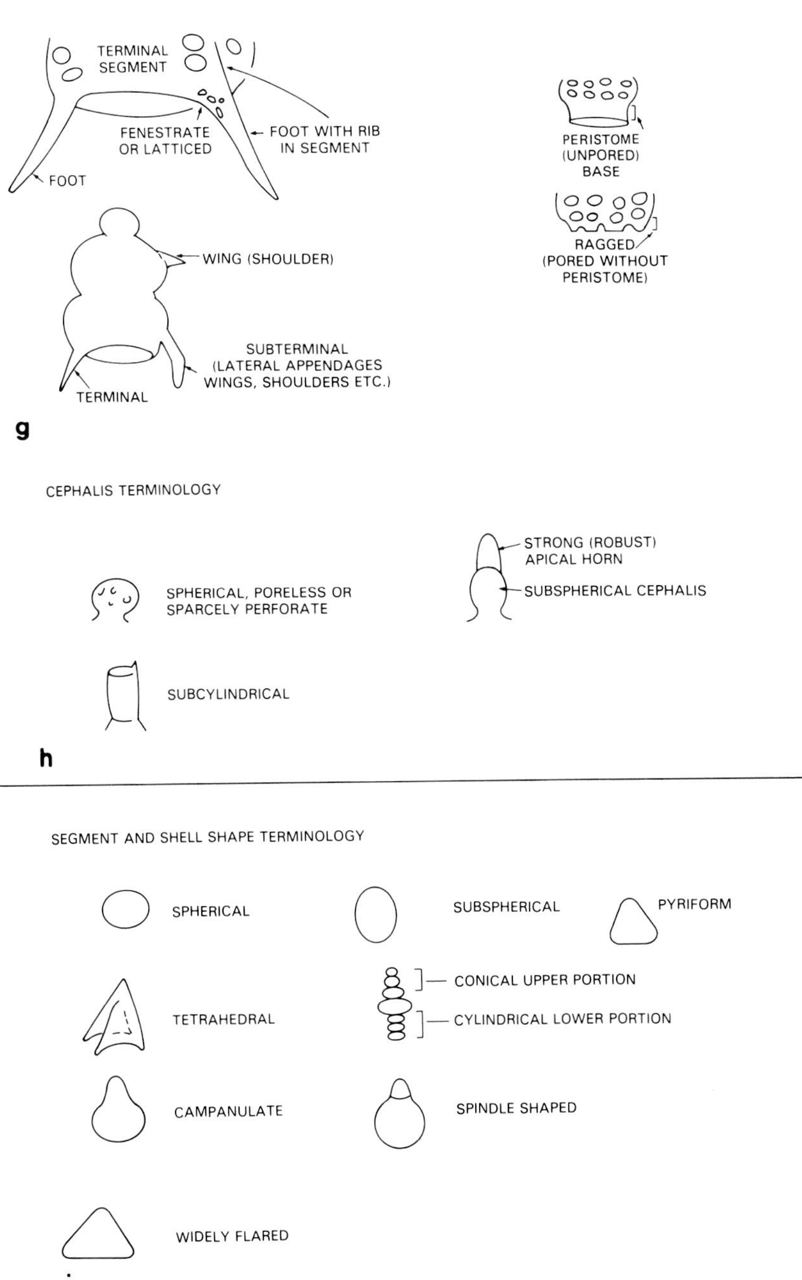

FIGURE 2.29 *(continued)*

tuberance of a skeleton. The outermost shell of a multishelled or single-shelled form is referred to as the cortical shell (Fig. 2.28a). Internal shells are referred to as medullary (Fig. 2.28a). Connections between shells and structures are referred to as bars or beams (Fig. 2.28a). External projections supported by an internal bar (a radial bar) are main (or polar or principal) spines, whereas those not supported by bars are by-spines (by-spines are usually much smaller than main spines) (Fig. 2.28a). The pores on the shells may take on many forms or arrangements (Fig. 2.28c). The overall shape of spumellarian skeletons plus accessory structures may be seen in Fig. 2.28d. These general shapes may range from spherical (sometimes domed or spiraled) to discoidal to various extensions from the spherical or discoidal base plan such as ellipsoidal to armed. There also can be a very complicated coiled basic structure referred to as larcoid with gates (holes) and girdles (latticed structure). A number of structures are shown in Fig. 2.28, such as the patagium, which connects arms, and the pylome, which is a large pore or passageway into the skeleton (many times with a protruding external tube). Nassellarian skeletal terminology (Fig. 2.29) is even more complex than spumellarian terminology. Of primary importance in some nassellarian classifications is the nature of the bars in the cephalis (head or first chamber) and upper thorax (second chamber) (Fig. 2.29a–d). The relation of the latticed skeleton to these cephalic bars is also taxonomically important, such as the eucephalic lobe of the cephalis having the median bar as its base. The chamber (or segment or joint) following the thorax is termed the *abdomen* and all further chambers are referred to as *postabdominal* (Fig. 2.29c). A stricture, sometimes reinforced by an internal ring, occurs between chambers (Fig. 2.29c). Entrances larger than pores are referred to as tubes and apertures (Fig. 2.29d). Tubes are openings in the cephalis, usually supported by spines. Apertures are terminal openings and may be either constricted or open (Fig. 2.29c). The pores may be of various shapes and arrangements (Fig. 2.29e–f). Various structures occur at the termination, such as feet, which may or may not be latticed (with pores) (Fig. 2.29g). These feet may possess a rib that is a raised structure within the terminal segment, and they may be terminal (at the base of the terminal segment) or subterminal (Fig. 2.29g). There also may be no feet and the terminal chamber may end in a ragged fashion or it may have an unpored (and sometimes extended) base termed the peristome (Fig. 2.29g). Various terminologies relative to the cephalis, segment, and general shell shape are also given (Fig. 2.29h–i).

Other sources reviewing polycystin skeletal morphology and terminology are Anderson (1983), Campbell (1954), Petrushevskaya (1975), and Petrushevskaya and Kozlova (1972).

Function of the Test

Although there has been speculation as to the function of the radiolarian test (skeleton) and Anderson (1983) does show that it is important in feeding, the functions are probably diverse, reflected in the diversity of skeletal types, including support for the rhizopodia, protection against predation, increased surface area for food capture, and perhaps ballast. The skeletons of spumellarians many times are surrounded by cytoplasm and soft parts of their own making so that the outline of those skeletons in no way is refected in the living (with soft parts) form. However, the skeletons of some spumellarians and apparently most nassellarians are reflected in their living form. In these forms skeletal uses such as those suggested by Sloan (1981) for a nassellarian with cephalis and apical spine providing a protected (terminal spines) basket for the soft parts, symbiotic algae, and so on, may indeed be valid. The junior author of this chapter has collected a number of organic aggregates at depth that have elongated conical nassellarians adhering to their periphery. It may be that this type of elongated cone morphology, so common to the deep sea, is a morphological adaptation for attaching to sinking organic detrital aggregates.

Classification

Meyen (1834) first described radiolarians from plankton tows taken from the Atlantic Ocean and China Sea. Ehrenberg (1847), after publishing a number of short papers (mainly on fossil radiolarians), produced the first classification of radiolarians, which he called Polycystina. Muller (1858) first called the group Radiolaria. Hertwig (1879) classified the radiolarians based on the nature of their unique feature, the central capsule (that divides the protoplasm into endo- and ectoplasm), and he appears to be the first truly to recognize the radiolarians as protozoans. Haeckel, a contemporary and student of Muller and Hertwig, published a monograph on Radiolaria that essentially documented all known fossil radiolarians of the time and included a new radiolarian classification (Haeckel, 1862). Haeckel (1881) further developed this classification with the use of material from the H.M.S. *Challenger* expedition into a form that at least for the major subdivisions greatly resembled the radiolarian classification of Hertwig (1879). However, Haeckel's major radiolarian work, and classification, is included as a number of volumes in the *H.M.S. Challenger Reports* (Haeckel, 1887), in which he not only reported on almost 3,000 new species but included a monographic study of all radiolarians (living and fossil) known to that date. Haeckel's classification, although "artificial" and not "natural," remained the classification for almost a century. In the 1960s Riedel began to revise Haeckel's radiolarian classification in an attempt to make it more "natural" and illustrated the geologic ranges of his radiolarian families (Riedel, 1967). A major radiolarian classification was produced by Riedel (1971) that included a number of revisions at the generic level. Petrushevskaya (1971) produced another new radiolarian classification, which she further developed in two volumes of the *Initial Reports of the Deep Sea Drilling Project* (Petrushevskaya and Kozlova, 1972; Petrushevskaya, 1975).

Today most radiolarian workers use a combination of Riedel's and Petrushevskaya's schemes; in fact most use the combined schemes of Nigrini and Moore (1978) and Nigrini and Lombari (1984), which are used in catalogues (guides) of modern and Miocene radiolarians, respectively. Table 2.11 shows the combined classification scheme of Nigrini and Moore (1978). These schemes were primarily devised as aids to the Climate Long Range Investigation Mapping and Planning (CLIMAP) program, and the Cenozoic Paleoceanography (CENOP) program. It must be mentioned that these are classifications that are based on the skeletons of mainly recent and Cenozoic radiolarians and are conceived by micropaleontologists; this presents at least two major biases, one toward the Cenozoic and another toward fossilized characteristics. Other classificatory schemes of a more biological nature, as well as excellent reviews of the mentioned schemes, may be found in Anderson (1983). However, since most fossil radiolarian workers are either Cenozoic, Mesozoic, or Paleozoic specialists, little effort has been expended on combining the systematic developments between each of the eras. Since this is not the place for such a synthesizing effort, the higher radiolarian taxa will be referred to on an informal basis. Table 2.12 gives the informal higher taxonomic units used here. These units are close to family level, and brief descriptions are given, as are comments on the environment and geologic range of each informal taxon. This informal scheme is similar to that of Kling (1978) and extensive unpublished notes produced by Riedel.

TABLE 2.11 Polycystin Radiolarian Classification Scheme to Family of Nigrini and Moore (1979)

Subclass: *Radiolria* (Muller, 1858)
Order: *Polycystina* (Ehrenberg, 1838; *emend.* Riedel, 1967)
Suborder: *Spumellaria* (Ehrenberg, 1875)
Family: *Collosphaeridae* (Muller, 1858)
Family: *Actinommidae* (Haeckel, 1862; *emend.* Petrushevskaya, 1975)
Family: *Phacodiscidae* (Haeckel, 1881)
Family: *Spongodiscidae* (Haeckel, 1862; *emend.* Riedel, 1967)
Family: *Pyloniidae* (Haeckel, 1881)
Family: *Litheliidae* (Haeckel, 1862)
Suborder: *Nassellaria* (Ehrenberg, 1875)
Family: *Plagoniidae* (Haeckel, 1881; *emend.* Riedel, 1967)
Family: *Trissocyclidae* (Haeckel, 1881; *emend.* Goll, 1968)
(= *Acanthodesmiidae* (Haeckel, 1862; in Riedel, 1971)
Family: *Carpocaniidae* (Haeckel, 1881; *emend.* Riedel, 1967)
Family: *Theoperidae* (Haeckel, 1881; *emend.* Riedel, 1967)
Family: *Pterocoryidae* (Haeckel, 1881; *emend.* Riedel, 1967)
Family: *Artostrobiidae* (Riedel, 1967; *emend.* Foreman, 1973)
Family: *Cannobotryidae* (Haeckel, 1881; *emend.* Riedel, 1967)

Ecology and Paleoecology of Radiolarians

Biology of Radiolarians

Only those aspects of radiolarian biology especially pertinent to paleobiology are dealt with here. An excellent compilation of other aspects of radiolarian biology can be found in a book on radiolarians by Anderson (1983).

Life Spans and Turnover Rates

By comparing radiolarian standing crops to their flux to the seafloor and preservation in marine varved sediments (so years could be counted), Casey et al. (1971) calculated polycystin turnover rates of a few days to months and suggested that radiolarians within their normal zoogeographic range live for about a month. This "average life span of a month" has been supported by the length of time radiolarians live in culture (Anderson, 1983) and by comparison of standing crops and fluxes to sediment traps (Takahashi, 1984).

Niches, Associations, and Exclusions

Radiolarians have been suggested to occupy the following niches: nannoherbivore, bacterivore, detritivore, symbiont bearer, and absorber of dissolved organic matter (Casey et al., 1979a). Anderson (1983) used similar terms in describing the niches of radiolarians. His terms were bacterivore, herbivore, omnivore, symbiotroph, detritivore, and maybe osmotroph. The larger phaeodarian radiolarians were suspected to belong to the detritivore and bacterivore niches by Casey et al. (1979a). Using water bottle samplings from the southern Sargasso Sea and Gulf Stream, Casey et al. (1979a) found that at the depths where high concentrations of acantharian radiolarians occur, polycystin radiolarians were absent; it was also noted that where tintinnids (only shelled tintinnids were counted), phaeodarians, planktonic foraminiferans, and pteropods peaked in abundance, the standing crops of both nassellarian and spumellarian radiolarians decreased. Polycystin radiolarian densities and diversities also appear to decline because of the

TABLE 2.12 Common Name Classification of Radiolarians

POLYCYSTINE Radiolarians (Natural Groups)

Polycystines (skeleton of opaline silica)

Spumellarians (sphere) (polycystines that are generally spherical or based on sphere)

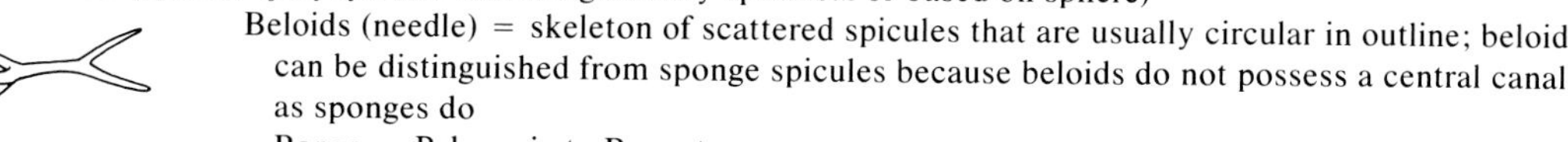

Beloids (needle) = skeleton of scattered spicules that are usually circular in outline; beloids can be distinguished from sponge spicules because beloids do not possess a central canal as sponges do
Range = Paleozoic to Recent.
Environment = shallow cosmopolitan.

Entactinids (ray spine within) = skeleton spherical to ellipsoidal with a latticed wall structure, distinguished from other groups by bars running to the center of the skeleton
Range = Cambrian to Carboniferous (Biostrat. Imp.).
Environment = warm-water sphere.

Orosphaerids (mountainous sphere) = a large spherical or cup-shaped skeleton with a coarse, polygonal lattice and commonly club-shaped spines, distinguished by large size of complete specimens (a few millimeters) but in sediments usually find only pieces of lattice or whole spines (some Paleozoic forms resemble orosphaerids, some paleoactinomonids, and especially the rotasphaerids)
Range = Eocene to Recent (could be Biostrat. Imp.).
Environment = probably all deep-sea or cold-water sphere, occur in red clays as one of the few preserved fossils.

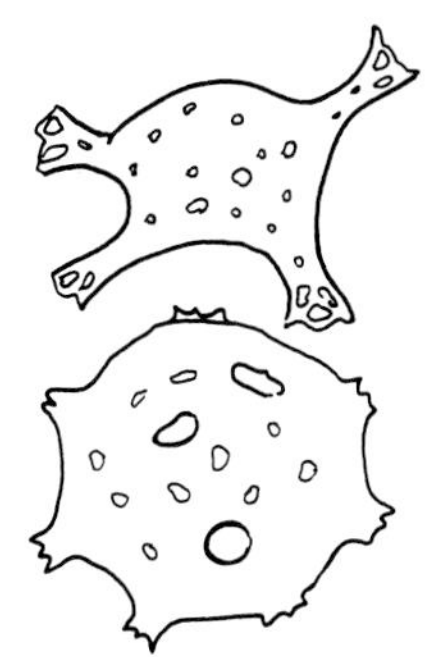

Collosphaerids (neck sphere) = single spheres with usually more interpore area than pore area, many times with internal or external tubular or spiney projections, can be distinguished from similar forms by the preceding characteristics and usually frail (not robust nature).
Range = Lower Miocene (or Upper Oligocene) to Recent (Biostrat. Imp.).
Environment = warm water sphere; dominate the oligotrophic anticyclonic gyres; colonial and possessing symbiotic algae (zooxanthellae); a few more robust forms appear to be cold water forms; shelfal tolerant.

Actinommids (ray spined) = a catch-all polyphyletic group containing generally spherical forms (some rough subgroups follow)
Range of "group" = Paleozoic?, Triassic to Recent.

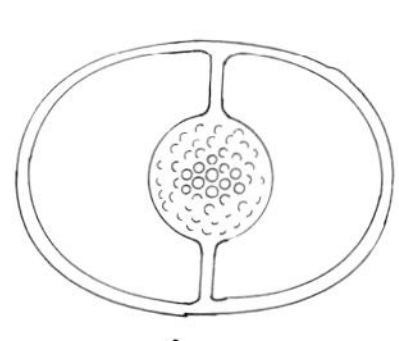

Saturnalins (like Saturn) = spherical latticed skeleton possessing a circular outer ring connected to a latticed or sometimes spongy shell.
Range = Triassic to Recent; forms with ornamented outer ring appear to be restricted to the Mesozoic.
Environment = warm-water sphere.

Stylosphaerids (pillar sphere) = latticed shells with polar spines, one or two (the saturnalins are probably a subgroup of this group).
Range = Paleozoic? to Recent.
Environment = cosmopolitan.

Cubosphaerids (six-spined sphere, in cube pattern) = latticed shells with polar and equatorial spines (six) (tripospherids = three spines (*Trilonche* Devonian to Mesozoic), staurosphaerids = four spines (staurolonchids, J to K), pentasphaerids = five spines).
Range = Paleozoic to Recent.
Environment = cosmopolitan.

Astrosphaerids (star sphere) = latticed shells with numerous spines.
Range = Paleozoic to Recent.
Environment = cosmpolitan.

Cenodiscids (empty disk) = lattice shell (single) without obvious spination; can usually be distinguished from collosphaerids by having less interpore area, thicker skeleton, and no tubes.
Range = Cambrian to Recent.
Environment = wide range of environments but great abundance in sediment samples usually means the sample reflects cold surface waters or has undergone considerable dissolution.

Liospherids (layered sphere) = lattice shell (multiple) without obvious spines.
Range = Paleozoic to Recent.
Environment = cosmopolitan.

TABLE 2.12 *(continued)*

Artiscins (loaf of bread) = elliptical outer (cortical) latticed shell many times with an equatorial constriction and caps and or spongy columns.
Range = Oligocene to Recent.
Environment = warm-water sphere, symbiotic algae, shelfal tolerant.

Phacodiscids (lens disk) = discoidal, biconvex latticed outer shell
Range = Mesozoic to Recent.
Environment = warm-water sphere.

Coccodiscids (kernel disk) = lens shaped with a latticed center and spongy chambered girdle or arms (orbiculiformids J–K and maybe cavaspongidae LK included here).
Range = Mesozoic to Oligocene.
Environment = warm-water sphere.

Spongodiscids (spongy disk) = polyphyletic group of generally discoidal and spongy nature, group range Devonian to Recent (subgroups follow)

Spongotrochins (spongy wheel) = spongy lens (sometimes with evidence of "rings") and many times with large, numerous spines.
Range = Mesozoic (maybe Paleozoic) to Recent (includes orbicuformids?).
Environment = commonly cold-water sphere, good indicators of upwelling.

Spongopylins (spongy with tube) = spongy lens without spines but with a pylome.
Range = Mesozoic to Recent.
Environment = cold-water sphere.

Spongasterins (spongy stomach) = spongy disk with margins faceted and radiating.
Range = Paleocene to Recent (K?) thickenings internally (euchitonids) (Biostrat. Imp.).
Environment = warm-water sphere, tolerant of near-shore conditions (indicator of such), with symbiotic algae.

Dictyocorins (netted head) = spongy disc with three arms (chambered or spongy).
Range = Mesozoic to Recent.
Environment = mainly warm-water sphere, common (dictyocorins not euchitonids) in eutrophic environments (indicative of such), some with symbiotic zooxanthellae (there are cool and warm morphotypes).

Stylodictids or Porodiscids (chambered or pored net or disk) = disk with rings of lattice, may be a subgroup of spongodiscids.
Range = Permotriassic to Recent.
Environment = appears to be mainly subsurface and cold-water forms, high latitude and upwelling.

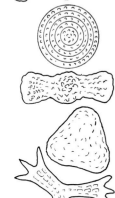

Spongurins (spongy) = spongy cylinder (*Amphirhopalum* and prunobrachids included here).
Range = Mesozoic to Recent.
Environment = cosmopolitan, cold and warm morphotypes. (*Archaeospongoprunum* = spongurin with polar spines Jurassic–Cretaceous).

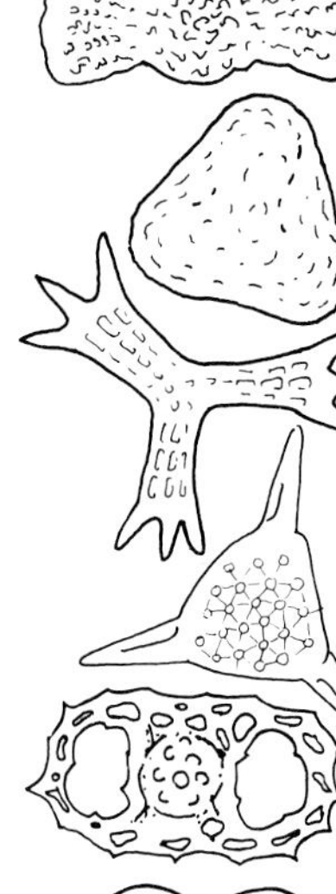

Phaseliformins (pouch form) = subellipsoidal spongy.
Range = Upper Cretaceous.
Environment = ?

Hagiastrids (holy star) = flat "disk" with spongy-rectangular mesh and two to four (usually three) radial arms (large) that may possess spines.
Range = Mesozoic (Import. Biostrat.)
Environment = warm-water sphere.

Pseudoaulophacids (false tent lens) = lenslike or more commonly triangular spongy and triangular mesh, usually with a few prominent marginal spines.
Range = Mesozoic.
Environment = ?

Pylonids (gateways) = shell an ellipsoid of girdles with gates (holes).
Range = Eocene to Recent.
Environment = mainly warm-water sphere.

Tholonids (domes) = outer (cortical) shell elliptical with bulblike extensions of the shell and some spines.
Range = Pliocene to Recent.
Environment = deep- and or cold-water sphere (in intermediate waters).

TABLE 2.12 *(continued)*

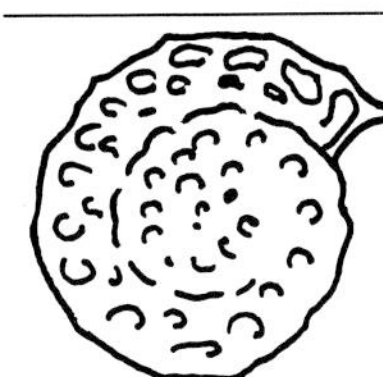

Lithelids (sphered) = coiled latticed shell (includes larcospirins and sphaeropylins).
Range = Carboniferous to Recent.
Environment = cosmopolitan with cold-water morphotype tightly coiled and warm-water sphere morphotype loosely (more open) coiled.

Radiolaria Incertae Sedis (probably spumellarianlike).
Paleoactinommids (old spiney one) = all the Paleozoic Spumellaria with single latticed shell or two or more concentric shells and lacking internal spicular system of Entactinids and characteristic lattice structure of Rotasphaerids (Paleoactinommids = hollow spines and actinommids = solid spines, warm-water sphere).

Rotasphaerids (wheel sphere) = Paleozoic Spumlellaria with single spherical, latticed shell with angular meshes, lacking the internal spicular system of Entactinids and displaying a point or points upon the shell from which radiate five or more strong, straight lattice bars (Rotasphaerids resemble the deep-living orosphaerids, warm-water sphere).

Nassellarians (nose) = polycystines that are generally cone shaped.
Plagonids (snares) = polyphyletic group of simple forms, spicule to two joints.

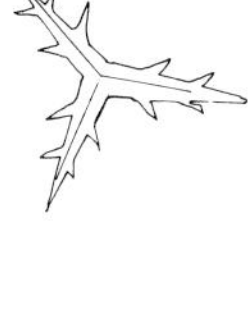

Plagiacanthins (thorn snare) = thorax and sometimes walls of cephalis reduced, sometimes to the extreme of only spicules remaining; these "spicules" can be distinguished from beloids in that the plagiacanthins usually possess faceted spines and the beloids, round spines.
Range = Miocene to Recent (but "spines" in Mesozoic and paleozoic).
Environment = cosmopolitan in surface and subsurface.

Lophophaenins (open neck) = cephalis and thorax about equally developed, usually two lobes (or three) in cephalis; apical, dorsal, and primary spines, usually obvious and extending beyond lattice shell.
Range = Cretaceous to Recent.
Environment = cosmpolitan, mainly shallow but also deep members (the antarcticins appear to be good shallow Antarctic forms and are so far the only radiolarians recognized from Antarctic Bottom Water in the Northern Hemisphere) (*Peridinium spinipes* may be a good temperate eutrophic indicator).

Sethoperins (sieve bearer) = cephalis and six latticed plates extending from cephalis, or modified as a basketlike thorax.
Range = Cenozoic.
Environment = warm-water sphere.

Sethophormins (sieve basket) = cephalis large, hemispherical with a flattened umbrella-shaped thorax.
Range = Cretaceous to Recent.
Environment = cosmopolitan.

Cyrtentactins (cage with spine within) = looks like "cross" between entactinid (spumellarian) and lophophaenin (nassellarian); may be earliest nassellarian (Pylenonemids included here).
Range = Devonian to Carboniferous.
Environment = ?

Carpocanids (fruit basket) = cephalis small and recessed into the thorax; thorax usually elongate and usually with longitudinally arranged pores (Williriedellids Mesozoic. *Diacanthacapsa* K, *Sethocapsa* J to K, *Gongylothorax* J to K, *Heliocryptocapsa* K and *Kozurium* K included here).
Range = Eocene to Recent (J?).
Environment = warm-water sphere mainly.

Theoperids (devine pouch) = small, spherical, usually poreless cephalis, with one or more postcephalic chambers, usually only one small apical spine on top of cephalis.

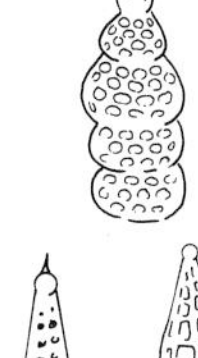

Eucyrtidins (true cage) = usually more than one postcephalic chamber (joint), and pores not quadrangular (Dictyomitrids J to K, Neosciadiocapsids K to E, Parvacingulids J to K, Spongocapsulids J to K, Syringocapsids, J to K, Ultranaporaids J to K, Xitids J to K).
Range = Mesozoic to Recent (Biostrat. Imp.).
Environment = cosmopolitan but mainly in warm-water sphere.

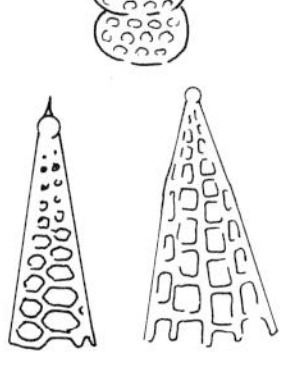

Plectopyramins (woven pyramid) = usually only one postcephalic joint, and this joint cone shaped and usually with longitudinally arranged quadrangular pores.
Range = Mesozoic to Recent.
Environment = cosmopolitan but mainly deep- or cold-water sphere (enhanced under high prod. and divergences).

TABLE 2.12 *(continued)*

Pterocorids or Pterocorythids (winged helmet) = large, elongate cephalis with one or more postcephalic chambers; cephalis usually consists of a main lobe (eucephalic) and two lateral lobes separated by furrows; the cephalis is pored and usually bears a large vertical spine (apical) that emerges from the side of the cephalis.
Range = late Cretaceous to Recent (Biostrat. Imp.).
Environment = warm-water sphere.

Amphipyndacids (double) = small, spherical, usually poreless cephalis with several postcephalic joints; cephalis divided into two vertical segments by transverse internal ledge (probably a Eucyrtidin).
Range = Cretaceous to lower Tertiary (Biostrat. Imp.).
Environment = warm-water sphere, (a few living "similar" forms are deep-sea).

Artostrobids (loaf top) = cephalis of eucephalic lobe, other lobes, and sometimes tubes; many postcephalic joints usually with latitudinally arranged pores; spines may sometimes be seen protruding from the collar structure into the postcephalic chamber inner area.
Range = Cretaceous to Recent (may be good for cosmopolitan biostrat.).
Environment = cosmopolitan, warm-water and cold-water (and deep-water) morphotypes can be distinguished (robust artostrobids appear to be upwelling enhanced).

Cannobotryids (tube cluster) = cephalis of eucephalic lobe, other lobes that may extend as tubes; usually only one or two postcephalic joints; pores commonly appear to be randomly arranged.
Range = Eocene (K?) to Recent.
Environment = cosmopolitan but more diverse in warm-water sphere; cold (deep) and warm-water morphotypes can be distinguished

Acanthodesmiids (thorn bundle) = D-shaped ring or latticed, bilobed chamber with internal D-shaped sagittal ring.

Acanthodesmins = D-shaped ring or modification of such, or sagittal ring (D-ring) surrounded by two latticed lateral lobes; apical and basal poles not obvious.
Range = Eocene to Recent.
Environment = mainly warm-water sphere, usually shallow and apparently with symbiotic algae.

Triospyrins (three wraps) = sagittal ring (D-ring) surrounded by two latticed lateral lobes; apical and basal poles obvious; commonly with basal feet and apical spines (sometimes very long).
Range = Paleocene to Recent (Biostrat. Imp.).
Environment = mainly warm-water sphere.

Rotaformids (wheel form) = lens-shaped, central area enclosing what appears to be nassellarian cephalis; this central area is connected to an outer ring by radial bars.
Range = Cretaceous (Biostrat. Imp.).
Environment = warm-water sphere.

Radiolaria Incenrtae Sedis (Probably nassellarianlike)

Albaillellids (slim ring) = skeleton consisting of three spines in a plane forming a triangle; riblike elements may protrude from the spines, giving the skeleton a jointed appearance; may be oldest nassellarian.
Range = Silurian? to Carboniferous (maybe Triassic) (Imp. Biost.).
Environment = ? (open ocean, perhaps deep-living).

Palaeoscenids (old tent) = skeleton consisting of four diverging basal spines with riblike elements protruding, topped by two or four "apical" spines.
Range = Devonian to Carboniferous (Biostrat. Imp.).
Environment = ?

presence of high concentrations of diatoms. This appears to be the case in high-latitude shallow waters and in the eastern boundary currents tapping these waters, where diatoms are dominant. Perhaps radiolarians decline in abundance owing to direct competition for food sources (or are themselves a food source) with the presence of tintinnids, phaeodarians, planktonic foraminiferans, and/or pteropods. Perhaps radiolarians lose when diatoms compete for dissolved silica. Perhaps acantharians, under bloom conditions, release an ex-metabolite that inhibits polycystin radiolarians. Casey et al. (1979a) suggested some general distributional patterns for radiolarians occupying different niches. Nannoherbivores are considered to be generally restricted to the upper 200 m of the water column. Those radiolarians with symbiotic algal associations dominate the oligotrophic subtropical anticyclonic gyres and warm shelfal and neritic-influenced regions (such as the eastern tropical Pacific). Both detritivores and bacterivores are found at depth and in shallow subsurface high latitudes.

Densities

Polycystin radiolarian surface water densities (standing crops) appear to be highest in the subtropical anticyclonic gyres. The dominant radiolarians in these waters appear to be symbiotrophs. Lowest polycystin radiolarian surface water densities appear to occur in shelfal waters, followed by subpolar and polar waters, and the eastern boundary currents that tap these waters. Polycystin radiolarian standing crops fluctuate greatly in shelfal waters, usually decreasing at times of high phytoplanktonic standing crops and increasing during times of open-ocean water intrusion (Casey et al., 1981). High-speed surface net tows across the California Current into the anticyclonic gyre illustrate a drastic increase in polycystin radiolarian standing crops as gyre waters are entered. Radiolarian density usually peaks at subsurface depths of 100 or 200 m, many times just below the pigment depth and/or at the peak in bacterial standing crop. The highest deep-water standing crops (living polycystin radiolarians from stained vertical plankton tows) appear to be from under higher-productivity regions (Casey et al., 1979a). Apparently, a higher flux of organic detritus from these more eutrophic waters sustains a high polycystin radiolarian standing crop of detritivores and bacterivores at depth.

Diversities

Polycystin radiolarian diversity in surface waters appears to be highest in the subtropical anticylonic gyres (at least in the North Pacific gyre), and lowest in the surface waters of shelfal and high latitudes. Polycystin radiolarian diversity fluctuates greatly in shelfal waters, with lowest value during periods of high phytoplankton standing crop and highest values during periods of open-ocean water intrusion (Casey et al., 1981). Radiolarian diversity with depth usually peaks at some depth below the surface.

Distribution and Ecology

Polycystin radiolarians are geographically the greatest ranging (pole to pole, surface to abyss) and taxonomically the most diverse of the well-preserved marine planktonic protozoa. There are about 400–500 relatively common polycystin radiolarian species living in the present oceans (Casey and others, 1983). Approximately 200 of these species live in shallow (0–200 m) central and tropical waters of the ocean (warm-water sphere), 40–50 in high-latitude (poleward of the subtropical or polar convergences) shallow (0–400 m) waters, 150–200 in deep (greater than 200 m) water, most of which appear to be tropical submergent forms (shallow in high latitudes and diving as tropical submergent or tropical avoidant forms), and about 40–50 eurybathyal forms. Most of these deeper-living forms are vertically stratified (at the subspecific, specific, and high taxonomic levels), as are the shallower-water forms.

Only research of the last 10 years or so will be reviewed (and in some cases expanded upon) here; a review of the earlier work on radiolarian distribution and ecology can be found in Casey (1971, 1977).

Plankton Tow Studies

Since most radiolarians in the water column below about 100 m are dead (empty) skeletons, it is best to use living (stained) specimens, especially in studies dealing with vertical distributions. Using stained plankton material and *R*-mode cluster analysis, McMillen and Casey (1978) were able to describe six depth-stratified radiolarian groups from the Gulf of Mexico and Caribbean Sea. They also found that water masses in the Gulf of Mexico and Caribbean can be distinguished by their radiolarian assemblages, and they designated certain species as good indicators of North Atlantic Deep Water, Antarctic Intermediate Water, Subtropical Underwater, and surface water. Using stained plankton material, Spaw (1979) distinguished five depth-restricted radiolarian assemblages and one eurybathic assemblage from the southern Sargasso Sea. She found that most spumellarian species were found in samples from the upper 2,000 m and that the artiscins and pylonids were restricted to the upper 200 m. She found that nassellarians occur throughout the water column (peaking in diversity between 200 and 2,000 m) with the most diverse nassellarian assemblage in the surface waters (0–200 m) and with plectopyramids and sethophormins living only below 2,000 m. Major assemblage breaks coincided with changes in the biological and physical structure of the water column, including the base of the warm-water sphere, the oxygen minimum, the change to intermediate water masses, and the change to deeper-water masses (Spaw, 1979).

The vertical distributions of radiolarian standing crops (densities) using stained material from different environments were compared by Casey et al. (1979a). The more eutrophic Gulf Stream set exhibited higher standing crops in both shallow and deep samples than the more oligotrophic southern Sargasso Sea or open-ocean Gulf of Mexico samples. Seasonal distributions of radiolarians (using stained material) were studied from the south Texas outer continental shelf (Leavesley et al., 1978; Casey et al., 1979a; Casey et al., 1981).

Plankton studies, especially using stained material, are very useful as aids in defining which radiolarians belong in which ecological groups, but careful analyses of the trophic activity and environmental tolerances of living organisms needs to be done to complement the distributional data.

Casey et al. (1982, 1983) supplemented information for McMillen and Casey (1978) and Spaw (1979) in developing a model of the dominant distribution patterns of spumellarian and nasellarian groups. This was intended to be a robust scheme illustrating the dominant distribution patterns of informal (approximately family-level) groups and is expanded upon in the following section.

Model of Modern Polycystin Zoogeography

Casey et al. (1982, 1983) developed and utilized a hypothetical model of a modern-day ocean for plotting radiolarian distributions. The main terminology used in that hypothetical ocean was that used by Reid et al. (1978). Here the shallow portions of this hypothetical ocean (Casey et al., 1982, 1983), which contains generic currents, convergences, divergences, and so on, characteristic of all shallow modern oceans, are overlayed by McGowan's (1974) basic biotic provinces of the oceanic Pacific (Fig. 2.30). Radiolarian provinces appear to mirror these provinces of other planktonic groups. Table 2.13 is a modification of McGowan's table comparing some of the characteristics of his biotic provinces of the North Pacific, and Casey and others (1986) table. Referring to both Fig. 2.30 and Table 2.13, radiolarian diversity in the subpolar cyclonic gyres appears to be low—dominated by cenodiscids, spongodiscids, spongotrochins, lithelids, and lophophaenins.

Endemism in the subpolar cyclonic gyre is moderate in the southern gyre, housing such endemic forms as antacticins. The transition zone contains a moderate diversity dominated by cold-water morphotypes of warm-water-sphere taxa and a few endemics. The transition zone appears to be a region of radiolarian isolation and evolution (see Casey et al., 1983). The subtropical anticyclonic gyre radiolarian fauna exhibits the highest radiolarian diversities, densities, and endemism of any of the zones. This zone is dominated by radiolarians hosting symbiotic zooxanthellae, and this association has been suggested to be the reason that these symbiont bearers dominate in these oligotrophic regions: They are able to sustain a substantial part of their nutrition by symbiotic photosynthesis (Casey et al., 1981). The equatorial zone is very similar to the subtropical anticyclonic gyre zone, but is not as high in diversity, density, or endemism. The eastern tropical zone is dominated by symbiont–bearing radiolarians, as is the subtropical anticyclonic gyre; however, unlike the gyre, there are few endemics and only low to moderate diversities and densities. The boundary current "zones" are really just tappings of either the equatorial or subpolar and transition zones and reflect these parental provinces in their radiolarian diversities, densities, and degree of endemism.

Ecological Groups of Radiolarians

Polycystin radiolarian groups were zoogeographically related to a hypothetical open-ocean environment by Casey et al. (1982, 1983). These radiolarian groups are essentially informally named "ecological" groups that are of family or lower taxonomic levels (see Table 2.12 for brief morphological, distributional, ecological, and geologic range descriptions of each group). Informal groups were used because of the current fluctuating state of radiolarian systematics, and because a formal system did not allow the flexibility needed at this time. These groups were related to a hypothetical ocean that represented present-day world open-oceanic parameters (Fig. 2.31a). Distributions of ecological groups were determined from plankton and well-preserved deep-sea surface sediment distributions. Figures 2.31b and 2.31c illustrate the dominant distributions of spumellarian and nassellarian groups respectively in this hypothetical ocean (from Casey et al., 1982). The percent of all warm-water-sphere radiolarians (both spumellarians and nassellarians, Fig. 2.32a) from well-preserved deep-sea oceanic sediments illustrates the distributional pattern of warm-water-sphere radiolarians with major poleward boundaries at the subtropical convergences and extensions of warm forms via the western boundary currents and "exclusions" from the cold eastern boundary currents. Although this pattern (Fig. 2.32a) is similar to that of most warm-water-sphere radiolarians some individual ecologic groups exhibit important differences. For example, the percentage of warm-water collosphaerids, although exhibiting the same basic patterns, shows dominance in the subtropical anticyclonic gyre and eastern tropical zones (Fig. 2.32b). Casey et al (1982) suggested that collosphaerids (colonial radiolarians with symbiotic algae) possess the ability to remain at or near the surface or drop below the surface to the nutricline, allowing them to dominate the warm-water sphere of the oligotrophic gyres. This association with symbiotic algae and their ability to

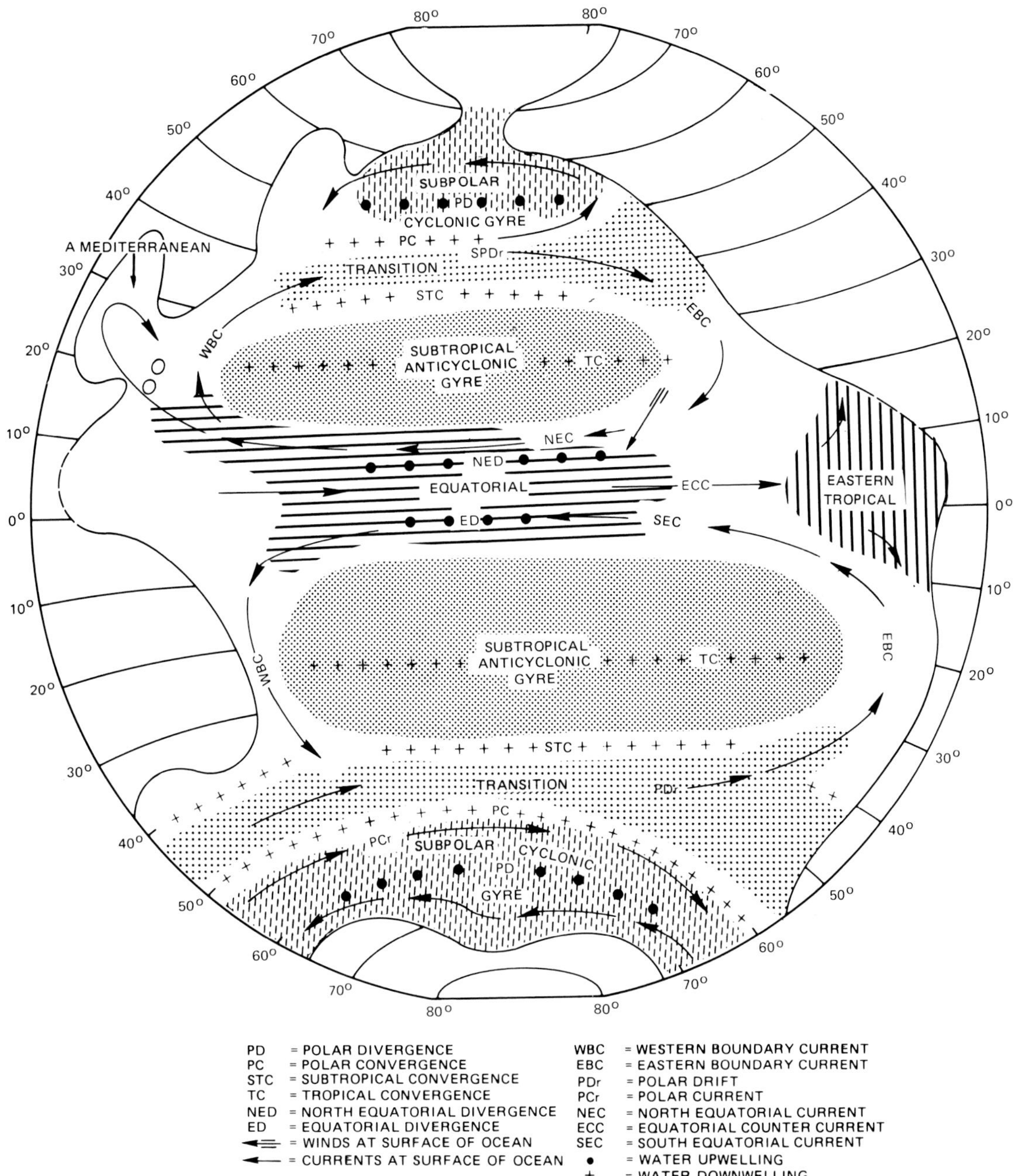

FIGURE 2.30 Model of modern polycystin zoogeography on a hypothetical ocean. Hypothetical ocean of Casey et al. (1982, 1983) and biotic provinces similar to those of McGowan (1974). Dots represent water movement into page. Crosses represent water movement out of page. Divergences, convergences, and currents are designated on the figure.

act as plants may also account for the dominance of collosphaerids in the eutrophic eastern tropical zone (Casey et al., 1982). The percentage of all cold-water-sphere radiolarians (both spulmellarians and nassellarians, Fig. 2.33a) from well-preserved deep-sea oceanic sediments illustrates the distributional patterns of cold-water-sphere radiolarians with especially high percentages poleward of the polar convergences and into the eastern boundary currents, with extensions along the open-ocean, low-latitude divergences. The percentage of all intermediate and deep-water forms (Fig. 2.33b) from

TABLE 2.13 Oceanographic and Radiolarian Characteristics of the Shallow Zones Plotted on Fig. 30.*

Zone of the hypothetical ocean	Primary productivity (mg C m^{-3} d^{-1})	Degree stratified (layering)	Leakage (loss to other zones)	Neritic influence	
Subpolar cyclonic gyre					
Northern	High (500+–250)	Low	Great	Great	
Southern	High (250–150)	Low	Great	Little	
Transition, Northern and southern	Moderate (150)	Low	Great	Little	
Subtropical anticyclonic gyre, northern and southern	Low (<100)	High	Very low	None	
Equatorial	High to moderate (250–150)	High to low	Moderate	Little	
Eastern tropical	High (500+–250)	Usually low	Moderate	Moderate to great	
Western boundary currents	High (250–150)	Low	Moderate	Moderate	
Eastern boundary currents	High (500+–250)	Low	Great	Moderate to great	
Subpolar cyclonic gyre					
Northern	Low	Low	Few, if any	Cenodiscids, spongodiscids, spongotrochins, robust lithellids, and lophophaeinids	Pliocene?
Southern	Low	Low	Moderate	Same as in northern	Pliocene
Transition northern & so.	Moderate	Moderate	Few	Cold-water morphotypes of warm taxa	Disappears and then reappears
Subtropical anti-cyclonic gyre					
Northern and southern	Very high	Very high	High	Symbiont-bearing and warm-water sphere such as collosphaerids, artiscins, and spongasters	Mid-Miocene
Equatorial	High	High	Moderate	Similar to gyre but not as many symbiont bearers	Mid-Miocene
Eastern tropical	Low to moderate	Low to moderate	Few	Symbiont-bearing	Pliocene
Western boundary current	High	Moderate	None	Expatriates from equatorial zone	Mid-Miocene
Eastern boundary current	Moderate	Low to moderate	None	Expatriates from subarctic and transition zones	Mid-Miocene

*Both Fig. 30 and this table draw on oceanographic information and biological information from McGowan (1974) and Reid et al. (1978).

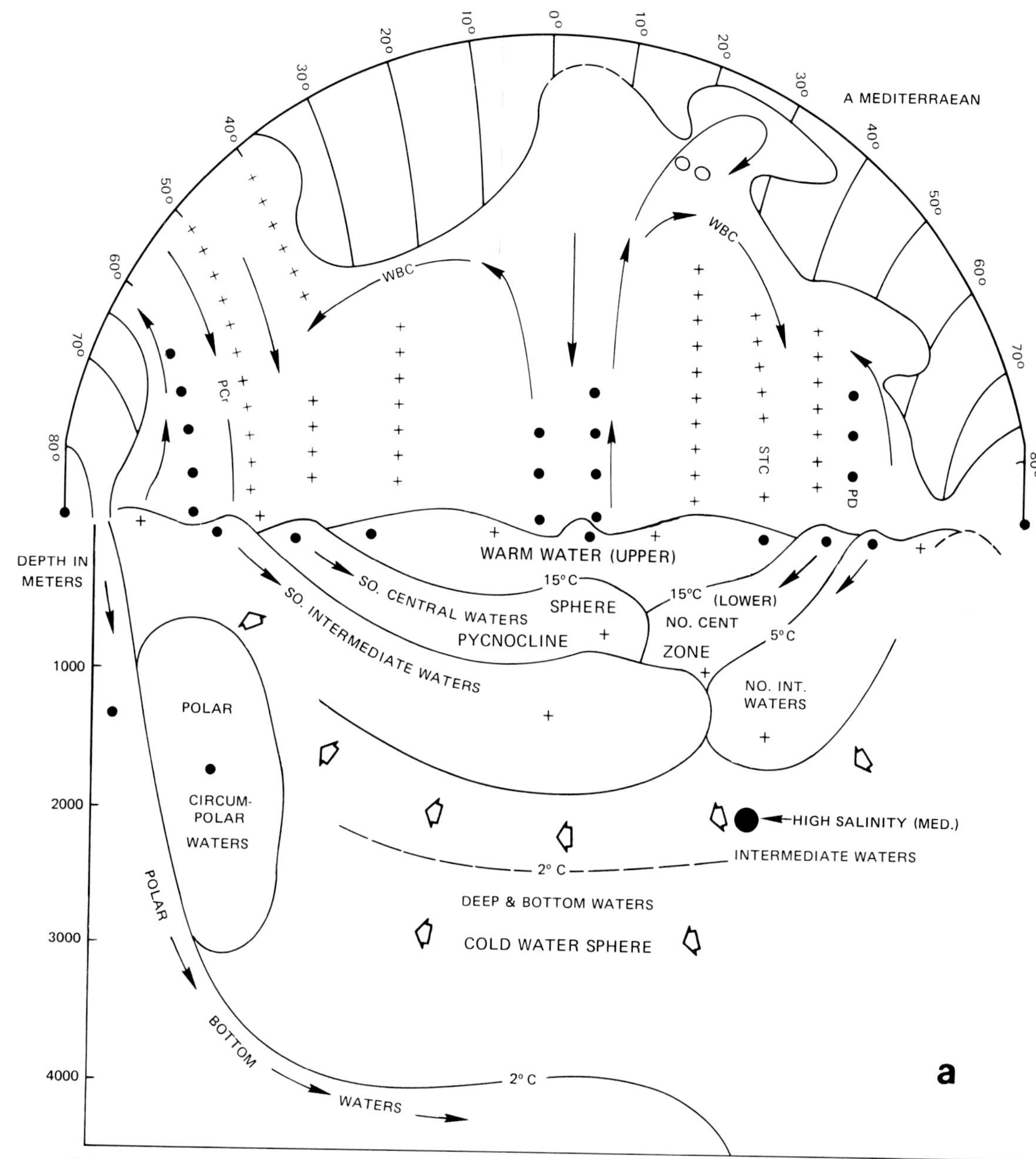

FIGURE 2.31 (a) Deep-sea circulation and water masses related to surface currents and convergences and divergences in a hypothetical ocean. (From Casey et al., 1982.) Currents, divergences, and convergences are designated as in Figure 2.30.

well-preserved deep-sea oceanic sediments is similar to the pattern for cold-water forms; however, of special paleoceanographic interest is the enhancement of intermediate and deep-water forms under the boundary currents, especially the eastern boundary currents, and the oceanic divergences and subtropical and polar convergences.

Papers placing living radiolarians into these ecological groups are Carson (1986), Casey (1982), Casey et al. (1983), Weinheimer (1986), Weinheimer et al. (1986), and Wigley (1982). Papers extrapolating into the fossil record and placing fossil radiolarians into these, or similar, ecological groups are Casey (1982), Casey et al. (1983), Domack (1986), Perez-Guzman and Casey (1986), and Weinheimer et al. (1986).

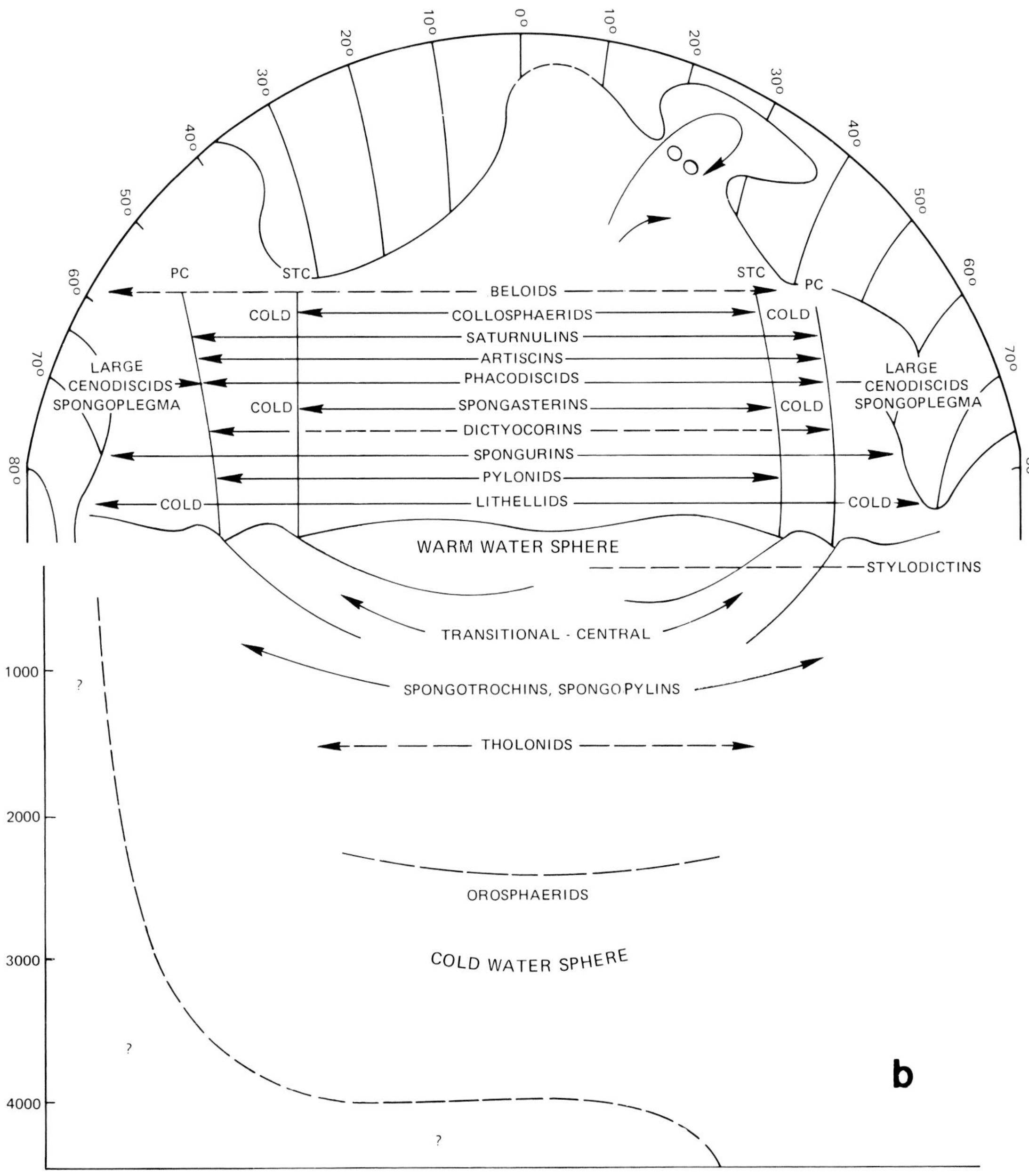

FIGURE 2.31 *(continued)* (b) Dominant distribution of spumellarian radiolarians in a hypothetical ocean. (From Casey et al., 1982.)

Taphonomy

As stated earlier, taphonomy is the study of the processes that affect organic remains after death and prior to final burial (Behrensmeyer and Kidwell, 1985). The ocean is undersaturated in silica; therefore, radiolarian preservation is not the norm. Radiolarians are usually preserved in sediments under regions of high productivity where there is a high rate of silica sedimentation and the tendency toward more acidic conditions in bottom waters because of the rotting of transported organic matter. Todays radiolarian oozes tend to develop on the margins of the tropical zones of calcareous–siliceous oozes where the calcareous oozes begin to dissolve away leaving radiolarian oozes. This is illustrated in Figure 2.34, where radiolarian and

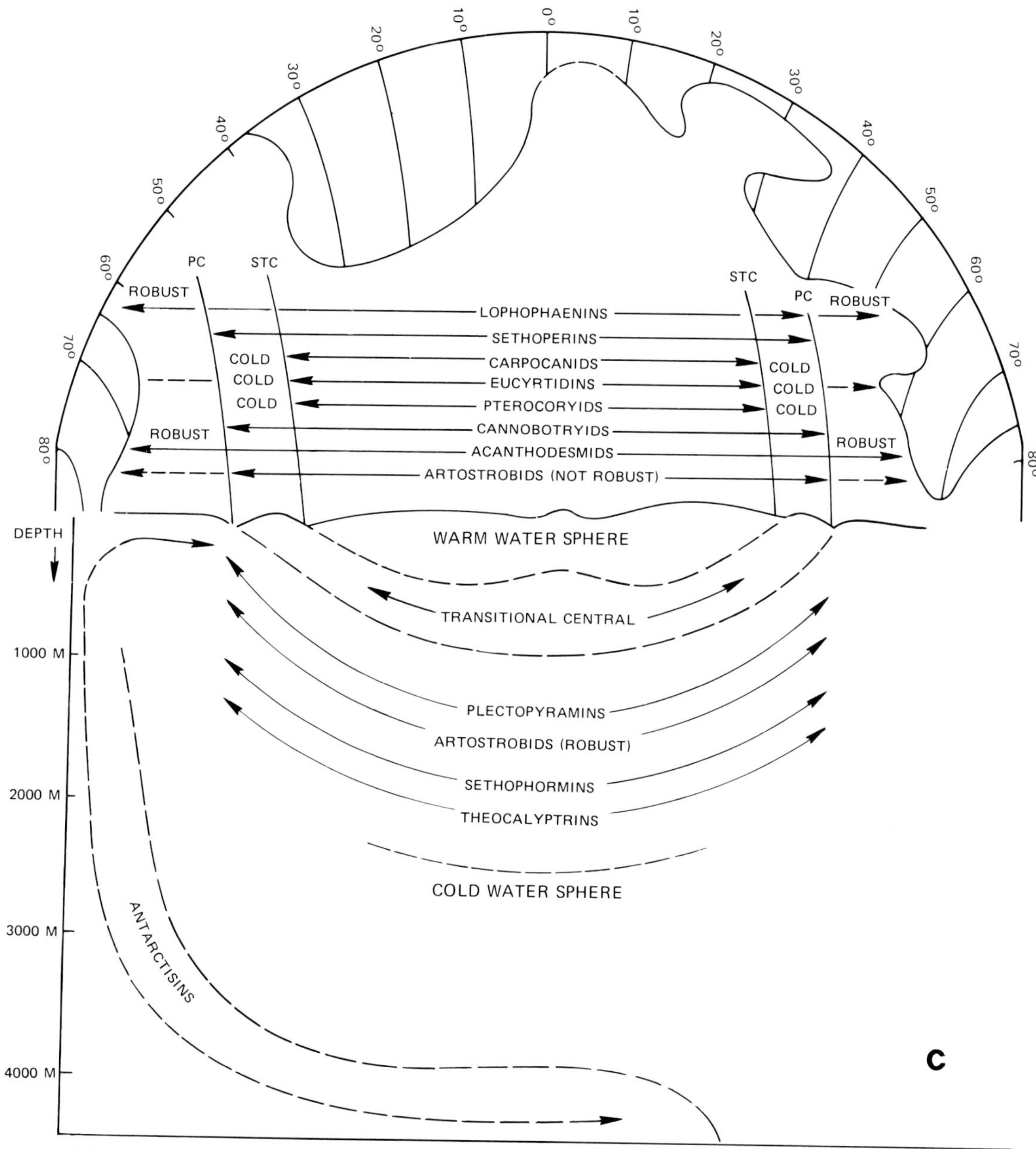

FIGURE 2.31 *(continued)* (c) Dominant distribution of nassellarian radiolarians in a hypothetical ocean. (From Casey et al., 1982.)

other deep-sea sediments are related to overlying oceanographic conditions. In the high-productivity equatorial oceans radiolarian oozes (equal to siliceous oozes (R > D) (radiolarian > diatom) develop; however, in high-latitude oceans, radiolarians usually do not form oozes and are masked by either diatoms or glacial marine sediments. As can be seen at the bottom of Figure 2.34, oceanic divergences and convergences appear to be important controlling factors in present-day distributions of radiolarian-bearing sediments.

Biostratigraphy

Dating of sediments and sedimentary rocks is an important endeavor for most aspects of historical geology and a prerequisite to detailed studies of paleoceanography, evolution, and so on. Riedel (1957)

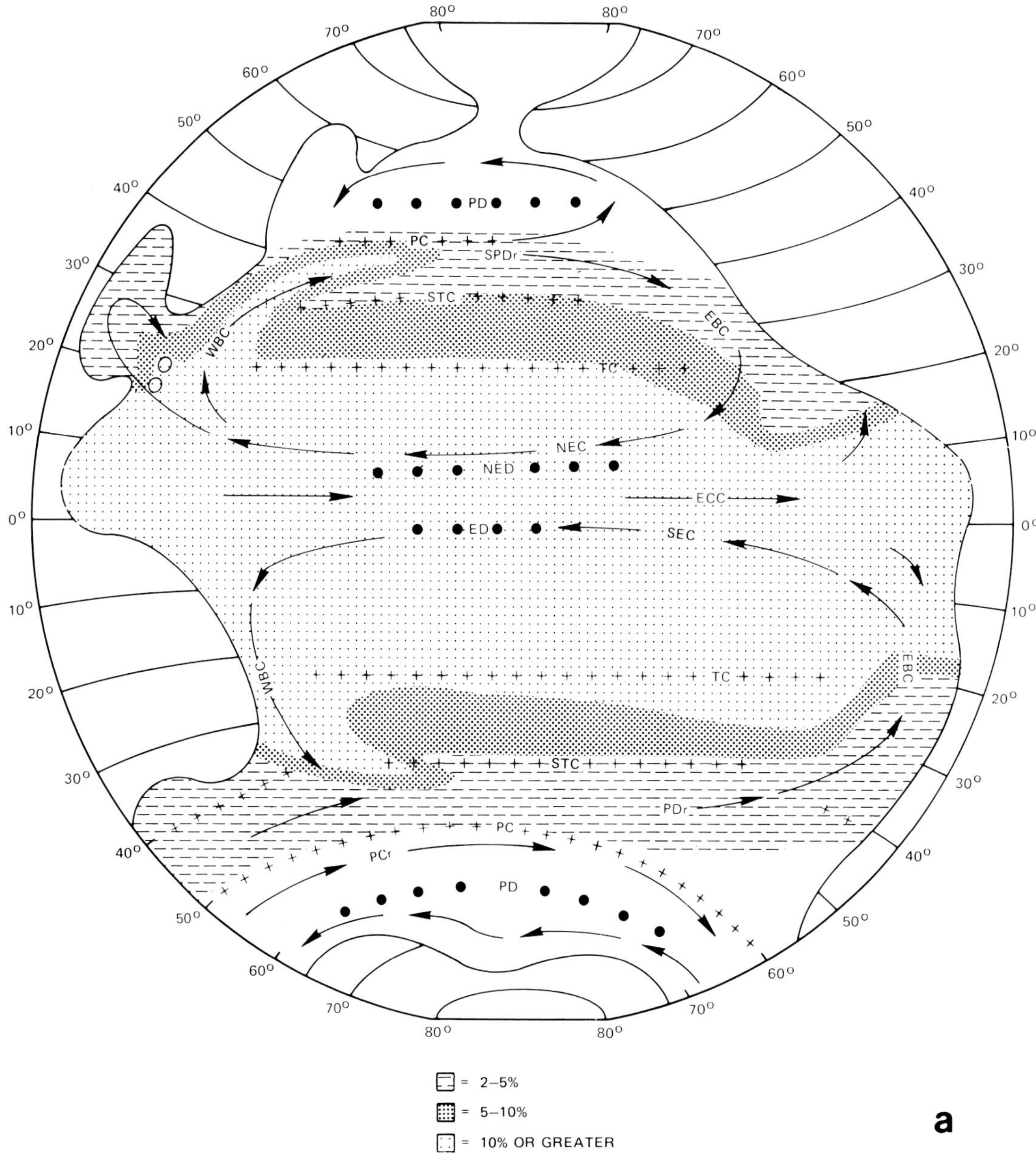

FIGURE 2.32 (a) Percent warm-water forms of entire radiolarian fauna in well-preserved (radiolarians) sediments in a hypothetical ocean. (From Casey et al., 1982.) Current, divergences, and convergences are designated as in Figure 2.30.

published the first radiolarian biostratigraphy using Swedish Deep-Sea Expedition cores from the western Pacific. The resolution of this study was the early, middle, and later Tertiary and Quaternary. Riedel was and has been the main force in developing radiolarian biostratigraphies, and many of the current American radiolarian micropaleontologists have studied in his laboratory. Riedel and Sanfilippo (1970) first developed their tropical radiolarian zonation using DSDP Leg 4 material (Sites 27–29) from the western tropical Atlantic and Caribbean, the land-based Oceanic Formation on Barbados, and a few short-piston cores from the Pacific for their Paleogene zones; and DSDP Caribbean Sites 30 and 31, the Experimental Mohole Site off Guadalupe Island in the Pacific, and Pacific short-piston cores for their Neogene zones. They described 19 radiolarian zones from the middle Eocene through Pliocene. Using DSDP Leg 7 material from the central and eastern tropical Pacific,

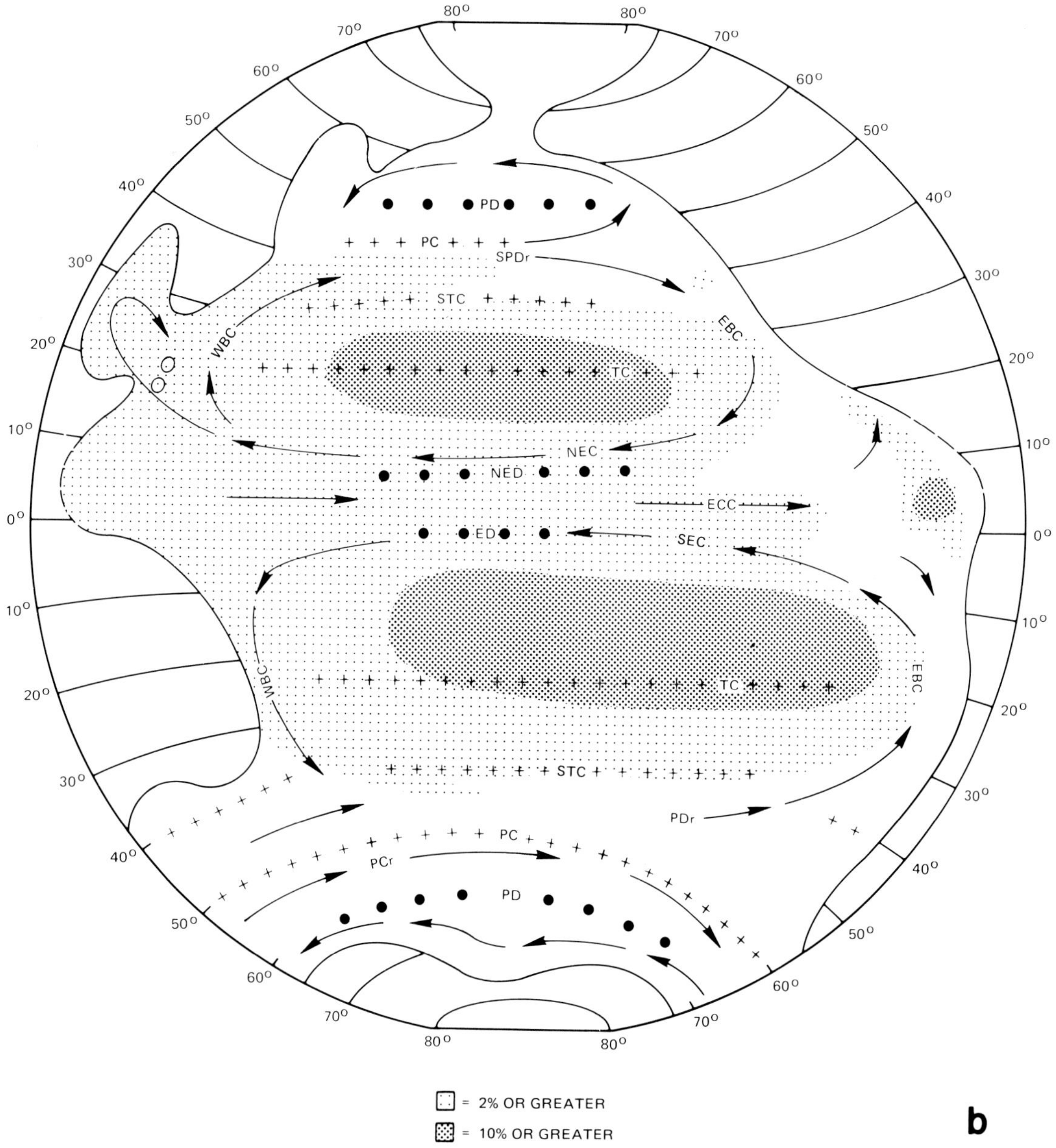

FIGURE 2.32 *(continued)* (b) Percent warm-water collosphaerids of entire radiolarian fauna in well-preserved (radiolarians) sediments in a hypothetical ocean. (From Casey et al., 1982.)

Riedel and Sanfilippo (1971) revised their earlier zonation (Riedel and Sanfilippo, 1970) and placed their Neogene zones on a much firmer basis with the use of continuous long DSDP sections instead of the previously used short-gravity cores. This latest zonation was also correlated to land-based sections from Europe and the Caribbean, which also allowed for a correlation to planktonic foraminiferan zonations. Various authors have modified this tropical radiolarian zonation of Riedel and Sanfilippo (1971) so that its most currently used form (Riedel and Sanfilippo, 1978) contains 26 zones ranging from late Paleocene to present. Most of the biostratigraphically important Neogene radiolarian first and last occurrences, and most of the Neogene portions of that zonation were correlated to paleomagnetically dated equatorial Pacific overlapping piston cores by Theyer et al. (1978). This was a significant advance, for datums (such as first or last occurrences), zones, and so on, could now be related to geologic time in fractions of millions of years. However, this was not the first time that a

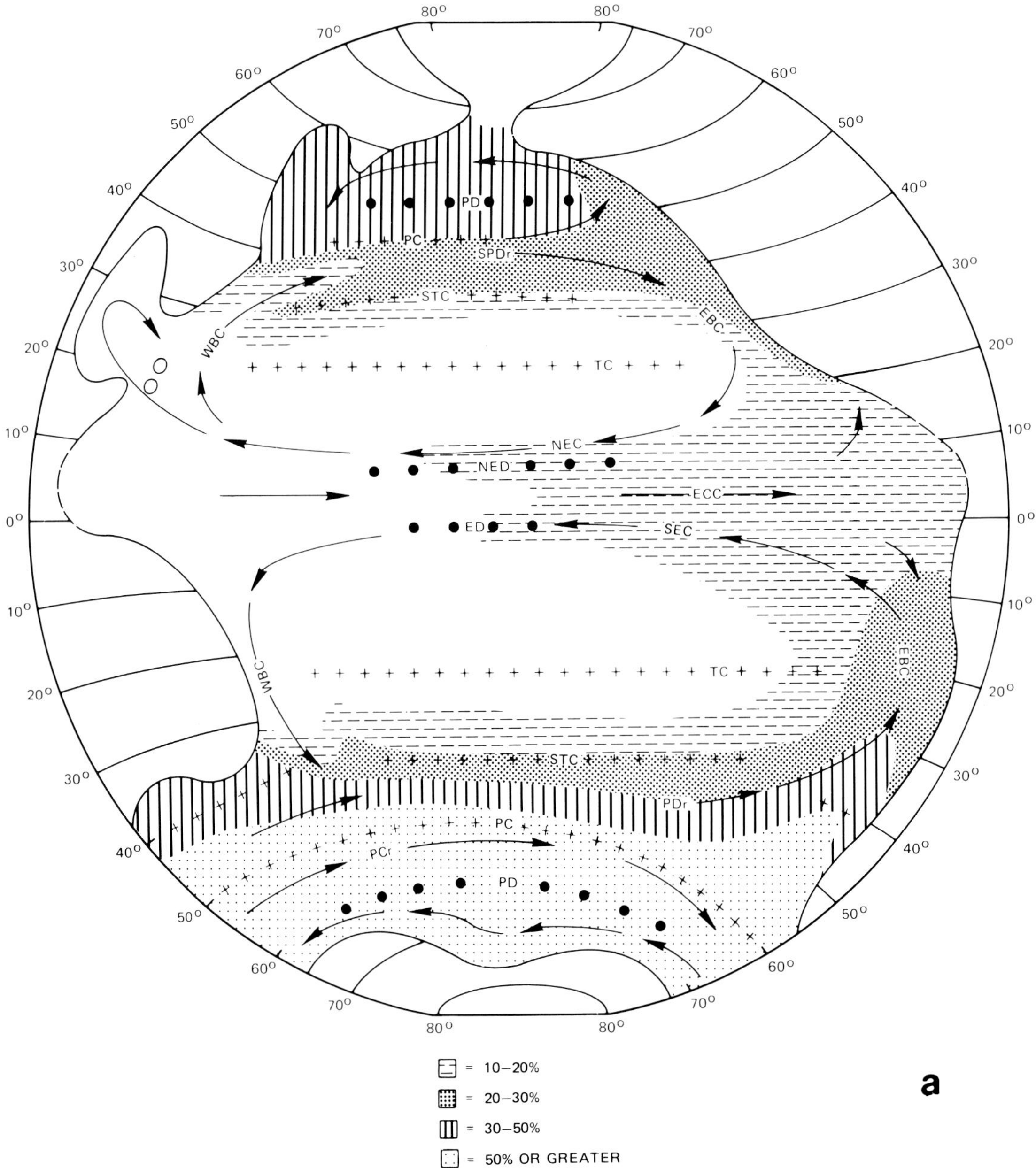

FIGURE 2.33 (a) Percent cold-water forms of entire radiolarian fauna in well-preserved (radiolarians) sediments in a hypothetical ocean. (From Casey et al., 1982). Currents, divergences, and convergences are designated as in Figure 2.30.

radiolarian zonation had been correlated paleomagnetically. Opdyke et al. (1966) related the late Neogene Antarctic radiolarian zones of Hays (1965) [which Sanfilippo et al. (1985) refer to as the first radiolarian zonation] to the magnetostratigraphic signature from the same cores Hays had used for his zonation. Sanfilippo et al. (1985) have modified their own zonation recently, and this new modification with 29 zones using the zonal concepts of the *International Stratigraphic Guide* (Hedberg, 1976) should be the most used low-latitude Cenozoic radiolarian zonation.

There have been a number of biostratigraphic zonations other than the Riedel and Sanfilippo tropical radiolarian zonation, such as the zonation just mentioned for the late Neogene of the Antarctic (Hays, 1965). Most of these are regional zonations, such as Hays (1965); Bandy et al. (1971); Petrushevskaya (1975); and Chen (1975), mainly for the Neogene of the Antarctic; Hays (1970); Kling

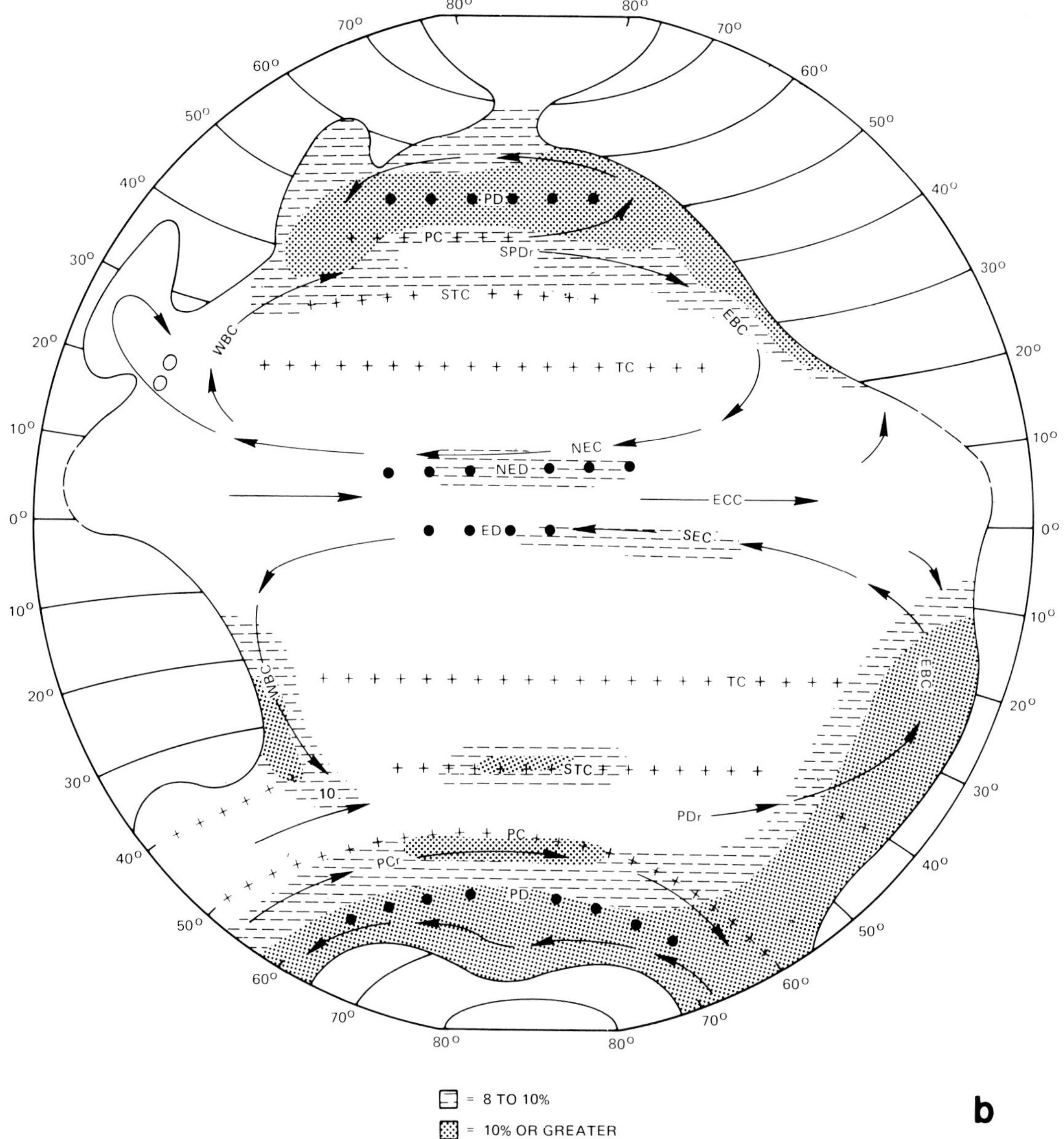

FIGURE 2.33 *(continued)* (b) Percent intermediate to deep-water forms of entire radiolarian fauna in well-preserved (radiolarians) sediments in a hypothetical ocean. (From Casey et al., 1982.)

(1973); Reynolds (1980); and Weaver et al. (1981), for the Neogene of the midlatitude North Pacific and borderlands; and Nigrini (1971), for a Quaternary zonation for the equatorial Pacific. Since these zonations are based mainly on shallow-water forms (either shallow cold- or warm-water-sphere forms), their biostratigraphic usefulness is of a regional nature. Attempts at developing more cosmopolitan zonations via the use of deep-water widespread radiolarians are those of Reynolds (1978), Casey and Reynolds (1980), and Casey and Wigley (1985). Casey and Wigley (1985) proposed a Neogene deep-living radiolarian zonation that consisted of six radiolarian zones with the nominant taxa of each zone first occurring at the base of its zone and that nominant taxa running throughout its zone. Although this is not a high-resolution zonation, its major advantage is that it is a cosmopolitan zonation that can be used for worldwide datums, and for the comparison and correlation of other more regional zonations.

There are few pre-Cenozoic radiolarian zonations. Using mainly land-based sections, Pessagno and co-workers have zoned portions of the Meso-

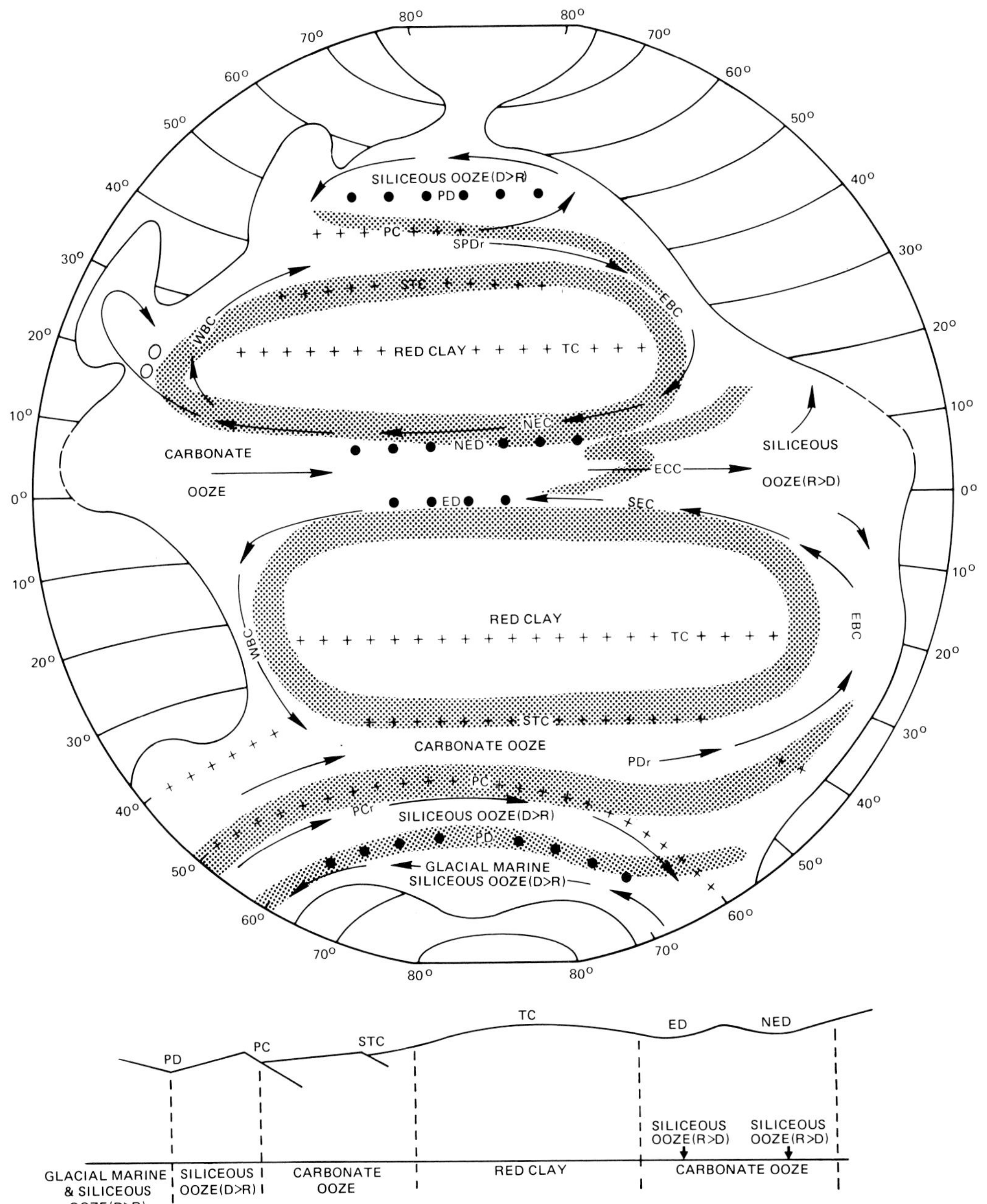

FIGURE 2.34 Dominant deep-sea sediment types related to surface currents, divergences, and convergences in a hypothetical ocean. Currents, divergences, and convergences are designated as in Figure 2.30. R = radiolarians, D = diatoms. All oceanic depths are 4,000 m.

zoic, and Pessagno (1977) established a Mesozoic radiolarian stratigraphy. Using mainly DSDP material, Sanfilippo and Riedel (1985) established a late Mesozoic radiolarian zonation. Recent papers by Schaff (1984, 1985) deal with radiolarian zonations of the Cretaceous. Baumgartner (1984) developed a low-latitude radiolarian zonation for the middle Jurassic to early Cretaceous; and Dumitrica et al. (1980) describe radiolarians useful for age determinations from the middle Triassic of the southern Alps. Holdsworth and Jones (1980) developed a preliminary radiolarian zonation for the late Devonian through Permian.

Paleoceanography

The radiolarian subgroups Polycystina and Phaeodarina are preserved in the fossil record and useful for paleoceanographic reconstructions. The polycystin radiolarians are chronologically the longest ranging (Cambrian to Holocene), geographically the greatest ranging (pole to pole, surface to abyss), and taxonomically the most diverse of the well-preserved microzooplankton. The Phaeodarina occur sporadically in sediments from at least the Cretaceous to the Holocene. Because of this considerable diversity, wide yet distinct horizontal and vertical distributions, and long geologic range, radiolarians have been used as tools to determine such paleoceanographic conditions as existence and initiation of specific water masses, paleotemperatures, paleocirculations, paleodepth, distance from shore, paleoanoxia, paleoupwelling and paleoproductivity. Selected examples will be given here to illustrate the usefulness of radiolarians for paleoceanographic reconstructions. These examples of specific paleoceanographic indices will be followed by a case study of paleoceanographic reconstruction using these specific indices to reconstruct the paleo-California Current system.

Two major multi-institutional paleoceanographic reconstruction efforts should be mentioned at the onset. These are the Climate Long Range Investigation Mapping and Planning (CLIMAP) program, which investigated paleoceanographic conditions of the Quaternary, and a number of papers generated by this program may be found in Cline and Hays (1976); and the Cenozoic Paleoceanography Project (CENOP), which investigated paleoceanographic conditions of the Neogene, and a number of papers generated by this project are reviewed in Kennett (1982). These projects have used computer-assisted techniques to suggest many paleoceanographic constructions for many different phenomena (paleotemperature, paleocirculation, etc.). The basic analysis for these interprepations was developed by Imbrie and Kipp (1971). This analysis uses the radiolarian component (or other microfossil groups) from core tops that are placed in several assemblages by factor analysis. These assemblages are related to observed oceanographic parameters from overlying waters (currents, convergences, surface water temperatures of differing seasons, etc.), and these parameters are taken down-core via that assemblage relationship.

Two other matters should also be mentioned at the onset, and the first deals with the use of the entire radiolarian fauna in recent or fossil studies. Recent studies by Riedel et al. (1985), Mullineaux and Westberg-Smith (1986), Empson-Morin (1984), and others have utilized the entire radiolarian fauna for paleoceanographic interpretations. Although more laborious, this use of the entire preserved fauna appears to generate the most reliable paleoceanographic reconstructions. The other matter is the importance of studying recent radiolarians from both plankton and sediment samples as aids for developing radiolarians as good paleoceanographic indicators. Although this latter matter could be discussed in any of the following sections, it is illustrated in the section on circulation, where first the use of radiolarians as indicators of present day-circulation is given, followed by their use as indicators of ancient circulation.

Paleoceanographic Phenomena Indicated by Radiolarians

Paleowater Masses

The time of evolution of radiolarians indicative of a specific water mass has been used to suggest the time of initiation of that specific water mass (Casey, 1973). Casey et al (1982, 1983), as mentioned earlier, using worldwide Holocene marine sediments, developed and presented a model of radiolarian distribution related to a hypothetical ocean that displayed generic water mass, oceanographic front, current, and other physical oceanographic data (Fig. 2.31a). The major radiolarian distributional patterns, such as the dominant distributional patterns of the main groups of spumellarians (Fig. 2.31b), could be plotted on this model (including their depth distributions) from the sedimentary record. These same patterns may also be reconstructed for fossil sediments, as had been done by Perez-Guzman and Casey (1986). Basically these fossil reconstructions are extrapolations of the modern models. For example, the subtropical convergences may be reconstructed as being at the poleward extent of radiolarian assemblages where the collosphaerid group is represented as greater than 2 percent of the total radiolarian fauna. Using these criteria, radiolarians occupying the region between the present and/or fossil subtropical convergences can then be placed in the upper warm-water-sphere environmental group. Since the collosphaerids are well represented in Neogene but not in pre-Neogene sediments, this specific reconstruction is limited to the Neogene. However, on a broader scale it appears that robust radiolarian environments (such as just the warm- or cold-water sphere) may be deciphered as far back as the earliest radiolarian record, where some of the radiolarians illustrated appear to represent the cold- and some the warm-water sphere (Fig. 2.35).

Paleotemperatures

Perhaps the first significant study that used polycystin radiolarians as paleoceanographic indices was that of Hays (1965). In this study Hays divided the radiolarians into assemblages (sub-Antarctic, Antarctic, etc.), correlated these assemblages to surface-water conditions such as convergences, and then extrapolated these conditions down-core,

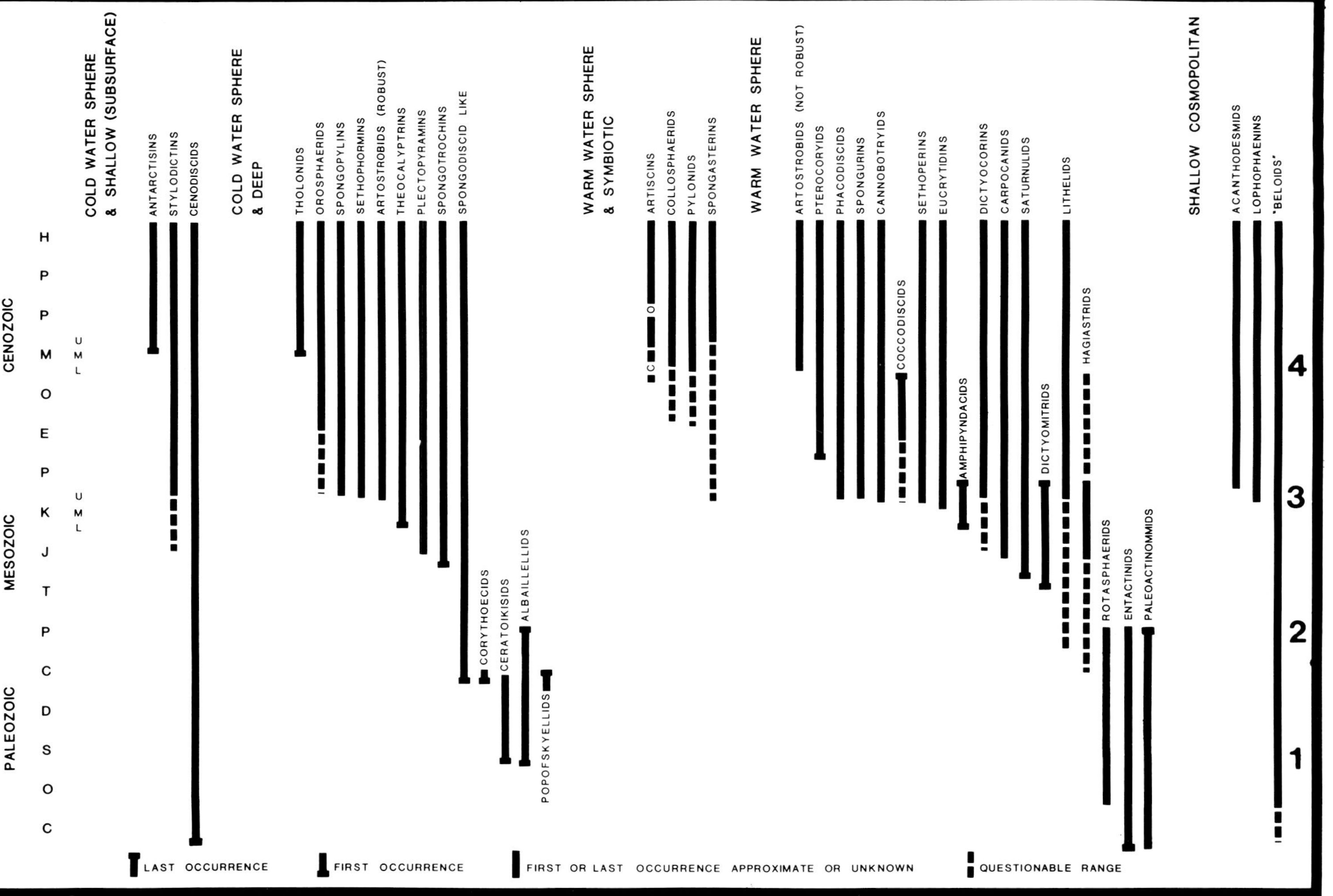

FIGURE 2.35 Geologic ranges of radiolarian "family level" groups, arranged by environmental group. Modified from Casey et al. (1983) with additions from Riedel and Sanfilippo (1981) and Lipps (1985).

finding that down-core these radiolarians indicated a cooling of Antarctic waters since late Tertiary times. Some of these same Antarctic cores were used by Bandy et al. (1971) in an attempt to document the magnitude of these changes in relation to paleotemperature. The technique used by Bandy et al. (1971) was to establish a radiolarian temperature index from the radiolarians contained in surface sediments compared to present sea surface temperatures and to extrapolate this index down-core. This same technique was used by Casey (1972) to determine paleotemperatures from southern Californian and adjacent Neogene sediments and sedimentary rocks. The CLIMAP and later CENOP projects used computer-aided techniques (see discussion at beginning of paleoceanography section herein for a brief review of these techniques) to reconstruct paleotemperatures for entire oceans at specific time planes. Moore (1978), using radiolarians in a CLIMAP-supported study, compiled (and cites) a number of these studies in reconstructing Pleistocene paleotemperatures for the Pacific Ocean. Romine (1985), using radiolarians in a CENOP-supported study, reconstructed the paleotemperatures of the North Pacific at 8 Ma. Weaver et al. (1981) reconstructed paleotemperatures from the late Neogene of the northeastern Pacific by establishing environmental groups (as mentioned previously in the section on paleowater masses) and then using these groups for paleotemperature analysis.

Another type of paleotemperature is the maximum paleotemperature a sedimentary basin was subjected to, an important paleotemperature for oil and gas exploration. A common method of determining such a paleotemperature is to use the color changes of organic-rich microfossils, which change from light to dark during exposure. Common fossils used in this way are palynomorphs and conodonts. Casey (1986) suggests that phaeodarian radiolarians (which possess organic-rich skeletons) may also be used in this manner. He conducted pyrolysis experiments using recently dead and fossil phaeodarians and noted a color change similar to that described for palynomorphs.

Circulation

This section precedes the section on paleocirculation and deals with the use of radiolarians in determinations of present-day circulation. It emphasizes the need to study the recent environment and recent radiolarians when attempting to use radiolarians as paleoceanographic indicators. Radiolarians from plankton tows have been used as indicators of the movements and provenances of oceanic waters. Some of the older works of this nature may be found in Casey (1977). Radiolarian taxa and radiolarian densities and diversities were used in conjunction with physical oceanographic measurements to characterize the circulation and provenance of waters overlying the south Texas outer continental shelf (Leavesley et al., 1978). Weinheimer et al. (1986) used radiolarians from plankton tows to characterize the circulatory changes within the California Current System off southern California during El Niño and anti–El Niño conditions. During El Niño conditions the California Current off southern California appears to diminish in size and move over the southern California continental borderland. During this time high-standing crops of gyre and/or eastern tropical Pacific radiolarians invade. A number of radiolarian taxa may be used to indicate these fluctuations, but perhaps the best are members of the radiolarian genus *Spongaster. Spongaster tetras irregularis* is a characteristic radiolarian component of the California Current (CC in Fig. 2.36 and Fig. 18 of Table 2.12), *Spongaster tetras tetras* is characteristic of the gyre waters (CG in Fig. 2.36 and Fig. 17 of Table 2.12), and *Spongaster pentas* is characteristic of eastern tropical Pacific waters (ETP in Fig. 2.36 and Fig. 19 of Table 2.12).

During anti–El Niño (or what most would consider normal oceanographic) conditions these subspecies and species are distributed as shown in Fig. 2.36. However, at the initiation of the 1983 El Niño cycle *S. tetras tetras* was the only *Spongaster* found in southern California waters, indicating that the California Current had slowed and gyre waters had invaded the area off southern California. During the peak of the 1983 El Niño the dominant *Spongaster* was *S. pentas,* indicating that eastern tropical Pacific waters had invaded the area. In the winter following the 1983 El Niño, conditions were apparently returning to normal, with the dominant *Spongaster* being *S. tetras irregularis,* although there were traces of both *S. pentas* and *S. tetras tetras,* indicating that completely anti–El Niño–like conditions had not returned. Although these El Niño conditions are not the norm, the main radiolarian imprint on the underlying sediments appears to be imposed during El Niño conditions and mainly reflects El Niño conditions (Cleveland and Casey, 1986). As the California Current diminishes, its main stream (or a main branch) swings over the southern California borderland and higher than normal (i.e., higher than anti–El Niño) standing crops of cold, intermediate, and deep- (all characteristic of the California Current) as well as warm-water radiolarians occur in the area (Weinheimer et al., 1986). These higher-standing crops of warm, cold, and intermediate and deep peaked during the 1964

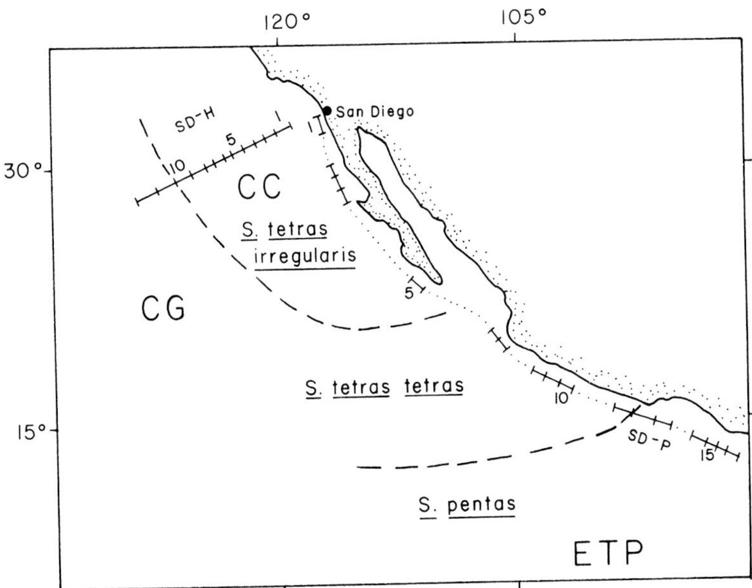

FIGURE 2.36 Dominance of subunits of the genus *Spongaster* as indicators of water masses and provenance of waters. (From Weinheimer et al., 1986.)

El Niño, as shown in plankton tows taken before and at the initiation of that El Niño (Weinheimer et al., 1986). Sediments deposited as varves in the Santa Barbara Basin exhibit these same radiolarian peaks for the 1964 El Niño (64–66 interval in Fig. 2.37), and especially so for the very strong 1957 El Niño (56–58 interval in Fig. 2.37) (Weinheimer et al., 1986). Since El Niños are so strongly imprinted in the sediments, it should not be surprising that the main radiolarian imprint on the southern Californian borderland (and perhaps for much of the California Current system) is one of El Niño. The main radiolarian imprint of the California Current fauna on southern California continental borderland sediments is one of the current running through the borderland (El Niño–like) rather than running in a southerly direction along the borderland (anti–El Niño–like) (Cleveland and Casey, 1986). Weinheimer (1986) reports the 1983 El Niño is reflected in the sediments of the Santa Barbara Basin in essentially the same sequence of events as reported from the plankton of that year by Weinheimer et al. (1986).

Paleocirculation

Many of the papers dealing with paleotemperatures also deal with aspects of paleocirculation. Romine and Lombari (1985), using CENOP techniques, reconstructed the paleocirculation of the North Pacific at 8 Ma and compared it with present and Pleistocene circulations. One of the most innovative paleocirculation reconstructions was that of Pisias (1979), who reconstructed paleocirculations for the California Current during the last 8,000 years. He constructed paleodynamic height circulation maps using techniques similar to those used by CLIMAP and CENOP. The paleo-California Current was also studied by Weinheimer et al. (1986) using Neogene land-based and Deep Sea Drilling (DSDP) samples. The dimensions of the paleo–California Current were outlined by the 30 percent cold-water radiolarian contour for the 5, 8, and 10 Ma time slices (Fig. 2.38). During known oceanic warm periods (8 and 10 Ma) the paleo–California Current so defined was narrower and did not extend as far south as during a known oceanic cold period (5 Ma).

Paleodepth

Few papers have attempted to use radiolarians as paleodepth indices. The reasons for this are that foraminiferans usually handle this aspect well, few people work on radiolarian paleoecology, and in general radiolarian-rich sediments or rocks are assumed to be deposited in deep-sea environments. Usually an increase in radiolarian number (number of radiolarians per gram sediment) or abundance in land-based sections has been used to indicate increased paleodepth. Wigley (1982) noted that the number of radiolarian species and higher-order taxa showed some general trends with depth. Cleveland and Casey (1986) plotted the radiolarian number for shelfal, slopal, and basinal recent sediments from the continental borderland off southern California. They found that the radiolarian number was very low for shelfal sediments (usually less than 50/g)

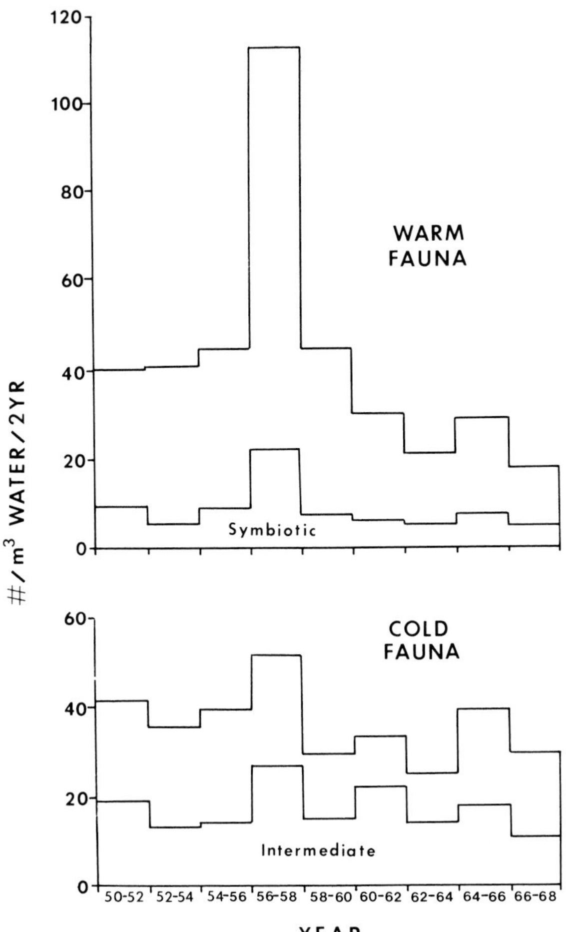

FIGURE 2.37 Radiolarian flux rates into the Santa Barbara Basin during anti–El Niño and El Niño conditions. (From Weinheimer et al., 1986.)

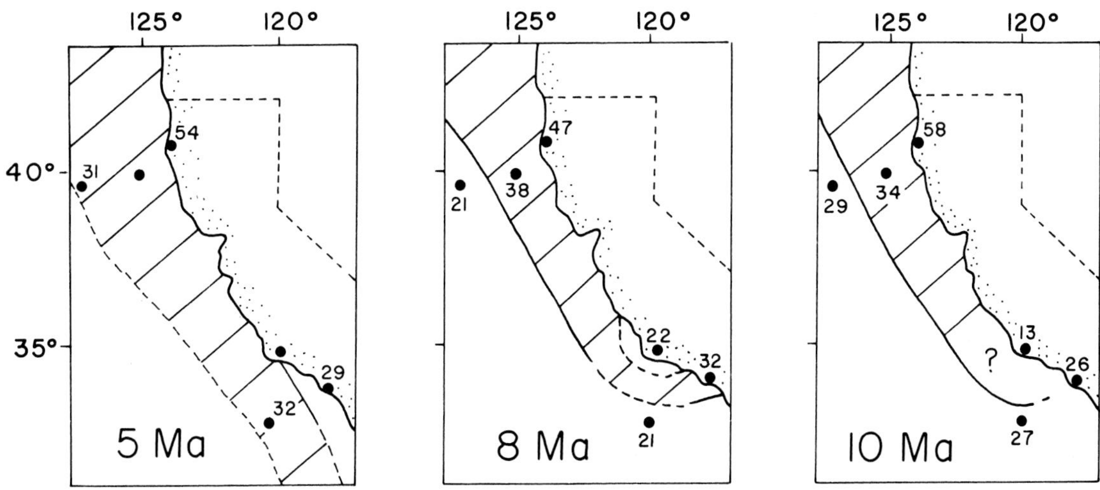

FIGURE 2.38 Extent of the California Current during 5, 8 and 10 Ma. (From Weinheimer et al., 1986.)

and that there was fairly consistent increase in radiolarian number with increased depth (to about 800/g at about 1,300 m). In an area of known upwelling (off San Miguel Island) a peak in radiolarian numbers occurred at shelf break depth when compared with shallower and deeper depths (Cleveland and Casey, 1986). This peak is believed to represent a mortality of forms upwelled to the shelf since this assemblage contains many deep-living forms. We believe that in a shallow section this may be used as an index of shelf break depths. Empson-Morin (1984) used a Fisher Alpha Diversity Index on radiolarians extracted from Mesozoic sediments to develop a neritic versus deep-water index. Mesozoic Deep Sea Drilling Project cores exhibited the highest indices, and neritic Californian and Romanian, the lowest.

Neritic Versus Open-Ocean Waters and Sea-Level Fluctuations

Casey et al. (1981) noted that spumellarian radiolarians dominate shelf waters, whereas nassellarian radiolarians are more common to open-ocean waters. McMillen (1979) used the ratio of spumellarians to nassellarians to interpret near-shore versus offshore conditions; he then used this index (McMillen, 1979) to suggest fluctuations in sea level. Holdsworth (1977) used low radiolarian diversity to indicate near-shore and high radiolarian diversity to indicate open-ocean enviroments in Paleozoic sediments.

Paleoupwelling and Paleoeutrophism

Radiolarians from plankton tows have been used to distinguish upwelling conditions. The appearance of radiolarians usually restricted to depths deeper than 200 m in the upper waters off the California coast was suggested by Casey (1977) to reflect upwelling conditions. Leavesley et al. (1978) suggested that the presence of the radiolarian *Spongotrochus glacialis* in the shelfal waters off the south Texas coast indicated upwelling conditions. For those same waters Casey et al. (1981) also considered the radiolarian species *Helotholus* spp., *Spongopyle osculosa*, and *Lithelius minor* as upwelling indicators.

Unusually high accumulation rates of biogenic silica in general have been used to suggest regions of high fertility (Berger and Roth, 1975). More specifically to the present topic, the accumulation and good state of preservation of radiolarians in sediments has also been associated with regions of high productivity (Riedel and Sanfilippo, 1977). In comparing eutrophic southern California borderland and more oligotrophic Gulf of Mexico and Caribbean radiolarian assemblages, Wigley (1982) concluded that the greater percentage and diversity of known deep-water polycystin radiolarians in the southern California samples was due to the more dynamic upwelling conditions off California, which caused an enhancement of deep-water radiolarian diversity and number in the sediments. On a greater geographic scale, Casey et al. (1982) documented that intermediate- and deep-water radiolarians are enhanced in recent sediments under the polar cyclonic gyres, eastern boundary currents, and oceanic convergences and divergences—all regions of upwelling. The junior author of this paper considers that a number of the radiolarian assemblages distinguished by CLIMAP and CENOP researchers owe their uniqueness to an upwelling factor.

Increased percentages of intermediate- and deep-water radiolarians in Neogene land-based sections from northern to Baja California were used by Weaver et al. (1981) as an index of increased upwelling. Using this technique, they documented a significant upwelling period between 10 and 6 Ma.

Paleoanoxia

Phaeodarian radiolarians are radiolarians whose skeletons contain a significant amount of organic matter, so much organic matter, in fact, that the skeletons are usually preserved only in regions of low oxygen concentration. Driskill (1986) documented the phaeodarian component of Holocene sediments of the southern California continental borderland and developed a number of phaeodarian indices of anoxia. Her best index to date appears to be the ratio of the preserved phaeodarian surface area to the dry weight of sediment (a sort of phaeodarian number). Using this index she has been able to distinguish low oxygen environments in the Neogene section at Centerville Beach, California. She can distinguish three different low-oxygen environments: an anoxic basin environment, a main oxygen minimum zone impingement on the continental slope environment, and a shallow oxygen minimum zone impingement on the shelf break or uppermost slope. Geochemical data she presents support these paleoenvironmental reconstructions (Driskill, 1986).

Case Study of Paleooceanographic Reconstruction; the Reconstruction of the Paleo–California Current System

The preceding discussion may be supplemented by a more extensive review of radiolarian paleoceanographic studies prior to 1983 by Anderson (1983). Usually the previously described paleoindicators are used in conjunction with one another, and other information, to develop as complete a reconstruction of the past environment as possible. Many of these previously mentioned techniques were developed or refined for an ongoing study by the junior author and his colleagues. This long-range and ongoing study is an attempt to reconstruct the paleoceanography of the California Current system by using mainly radiolarian paleoindicators. First, information from living radiolarians was acquired, such as that mentioned in the section on circulation. This information lets one know where the radiolarians actually live (in the water column, their seasonality, their provenance as in the California Current, the eastern tropical Pacific, etc.). To ease into the fossil record, radiolarians from recent sediments were then studied. Here it was noted that in some areas the main sedimentary imprint was determined by El Niño events (Cleveland and Casey, 1986) and that these El Niño events were recorded in some recent sediments as distinct and datable events (Weinheimer, 1986; Weinheimer et al., 1986). Extrapolating into the more distant past (millions of years) the shape, size, and dynamics (degree of upwelling, El Niño–like or anti–El Niño–like, etc.) of the California Current system was diagramed at three time planes (Domack, 1986; Weinheimer et al., 1986). In the most recently completed, but as yet unpublished, extension of this research Nelson (1986) has reconstructed the paleoceanographic history of the Neogene Humboldt Basin, northern California. Using radiolarians for biostratigraphic and paleoceanographic indicators, along with organic carbon in sediments, general lithology, associated micro- and megafossils, and previously published isotopic information, he has been able to reconstruct the paleoceanography of the basin from latest middle Miocene to Pliocene. At 11.5 and 8 Ma (million years) he reconstructs a strong California Current (CC in Fig. 2.39) and Davidson Countercurrent (DC in Fig. 2.39). At 6.5 Ma he reconstructs a very strong California Current exhibiting exceptionally strong boundary current upwelling resulting in an expansion of the oxygen minimum zone, a condition he designates as strong anti–El Niño–like (Fig. 2.39). At 5.5 and 5.0 Ma he reconstructs an environment that he believes is similar to today's strong El Niño–like conditions and that he believes is related to the uplift of Panama to a depth that allows such an environment to exist (Fig. 2.39). These types (e.g., Nelson, 1986) of reconstructions are usually what the marine micropaleontologist or paleoceanographer is striving for. To achieve them he or she must use not only microfossil evidence such as the paleoceanographic indices mentioned, but all other available data. Many of these studies could be referred to as combined studies, since they utilize a combination of various microfossil groups and other information. Some of the important combined studies involving radiolarians are mentioned in the following section on combined studies.

Combined Studies

In 1971 a consortium of scientists from many institutions was formed to study the history of global climate over the past million years, particularly the elements of that history recorded in deep-sea sediments (CLIMAP Project Members, 1976). Radiolarians along with diatoms, coccolithophorids, and planktonic foraminiferans were used to reconstruct marine paleoenvironments and to help infer terrestrial paleoenvironments. The basic methods used were similar to many previous studies in that faunal assemblages from recent sediments were correlated to various current oceanographic parameters (such as August and February sea-surface temperatures, 100-m isohalines, etc.) and then extrapolated downcore (at time planes such as 18,000 ybp, considered to be the last glacial maxima). The actual mechanics were accomplished by a computer-assisted factor-analytic approach developed by Imbrie and Kipp (1971), briefly explained here at the beginning of the section on paleoceanography and nicely explained by Anderson (1983, p. 277). In a volume dealing regionally with the Atlantic, Antarctic, and Pacific many of the CLIMAP results are published (Cline and Hays, 1976). An especially well-known publication authored by CLIMAP project members (1976) reconstructed marine and terrestrial paleoclimates and paleoenvironments (including terrestrial biomes) worldwide for 18,000 years ago.

Combining various microfossil groups for developing high-resolution biostratigraphies and correlating individual group biostratigraphies has been a major and worthwhile endeavor. An early attempt at this was made by Hays et al (1969). Here 15 equatorial Pacific cores containing planktonic foraminiferan, radiolarian, calcareous nannofossil, diatom, and silicoflagellate faunas and floras and good paleomagnetic signatures were correlated for the last 4.5 Ma. The biostratigraphies were correlated and along with sedimentologic data paleoclimates were also investigated. Land-based sections from type section regions of the Neogene were biostratigraphically correlated using radiolarians, diatoms, silicoflagellates, and calcareous nannofossils (Sanfilippo et al., 1973). Aside from the useful biostratigraphic correlations it became apparent that there were significant discrepancies between first and last occurrences of biostratigraphically important radiolarians between the European samples and the earlier-developed radiolarian biostratigraphy from the Pacific. An excellent treatment of multigroup microfossil correlation (related to the paleomagnetic scale) and paleoclimatology for the late Neogene may be found in Berggren and van Couvering (1974).

Very good summaries of paleoceanographic stud-

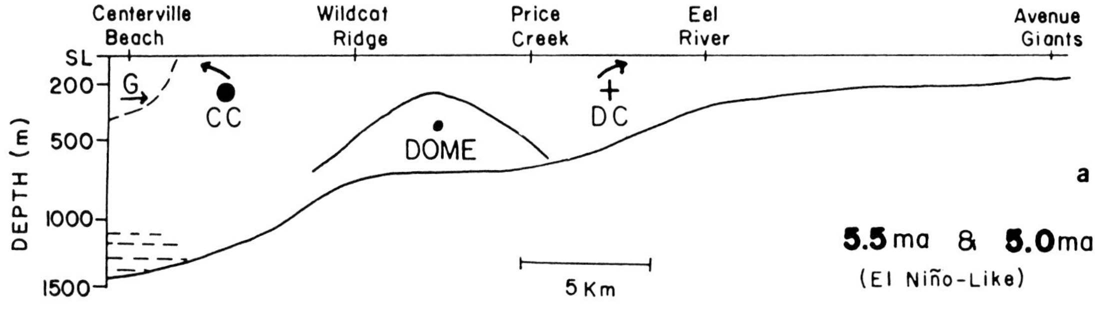

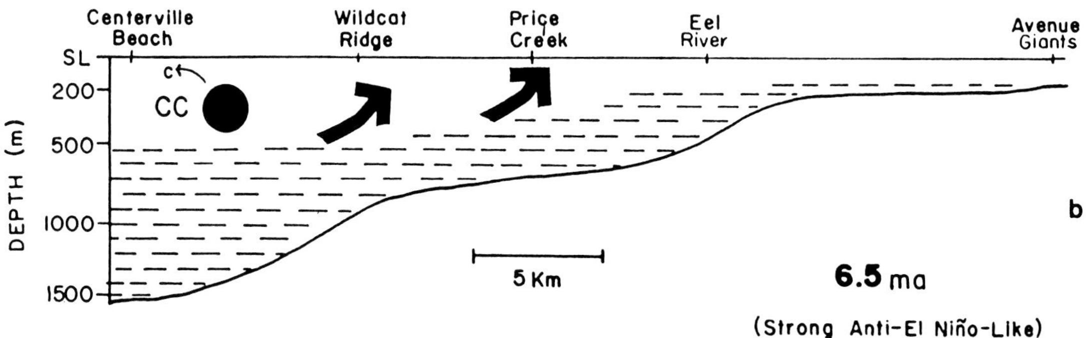

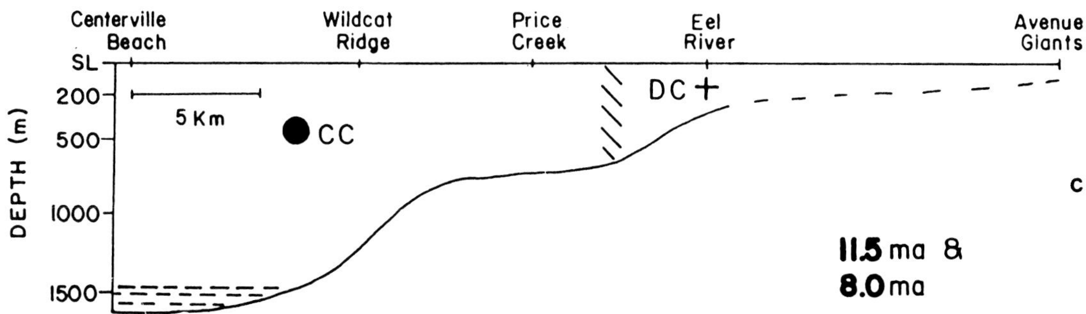

FIGURE 2.39 Humboldt Basin paleoceanographic reconstructions. Dots represent water movement out of page and crosses represent water movement into page. The size of the dot indicates relative strength and volume of current; the larger the dot, the stronger the current and the more water transported. The CC means California Current; DC means Davidson Countercurrent. The horizontal dashed lines represent oxygen minimum waters, and the locations at the reconstructed sea surface (SL), such as Centerville Beach, represent the paleolocations of these present land-based sections. (a) Humboldt Basin paleoceanographic reconstruction at 11.5 and 8.0 Ma. (b) Humboldt Basin paleoceanographic reconstruction at 6.5 Ma. (c) Humboldt Basin paleoceanographic reconstruction at 5.5 and 5.0 Ma.

ies using combinations of microfossil groups may be found in Berger (1981) and Kennett (1982).

Evolution and Extinction of Radiolaria

The oldest well-preserved and carefully documented radiolarians are from the Lower Ordovician of Spitsbergen (Fortey and Holdsworth, 1971); although paleoactinomonids and others appear to be present in Lower Cambrian rocks. The Lower Ordovician radiolarian fauna exhibits the three basic radiolarian skeletal types: the unshelled spicule, closed sphere, and cone (Fortey and Holdsworth, 1971). This fauna was apparently deposited in a comparatively shallow environment (Fortely and Holdsworth, 1971). Apparently none of the Paleozoic radiolarian faunas described by and according to Holdsworth (1977) can be considered to have been deposited under truly oceanic conditions; however, he suggested that albaillellids and pylentonemids may indeed have been deeper forms that invaded these shallower environments during times of open-ocean intrusion. These comments on shallow versus deep radiolarians by Holdsworth (1977) and inferences using his logic by Casey et al. (1983) allowed for the development of the Paleozoic portion of a paleogeographic radiolarian range chart (Casey et al., 1983). A modification of this chart (Fig. 2.35) suggesting that cold-water- and warm-water-sphere radiolarians are distinguishable in the Cambrian, is the junior author's interpretation. A deep radiolarian fauna had developed by at least Silurian times. It might well be that the initiation of the mid-Paleozoic cold-water sphere (1 in Fig. 2.35) set the stage for the evolution of the first deep, cold-water-sphere forms (albaillellids and perhaps ceratoikisids) (Casey et al., 1983). The initiation of the late Paleozoic cold-water sphere was suggested to have resulted in a major radiolarian reorganization of deep, cold-water-sphere radiolarians with extinctions of albailleleids and popofskyellids, and the evolution of deep, cold-water-sphere spongodiscidlike forms and corythoecids (Casey et al., 1983); however, in a diverse Permian radiolarian assemblage from west Texas, Cornell (1983) reports occurrences of paleoactinommids, entactiniids, rotaphaerids, and albaillellids, so Figure 2.35 has been modified to include these occurrences and extend the ranges of these groups. It now appears that the previously considered depauperate nature of Permian radiolarian faunas (Tappan and Leoblich, 1973) is not an appropriate evaluation at least for the early Permian (Cornell, 1983; Nazarov and Ormiston, 1985). However, the late Paleozoic and Paleozoic–Mesozoic transition appears to represent a major radiolarian reorganization with the extinctions of albaillellids, rotasphaerids, entactinids, and paleoactinommids (2 in Figure 2.35). This reorganization was suggested to be caused by the reduction in the number of water masses as a result of the suturing of the continents and eutrophication due to Permian glaciation (Casey et al., 1983).

The first unquestionable nassellarians appear in the Triassic, although Riedel and Sanfillipo (1981) suggest that a nassellarian group, the theoperids, may occur in the Permian. Approximately half of the extant groups shown in Fig. 2.35 evolved during the Mesozoic, and there are obvious generic affinities with extant radiolarians in the late Cretaceous. Characteristic Mesozoic (especially late Mesozoic) forms are amphipyndacids, dictyomitrids and hagiastrids. Although most other planktonic fossil groups undergo a drastic taxonomic reduction at the Mesozoic–Cenozoic transition, it appears that the major radiolarian faunal reorganization occurred within the early Cenozoic. In an exceptionally well-preserved Paleocene radiolarian assemblage from a DSDP core in the Tasman Sea, Dumitrica (1973) found amphipynacids and dictyomitrids. Hagiastrids, so characteristic of the Mesozoic, are commonly found in early Paleogene and to some extent in later Paleogene sediments (Fig. 2.35). Therefore, it appears that the major radiolarian events during the late Mesozoic, Mesozoic–Cenozoic transition, and shortly thereafter are not mass extinctions of "Mesozoic" groups but the evolution of Cenozoic groups (however, extinctions at the species level during this time may have been very significant). "Cenozoic" groups evolving at this time were the stylodictins, orosphaerids, spongopylins, sethophormins, robust arto strobids, spongasterins, pterocoryids, phacodiscids, spongurins, cannobotryids, coccodiscids, sethoperins, eucrytidins, dictyocorins, acanthodesmids, and lophophaenins (3 in Fig. 2.35). The environmental pressure related to this rediversification in radiolarians might have been the rediversification of planktonic foraminiferans and other planktonic groups after the terminal Cretaceous extinctions within those groups (Casey et al., 1983). From the Cretaceous on, radilarians have had to share their nannoherbivore and symbiotic niches with the planktonic foraminiferans. The evolution of diatoms, and their rise to dominance in the Cretaceous, must have had a major impact on radiolarian evolution. Apparently with the development of diatoms a "competition" for dissolved silica resulted that radiolarians had not previously experienced. Moore (1969) exhibited that the average weight of radiolarian skeletons decreased during the Cenozoic, and Harper and Knoll (1975) suggested that the reason for this decrease was due to competition from diatoms. Casey et al. (1978) showed that warm-water-sphere radiolarians exhibited a loss of skeletal silica through the Cenozoic, but deep, cold-water-sphere forms did not suggesting that only those radiolarians in "competition" with diatoms were affected. The last major radiolarian reorganization occurred at the Paleogene–Neogene transition (4 in Fig. 2.35). Tholonids, antarctisins, collosphaerids, artiscins, spongasterins, and pylonids (and perhaps not robust artostrobids) all either evolved or increased in dominance and diversity at this transition so that they can be considered Neogene groups. The tholonids appear to be restricted to present-day intermediate water masses. The antarctisins appear to exist mainly as polar subsurface forms with a few diving with Antarctic Bottom Water. The evolution of these two cold-water-sphere groups in the mid-Miocene was considered to be at least in part a response to the initiation of the Neogene water mass regime at that time and the establishment of the new intermediate and circumpolar water masses that they invaded (Casey et al., 1983). The collosphaerids, artiscins (at least those referred to in the past as the genus *Ommatartus* and designated in Fig. 2.35 with an O), spongasterins, and pylonids are all known to be warm-watersphere symbiont-bearing forms. Therefore, it was suggested that the development of the Neogene water mass regime with oligotrophic subtropical gyres opened the necessary niche for the evolution of these symbiotic groups (Casey et al., 1983). Similar oligotrophic niches were probably available and occupied by coccodiscids and perhaps hagiastrids in the Paleogene; however, it is suggested that the Neogene reorganization of water masses eliminated these forms.

Riedel and Sanfillipo (1981) referred to the distinctness of the three phases in radiolarian history (essentially the three eras of the Phanerozoic), and similarities in transitions from Paleozoic to Mesozoic and Mesozoic to Cenozoic faunas with a reduction in families at the boundaries, a relatively small number of families immediately above these boundaries, and an increase in families as the era progresses. They also suggest that the very incomplete available records of the Mesozoic and Paleozoic suggest that these eras (phases) were as diverse morphologically as the Cenozoic.

Speciation and Microevolution

In recent years radiolarians have been used quite extensively in studies concerned with speciation and microevolution. Radiolarians are especially suited for such studies because they are abundant, well preserved, and diverse; they exhibit a range of morphological characteristics; and their present-day distributions and ecologies are fairly well documented. Some of the drawbacks to such studies, however, are that their present-day distributions and ecologies are not known to the extent to which they should be, radiolarian genetics and reproduction are still relatively unknown, and most workers on radiolarian speciation and microevolution come from backgrounds other than biology.

The main impetus for this research appears to have been the papers of Eldredge and Gould (1972) and Gould and Eldredge (1977), where they suggested and expanded upon their theory of punctuated equilibria, which suggests that speciation occurs in jumps (saltation) between periods of "nonevolution" (stasis). Addressing this theory directly, papers by Bjorklund and Goll (1979), Kellogg (1983), and Lazarus et al. (1985) suggest that other modes of evolution appear to explain best their radiolarian studies. Bjorklund and Goll (1979) studied the collosphaerid radiolarian lineages *Acrosphaera, Collophaera,* and *Trisolenia* from the equatorial Pacific, which exhibited evidence for both phyletic gradualism and punctuated equilibria. Kellogg (1983) expanded on the work of Hays (1970) and Kellogg (1976), suggesting that the evolution of the radiolarian *Eucyrtidium matuyami* from *E. calvertense* was the product not of a single or a few large steps early in the speciation process but rather of the disproportionate occurrence of many small steps in a particular direction during neosympatry (character displacement in the size of the two species). Lazarus et al. (1985) conducted a global taxonomic and biometric study of species of the radiolarian genus *Pterocanium* and came to the conclusion that over the last 6 Ma the five major lineages studied exhibited gradual phyletic evolution, examples of symptric or parapatric speciation, and the possibility of hybridization.

Goll (1976) was the first to consider hybridization as an important factor in radiolarian evolution. Using species of acanthodesmids on a global scale, he noted evidence of hybridization and backcrossing. Hybridization was also suggested as a mode of speciation for species of collosphaerids studied by Bjorklund and Goll (1979). It is interesting to note that most living members of these groups (acanthodesmids and collosphaerids) are symbiotrophs, perhaps suggesting that their "plantlike" nature may in some way be related to their proposed speciation by hybridization. Hybridization appears to be a much more common method of speciation in plants than in animals.

Casey (1982) used the geographical occurrences of radiolarians in plankton, recent and fossil sediments, to develop models for evolution within the *Lamprocyrtis* and *Stichocorys* lineages. For extant forms distributions from plankton tows were used, and for extinct forms their paleodistributions were reconstructed using the same methods as described here previously (Casey et al., 1983). Speciation by branching (allopatric speciation) with the ancestor coexisting in time for a significant period occurred when the new species occupied an environment different from that of its ancestor. Speciation by replacement (maybe sympatric), with the ancestor–descendant overlap brief, occurred when the new species remained in the geographic environment of the ancestral species. It was concluded that these long overlaps in the sedimentary record might suggest gradualism and the brief overlaps in the sedimentary record punctuation; however, both appear to occur almost instantaneously (except perhaps for the cases of allopatry described) when the paleogeographies, especially the vertical paleogeographies, are reconstructed. It was also noted that it appears to require some significant amount of time for an allopatrically evolved species to inhabit it maximum geographic range but very little time for a supposed sympatrically evolved species to inhabit its maximum geographic range. Shortly after the Casey (1982) paper was submitted, it became apparent to the author that one of the species dealt with in that paper and thought to be extinct might well be alive in the North Atlantic Central Water Mass and finding it in that water mass would support the reconstruction of that species being a central water form. Plankton tows from the North Atlantic Central Water Mass were examined and the species was found (Casey et al., 1983).

Riedel and Sanfilippo (1980) illustrate nine radiolarian lineages, most of which they consider to exhibit gradual changes. Also, the terminal species of many extinct lineages have bizarre forms, unusual for radiolarian skeletons; these observations led the authors earlier (Riedel and Sanfilippo, 1978) to invoke the concept of orthogenesis in interpretating these apparently inertial evolutionary changes. In contrast to the more straightforward successions described (Riedel and Sanfilippo, 1981), the evolutionary sequence of the radiolarian genus *Spongaster* exhibited more complex patterns of apparently contrary tendencies of morphological changes (tendencies toward biopolarity and polygonality).

Spongaster pentas (the pentagonal form, Fig. 19 of Table 2.12) is an important biostratigraphic indicator (Riedel and Sanfilippo, 1978) whose first and last occurrences (evolution and extinction) have been dated in the equatorial Pacific paleomagnetically as 4.6 and 3.4 Ma respectively (Theyer et al., 1978). However, it was found to be living in the Gulf of Mexico and suggested to be a relict left when the Isthmus of Panama separated the previously connected equatorial Atlantic and Pacific (Casey et al., 1979b). Casey et al. (1979b) suggested the *S. pentas* died out in the Pacific when Panama emerged and therefore suggested an age of about the proposed extinction of *S. pentas* (extinction in the Pacific anyway) for that emergence. Carson and Casey (1986) found *S. pentas* living in the eastern tropical Pacific and apparently restricted to that environment (Fig. 2.36), an environment the same as the eastern tropical environment of Figure 2.30 and Table 2.13. After this initial finding of exant *S. pentas* in the Pacific, Carson and Casey (1986) compiled the first and last occurrences of this species in North Pacific DSDP cores. *S. pentas* exhibited a broad distribution at about 5 Ma. At about 4 Ma its range began to contract to the tropical regions and ultimately to the eastern tropical Pacific, where it has been reported in both Quaternary sediments (listed as reworked) and now plankton samples. They suggested that the evolution of *S. tetras tetras* (Fig. 17 of Table 2.12, a quadrangular form) at approximately 3.4 Ma and *S. tetras irregularis* (similar to *S. tetras tetras* except that it is rectangular rather than squarish in outline; Fig.17 of Table 2.12) sometime later may have resulted from a progressive opening of middle- and high-latitude niches vacated by the parent *S. pentas* stock due to climatic deterioration. This evolution would most likely be speciation (or "subspeciation") by replacement (maybe sympatric) in the terminology of Casey (1982).

Future Radiolarian Studies

The studies of Radiolaria are still at very immature stages. Many of the aspects of greatest importance to micropaleontologists are at more advanced stages, such as radiolarian biostratigraphy. However, there is much to be done and a few studies that appear to be most pressing will be briefly discussed here.

Radiolarian systematics is still in a flux. Most of the effort has been toward developing natural classifications using extant and Cenozoic radiolarian assemblages. It will be extremely important to incorporate Cenozoic, Mesozoic, and Paleozoic radiolarian assemblages into a comprehensive natural radiolarian classification. A concentrated combined effort clarifying the affinities of Cenozoic, Mesozoic, and Paleozoic taxa is our greatest systematic need in order to understand these affinities and radiolarian evolution, and to extend Neogene paleoceanographic reconstruction techniques and indicators into the Mesozoic and Paleozoic.

Radiolarian zoogeographies and paleozoogeographies are only superficially known. It is extremely important to document the distributions of at least the dominant extant radiolarians and to reconstruct (using this knowledge of extant radiolarians) paleozoogeographies. This is sorely needed in evolution studies and a necessity for studies concerned with paleoceanography. Similarly needed are studies concerned with radiolarian ecology.

Radiolarian paleoceanographic studies will benefit from advances in radiolarian classifications and zoogeographies. The culturing of radiolarians would add greatly to our knowledge, as would the development of better extraction techniques for some of the older rocks. However, some of the major benefits may have to await technological advances. Advances such as the ability to extract stable isotopic information from a very small number of (or individual) radiolarians would be extremely valuable in such paleoceanographic studies as those involved with paleotemperatures, paleosalinities, and so on. The use of amino acid racemization as a dating tool would be extremely valuable, as would the use of amino acids in developing a more natural classification (King, 1975, 1977). The development of computer-assisted taxonomic schemes and machines should enhance our ability to handle more information (Budai et al., 1980). Joint research efforts using a combination of microfossil groups and geochemical and mathematical techniques should provide at least higher-resolution paleoceanographic reconstructions if not entirely new paleoceanographic insights.

ACKNOWLEDGMENTS

The authors gratefully acknowledge the invitation of the editor to participate in this volume. P. L. Steineck is indebted to the over 70 foraminiferal specialists who generously forwarded copies of their publications; regrettably, space does not allow a full listing of their names. It is unfortunate that much of this interesting and important work could not be included in a single chapter of limited size. We are also indebted to John Haynes, James Kennett, Anne Miller, Pamela Hallock-Muller, C. Wyllie Poag, Fay Woodruff, and David Yozzo for their donation of scanning electron micrographs and original graphic materials. Ms. Nancy Dizzine and Dr. Barbara Dexter edited and typed portions of the manuscript. Ms. Gina Federico prepared the illustrations containing micrographs of Foraminifera. Martin Buzas and Jere Lipps kindly read portions of the manuscript and offered many helpful suggestions.

Support for the radiolarian portion of this chapter was partially provided by the National Science Foundation, Marine Geology and Geophysics, Grants OCE-74-21805, OCE-84-08852, and OCE-86-20446; acknowledgment is also made to the donors of the Petroleum Research Fund administered by the American Chemical Society and to the W.M. Keck Foundation for partial support of this study. Ms. Rose Graves of the University of San Diego aided in the editing and typing; Ms. Lisa Harrington and Ms. Rebecca Stanley of the Marine Studies Program of the University of San Diego aided in the drafting.

REFERENCES

Adams, C.G. (1976) Larger Foraminifera and the late Cenozoic history of the Mediterranean region. *Paleogeogr. Paleoclimatol. Paleoecol.* 20:47–61.

Adams, C.G. (1983) Speciation, phylogenesis, tectonism, climate and eustacy: factors in the evolution of Cenozoic larger foraminiferal bioprovinces. In Sims, R.W., Price, J.H., and Whalley, P.E.S. (eds.), *Evolution, Time and Space: The Evolution of the Biosphere.* Systematics Association, special Vol. 23. New York: Academic Press, pp. 255–289.

Adelseck, C.G., Jr. (1977) Dissolution of deep-sea carbonate: preliminary calibration of preservational and morphologic aspects. *Deep-Sea Res.* 24:1167–1185.

Adelseck, C.G., and Berger, W.H. (1975) On the dissolution of planktonic Foraminifera and associated microfossils during settling and on the sea floor. In Sliter, W.V., Bé, A.W.H. and Berger, W.H. (eds.), *Dissolution of Deep-Sea Carbonates.* Lawrence, Kan.: Allen Press, Inc. Cushman Found. Foram. Res. Spec. Publ. 13, pp. 70–81.

Alexander, S.P., and Banner, F.T. (1984) The functional relationship between skeleton and ectoplasm in *Haynesina germanica* (Ehrenberg), *J. Foram. Res.* 14:159–170.

Alvarez, W. (1984) The end of the Cretaceous: sharp boundary or gradual transition? *Science,* 223:1183–1186.

Alvarez, W., Alvarez, L.W., Asaro, F., and Michel, H.V. (1982) Current status of the impact theory for the terminal Cretaceous event. In Silver, L.T., and Schultz, P.H. (eds.), Geological Implications of Impacts of Large Comets and Asteroids on the Earth. *Geol. Soc. Amer.,* Spec. Paper 190, pp. 305–316.

Alvarez, W., Kauffman, E.G., Alvarez, L.W., Asaro, F., and Michel, H.V. (1984) Impact theory of mass extinctions and the invertebrate fossil record. *Science* 223:1135–1140.

Andersen, H.V. (1984) The wall structure of the superfamily Miliolacea and its lineages in the Gulf Coastal Plain. *Tulane Stud. Geol. Paleontol.* 18:1–19.

Anderson, E.J., Goodwin, P.W., and Sobieski, P. (1984) Episodic accumulation and the origin of formation boundaries in the Heldeberg Group of New York State. *Geology* 12:120–123.

Anderson, O.R. (1983) *Radiolaria.* New York: Springer-Verlag, 355 pp.

Anderson, O.R., and Tuntivate-Choy, S. (1984) Cytochemical evidence for peroxisomes in planktonic foraminifers. *J. Foram. Res.* 14:203–205.

Angell, R.W. (1980) Test morphogenesis (chamber formation) in the foraminifer *Spiroloculina hyalina* Schulze. *J. Foram. Res.* 10:89–101.

Arnold, A.J. (1982a) Techniques for biometric analysis of Foraminifera. *Proc. Third North American Paleontological Convention,* Vol. 1, Geol. Sur. Canada, Ottawa, pp. 13–15.

Arnold, A.J., (1982b) Species survivorship in the Cenozoic Globigerinida. *Proc. Third North American Paleontological Convention,* Vol. 1, Geol. Sur., Canada, Ottawa, pp. 9–12.

Arnold, A.J. (1983a) Foraminiferal thanatocoenoses on the continental slope off Georgia and South Carolina. *J. Foram. Res.* 13:79–90.

Arnold, A.J. (1983b) Phyletic evolution in the *Globorotalia crassaformis* (Galloway and Wissler) lineage: a preliminary report. *Paleobiology* 9(4):390–397.

Arnold, A.J., and SenGupta, B.K. (1981) Diversity changes in the foraminiferal thanatocoenoses of the Georgia–South Carolina continental slope. *J. Foram. Res.* 11:268–276.

Arnold, Z.M. (1978) An allogromiid ancestor of the milioliacean foraminifera. *J. Foram. Res.* 8:83–96.

Arnold, Z.M. (1979) Biological clues to the origin of miliolacean Foraminifera. *J. Foram. Res.* 9:302–321.

Arnold, Z.M. (1984) The gamontic karyology of the saccamminid foraminifer *Psammophaga simplora* Arnold. *J. Foram. Res.* 14:171–186.

Bandy, O.L. (1972) Origin and development of *Globorotalia (Turborotalia) pachyderma* (Ehreneberg). *Micropaleontology,* 18:294–318.

Bandy, O.L., Casey, R.E., and Wright, R.C. (1971) Late Neogene planktonic zonation, magnetic reversals, and radiometric dates, Antarctic to the tropics. Antarctic Res. Series, Vol. 15, Antarctic Oceanology I. Washington, D.C.: American Geophysical Union, pp. 1–26.

Banner, F.T. (1982) A classification and introduction to the Globigerinacea. In Banner, F.T., and Lord, A.R. (eds.). *Aspects of Micropaleontology,* London: George Allen and Unwin, pp. 142–239.

Banner, F.T., and Desai, D. (1985) The genus *Clavulinoides* Cushman emended and the new Cretaceous genus *Clavulinopsis*. *J. Foram. Res.* 15:79–81.

Banner, F.T., and Pereira, C.P.G. (1981) Some biserial and triserial agglutinated smaller Foraminifera: their wall structure and its significance. *J. Foram. Res.* 11:85–117.

Banner, F.T., Pereira, C.P.G., and Desai, D. (1985) "Tretomphaloid" float chambers in the Discorbidae and Cymbaloporidae. *J. Foram. Res.* 15:159–178.

Banner, F.T., Sheehan, R., and Williams, E. (1973) The organic skeleton of rotaline Foraminifera: a review. *J. Foram. Res.* 3:30–42.

Baumgartner, P.O. (1984) A Middle Jurassic–Early Cretaceous low-latitude radiolarian zonation based on Unitary Associations and age of Tethyan radiolarites. *Ecol. Geol. Helv.* 77:729–837.

Bé, A.W.H. (1982) Biology of planktonic Foraminifera. In Broadhead, T.W. (ed.). *Foraminifera: Notes for a Short Course,* Univ. Tennessee Dept. of Geological Sciences, Studies in Geology, Vol. 6, pp. 51–89.

Bé, A.W.H., Hemleben, C., Anderson, O.R., and Spindler, M. (1980) Pore structures in planktonic Foraminifera. *J. Foram. Res.* 10:117–128.

Behrensmeyer, A.K., and Kidwell, S.M. (1985) Taphonomy's contributions to paleobiology. *Paleobiology* 11:105–119.

Belanger, P.E. (1982) Paleo-oceanography of the Norwegian Sea during the past 130,000 years: coccolithophorid and foraminiferal data. *Boreas* 11:29–36.

Belanger, P.E., Curry, W.B., and Matthews, R.K. (1981) Core-top evaluation of benthic foraminiferal isotopic ratios for paleo-oceanographic interpretations. *Paleogeogr. Paleoclimatol. Paleoecol.* 33:205–220.

Belanger, P.E., and Streeter, S.S. (1980) Distribution and ecology of benthic Foraminifera in the Norwegian–Greenland Sea. *Mar. Micropaleontol.* 5:401–428.

Bellemo, S. (1974a) The compound and intermediate wall structures in Cibicidinae (Foraminifera) with remarks on the radial and granular wall structures. *Bull. Geol. Inst. University of Uppsala,* N.S., 6:1–11.

Bellemo, S. (1974b) Ultrastructure in Recent radial and granular calcareous foraminifers. *Bull. Geol. Inst. University of Uppsala,* N.S., Vol. 6, pp. 117–122.

Bellemo, S. (1976) Wall ultramicrostructure in the foraminifer *Cibicides floridanus* (Cushman). *Micropaleontology* 22:352–362.

Bellemo, S. (1979) Test wall structures in some Spirillinacea (Foraminifera). *Bull. Geol. Inst.* University of Uppsala, N.S., Vol. 8, pp. 77–82.

Belyaeva, N.V., and Burmistrova, I.I. (1985) Critical carbonate levels in the Indian Ocean. *J. Foram. Res.* 15:337–340.

Berger, W.A. (1981) Paleoceanography: The deep-sea record. In Emiliani, C. (ed.), *The Oceanic Lithosphere: The Sea,* Vol. 7. New York: Wiley, pp. 1437–1519.

Berger, W.H. and Roth, P.H. (1975) Oceanic micropaleontology: progress and prospect. *Rev. Geophys. Space Phys.* 13:561–585.

Berggren, W.A. (1968) Phylogenetic and taxonomic problems in some Tertiary planktonic foraminiferal lineages. *Tulane Stud. Geol.* 6:1–12.

Berggren, W.A. (1984a) Neogene planktonic foraminiferal biostratigraphy and biogeography: Atlantic, Mediterranean and Indo-Pacific regions. In Tsuchi, R., and Ikebe, N. (eds.). *Pacific Neogene Datum Planes—Contribution to Biostratigraphy and Chronology.* Tokyo, Japan: University of Tokyo Press, pp. 111–161.

Berggren, W.A. (1984b) Correlation of Atlantic, Mediterranean and Indo-Pacific Neogene stratigraphies: geochronology and chronostratigraphy. In Tsuchi, and Ikebe, N. (eds.). Pacific Neogene Datum Planes—Contribution to Biostratigraphy and Chronology, Tokyo, Japan: University of Tokyo Press, pp. 93–110.

Berggren, W.A., Aubrey, M.P., and Hamilton, N. (1983) Neogene magnetobiostratigraphy of deep sea drilling project site 516 (Rio Grande Rise, south Atlantic). In Barker, P.F., Carlson, R.L., and Johnson, D.A. (eds.), *Initial Reports of the Deep Sea Drilling Project,* Vol. 72, pp. 675–692. U.S. Govt. Printing Office Washington, D.C.

Berggren, W.A., and Hollister, C.D. (1977) Plate tectonics and paleocirculation—commotion in the ocean. *Tectonophysics* 38:11–48.

Berggren, W.A., Kent, D.V., Flynn, J.J., and van Couvering, J.A. (1985) Cenozoic geochronology. *Geol. Soc. Amer. Bull.* 96:1407–1418.

Berggren, W.A., et al. (1980) Towards a Quaternary time scale. *Quaternary Res.* 13:277–302.

Berggren, W.A., and D. Schnitker (1983) Cenozoic marine environments in the North Atlantic and Norwegian–Greenland Sea. In Bott, M., Saxou, A., Talwani, M., and Thiede, J. (eds.), *Structure and Development of the Greenland-Scotland Ridge.* New York: Plenum, pp. 495–547.

Berggren, W.A., and van Couvering, J. (1984) *Catastrophes and Earth History: The New Uniformitarianism,* Princeton, N.J.: Princeton University Press, 250 pp.

Berggren, W.A., and van Couvering, J.A. (1974) The Late Neogene biostratigraphy, geochronology, and paleoclimatology of the last 15 m.y. *Paleogeogr. Paleoclimatol. Paleoecol.* 16:1–216.

Berthold, W.O. (1976) Ultrastructure and function of wall perforations in *Patellina corrugata* Williamson (Foraminifera). *J. Foram. Res.* 6:22–29.

Bjorklund, R., and Goll, R.M. (1979) Internal skeletal structures of *Collosphaera* and *Trisolenia*: a case of repetitive evolution in the Collosphaeridae. *J. Paleontol.* 53:1293–1326.

Blake, G.H., and Douglas, R.G. (1980) Pleistocene occurrence of *Melonis pompiliodes* in the California borderland and its implication for foraminiferal paleoecology. Cushman Found. Foram. Res. Spec. Publ. 19, pp. 40–59.

Bloeser, B. (1985) *Melanocyrillium*, a new genus of structurally complex Late Proterozoic microfossils from the Kwagunt Formation (Chuar Group), Grand Canyon, Arizona. *J. Paleontol.* 59:741–765.

Bloeser, B., Schopf, M.W., Horodyski, R.J., and Breed, W.J. (1977) Chitinozoans from the Late Precambrian Chuar group of the Grand Canyon, Arizona. *Science* 195:676–679.

Blow, W.H. (1969) Late Middle Eocene to Recent planktonic foraminiferal biostratigraphy. *Proc. 1st Inter. Conf. Planktonic Microfossils*, Geneva, pp. 19–202.

Blow, W.H. (1979) The Cainozoic Globigerinida. A study of the morphology, taxonomy, evolutionary relationships and the stratigraphical distribution of some Globigerinids (mainly Globigerinacea). Leiden: E.J. Brill, 1413 pp.

Boersma, Ann (1984) Campanian through Paleocene paleotemperature and carbon isotope sequence and Cretaceous–Tertiary boundary in the Atlantic Ocean. In Berggren, W.A., and van Couvering, John (eds.), *Catastrophes and Earth History: The New Uniformitarianism*. Princeton, N.J.: Princeton University Press, pp. 247–278.

Boersma, A., and Premoli-Silva, I. (1983) Paleocene planktonic foraminiferal biogeography and the paleoceanography of the Atlantic Ocean. *Micropaleontology*, 29:355–381.

Boltovsky,, E., and Lena, H. (1971) The Foraminifera (except family Allogromiidae) which dwell in fresh water. *J. Foram. Res.* 1:71–76.

Bouvier-Soumagnac, Y., and Duplessy, J.C. (1985) Carbon and oxygen isotopic composition of planktonic Foraminifera from laboratory culture, plankton tows and recent sediment: implications for the reconstruction of paleoclimatic conditions and of the global carbon cycle. *J. Foram. Res.* 15:302–320.

Brady, H.B. (1884) Report on the Foraminifera dredged by H.M.S. *Challenger* during the years 1873–1876. *Repts. Scientific Results of the Voyage of H.M.S. Challenger* Vol. 9 (*Zoology*), 814 pp.

Brasier, M.D. (1975a) Morphology and habitat of living benthonic foraminiferids from Caribbean carbonate environments. *Rev. Espanol. Micropaleontol.* 7:567–578.

Brasier, M.D. (1975b) An outline history of seagrass communities. *Paleontology* 18:681–702.

Brasier, M.D. (1980) *Microfossils*. London: George Allen and Unwin, 193 pp.

Brasier, M.D. (1982a) Architecture and evolution of the foraminiferid test. A theoretical approach. In Banner, F.T., and Lord, A.R. (eds.), *Aspects of Micropaleontology*. London: George Allen and Unwin, pp. 1–41.

Brasier, M.D. (1982b) Foraminiferid architectural history: review using the Min LOC and PI methods. *J. Micropaleontol.* 1:95–105.

Brasier, M.D. (1984) Some geometrical aspects of fusiform planispiral shape in larger Foraminifera. *J. Micropalaeontol.*, 3:11–15.

Bremer, M.L., and Lohmann, G.P. (1982) Evidence for primary control of the distribution of certain Atlantic Ocean benthonic Foraminifera by degree of carbonate saturation. *Deep-Sea Res.* 29:987–998.

Brouwer, J. (1965) Agglutinated foraminiferal faunas from some turbiditic sequences. *K. Nederl. Akad. Wetensch. Proc.*, Ser. B, 68:309–334.

Budai, A., Riedel, W.R., and Westberg, M.J. (1980) A general-purpose paleontologic information device. *J. Paleontol.* 54:259–262.

Burke, S.C. (1981) Recent benthic Foraminifera of the Ontong Java Plateau. *J. Foram. Res.* 11:1–20.

Buzas, M.A. (1969) Foraminiferal species densities and environmental variables in an estuary. *Limnol. Oceanogr.* 14:411–422.

Buzas, M.A. (1974) Vertical distribution of *Ammobaculites* in the Rhode River, Maryland. *J. Foram. Res.* 4:144–147.

Buzas, M.A. (1977) Vertical distribution of Foraminifera in the Indian River, Florida. *J. Foram. Res.* 7:234–237.

Buzas, M.A. (1982) Regulation of foraminiferal densities by predation in the Indian River, Florida. *J. Foram. Res.* 12:66–71.

Buzas, M.A., and Culver, S.J. (1980) Foraminifera: distribution of provinces in the western North Atlantic. *Science* 209:687–689.

Buzas, M.A., and Culver, S.J. (1984) Species duration and evolution: benthic Foraminifera on the Atlantic continental margin of North America. *Science* 225:829–830.

Buzas, M.A., Culver, S.J., and Isham, L.B. (1985) A comparison of fourteen elphidid (Foraminiferida) taxa. *J. Paleontol.* 59:1075–1090.

Buzas, M.A., and Severin, K.P. (1982) Distribution and systematics of Foraminifera in the Indian River, Florida. *Smithsonian Contrib. Mar. Sci.* No. 16, 73 pp.

Buzas, M.A., Smith, R.K., and Beem, K.A. (1977) Ecology and systematics of Foraminifera in two *Thalassia* habitats, Jamaica, West Indies. *Smithsonian Contrib. Paleobiol.* No. 31, 139 pp.

Campbell, A.S. (1954) Radiolaria. Treatise on Invertebrate Paleontology, Part D, Protista 3. *Geol. Soc. Amer.*, pp. 11–163.

Caron, M., and Homewood, P. (1983) Evolution of early planktonic foraminifers. *Mar. Micropaleontol.* 7:453–462.

Carson, T.L. (1986) Radiolarian response to the 1983 California El Niño. In Casey, R.E., and Barron, J.A. (eds.) *Siliceous Microfossil and Microplankton Studies of the Monterey Formation and Modern Analogs*. Soc. Econ. Paleontol. Mineral., Pacific Coast Section, Los Angeles, pp. 9–19.

Carson, T.L., and Casey, R.E. (1986) Zoogeography, paleozoogeography, and evolution of the radiolarian genus *Spongaster* in the North Pacific. In Casey, R.E. and Barron, J.A. (eds.), *Siliceous Microfossil and Microplankton Studies of the Monterey Formation and Modern Analogs*. Soc. Econ. Paleontol. Mineral., Pacific Coast Section, Los Angeles, pp. 97–103.

Casey, R.E. (1971) Distribution of polycystine Radiolaria in the oceans in relation to physical and chemical conditions. In Funnell, B.M., and Riedel, W.R. (eds.), *The Micropalaeontology of Oceans*. New York: Cambridge University Press, pp. 151–159.

Casey, R.E. (1972) Neogene radiolarian biostratigraphy and paleotemperatures; southern California, the Experimental Mohole, Antarctic Core E 14-8. *Paleogeogr. Paleoclimatol. Paleoecol.*, 12:115–130.

Casey, R.E. (1973) Radiolarian evidence for the initiation and development of Neogene glaciations and the Neogene water mass regimes. *Abstract with Programs,* Vol. 5. Geol. Soc. Amer. Meet., Dallas, Texas, pp. 570–571.

Casey, R.E. (1977) The ecology and distribution of recent Radiolaria. In Ramsey, A.T.S. (ed.), *Oceanic Micropalaeontology,* New York: Academic Press, pp. 809–843.

Casey, R.E. (1982) *Lamprocyrtis* and *Stichocorys* lineages: biogeographical and ecological perspectives relating to the tempo and mode of polycystine radiolarian evolution. In Mamet, B., and Copeland, M.J. (eds.), *Proc. Third N. Amer. Paleontol. Conv.*, Vol. 1, pp. 77–82.

Casey, R.E. (1986) Phaeodarian radiolarians as potential indicators of thermal maturation. In Casey, R.E. and Barron, J.A. (eds.), *Siliceous Microfossil and Microplankton Studies of the Monterey Formation and Modern Analogs*. Soc. Econ. Paleontol. Mineral., Pacific Section, Los Angeles, pp. 87–89.

Casey, R.E., Carson, T.L., and Weinheimer, A.L. (1986) The modern California Current System and radiolarian responses to "normal" (anti–El Niño) conditions. In Casey, R.E., and Barron, J.A. (eds.), *Siliceous Microfossil and Microplankton Studies of the Monterey Formation and Modern Analogs*. Soc. Econ. Paleontol. Mineral., Pacific Section, Los Angeles, pp. 1–7.

Casey, R.E., Drickman, D., Kunze, F., Reynolds, R., Schafersman, S., and Spaw, J.M. (1978) Deep-living polycystine radiolarian ecology, paleoecology, evolution, biostratigraphy, and preservation. *Abstract with Programs,* Vol. 10. Geol. Soc. Amer., No. 7, pp. 378.

Casey, R., Gust, L., Leavesley, A., Williams, D., Reynolds, R., Duis, T., and Spaw, J.M. (1979a) Ecological niches of radiolarians, planktonic foraminiferans and pteropods inferred from studies on living forms in the Gulf of Mexico and adjacent waters. *Trans. Gulf Coast Assoc. Geol. Soc.* 29:216–223.

Casey, R.E., Leavesley, A., Spaw, J.M., McMillen, K., and Sloan, J. (1981) Radiolarian species composition, density and diversity as indices of the structure and circulation of waters overlying the South Texas shelf. *Trans. Gulf Coast Assoc. Geol. Soc.* 31:257–263.

Casey, R., McMillen, K., Reynolds, R., Spaw, J.M., Schwarzer, R., Gervirtz, J., and Bauer, M. (1979) Relict and expatriated radiolarian fauna in the Gulf of Mexico and its implications. *Trans. Gulf Coast Assoc. Geol. Soc.* 29:224–227.

Casey, R.E., Partridge, T.M., and Sloan, J.R. (1971) Radiolarian life spans, mortality rates and seasonality gained from Recent sediment and plankton samples. In Farinacci, A. (ed.), Proc. Second Planktonic Conference, Rome, 1970, pp. 159–165.

Casey, R.E., and Reynolds, R.A. (1980) Late Neogene radiolarian biostratigraphy related to magnetostratigraphy and paleoceanography: suggested cosmopolitan radiolarian datums. In Sliter, W.V., (ed.), *Studies in Marine Micropaleontology and Paleoecology: a memorial volume to Orville L. Bandy,* Cushman Foundation Special Publication No. 19, pp. 287–300.

Casey, R.E., Spaw, J.M., and Kunze, F.R. (1982) Polycystine radiolarian distributions and enhancements related to oceanographic conditions in a hypothetical ocean. *Trans. Gulf Coast Assoc. Geol. Soc.* 32:228–237.

Casey, R.E., and Wigley, C.R. (1985) Neogene deep-living radiolarian zonation and its uses. *Abst. Amer. Assoc. Petrol. Geol. Nat. Meeting,* New Orleans, p. 275.

Casey, R.E., Wigley, C.G., and Perez-Guzman, A.M. (1983) Biogeographic and ecologic perspective on polycystine radiolarian evolution. *Paleobiology* 9:363–376.

Chen, P. (1975) Antarctic radiolaria, Leg 28 Deep Sea Drilling Project. In Frakes, L.A., and Hayes, D.E. (eds.), *Initial Reports of the Deep Sea Drilling Project,* Vol. 28, Washington, D.C.: U.S. Govt. Printing Office, pp. 437–513.

Cifelli, R. (1982) Early occurrences and some phylogenetic implications of spiny, honeycomb-textured planktonic Foraminifera. *J. Foram. Res.* 12:105–115.

Cifelli, R., and Scott, R.H. (1983) The New Zealand early Miocene globorotallids *Globorotalia incognita* Walters and *Globorotalia zealandica* Hornibrook. *J. Foram. Res.* 13:163–166.

Cleveland, M.N., and Casey, R.E. (1986) Radiolarian indices of physical and chemical oceanographic phenomena in Recent sediments of the Southern California Continental Borderland. In Casey, R.E., and Barron, J.A. (eds.), *Siliceous Microfossil and Microplankton Studies of the Monterey Formation and Modern Analogs*. Soc. Econ. Paleontol. Mineral., Pacific Section, Los Angeles, pp. 21–30.

CLIMAP Project Members (1976) The surface of the Ice Age Earth. *Science* 191:1131–1137.

Cline, R.M., and Hayes, J.D., eds. (1976) Investigations of Late Quaternary paleoceanography and paleoclimatology, *Geol. Soc. Amer. Mem.* 145, Boulder, Colorado.

Coleman, A.R. (1980) Test structure and function of the agglutinated foraminifer *Clavulina*. *J. Foram. Res.* 10:143–152.

Collison, P. (1980) Vertical distribution of Foraminifera off the coast of Northlumberland, England. *J. Foram. Res.* 10:75–78.

Conger, S.D., Green, H.W., and Lipps, J.H. (1977) Test ultrastructure of some calcareous Foraminifera. *J. Foram. Res.* 7:279–321.

Conkin, J.E., and Conkin, B.A. (1982) North American Paleozoic agglutinate Foraminifera. In Broadhead, T.W. (ed.), *Foraminifera: notes for a short course*. University of Tennessee, Dept. of Geology, Studies in Geology No. 6, pp. 177–219.

Conway, Morris, S., and Fritz, W.H. (1980) Shelly microfossils near the Precambrian/Cambrian boundary, Mackenzie Mountains, North-West Canada. *Nature* 286:381–384.

Corliss, B.H. (1979) Size variation in the deep-sea benthonic foraminifer *Globocassidulina subglobosa* (Brady) in the southeast Indian Ocean. *J. Foram. Res.* 9:50–60.

Corliss, B.H. (1983) Quaternary circulation of the Antarctic Circumpolar Current. *Deep-Sea Res.* 30:47–63.

Corliss, B.H., Aubrey, M.P., Berggren, W.A., Fenner, J.M., Keigwin, L.D., Jr., and Keller, G. (1984) The Eocene/Oligocene boundary event in the deep sea. *Science* 226:806–810.

Corliss, B.H., and Honjo, S. (1981) Dissolution of deep-sea benthic Foraminifera. *Micropaleontology* 27:356–378.

Cornell, W.C. (1983) Some Permian (Leonardian) radiolarians from Bone Springs Limestone, Delaware Basin, West Texas. *Bull. Amer. Assoc. Petrol. Geol.* 67:444.

Cracraft, J., and Eldredge, N. (1979) *Phylogenetic Analysis and Paleontology*, New York: Columbia University Press, 310 pp.

Crowley, T.J. (1985) Late Quaternary carbonate changes in the North Atlantic and Atlantic/Pacific comparisons. In Sundquist, E.T., and Broecker, W.S. (eds.), *The Carbon Cycle and Atmospheric CO_2: Natural Variation, Archean to Present*, Washington, D.C.: Amer. Geophys. Union, pp. 271–284.

Crowley, T.J., and Matthews, R.K. (1983) Isotope-plankton comparisons in a late Quaternary core with a stable temperature history. *Geology* 11:275–278.

Cullen, J.L., and Prell, W.L. (1984) Planktonic Foraminifera of the northern Indian Ocean: distribution and preservation in surface sediments. *Mar. Micropaleontol.* 9:1–52.

Culver, S.J., and Buzas, M.A. (1981) Recent benthic foraminiferal provinces on the Atlantic continental margin of North America. *J. Foram. Res.* 11:217–240.

Culver, S.J., and Buzas, M.A. (1982) Recent benthic foraminiferal provinces between Newfoundland and Yucatan. *Geol. Soc. Amer. Bull.* 93:269–277.

Culver, S.J., and Buzas, M.A. (1983a) Recent benthic foraminiferal provinces in the Gulf of Mexico. *J. Foram. Res.* 13:21–31.

Culver, S.J., and Buzas, M.A. (1983b) Benthic Foraminifera at the shelfbreak: North American Atlantic and Gulf margins. *Society of Economic Paleontologists and Mineralogists*, Spec. Publ. No. 33, pp. 359–371.

Curry, D. (1982) Differential preservation of foraminiferids in the English Upper Cretaceous. In Banner, F.T., and Lord, A.R. (eds.) *Aspects of Micropaleontology*. London: George Allen and Unwin, pp. 240–261.

Curry, W.B., and Lohmann, G.P. (1982) Carbon isotopic changes in benthic Foraminifera from the Western South Atlantic: reconstruction of glacial abyssal circulation patterns. *Quaternary Res.* 18:218–235.

Curry, W.B., and Lohmann, G.P. (1984) Reduced advection into Atlantic Ocean deep eastern basins during last glaciation maximum. *Nature* 306:577–580.

Curry, W.B., and Matthews, R.K. (1981) Equilibrium ^{18}O fractionation in small size fraction planktonic Foraminifera. Evidence from Recent Indian Ocean sediments. *Mar. Micropaleontol.* 6:327–337.

Dailey, D.H. (1983) Late Cretaceous and Paleocene benthic Foraminifera from Deep-Sea Drilling Project site 516, Rio Grande Rise, Western South Atlantic Ocean. In Barker, F.P., et al., Initial Reports of the Deep-Sea Drilling Project, Vol. 72. Washington, D.C.: U.S. Govt. Printing Office, pp. 775–782.

Delaca, T.E., Karl, D.M., and Lipps, J.H. (1981) Direct use of dissolved organic carbon by agglutinated benthic Foraminifera. *Nature* 289:287–289.

Delaca, T.E., Lipps, J.H., and Hessler, R.R. (1980) The morphology and ecology of a new large agglutinated Antarctic foraminifer (Textulariina: Notodendrodidae *nov.*). *Zool. J.* Linnean Soc. 69:205–224.

Dingus, L., and Sadler, P.M. (1982) The effects of stratigraphic completeness on estimates of evolutionary rates. *Syst. Zool.* 31:400–412.

Dodson, M.M., and Reyment, R.A. (1980) Analysis of the extinction of the late Cretaceous foraminifer *Afrobolivina afra*. *Cret. Res.* 1:143–164.

Domack, C.R. (1986) Reconstruction of the California Current at 5, 8, and 10 million years B.P. using radiolarian indicators. In Casey, R.E., and Barron, J.A. (eds.), *Siliceous Microfossil and Microplankton Studies of the Monterey Formation and Modern Analogs*. Soc. Econ. Paleontol. Mineral., Pacific Section, Los Angeles, pp. 39–54.

Douglass, R.C. (1967) Permian Tethyan fusulinids from California. U.S. Geological Survey Professional Paper 593A, Washington, D.C., 25 pp.

Douglass, R.C. (1977) The development of fusulinid biostratigraphy. In Kauffman, E.G., and Hazel, J.E. (eds.), *Concepts and Methods of Biostratigraphy*, Stroudsburg, Pa., pp. 463–481. Dowden Hutchinson and Ross, Inc.

Douglass, R.C., and Nestell, M.K. (1984) Fusulinids of the Atoka Formation, lower-middle Pennsylvanian, south-central Oklahoma. In Sutherland, P.K., and Manger, W.L. (eds.), *The Atokan Series (Pennsylvanian) and Its Boundaries—A Symposium*, Oklahoma Geol. Survey Bull. No. 136, Tulsa, pp. 19–39.

Douglas, R.G. (1981) Paleoecology of continental margin basins: a modern case history from the borderland of Southern California. In Douglas, R.G., Colburn, I.P., and Gorsline, D.S. (eds.), *Depositional Systems of Active Continental Margin Basins: Short Course Notes*. Soc. Econ. Paleontol. Mineral., Pacific Section, Los Angeles, pp. 121–156.

Douglas, R.G., and Heitman, H. (1979) Slope and basin benthic Foraminifera of the California borderland. In Doyle, L., and Pilkey, O. (eds.), *Geology of Continental Slopes*. Soc. Econ. Pal. Mineral., Spec. Pub. 27, Tulsa, p. 231–246.

Douglas, R.G., Liestman, J., Walch, C., Blake, G., and Cotton, M.L. (1980) The transition from live to sediment assemblage in benthic Foraminifera from the Southern California borderland. In Field, M., Bouma, A., Colburn, I., Douglas, R. (eds.), *Quaternary Depositional Environment of the Pacific coast*, Pacific Coast Paleogeography Symposium, Vol. 4, Los Angeles, pp. 257–280.

Douglas, R.G., and Savin, S.M. (1978) Oxygen isotopic evidence for the depth stratification of Tertiary and Cretaceous planktonic Foraminifera. *Mar. Micropaleontol.* 3:175–194.

Douglas, R.G., and Woodruff, F. (1981) Deep-sea benthic Foraminifera. In Emiliani, C. (ed.), *The Oceanic Lithosphere, The Sea*, Vol. 7. New York: Wiley-Interscience, pp. 1233–1327.

Driskill, L.E. (1986) Preliminary report: phaeodarian radiolarians as indicators of Recent and ancient (Monterey) anoxic events in California. In Casey, R.E., and Barron, J.A. (eds.), *Siliceous Microfossils and Microplankton Studies of the Monterey Formation and Modern Analogs*. Soc. Econ. Paleontol. Mineral., Pacific Section, Los Angeles, pp. 7–85.

Drooger, C.W. (1984) Evolutionary patterns in lineages of orbitoidal foraminifera. *K. Nederl. Akad. Wetensch. Proc.*, Ser. B, 87:103–130.

Dumitrica, P. (1973) Paleocene Radiolaria, DSDP Leg 21. In Burns, R.E., and Andrews, J.E. et al. (eds.), Initial Reports of the Deep Sea Drilling Project, Vol. 21. Washington, D.C.: U.S. Govt. Printing Office, pp. 787–817.

Dumitrica, P., Kozur, H., and Mostler, H. (1980) Contribution to the radiolarian fauna of the Middle Triassic of the Southern Alps. *Geol. Palaeontol. Mitt. Innsbruck*, 10:1–46.

Dunbar, R.B. (1983) Stable isotope record of upwelling and climate from Santa Barbara Basin, California. In Suess, E., and Thiede, J. (eds.) *Coastal Upwelling: Its Sedimentary Record*, Part A. New York: Plenum, pp. 217–246.

Dunbar, R.B., and Wefer, G. (1984) Stable isotope fractionation in benthic Foraminifera from the Peruvian continental margin, *Mar. Geol.* 59:215–225.

Echols, R.J., and Fowler, G.S. (1973) Agglutinated tintinnid loricae from some Recent and Late Pleistocene shelf sediments. *Micropaleontology* 19:431–443.

Ehrenberg, C.G. (1874) Uber einer halibiolithische, von Herrn R. Schomburgk entdeckte, vorherrschend aus microskopischen Polycystinen gebildete. Gebirgsmasse von Barbados, Monatsber, *Klg. Preuss. Akad. Wiss. Berliln*, Jahrg. 1846, pp. 382–385.

Ekdale, A.A., and Bromley, R.G. (1984) Sedimentology and ichnology of the Cretaceous–Tertiary boundary in Denmark: implications for the causes of the terminal Cretaceous extinction. *J. Sedimentary Petrol.* 54:681–703.

Eldredge, N., and Cracraft, J. (1980) *Phylogenetic Patterns and the Evolutionary Process: Method and Theory in Comparative Biology*. New York: Columbia University Press, 360 pp.

Eldredge, N., and Gould, S.J. (1972) Punctuated equilibria: an alternative to phyletic gradualism. In Schopf, T.J.M. (ed.), *Models in Paleobiology*. San Francisco: Freeman, Cooper, pp. 82–115.

Eldredge, N., and Novacek, J.W. (1985) Systematics and paleobiology. *Paleobiology* 11:65–74.

Ellison, R.L., and Peck, G.E. (1983) Foraminiferal recolonization on the continental shelf. *J. Foram. Res.* 13:231–241.

Empson-Morin, K.M. (1984) Depth and latitude distribution of Radiolaria in Campanian (Late Cretaceous) tropical and subtropical oceans. *Micropaleontology* 30:87–115.

Eva, A.N. (1980) Pre-Miocene seagrass communities in the Caribbean. *Paleontology* 23:231–236.

Fairbanks, R.G., et al. (1982) Vertical distribution and isotopic fractionation of living planktonic Foraminifera from the Panama Basin. *Nature* 298:841–844.

Fermont, W.J.J. (1977) Biometrical investigation of the genus *Operculina* in Recent sediments of the Gulf of Elat, Red Sea. *Utrecht Micropaleontol. Bull.* 15:111–147.

Fermont, W.J.J. (1982) Discocyclinidae from Ein Avedat (Israel). *Utrecht Micropaleontol Bull.* 27:1–273.

Fermont, W.J.J., Krevler, R., and van der Zwaan, G.J. (1983) Morphology and stable isotopes as indicators of productivity and feeding patterns in Recent *Operculina ammonoidiea* (Gronovius). *J. Foram. Res.* 13:122–128.

Feyling-Hanssen, R.W. (1972) The foraminifer *Elphidium excavatum* (Terquem) and its variant forms. *Micropaleontology* 18:337–354.

Fischer, A.G. (1984) The two Phanerozoic supercycles. In Berggren, W.A., and van Couvering, John (eds.), *Catastrophes and Earth History: The New Uniformitarianism*. Princeton, N.J.: Princeton University Press, pp. 129–150.

Fischer, A.G., and Arthur, M.A. (1977) Secular variations in the pelagic realm. Soc. Econ. Paleontol. Mineral. Spec. Pub. 25, pp. 19–50.

Fleisher, R.L. (1974) Cenozoic planktonic Foraminifera, Arabian Sea (Deep Sea Drilling Project, leg 23A) In Whitmarsh, R.B., Rose, D.A., et al., Initial Reports of the Deep Sea Drilling Project, Vol. 23 Washington, D.C.: U.S. Government Printing Office, pp. 1001–1071.

Fleisher, R.L. (1975) Oligocene planktonic foraminiferal biostratigraphy, central North Pacific Ocean (Deep Sea Drilling Project, leg 21). In Larson, R.L., Moberly, R., et al., Initial Reports of the Deep Sea Drilling Project, Vol. 32 Washington, D.C.: U.S. Government Printing Office, pp. 752–763.

Fortey, R.A., and Holdworth, B.K., (1971) The oldest known well-preserved Radiolaria. *Boll. Soc. Paleontol. Ital.*, 10:35–41.

Frankel, L. (1975) Pseudopodia of surface and subsurface dwelling *Miliammina fusca* (Brady). *J. Foram. Res.* 5:211–219.

Ganssen, G., and Sarnthein, M. (1983) Stable-isotope composition of foraminifers: the surface and bottom water record of coastal upwelling. In Seuss, E. and Thiede, J. *Coastal Upwelling: Its Sedimentary Record*, Part A. New York: Plenum, pp. 99–123.

Gibson, T., and Buzas, M. (1973) Species diversity patterns in modern and Miocene Foraminifera of the eastern margin of North America. *Geol. Soc. Amer. Bull.* 84:217–238.

Goll, R.M. (1976) Morphological intergradation between modern populatons of *Lophospyris* and *Phormospyris* (Trissocyclidae: Radiolaria). *Micropaleontology* 22:379–418.

Gould, S.J. (1980a) Is a new and general theory of evolution emerging? *Paleobiology* 6:119–130.

Gould, S.J. (1980b) The promise of paleobiology as a nomothetic, evolutionary discipline. *Paleobiology* 6:96–118.

Gould, S.J. (1985) The paradox of the firt tier: an agenda for paleobiology. *Paleobiology* 11:2–12.

Gould, S.J., and Eldredge, N. (1977) Punctuated equilibria: the tempo and mode of evolution reconsidered. *Paleobiology* 3:115–151.

Gradstein, F.M., and Agterberg, F.P. (1982) Models of Cenozoic foraminiferal stratigraphy—Northwestern Atlantic margin. In Cubitt, J.M. and Reyment, R.A. (eds.), *Quantitative Stratigraphic Correlation*. New York: Wiley, pp. 119–170.

Gradstein, F.M., and Berggren, W.A. (1981) Flysch-type agglutinated Foraminifera and the Maestrichtian to Paleogene history of the Labrador and North Seas. *Mar. Micropaleontol.* 6:211–268.

Graham, D.W., Corliss, B.H., Bender, M.L., and Keigwin, L.D. Jr. (1981) Carbon and oxygen isotopic disequilibria of Recent deep-sea benthic Foraminifera. *Mar. Micropaleontol.* 6:483–498.

Green, H.W., II, Lipps, J.H., and Showers, W.J. (1980) Test ultrastructure of fusulinid Foraminifera. *Nature* 283:853–855.

Grell, K.G. (1979) Cytogenetic systems and evolution in Foraminifera. *J. Foram. Res.* 9:1–13.

Grigelis, A., and Gorbatchik, T. (1980) Morphology and taxonomy of Jurassic and early Cretaceous representatives of the superfamily Globigerinacea (Favusellidae). *J. Foram. Res.* 10:180–190.

Grossman, E.L. (1984) Stable isotope fractionation in live benthic Foraminifera from the Southern California Borderland. *Paleogeogr. Palaeclimatol. Palaeoecol.* 47:301–327.

Haeckel, E. (1862) Die Radiolarien (Rhizopoda Radiolaria). Eine Monographie. Berlin: Reimer, 572 pp.

Haeckel, E. (1881) Uever die Tiefsee Radiolarien der Challenger Expedition: Sitzzungsber. *Med. Naturw. Gesell. Jena, Jahrg.*, pp. 35–36.

Haeckel, E. (1887) Report on the Radiolaria collected by H.M.S. *Challenger* during the years 1873–1875: Rept. Voyage "Challenger," Zool. (London) 18, 1803 pp.

Hallock, P. (1979) Trends in test shape with depth in large, symbiont-bearing Foraminifera. *J. Foram. Res.* 9:61–69.

Hallock, P. (1981a) Light dependence in *Amphistegina*. *J. Foram. Res.* 11:40–46.

Hallock, P. (1981b) Algal symbiosis: a mathematical analysis. *Mar. Biol.* 62:249–255.

Hallock, P. (1982) Evolution and extinction in larger Foraminifera. Proc. Third North American Paleonotological Convention, Geol. Sur., Canada, Ottawa, v. 1, pp. 221–225.

Hallock P. (1984) Distribution of selected species of living algal-symbiont bearing Foraminifera on two Pacific coral reefs. *J. Foram. Res.* 14:250–261.

Hallock, P. (1985) Why are larger Foraminifera large? *Paleobiology* 11:195–208.

Hallock, P., and Hansen, H.J. (1979) Depth adaption in *Amphistegina:* change in lamellar thickness. *Bull. Geol. Soc. Denmark* 27:99–104.

Haman, D. (1983) Modern Textulariina (Foraminiferida) from the Balize Delta, Louisiana. In Verdenius, J.G., vanHinte, J.E., Fortuis, A.R. (eds.), Proceedings of the First Workshop on Arenaceous Foraminifera, 7–9 September 1981. Continental Shelf Institute, Norway, IKU Publication No. 108, pp. 59–87.

Hansen, H.J. (1979) Test structure and evolution in the Foraminifera. *Lethaia* 12:173–182.

Hansen, H.J., and Dahlberg, P. (1979) Symbiotic algae in milioline Foraminifer: CO2 uptake and shell adaptations. *Bull. Geol. Soc. Denmark* 28:47–55.

Harland, W.B., Cox, A.V., Llewellyn, P.G., Pickton, C.A.G., Smith, A.G., and Walters, R. (1982). A geologic time scale. New York: Cambridge University Press, 131 pp.

Harper, H.E., and Knoll, A.H. (1975) Silica, diatoms, and Cenozoic radiolarian evolution. *Geology* 3:175–177.

Hart, M.B. (1980) A water-depth model for the evolution of the planktonic Foraminiferida. *Nature* 286:252–254.

Hay, W.W. (1983) The global significance of regional Mediterranean Neogene paleoenvironmental studies. *Utrecht Micropaleont. Bull.* 30:9–24.

Haynes, J.R. (1981) *Foraminifera*. New York: Wiley, 931 pp.

Hays, J.D. (1965) Radiolaria and late Tertiary and Quaternary history of Antarctic Seas. In Llano, G.A. (ed.), *Biology of Antarctic Seas*, Vol. II. Amer. Geophys. Union, Antarct. Res. Ser. 5, pp. 125–184.

Hays, J.D. (1970) The stratigraphy and evolutionary trends of Radiolaria in North Pacific deep-sea sediments. In Hays, J.D. (ed.), *Geologic Investigations of the North Pacific*, Geol. Soc. Amer. Mem. 126, pp. 186–218.

Hays, J.D., Saito, T., Opdyke, N.D., and Burckle, L.H. (1969) Pliocene–Pleistocene sediments of the equatorial Pacific—their paleomagnetic, biostratigraphic and climatic record. *Geol. Soc. Amer. Bull.* 80:1481–1514.

Healy-Williams, N. (1983) Fourier shape analysis of *Globorotalia truncatulinoides* from late Quaternary sediments in the southern Indian Ocean. *Mar. Micropaleontol.* 8:1–16.

Healy-Williams, N., Ehrlich, R., and Williams, D.F. (1985) Morphometric and stable isotopic evidence for subpopulations of *Globorotalia truncatulinoides*. *J. Foram. Res.* 15:242–253.

Healy-Williams, N., and Williams, D.F. (1981) Fourier analysis of test shape of planktonic Foraminifera. *Nature* 289:485–487.

Hedberg, H.D. (1976) *International Stratigraphic Guide: A Guide to Stratigraphic Classification, Terminology, and Procedure*. New York: Wiley, 200 pp.

Hemleben, C. (1975) Spine and pustule relationships in some Recent planktonic Foraminifera. *Micropaleontology* 21:334–340.

Hemleben, C., Bé, A.W.H., Anderson, O.R., and Tuntivate, S. (1977) Test morphology, organic layers and chamber formation of the planktonic foraminifer *Globorotalia menardii* (d'Orbigny). *J. Foram. Res.* 7:1–25.

Hemleben, C., and Spindler, M. (1983) Recent advances in research on living planktonic Foraminifera. *Utrecht Micropaleontol. Bull.* 30:141–170.

Hemleben, C., Spindler, M., Breitinger, I., and Deuser, W.G. (1985) Field and laboratory studies on the ontogeny and ecology of some globorotaliid species from the Sargasso Sea off Bermuda. *J. Foram. Res.* 15:254–272.

Herman, Y. (1981) Causes of massive biotic extinctions and explosive evolutionary diversification throughout Phanerozoic time. *Geology* 9:104–108.

Hermelin, J.O.R. (1983) Biogeographic patterns of modern *Reophax dentaliniformis* Brady (arenaceous benthic Forminifera) from the Baltic Sea. *J. Foram. Res.* 13:155–162.

Hermelin, J.O.R., and Malmgren, B.A. (1980) Multivariate analysis of environmentally controlled variation in *Lagena,* Late Maastrichtian, Sweden. *Cretaceous Res.* 1:193–206.

Hertwig, R. (1879) *Der Organismus der Radiolarien.* Jena: G. Fischer, 149 pp.

Hoffman, A., and Kitchell, J.A. (1984) Evolution in a pelagic planktic system: a paleobiologic test of models of multispecies evolution. *Paleobiology* 10:9–34.

Hohenegger, J. (1978) A population genetic interpretation of the morphological transformation of Triassic Foraminifera. *Lethaia* 11:199–215.

Holdsworth, B.K. (1977) Paleozoic Radiolaria: stratigraphic distribution in Atlantic borderlands. In Swain, F.M. (ed.), *Stratigraphic Micropaleontology of Atlantic Basins and Borderlands.* Amsteerdam: Elsevier, pp. 167–184.

Holdsworth, B.K., and Jones, D.L. (1980) Preliminary radiolarian zonation for Late Devonian through Permian time. *Geology* 8:281–285.

Hottinger, L. (1982) Larger Foraminifera, giant cells with a historical background. *Naturwissenshaften* 69:361–371.

Hottinger, L. (1983) Processes determining the distribution of larger Foraminifera in space and time. *Utrecht Micropaleontol. Bull.* 30:239–254.

Huang, C. (1981) Observations on the interior of some late Neogene planktonic Foraminifera. *J. Foram. Res.* 11:173–190.

Hull, D.L. (1976) Are species really individuals. *Syst. Zool.* 25:174–191.

Hutson, W.H. (1977) Transfer functions under no-analog conditions: experiments with Indian Ocean planktonic Foraminifera. *Quaternary Res.* 8:355–367.

Hutson, W.H., and Prell, W.L. (1980) A paleocological transfer function FI-2 for Indian Ocean planktonic Foraminifera. *J. Paleontol.* 54:381–399.

Imbrie, J. (1985) A theoretical framework for the Pleistocene ice ages. *J. Geol. Soc.* (Lond.) 142:417–432.

Imbrie, J., and Kipp, N.G. (1971) A new micropaleontological method for a quantitative paleoclimatology: application to a late Pleistocene Caribbean core. In Turekian, K.K. (ed.), *The Late Cenozoic Glacial Ages.* New Haven, Conn.: Yale University Press, pp. 71–81.

Ingle, J.C. (1980) Cenozoic paleobathymetry and depositional history of selected sequences within the southern California continental borderland. Cushman Found. Foram. Res. Spec. Publ. 19, pp. 163–195.

Ingle, J.C., and Keller, G. (1980) Benthic foraminiferal biofaces of the Eastern Pacific margin between 40°S and 32°N. In *Quaternary Depositional Environments of the Pacific Coast.* Pacific Coast Paleogeography Symposium 4. Pacific Section, Soc. Econ. Paleontologists and Mineralogists, Los Angeles, pp. 341–355.

Ingle, J, Keller, G., and Kolpack, R. (1980) Benthic foraminiferal biofacies, sediments and water-masses of the southern Peru–Chile Trench area, Southeastern Pacific Ocean. *Micropaleontology* 26:113–150.

Ivert, H. (1980) Relationship between stratigraphical variation in the morphology of *Gabonella elongata* and geochemical composition of the host sediment. *Cret. Res.* 1:223–233.

Jablonski, D., and Bottjer, D.J. (1983) Soft-bottom epifaunal suspension-feeding assemblages in the late Cretaceous: implications for the evolution of benthic paleocommunities. In Tevesz, M.J.S. and McCall, P.L. (eds.), *Biotic Interactions in Recent and Fossil Benthic Communities.* New York: Plenum, pp. 747–812.

Jablonski, D., Flessa, K.W., and Valentine, J.W. (1985) Biogeography and paleobiology. *Paleobiology* 11:75–90.

Kahn, M.I. (1979) Non-equilibrium oxygen and carbon isotopic fractionation in tests of living planktonic Foraminifera. *Oceanol. Acta* 2:195–208.

Kahn, M.I. (1981) Ecological and palaeoecological implications of the phenotypic variation in three species of living planktonic Foraminifera from the North Eastern Pacific Ocean (50°N, 145°W). *J. Foram. Res.* 11:203–211.

Keigwin, L.D., Bender, M.L., and Kennett, J.P. (1979) Thermal structure of the Deep Pacific Ocean in the early Pliocene. *Science* 205:1386–1388.

Keigwin, L.D., and Boyle, E.A. (1985) Carbon isotopes in deep-sea benthic Foraminifera: precession and changes in low-latitude biomass. In Sundquist, E.T., and Broecker, W.S. (eds.) *The Carbon Cycle and Atmospheric* CO_2*: Natural Variations, Archean to Present,* Washington, D.C.: American Geophysical Union, pp. 319–328.

Keir, R.S., and Hurd, D.C. (1983) The effect of encapsulated fine grain sediment and test morphology on the resistance of planktonic Foraminifera to dissolution. *Mar. Micropaleontol.* 8:193–214.

Keller, G. (1981a) Origin and evolution of the genus *Globigerinoides* in early Miocene of the Northwestern Pacific, DSDP Site 292. *Micropaleontology* 27:293–304.

Keller, G. (1981b) The genus *Globorotalia* in the early Miocene of the equatorial and northwestern Pacific. *J. Foram. Res.* 11:118–132.

Keller, G. (1982) Biochronology and paleoclimatic implications of middle Eocene to Oligocene planktonic foraminiferal faunas. *Mar. Micropaleontol.* 7:463–486.

Keller, G. (1983) Paleoclimatic analyses of middle Eocene through Oligocene planktonic foraminiferal faunas. *Paleogeogr. Paleoclimatol. Paleoecol.* 43:73–94.

Keller, G., Barron, J.A., and Burkle, L.H. (1982) North Pacific late Miocene correlations using microfossils, stable isotopes, percent Ca_2CO_3 and magnetostratigraphy. *Mar. Micropaleontol.* 7:327–359.

Keller, G., D'Hondt, S., and Vallier, T.L. (1983) Multiple microtektite horizons in Upper Eocene marine sediments: no evidence for mass extinctions. *Science* 221:150–152.

Kellogg, D.E. (1976) Character displacement in the radiolarian genus *Eucyrtidium*. *Evolution* 29:736–749.

Kellogg, D.E. (1983) Phenology of morphologic change in radiolarian lineages from deep-sea cores: implications for macroevolution. *Paleobiology* 9:355–362.

Kellogg, T.B. (1977) Paleoclimatology and paleo-oceanography of the Norwegian and Greenland Seas: the last 450,000 years. *Mar. Micropaleontol.* 2:235–249.

Kellogg, T.B. (1980) Paleoclimatology and paleo-oceanography of the Norwegian and Greenland Seas: glacial–interglacial contrasts. *Boreas* 9:115–137.

Kellogg, T.B. (1984a) Late Glacial-Holocene high-frequency climatic changes in deep-sea cores from the Denmark Strait. In Morner, N.A., and Karlen, W. (eds.) *Climatic Changes on a Yearly to Millennial Basis*. Boston: D. Reidel, pp. 123–133.

Kellogg, T.B. (1984b) Paleoclimatic significance of subpolar Foraminifera in high-latitude marine sediments. *Can. J. Earth Sci.* 21:189–193.

Kennet, J.P. (1976) Phenotypic variation in some Recent and late Cenozoic planktonic Foraminifera. In Hedly, R.H., and Adams, C.G. (eds.), *Foraminifera*, Vol. 2. London: Academic Press, pp. 111–170.

Kennett, J.P. (1978) The development of planktonic biogeography in the southern ocean during the Cenozoic. *Mar. Micropaleontol.* 3:301–345.

Kennett, J.P. (1980) Paleoceanographic and biogeographic evolution of the Southern Ocean during the Cenozoic, and Cenozoic microfossil datums. *Paleogeogr. Paleoclimatol. Paleoecol.* 31:123–152.

Kennett, J.P. (1982) *Marine Geology*. Englewood Cliffs, N.J.: Prentice-Hall, 813 pp.

Kennett, J.P. (1983) Paleo-oceanography: global ocean evolution. *Reviews of Geophysics and Space Physics*, U.S. National Report to International Union of Geodesy and Geophysics, Vol. 21, pp. 1258–1274.

Kennett, J.P., Malmgren, B.A., and Srinivasan, M.S. (1982) Evolutionary biology of planktonic Foraminifera. In *Proc. Third North American Paleontological Convention*, Vol. 1 Geol. Sur. Canada, Ottawa, p. 291.

Kennett, J.P., and Srinivasan, M.S. (1980) Surface ultrastructural variation in *Neogloboquadrina pachyderma* (Ehrenberg): phenotypic variation and phylogeny in the late Cenozoic. Cushman Found. Foram. Res. Spec. Publ. 19, pp. 134–162.

Kennett, J.P., and Srinivasan, M.S. (1983) *Neogene Planktonic Foraminifera: A Phylogenetic Atlas*. Stroudsburg, Pa.: Hutchinson Ross, 263 pp.

Kennett, J.P., and Srinivasan, M.S. (1984) Neogene planktonic foraminiferal datum planes of the South Pacific: Mid to equatorial latitudes. In Tsuchi, R., and Ikebe, N. (eds.), *Pacific Neogene Datum Planes: Contribution to Biostratigraphy and Chronology*, Tokyo, Japan: University of Tokyo Press, pp. 11–25.

Kent, D.V., and Gradstein, F.M. (1985) A Cretaceous and Jurassic geochronology. *Bull. Geol. Soc. Amer.* 96:1419–1427.

Killingsley, J.S., Johnson, R.F., and Berger, W.H. (1981) Oxygen and carbon isotopes of individual shells of planktonic Foraminifera from the Ontong-Java Plateau, equatorial Pacific. *Paleogeogr. Paleoclimatol. Paleoecol.* 33:193–204.

King, K (1975) Amino acid composition of the silicified organic matrix in fossil polycystine Radiolaria. *Micropaleontology* 21:215–226.

King, K. (1977) Amino acid survey of Recent calcareous and siliceous deep-sea microfossils. *Micropaleontology* 23:180–202.

King, K., Jr., and Hare. P.E. (1972) Amino acid composition of the test as a taxonomic character for living and fossil planktonic Foraminifera. *Micropaleontology* 18:285–293.

Kitchell, J.A. (1985) Evolutionary paleoecology: recent contributions to evolutionary theory. *Paleobiology* 11:91–104.

Kling, S.A. (1973) Radiolaria from the eastern North Pacific, Deep Sea Drilling Project Leg 18. In Mosich, L.F., and Weser, O.E. (eds.), *Initial Reports of the Deep Sea Drilling Project*, Vol. 18. Washington, D.C.: U.S. Government Printing Office, pp. 617–671.

Kling, S.A. (1978) Radiolaria. In Haq, B.U., and Boersma, A. (eds.), *Introduction to Marine Micropaleontology*, Amsterdam: Elsevier, pp. 203–244.

Kobluk, D.R., and James, N.P. (1979) Cavity dwelling organisms in lower Cambrian patch reefs from southern Labrador. *Lethaia* 12:193–218.

Lamb, J.L., and Miller, T.H. (1984) *Stratigraphic Significance of Uvigerinid Foraminifers in the Western Hemisphere*. Lawrence, Kansas: University of Kansas Paleontological Contributions, Article 66, 99 pp.

Lazarus, David (1983) Speciation in pelagic Protista and its study in the planktonic microfossil record: a review. *Paleobiology* 9:327–340.

Lazarus, D., Scherer, R.P., and Prothero, D.R. (1985) Evolution of the radiolarian species-complex *Pterocanium:* a preliminary survey. *J. Paleontol.* 59:183–220.

Leavesley, A., Bauer, M., McMillen, K.J., and Casey, R.E. (1978) Living shelled microzooplankton (radiolarians, foraminiferans and pteropods) as indicators of oceanographic processes in water over the outer continental shelf of south Texas. *Trans. Gulf Coast Assoc. Geol. Soc.* 28:229–238.

Leckie, R.M., and Webb, P.N. (1985) *Candeina antarctica*, n. sp., and the phylogenetic history and distribution of *Candeina* spp. in the Paleogene–early Neogene of the Southern Ocean. *J. Foram. Res.* 15:65–78.

Leutenegger, S. (1977) Ultrastructure de Foraminifères perforés et imperforés ainsi que de leurs symbiotes. *Cah. Micropaleont.* (CNRS, Paris) 3:1–52.

Leutenegger, S. (1984) Symbiosis in benthic Foraminifera: specificity and host adaptations. *J. Foram. Res.* 14:16–35.

Leutenegger, S., and Hansen, H.J. (1979) Ultrastructural and radiotracer studies of pore function in Foraminifera. *Mar. Biol.* 54:11–16.

Leventer, A, Williams, D.F., and Kennett, J.P. (1982) Dynamics of the Laurentide ice sheet during the last deglaciation: evidence from the Gulf of Mexico. *Earth Planet. Sci. Lett.* 59:11–17.

Levinton, J.S., and Ginzburg, L. (1984) Repeatability of taxon longevity in successive Foraminifera radiations and a theory of random appearance and extinction. *Proc. Nat. Acad. Sci.* 81:5478–5481.

Lipps, J.H. (1966) Wall structure, systematics and phylogeny of Cenozoic planktonic Foraminifera. *J. Paleontol.* 40:1257–1263.

Lipps, J.H. (1973) Test structure in Foraminifera. *Ann. Rev. Microbiol.* 27:471–488.

Lipps, J.H. (1975) Feeding strategies of test function in Foraminifera. Abstracts, First International Conference on Benthic Foraminifera of Continental Margins. Dalhousie University, Halifax, p. 26.

Lipps. J.H. (1979) The ecology and paleoecology of planktic Foraminifera. In *Foraminiferal Ecology and Paleoecology,* Society of Economic Paleontologists and Mineralogists. Short Course No. 6, Houston, Texas, pp. 62–104.

Lipps, J.H. (1981) What, if anything, is micropaleontology. *Paleobiology* 7:167–199.

Lipps, J.H. (1982) Biology/paleobiology of Foraminifera. In Broadhead, T.W., (ed.), *Foraminifera: Notes for a Short Course,* University of Tennessee Dept. of Geological Sciences Studies in Geology No. 6, pp. 1–21, University of Tennessee, Knoxville.

Lipps, J.H. (1983) Biotic interactions in benthic Foraminifera. In Tevesz, M.J.S. and McCall, P.L. (eds.), *Biotic Interactions in Recent and Fossil Benthic Communities.* New York: Plenum, pp. 331–376.

Lipps, J.H., and Valentine, J.W. (1970) The role of Foraminifera in the trophic structure of marine communities. *Lethaia* 3:279–286.

Loeblich, A.R., and Tappan, H. (1964) *Sarcodina, Chiefly Thecamoebians and Foraminiferida. Treatise on Invertebrate Paleontology,* Part C, *Protista,* Vol. 2, Lawrence, Kansas: Geol. Soc. Amer., 900 pp.

Loeblich, A.R. Jr., and Tappan, H. (1981) Suprageneric revisions of some calcareous Foraminiferida. *J. Foram. Res.* 11:159–164.

Loeblich, A.R., and Tappan, H. (1982) Classification of the Foraminiferida. In Broadhead, T.W. (ed.), *Foraminifera: Notes for a Short Course.* Univ. Tennessee, Dept. of Geol. Sciences Stud. Geol. 6, Knoxville: University of Tennessee, pp. 22–37.

Loeblich, A.R., and Tappan, H. (1984) Suprageneric classification of the Foraminiferida (Protozoa). *Micropaleontology* 30:1–70.

Loeblich, A.R., Jr., and Tappan, H. (1985) Some new and redefined genera and families of agglutinated Foraminifera II. *J. Foram. Res.* 15:175–217.

Lohmann, G.P. (1978) Abyssal benthonic Foraminifera as hydrographic indicators in the western south Atlantic Ocean. *J. Foram. Res.* 8:6–34.

Lohmann, G.P. (1983) Eigenshape analysis of microfossils: a general morphometric procedure for describing changes in shape. *Math. Geol.* 15:659–672.

Lohmann, G.P., and Malmgren, B.A. (1983) Equatorward migration of *Globorotalia truncatulinoides* ecophenotypes through the late Pleistocene: gradual evolution or ocean change? *Paleobiology* 9:414–421.

Loubere, P. (1981) Oceanographic parameters reflected in the seabed distribution of planktonic Foraminifera from the North Atlantic and Mediterranean Sea. *J. Foram. Res.* 11:137–158.

Loubere, P. (1982) The western Mediterranean during the last glacial: attacking a no-analog problem. *Mar. Micropaleontol.* 7:311–326.

Loutit, T.S., and Kennett, J.P. (1979) Application of carbon isotope stratigraphy to late Miocene shallow marine sediments, New Zealand. *Science* 204:1196–1199.

Loutit, T.S., Kennett, J.P., and Savin, S.M. (1983) Miocene equatorial and south-west Pacific paleoceanography from stable isotope evidence. *Mar. Micropaleontol.* 8:215–234.

Loutit, T.S., and Kiegwin, L.D., Jr., (1982) Stable isotopic evidence for Miocene sea-level fall in the Mediterranean region. *Nature* 300:163–166.

Lowenstam, H.A. (1981) Minerals formed by organisms. *Science* 211:1126–1131.

Lowenstam, H.A., and Weiner, S. (1983) Mineralization by organisms and the evolution of biomineralization. In Westbroek, P., and de Jonge, E.W. (eds.), *Biomineralization and Biological Metal Accumulation.* Reidel, pp. 191–203.

Lutze, G.F., and Coubourn, W.T. (1984) Recent benthic Foraminifera from the continental margin of Northwest Africa: community structure and distribution. *Mar. Micropaleontol.* 8:361–402.

Luz, B., and Reiss, Z. (1983) Stable carbon isotopes in Quaternary Foraminifera from the Gulf of Aqaba (Elat), Red Sea. *Utrecht Micropaleontol. Bull.* 30:129–140.

Malmgren, B.A. (1983) Ranking of dissolution susceptibility of planktonic Foraminifera at high latitudes of the South Atlantic Ocean. *Mar. Micropaleontol.* 8:183–192.

Malmgren, B.A., Berggren, W.A., and Lohmann, G.P. (1983) Evidence for punctuated gradualism in the late Neogene *Globorotalia tumida* lineage of planktonic Foraminifera. Paleobiology 9:377–389.

Malmgren, B.A., Berggren, W.A., and Lohmann, G.P. (1984) Species formation through punctuated gradualism in planktonic Foraminifera. *Science* 225:317–319.

Malmgren, B.A., and Kennett, J.P. (1976) Biometric analysis of phenotypic variation in recent *Globigerina bulloides* d'Orbigny in the Southern Indian Ocean. *Mar. Micropaleontol.* 1:3–25.

Malmgren, B.A., and Kennett, J.P. (1977) Biometric differentiation between recent *Globigerina bulloides* and *Globigerina falconensis* in the Southern Indian Ocean. *J. Foram. Res.* 7:130–148.

Malmgren, B.A., and Kennett, J.P. (1978) Late Quaternary paleoclimatic application of mean size variations in *Globigerina bulloides* d'Orbigny in the Southern Indian Ocean. *J. Paleontol.* 52:1195–1207.

Malmgren, B.A., and Kennett, J.P. (1981) Phyletic

gradualism in a late Cenozoic planktonic foraminiferal lineage; DSDP Site 284, Southwest Pacific. *Paleobiology* 7:230–240.

Malmgren, B.A., and Kennett, J.P. (1982) The potential of morphometrically based phylo-zonation: application of a late Cenozoic planktonic foraminiferal lineage *Mar. Micropaleontol.* 7:285–296.

Malmgren, B.A., and Kennett, J.P. (1983) Phyletic gradualism in the *Globorotalia inflata* lineage vindicated. *Paleobiology* 9:427–428.

Marszalek, D.S., Wright, R.C., and Hay, W.W. (1969) Function of the test in Foraminifera. *Trans. Gulf Coast Assoc. Geol. Soc.* 19:341–352.

Matthews, R.K., and Poore, R.Z. (1980) Tertiary O^{18} record and glacioeustatic sea-level fluctuations. *Geology* 8:501–504.

McGowran, B. (1968) Reclassification of early Tertiary *Globorotalia. Micropaleontology* 14:61–80.

McGowan, J.A. (1974) The nature of oceanic ecosystems. In Miller, C.B. (ed.), *The Biology of the Oceanic Pacific.* Corvallis, Oreg.: Oregon State University Press, pp. 9–28.

McIntyre, A., Kipp, N.G., Bé, A.W.H., Crowley, T., Kellogg, T., Gardner, J.V., Prell, W., and Ruddiman, W.F. (1976) Glacial North Atlantic 18,000 years ago: a CLIMAP reconstruction. *Geol. Soc. Amer. Mem.* 45:43–76.

McKinney, M.L. (1985) Distinguishing patterns of evolution from patterns of deposition. *J. Paleontol.* 59:561–567.

McMillen, K.J. (1979) Radiolarian ratios and the Pleistocene-Holocene boundary. In *Trans. Gulf Coast Assoc. Geol. Soc.* 29:298–301.

McMillen, K.J., and Casey, R.E. (1978) Distribution of living polycystine radiolarians in the Gulf of Mexico and Caribbean Sea, and comparison with the sedimentary record. *Mar. Micropaleontol.* 3:121–145.

Meyen, F.J.F. (1834) Uber das Leuchten des Meeres und Beschreibung einger Polypen und anderer niederer Theire; In *Beitrage zur Zoologie, gesammelt auf einer Reise um die Erde:* Nova Acta Acad. Caesar. Leop. Carol, 16(8) suppl. 1, abh. 5:125–216.

Milam, R.W., and Anderson, J.B. (1981) Distribution and ecology of recent benthic Foraminifera of the Adelie-George V continental shelf and slope, Antarctica. *Mar. Micropaleontol.* 6:297–325.

Miller, A.A.L., Scott, D.B., and Medioli, F.S. (1982) *Elphidium excavatum* (Terquem): ecophenotypic versus subspecific variation. *J. Foram. Res.* 12:116–144.

Miller, K.G. (1982) Cenozoic benthic Foraminifera: case histories of paleoceanographic and sea-level changes. In Broadhead, T.W. (ed.), *Foraminifera: Notes for a Short Course.* Univ. Tenn. Dept. of Geol. Sci., Studies in Geology 6, pp. 107–126.

Miller, K.G. (1983) Eocene–Oligocene paleoceanography of the deep Bay of Biscay: benthic foraminiferal evidence. *Mar. Micropaleontol.* 7:403–440.

Miller, K.G., Aubry, M.P., Kahn, M.J., Melillo, A.J., Kent, D.V., and Berggren, W.A. (1985) Oligocene–Miocene biostratigraphy of the Western North Atlantic. *Geology* 13:257–261.

Miller, K.G., Curry, W.B., and Ostermann, D.R. (1984) Late Paleogene (Eocene to Oligocene) benthic foraminiferal oceanography of the Goban Spur region, Deep Sea Drilling Project Leg 80. In de Graciansky, P.C. and Pag, C.W. et al. (eds.), *Initial Reports of the Deep Sea Drilling Project,* Washington, D.C.: Vol. LXXX, U.S. Government Printing Office, pp. 505–538.

Miller, K.G., and Fairbanks, R.G. (1985) Oligocene to Miocene carbon isotope cycles and abyssal circulation changes. In Sundquist, E.T., and Broecker, W.W. (eds.), *The Carbon Cycle and Atmospheric CO_2: Natural Variations Archean to Present.* Washington, D.C.: Amer. Geophys. Union, pp. 469–486.

Miller, K.G., Gradstein, F.M., and Berggren, W.A. (1982) Late Cretaceous to early Tertiary agglutinated benthic Foraminifera in the Labrador Sea. *Micropaleontology* 28:1–30.

Miller, K.G., and Lohmann, G.P. (1982) Environmental distribution of recent benthic Foraminifera on the north-east United States continental slope. *Geol. Soc. Amer. Bull.* 93:200–206.

Mix, A.C., and Fairbanks, R.G. (1985) North Atlantic surface-ocean control of Pleistocene deep-ocean circulation. *Earth Planet. Sci. Lett.* 73:231–243.

Molfino, B., Kipp, N.G., and Morley, J.J. (1983) Comparison of foraminiferal, coccolithophorid, and radiolarian paleotemperature equations: assemblage coherency and estimate concordancy. *Quaternary Res.* 7:279–313.

Monger, J.W.H., and Ross, C.A. (1984) Upper Paleozoic volcano-sedimentary assemblages of the western North American Cordillera. In Nassichock, W.W. (ed.) *Neuvieme Congres International de Stratigraphie et de Geologie du Carbonifere; Compte Rendu,* Vol. 3, Paleogeography and Paleotectonics, Carbondale: Southern Illinois University Press, pp. 219–228.

Moore, T.C. (1969) Radiolaria: change in skeletal weight and resistance to solution. *Bull. Geol. Soc. Amer.* 80:2103–2108.

Moore, T.C. (1978) The distribution of radiolarian assemblages in the modern and ice-age Pacific. *Mar. Micropaleontol.* 3:229–266.

Moore, T.C., Jr., Burckle, L.H., Geitzenauer, K., Lue, B., Molina-Cruz, A., Robertson, J.H., Sachs, H., Sancetta, C. (1980) The reconstruction of sea surface temperatures in the Pacific Ocean of 18,000 B.P. *Mar. Micropaleontol.* 5:215–248.

Moorkens, T.L. (1976) Palokologische bedeuteng einiger vergesellschaft-ungen on sandschaligen Foraminiferin ausdem NW europaischen Alttertiar und ihre Beziehung zu Muttergestein-en. *Erdol. Kohle* 75:77–95.

Muller, J. (1858) Uber die Thalassicollen, Polycystinen und Acanthometren des Mittelmeeres. *Abr. Pruess. Akad. Wiss. Jahg.* pp. 1–62.

Mullineaux, L.S., and Westberg-Smith, M.J. (1986) Radiolarians as paleoceanographic indicators in the Miocene Monterey Formation, upper Newport Bay, CA. *Micropaleontology* 32:48–71.

Murray, J.W. (1973) *Distribution and ecology of living benthic foraminiferids.* London: Heinemann, 476 pp.

Murray, J.W. (1982) Benthic Foraminifera: the validity of living, dead or total assemblages for the interpretation of paleoecology. *Micropaleontol.* 1:137–140.

Murray, J.W. (1984a) Benthic foraminifera: some relationship between ecological observations and palaeoecological interpretations. *Benthos '83: 2nd International Symposium on Benthic Foraminifera*, pp. 465–469.

Murray, J.W. (1984b) Paleogene and Neogene benthic foraminifers from Rockall Plateau. In Roberts, D.G. and Schnitker, D. (eds.), *Initial Reports of the Deep-Sea Drilling Project*, Vol. LXXXI. Washington, D.C.: U.S. Government Printing Office, pp. 503–534.

Murray, J.W. (1985) Recent Foraminifera from the North Sea (Forties and Ekofisk areas) and the continental shelf west of Scotland. *J. Micropaleontol.* 4:117–126.

Murray, J.W., Sturrock, S., and Weston, J. (1982) Suspended load transport of foraminiferal tests in a tide- and wave-swept sea. *J. Foram. Res.* 12:51–65.

Nazarov, B.B., and Ormiston, A.R. (1985) Radiolaria from the Late Paleozoic of the southern Urals, USSR and West Texas, USA. *Micropaleontology* 31:1–54.

Nelson, C.O. (1986) Radiolarian biostratigraphic and paleoceanographic studies of Monterey-like rocks of the Humboldt Basin, northern California. Ph.D. thesis, Rice University, Houston, 198 pp.

Nigrini, C. (1971) Radiolarian zones for the Quaternary of the equatorial Pacific Ocean. In Funnell, R.M. and Riedel, W.R. (eds.) *The Micropalaeontology of Oceans*. New York: Cambridge University Press, pp. 443–461.

Nigrini, C., and Lombardi, G. (1984) *A Guide to Miocene Radiolaria*. Cushman Found. Foram. Res. Spec. Publ. 22, 422 pp.

Nigrini, C., and Moore, T.C. (1979) *A Guide to Modern Radiolaria*. Cushman Found. Foram. Res. Spec. Publ. 16, 342 pp., Washington, D.C.

Nitecki, M.H., ed. (1984) *Extinctions*. Chicago: University of Chicago Press, 389 pp.

Okimura, Y., Ishii, K., and Ross, C.A. (1985) Biostratigraphical significance and faunal provinces of Tethyan late Permian smaller Foraminifera. In Nakazawa, K., and Dickins, J.M. (eds.), *The Tethys: Her Paleogeography and Paleobiogeography from Paleozoic to Mesozoic*. Tokai University Press, pp. 115–138.

Olsson, R.K. (1982) Cenozoic planktonic Foraminifera: a paleobiogeographic summary. In Broadhead, T.W. (ed.), *Foraminifera: Notes for a Short Course*. Univ. Tenn. Dept. Geol. Sci., Studies in Geol. 6, Knoxville: University of Tennessee, pp. 127–147.

Opdyke, N.D., Glass, B., Hays, J.D. and Foster, J. (1966) Paleomagnetic study of Antarctic deep-sea cores. *Science* 154:349–357.

Osterman, L.E., and Kellogg, T.B. (1979) Recent benthic foraminiferal distributions from the Ross Sea, Antarctica: reaction to ecologic and oceanographic conditions. *J. Foram. Res.* 9:250–269.

Ozawa, Tomowo (1975) Evolution of *Lepidolina multiseptata* (Permian, Foraminifera) in East Asia. *Mem. Fac. Sci. Kyushu Univ., Ser. D., Geol.* 23:117–164.

Painter, P.K., and Spencer, R.S. (1984) A statistical analysis of variants of *Elphidium excavatum* and their ecological control in Southern Chesapeake Bay, Virginia. *J. Foram. Res.* 14:120–128.

Parker, F.L. (1962) Planktonic foraminiferal species in Pacific sediments. *Micropaleontology* 8:219–254.

Parker, F.L. (1967) Late Tertiary biostratigraphy (planktonic Foraminifera) of tropical Indo-Pacific deep-sea cores. *Bull. Amer. Paleontol.* 52(235):115–187.

Perch-Nielsen, K., McKenzie, J., and Quziang, H. (1982) Biostratigraphy and isotope stratigraphy and the "catastrophic" extinction of calcareous nannoplankton at the Cretaceous/Tertiary boundary. In Silver, L.T., and Schultz, P.H. *Geological Implications of Impacts of Large Comets and Asteroids on the Earth*. Boulder, Colorado: Geol. Soc. Amer. Spec. Pap. 190, pp. 353–372.

Perez-Guzman, A.M., and Casey, R.E. (1986) Paleoceanographic reconstructions from radiolarian-bearing Baja California and adjacent sections. In Casey, R.E., and Barron, J.A. (eds.), *Siliceous Microfossil and Microplankton Studies of the Monterey Formation and Modern Analogs*. Soc. Econ. Paleontol. Mineral., Pacific Section, Los Angeles, pp. 55–68.

Pessagno, E.A. (1977) Radiolaria in Mesozoic stratigraphy. In Ramsay, A.T.S. (ed.), *Oceanic Micropalaeontology*. New York: Academic Press, pp. 913–950.

Peterson, L.C. (1984) Recent abyssal benthic foraminiferal biofacies of the eastern equatorial Indian Ocean. *Mar. Micropaleontol.* 8:479–520.

Peterson, L.C., and Prell, W.L. (1985) Carbonate preservation and rates of climatic change: an 800 kyr record from the Indian Ocean. In Sundquist, E.T., and Broecker, W.S. (eds.), *The Carbon Cycle and Atmospheric* CO_2*: Natural Variations, Archean to Present*. Washington, D.C.: Amer. Geophys. Union, pp. 251–270.

Petrushevskaya, M.G. (1971) On the natural system of polycystine Radiolaria (Class Sarcodina). In Farinacci, A. (ed.), *Proc. II Planktonic Conf.*, 1970, Romae, pp. 981–992.

Petrushevskaya, M.G. (1975) Cenozoic radiolarians of the Antarctic, Leg 29, DSDP. In Kennett, J.P., Houtz, R.E., et al. (eds.), *Initial Reports of the Deep Sea Drilling Project*, Vol. 29. Washington, D.C.: U.S. Government Printing Office, pp. 541–675.

Petrushevskaya, M.G., and Kozlova, G.E. (1972) Radiolaria. In J.D. Hays et al, Initial Reports of the Deep Sea Drilling Project, Vol. 14. Washington, D.C.: U.S. Government Printing Office, pp. 630–648.

Pisias, N.G. (1979) Model for paleoceanographic reconstructions of the California Current during the last 8000 years. *Quaternary Res.* 11:373–386.

Poag, C.W. (1978) Paired foraminiferal ecophenotypes in Gulf Coast estuaries: ecological and paleoecological implications. *Trans. Gulf Coast Assoc. Geol. Soc.* 28:395–421.

Poore, R.Z., and Matthews, R.K. (1984a) Oxygen isotope ranking of late Eocene and Oligocene planktonic foraminifers: implications for Oligocene sea-surface temperatures and global ice-volume. *Mar. Micropaleontol.* 9:111–134.

Poore, R.Z., and Matthews, R.K. (1984b) Late Eocene–Oligocene oxygen and carbon-isotope record from south Atlantic ocean, Deep Sea Drilling Project Site

522. In Hsu, K.J., La Brecque, J.L. et al. (eds.), *Initial Reports of the Deep Sea Drilling Project,* Vol. 73. Washington, D.C.: U.S.. Government Printing Office, pp. 725–735.

Prell, W.L., and Curry, W.B. (1981) Faunal and isotopic indices of monsoonal upwelling, Western Arabian Sea. *Oceanol. Acta* 4:91–98.

Prell, W.L., Hutson, W.H., and Williams, D.F. (1979) The subtropical convergence and late Quaternary circulation in the southern Indian Ocean, *Mar. Micropaleontol.* 4:225–234.

Prell, W.L., et al. (1980) Surface circulation of the Indian Ocean during the last glacial maximum, approximately 18,000 YBP. *Quaternary Res.* 14:309–336.

Price, M.V. (1980) On the significance of test form in benthic salt-marsh Foraminifera. *J. Foram. Res.* 10:129–135.

Reid, J.L., Brinton, E., Fleminger, A., Venrich, E.L., and McGowan, J.A. (1978) Oceanic circulation and marine life. In Charnock, H., and Deacon, G. (eds.), *Advances in Oceanography.* New York: Plenum, pp. 65–130.

Resig, J.M. (1974) Recent Foraminifera from a landlocked Hawaiian lake. *J. Foram. Res.* 4:69–76.

Reyment, R.A. (1982) Phenotypic evolution in a Cretaceous foraminifer. *Evolution* 36:1182–1199.

Reyment, R.A. (1983) Phenotypic evolution in microfossils. *Evol. Biol.* 16:209–254.

Reynolds, L., and Thunell, R.C. (1985) Seasonal succession of planktonic Foraminifera in the subpolar North Pacific. *J. Foram. Res.* 15:282–301.

Reynolds, R.A. (1978) Cosmopolitan biozonation for late Cenozoic radiolarians and paleoceanography from Deep Sea Drilling Project Core 77B of Leg 9. *Trans. Gulf Coast Assoc. Geol. Soc.* 28:423–431.

Reynolds, R.A. (1980) Radiolarians from the western North Pacific, Leg 57, Deep Sea Drilling Project. In Lee, M., and Stout, L.N. (eds.), *Initial Reports of the Deep Sea Drilling Project,* Vol. 57. Washington, D.C.: U.S. Government Printing Office, pp. 735–769.

Riding, R., and Brasier, M. (1975) Earliest calcareous Foraminifera. *Nature* 257:208–210.

Riedel, W.R. (1957) Radiolaria: a preliminary stratigraphy. *Rept. Swed. Deep Sea Expedition* 6(3):59–96.

Riedel, W.R. (1967) Subclass Radiolaria. In Harland, W.B., et al. (eds.) *The Fossil Record.* London: Geological Society of London, pp. 291–298.

Riedel, W.R. (1971) Systematic classification of polycystine Radiolaria. In Funnell, B.M. and Riedel, W.R. (eds.), *The Micropalaeontology of Oceans.* New York: Cambridge University Press, pp. 649–661.

Riedel, W.R., and Sanfilippo, A. (1970) Radiolaria, Leg 4, Deep Sea Drilling Project. In Bader, R.G., et al. (eds.), *Initial Reports of the Deep Sea Drilling Project.* Vol. 4. Washington, D.C.: U.S. Government Printing Office, pp. 503–575.

Riedel, W.R., and Sanfilippo, A. (1971) Cenozoic Radiolaria from the western tropical Pacific, Leg 7, In Winterer, E.L., et al. (eds.)., *Initial Reports of the Deep Sea Drilling Project,* Vol. 7. Washington, D.C.: U.S. Government Printing Office, pp. 1529–1672.

Riedel, W.R., and Sanfilippo, A. (1977) Cainozoic Radiolaria. In Ramsay, A.T.S. (ed.), *Oceanic Micropaleontology.* Vol. 2. New York: Academic Press, pp. 847–912.

Riedel, W.R., and Sanfilippo, A. (1978) Stratigraphy and evolution of tropical Cenozoic radiolarians. *Micropaleontology* 23:61–96.

Riedel, W.R., and Sanfilippo, A. (1981) Evolution and diversity of form in radiolaria. In Simpson, T.L., and Volcani, B.E. (eds.), *Silicon and Siliceous Structures in Biological Systems.* New York: Springer-Verlag, pp. 323–346.

Riedel, W.R., Westberg-Smith, M.J., and Budai, A. (1985) Late Neogene Mediterranean Radiolaria from the point of view of paleoenvironment. In Stanley, D.J., and Wetzel, F. (eds.), *Geologic Evolution of the Mediterranean Basin.* New York: Plenum, pp. 487–523.

Romine, K. (1985) Radiolarian biogeography and paleoceanography of the North Pacific at 8 Ma. In Kennett, J.P. (ed.) *The Miocene Ocean: Paleoceanography and Biogeography.* Boulder: Geological Society of America Memoir 163, pp. 237–272.

Romine, K., and Ombari, G. (1985) Evolution of Pacific circulation in the Miocene: radiolarian evidence from DSDP Site 289. In Kennett, J.P. (ed.) *The Miocene Ocean: Paleoceanography and Biogeography,* Boulder: Geological Society of America Memoir 163, pp. 273–290.

Ross, C.A. (1967) Development of fusulinid (Foraminiferida) faunal realms. *J. Paleontol.* 41:1341–1354.

Ross, C.A. (1971) Palaeoecology of late Pennsylvanian fusulinids (Foraminiferida), Texas. *C. R. Congr. Intern. Strat. Geol. Carbonif.* (Sheffield 1967) 4:1429–1440.

Ross, C.A. (1972) Paleobiological analysis of fusulinacean (Foraminiferida) shell morphology. *J. Paleontol.* 46:719–728.

Ross, C.A. (1982a) Paleobiology of fusulinaceans. *Proc. Third North American Paleontological Convention,* Vol. 2. Geol. Sur. Canada, Ottawa, pp. 441–445.

Ross, C.A. (1982b) Paleozoic Foraminifera-fusulinids. In Broadhead, T..W. (ed.), *Foraminifer: Notes for a Short Course* Univ. Tenn. Dept. Geol. Sci., Studies in Geology 6, p. 163–176.

Ross, C.A. (1984) Fusulinacean biostratigraphy near the Carboniferous–Permian boundary in North America. In Sutherland, P.K., and Monger, W.L. (eds.) *Neuvieme Congres International de Strahgraphie et de Geologie du Carbonifere: Compte Rendu,* Vol. 2, *Biostratigraphy,* Carbondale: Southern Illinois University Press, pp. 535–542.

Ross, C.A., and Ross, J.R.P. (1983) Late Paleozoic accreted terranes of western North America. In Stevens, C.H. (ed.), Pre-Jurassic rocks in western North America suspect terranes. Soc. Econ. Paleontol. Mineral., Pacific Section, Los Angeles, pp. 7–22.

Ross, C.A., and Ross, J.R.P. (1985a) Carboniferous and early Permian biogeography. *Geology* 13:27–30.

Ross, C.A., and Ross, J.R.P. (1985b) Late Paleozoic depositional sequences are synchronous and worldwide. *Geology* 13:194–197.

Röttger, R., and Hallock, P. (1982) Shape trends in *Heterostegina depressa* (Protozoa, Foraminiferida). *J. Foram. Res.* 12:197–204.

Ruddiman, W.F., and McIntyre, A. (1984) Ice-age thermal resonse and climatic role of the surface Atlantic Ocean, 40°N to 63°N. *Geol. Soc. Amer. Bul.* 95:381–396.

Saito, T., Thompson, R., and Breger, D. (1976) Skeletal ultramicrostructure of some elongate-chambered planktonic Foraminifera and related species. In Takayanagi, Y., and Saito, T. (eds.) *Progr.* Micropaleontology. New York: Amer. Mus. Nat. Hist., Micropaleontology Press, pp. 278–304.

Sanfilippo, A., Burckle, L.H., Martini, E., and Riedel, W.R. (1973) Radiolarians, diatoms, silicoflagellates and calcareous nannofossils in the Mediterranean Neogene. *Micropaleontology* 19:209–234.

Sanfilippo, A., and Riedel, W.R. (1985) Cretaceous Radiolaria. In Bolli, H.M., Saunders, J.B., and Perch-Nielsen, K. (eds.), *Plankton Stratigraphy.* Cambridge: Cambridge University Press, p. 573–630.

Sanfilippo, A., Westberg-Smith, J., and Riedel, W.R. (1985) Cenozoic Radiolaria. In Bolli, H.M., Saunders, J.B., and Perch-Nielsen, (eds.), *Plankton Stratigraphy.* Cambridge: Cambridge University Press, 631–712.

Savin, S.M., Abel, L., Barrera, E., Hodell, D., Kennett, J.P., Murphy, M., Keller, G., Killingley, J., and Vincent, E. (1985) The evolution of Miocene surface and near-surface marine temperatures: oxygen isotopic evidence. *Geol. Soc. Amer. Mem.* 163:49–82.

Savin, S.M., Douglas, R.G., Keller, G., Killingley, J.S., Shaughnessy, L., Sommer, M.A., Vincent, E., and Woodruff, F. (1981) Miocene benthic foraminiferal isotope records: a sysnthesis. *Mar. Micropaleontol.* 6:423–450.

Schaaf, A. (1984) Les radiolaries du Crétacé Inferieur et Moyen: biologie et systématique. *Sci. Geol. Mem.* 75:1–189.

Schaaf, A. (1985) Un nouveau canevas biochronologiques de Crétacé Inferieur et Moyen: les biozones à radiolaires. *Sci. Geol. Bul.* 38:227–269.

Schafer, C.T., and Cole, F.E., (1982) Living benthic Foraminifera distributions of the contintental slope and rise east of Newfoundland, Canada. *Geol. Soc. Amer. Bull.* 93:207–217.

Schafer, C.T., Cole, F.E., and Carter, L. (1981) Bathyal zone benthic foraminiferal genera off northeast Newfoundland. *J. Foram. Res.* 11:296–313.

Schallreuter, R.E.L. (1983) Calcareous foraminifers from the Ordovician of Baltoscandia. *J. Micropaleontol.* 2:1–6.

Schindel, D.E. (1980) Microstratigraphic sampling and the limits of paleontologic resolution. *Paleobiology* 6:408–426.

Schnitker, D. (1979) The deep waters of the Western North Atlantic during the past 24,000 years, and the re-initiation of the Western Boundary Undercurrent. *Mar. Micropaleontol.* 4:235–264.

Schnitker, D. (1980) Quaternary deep-sea benthic foraminifers and bottom water-masses. *Ann. Rev. Earth Plant. Sci.* 8:343–370.

Schnitker, D. (1982) Climatic variability and deep ocean circulation: evidence from the North Atlantic. *Paleogeogr. Paleoclimatol. Paleoecol.* 40:213–234.

Schnitker, D. (1984) High resolution records of benthic foraminifers in the late Neogene of the Northeastern Atlantic. In Roberts, D.G., and Schnitker, D. (eds.), *Initial Reports of the Deep Sea Drilling Project,* Vol. 81. Washington, D.C.: U.S. Government Printing Office, pp. 611–622.

Schopf, T.J.M. (1980) *Paleoceanography.* Cambridge, Mass.: Harvard University Press, 350 pp.

Schwab, D., and Plapp, R. (1983) Quantitative chemical analysis of the shell of the monthalamous foraminifer *Allogromia laticollaris* Arnold. *J. Foram. Res.* 13:69–71.

Scott, D.B. (1976) Quantitative studies of marsh foraminiferal patterns in Southern California and their application to Holocene stratigraphic problems. *First International Symposium on Benthonic Foraminifer of Continental Margins, Part A, Ecology and Biology.* Maritime Sediments Spec. Publ. 1, pp. 153–170.

Scott, D.B., Gradstein, F., Schaffer, C., Miller, A., and Williamson, M. (1983) The recent as a key to the past: does it apply to agglutinated foraminiferal assemblages? In Verdenius, V.G., van Hinte, J.E., and Fortuin, A.R. (eds.) *Proceedings of the First Workshop on Arenaceous Foraminifera 7-9, September 1981,* Trondheim, Norway, pp. 147–158.

Scott, D.B., and Medioli, F.S. (1980) Quantitative Studies of Marsh Foraminiferal Distributions in Nova Scotia. Implications for Sea Level Studies. *Cushman Found. Foram. Res. Spec. Publ.* 17, 58 pp.

Scott, D.B., Schafer, C.T., and Medioli, F.S. (1980) Eastern Canadian estuarine Foraminifera: a framework for comparison. *J. Foram. Res.* 10:205–234.

Scott, G.H. (1973a) Peripheral structures in chambers of *Globorotalia scitula praescitula* and some descendants. *Rev. Espanola Micropaleontol.* 5:235–246.

Scott, G.H. (1973b) Ontogeny and shape in *Globorotalia menardii. J. Foram. Res.* 3:142–146.

Scott, G.H. (1974) Biometry of the foraminiferal shell. In Hedley, R.H., and Adams, C.G. (eds.), *Foraminifera,* Vol. 1. New York: Academic Press, pp. 55–152.

Scott, G.H. (1976) Estimation of ancestry in planktonic Foraminifera: *Globoquardrina deniscens.* New Zealand *J. Geol. Geophys* 19:311–325.

Scott, G.H. (1979) The Late Miocene to early Pliocene history of the *Globorotalia miozea* plexus from Blind River, New Zealand, *Mar. Micropaleontol.* 4:341–361.

Scott, G.H. (1982) Tempo and stratigraphic record of speciation in *Globorotalia puncticulata. J. Foram. Res.* 12:1–12.

Scott, G.H. (1983) Gradual evolution in a planktonic foraminiferal lineage reconsidered. Divergence and phyletic transformations in the history of the *Globorotalia inflata* lineage. *Paleobiology* 9:422–426.

Sen Gupta, B.K., and Strickert, D.P. (1982) Living benthic Foraminifera of the Florida–Hatteras slope: distribution trends and anomalies. *Geol. Soc. Amer. Bull.* 93:218–224.

Shackleton, N.J., Corfield, R.M., and Hall, M.A. (1985) Stable isotope data and the ontogeny of Paleocene planktonic Foraminifera. *J. Foram. Res.* 15:321–336.

Shackleton, N.J., Hall, M.A., and Boersma, A. (1984) Oxygen and carbon isotope data from Leg 74 foraminifers. In Moore, T.C., Jr., and Rabinowitz, P.D. (eds.), *Initial Reports of the Deep Sea Drilling Project*, Vol. 74. Washington, D.C.: U.S. Government Printing Office, pp. 599–612.

Shackleton, N.J., Imbrie, J. and Hall, M.A. (1983) Oxygen and carbon isotope record of east Pacific core V19-30: implications for the formation of deep water in the late Pleistocene North Atlantic. *Earth Plane. Sci. Lett.* 65:233–244.

Shackleton, N.J., and Pisias, N.G. (1985) Atmospheric carbon dioxide, orbital forcing, and climate. In Sundquist, E.T., and Broecker, W.S. (eds.) *The Carbon Cycle and Atmospheric CO_2: Natural Variation Archean to Present.* Washington, D.C.: Amer. Geophys. Union, pp. 303–318.

Shackleton, N.J., and Vincent, E. (1978) Oxygen and carbon isotope studies in Recent foraminifers from the southwest Indian Ocean. *Mar. Micropaleontol.* 3:3–13.

Silver, L.T., and Schultz, P.H., eds. (1982) *Geological Implications of Impacts of Large Asteroids and Comets on the Earth.* Boulder, Colorado: Geol. Soc. Amer. Spec. Paper 190, 527 pp.

Sloan, J.R. (1981) Radiolarians of the North Philippine Sea: their biostratigraphy, preservation, and paleoecology. Ph.D. thesis, University of California, Davis, 154 pp.

Smit, J. (1982) Extinction and evolution of planktonic Foraminifera at the Cretaceous/Tertiary boundary after a major impact. In Silver, L.T., and Schultz, P.H. (eds.), *Geological Implications of Impacts of Large Comets and Asteroids on the Earth.* Boulder, Colorado: Geol. Soc. Amer. Spec. Pap. 190, pp. 329–352.

Snyder, S.W., Muller, C., and Miller, K.G. (1984) Eocene–Oligocene boundary: biostratigraphic recognition and gradual paleoceanographic change at DSDP site 549. *Geology*, 12:112–115.

Spaw, J.M. (1979) Vertical distribution, ecology and preservation of recent polycystine Radiolaria of the north Atlantic Ocean (southern Sargaso Sea region). Ph.D. thesis, Rice University, Houston, 185 pp.

Srinivasan, M.S., and Kennett, J.P. (1974) Secondary calcification of the planktonic foraminifer *Neogloboquadrina pachyderma* as a climatic index. *Science* 186:630–632.

Srinivasan, M.S., and Kennett, J.P. (1975) The status of *Bolliella, Beella, Protentella* and related planktonic Forminifera based on surface ultrastructure. *J. Foram. Res.* 5:155–165.

Srinivasan, M.S., and Kennett, J.P. (1981a) A review of Neogene planktonic foraminiferal biostratigraphy: applications in the equatorial and South Pacific. Tulsa, Oklahoma: The Society of Economic Paleontologists and Mineralogists Spec. Publ. 32, pp. 395–432.

Srinivasan, M.S., and Kennett, J.P. (1981b) Neogene planktonic foraminiferal biostratigraphy and evolution: equatorial to subantarctic, South Pacific. *Mar. Micropaleontol.* 6:499–533.

Stanley, S.M. (1982) Macroevolution and the fossil record. *Evolution* 36:460–473.

Stanley, S.M. (1984a) Temperature and biotic crisis in the marine realm. *Geology* 12:205–208.

Stanley, S.M. (1984b) Marine mass extinctions: a dominant role for temperatures. In Nitecki, M.H. (ed.), *Extinctions.* Chicago: University of Chicago Press, pp. 69–118.

Stanley, S.M. (1985) Rates of evolution. *Paleobiology* 11:13–26.

Steineck, P.L. (1971) Phylogenetic reclassification of Paleogene planktonic Foraminifera. *Texas J. Sci.* 23:167–178.

Steineck, P.L., and Bergstein, J. (1979) Foraminifera from Hommocks salt-marsh, Larchmont Harbor, New York. *J. Foram. Res.* 9:147–158.

Steineck, P.L., and Fleisher, R.L. (1978) Towards the classical evolutionary reclassification of Cenozoic Globigerinacea (Foraminiferida). *J. Paleontol.* 52:618–635.

Stenseth, N.C., and Maynard-Smith, J. (1984) Coevolution in ecosystems: Red Queen or stasis? *Evolution* 38:870–880.

Streeter, S.S., and Lavery, S.A. (1982) Holocene and latest glacial benthic Foraminifera from the slope and rise off eastern North America. *Geol. Soc. Amer. Bull.* 93:190–199.

Streeter, S.S., and Shackleton, N.J. (1979) Paleocirculation of the Deep North Atlantic: 150,000-year record of benthic Foraminifera and oxygen-18. *Science* 203:168–171.

Takahashi, K. (1984) Radiolaria: sinking population, standing stock, and production rate. *Mar. Micropaleontol.* 8:171–181.

Tappan, H. (1971) Foraminiferida. *McGraw-Hill Encyclopedia of Science and Technology,* Vol. 5. New York: McGraw-Hill, pp. 467–475.

Tappan, H. (1976) Systematics and the species concept in benthonic foraminiferal taxonomy. *First International Symposium on Benthic Foraminifera of Continental Margins, Part A., Ecology and Biology,* Maritime Sediments *Spec. Publ.* 1, pp. 301–313.

Tappan, H. (1982) Extinction or survival: selectivity and causes of Phanerozoic crises. In Silver, L.T., and Schultz, P.H. (eds.) *Geological Implicatons of Impacts of Large Asteroids and Comets on the Earth.* Geol. Soc. Amer. Spec. Paper 190, pp. 265–276.

Tappan, H., and Leoblich, A.R. (1968) Lorica composition of modern and fossil Tintinnida (ciliate Protozoa), systematics, geologic distribution, and some new Tertiary taxa. *J. Paleontol.* 42:1378–1394.

Tappan, H., and Leoblich, A.R. (1973) Evolution of the oceanic plankton. *Earth Sci. Rev.* 9:207–240.

Theyer, F., Mato, C.Y., and Hammond, S.R. (1978) Paleontologic and geochronologic calibration of latest Oligocene to Pliocene radiolarian events, equatorial Pacific. *Mar. Micropaleonotol.* 3:377–395.

Thiede, J. (1983) Skeletal plankton and nekton in upwelling water masses off northwestern South America and northwest Africa. In Seuss, E., and Thiede, J. (eds.), *Coastal Upwelling: Its Sediment Record,* Part B. New York: Plenum, pp. 183–207.

Thomas, E. (1985) Late Eocene to Recent deep-sea benthic foraminifers from the central equatorial Pacific Ocean. In Mayer, L., Thayer, R., et al. (eds.), *Initial Reports of the Deep Sea Drilling Project,* Vol. 85. Washington, D.C.: U.S. Government Printing Office, pp. 655–693.

Thomas, F.C., and Schafer, C.T. (1982) Distribution and transport of some common foraminiferal species in the Minas Basin, eastern Canada. *J. Foram. Res.* 12:24–38.

Thompson, P.R. (1976) Planktonic foraminiferal dissolution and the progress towards a Pleistocene equatorial Pacific transfer function. *J. Foram. Res.* 6:208–229.

Thunnell, R.C. (1981) Cenozoic paleotemperature changes and planktonic foraminiferal speciation. *Nature* 289:670–672.

Thunell, R.C. and Honjo, S. (1981) Calcite dissolution and the modification of planktonic foraminiferal assemblages. *Mar. Micropaleontol.* 6:169–182.

Tjalsma, R.C., and Lohmann, G.P. (1983) Paleocene-Eocene bathyal and abyssal benthic Foraminifera from the Atlantic Ocean. Micropaleontology Spec. Publ. 4, New York: Micropaleontology Press, The American Museum of Natural History, 90p.

Towe, K.M. (1967) Wall structure and cementation in *Haplophragmoides canariensis*. Contributions from the Cushman Foundation for Foraminiferal Research, Vol. 18, pp. 147–151.

Towe, K.M. (1972) Invertebrate shell structure and the organic matrix concept. *Biomineralization Res. Rep.* pp. 2–14.

Towe, K.M., Berthold, W.V., and Appleman, D.E. (1977) The crystallography of *Patellina corrogata* Williamson: a-axis preferred orientation. *J. Foram. Res.* 7:58–61.

Towe, K.M., and Cifelli, R. (1967) Wall ultrastructure in the calcareous Forminifera: crystallographic aspects and a model for calcification. *J. Paleontol.* 41:742–762.

Van Hinte, J.E. (1978) Geohistory analysis—application of micropaleontology in exploration geology. *Amer. Assoc.* Petroleum Geol. Bull. 62:201–222.

Van Valen, L.M. (1984a) The case against impact extinctions. *Nature* 311:17–18.

Van Valen, L.M. (1984b) A resetting of Phanerozoic community evolution. *Nature* 307:660–662.

Van Valen, L.M. (1985a) Why and how do mammals evolve unusually rapidly? *Evolutionary Theory* 7:127–132.

Van Valen, L.M. (1985b) A theory of origination and extinction. *Evolutionary Theory* 7:133–142.

Van Valen, L.M. (1985c) How constant is extinction. *Evolutionary Thoery* 7:93–106.

Verdenius, J.G., and Van Hinte, J. (1983) Central Norwegian-Greenland Sea: Tertiary arenaceous Foraminifera, biostratigraphy and environment. In Verdenius, J.G., Van Hinte, J.E., and Fortuin, A.R. (eds.), *Proceedings of the First Workshop on Arenaceous Foraminifera 7–9 September 1981*. Trondkeim, Norway, pp. 173–224.

Vergnaud-Grazzini, C. (1983) Reconstruction of Mediterranean late Cenozoic hydrography by means of carbon isotope analyses. *Utrecht Micropaleontol. Bull.* 30:25–48.

Vincent, E., and Berger, W.H. (1981) Planktonic Foraminifera and their use in paleoceanography. In Emiliani, C. (ed.), *The Oceanic Lithosphere, The Sea*, Vol. 7. New York: Wiley, pp. 1025–1119.

Vincent, E., and Berger, W.H. (1985) Carbon dioxide and polar cooling in the Miocene: the Monterey hypothesis. In Sundquist, E.T., and Broecker, W.S. (eds.), *The Carbon Cycle and Atmospheric CO_2: Natural Variations Archean to Present*. Amer. Geophys. Union, Washington, D.C., pp. 455–468.

Vincent, E., Killingley, J.S., and Berger, W.H. (1980) The magnetic epoch-6 carbon shift: a change in the ocean's $^{13}C/^{12}C$ ratio 6.2 milion years ago. *Mar. Micropaleontol.* 5:185–203.

Vrba, E.S. (1985) Environment and evolution: alternate causes of the temporal distribution of the evolutionary events. *Suid-Afrikaanse Tydsbiref Wet.* 81:229–236.

Vrba, E.S., and Eldredge, N. (1984) Individuals, heirarchies and processes: towards a more complete evolutionary theory. *Paleobiology* 10:139–164.

Weaver, F.M., Casey, R.E., and Perez, A.M. (1981) Stratigraphic and paleocenographic significance of early Pliocene to middle Miocene radiolarian assemblages from northern to Baja California. In Garrison, R.E., Douglas, R.G., et al., (eds.), *Monterey Formation and Related Siliceous Rocks of California*. Soc. Econ. Paleontol. Mineral., Pacific Section, Los Angeles, pp. 71–86.

Wefer, G., Dunbar, R.B., and Seuss, E. (1983) Stable isotopes of foraminifers off Peru recording high fertility and changes in upwelling history. In Seuss, E., and Thiede, J. (eds.), *Coastal Upwelling: Its sediment record.*, Part B. New York: Plenum, pp. 295–308.

Wei, K.Y., and Kennett, J.P. (1983) Nonconstant extinction rates of Neogene Planktonic Foraminifera. *Nature* 305:218–220.

Weiner, S., and Erez, J. (1984) Organic matrix of the shell of the foraminifer *Heterostegina depressa*. *J. Foram. Res.* 14:206–212.

Weiner, S., Traub, W., and Lowenstam. H.A. (1983) Organic matrix in calcified exoskeletons. In Westbroek, P. and deJong, E.W. (eds.), *Biomineralization and Biological Metal Accumulation*. D. Reidel, p. 205–224.

Weinheimer, A.L. (1986) Radiolarian indicators of El Niño and anti-El Niño events in the Recent sediments of the Santa Barbara Basin. In Casey, R.E., and Barron, J.A., (eds.), *Siliceous Microfossil and Microplankton Studies of the Monterey Formation and Modern Analogs*. Soc. Econ. Paleontol. Mineral., Pacific Section, Los Angeles, pp. 31–38.

Weinheimer, A.L., Carson, T.L., Wigley, C.R., and Casey, R.E. (1986) Radiolarian responses to Recent and Neogene California El Ninño and anti-El Niño events. *Paleogeogr. Paleoclimatol. Paleoecol.* 53:3–25.

Werdelin, L., and Hermelin, J.O.R. (1983) Testing for ecophenotypic variation in a benthic foraminifer. *Lethaia* 16:303–307.

Weston, J.F. (1984) Wall structure of the agglutinated foraminifers *Eggerella bradyi* and *Karreriella bradyi* (Cushman). *J. Micropaleontol.* 3:29–32.

Weston, J.F. (1985) Comparison between Recent benthic foraminiferal faunas of the Porcupine Seabight and Western Approaches continental slope. *J. Micropaleontol. 4:165–184.*

Weston, J.F., and Murray, J.W. (1984) Benthic Foraminifera as deep-sea water-mass indicators. Benthos '83, 2nd International Symposium on Benthic Foraminifera, pp. 605–610.

Wigley, C.R. (1982) Comparison of radiolarian thanatocoenosis and biocoenosis from the oligotrophic Gulf of Mexico and Caribbean, and the eutrophic Southern California Sea. In *Trans. Gulf Coast Assoc. Geol. Soc.* 32:309–317.

Wilkinson, I.P. (1979) The taxonomy, morphology and distribution of the Quaternary and Recent foraminifer *Elphidium clavatum Cushman J. Paleontol.* 53:628–716.

Williams, D.F., Healy-Williams, N., Thunnell, R.C., and Leventer, A. (1983) Detailed stable isotope and carbonate records from the Upper Maestrichtian–Lower Paleocene section of hole 516F (Leg 72) including the Cretacceous–Tertiary boundary, 1983. In Barker, F.T., et al. *Initial Reports of the Deep-Sea Drilling Project,* Vol. 27. Washington, D.C.: U.S. Government Printing Office, pp. 921–930.

Williamson, M.A. (1982) Distribution of Recent Foraminifera on the Nova Scotian Shelf and Slope. In Mamet, B., and Copeland, M.J. (eds.), *Proc. Third North American Paleotological Convention,* Vol. 2. Geol. Sur. Canada, Ottawa, pp. 579–584.

Williamson, M.A. (1985) Recent forminiferal diversity on the continental margin off Nova Scotia, Canada. *J. Foram. Res. 15:43–51.*

Williamson, M.A., Keen, C.E., and Mudie, P.J. (1984) Foraminiferal distribution on the continental margin off Nova Scotia. *Mar. Micropaleontol.* 9:219–240.

Winkler, W., and van Stuijvenberg, J. (1982) Flysch-type agglutinated Foraminifera and the Maestrichtian to Paleogene history of the Labrador and North Seas—Comments. *Mar. Micropaleontol.* 7:359–360.

Wood, K.C., Miller, K.G., and Lonmann, G.P. (1985) Middle Eocene to Oligocene benthic foraminifera from the Oceanic formation, Barbados. *Micropaleontology* 31:181–197.

Woodruff, F. (1985) Changes in deep-sea benthic foraminiferal distribution in the Pacific Ocean: relationship to paleoceanography. In Kennet, J.P. (ed.), *The Miocene Ocean: Paleoceanography and Biogeography,* Mem. Geol. Soc. Am. 163:131–177.

Woodruff, F., and Douglas, R.G. (1981) Response of deep-sea benthic Foraminifera to Miocene paleoclimatic events. DSDP Site 289. *Mar. Micropaleontol.* 6:617–632.

Woodruff, F., and Savin, S.M. (1985) ^{13}C values of Miocene Pacific benthic forminifera: Correlations with seal level and biological productivity. *Geology* 13:119–122.

Woodruff, F., Savin, S.M., and Douglas, R.G. (1981) Miocene stable isotope record: a detailed OSDP Pacific Ocean study and its paleoclimatic implications. *Science* 212:665–668.

Zimmerman, M.A., Williams, D.F., and Rottger, R.F. (1983) Symbiont-influenced isotopic disequilibrium in *Heterostegina depressa. J. Foram. Res.* 13:115–121.

3

Zoogeography of Marine Protozoa: An Overview Emphasizing Distribution of Planktonic Forms

BERNT ZEITZSCHEL

INTRODUCTION

According to van der Spoel and Heyman (1983), both the geographic distribution and the speciation of plankton, including marine protozoa, are products of the geologic history of the oceans, the continental barriers, the current patterns and the limitation of survival of individual species and populations by both biotic and abiotic environmental conditions.

The geographic distribution also depends on life history of species, the mobility of populations, the selective pressure, and the time during which a taxon has existed. Each taxon forms part of biological evolution and is therefore never sharply delimited in time. Environmental conditions exert a natural selective influence on each population, which results in a survival of the most adequately adapted pheno- and genotypes at a particular place and during a particular time period (van der Spoel and Heyman, 1983).

According to these authors speciation can be defined as the interaction of time, selective pressure, and distribution, resulting in a certain genetic configuration of the genome. This implies that the variability or stability of a genotype is directly linked with the variability or stability of distribution patterns.

METHODS

A thorough analysis of marine protozoan distribution in space and time requires a combined approach of water column sampling and sediment analyses. The former provides information exclusively about extant species present at the location during the sampling program. The latter data, obtained from sediment samples, permit a broader time span of analysis, depending on the age of the sediment stratum examined (Anderson, 1983).

Sampling techniques for qualitative and quantitative analysis of marine protozoa or microzooplankton in general, in the water column are the same as for phytoplankton. The size categories of plankton organisms are given in Table 3.1. Marine protozoa belong mainly to the size range of nano- and microplankton.

A large variety of sampling and analyzing systems has been published to describe distributions and size spectra of marine organisms in the sea. Figure 3.1, from Yentsch and Yentsch (1984), gives an overview of the optimal spatial scales that are sampled adequately by a selection of instruments, including measurements of such environmental determinants as temperature, salinity, and nutrients. It is obvious from this diagram that, apart from the very new method of flow cytometry, the lowest common sampling scales of few to several dozens of kilometers in the horizontal plane are usually sufficient in open ocean situations.

Plankton nets of various designs, including closing nets, have been widely used as sampling devices in microzooplankton investigations. The advantage of nets is that they allow us to concentrate organisms from a relatively large volume of water quite easily. In particular, nets have the advantage of eliminating the effects of marked vertical stratification. The main disadvantage of nets is the distorted species composition shown by net samples. Only a portion of the organisms entering the net are retained by the gauze. One can assume that only about 50 percent of the shell-bearing marine protozoa are retained by nets with a mesh size of 40 μm. Nets with very fine meshes (e.g., 5–10 μm) catch small organisms more efficiently than coarse nets. However, their filtering efficiency is limited, as these nets tend to clog easily, because of their small free sifting surface. At the same time, existing fine-mesh nets are difficult to operate in bad weather (Fraser, 1968; Tranter, 1968).

Because of their selective and unpredictable filtering properties, and the uncertainty with regards to relating the organisms caught to a specific volume of water, nets should not be employed in quantitative protozoan sampling. Sampling with nets may, however, be useful in providing ample material for qualitative purposes (Tangen, 1978).

TABLE 3.1 Size Classes of Marine Plankton.

Plankton	Size	Organisms
Picoplankton	<2 μm	Bacteria, cyanobacteria, smallest eukaryotic organisms
Nanoplankton	2–20 μm	Microflagellates, dinoflagellates, coccolithophores, small diatoms, small ciliates
Microplankton	20–200 μm	Diatoms, dinoflagellates, protozoa, nauplii
Mesoplankton	200 μm–20 mm	Large diatoms, dinoflagellates, copepods
Macroplankton	20 mm–20 cm	Euphausiids, pteropods, appendicularians, ctenophores
Megaplankton	>20 cm	Colonial appendicularians, medusae

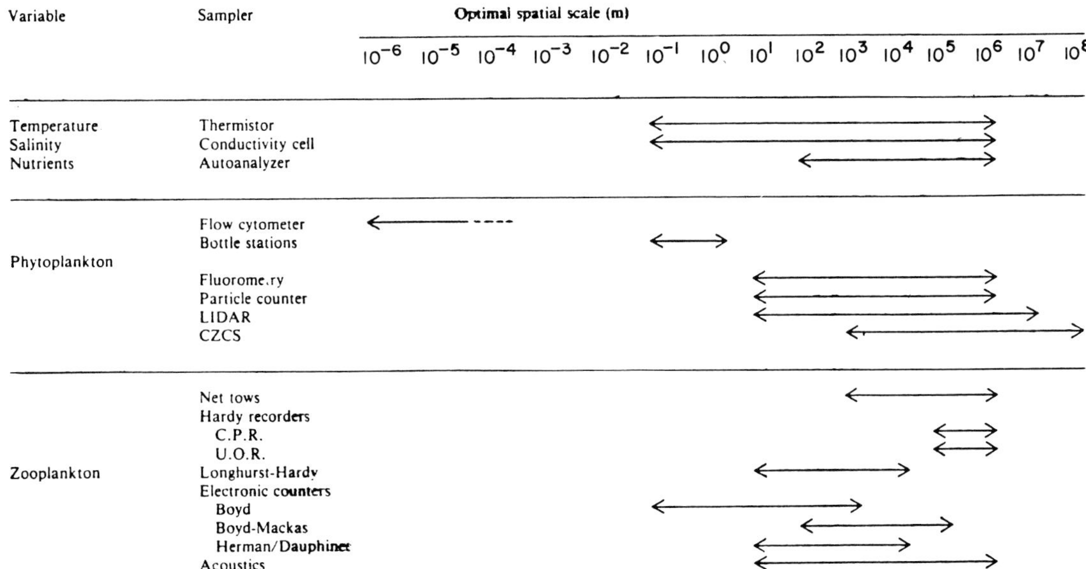

FIGURE 3.1 Normal spatial scales that can be sampled appropriately with various oceanographic instruments. (After Yentsch and Yentsch, 1984.)

Since 1930 the Continuous Plankton Recorder has been deployed from ships of opportunity in the North Sea, and later in the North Atlantic, to describe and analyze the variability of the plankton in space and time. Although the instrument is not designed to sample microzooplankton quantitatively, some biogeographic descriptions of marine protozoa have been achieved (Zeitzschel, 1966; Lindley, 1975; Reid, personal communication). Robinson (1978) considers that the Plankton Recorder survey provides an index of the abundance of species that is sensitive to changes on a monthly time scale and provides the only available information by a uniform method of sampling over a wide area, over a long period of years.

Plankton sampling data, obtained in open ocean locations at various depths in the water column using conventional plankton nets, are the major source of information for biogeographic analysis of marine protozoa.

Water bottles or pumps provide an alternative to plankton nets. Any sampling by water bottles, however, would generally be restricted to the larger bottles (10–30 L) in order to obtain sufficient numbers of the bigger microzooplankton for enumeration. The concentration of the organisms from large water bottles using fine screens at sea is, however, somewhat cumbersome. The main disadvantage of water bottles is that many samples must be taken to cover different depth ranges.

The use of pumps for sampling microzooplankton provides the potential for obtaining quantitative samples of the entire size spectrum of marine protozoa (Beers et al., 1967; Lenz, 1972). Furthermore, pump systems can allow for sampling any volume of material desired either for use at discrete depths or for integration over various horizontal, vertical, or oblique intervals. There are, however, mechanical restrictions with pumping devices if the water depth exceeds about 200 m.

The pumped water can be retained unconcentrated or preconcentrated in relatively small settling chambers for study of the smaller, generally more abundant microzooplankters, or it can be passed through various mesh-sized screens in order to concentrate the larger, less abundant organisms in the bigger size classes (Beers, 1978).

Sorokin (1981) points out that even the most gentle method possible of concentrating water samples to count zooflagellates and ciliates with the use of 5-μm nylon gauze, of 1–5-μm nucleopore filters or membrane filters, were found to be useless. During such concentration procedures a large part of the total ciliate population, observable live in the intact sample, dissappears. This loss averages more than 95 percent for both naked and loricated forms. Thus, even the most gentle concentration methods can be used only for examination of species composition of tintinnid populations. Sorokin (1981) recommends counting zooflagellates and ciliates in the living state, as soon as possible after sampling, in specially designed counting chambers. Dale and Burkill (1982) described a quick and simple technique for enumeration of living pelagic ciliates. Table 3.2 gives an impression of the abundance of zooflagellates, ciliates, and total microzooplankton in various parts of the oceans (after Sorokin, 1981).

Open ocean surface sediment samples are commonly obtained by grabs, box-cores, and corers. The disadvantage of all this sampling gear is that the "fluff," the newly settled fine material, is often washed out. Good results in the collection of settling material, including protozoan skeletons, loricae, and cysts in the water column and close to the sea bottom, have been achieved in recent years using sediment traps in moored and drifting mode (Takahashi and Honjo, 1981; v. Bodungen et al., 1985).

INTERPRETATION OF OBSERVATIONS

Attempts to arrive at biogeographic meaning are mainly efforts to use the presence or absence of one or more species or stages in their life cycles, and sometimes their morphological appearance, as

TABLE 3.2 Number (N) and Biomass (B, mg/m^3) of Zooflagellates and Ciliates (maximum values), Total Microzooplankton Biomass (mg/m^3) in Different Marine Habitats (Maximum Values), and Integral Biomass of Protozoa Within the Water Column.

Habitat	Zooflagellates ZN 10^6/L	Zooflagellates B	Ciliates N 10^3/L	Ciliates B	Total microzooplankton, B, mg/m^3	Integral biomass of protozoa in the euphotic zone, g/m^2 (wet wt)
Brackish fish pond, lagoons Valpisani (Italy)	4.8	450	370	11,700	13,000	8.0
Meromictic Faro Lagoon (Sicily), redox zone	23.0	1,500	40	840	2,400	6.0
Lagoon of Venice (Italy), April	4.0	240	17	430	750	0.8
De Castri Bay, Japan Sea, N.E. coast, July	2.0	120	19	440	650	3.3
Japan Sea, pelagic part, June	9.5	450	9	190	700	10.2
Peruvian upwelling, patch of diatom bloom, late phase of succession	6.0	320	86	800	1,200	14.0
Peruvian upwelling, area of "red tide"	13.0	720	145	1,800	2,500	20.0
Peruvian upwelling frontal area	0.7	40	22	810	950	6.8
Equatorial upwelling Eastern Pacific Ocean, 97°W	0.4	16	30	50	250	2.5
Peruvian upwelling area of strong upwelling near San Juan	9.0	500	21	650	1,300	12
Equatorial divergence, Central Pacific Ocean, 154°W	0.5	20	7	10	40	1.6
Trade wind area of South Pacific Ocean	3.0	100	1.5	30	150	1.3
Black Sea, coastal area	4.0	200	3.7	140	600	3.5
Black Sea, open area	1.8	100	0.45	20	100	2.8
Black Sea, upper boundary of redox zone, 130 m depth	0.7	40	0.04	8	50	—
Antarctic Ocean, midsummer	—	—	29	113	—	2.8

After Sorokin, 1981.

indicators of certain environmental conditions, including hydrographic events, eutrophication or warming trends, or long-term changes symptomatic of environmental disturbances (Smayda, 1978).

A thorough analysis, including meaningful statistical treatment of distributions of marine organisms, depends entirely on a logical sampling design. Venrick (1978) has discussed sampling strategies for phytoplankton that apply also for marine protozoa. The first step in sampling design is to state the problem in a clear and logical manner, for the nature of the problem will define the scales of time and space that should be contemplated. A sampling program has to result in an acceptable allocation of the effort that goes into counting from 100 to 1,000 samples, a typical range. The universe to be sampled cannot be considered statistically uniform; neither can the samples be taken at random. This poses difficulties in the statistical treatment of the data and the need for simplification. Limitations of equipment, time, and manpower also make it difficult to project or adhere strictly to a program involving an ideal distribution of sampling points (Margalef, 1978a).

Venrick (1978) points out that in reality, the conditions demanded by classical statistical methods are rarely, if ever, realized in the planktonic environment, because the distributional complexities of the plankton, confounded by their three- or four-dimensional nature, are far removed from the theoretical distributions underlying parametric statistical procedures. Furthermore, the difficulties of imposing a specific sampling strategy on an invisible, mobile population are enormous. Rarely, if ever, can one be assured of sampling the same population on subsequent attempts. The inability to sample the same population repeatedly (in the case of temporal analysis) or to sample locations simultaneously (in case of spatial studies) imposes an element of spatial–temporal interactions on the data that is difficult to extract. According to Venrick (1978), many studies have attempted to describe the complexities of planktonic distributions and to adapt them to statistical treatments. At the present time, there is no evidence that the specific results from one investigation can be extended to different organisms (i.e., from macrozooplankton to phytoplankton) or to different environments (i.e., from neritic to oceanic) or to different scales of sampling (i.e., from net tows to water bottle samples).

Venrick (1978) concludes that in the interest of gaining biological information it may be necessary to relax the statistical requirements to a greater or lesser degree and to use statistical procedures as qualitative rather than as probabilistic tools. But statistical requirements must be understood before they can be ignored. Only if the scientist has a firm understanding of the principles of classical statistics can he violate them without risk of fooling himself or his reader.

Despite this fact numerous attempts have been made to use mathematical treatments (e.g., to define groups or associations of plankton organisms). Lengendre and Legendre (1978) evaluate the different methods used in phytoplankton research.

The process of grouping usually involves two main steps: first, one establishes the degree of similarity (or distance) between the organisms to be grouped, through some coefficient or other evaluation method that seems to be appropriate to the data at hand. Then one proceeds to find the groups by applying a cluster rule on the association matrix and by deciding whether or not a given pair of organisms obey the clustering rule at the stated clustering level.

The biogeographic approach to association identification is usually a study of the relationships between samples in terms of their species composition. This is referred to as a Q-study and it involves the calculation of a coefficient of association (similarity or distance) between all pairs of samples. The Q-study of the affinity between sampling localities is often carried out on a large scale. When groups of localities are established, the species inhabiting them may be described as biogeographic communities.

Legendre and Legendre (1978) provide a practical guide to the choice of a proper association procedure (Table 3.3). In their conclusions they warn, however, against excessive enthusiasm about numerical methods in biogeographic research.

The recurrent group analysis (Fager, 1957, 1963; Fager and McGowan, 1963), the cluster analysis (Parker and Berger, 1971), and the principal component analysis (Williamson, 1961; Colebrook, 1964) have been successfully applied in biogeographical studies of marine protozoa.

The ecological situation under investigation is of importance in choosing a procedure to identify associations. For example, phytoplankton populations in areas of the oceans dominated by strong environmental gradients should not be approached in the same way as populations from mere uniform environments. Strong environmental gradients generally lead to better-defined population patterns, which do not require highly sophisticated mathematical analysis. In other words, there is no point in overidentifying associations; a straightforward procedure, when applicable, is better than a very involved one (Legendre and Legendre, 1978).

The shortcoming of oceanic biogeography in general and of biogeography of marine protozoa in particular is the lack of adequate data sets. Especially

TABLE 3.3 Characterization of Associative Methods.

Method		Discriminative characteristics
BINARY DATA (presence-absence)	*1. Similarity measures*	
	Jaccard	For single-linkage clustering
	Fager and McGowan	For complete-linkage clustering
	Krylov (χ^2)	For complete-linkage clustering with probabilistic threshold
	2. Clustering procedures	
	Single linkage	Associations at unspecified level
	Complete linkage (Fager)	Clusters of highly associated species with clouds of related species
QUANTITATIVE DATA (cell counts)	*1. Similarity measures*	
	Correlation coefficients:	
	Kendall (τ)	Semiquantitative data
	Pearson (r)	Quantitative data; well-structured environment
	Probabilistic index (Goodall, Orlóci)	Any type of data; for probabilistic agglomerative clustering
	2. Clustering procedures	
	Clustering the correlation matrix	Associations at unspecified level (single-linkage clustering)
	Probabilistic agglomeration	Densely associated species, clustered according to a probability criterion
	Factorial methods	Of questionable value in identifying phytoplankton associations

After Legendre and Legendre, 1978.

for oceanwide distributions, the records are a compilation of data from any available sampling program over a very large time period using a variety of sampling techniques, applying various methods of identification and enumeration, including graphic documentation in different scales and projections.

Despite this fact some useful information on the biogeography of some marine protozoa has been published mainly by palaeontologists. One outstanding example of unified presentation of very heterogeneous material has been achieved by van der Spoel and Heyman in their *Comparative Atlas of Zooplankton,* published in 1983. Some examples of this work will be given later in this chapter.

PHYSICAL ENVIRONMENT

The complex movements of water in the sea can be separated formally into transport (advection) and mixing processes (Fig. 3.2). The large-scale transport of water in currents, such as the equatorial current system or the Gulf Stream, is the best-known form of lateral movement.

The largest vertical motion of water is found in those regions, mainly on the western sides of the continents, where upwelling occurs. Vertical velocities in such areas are of the order of 10–3 cm/s, several orders of magnitude smaller than average horizontal transport in the main currents.

Turbulent mixing, both lateral and vertical, is always present, especially in the surface layer of the sea. Thus, although lateral movements disperse and intermix plankton populations, turbulent mixing generally controls the production of phytoplankton.

Horizontal movements in the sea are most pronounced in surface currents, undercurrents and deep currents, tides and internal waves, whereas vertical movements of water occur in zones of mixing, upwelling regions, and zones of convergence and divergence. Special circulation patterns, like the Langmuir circulation induced by wind (Langmuir, 1938), the dial thermal surface layer convection (Woods and Onken, 1982), and small-scale horizontal vortices caused by thermohaline driving forces, also influence the patterns of production and distribution of planktonic organisms.

Owen (1981) has summarized the ecological effects of combinations of physical characteristics of fronts and eddies with interactive characteristics of organisms. Fronts and eddies can mechanically affect local concentrations of organisms, juxtapose populations that would not otherwise interact, create new communities, conserve and translocate selected species ensembles and concentrations, attract and sustain motile animals, serve as repro-

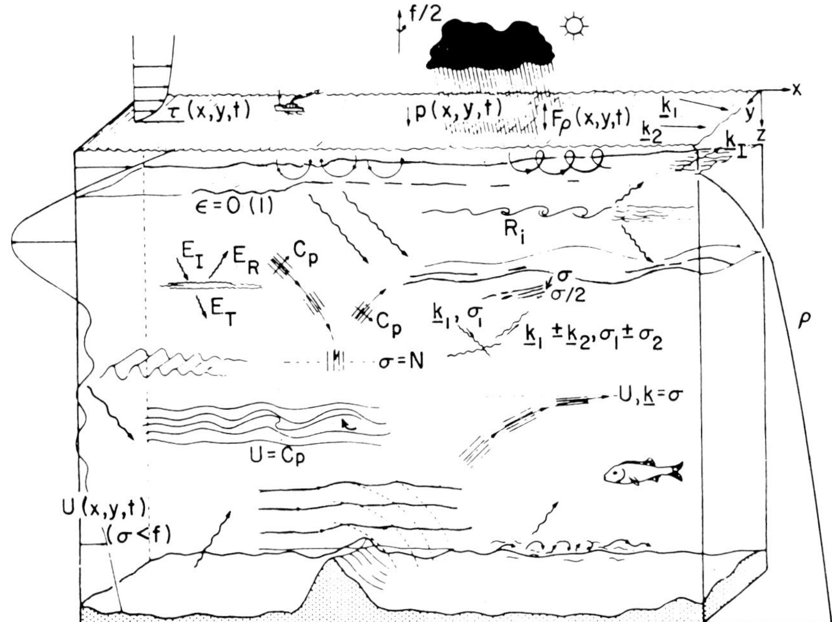

FIGURE 3.2 Physical processes in the ocean. (After Thorpe, 1975.) The figure indicates vertical profiles of wind stress (up left), current velocity (left), and density distribution (right). From top to bottom indications are given (e.g., of earth rotation, precipitation, atmospheric pressure, Langmuir circulation, shear, energy reflection, rays, Rossby waves, and bottom friction).

duction refuges, mechanically limit dispersal of meroplankton or neritic populations, influence selectivity pattern components of populations or communities (quasi-ordered patchiness), induce/sustain higher local production of organisms, modify migration patterns as diverse as annual, transoceanic fish movements and diel vertical migration of motile phytoplankton and microzooplankton, and collect surface-active and particulate substances.

Gagnon and Lacroix (1982) discuss the effect of tidal advection and mixing on the statistical dispersion of zooplankton. They conclude that in tidal estuaries, advection phenomena are more easily recognizable than turbulence effects.

One of the most significant factors contributing complexity to the interpretation of plankton distributional and time-series data is the great variability of abundance estimates, on all scales in space and time caused by patchiness (Steele, 1976, 1978; Longhurst, 1981).

Figure 3.3 depicts a conceptual model of the time–space scales of zooplankton biomass variability and the factors contributing to these scales. I, J, and K are bands centered about thousands, hundreds, and several kilometers in space scales, with time variations between weeks and geologic time scales (Haury et al., 1978).

Munk (1950) published a generalized diagram of the wind-driven circulation in an ocean extending from the equator to polar latitudes (Fig. 3.4). The sun is ultimately responsible for ocean circulation. It acts indirectly through the differential heating of the atmosphere, which causes the winds to blow and acts directly by heating the ocean with solar radiation. The mechanisms and the resulting motions are coupled and interactive, but it has been conceptually useful to separate the circulation of the ocean into the wind-driven and the thermohaline (Smith, 1976). The wind is the predominant force driving the circulation in the upper kilometers of the ocean. The thermohaline circulation is driven by density differences resulting from the heat fluxes and salt fluxes, resulting from evaporation and precipitation at the surface. The regional physical classification of the world ocean is given in Figure 3.5. Figure 3.6 shows the currents of the surface of the world ocean. Knowledge of the average surface circulation is derived largely from ships' drift observations and from geostrophic calculations based on the observed density distribution.

Richardson (1976) has reported that actual circulation (e.g., of the Gulf Stream) is rather complex, as shown by the distribution and movements of ring eddies several hundreds of kilometers in diameter in the North Atlantic (Fig. 3.7). Magaard et al. (1983) published a detailed account on the role of eddies in general ocean circulation.

An example of smaller eddies is given in Figure 3.8. The Landsat spectral image reflects the mass

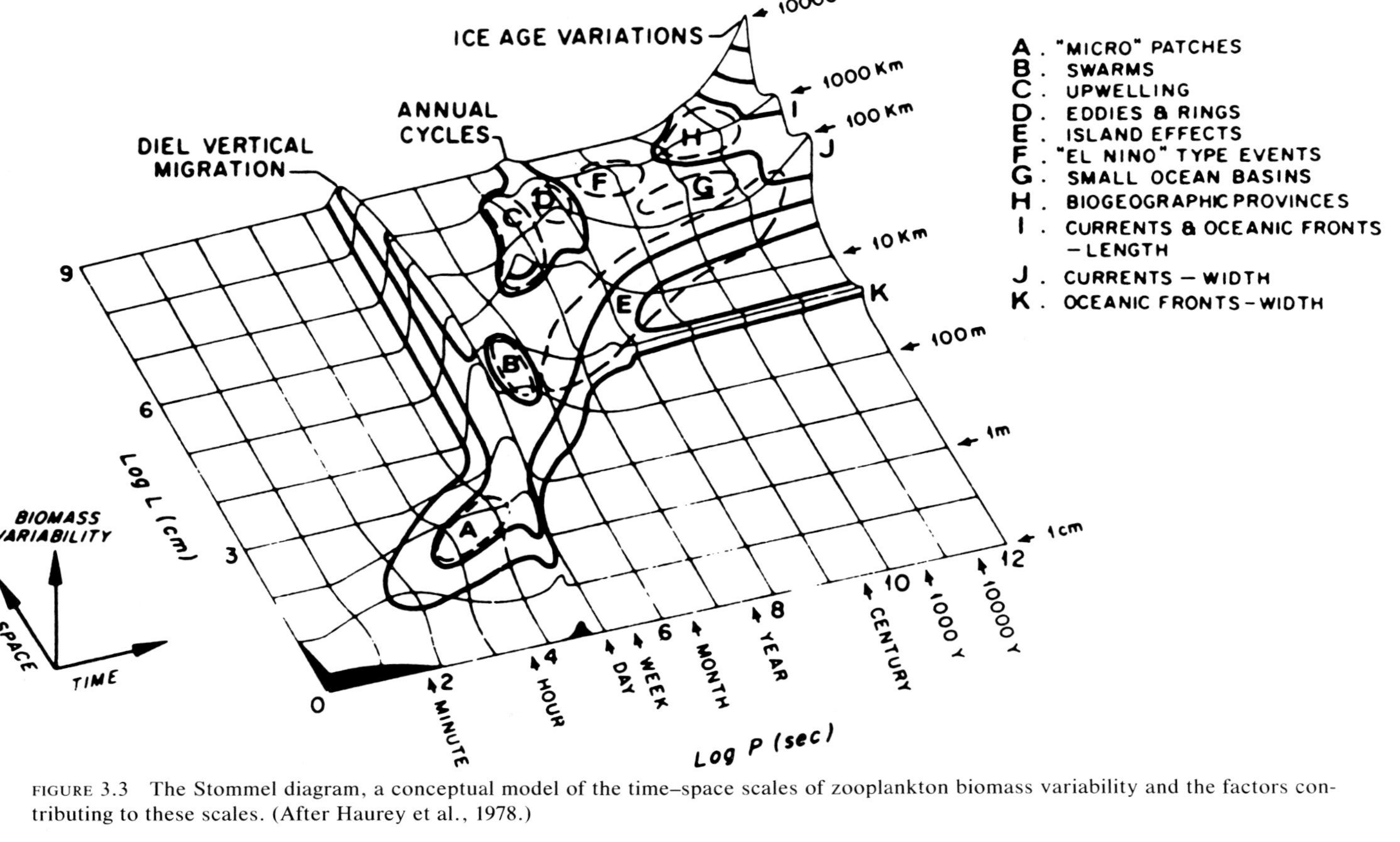

FIGURE 3.3 The Stommel diagram, a conceptual model of the time–space scales of zooplankton biomass variability and the factors contributing to these scales. (After Haurey et al., 1978.)

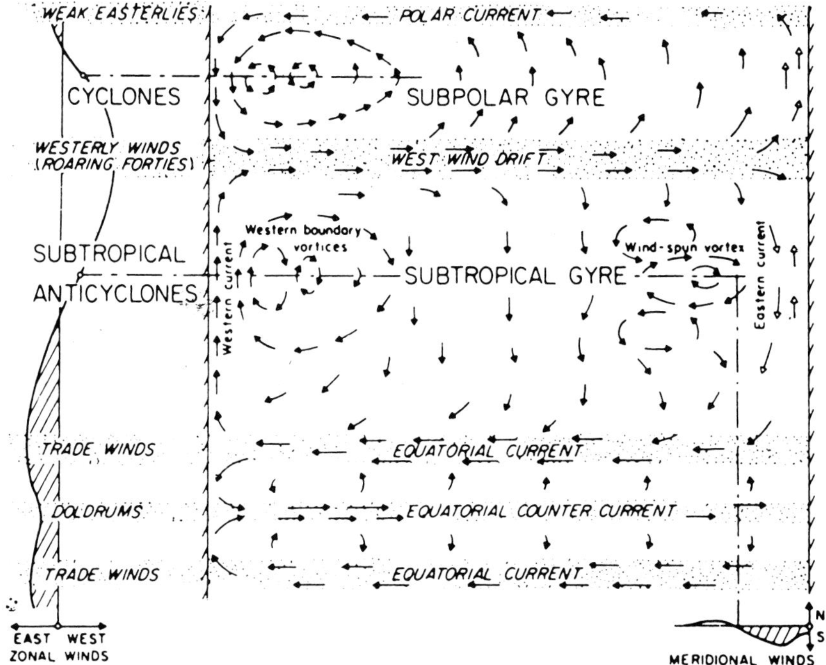

FIGURE 3.4 Generalized schematic of the wind-driven circulation in the ocean extending from equator to polar latitudes. (After Munk, 1950.)

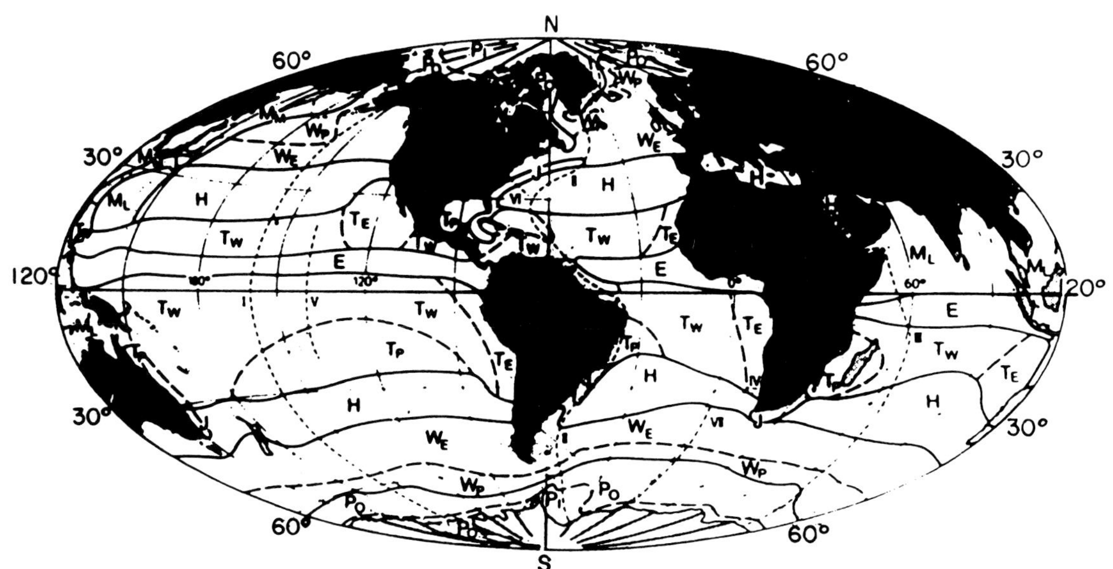

FIGURE 3.5 Regional classification of the world ocean. T: Region of trade wind currents. E: Region of equatorial countercurrents. M: Region of monsoon currents. H: Region of horse latitudes. J: Jet stream region. W: Region of west wind drift. P: Polar region. (After Dietrich et al., 1980.)

occurrence of blue-green algae in the southwestern Baltic Sea. The scale of these eddies is some 10 km.

Table 3.4 summarizes the major physical factors that influence primary productivity of phytoplankton and the distribution of plankton, including marine protozoa (after Zeitzschel, 1978).

MARINE BIOGEOGRAPHY

Marine biogeography, as a comprehensive and independent subject, is somewhat more than 125 years old. The first account was given by Forbes and Godwin-Austen, who published "The natural

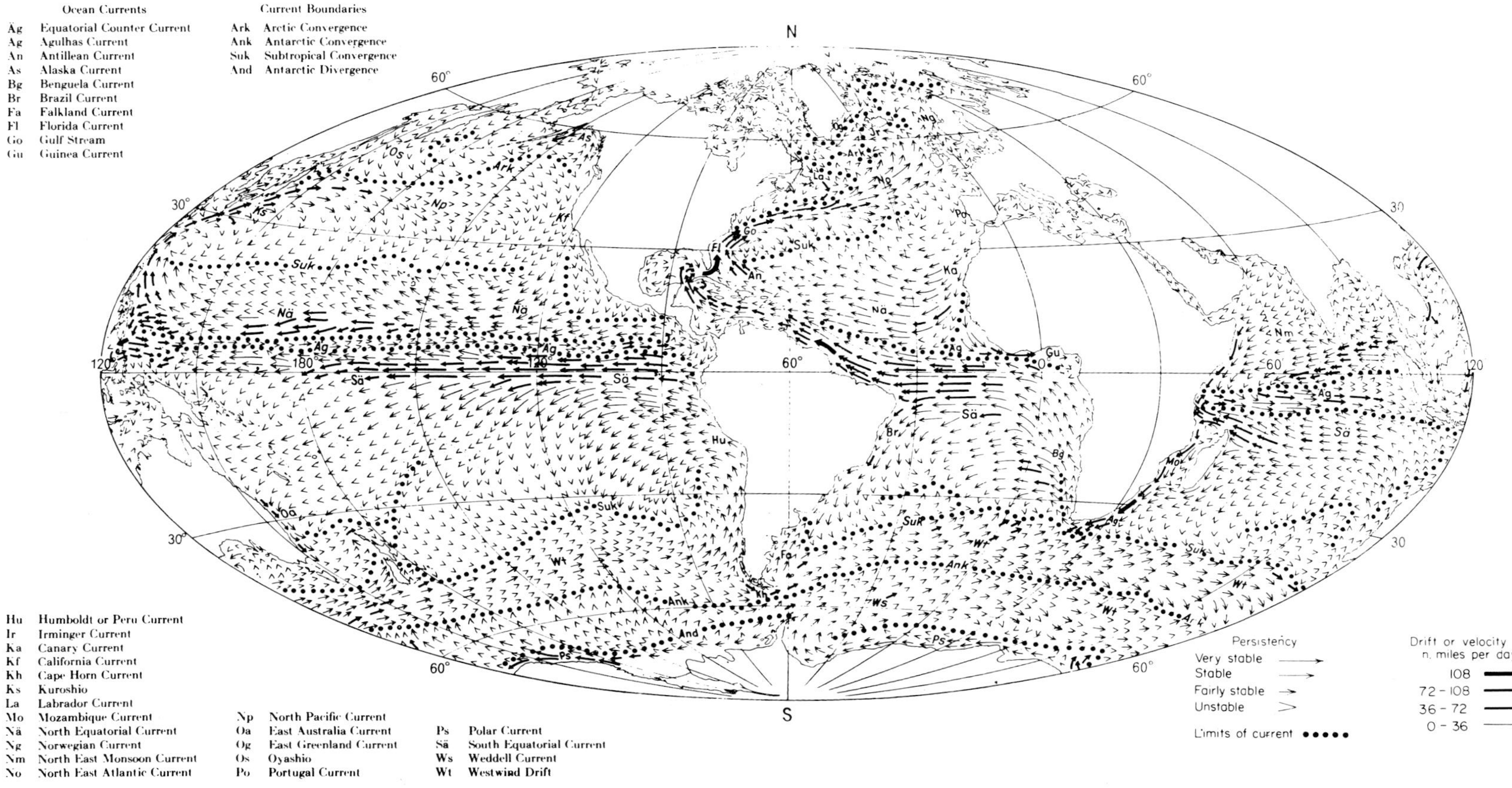

FIGURE 3.6 Currents of the surface of the world ocean in northern winter. (After Dietrich et al., 1980.)

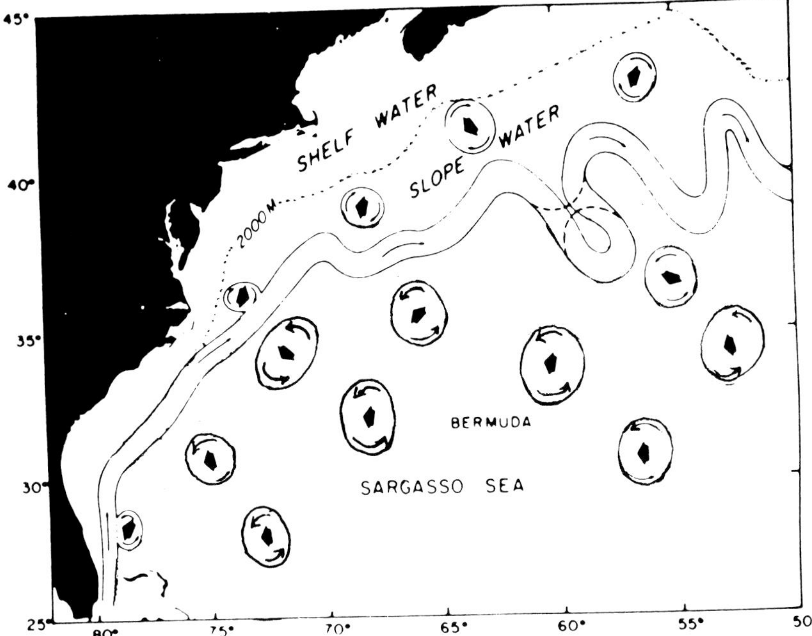

FIGURE 3.7 Schematic representation of the path of the Gulf Stream and the distribution and movement of ring eddies. (After Richardson, 1976.)

history of European Seas" in 1859. Many plankton investigators—such as Haeckel (1887), Giesbrecht (1892), Meisenheimer (1905), Brandt (1906, 1907, 1910), Steuer (1933), Russell (1935, 1939), Ekman (1953), Hedgpeth (1957), Johnson (1957), Balech (1958, 1960), Bradshaw (1959), Fraser (1962), Boltovskoy (1962, 1969, 1979, 1982, 1985), Johnson and Brinton (1963), Zeitzschel (1966, 1969, 1971, 1982), McGowan (1971), Bé and Tolderlund (1971), Fleminger and Hülsemann (1973), Bé (1977), Bé and Hutson (1977), van Der Spoel and Pierrot-Bults (1979), Buzas and Culver (1980), Beklemishev (1981), Peres (1982), Anderson (1983), Robinson (1983), van der Spoel (1983), van der Spoel and Heymann (1983), and others—have noted a close relationship between faunistic and hydrographic features of the major oceanic water masses and the fact that their geographic variations are particularly strong in the epipelagic rather than in the meso- or bathypelagic zone.

One of the earliest attempts to establish the pattern of species distribution of a group of plankton organisms was that of Giesbrecht (1892), who, on the basis of studies on copepods, divided the pelagic fauna into three "main regions": a warm-water region and a northern and a southern cold-water region.

Steuer (1933), who also studied copepods, came to somewhat different conclusions. In a summary of the results of his own studies and those of other authors, he proposed 12 distributional regions and subregions (Fig. 3.9).

In broad terms, three major biogeographical divisions can be distinguished for marine plankton communities: a Circumglobal Warm-water region, and the Northern and Southern Cold-water regions. The latter two can be further subdivided into Polar (= Arctic and Antarctic) and Subpolar (= Subarctic and Subantarctic) provinces. The warm-water region comprises two subtropical provinces with a single tropical province in between. Each province has its reciprocal in the other hemisphere, so that, although a total of nine geographic subdivisions can be distinguished, there are faunistically only five discrete provinces. These faunal provinces correspond in general to the major hydrographic regions in the world ocean.

The primary biogeographic classification of phytoplankton is into regional groups that describe their distributional range in terms of near-shore versus offshore occurrences, as well as latidudinally (Smayda, 1978). A species may thus be described as being neritic (sometimes as brackish, estuarine or coastal) or oceanic; and may also be described by a latitudinal descriptor, such as arctic, boreal, temperate, subtropical, or tropical. The ecologist's interest in such biogeographic classifications is in the associated descriptions of environmental parameters. The latitudinal descriptor suggests something about the species tolerance (preference) to temperature and light, whereas the onshore versus offshore classifier gives some indication about its possible requirements or tolerance for nutrients and its osmotic needs. This concept is

FIGURE 3.8 Landsat spectral channels MSS 4,5,6 (August 1975) showing mass occurrence of blue green algae in the southwestern Baltic Sea. Note the appearance of small eddies of a diameter of 20–30 km. (After Ulbricht and Horstmann, 1980.)

illustrated in Figure 3.10, where 12 geographic types are proposed from a compilation of the records of 379 phytoplankton species (Smayda, 1978).

Margalef (1978b) states that the best predictor of primary production and of dominant life forms in phytoplankton is the available external energy on which advection and turbulence depend. The combination of sedimentation with turbulence or variance in the components of velocity is, according to this author, the most important factor in the biology of phytoplankton.

Colebrook (1972) proposed a classification of the geographic distribution of plankton organisms of the North Atlantic and the North Sea, based on an analysis of charts of distributions of 64 species of phytoplankton and zooplankton from long-term data of the Continuous Plankton Recorder survey. Figure 3.11 shows the principal types of distribution for this most intensively investigated area of the world ocean.

Beklemishev et al. (1977) published maps of general phytoplankton geography and zooplankton distribution patterns (Figs. 3.12, 3.13).

In comparing the phytoplankton geography of the world ocean with the zooplankton dispersal in these distributional maps some remarkable differences are obvious. Most striking is the strong tapering of the Northern Hemisphere belts in the west, the narrowness of the transitional belts and the particular subdivision in the tropical areas. For

TABLE 3.4 Major Physical Factors Influencing Primary Productivity and Distribution of Plankton. Symbols in the Fourth and Fifth Columns Indicate Different Categories of Influence. Parentheses Around the + Signs Imply Some Degree of Reservation.

Type of motion	Forms of water motions	Special phenomena	Influence on Productivity	Influence on Distribution
Motions with large-scale effects	Permanent (gradient) flow	Meanders	(+)	+ +
	Wind currents	Eddies	(+)	+ +
	Divergence/convergence	Fronts	+	(+)
	(Upwelling)	Patches	+ +	(+)
	(Convective circulation)	—	(+)	(+)
Motions with small-scale effects	Tidal currents	Eddies, fronts	(+)	+
	Surface wave action	Foam, bubbles	(+)	—
	Wind mixing	Langmuir circulation	(+)	(+)
	Internal wave action	Slicks	(+)	(+)
	Upwelling	Patches	+ +	(+)
	Divergence/convergence	Fronts, eddies	+	(+)
	Molecular diffusion	Step structure	(+)	—
	Convective circulation	—	(+)	(+)
	Shear instabilities	Fronts	(+)	—
Motions caused by morphological features (small-scale)	Island circulation	Eddies	+	+
	Land promontories	Eddies	+	(+)
	Topography (ridges, submarine canyons)	Meanders	+	(+)
	Run-off and land drainage	Plumes	+	+

After Zeitzschel, 1978.

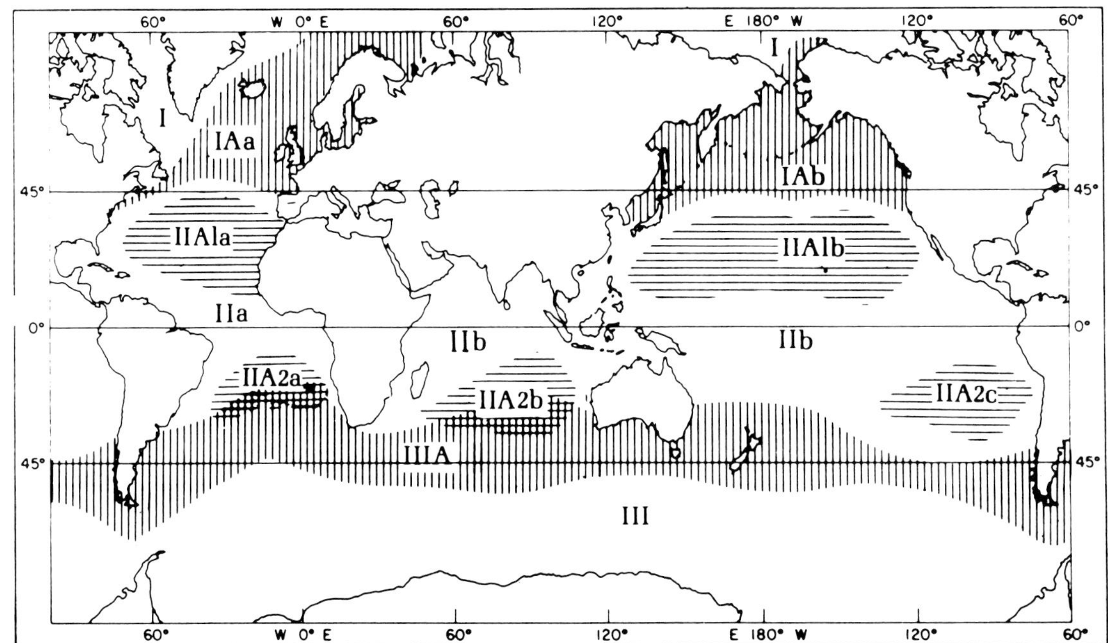

FIGURE 3.9 Distributional regions and subregions of the world oceans. (After Steuer, 1933.)

I	Circumpolar Arctic region
I A a	Subarctic Atlantic subregion
I A b	Subarctic Pacific subregion
II a	Tropical Atlantic region
II b	Tropical Indo-Pacific region
II A 1a	Northern subtropical Atlantic subregion
II A 1b	Northern subtropical Pacific subregion
II A 2a	Southern subtropical Atlantic subregion
II A 2b	Southern subtropical Indian subregion
II A 2c	Southern subtropical Pacific subregion
III	Circumpolar Antarctic region
III A	Circumpolar subantarctic subregion

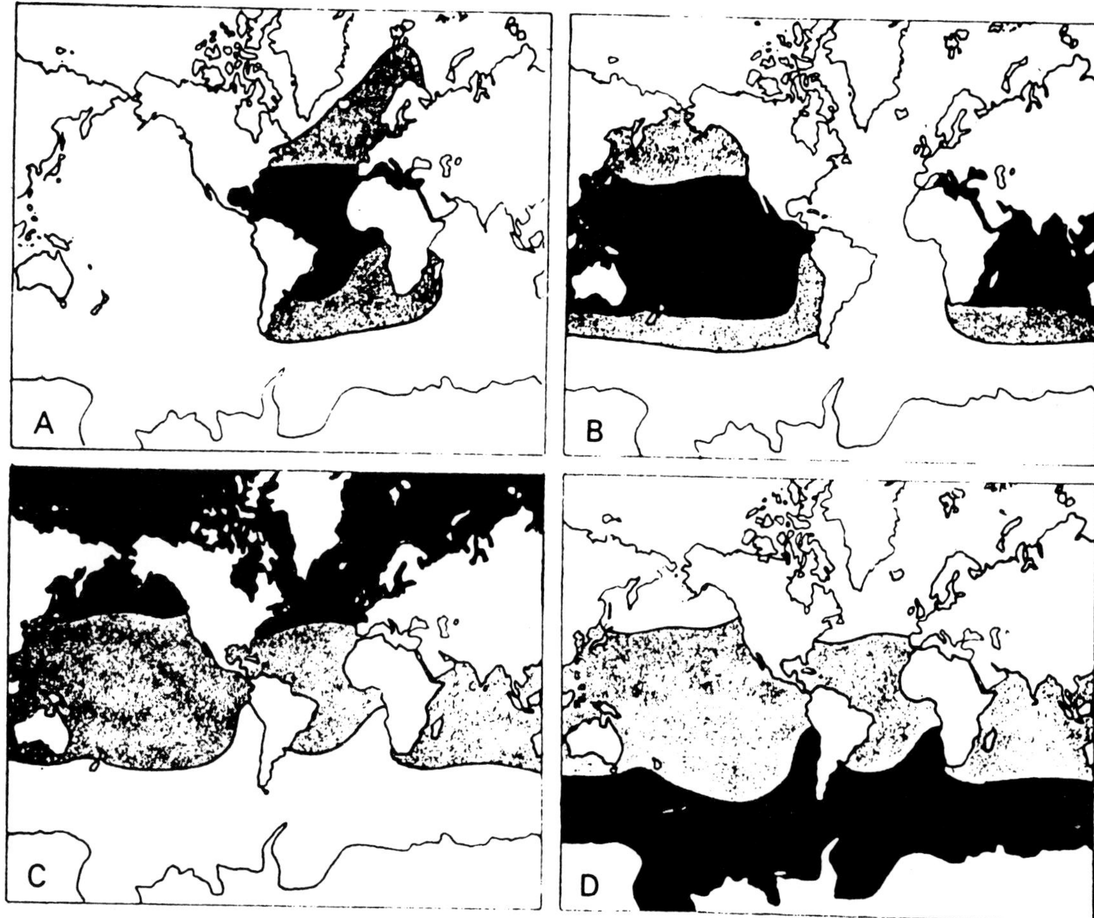

FIGURE 3.10 An attempt to delimit geographic types of phytoplankton distribtuion. From a compilation of the records available for 379 species of planktonic algae.

(I) Cosmopolitan, "euryoic" (i.e., tolerant to a large set of environmental variables): combination of C plus D.
(II) Cosmopolitan, temperate and warm waters: A + B.
(III) Cosmopolitan, warm waters: A + B, black areas only.
(IV) Cosmopolitan, cold waters: C + D, black areas only.
(V) Atlantic, euryoic: A.
(VI) Atlantic, warm waters: A, black area only.
(VII) Indo-Pacific, euryoic: B.
(VIII) Indo-Pacific, warm waters: B, black area only.
(IX) Boreal, euryoic: C.
(X) Boreal, cold waters: C, black area only.
(XI) Austral, euryoic: D.
(XII) Austral, cold waters: D, black area only. (After Margalef, 1961; rearranged by Sournia, 1978, from Smayda, 1979).

zooplankton, the supply of nutrient salts has only a secondary indirect effect, which, moreover, fades out because of the influence of currents and of the length of the food chain. For phytoplankton, these influences are direct, so that in fertile areas and regions of strong mixing, other patterns are found than those for zooplankton (van der Spoel and Heyman, 1983). Fronts and mesoscale eddy fields are important in determining the distribution of phytoplankton. The nature of fronts as boundaries between water masses and the enhancement of mixing along them leads to increased productivity and higher species diversity in their proximity (Pingree and Mardell, 1981; Zeitzschel, 1985).

McGowan (1971) compiled a list of 108 zooplankton species (including 17 species of foraminifera and 10 of radiolarians) with an indication of their distribution in the Pacific. On the basis of the com-

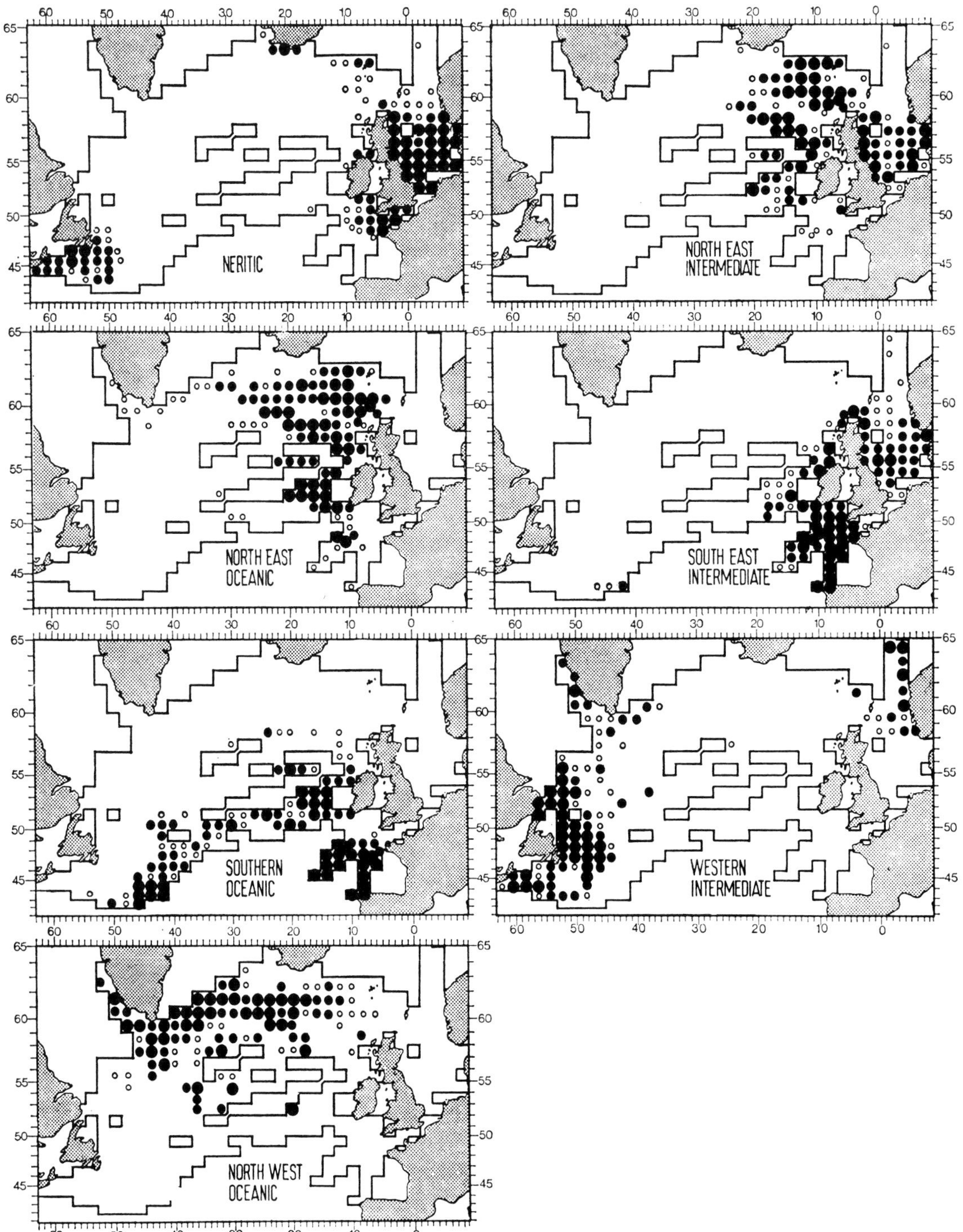

FIGURE 3.11 Charts showing the geographic distributions of classified groups of plankton organisms from the Continuous Plankton Recorder survey. (After Colebrook, 1972.)

piled information he proposed a number of generalizations about the patterns of zooplankton species distribution in the Pacific. These are

1. There is a remarkable amount of agreement between most of the species of all the taxa as to the position and shape of their distributional boundaries.
2. A large number of species in each group shows patterns of distribution the boundaries of which are almost identical with the boundaries of physical water masses.

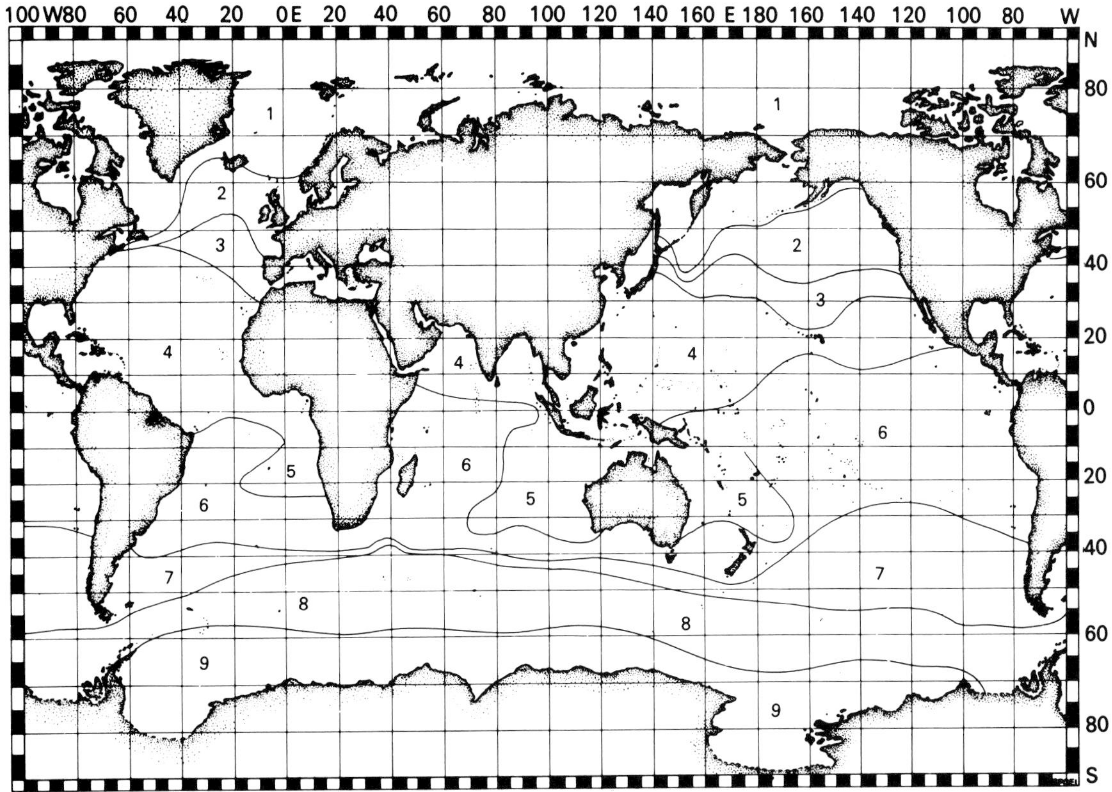

FIGURE 3.12 Phytoplankton geography of the world oceans. 1 = arctic area, 2 = arctic–boreal area, 3 = N. transitional area, 4 = tropical area, 5 = mixed tropical area, 6 = equatorial area, 7 = S. transitional area, 8 = subantarctic area, 9 = antarctic area. (After Beklemishev et al., 1977, from van der Spoel and Heyman, 1983).

3. A number of species have areas of the highest level of abundance within a water mass but their boundaries extend somewhat beyond the boundary of the water mass.
4. Many species are found throughout several water masses.
5. Some species are limited to certain large parts of some water masses.

Van der Spoel and Heyman (1983) provide a map of the important faunal centers in the world ocean (Fig. 3.14). This chart is included in their atlas to enable one to compare ranges that have a different geographic shape but are grouped around a fixed center or area. According to these authors, terrestrial and marine dispersal mechanisms are quite different. For terrestrial organisms, the faunal centers, from which after regressions taxa spread over larger distances, are geographically fixed areas, each with its distinct geologic and climatologic history. For planktonic animals, a faunal center has to be situated in a water mass and is therefore not geographically fixed. In the geologic past it may have shifted so that its history is uncertain. Moreover, real regression of planktonic animals will have been slight, as most planktonic taxa escape bad conditions by shifting their entire ranges.

Van der Spoel and Heyman (1983) define a center as an area with conditions more or less characteristic of the faunal elements of that center, and the taxa share the same ecological preferences or the same ancestral form. So, not all the centers included in Figure 3.14 are centers of speciation or dispersal; a number of them are merely ecologically defined areas.

Figure 3.15 depicts a diagrammatic areal subdivision of the world ocean as proposed by van der Spoel and Heyman (1983).

Only 14 main areas are depicted; the boundaries are based on the averages compiled from the literature. Surface waters forming the epipelagic realm are considered to reach from the surface down to 100 m. The mesopelagic zone is defined as the depth from 200 to 1,000 m, whereas the bathypelagic zone extends 1,000 m. Definitions of some of the terms frequently used by van der Spoel and Heyman (1983) are as follows: Central waters are the gyrally flowing waters enclosed by eastern and western boundary currents, equatorial currents, and the westward-flowing currents at about 40 de-

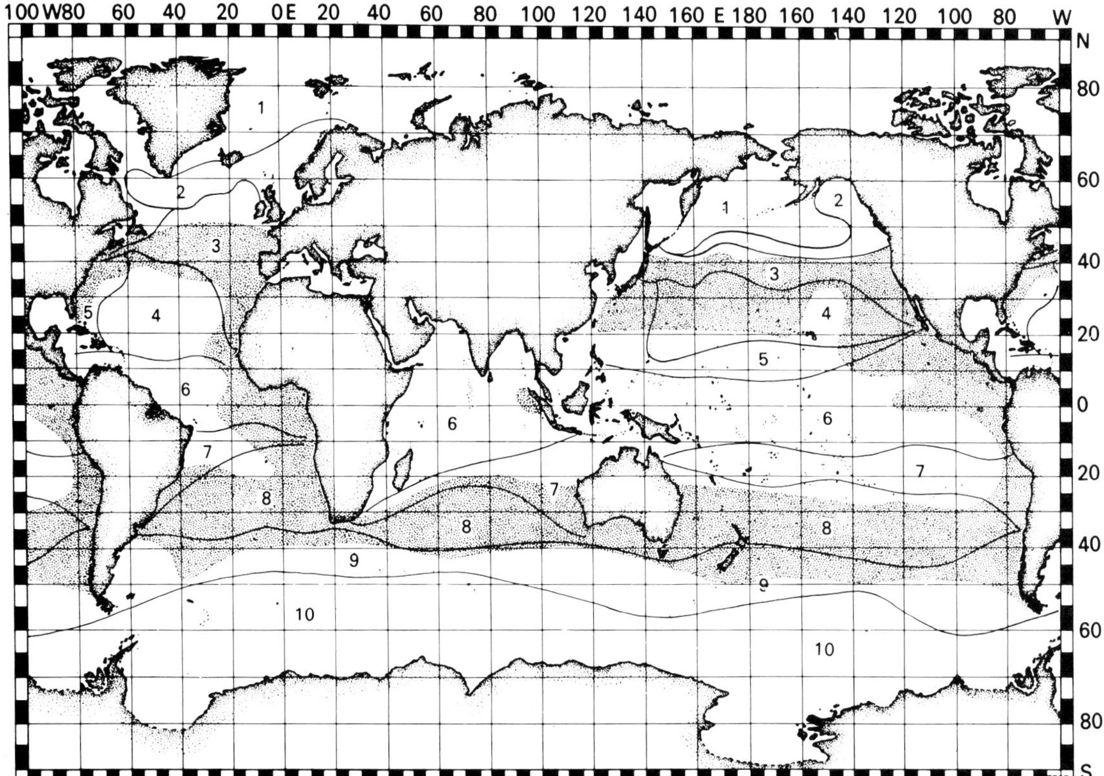

FIGURE 3.13 Zooplankton distribution pattern. 1 = arctic area; 2 = subarctic area; 3 = N. temperate area; 4 = N. subtropical area; 5 = N. subtropical area - terminal area; 6 = equatorial area; 7 = S. subtropical - terminal area, 8 = S. subtropical - central area; 9 = S. temperate area; 10 = antarctic area, dotted = areas of strongly mixing water masses. (After Beklemishev et al., 1977, from van der Spoel and Heyman, 1983.)

grees latitude. Boundary currents are divided into western boundary currents, which flow from equatorial to temperate areas, and eastern boundary currents, which flow from temperate to equatorial areas. Examples of the former are the Gulf Stream and the Kuroshio and Brazil Currents. Examples of the latter are the Canary, Benguela, and Humboldt or Peru Currents. Neritic areas are the areas over the continental shelf, theoretically bordered by the 200-m depth line, with great fluctuations in temperature and salinity, usually with relatively high production.

Terminal water is a term often used also in publications by Soviet scientists for the mixed waters of a latitudinal current between pairs of large gyres. The term *cyclical* is then applied to the large gyres themselves. Transitional waters are the waters of a strongly mixed character between subpolar waters and subtropical waters.

Concerning the zoogeography of marine protozoa, sufficient data are only available for three shell- or lorica-bearing groups: foraminifera, radiolaria, and tintinnids mainly from the surface or from integrated net samples. Some vertical migration has been reported from these groups of protozoa, but the active migration seems to be negligible compared with the vertical migration of the meso- and macroplankton.

BIOGEOGRAPHY OF FORAMINIFERA

Recent benthic foraminifera are perhaps the best known of the marine protozoa with respect to their distribution and ecology. Since this topic has been monographed many times (e.g., Phleger, 1960; Murray, 1973; Boltovskoy and Wright, 1976) and is being reviewed by a specialist in a book on foraminifera (Murray, in Lee, J.J., and Anderson, O.R. [eds.] *Biology of Foraminifera*), this protozoan zoogeographic topic has largely been excluded from consideration in this chapter.

Bé (1977) reviewed the biogeographic distribution of planktonic foraminifera in the epipelagic zone of the world ocean. One can recognize first-order patterns of a circumpolar nature, as well as second-order patterns, which appear to be the result of genetic isolation and provincialism.

According to Bé and Tolderlund (1971), many cold-water species have a bipolar distribution, in-

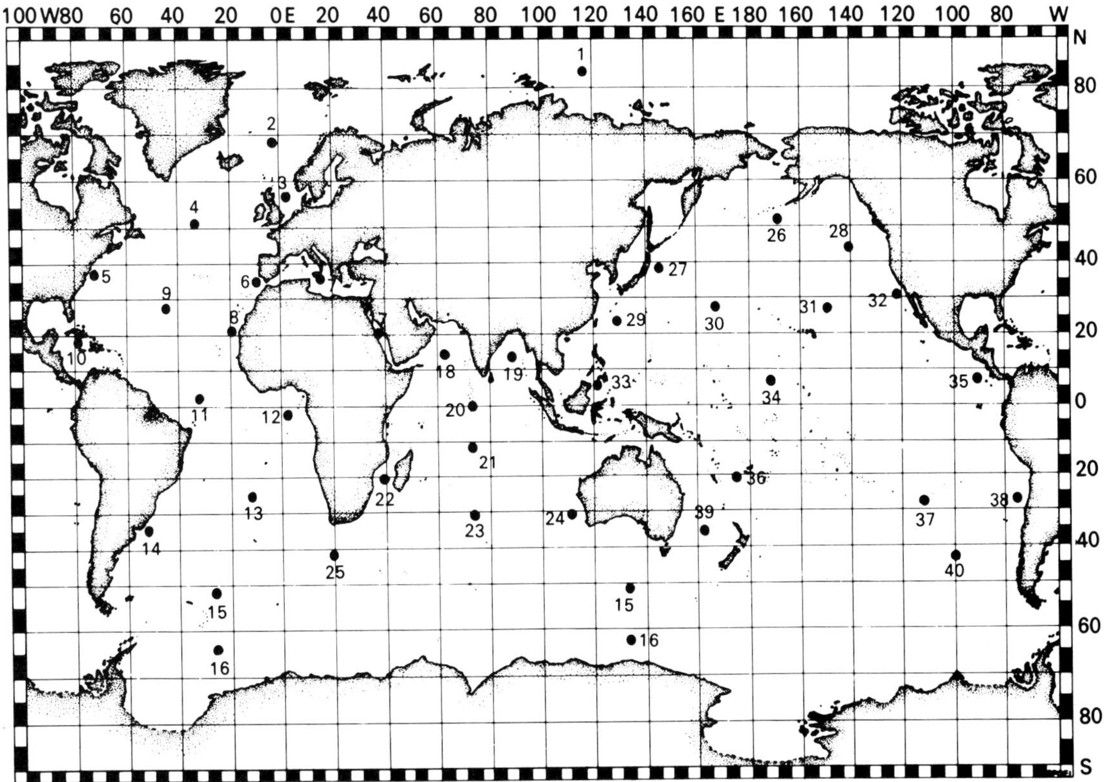

FIGURE 3.14 Faunal centers of the world ocean. 1 = Arctic C.; 2 = Subarctic Atlantic C.; 3 = North Sea C.; 4 = N. Transitional Atlantic C.; 5 = N.W. Atlantic C.; 6 = Atlanto-Mediterranean C.; 7 = Mediterranean C.; 8 = N.W. African C.; 9 = N. Subtropical Central Atlantic C.; 10 = Caribbean C.; 11 = Equatorial Atlantic C.; 12 = E. Equatorial Atlantic C.; 13 = S. Subtropical Central Atlantic C.; 14 = Argentina C.; 15 = Subantarctic C.; 16 = Antarctic C.; 17 = Red Sea C.; 18 = Arabian C.; 19 = Bay of Bengal C.; 20 = Equatorial Indian C.; 21 = S. Tropical Terminal Indian C.; 22 = Madagascar C.; 23 = S. Subtropical Central Indian C.; 24 = W. Australian C.; 25 = Agulhas C.; 26 = Subarctic Pacific C.; 27 = Japan C.; 28 = N.E. Transitional Pacific C.; 29 = Kuroshio C.; 30 = N.W. Subtropical Central Pacific C.; 31 = N.E. Subtropical Central Pacific C.; 32 = California Current C.; 33 = Indo-Malayan C.; 34 = Equatorial Pacific C.; 35 = E. Equatorial Pacific C.; 36 = S.W. Subtropical Central Pacific C.; 37 = S.E. Subtropical Central Pacific C.; 38 = Peru C.; 39 = Tasman Sea C.; 40 = S.E. Transitional Pacific C.

habiting both the northern and the southern cold-water provinces. Their present disjunct distribution has been variously explained. One of the most plausible explanations is that from a former continuous distribution the recent isolation of bipolar species is caused by a warming of the ocean waters since the Würm glacial epoch. This has consequently resulted in the reduction of this geographic range (Fig. 3.16).

Stenothermal cold-water species, such as the chaetognath *Eukrohnia hamata* (Alvarino, 1965), which have continuous distribution by "submerging" below the tropical–subtropical waters, are very rare. None of the planktonic foraminiferal species is known to exhibit deep tropical submergence. Even rarer are the eurythermal, cosmopolitan species, which have a worldwide distribution in all water types. *Globigerinita glutinata* is a rare example in this category.

Bé and Tolderlund (1971) categorized species assemblanges of foraminifera according to their distributional pattern in the world ocean (Fig. 3.17). Figure 3.18 shows the total and optimum surface temperature ranges of individual species of planktonic foraminifera. They are based on correlations of surface temperatures and relative abundance estimates of the different species.

Figure 3.19, shows, as an example of results obtained by principal component analysis, a composite map of regional dominance of nine life assemblages of foraminifera in the Indian Ocean. A total of 154 vertical plankton tows, down to a depth of 300 m, have been considered. The dominant species in each assemblage are illustrated (after Bé and Hutson, 1977).

In warm-water areas *Pulleniatina obliquiloculata* is distributed between 30 degrees north and 30 degrees south (Fig. 3.20, Bé, 1977). This species is

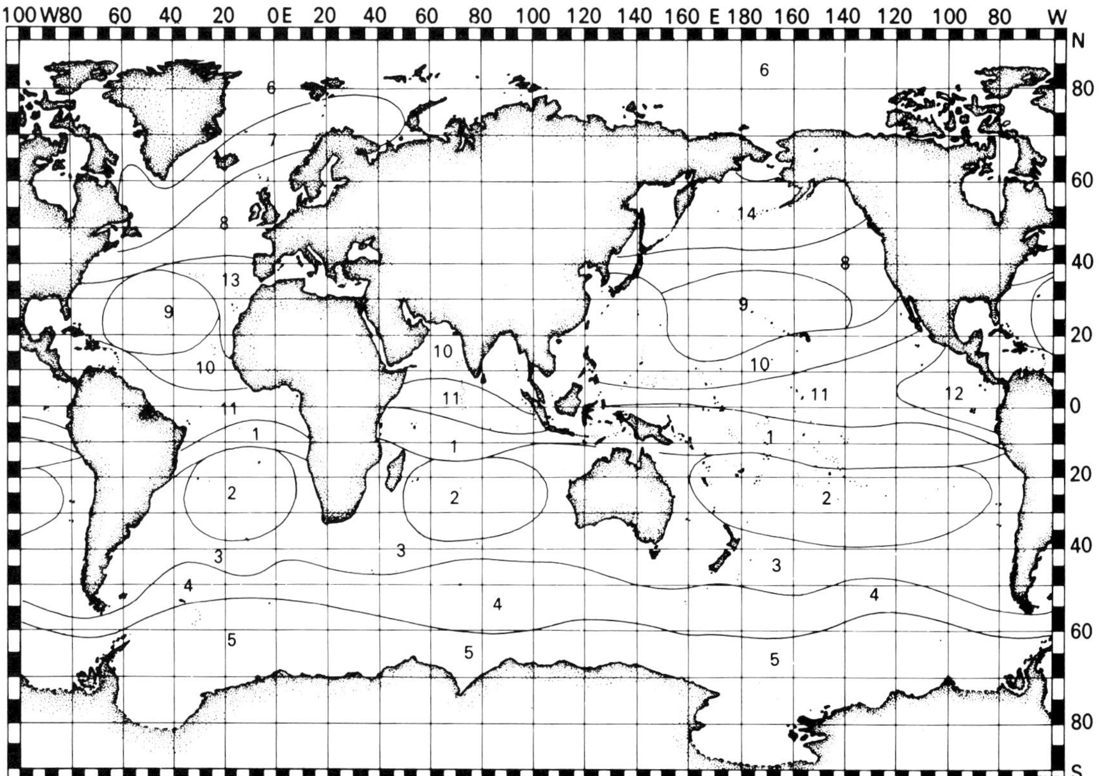

FIGURE 3.15 Diagrammatic areal subdivision of the world ocean. 1 = tropical - terminal water; 2 = S. Tropical - central water; 3 = transitional water; 4 = subantarctic water; 5 = antarctic water; 6 = arctic water; 7 = subarctic water; 8 = N. transitional water; 9 = N. subtropical - central water; 10 = N. tropical - terminal water; 11 = equatorial water; 12 = E. Pacific water; 13 = Atlanto–Mediterranean water; 14 = cool - temperate water. All transitional, subtropical, and tropical waters bordering continents can be considered boundary current waters. (After van der Spoel and Heyman, 1983.)

also reflected in the sediment records. The highest abundance is adjacent to the equator. The possibility of a passage south of South Africa, but reflected in the surface sediments, makes it acceptable that for some taxa, a circumglobal range may be found, although the normal boundaries of this foraminifera species are in the previously mentioned limits.

The distribution of living *Globoquadrina pachyderma* is an example of bipolarity (Fig. 3.21; Bé, 1977). In the surface sediments, however, distribution with a full north–south continuity in the tropical East Atlantic and almost a connection in the tropical East Pacific is found. According to van der Spoel and Heyman (1983) the question remains of whether real monotypic bipolar taxa exist and of whether such taxa do indeed entirely lack north–south connections.

BIOGEOGRAPHY OF RADIOLARIA

Anderson (1983) published a comprehensive account on the biology of radiolaria, including their geographic distribution. Much of this section on radiolarian biogeography is drawn from this work.

One of the earliest programs of open-ocean sampling has been the British *Challenger* expedition conducted from 1873 to 1876. The results of the analysis of radiolarian samples were reported by Haeckel (1887). He concluded that radiolaria occur in all the seas, in all climatic zones, and from the surface layers to great depths in the water column. His samples were taken in plankton nets and from sediment cores. Among the approximately 4,000 species that he examined, many were undoubtedly fossil forms obtained from the sediments. Perhaps only 400 or 500 of the more common polycystine radiolaria are living in the oceans today (Casey et al., 1979). Haeckel determined that some radiolarian species are limited to certain bathymetric faunal zones. He recognized three zones: Pelagic Faunal, Zonarial Faunal, and Abyssal Faunal. The Pelagic Faunal zone occurs from the surface to about 46 m. The radiolaria found here consist largely of Spumellaria and Acantharia with a few members of Nassellaria and Phaeodaria. The Zonarial Fauna

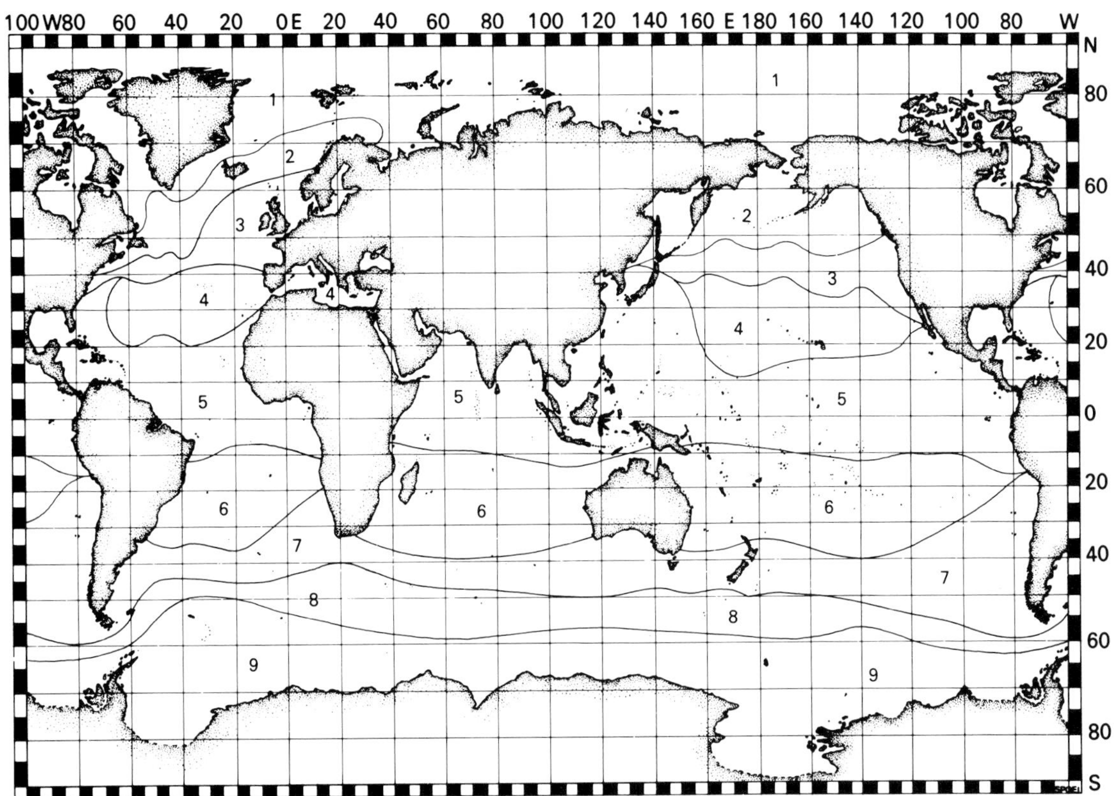

FIGURE 3.16 World distribution zones of planktonic foraminifera. 1 = arctic zone; 2 = subarctic zone; 3 = N. transitional zone; 4 = N. subtropical zone, 5 = tropical zone, 6 = S. subtropical zone. 7 = S. transitional zone, 8 = subantarctic zone, 9 = antarctic zone. (After Bé and Tolderlund, 1971, from van der Spoel and Heyman, 1983.)

occurs in strata at various bathymetric depths between the Pelagic Fauna and the Abyssal Fauna. In the upper portions of the Zonarial Faunal zone (46–3,656 m), Spumellaria predominate but are replaced gradually by Nassellaria and Phaeodaria in the deeper strata (3,656 m to just above the ocean floor). The Abyssal Fauna encompass largely Phaeodaria and Nassellaria that float very near to the ocean floor. Geographically, Haeckel reported that the greatest diversity and largest number of radiolarian species occur in the tropics. The abundance of species gradually declines toward the poles. This gradient is steeper in the Northern Hemisphere than in the Southern Hemisphere, and the Southern Hemisphere appears to have more species than the Northern Hemisphere. The richest diversity and greatest abundance of radiolaria were found in the Pacific (Anderson, 1983).

Subsequent research in the early twentieth century on general patterns of radiolarian distribution has been concisely reviewed by Casey (1971a, 1971b, 1977), who proposed a convenient model of biogeographic zones suitable for use with polycystine radiolaria. Casey (1971a) proposed seven geographic zones distributed from north to south and further refined on the basis of depth into shallow zones (surface to 100 or 200 m, and perhaps 400 m in the lower latitudes) and deeper zones (Fig. 3.22). The hydrographic conditions correlated with each of the seven shallow-water zones are given from Casey (1971a):

1. *Subarctic Faunal Zone*. Waters north of the North Pacific drift and the Subarctic Convergence (also sometimes termed Arctic or Polar Convergence).
2. *Transition Faunal Zone*. The North Pacific drift waters bounded on the north by the Subarctic Convergence and on the south by the Subtropical Convergence.
3. *North Central "shallow" Faunal Zone*. Waters within the large anticyclonic circulation pattern of the North Pacific, which could be divided easily into two parts (east and west) as the circulation and Central water masses are.
4. *Equatorial Faunal Zone*. The regions occupied by the North and South Equatorial Current systems.
5. *South Central "shallow" Faunal Zone*. Waters within the large anticyclonic circulation pattern

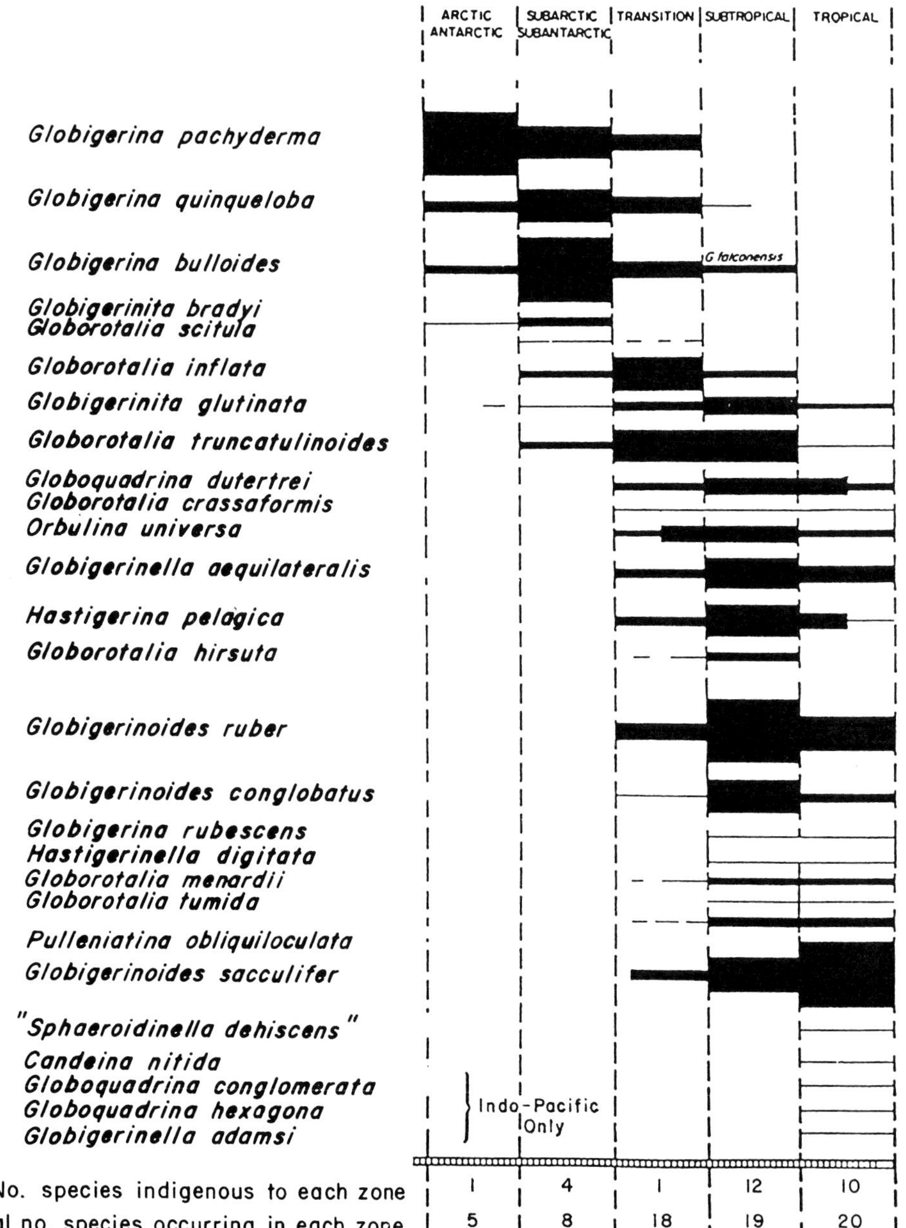

FIGURE 3.17 Species assemblages and ranges in the five world distributional zones. Varying thickness represents relative abundance within each zone. (After Bé and Tolderlund, 1971.)

of the South Pacific, which could be divided easily into two parts (east and west) as the circulation and Central water masses are.

6. *Subantarctic Faunal Zone.* Waters bounded to the north by the Subtropical Convergence (the Subantarctic Convergence) and to the south by the Polar Convergence (the Antarctic Convergence).
7. *Antarctic Faunal Zone.* Waters bounded by the Polar Convergence on the north and the Antarctic Continent on the south.

Some species of radiolaria may be endemic to or characteristic of certain water masses below the shallow-water zone. Casey suggests the following classification for these indicator fauna:

1. The North Transition–Central fauna, which is endemic to waters of the North Pacific Central water mass (or masses, east and west), is shallow at the areas of formation and dives with the water mass at the North Pacific Subtropical Convergence.

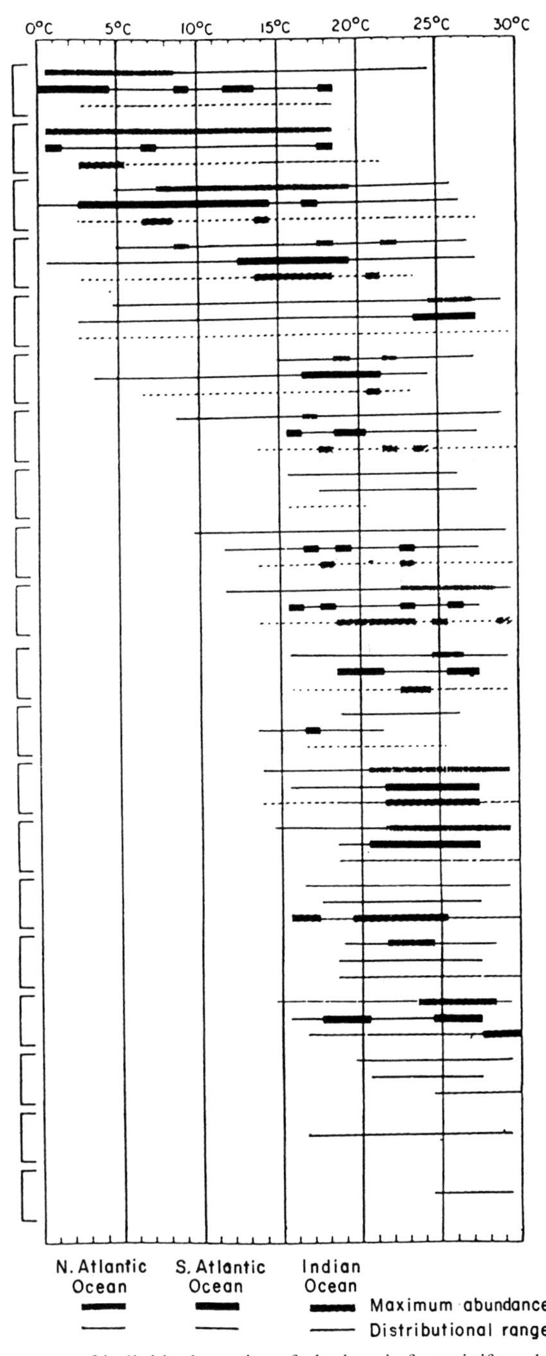

FIGURE 3.18 Total and optimum surface temperature ranges of individual species of planktonic foraminifera, based on correlation of surface temperatures and species relative abundance data. (After Bé and Tolderlund, 1971.)

2. The Subarctic–Intermediate fauna endemic to waters of the North Pacific Intermediate water mass, which is shallow north of the Subarctic Convergence and dives with the waters at the Convergence.

A general distributional scheme for polycystine radiolaria that is applicable to the world ocean and adjacent seas and incorporates the foregoing hydrographic and biogeographic data has been proposed by Casey (1971a, p. 156) as follows:

I. Shallow Water Faunal Zones (approximately 0–200 m)
   1. Polar (60 degrees north or south and greater)

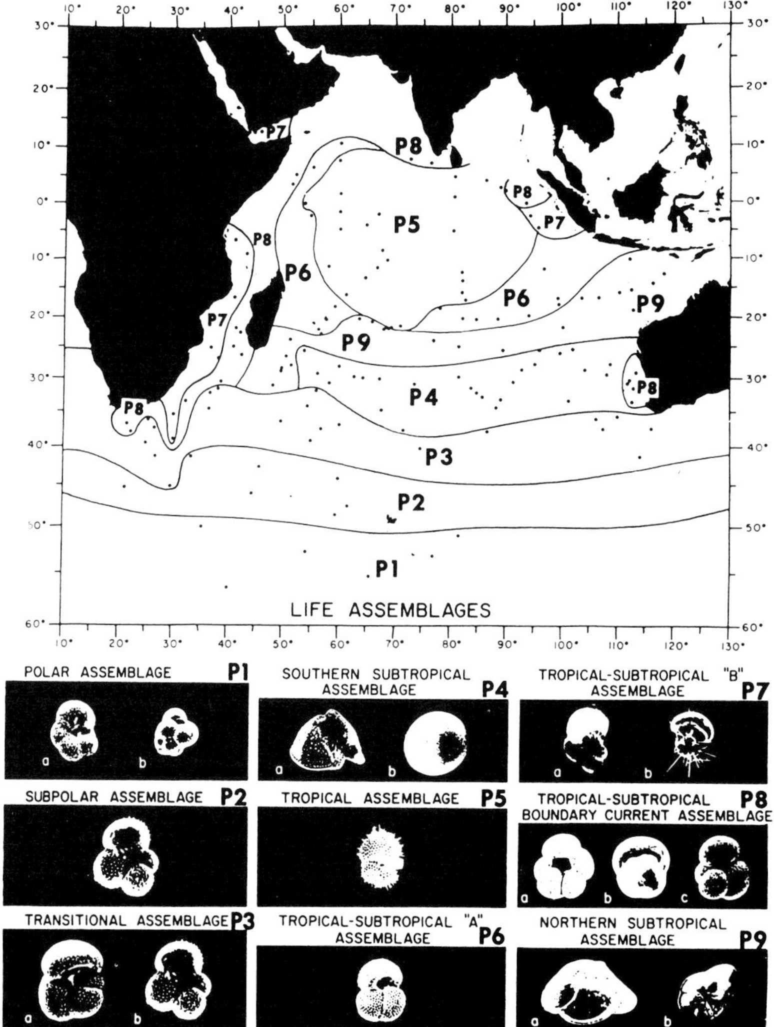

FIGURE 3.19 Composite map of regional dominance of nine life assemblages of plankton foraminifera, based on 154 vertical tows in the upper 300 m of water. Dominant species in each assemblage are illustrated. (After Bé and Hutson, 1977.) P1: *Globoquadrina pachyderma, Globigerinita uvula*; P2: *Globigerina bulloides*; P3: *Globigerina bulloides, Globorotalia inflata*; P4: *Globorotalia truncatulinoides, Orbulina universa*; P5: *Globigerinoides sacculifer*; P6: *Globigerinoides ruber*; P7: *Globigerinella aequilateralis, Hastigerina pelagica*; P8: *Globoquadrina dudertrei, Pulleniatina obliquiloculata*; P9: *Globorotalia menardii, Globorotalia tumida*.

    (1) Subarctic
    (2) Antarctic
2. Subpolar (50–60 degrees north or south)
    (1) Transition (North Pacific and North Atlantic)
    (2) Subantarctic
3. Central (10–50 degrees north or south)
    (1) North (may be divided into east and west where appropriate)
    (2) South (may be divided into east and west where appropriate)
4. Equatorial (0–10 degrees north or south)
5. Special (adjacent seas such as Arctic and Mediterranean shallow-water faunas)

II. Deep-Water Faunal Zones (diving below or existing below Shallow-Water faunas)
1. Central
    (1) Transition-Central (0–40 degrees south, 200–700 m)
    (2) Subantarctic-Central (0–40 degrees north, 200–700 m)
2. Intermediate
    (1) Subarctic-Intermediate (equator > 900 m to Arctic > 200 m)
    (2) Antarctic-Intermediate (equator > 900 m to Antarctic > 200 m)
3. Also Common, Deep, and Bottom, and Bottom Faunal Zones (in oceans where appropriate)

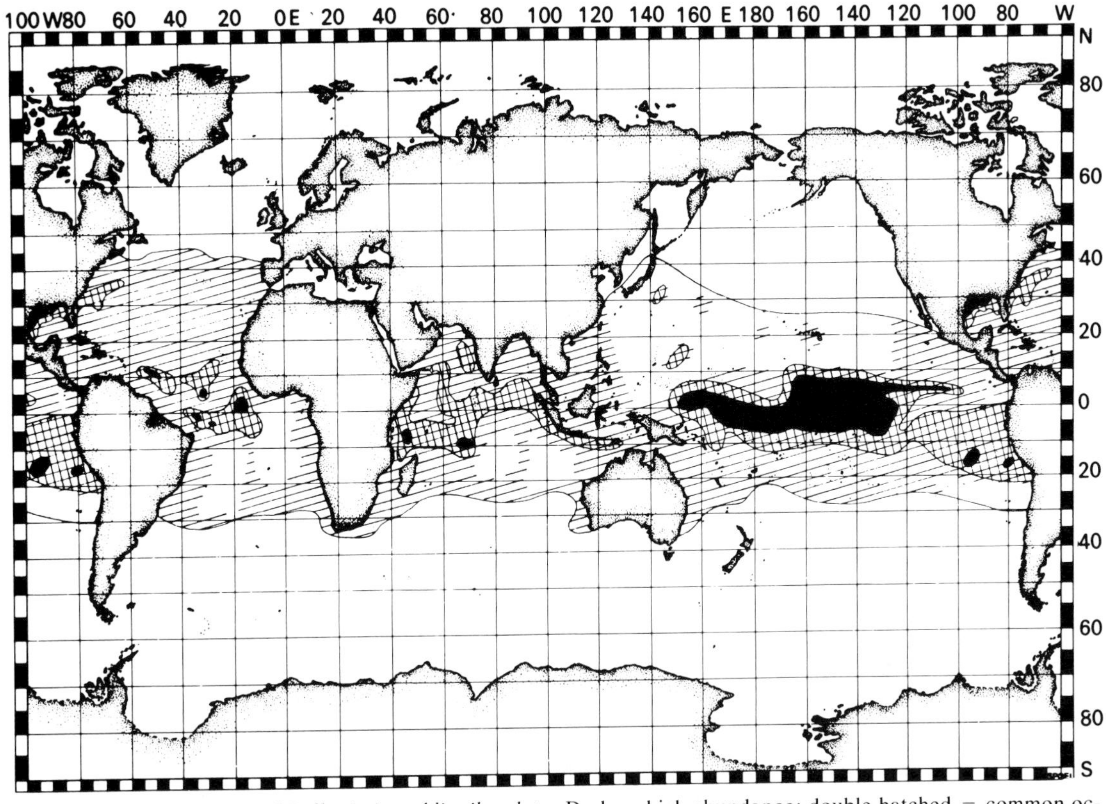

FIGURE 3.20 Distribution of *Pulleniatina obliquiloculata*. Dark = high abundance; double hatched = common occurrence; hatched = only recorded from surface sediments. (After Bé, 1977, from van der Spoel and Heyman, 1983.)

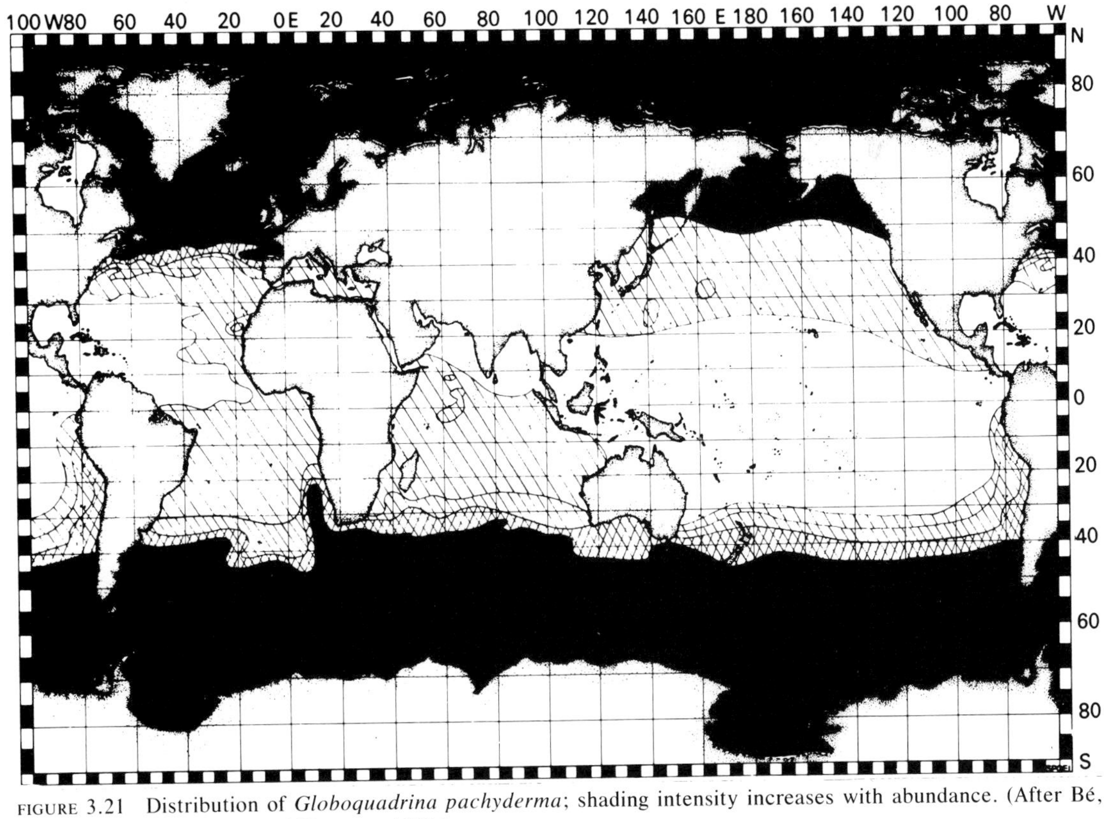

FIGURE 3.21 Distribution of *Globoquadrina pachyderma*; shading intensity increases with abundance. (After Bé, 1977, from van der Spoel and Heyman, 1983.)

4. Special (adjacent seas, such as Arctic and Mediterranean deep-water faunas)

III. Cosmopolitan Faunal Zone (cutting across other faunal zones)
  1. Shallow
  2. Deep

IV. Narrow Endemics (within a part of any previous zone)

The intermediate water zones in section II-2 are water masses that occur at great depths (> 900 m) at the equator but gradually extend toward the surface near the poles, where they encompass all the water columns beneath 200 m, as illustrated for the Pacific Ocean in Figure 3.22.

The major geographic distribution of some commonly occurring extant polycystine and phaeodarian radiolaria published by Haeckel (1887) is presented in Table 3.5. These data have been synthesized from several sections of Haeckel's report, including distributional data presented at the end of species descriptions and in the narrative. Where possible, the habitats have been coded to correspond to Casey's biogeographic model (1971a).

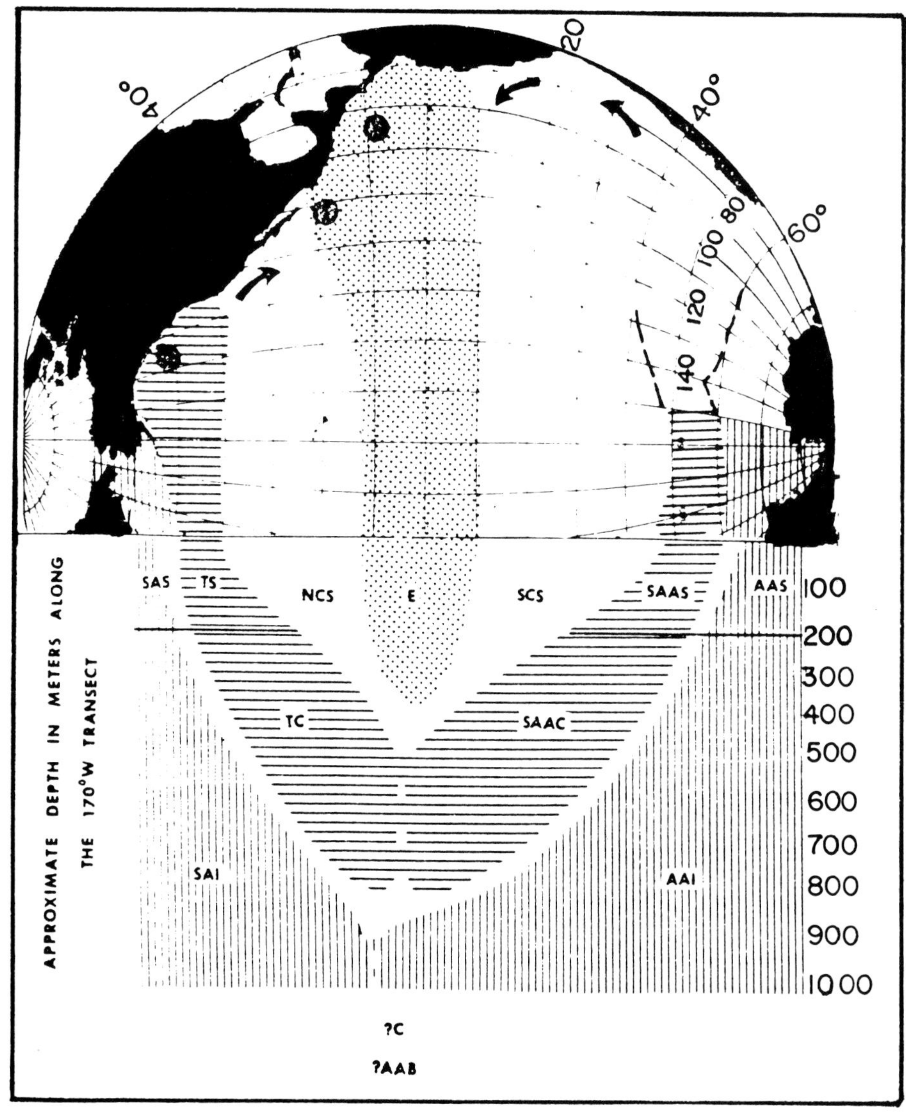

FIGURE 3.22 Biogeographic zones for polycystine radiolarians in the Pacific Ocean. ASS = Antarctic shallow faunal zone; AAI = Antarctic intermediate faunal zone; SAAS = Subantarctic shallow faunal zone; SAAC = Subantarctic central faunal zone; SCS = South central shallow faunal zone; E = Equatorial faunal zone; NCS = North central shallow faunal zone; TC = Transition central faunal zone; TS = Transition shallow faunal zone; SAI = Subarctic intermediate faunal zone; SAS = Subarctic shallow faunal zone. Perhaps a Common faunal zone (C) exists below the Intermediate zones and an Antarctic bottom faunal zone (AAB) exists below the Common zone. (After Casey, 1971, from Anderson, 1983.)

TABLE 3.5 Distribution of Some Solitary Radiolaria.

Species	Location		
	Atlantic Ocean	Pacific Ocean	Indian Ocean
Spumellaria			
Actissa princeps		Surface [E]	Ceylon, Surface [E]
Thalassolampe margarodes	Med., Canary Is.—surface [E]		
Thalassicolla nucleata	Mediterranean and cosmopolitan in all oceans between 40°N and 40°S, Surface [NCS, E, SCS]		
Cenosphaera primordialis		Central, surface [E]	Surface [E]
C. coronata		Central, 4,800 m	
C. gigantea		Central, 5,300 m	
C. reticulata	Med. (Messina), surface [E]		
C. tenerrima		Central, surface [E]	
Corposphaera infundibulum	North		
C. melissa		Central, 5,300 m	
C. prunulum	South, surface [SCS]		
C. nobilis	Cosmopolitan in all oceans of various depths		
Plegmosphaera pachylplegma		Central, surface [E]	
P. leptoplegma	North, surface [NCS]		
Saturnalis circularis	South, 4,000 m		
S. circoides			Zanzibar, 4,000 m [AAB]
S. rotula		North, surface [NCS]	
Stylosphaera musa	Tropical, 4,100 m		
S. polyhymnia	Cosmopolitan in all oceans, surfaces		
Hexostylus phoenaxonius		Central, 4,700 m	
H. sapientum	North, surface [NCS]		
H. maximus		Central, 5,300 m	
H. marginatus		South, 2,700 m	
Hexacontium hexagonale			Ceylon, surface [E]
Clodococcus arborescens	Med., North, Canary Isl.—surface [E to NCS]		
C. antarcticus			Antarctic, 3,600 m [SAI]
C. quadricuspis		Central, 5,300 m	
C. japonicus		North, surface [NCS]	
Elaphococcus furcatus	Tropical, surface [E]		
E. dichotimus	Arctic, surface		
E. umbellatus		Southeast, surface [SCS]	
Spongiomma radiatum		Central, surface [E]	
S. asteroides	South, surface [SCS]		
Spongopila dichotoma	Tropical, surface [E]		
S. verticillata		Tropical, surface [E]	
Centrocubus octostylus		Central, surface [E]	
Octodendron verticullatum		South, surface [SCS]	

TABLE 3.5 *(continued)*

Species	Location		
	Atlantic Ocean	Pacific Ocean	Indian Ocean
Spongosphaera polyacantha	Med. (Nice), surface [E]		
Spongodrymus elaphococcus	Tropical, surface [E]		
S. quadricuspis		Central, surface [E]	
Cannartus violina		Central, 5,300 m	
Cyphonium coscinoides		North, surface [NCS]	
C. ethmarium	Equatorial, surface [E]		
C. diattus			Western, Zanzibar, 4,000 m
Panartus tetraplus		Central, 4,300–5,300 m	
P. tetracolus	Equatorial, 4,410 m		
P. tetrathalamus	Cosmopolitan, various depths in all oceans		
Porodiscus orbiculatus	Cosmopolitan, surface		
Spongodiscus mediterraneus	Med.		
S. radiatus		Central, 5,300 m	
S. favus	North (Greenland) surface [SAS]		
Tholospira nautiloides			Between Ceylon & Socotra, Surface [E]
T. spinosa		South, surface [SCS]	
Phorticium pylonium	Cosmopolitan, common in all oceans, surface and various depths		
Soreuma irregulare		North, 5,300 m	
S. acervulina		South, 2,700 m	
S. setosum		Central, 4,400 m	
Nassellaria			
Plagonium sphaerozoum	Equatorial, surface [E]		
P. lampoxanthium		North, Surface [NCS]	
P. arborescens			Madagascar, Surface [E]
P. trigeminum		Central, 5,300 m	
P. distractis		South, Surface [SCS]	
Cortina tripus	Cosmopolitan, common in all oceans, surface, and various depths		
C. typus	Tropical, 4,500 m		
C. dendroides		Central, 4,300–5,300 m	
Semantis biforis		Central, 4,900–5,300 m	
S. distephanus	Tropical, surface [E]		
Dorcadospuris denata		Central, 4,300–5,300	
Carpocanium diodema	Cosmopolitan in many locations in all oceans, surface		
C. laeve	Med., tropical, 4,470 m		
C. cylindricum		Central, 4,400 m	
Pterocanium gravidum	South, 4,000 m [AAB]		
P. pyramis		Central, 5,000 m	

TABLE 3.5 *(continued)*

Species	Location		
	Atlantic Ocean	Pacific Ocean	Indian Ocean
P. depressum			Zanzibar, 4,000 m [AAB]
P. trilobum	Med. (Messina), surface		
Podocyrtis tripodiscus		Central, 4,400–5,400 m [AAB]	
P. conica	Cosmopolitan, tropical Atlantic and Pacific, 4,300–5,300 m [AAB]		
P. tridactyla			Madagascar [E]
P. ovata	Med., Surface [E]		
P. argulus	Tropical, 4,600 m		
Theocorys turgidula		Tropical, surface [E]	
T. veneris	Cosmopolitan, abundant, surface		
T. obliqua		Central, 5,300 m	
T. martis		South, 2,700 m	
Lithochytris cortina		Central, 5,000 m	
L. pyriformis	Tropical, 4,470 m		
L. lucerna		South, 3,200 m	
Callimitra carolotae		Central, 5,300 m	
C. emmae		Central, 5,000 m	
Cyrtocapsa tetrapera		Western, Tropical, 8,200 m	
C. cornuta		Central, 4,700 m	
C. diploconus	Tropical, 4,100 m		
C. fusulus		South, 2,700 m	
Phaeodaria			
Cannoraphis spinulosa		North, Surface	
C. lamphoxanthium		South, 4,700 m	
C. spathillata			Cocos Islands, surface [E]
Aulacantha scolymantha	Cosmopolitan, common, surface and at various depths		
A. cannulata		South, surface	
A. clavata	South, 3,723 m		
Aulographis pandore	Cosmopolitan, in all oceans, surface to various depths		
A. bovicornis	South, surface [SCS]		
A. triangulum		South, 4,700 m	
A. stellata			Madagascar, surface
Aulosphaera trigonopa	Cosmopolitan, all oceans, surface		
A. flexuosa	North (Faeröe Channel), surface [TS to SAS]		
A. verticillata		South, surface [SCS]	
A. dendrophora		Central, 4,400 m	
Challengeria naresii	Cosmopolitan, 1,800–5,500 [AAB]		
C. sigmodon		North, 4,100 m	
C. pyramidalis	South, 3,700 m		

TABLE 3.5 *(continued)*

Species	Location		
	Atlantic Ocean	Pacific Ocean	Indian Ocean
C. elephas			Cocos Islands, surface [E]
C. murrayi		Northwestern, 4,100 m	
Tuscarora tubulosa		North, 3,700–5,600 m [AAB]	
T. tetrahedra	Tropical, 4,470 m		
T. belknapii		South, 3,700 m	
Coelodendrum ramosissimum	Cosmopolitan, surface to various depths		
C. lappaceum		South, 2,700–4,700 m [AAB]	
C. digitatum			Madagascar, surface [E]
Coelotholus regina		Southeastern, 2,500 m	
Coelographis regina		southeastern, 3,300 m	
C. hexostyla		North, 4,100 m	
C. gracillima	Med., Surface [E]		

*Table taken from Anderson (1983) which was based on data from Haeckel (1887). Abbreviations used: Med., Mediterranean; Isl., Island. Depth of station is given in meters (m). Symbols in parentheses are codes for Biogeographical Zones (Casey, 1971a); AAB, Antarctic Bottom Faunal Zone; E, Equatorial, NCS, North Central Shallow Zone; SAI, Subarctic Intermediate Faunal Zone; SAS, Subarctic Shallow Faunal Zone; and SCS, South Central Shallow Faunal Zone. These are assigned based on the best estimate from Haeckel's information. No assignment is made when the information is too limited to do so.

Among these few species chosen from each of the major radiolarian groups, there are clearly diverse faunal zones. Some dwell in near-surface waters, whereas others occur at great depths. They occur in all major oceans, and some species are cosmopolitan, widely distributed throughout the world's oceans. More recent studies of solitary polycystine radiolarian distributions have begun to elucidate abundance patterns in relation to physical, chemical, and biological factors in the environment.

Figure 3.23 gives an example of the worldwide distribution of two species of radiolaria from Goll (1976). *Phormospyris stabilis* and *Lophospyris pentagona* have centers of distributions in the Eastern Tropical Pacific, the Tropical Indian Ocean, the Eastern Atlantic Boundary waters, and the Atlantic Transitional waters.

BIOGEOGRAPHY OF TINTINNIDA

Publications on the regional distribution of tintinnids are relatively rare as compared with worldwide distributional maps of foraminifera and radiolaria. Zeitzschel (1966) used data from the Continuous Plankton Recorder survey to describe the distribution pattern for five of the more common tintinnid species in the North Atlantic. In addition, data from the following expeditions and publications were incorporated: Brandt (1906, 1907, 1910), Campbell (1942), Candeias (1930), Gaarder (1946), Grøntved and Seidenfaden (1938), Jørgensen (1899), Margalef and Duran (1953), Ostenfeld (1899, 1900), and Silva (1950). The genus *Acanthostomella*, for example, has two sharp divisions in distribution, one in distinctly tropical areas and the other in cold-water regions. The cold-water species *Acanthostomella norvegica* is widely distributed in the Northwest Atlantic Ocean and is seldom found south of the northerly side of the Gulf Stream. It can be very abundant over and near the Newfoundland Banks and not so common in the waters near the Shetland Islands. According to Grøntved and Seidenfaden (1938), this species is absent in regions with drift ice near Greenland and in actual Arctic waters. *A. norvegica* is a typical boreal species. The seasonal and vertical distribution of this species from two fjordlike bays in insular Newfoundland has been described by Davis (1985).

These observations were extended by a more detailed study by Lindley (1975), who presented distributional maps for 38 species (or entities) of tintinnids in the North Atlantic on the basis of data of the year 1965 from the Plankton Recorder survey. The distributions could be assigned to the geographic classification for this area, derived by Colebrook (1972). Figure 3.24 gives examples for the distribution of five species of tintinnids in relation to the seven proposed geographic distributions of classified groups of plankton depicted in Figure 3.11 after Colebrook (1972). In the "Southeast In-

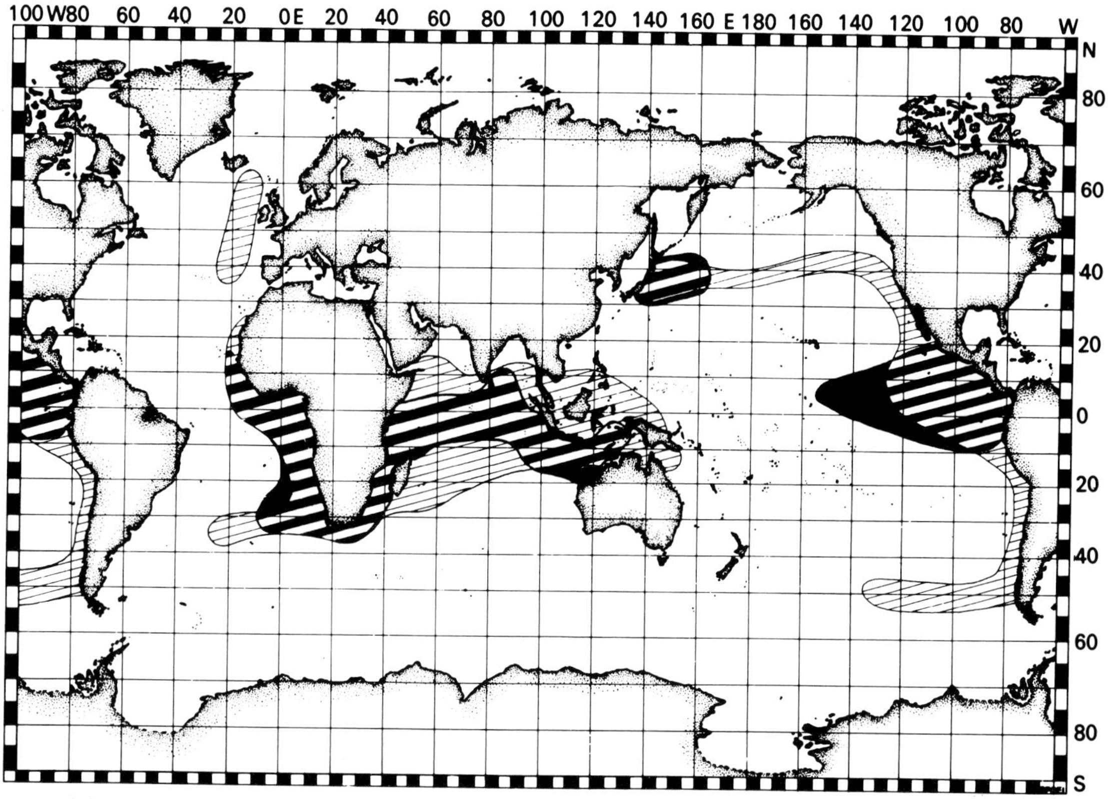

FIGURE 3.23 Distribution of *Phormospyris stabilis* (dark) and *Lophospyris pentagona* (hatched). (After Goll, 1976, from van der Spoel and Heyman, 1983.)

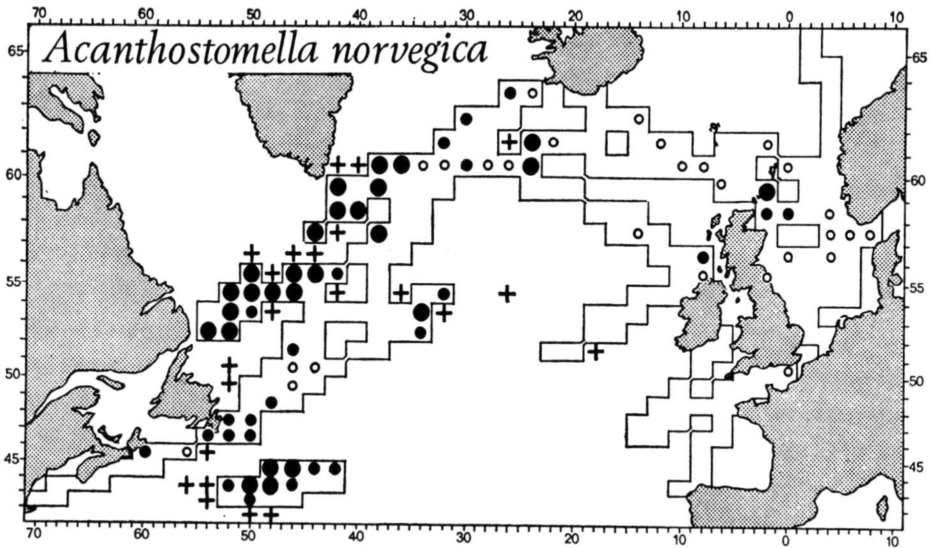

FIGURE 3.24 Geographic distribution of five selected species of tintinnids from the analysis of the Continuous Plankton Recorder in 1965. The area outlined in the charts is that which was sampled in at least three months of the year. Within this area, in most of the charts, the occurrence of a species within a standard rectangle (1° lat. by 2° long.) is shown by a dot. As a measure of abundance, three categories are given with an equal number of rectangles in each category. The category levels (in mean log) are as follows: *Acanthostomella norvegica*, 0.24–0.60; *Codonellopsis lagenula*, 0.30–1.00; *Dictyocysta elegans*, 0.50–1.60; *Helicostomella subulata*, 0.27–0.50; *Ptychocylis urnula*, 0.20–0.50. (Example: *A. norvegica*: large dots, 0.60; small dots, 0.24–0.60; open circles < 0.24 per sample). Tintinnids found in the less-well-sampled rectangles (outside the outlined area) are shown as plus symbols, without regard to abundance. (After Lindley, 1975.)

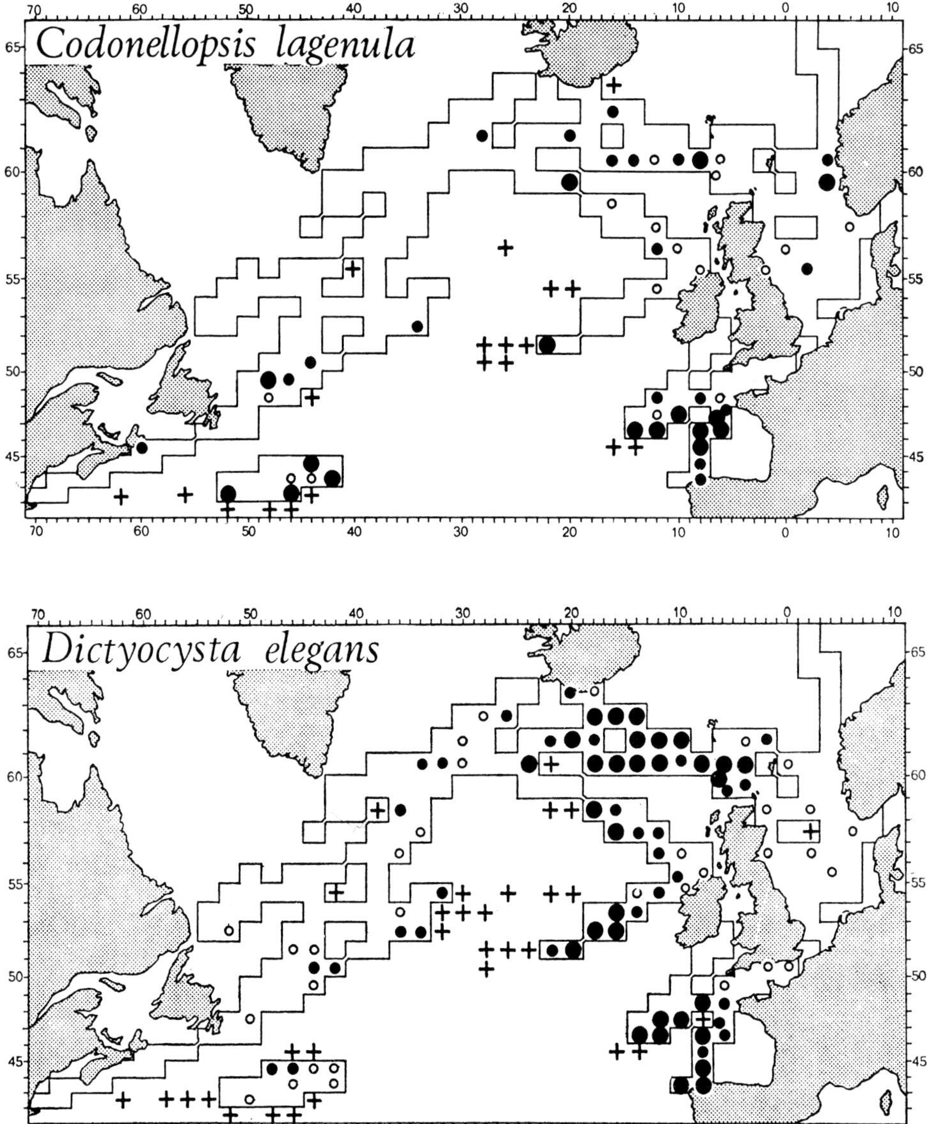

FIGURE 3.24 *(continued)*

termediate" zone and in the "Northeast Intermediate" zone no tintinnids were found during the period of investigation. In Table 3.6 the geographic classification of all analyzed species of tintinnids are given.

Zeitzschel (1969) described the distribution of 27 major tintinnid species from his own findings from material of the German "Meteor" Expedition of 1964–1965 and relevant literature. He used a somewhat more specific concept of Hela and Laevastu (1962), who described the natural regions of the oceans according to the surface current system and geographically distinct areas of the world ocean.

Taniguchi (1984) investigated the microzooplankton from 1-L water samples in the upper water column at 15 stations along a north–south section from the Chukchi Sea to the northern North Pacific Ocean through the Bering Sea (70 degrees north–45 degrees north). The result of this work is given in Figure 3.25. The dominant taxa were ciliates other than tintinnids, tintinnid ciliates, and copepod nauplii. Although the dominance in individual number was in the order depicted in Figure 3.25, nauplii were the most important by weight. Regional distribution of ciliates other than tintinnids coincided with that of chlorophyll stocks, being small in oceanic waters in the south, large in shelf waters in the north, and largest at the shelf break in the cen-

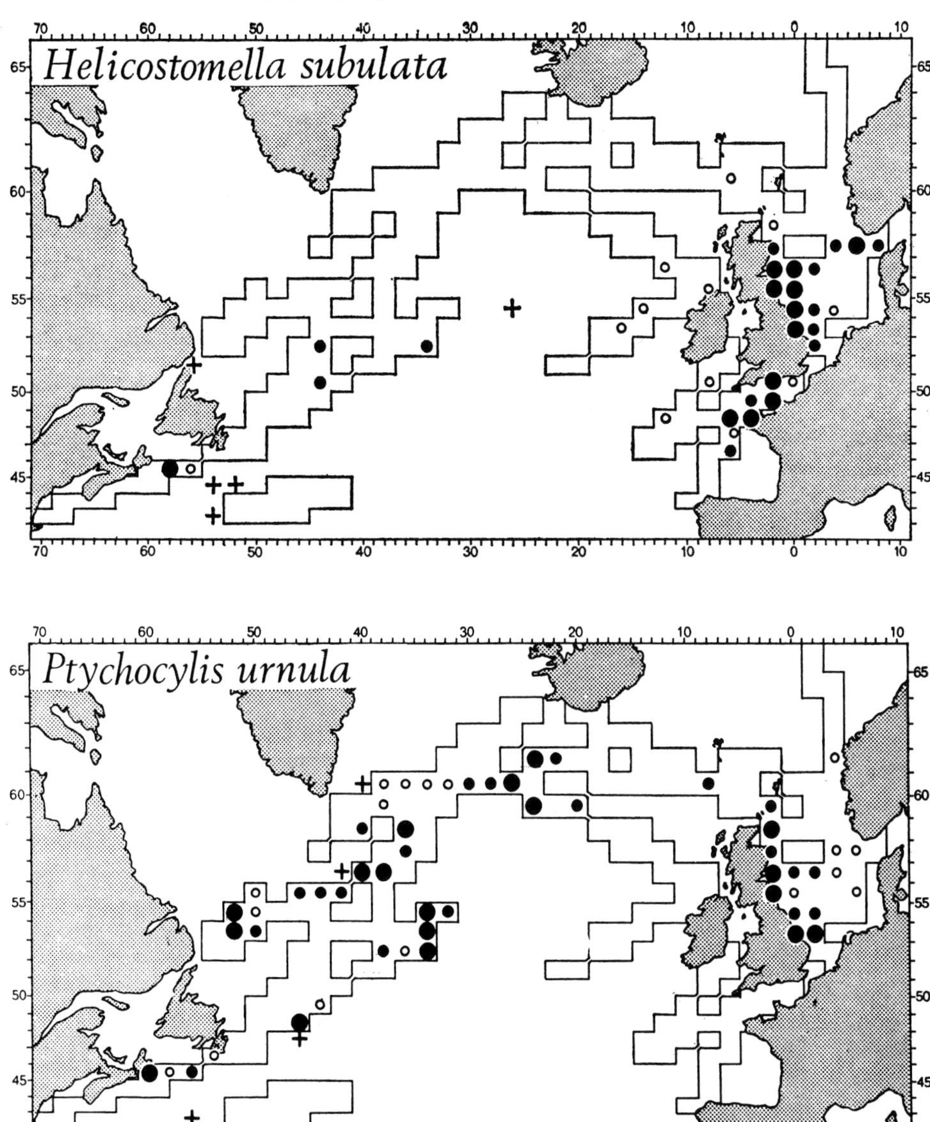

FIGURE 3.24 *(continued)*

ter of the investigated areas. Although the same positive relationship to chlorophyll was also observed roughly in tintinnids, it was not always evident in copepod nauplii. Foraminifera and radiolaria formed minor constituents of the microzooplankton during this cruise in July of 1978.

Data on the seasonal distribution of marine protozoa in the open ocean are very scarce. Zeitzschel (1967) investigated the seasonal cycle of tintinnids from samples of the weather ships *India* (? = 60 degrees north, ? = 20 degrees west) and *Juliett* (? = 53 degrees, 0 minutes north, ? = 18 degrees, 40 minutes west) in the North Atlantic. The tintinnid maximum at station *India* was in July; from August to March only few specimens were recorded.

At station *Juliett* no distinct seasonal trend of the total tintinnids could be found.

Lindley (1975) reports the seasonal cycle of tintinnids from seven areas of the North Atlantic and the North Sea (Fig. 3.26). In Figure 3.27 the seasonal cycle of 10 species of tintinnids is depicted. Figure 3.28 shows the seasonal variation of total tintinnids for the seven areas for the year 1965. Consistent patterns can be recognized. More species were abundant and continued to be so over a longer period in the neritic and southern ocean

TABLE 3.6 Geographic Classification of Tintinnid Distribution in the North Atlantic and the North Sea.

Species
Neritic
Favella serrata (Mobius) Brandt.
Helicostomella subulata (Ehrb.) Jörg.
Parafavella gigantea group
Stenosemella spp.
Tintinnidium mucicola (Claparède and Lachmann) Daday
Tintinnopsis spp.
Northeast Intermediate
No tintinnids were found with this type of distribution.
Northeast Oceanic
Dictyocysta elegans (Ehrb.)
Southeast Intermediate
No tintinnids were found with this type of distribution.
Southern Oceanic
Amphorides quadrilineata (Claparède and Lachmann)
Codonaria cistellula (Fol) Brandt.
Codonella galea Haekel
Codonellopsis lagenula (Claparède and Lachmann) Jörg
Codonellopsis orthoceras Haekel
Dadayiella ganymedes (Entz. Sr.) Kofoid and Campbell
Dictyocysta mitra Haekel
Epipocylis undella (Ostenfeld and Schmidt) Jörg.
Epipocyloides reticulata (Ostenfeld and Schmidt) Jörg.
Eutintinnus fraknói (Daday) Jörg.
Eutintinnus lusus-undae (Entz. Sr.) Jörg.
Favella azorica (Cleve) Jörg.
Parundella caudata (Ostenfeld) Jörg.
Petalotricha ampulla (Fol) Kent
Proplectella claparèdei (Entz Sr.) Kofoid and Campbell
Proplectella subacuta (Cleve) Jörg.
Protorhabdonella curta (Cleve) Jörg.
Rhabdonella amor (Cleve) Brandt
Rhabdonella spiralis (Fol) Brandt
Salpingella lineata (Entz. Sr.) Kofoid and Campbell
Steenstrupiella steenstrupii (Claparède and Lachmann) Kofoid and Campbell
Undella dohrnii (Daday) Jörg.
Western Intermediate
Parafavella denticulata (Ehrb.) Kofoid and Campbell and *P. edentata* Brandt
Ptychocylis obtusa Brandt
Ptychocylis urnula (Claparède and Lachmann)
Northwest Oceanic
Acanthostomella novegica (Daday) Jörg.
Unclassified
Coxliella pseudannulata Jörg.
Cymatocylis spp.
Epipocylis acuminata (Daday) Jörg.
Salpingella acuminata (Claparède and Lachmann)
Tintinnus bursa (Cleve) Kofoid and Campbell

After Lindley, 1975.

areas than in the colder northern areas. Lindley (1975) identified a marked seasonal succession in the neritic and warm-water areas.

It is notable that the seasonal cycles of total tintinnids started earliest and lasted longest in Areas 5, 6, and 7, where there was also the strongest bimodality of the seasonal cycle. In the other areas the seasons tended to be unimodal.

PELAGIC SEDIMENTATION

Over geologic time periods the accumulation of shell-bearing organisms in the sediment is a marked phenomenon for large parts of the ocean basins (Fig. 3.29). Diatom ooze occurs in the high productive polar and upwelling areas, whereas foraminifera ooze covers large parts of the Atlantic, Pacific, and Indian Oceans. Ooze containing high amounts of radiolaria is restricted mainly to a longitudinal belt at about 10 degrees north in the Pacific Ocean.

Many zoogeographic studies of marine plankton organisms incorporate distributional data from surface sediment material. Distributional ranges in sediments mostly differ from living plankton ranges because sedimentation is of a variable intensity in different areas and time periods because of differences of a variety of physical, morphological, and biological factors. Assuming "normal" sedimentation rates of several dozens of meters per day allows, for example, skeletons of radiolaria or foraminifera to drift considerable distances from their place of origin.

Only skeleton-bearing species that have relatively stable shells are preserved in the sedimentary record. The large gelatinous *Collodaria* and some spicule-bearing colonial radiolaria are not represented in sediment samples.

Many Phaeodaria do not leave a fossil record, because their organic-rich skeletons are porous or granular in texture and do not persist. It is essential to recognize that the deep-water zones are marked by cosmopolitan species dwelling within this rather

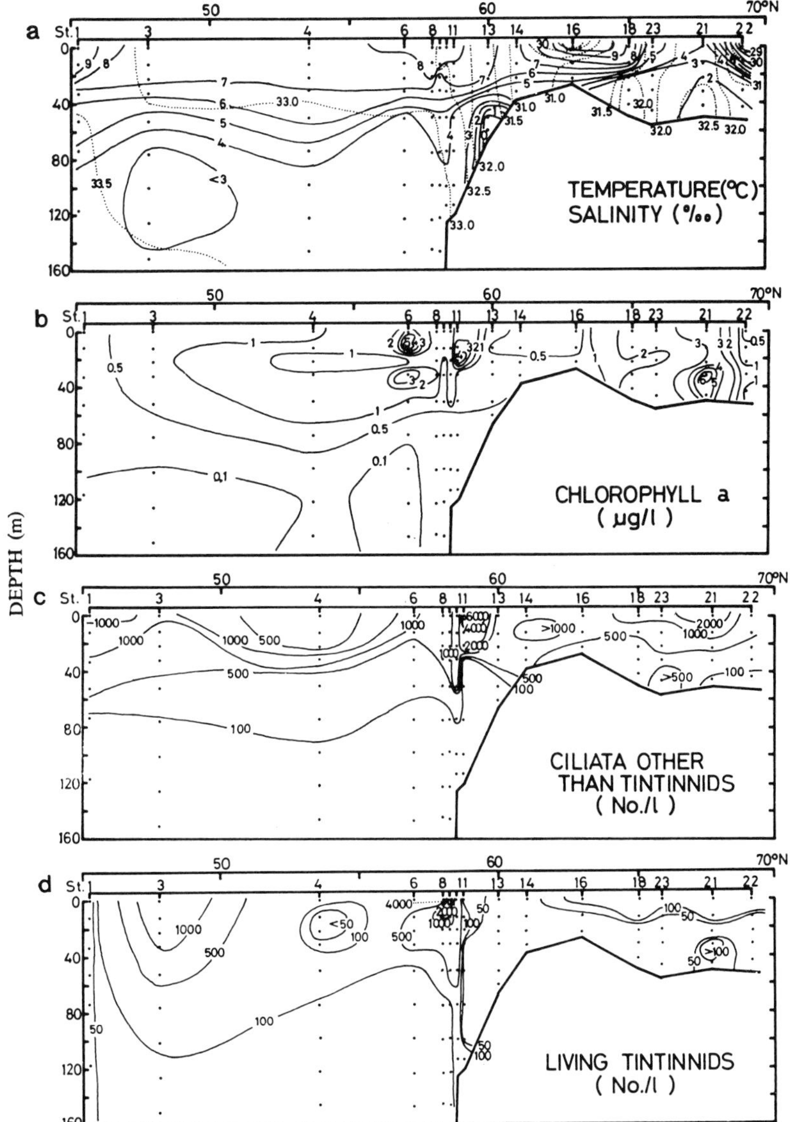

FIGURE 3.25 Vertical profiles of environmental parameters and population densities and biomasses of various microzooplankton in the Chukchi Sea, Bering Sea, and North Pacific. a. temperature (solid line) and salinity (dotted line). g. biomass in volume of holomicrozooplankton (naked ciliates, tintinnids, foraminifera, raddiolarians, rotifers). h. biomass in volume of total microzooplankton, including holomicrozooplankton and larvae of copepods, pteropods, and bivalves. (After Taniguchi, 1984.)

homogeneous environment. Transfer of species from one major geographic region to another within this zone is very likely. Thus, high-latitude forms are also tropically submergent, resulting in an exchange of deep-water forms beneath the tropics. An example is the exchange of fauna between the North Pacific Intermediate water and the Antarctic Intermediate water masses.

As an example, Figure 3.30 shows the differences between the larger sediment range hatched and the smaller living plankton distribution of the foraminifera *Globorotalia hirsuta*. The differences are most pronounced in the North Atlantic and Subtropical–Central Indian Ocean. These differences may be due to transport of living specimens or of specimens in the period between dying and sedimentation (Bé, 1977; Bé and Tolderlund, 1971; van der Spoel and Heyman, 1983).

Goll and Bjørklund (1971) examined polycystine radiolarian skeletons in the surface sediments of the North Atlantic Ocean and determined their abundance (as radiolaria per gram bulk sediment) and quality of preservation based on 334 samples taken at locations between latitudes 11 degrees

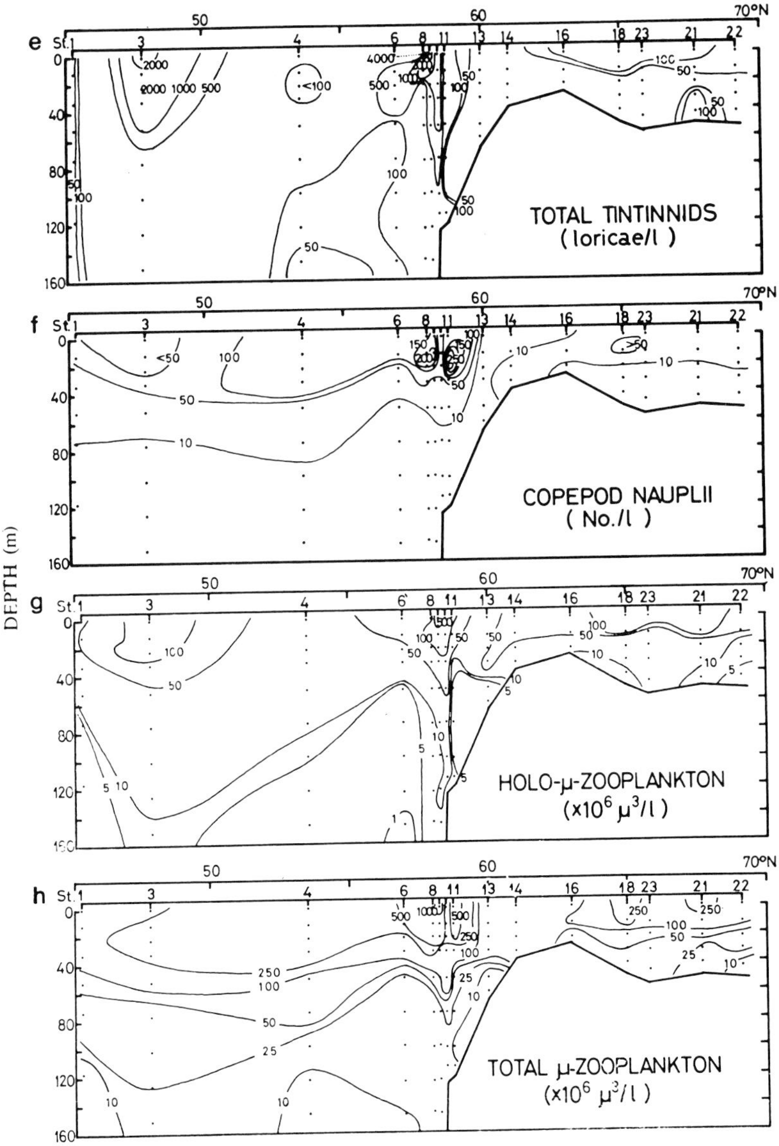

FIGURE 3.25 *(continued)*

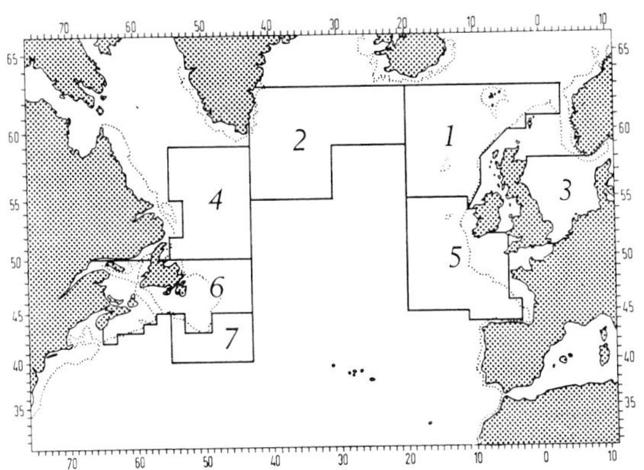

FIGURE 3.26 The area used for analysis of seasonal cycles of tintinnids. (After Lindley, 1975.)

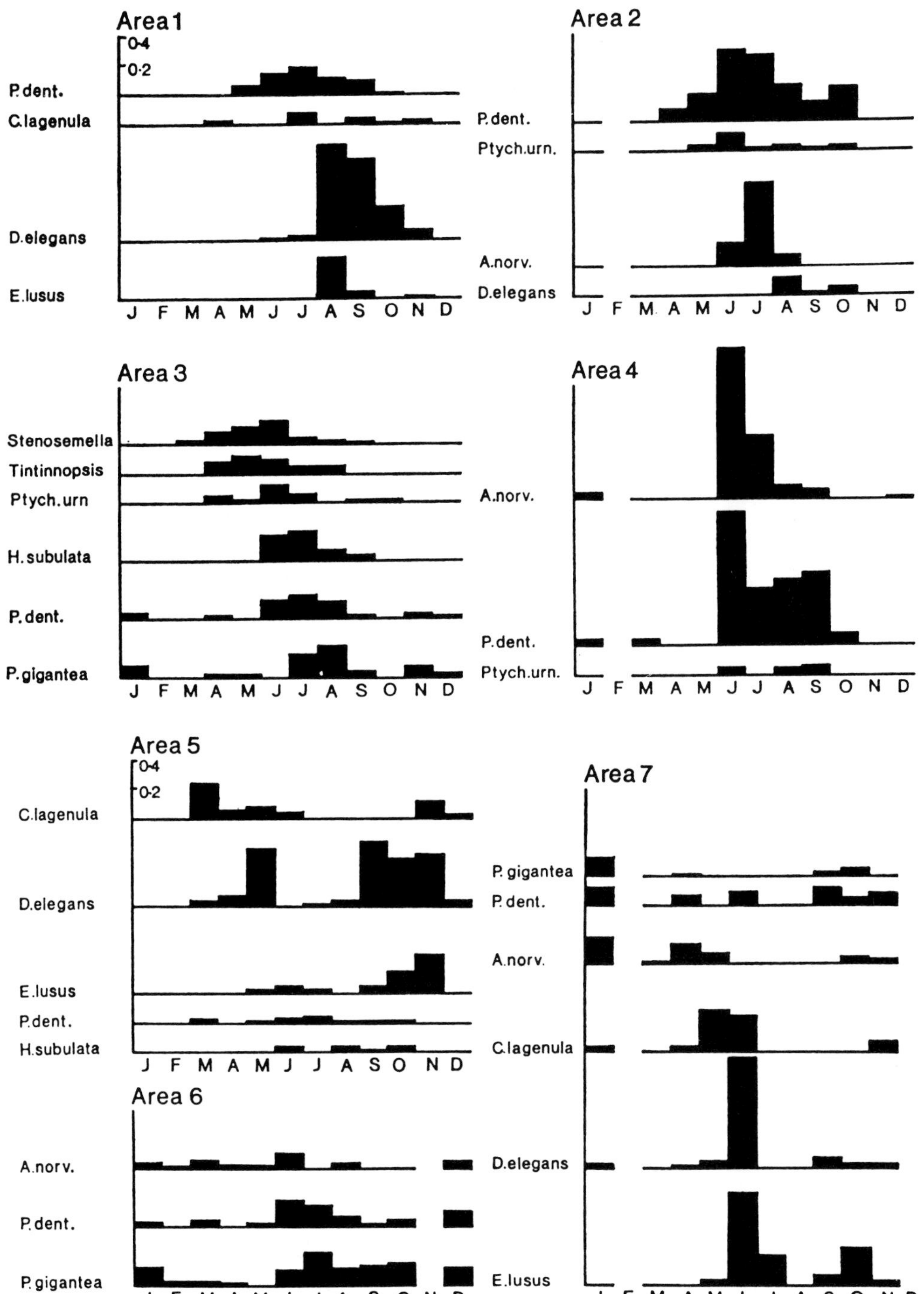

FIGURE 3.27 Seasonal cycles of tintinnid species in areas 1–7. The scale in mean log number per subsample is in the top left-hand corner. A gap in the base line indicates lack of sampling. The following abbreviations are used: *A. norv.* = *Acanthostomella norvegica*; *C. lagenula* = *Codonellopsis lagenula*; *D. elegans* = *Dictyocysta elegans*; *E. lusus* = *Eutintinnus lusus-undae*; *H. subulata* = *Helicostomella subulata*; *P. dent.* = *Parafavella denticulata* and *P. edentata* combined; *P. gigantea* = *Parafavella gigantea* group; *Ptych. urn.* = *Ptychocylis urnula*; *Stenosmella* = *Stenosmella* spp.; *Tintinnopsis* = *Tintinnopsis* spp. (After Lindley, 1975.)

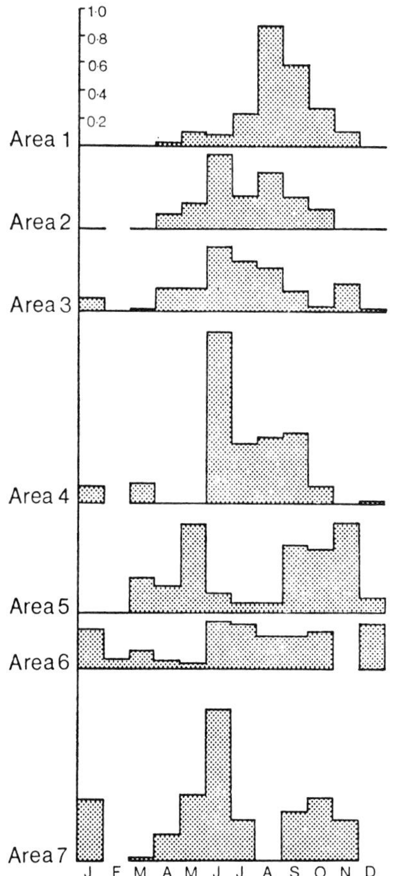

FIGURE 3.28 Seasonal cycles of total tintinnids in areas 1–7. A gap in the base line indicates absence of sampling. The scale, mean log number per subsample, is given at the top of the diagram. (After Lindley, 1975.)

south and 63 degrees north. The maximum concentration of radiolaria known in North Atlantic sediments is 118,000 specimens/g, or about 5 percent of the dry-bulk sediment weight. Several regions in the North Atlantic accumulate specimens in a concentration greater than 10,000 radiolaria/g. In general, the sedimentary accumulation of radiolaria coincides with current systems. Radiolaria are most abundant in sediments from east of the Mid-Atlantic Ridge and from the Caribbean Sea. A high abundance of radiolarian skeletons (species 44 μm or larger) was found in a zone occurring approximately between 15 degrees north and 10 degrees south and extending as a tongue into the Caribbean Sea. A similarly high-density region occurred further north, approximately between 30 and 60 degrees north, which was near the northern limit of their sampling range. Very large quantities of radiolarian skeletons (>10,000 radiolaria/g) occurred near the equator and in the northeastern Atlantic approximately between 45 and 60 degrees north and 15 and 35 degrees west. Sediment samples, underlying the central circulation gyre of the North Atlantic (approximately east of Florida), were low in abundance or barren of skeletons. Sediments underlying strong currents (Equatorial, Gulf Stream, and North Atlantic currents) generally yielded samples with more than 1,000 radiolaria/g, although abundance fluctuated greatly. Goll and Bjørklund recognize that abundance patterns of this kind cannot be explained simply; however, they suggest three possible categories of causal factors: (1) variations in production of living radiolaria, (2) the masking effect of other sedimentary constituents, and (3) opal solution. Unfortunately, they did not find that the sediment abundance of radiolarian skeletons could be related to such biologically significant factors as primary productivity or even to estimates of surface water abundance of radiolaria as assessed by other researchers. They also eliminated the masking effect of other constituents as a contributing factor and concluded that opal solution is probably a major factor in explaining variations in sediment abundance. Although low abundances may be attributed to opal solution, regions of high abundance of skeletons probably represent localized high productivity if current effects can be neglected.

Goll and Bjørklund (1971) have also mapped the distribution of several species of solitary radiolaria. Interestingly, they find a substantial difference between the species distribution of radiolaria and that of planktonic foraminifera. Most notable is the restricted nature of radiolarian species distribution compared with the broader distribution of most planktonic foraminiferal species. Some species of radiolaria were truly cosmopolitan (e.g., *Theocalyptra davisiana*). However, the majority of the species exhibited a much more limited distribution. Eight species of radiolaria were examined as typical representatives of solitary radiolarian distribution.

Takahashi and Honjo (1981) have presented some interesting taxon-quantitative data on radiolarian abundance and vertical flux of radiolarian skeletons in the western Tropical Atlantic Ocean using samples obtained from sediment traps. The sediment traps were deployed for 98 days from November 1977 through February 1978 at a station located at latitude 13 degrees, 33.2 minutes north, longitude 54 degrees, 1 minute west. Some of the most abundant species collected at the four sampling depths are presented in Table 3.7. The total number of skeletons accumulated per square meter per day are reported in this condensed table. A complete list of specimens collected is given in the paper by Takahashi and Honjo. Their data include polycystine and phaeodarian species and is therefore one of the few modern research studies to provide es-

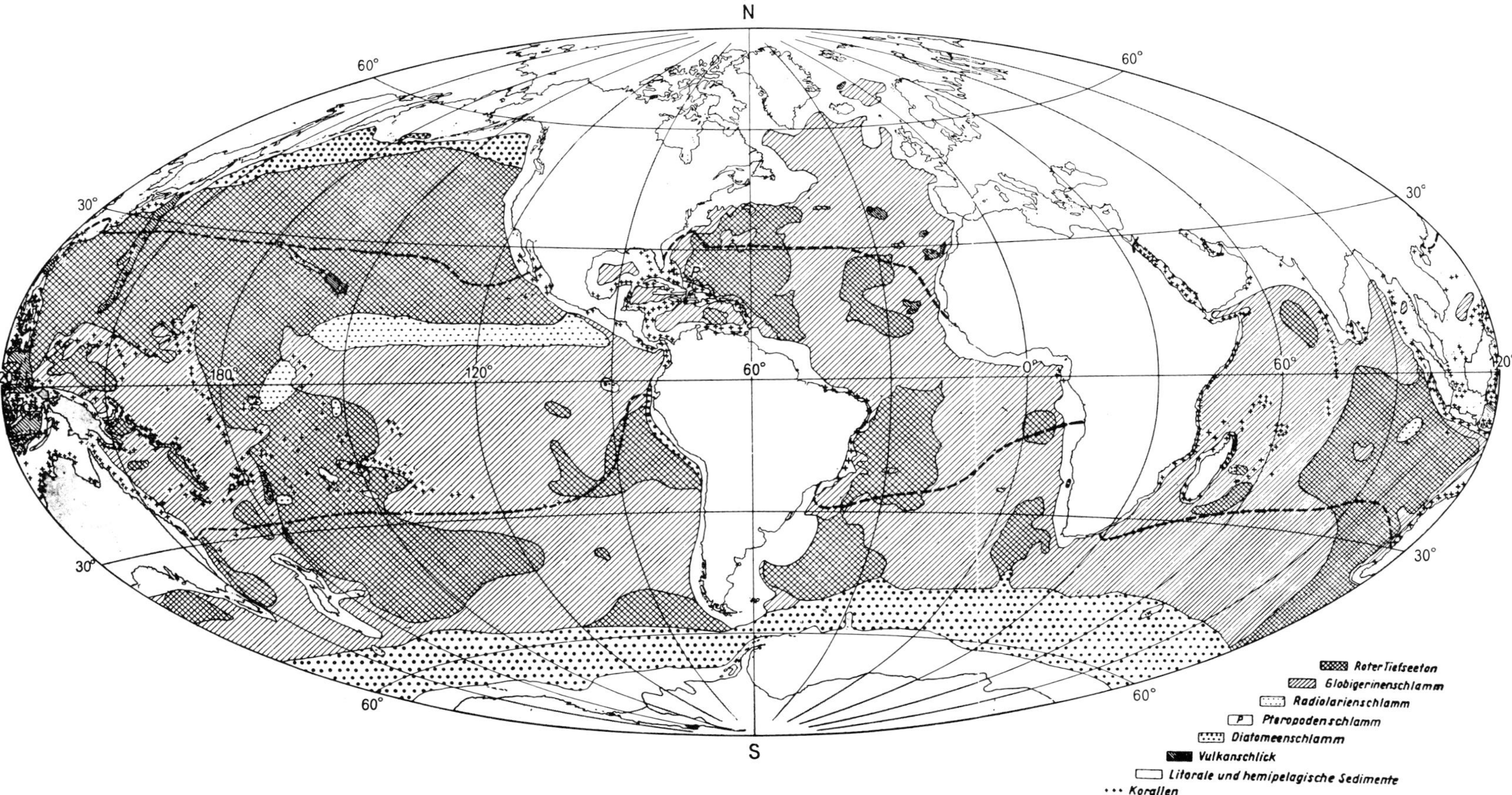

FIGURE 3.29 Distribution of deep-sea sediments. (After Dietrich et al., 1980.)

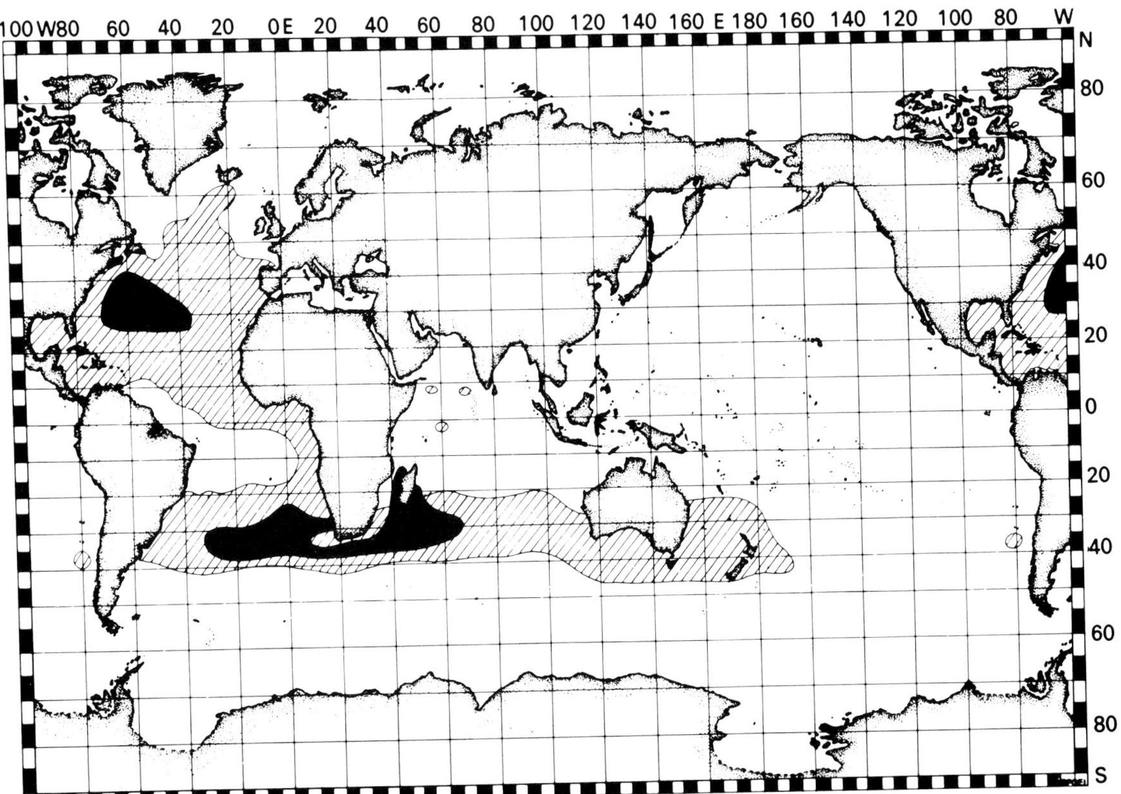

FIGURE 3.30 Distribution of *Globorotalia hirsuta*. Hatched = surface sediment distribution; dark = living plankton distribution. (After Bé and Tolderlund, 1971; Bé, 1977, from van der Spoel and Heyman, 1983.)

timates of phaeodarian abundance in the Tropical Atlantic Ocean. Among the samples collected at the four depths, Spumellaria comprised approximately 20–30 percent of the total sample; Nassellaria, 60–70 percent; and Phaeodaria, 6–8 percent. The number of species encountered in all samples in each group was 89 Spumellaria, 84 Nassellaria, 34 Phaeodaria, and 1 Sticholonchida. The diversity index of Nassellaria increased substantially from 389 to 988 m. This probably represents the contribution of deep-water species and tropical submergent species. Among the species reported by Takahashi and Honjo (1981) that conform to Casey's deep-water forms are *Cyrtopera languncula, Peripyramis circumtexta, Litharachnium tentorium,* and *Cornutella profunda*. It is interesting to note that several species of Phaeodaria occur appreciably in the upper water layer above 389 m. These include *Challengeron willemoesii, Protocystis xiphodon, Euphysetta pusilla, Borgertella caudata,* and *Conchidium caudatum*. It is difficult to make estimates about deep-dwelling forms, as the skeletons of Phaeodaria are more fragile and may be degraded during settling. As mentioned by Takahashi and Honjo, this also makes it difficult to make precise estimates of their vertical flux, as some of the shells occur only as pieces in the deep traps. The total radiolarian flux ranged from 16.0×10^3 to 23.7×10^3 shells $m^{-2}\ d^{-1}$. An earlier estimate by Honjo (1978) obtained in the Sargasso during winter was 14.0×10^3 shells $m^{-2}\ d^{-1}$, which further corroborates a lower productivity in this region when the water mass is colder (Anderson, 1983).

PLANKTON INDICATOR SPECIES

Physical and/or chemical and/or biological determinants may be used to characterize bodies of water in the ocean. Zoogeographers and taxonomists are often keen to give advice to physical oceanographers, recalling the idea of Johnson (1957) to use plankton organisms as tiny drift bottles.

The concept is quite simple. As water moves from one place to another, it gradually mixes with the surrounding water and becomes either cooler or warmer, as the case may be. Its physical and chemical characteristics also mix until its identity is lost

TABLE 3.7 Some Abundant Solitary Radiolaria Collected at Varying Depths in Sediment Traps in the Southwestern Atlantic Ocean

Taxon	Collection depth (m)			
	389	988	3,755	5,068
Spumellaria				
Family Liosphaeridae Haeckel				
Cenosphaera huxleyi Müller	5	46	107	38
Styptosphaera sp.	74	74	162	143
Plegomosphaera lepticali Renz	21	0	15	18
Total Liosphaeridae	100	120	284	199
Family Actinommidae Haeckel				
Theocosphaera inermis (?) (Haeckel)	251	58	409	194
Actinomma arcadophorum Haeckel	56	40	56	65
Trilobatum (?) acuferum Popofsky	238	58	673	437
Stylosphaera melpomene Haeckel	132	64	52	110
Ommatartus tetrathalamus (Haeckel) subsp. A	104	161	156	149
Total Actinommidae	1130	695	2077	1610
Family Phacodiscidae Haeckel				
Spongodiscus resugens Ehrenberg *Spongodiscus* sp. B	808	929	1337	1498
Spongotrochus glacialis (Popofsky)				
Total Phacodiscidae	920	1027	1497	1631
Family Porodiscidae Haeckel				
Euchitonia furcata Ehrenberg	106	32	128	46
Hymeniastrum euclidis Haeckel	81	72	126	102
Porodiscus micromma (Harting)	76	160	257	114
Stylochlamydium asteriscus Haeckel	121	72	55	91
Total Porodiscidae	410	375	677	429
Family Litheliidae Haeckel				
Tholospira cervicornis Haeckel group	191	365	176	371
Pylonena armata Haeckel group	163	85	186	111
Tetrapyle octacantha Müller	513	444	947	458
Octopyle stenozona Haeckel	*53*	*56*	*72*	*37*
Total Lithellidae	*954*	*956*	*1402*	*1013*
Nassellaria				
Family Plagoniidae Haeckel				
Cladoscenium anacoratum Haeckel	123	320	267	419
Obeliscus pseudocuboidea Popofsky	83	131	167	113
Phormacantha hystrix (Jørgensen)	150	141	330	260
Peridium spinipes Haeckel	1251	2184	3114	2173
Lophophaena cylindrica (Cleve) *Peromelissa phalacra* Haeckel	2445	1283	1925	1570
Acanthocorys cf. *variabilis* Popofsky	594	538	982	146
Helotholus histricosa Jørgensen	313	251	238	120
Total Plagoniidae	5112	5334	7464	5134
Family Acanthodesmiidae Haeckel				
Zygocircus capulosus Popofsky	248	596	795	285
Zygocircus productus (Hertwig)	453	436	634	431
Acanthodesmia vinculata (Müller)	108	91	69	54
Dictyospyris sp. A and B	109	54	36	9
Amphyspyris costata Haeckel	128	175	56	56
Total Acanthodesmiidae	1197	1570	1768	1068

TABLE 3.7 *(continued)*

Taxon	Collection depth (m)			
	389	988	3,755	5,068
Family Theoperidae Haeckel				
Cornutella profunda Ehrenberg	64	141	367	160
Pterocanium praetextum (Ehrenberg)	207	158	102	88
Eucyrtidium hexastichum (Haeckel)	120	45	133	40
Theocalyptra davisiana cornutoides (Petrushevskaya)	82	354	784	207
Theocalyptra davisiana davisiana (Ehrenberg)	83	122	197	75
Total Theoperidae	707	1244	2467	1137
Family Pterocorythiidae Haeckel				
Pterocorys zancleus (Müller) *Pterocorys campanula* (Haekel)	2966	2381	2698	1654
Total Pterocorythidae	2975	2418	2820	1733
Family Artostrobiidae Riedel				
Spirocyrtis scalaris Haeckel	91	79	67	101
Spirocyrtis sp. aff. *S. seriata*/*S. subscalaris*	1496	629	653	660
Carpocanarium papillosum (Ehrenberg)	0	7	2	56
Total Artostrobiidae	1587	748	874	860
Family Cannobotryidae Haeckel				
Acrobotrys sp. A, B, and C	123	76	126	68
Botryocyrtis scutum (Harting)	226	167	212	111
Total Cannobotryidae	349	243	338	178
Phaeodaria (Haeckel)				
Family Challengeriidae Murray				
Challengeron willemoesii Haeckel	128	110	62	36
Protocystis ziphodon Haeckel	135	64	36	40
Total Challengeridae	291	199	137	109
Family Medusettidae Haeckel				
Euphysetta pusilla Cleve	407	229	188	21
Medusetta ansata Borgert	73	15	59	33
Total Medusettidae	488	330	345	147
Family Lirellidae Ehrenberg				
Borgertella caudata (Wallich)	252	62	149	180
Lirella bullata (Stadum and Ling)	2	384	603	705
Total Lirellidae	254	453	1132	1084
Family Conchariidae Haeckel				
**Conchidium caudatum* (Haeckel)	103	2	25	30
**Conchopsis compressa* Haeckel	0	3	0	0
*Total Conchariidae	130	6	35	32

*After Takahashi and Honjo, 1981; From Anderson, 1983.

and the differences can no longer be measured. But although the plankton also becomes mixed, the actual individuals carried by the current cannot be changed and they remain recognizable until they die and disintegrate. If the chosen organisms are easily identified, the water mass can be labeled at sight simply by towing a plankton net and looking at the catch. This has been realized for centuries, but only in a vague way until 1935, when Russell pointed out how useful such labels were and how easily they could be obtained. Because certain species in the plankton could be used to indicate the origin of the water mass containing them, Russell coined the term *plankton indicator species*.

Not all plankters are good indicators of water masses. There are several criteria that should be respected:

1. Correct identification of particular species.
2. Knowledge of the life history (e.g., optimal conditions for reproduction).
3. Knowledge of the limiting environmental factors.
4. The relatively long life-span of species.

Organisms that are most suitable in this respect include the chaetognaths, pteropods, heteropods, euphausids, and copepods.

Marine protozoa do not fulfill most of the preceding criteria. The weakest point is their short life-span and their capacity to respond very quickly to favorable environmental conditions with respect to reproduction. In addition, they can form resting spores under certain circumstances. These spores might be important for seeding special water masses; it is difficult, however, to track drifting resting spores because often they cannot be identified.

Despite these restrictions, marine autotrophic and heterotrophic protista have been used successfully as indicator species (e.g., diatoms, dinoflagellates, coccolithophores, foraminifera, radiolarians, and tintinnids).

Clearly, attempts to arrive at biogeographic meaning are simply efforts to use the presence or absence of one or more species, or stages in their life cycles and sometimes their morphological appearance as indicators of certain environmental conditions, including hydrographic events, eutrophication or warming trends, or long-term changes symptomatic of environmental disturbances. According to Smayda (1978), the extent to which this is successful depends, *inter alia,* on the environmental condition being monitored and whether the appropriate indicator species is known, if it occurs at all. The information sought from the indicator species sometimes requires of the investigator minimal autoecological insight, such as using the organism to indicate certain water mass incursions. At the other extreme, considerable autoecological knowledge is required to find and use an indicator organism that would be symptomatic of a very specific condition (e.g., a certain pollutant). In the latter case the organism is used similar to a miner's canary, that is, to bioassay a unique environmental disturbance.

Zeitzschel (1969) tried the approach introduced by Bary (1959) to identify indicator species of tintinnids in the western Arabian Sea. Figure 3.31 gives a classical temperature/salinity diagram from 64 "Meteor" stations. Four main water bodies may be identified during the northeastern monsoon period.

1. The origin of the Somali current in the northwestern part of the Arabian Sea (cluster of station right corner) with low temperature and high salinities.
2. Stations off the East African coast between 3 degrees north and 1 degree, 30 minutes south, representing the Somali current (Plankton station 132, 134, 149, 155, 157).
3. Stations representing the South Equatorial current (top left) with low salinities and high temperatures.
4. An intermediate mixing zone in the open ocean, especially stations 168 and 169 close to 4 degrees south and 500 nm off the coast.

Figure 3.32 gives examples of some typical temperature–salinity–tintinnid diagrams for four species. These species show distinct distributional patterns. *Epiplocylis undella* is especially abundant in the northern area and at station 169 in the mixing zone of the Somali and East African coastal currents. *Rhabdonella spiralis* is only abundant in the high saline waters of the northern area. *Dadayiella ganymedes* was mainly identified in the colder northern waters, as well as in the mixing zone at station 151. There is no record of this species at stations 165 and 167. It was very abundant in the mixing zone at stations 149 and 169 at salinities of 35.3 percent.

Boltovskoy (1969) analyzed plankton samples that were taken during four expeditions of Equalant I in the Tropical West Atlantic. He came to the conclusion that the upper layer (100–0 m) can be divided, according to its foraminiferal fauna, into two zones that are rather well pronounced and differ clearly from one another (Fig. 3.33). These two zones are

1. South–southeastern zone.
2. North–northwestern zone.

The first zone differs from the second in the following ways:

1. It is poorer qualitatively. The species encountered are *Globigerina dudertrei, G. rubescens, Globigerinella aequilateralis, Globigerinita glutinata, Globigerinoides ruber, G. trilobus,* and *Orbulina universa*. In the north–northwestern zone, apart from those listed previously, the following species have also been found: *Globigerinoides conglobatus, G. trilobus* (f. *succuli-*

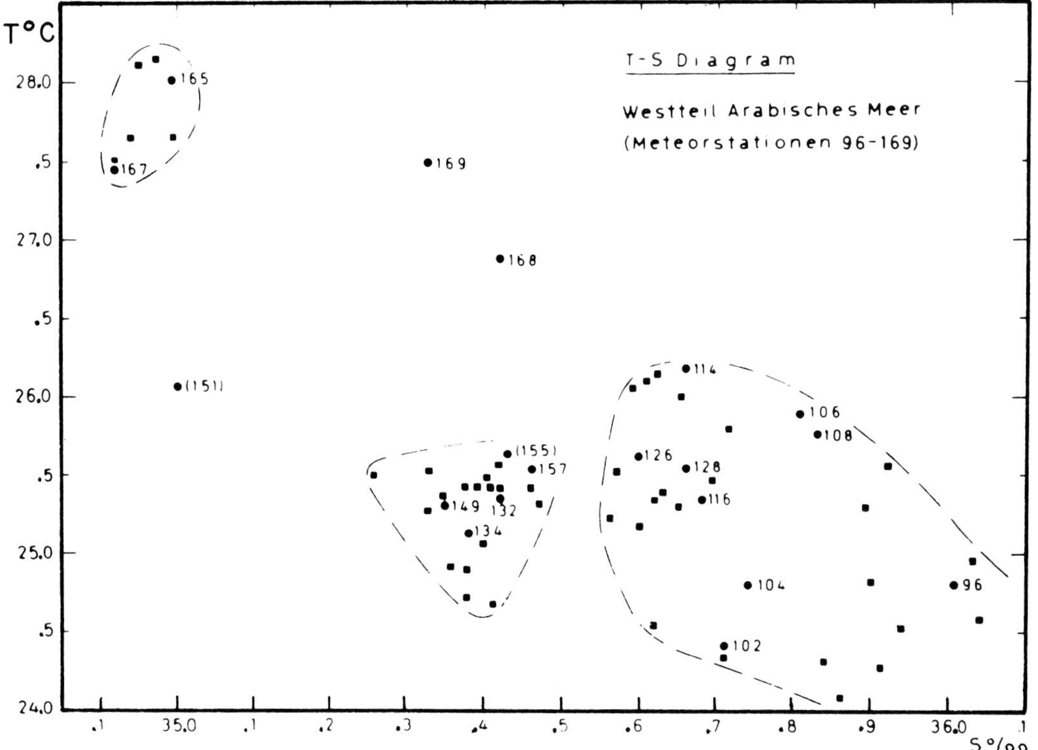

FIGURE 3.31 Abundance and preservation of radiolaria in the surface sediments of the North Atlantic Ocean. (After Goll and Bjorklund, 1971.)

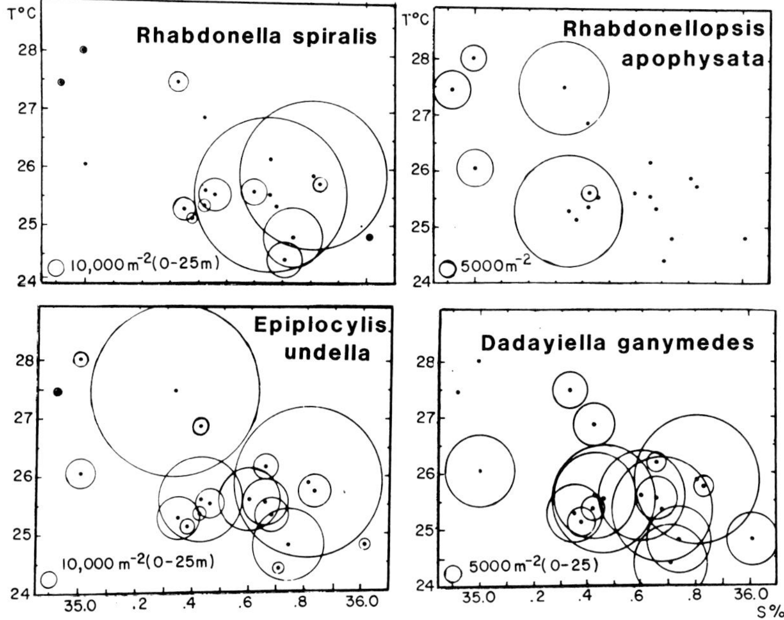

FIGURE 3.32 Temperature–salinity diagram for 64 "Meteor" stations in the western Arabian sea. (After Zeitzschel, 1969.)

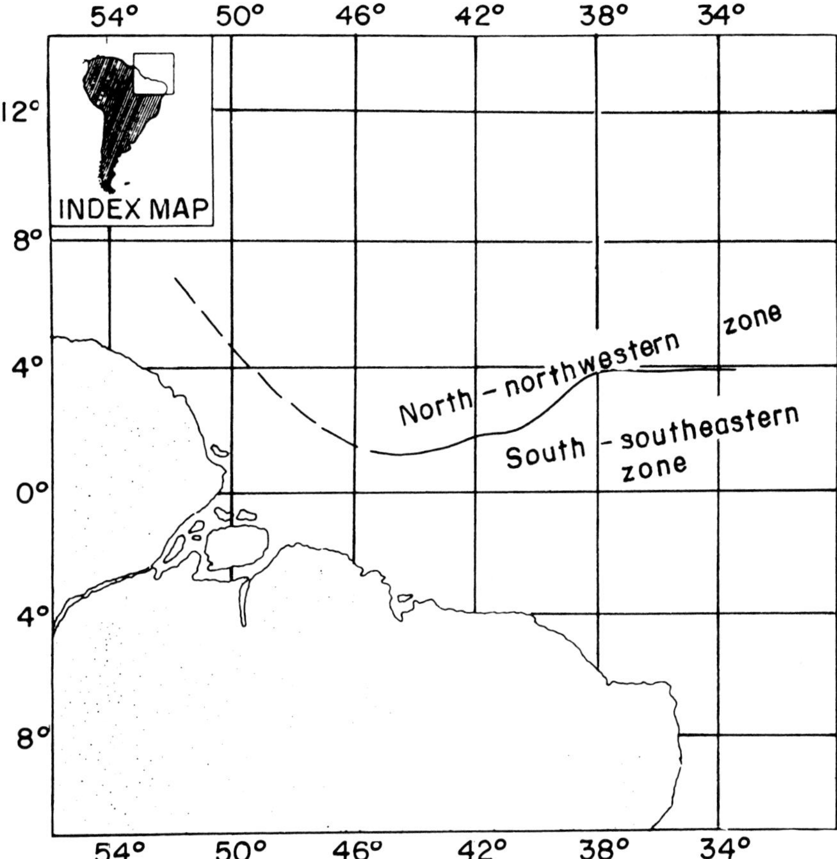

FIGURE 3.33 Temperature–salinity–tintinnid diagram for four species in the western Arabian Sea. For "Meteor" station numbers see Figure 2.32. (After Zeitzschel, 1969.)

fera), *Candeina nitida*, *Pulleniatina obliquiloculata*.

2. It is poorer quantitatively. The comparison of the number of shells found in the samples of the same type shows that on average the south–southeastern zone has only one third of the population of the north–northwestern zone.

According to Boltovskoy (1969) the separation of the two zones is faunistically well pronounced and the difference between them can be noticed immediately by observation of the corresponding samples by any foraminifera specialist.

A more general account of the biogeography of the southwestern Atlantic is given by the same author (Boltovskoy, 1979, 1982, 1985).

CONCLUSIONS

According to Anderson (1983), few documented generalizations can be made about the ecology of radiolaria and other marine protozoa, owing to the broad scope of species studied, the varied and sometimes occasional regions examined, and often the lack of simultaneous physical and chemical monitoring of environmental variables during microplankton sampling.

Some reasonable perspectives on marine protozoan patterns of interaction with the biological and physical environment emerge, however, from the sum total of research on shell-bearing protozoan life history, patterns of geographic distribution, and environmental correlation with faunal assemblages.

There is clear evidence from much of the research literature that many species of foraminifera, radiolaria, and tintinnina inhabit identifiable masses of ocean water and occupy faunal niches or biogeographic zones comparable with meso- and macrozooplankton.

In recent years, however, biogeographic research has often been criticized, mainly because a unified theory of pelagic biogeography seemed not to be evident (Frost, 1979). The field of marine biogeography was often looked down upon; actual research in biological oceanography was mainly centered on dynamic aspects of ecosystem research.

There was a growing interest and need to obtain meaningful physiological rate measurements and estimates of ecological fluxes rather than to describe distributions of plankton organisms in the world ocean (Platt et al., 1981). Besides, physical oceanographers appear not to be much interested in advice from planktologists concerning the identification of water masses through the use of indicator species. Rather, they use such new techniques as satellite-tracked buoys, or they cooperate with marine chemists who provide data (e.g., on relevant isotopes).

Finally, it was apparent that younger invesitgators had rediscovered the classical field of investigating the life history or life cycle strategies of planktonic organisms. It is felt that although environmental factors or external energy may govern important processes in the ocean, some "biology" might be involved with respect to the functioning of complex marine systems (Steidinger and Walker, 1985). These ideas were also considered by marine biologists interested in biogeography and have resulted in the formulation of some new concepts. Three hypotheses, for example, were proposed by van der Spoel (1985) at an International Conference on pelagic biogeography:

1. Within plankton communities it is not the influences of the environment but the characteristics of the organisms that primarily determine the range of the species.
2. Evolution in the pelagic environment is too slow for organisms to achieve an optimal, adaptive, ecological balance within the larger available ocean habitat.
3. It is likely that population dynamics and balanced interactions among organisms will also restrict their range.

REFERENCES

Alvarino, A. (1965) Chaetognaths. In Barnes, H. (ed.), *Annual Review of Oceanography and Marine Biology*. London: Allen and Unwin, pp. 115–194.

Anderson, O.R. (1983) *Radiolaria*. New York: Springer-Verlag, 355 pp.

Balech, E. (1958) Los dinoflagelados y tintinnoineas como indicatores oceanograficos. In *Symposium Sobre plankton*. Sao Paulo: Centro de Coop. Cient. de UNESCO para America Latina, pp. 33–36.

Balech, E. (1960) The changes in the phytoplankton population off the California coast. *Calif. Coop. Ocean. Fish. Invest.* 7:127–132.

Bary, B.M. (1959) Species of zooplankton as a means of identifying different surface waters and demonstrating their movements and mixing. *Pac. Sci.* 13:14–54.

Bé, A.W.H. (1977) An ecological, zoogeographic and taxonomic review of recent planktonic foraminifera. In Ramsay, A.T.S. (ed.), *Oceanic Micropaleontology*. London: Academic Press, pp. 1–100.

Bé, A.W.H., and Tolderlund, D.S. (1971). Distribution and ecology of living planktonic foraminifera in surface waters of the Atlantic and Indian Ocean. In Funnell, B.W., and Riedel, W.R. (eds.), *Micropaleontology of Oceans*. London: Cambridge University Press, pp. 105–149.

Bé, A.W.H., and W.H. Hutson (1977) Ecology of planktonic foraminifera and biogeographic patterns of life and fossil assemblages in the Indian Ocean. *Micropaleontology* 23:369–414.

Beers, J.R. (1978) About microzooplankton. In Sournia, A. (ed.), *Phytoplankton Manual*. 6. Paris: UNESCO, pp. 288–296.

Beers, J.R., Stewart, G.L. and Strickland, J.D.H. (1967) A pumping system for sampling small plankton. *J. Fish. Res. Bd. Canada*24:1811–1818.

Beklemishev, C.W. (1981) Biological structure of the Pacific Ocean as compared with two other oceans. *J. Plankton Res.* 3:531–549.

Beklemishev, C.W., Parin, N.B., and Semina, G.N. (1977) Pelagial. In Vinogradov, M. (ed.), *Biogeographical Structure of the Ocean* (Ocean Biogeography). Moscow (in Russian), pp. 219–261.

Bodungen, B. v., Smetacek, V., Tilzer, M.M., and Zeitzschel, B. (1985) Primary production and sedimentation during austral spring in the Antarctic peninsula region. *Deep-Sea Res.*.

Boltovskoy, E. (1962) Planktonic foraminifera as indicators of different water masses in the South Atlantic. *Micropaleontology* 8:403–408.

Boltovskoy, E. (1969) Distribution of planktonic foraminifera as indicators of water masses in the western part of the Tropical Atlantic. In *Proceedings of the Symposium on the oceanography and fisheries resources of the Tropical Atlantic*. Results of ICITA and GTS, Paris: UNESCO, pp.45–55.

Boltovskoy, E., and Wright, R. (1976) *Recent Foraminifera*. The Hague: Dr. W. Junk b.v. Pub., 515 pp.

Boltovskoy, E. (1979) Zooplankton in the southwestern Atlantic. *S. Afr. J. Sci.* 75:541–544.

Boltovskoy, E. (1982) North-south zooplanktonic diversity variations in the south-western Atlantic Ocean. *Physis (A)* 44:1–6.

Boltovskoy, E. (1985) Biogeography of the southwestern Atlantic: overview, current problems and prospects. International Conference on Pelagic Biogeography, Noordwijkerhout, abstract, p. 6.

Bradshaw, I.S. (1959) Ecology of living planktonic foraminifera of the North and Equatorial Pacific Ocean. *Cushman Found. Foram. Res. Contr.* 10:26–64.

Brandt, K. (1906) Die Tintinnodeen der Plankton-Expedition. Tafelerklärungen nebst kurzer Diagnose der neuen Arten. *Ergebn. Atl. Planktonexped.* 3:1–33.

Brandt, K. (1907) Die Tintinnodeen der Plankton-Expedition. *System. Teil Ergebn. Atl. Planktonexped.* 3:1–499.

Brandt, K. (1910) Rapports sur les espèces du plankton.

Cons. Perman. Intern. Explor. Mer. Bull. Trim. Res. acquis pendant trois. period et dans per. intermed. 1:1–19.

Buzas, M.A., and Culver, S.J. (1980) Foraminifera: distribution of provinces in the western North Atlantic. *Science* 209:687–689.

Candeias, A. (1930) Estudos de plancton na Baia de Sesimbra. *Bull. Soc. Port. Sci. Nat.* 11:12–14.

Casey, R.E. (1971a) Distribution of polycystine radiolaria in the oceans in relation to physical and chemical conditions. In Funnel, B.M., and Riedel, W.R. (eds.), *The Micropaleontology of Oceans*. Cambridge: Cambridge University Press, pp. 151–159.

Casey, R.E. (1971b) Radiolarians as indicators of past and present water-masses. In Funnel, B.M., and Riedel, W.R. (eds.), *The Micropaleontology of Oceans*. Cambridge: Cambridge University Press, pp. 331–349.

Casey, R.E. (1977) The ecology and distribution of recent radiolaria. In Ramsey, A.T.S. (ed.), *Oceanic Micropaleontology,* Vol. 2. London: Academic Press, pp. 809–845.

Casey, R.E., Gust, L., Leavesley, A., Williams, D., Reynolds, R., Duis, T., and Span, J.M. (1978) Ecological niches of radiolarians, planktonic foraminiferians and pteropods inferred from studies on living forms in the Gulf of Mexico and adjacent waters. *Trans. Gulf Coast Assoc. Geol. Soc.* 29:216–223.

Colebrook, J.M. (1964) Continuous plankton records: a principal component analysis of the geographical distribution of zooplankton. *Bull. Mar. Ecol.* 6: 78–100.

Dale, T., and Burkill, P.H. (1982) Live counting—a quick and simple technique for enumerating pelagic ciliates. *Ann. Inst. Oceanogr.* 58 (Suppl.):267–276.

Davis, C.C. (1985) Acanthostomella norvegica (Daday) in insular Newfoundland waters, Canada (Protozoa: Tintinnina). *Int. Rev. Hydrobiol.* 70:21–26.

Dietrich, G., Kalle, K., Krauss, W., and Siedler, G. (1980) *General Oceanography*. New York: Wiley, 625 pp.

Ekman, S. (1953) *Zoogeography of the Sea*. London: Sedgwick and Jackson, 417 pp.

Fager, E.W. (1953) Determination and analysis of recurrent groups. *Ecology* 38:586–595.

Fager, E.W. (1963) Communities of organisms. In Hill, M.N. (ed.), *The Sea,* Vol. 2. London: Interscience, pp. 415–437.

Fager, E.W., and McGowan, J.A. (1963) Zooplankton species in the North Pacific. *Science* 140:453–460.

Fleminger, A., and Hülsemann, K. (1973) Relationships of Indian Ocean epiplankton calanoids to the world oceans. In Zietzschel, B. (ed.), *Biology of the Indian Ocean*. Heidelberg: Springer Verlag, pp. 339–347.

Forbes, E., and Godwin-Austen, R. (1859) *The natural history of European seas*. London: van Voorst, 306 pp.

Fraser, J. (1962) *Nature Adrift: The Story of Marine Planktons*. London: Foulis, 178 pp.

Fraser J. (1968) Standardization of zooplankton sampling methods at sea. In Fraser, J. (ed.), *Zooplankton Sampling*. Paris: UNESCO, pp. 147–169.

Frost, B.W. (1979) Problems in marine biogeography. *Science* 209:1112.

Gaarder, K.R. (1946) Tintinnoinea. Report on the scientific results of the "Michael Sars" North Atlantic deep-sea exped. 1910. 2(1):3–37.

Gagnon, M., and Lacrois, G. (1982) The effects of tidal advection and mixing on the statistical dispersion of zooplankton. *J. Exp. Mar. Biol. Ecol.* 56:9–22.

Giesbrecht, W. (1892) Systematic und Faunistic der pelagischen Copepoden des Golfes von Neapel und der angrenzenden Meeresabschnitte. *Fauna Flora Golfes von Neapel* 19:1–831.

Goll, R.H. (1976) Morphological intergradation between modern populations of Lophospyris and Phormaspyris (Trissocyclidae, Radiolaria). *Micropaleontology* 22:379–418.

Goll, R.H., and Bjørklund, K.R. (1971) Radiolaria in surface sediments of the North Atlantic Ocean. *Micropaleontology* 17:434–454.

Gran, H.H. (1929) Quantitative plankton investigations carried out during the expedition with the "Michael Sars," July–September 1924. *Rapp. Proc. Verbaux* 56:1–50.

Grøntved, J., and Seidenfaden, G. (1984) The Godthaab Expedition, 1928. The phytoplankton of the waters west of Greenland. *Medd. Grønland* 82(5):213–224.

Haeckel, E. (1887) Report on Radiolaria collected by H.M.S. Challenger during the years 1873–1876. In Thompson, C.M., and Murray, J. (eds.), *The Voyage of H.M.S. Challenger,* 18. London: Her Majesty's Stationery Office, pp. 1–1760.

Haury, L.R., McGowan, J.A., and Wiebe, P.H. (1978) Patterns and processes in the time-space scales of plankton distributions. In. Steele, J.H. (ed.), *Spatial Pattern in Plankton Communities*. New York: Plenum, pp. 277–327.

Hedgpeth, J.W. (1957) Marine biogeography. In Hedgpeth, J.W. (ed.), *Treatise on Marine Ecology and Paleoecology*. 1 Ecology. Mem. Geol. Soc. Amer. Mono. 67, pp. 359–382.

Hela, I., and Laevastu, T. (1962) *Fisheries Hydrography*. London: Fish. News (books), 137 pp.

Honjo, S. (1978) Sedimentation of materials in the Sargasso Sea at a 5,367 m deep station. *J. Mar. Res.* 36:469–492.

Johnson, M.W. (1957) Plankton. In Hedgpeth, J.W. (ed.), *Treatise on Marine Ecology and Paleoecology*. 1 Ecology. Mem. Geol. Soc. Amer. Mono. 67, pp. 443–460.

Johnson, M.W., and Brinton, E. (1963) Biological species, water masses, and currents. In Hill, M.N. (ed.), *The Sea,* Vol. 2. London: Interscience, pp. 381–414.

Jørgensen, E. (1899) Über die Tintinnodeen der norwegischen Westküste. *Bergen. Mus. Aarbog* 2:1–48.

Langmuir, I. 1938) Surface motion of water induced by wind. *Science* 87:119–123.

Legendre, L., and Legendre, P. (1978) Associations. In Sournia, A. (ed.), *Phytoplankton Manual*. Paris: UNESCO, pp. 261–272.

Lenz, J. (1972) A new type of plankton pump on the vacuum principle. *Deep-Sea Res.* 19:453–459.

Lindley, J.A. (1975) Continuous plankton records: a

plankton atlas of the North Atlantic and North Sea: Suppl.3-Tintinnida (Protozoa, Ciliophora) in 1965 Bull. Mar. Ecol. 8:201–213.

Longhurst, A.R. (1981) Significance of spatial variability. In Longhurst, A.R. (ed.), *Analysis of Marine Ecosystems*. London: Academic Press, pp. 415–441.

Magaard, L., Müller, P., and Pujalet, R., eds., (1983) The role of eddies in the general ocean circulation. Proc. Hawaiian Winter Workshop, Honolulu.

Margalef, R. (1978a) Sampling design—some examples. In Sournia, A. (ed.), *Phytoplankton Manual* Paris: UNESCO, pp. 17–31.

Margalef, R. (1978b) Life-forms of phytoplankton as survival alternatives in an unstable environment. *Oceanol. Acta* 1:493–509.

Margalef, R., and Duran, M. (1953) Microplancton de Vigo, de Octubre de 1951 a Septiembre de 1952. *Publ. Inst. Biol. Appl.* 13:5–78.

McGowan, J.A. (1971) Oceanic biogeography of the Pacific. In Funnell, B.M., and Riedel, W.R. (eds.). *The Micropaleontology of Oceans*. Cambridge: Cambridge University Press, pp. 3–74.

Meisenheimer, J. (1905) Die tiergeographischen Regionen des Pelagials auf Grund der Verbreitung der Pteropoden. *Zool. Anz.* 29:155–163.

Murray, J.W. (1973) *Distribution and Ecology of Living Benthic Foraminiferids*. Heinemann, 274 pp.

Munk, W.H. (1950) On the wind driven ocean circulation. *J. Meteorol.* 7:79–93.

Ostenfeld, C.H. (1899) "Plankton" in Iagttagelser over overfladevandets temperatur saltholdighed og plankton paa islandske og grønlandske skibsrouter i 1899, af C. F. Wandel bearbjdede af Martin Knudsen og C. Ostenfeld. *Kjøbenhavn. Gad.* 47:93.

Ostenfeld, C.H. (1900) "Plankton i 1899" in Iagttagelser over overfladevandets temperatur, saltholdighed og plankton paa islandske og grønlandske skipsrouter i 1899, af C. F. Wandel bearbjdede af Martin Knudsen og C. Ostenfeld. *Kjøbenhavn. Gad.* 43–93.

Owen, R.W. (1981) Fronts and eddies in the sea. In Longhurst, A.R. (ed.), *Analysis of Marine Ecosystems*. London: Academic Press, pp. 197–233.

Parker, F.L., and Berger, W.H. (1971) Faunal and solution patterns of planktonic foraminifera in surface sediments of the South Pacific. *Deep-Sea Res.* 18:73–107.

Peres, J.M. (1982) Major pelagic assemblages. In Kinne, O. (ed.), *Marine Ecology*, Vol. 5, Part I. New York: Wiley, pp. 187–311.

Phleger, F.B. (1960) Ecology and distribution of recent foraminifera. Baltimore: Johns Hopkins University Press, 297 pp.

Pingree, R.D., and Mardell, G.T. (1981) Slope turbulence, internal waves and phytoplankton growth at the Celtic Sea shelf-break. *Philos. Trans. Roy. Soc. Lond. A* 302:663–682.

Platt, T., Mann, K.H., Ulanowicz, R.E., eds. (1981) *Mathematical Models in Biological Oceanography*. Paris: UNESCO, 156 pp.

Richardson, P. (1976) Gulf Stream rings. *Oceanus* 65–68.

Robinson, G.A. (1983) Continuous plankton records: phytoplankton in the North Sea, 1958–1980, with special reference to 1980. *Br. Phycol. J.* 18:131–139.

Robinson, G.A., and Hiby, A.R. (1978) The continuous plankton recorder survey. In Sournia, A. (ed.), *Phytoplankton Manual*. Paris: UNESCO, pp. 59–63.

Russell, F.A. (1935) On the value of certain plankton animals as indicators of water movements in the English Channel and North Sea. *J. Mar. Biol. Assoc. U.K.* 20:309–332.

Russell, F.A. (1939) Hydrographical and biological conditions in the North Sea as indicated by plankton organisms. *J. Cons. Int. Explor. Mer* 14:171–192.

Silva, E.S. (1950) Les tintinnides de la bai de Cascais (Portugal). *Bull. Inst. Oceanogr.* 979:1–28.

Smayda, T.J. (1978) Biogeographical meaning: indicators. In Sounia, A. (ed.), *Phytoplankton Manual*. Paris: UNESCO, pp. 225–229.

Smith, R.L. (1976) Waters of the sea: the oceans' characteristics and circulations. In Cushing, D.H., and Walsh, J.J. (eds.), *The Ecology of the Seas*. Oxford: Blackwell, pp. 24–58.

Sorokin, Yu. I. (1981) Microheterotrophic organisms in marine ecosystems. In Longhurst, A.R. (ed.), *Analysis of Marine Ecosystems*. London: Academic Press, pp. 293–342.

Spoel, van der, S. (1983) Patterns in plankton distribution and their relation to speciation. The dawn of pelagic biogeography. In Sims, R.W., Price, J.H., and Whalley, P.E.S. (eds.), *Evolution, Time and Space: The Emergence of the Biosphere*. London: Academic Press.

Spoel, van der, S. (1985) What is unique about open-ocean biogeography: zooplankton? International conference on pelagic biogeography. *Noordwijkerhout* (abstract), p. 42.

Spoel, van der, S., and Heyman, R.P. (1983) *A Comparative Atlas of Zooplankton: Biological Patterns in the Oceans*. Berlin: Springer Verlag, 186 pp.

Steele, J. (1976) Patchiness. In Cushing, D.H., and Walsh, J.J. *The Ecology of the Seas*. Oxford: Blackwell, pp. 98–115.

Steele, J. (1978) Some comments on plankton patches. In Steele, J.H. (ed.), *Spatial Patterns in Plankton Communities*. New York: Plenum, pp. 1–20.

Steidinger, K.A., and Walker, L.M. (1984) *Marine Plankton Life Cycle Strategies*. Boca Raton: CRC Press, 158 pp.

Steuer, A. (1933) Zur planmäßigen Erforschung der geographischen Verbreitung des Haliplanktons, besonders der Copepoden. Zoogeographica; *Int. Rev. Comp. Causal Animal Geogr.* 1:269–302.

Takahashi, K., and Honjo, S. (1981) Vertical flux of Radiolaria: a taxon-quantitative sediment trap study from the western tropical Atlantic. *Micropaleontology* 27:140–190.

Tangen, K. (1978) Nets. In A Sournia (ed.), *Phytoplankton Manual*. Paris: UNESCO, pp. 50–58.

Taniguchi, A. (1984) Microzooplankton biomass in the arctic and subarctic Pacific Ocean in summer. *Mem. Nat. Inst. Polar Res.* (spec. issue) 32:63–76.

Thorpe, S.A. (1975) The excitation, dissipation, and interaction of internal waves in the Deep Ocean. *J. Geophys. Res.* 80:328–338.

Tranter, D.J., ed. (1968) Reviews on zooplankton sampling methods. In *Zooplankton sampling*. Paris: UNESCO, pp. 11–114.

Ulbricht, K.A., and Horstmann, U. (1980) Landsat imagery of phytoplankton development in the Baltic Sea. *Int. Arch. Photogramm.* 13:932–934.

Venrick, E.L. (1978) Sampling strategies. In Sournia, A. (ed.), *Phytoplankton Manual*. Paris: UNESCO, pp. 7–16.

Williamson, M.H. (1961) An ecological survey of a Scottish herring fishery. Part IV: Changes in the plankton during the period 1949–1959. Appendix: A method for studying the relation of plankton variations to hydrography. *Bull. Mar. Ecol.* 5:207–229.

Woods, J.D., Onken, R. (1982) Diurnal variation and primary production in the ocean—preliminary results of a Langrangian ensemble model. *J. Plankton Res.* 4:735–756.

Yentsch, C.M., and Yentsch, C.S. (1984) Emergence of optical instrumentation for measuring biological properties. *Oceanogr. Mar. Biol.* (Ann. Rev) 22:55–98.

Zeitzschel, B. (1966) Die Verbreitung der Tintinnen im Nordatlantik. *Veröff. Inst. Meeresforsch.* 2:293–300.

Zeitzschel, B. (1967) Die Bedeutung der Tintinnen als Glied der Nahrungskette. *Helgoländer Meeresunters.* 15:589–601.

Zeitzschel, B. (1969) Tintinnen des westlichen Arabischen Meeres, ihre Bedeutung als Indikatoren für Wasserkörper und Glied der Nahrungskette. *"Meteor" Forsch. Ergebn.* D(4):47–101.

Zeitzschel, B. (1971) Primärproduktion und Phytoplanktonökologie im östlichen tropischen Pazifik. Habilitationsschrift Universität, Kiel, 115 pp.

Zeitzschel, B. (1978) Oceanographic factors influencing the distribution of plankton in space and time. *Micropaleontology* 24:139–159.

Zeitzschel, B. (1982) Zoogeography of pelagic marine protozoa. *Ann. Inst. Oceanogr.* 58 (Suppl.):91–116.

Zeitzschel, B. (1985) The dynamic of organic production in the Rockall Channel area. *Proc. Roy. Soc. Edinburgh.*

4

Feeding-Related Ecology of Marine Protozoa

Gerard M. Capriulo

All heterotrophic protista require organic substances for maintenance, growth, and reproduction. It has long been known that the required organics may be obtained by uptake of dissolved nutrients of small molecular weight through the cell membrane, or part of it, by feeding directly on particles, both living and detrital, through the process of endocytosis, by metabolic exchange with endosymbionts, or through combinations of all the preceding. It has not been until the last two decades, however, that the extreme importance of these forms, which include flagellate, sarcodine, and ciliate groups, to the food webs of the ocean has been recognized (Fig. 4.1). This chapter considers the feeding-related ecology of marine protozoa, with emphasis on phagotrophic forms and their qualitative, quantitative, and behavioral trophodynamic importance in marine ecology.

GENERAL FOOD WEB CONSIDERATIONS

Heterotrophic Flagellates

Recent advances in technology, particularly those associated with fluorescence–epifluorescence microscopy, have led to a deeper understanding of the significance of bacteria and heterotrophic–mixotrophic flagellates to marine food webs. These advances have provided more accurate methods of determining bacterial (Hobbie et al., 1977; Porter and Feig, 1980) as well as heterotrophic–phototrophic protista (and various inclusions in their food vacuoles) numbers (Haas, 1982; Caron, 1983; Sherr and Sherr, 1983a,b; Davis and Sieburth, 1982, 1984). Recently, Wikner et al. (1986) developed a method in which genetically marked minicells are added to natural seawater to estimate grazing rates of microflagellates. Sherr et al. (1987) developed a method that employs monodispersed, fluorescently labeled bacteria to estimate protozoan bacterivory. Protozoan bacterivory, particularly by flagellates, has also been quantified by other methods, including those involving serial dilution (Landry and Hassett, 1982), the uptake of fluorescent particles (e.g., Børsheim, 1984; Nygaard et al., 1988; McManus and Fuhrman, 1986a) the differential use of eukaryotic and prokaryotic inhibitors (e.g., Sanders and Porter, 1986; Fuhrman and McManus, 1984; Newell et al., 1983; Sherr et al., 1986a; Campbell and Carpenter, 1986) and the rate of loss of DNA from cells pre-labeled with tritiated thymidine (Servais et al., 1985). It is becoming increasingly more apparent that potentially serious problems plague some of these methods. Sieracki et al. (1987) pointed out that poor particle retention is achieved in flagellates fixed in various aldehydes. In essence, different fixatives have different effects on particle retention, resulting in mild to severe underestimates of consumption rates. Similar findings were reported by Pace and Bailiff (1987). Tremaine and Mills (1987a) and Taylor and Pace (1987) reported that certain eukaryote inhibitors, such as cycloheximide, inhibit not only eukaryotic growth but also bacterial growth. Such effects could well invalidate data collected using the troublesome inhibitors. Sherr et al. (1986a) demonstrated that if eukaryotic–antibiotic mixtures are chosen carefully and their concentrations rigidly controlled, conditions can be achieved in which no measurable effects of the eukaryotic inhibitor on bacterial growth are encountered. A general review of what controls bacterioplankton populations and the various methods used to measure grazing on bacterioplankton can be found in McManus and Fuhrman (1988a), Pace (1988) and Capriulo et al. (in press).

Use of the preceding methods has enabled us to develop an increasingly more accurate picture of the marine microbial food web. In particular, water column bacterial concentrations have been found to be much higher than was once thought (Table 4.1), with values of 10^7–10^8/ml or higher in eutrophic waters, 10^6–10^7/ml in coastal waters, and 10^5–10^6/ml in pelagic oligotrophic environments.

Water column bacterial concentrations have been found to be very stable over long periods of time, even as they may vary temporally, perhaps because of flagellate grazing pressure, on a diel basis (Meyer-Reil et al., 1979; Meyer-Reil, 1977; Azam et al., 1983). Benthic habitats, particularly those

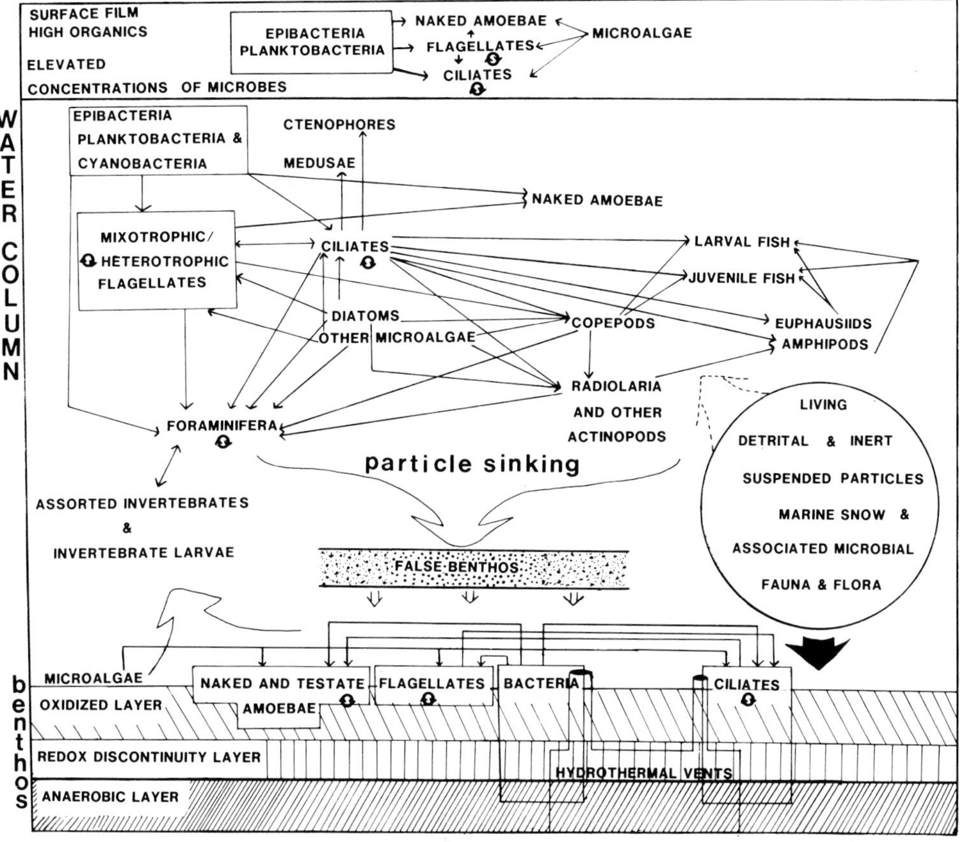

FIGURE 4.1 General overview of the marine food web depicting trophic interactions involving heterotrophic–mixotrophic protozoa. Arrows denote flow of biomass from prey to predator. Two-headed arrow indicates that biomass transfer is bidirectional. See text for details and discussions concerning marine snow, sea-surface film, hydrothermal vents, and false benthos. Microbial biomass associated with these specialized habitats is routinely cycled into the water column and benthic food webs. ♻ = within group cycling.

TABLE 4.1 Typical Bacterial Concentrations in Marine Systems

Bacterial concentration	Habitat	Source
5×10^5/ml	Southern California waters	Fuhrman et al. (1980)
$1–1.5 \times 10^6$/ml	Southern California waters	Fuhrman and Azam (1980)
$0.3–1 \times 10^7$/ml	Georgia coast	Newell and Fallon (1982)
1.2×10^3/ml	New York Bight	Kirchman et al. (1982)
1.3×10^3/ml	New York Bight	Ducklow and Kirchman (1983)
$0.8–5.7 \times 10^6$/ml	Kiel Bight (Germany)	Meyer-Reil (1977)
$0.8–4 \times 10^6$/ml	Peru upwelling	Sorokin (1978)
$0.2–1.4 \times 10^6$/ml	Caribbean Sea	Burney et al. (1982)
10^5/ml	North Carolina coastal waters	Ferguson and Rublee (1976)
$2–3.5 \times 10^5$/ml	Antarctic waters	Hanson et al. (1983)
	General review	Ducklow (1983)

underlying productive waters, exhibit bacterial concentrations in the 10^8/ml or higher levels, with 10^9 or more bacteria/g dry-weight detritus often found (Pomeroy, 1980). High bacterial concentrations are also found in the waters of the hydrothermal vent regions, where numerous flagellates and ciliates have been observed (Small and Gross, 1985; Sniezek, 1987) and associated with marine snow aggregates as well as with accumulations of detritus at density discontinuity layers in the water column (Caron et al., 1982; Capriulo, Small, and Angel, unpublished data; Silver et al., 1984; Small et al., 1983).

The bacteria are found free-floating in the water column (planktobacteria), attached to suspended or settled, living or inert particles or bottom sediments (epibacteria) and in tight "zones" around suspended and/or settled particles where possible attachment and detachment may routinely occur (pseudoepibacteria) (Sieburth, 1979).

The fate of this bacterial carbon has been considered in relationship to what is now termed the "microbial loop" in aquatic food webs (Pomeroy, 1974; Azam et al., 1983; Sherr et al., 1986b) in which the heterotrophic and mixotrophic flagellates, as well as various marine amoebae and ciliates, have emerged as central players.

The microflagellates, including both the Zoomastigophoreans and the Phytomastigophoreans (classification after Lee et al., 1985), are ubiquitous and often abundant in most marine ecosystems (Table 4.2), although they are noticeably absent in the marine caves of Bermuda (personal observation and E. B. Small, unpublished data). They have been found to ingest both bacteria and algae in pelagic environments (Goldman and Caron, 1985; Sherr et al., 1983; Sherr and Sherr, 1983a,b; Sieburth and Davis, 1982; Sieburth, 1976; Haas and Webb, 1979; Azam et al., 1983), although their importance to benthic ecosystems, where ciliates and certain amoebae are often the seemingly more important predators, has not yet been clearly elucidated.

Preferred Nutrition

It has been important to determine whether the heterotrophic–mixotrophic flagellates (henceforth referred to only as heterotrophic flagellates) live primarily as phagotrophs or osmotrophs. Data of Beers et al. (1980) suggest an osmotrophic nutritional mode, as does the work of Gold et al. (1970), where successful culture of flagellates on rich organic media was reported. The importance of dissolved organics was found to be less important by Kopylov et al. (1980), who studied the energy balance of *Parabodo attenuatus* and found a dual osmotrophic–phagotrophic nutritional mode with dissolved organics only accounting for 20–30 percent of that flagellate's caloric intake, and Haas and Webb (1979), who, working with five species of nonpigmented microflagellates 3–10 μm in size, from the York River Estuary, Virginia, found that none of their flagellates were able to utilize 11 organic compounds (nine amino acids, plus glucose and sodium acetate) at concentrations up to 0.75 mg/L. All five species did, however, ingest live bacteria and most likely have an obligate requirement for bacteria in their diets.

It is likely that, at the low concentrations of dissolved organics typical of most marine ecosystems,

TABLE 4.2 Some Typical Flagellate Concentrations with Associated Bacterial Densities

Flagellates (no./ml)	Bacteria (no./ml)	Habitat	Source
$<2 \times 10^2$ to $>3 \times 10^3$	1.5–3 1 × 10^6	Limfjorden, Denmark	Fenchel (1982d)
10^2–10^4	—	Marseille, France	Chretiennot (1974)
10^3–10^4	0.5–1 × 10^6 (thermocline) 0.05–0.08 × 10^6 (surface)	Japan Sea	Sorokin (1977)
2–6 × 10^3 (inshore) 0.3–2 × 10^3 (offshore)	— —	Georgia nearshore and estuarine	Sherr and Sherr (1983a)
5 × 10^3	2 × 10^6	Estuarine	Davis and Sieburth (1984)
3 × 10^3	1 × 10^6	Continental shelf	Davis and Sieburth (1984)
0.5 × 10^3	0.5 × 10^6	Sargasso Sea, Caribbean	Davis and Sieburth (1984)

uptake by flagellates is diffusion limited and proportional to the length (i.e., surface area) of the cell. This suggests that flagellates are poor competitors of bacteria for the uptake of dissolved organics and that uptake of dissolved organic material does not play a significant role in their nutritional life (Fenchel, 1982a–e). It can be concluded, then, that except for conditions of unrealistically high concentrations of dissolved organics, such as exist in culture media and at times in such specific microhabitats as marine snow or around sewage discharge areas, nonautotrophic (used synonymously with *nonphototrophic*) microflagellates are dependent on some level of particulate food, primarily bacteria, for growth and survival.

The preceding conclusion is supported by data of Sherr and Sherr (1983b), who found that about 30 percent of a population of Duplin River (Georgia, USA) estuarine heterotrophic flagellates had bacteria in their recently formed food vacuoles, and Sieburth and Davis (1982), who observed bacteria in the food vacuoles of heterotrophic microplankton that routinely release dissolved carbohydrates and other DOC compounds, perhaps to carry out microbial gardening. Additional supporting evidence of the importance of bacteria to the diet of many flagellate species also exists (Sieburth et al., 1976; Haas and Webb, 1979; Fenchel, 1982a–d; Azam et al., 1983; Wright and Coffin, 1984; Goldman and Caron, 1985; Davis and Sieburth, 1984b; Davis et al., 1985; Sherr et al., 1986b), and some earlier work (Laval, 1971; Leadbeater and Morton, 1975; Kudo, 1966) further documents the bacterivorous nature of choanoflagellates, bicoecids, and bodonids. Chroococcoid cyanobacteria, methane-oxidizing bacteria, and aggregated bacteria associated with excreted mucilage (some of which is released by the flagellates themselves), which acts as an adhesive combining organic debris and microbes together, are also used by flagellates and large protozoa (Johnson and Sieburth, 1979; Johnson et al., 1982; Caron et al., 1985; Harris and Mitchell, 1973; Goldman and Caron, 1985). Chroococcoid cyanobacteria were not digested by marine copepods (Johnson et al., 1982), although Roman (1978) demonstrated the importance of the cyanobacterium *Trichodesmium* trichomes to the diet of Sargasso Sea copepods.

Flagellate Feeding

Many of the flagellate feeding studies carried out to date (e.g., Daggett and Nerad, 1982; Fenchel, 1982b; Sherr et al., 1983) employed static bacterial populations as food. However, results using growing bacteria (Davis and Sieburth, 1984b; Kopylov et al., 1980) were comparable.

In general, flagellates employ two trophic strategies, including

1. Predation by filtration (as seen in bicoecids, actinomonads, paraphysomonads, monads, pseudobodonids), where a current is established, by flagella action passing water and bacteria over the cell surface, which leads to possible subsequent prey capture.
2. Predation by direct encounter during flagella and or prey movement (e.g., bodonids). Actual feeding mechanisms will be discussed in more detail later. It has been found that predation rates, at higher bacterial food concentrations, were greater for filtration feeders than for encounter feeders (Davis and Sieburth, 1984b). Also, many microflagellates adhere to particles. (Those that do not are rare.) Planktonic forms that do so likely graze surfaces, thereby picking up adhering epibacteria (Davis and Sieburth, 1984b).

Certain flagellates, such as *Pseudobodo* sp., can ingest picoplanktonic prasinomonads (Parslow et al., 1986) and coccoid forms of bacteria, but have trouble eating larger rods (Lucas et al., 1987). Also, at least certain heterotrophic microflagellates have been found to exhibit chemosensory behavior (Sibbald et al., 1987). General questions concerning microheterotrophic flagellate bacterivory and size-selective grazing have been considered by Anderson et al. (1986) and Anderson and Fenchel (1985). Fenchel (1986) provides a general review of the ecology of heterotrophic microflagellates and, along with Patterson (1986), reported on a new type of filter-feeding planktonic flagellate. Caron (1987) examined the ability of four species of heterotrophic microflagellates to graze attached versus unattached bacteria and found pronounced differences among species. *Monas* sp. and *Cryptobia* sp. efficiently grazed unattached bacteria and showed little or no ability to graze attached or aggregated cells, whereas *Rhynchomonas nasuta* and *Bodo* sp. were found to prefer attached and aggregated forms and possessed a limited ability to graze unattached forms. *R. nasuta* and *Bodo* sp. selected attached forms even when unattached bacteria were offered at higher concentrations than attached forms. Caron also noted that the existence of microflagellates in the plankton that feed on attached bacteria is linked to the distribution of suspended particles.

The impact of heterotrophic flagellate grazing on bacteria has been estimated to be as high as 50 percent (Sherr et al., 1984) and 66 percent (Linley et al., 1983) of the bacterial production. It is likely that in many marine ecosystems, bacterial populations are controlled by pico- and nano-sized flagellates and nano-sized ciliate populations. The picoflagellates themselves are likely controlled by nanociliate predators (Rassoulzadegan and Sheldon, 1986). Additionally, microflagellates are often found in microhabitats that support high concentrations of bacteria. Such microhabitats include patches of decaying organic matter such as those found in the benthos, suspended in the water column as marine snow, accumulated at density discontinuity layers in the open ocean or coastal water column (forming a type of "false benthos"), associated with floating or settled algae (e.g., seaweeds) or animal (e.g., appendicularian houses, remains of salps, doliolids, and the like) debris, or on floating neustonic or surface-water material, such as sargassum weeds, ship bottom fouling layers, whale skin, and the like (Sieburth, 1979; Sherr et al., 1983; Caron et al., 1982).

Sights such as those mentioned previously, rich in organic compounds, often foster bacterial growth. That growth is stimulated by flagellate grazing (Sherr et al., 1982 and 1983; Fenchel and Harrison, 1976). As one might expect, this kind of bacterial–flagellate interaction implies that flagellates should be important to various decomposition events. This connection was documented in the English Channel, where microheterotrophs were responsible for much of the decomposition of phytoplankton biomass (Newell and Linely, 1984); for the Sea of Japan (Sorokin, 1977), following the spring diatom bloom; and in Lake Kinneret, Israel, where high flagellate numbers were found associated with detritus derived from a bloom of the dinoflagellate *Peridinium cinctum* (Sherr et al., 1983). Typical patch sizes associated with flagellates are measured in the centimeters to meter range.

Ingestion and Clearance Rates

Flagellate clearance rates are proportional to filter porosity and efficiency as well as to food size, shape and motility, and flagellate mouth size (Fenchel, 1982b,d; 1986a,b; 1987). Clearance and ingestion are both functions of food concentration, with clearance being inverse hyperbolically related to bacterial food density (Sherr et al., 1983; Figure 4.2). It is additionally likely that clearance and ingestion rates, not only for microflagellates, but for all phagotrophic protozoa, are relatable by some constant of proportionality to vacuole and membrane recycling time as well as to concentrations of associated enzymes (Nisbet, 1984).

Table 4.3 provides some typical clearance and ingestion rates for some heterotrophic flagellates studied to date. Clearance rates generally vary between 0.0002 and 0.08 μl per flagellate per hour. This broad range of values, spanning two orders of magnitude, can likely be attributed to size differences among the flagellates studied (Sherr et al., 1983) as well as to differences in prey quantity and quality. The microflagellate *Actinomonas microbiles* has been found to be capable of clearing 10 times its own cell volume per hour of water, for food (Fenchel, 1982e). Reported clearance rates for Narragansett Bay (Rhode Island, USA) flagellates were higher than those of Sargasso Sea flagellates at low prey concentrations (less than 10^6 bacteria/ml), but this may be attributed to the fact that Narragansett Bay experiments were run at lower overall food concentrations than those employed in the Sargasso Sea experiments (Davis and Sieburth, 1984b). The clearance rates reported in Table 4.3 are about 100 times lower, on a cell-to-cell basis, than those reported for ciliates (see Table 4.12). Nygaard et al. (1988) also reported on the grazing rates of marine heterotrophic microflagellates feeding on bacteria.

Flagellate ingestion rates vary in the 100s to 1000s of bacteria ingested per flagellate per day with associated gross growth efficiencies in the 20–50 percent range, on average (see Table 4.3). In one study these rates were found to be linearly related

FIGURE 4.2 Clearance rates showing an inverse hyperbolic relationship with increeasing bacterial density (a) and ingestion rates showing a positive linear relationship to food concentration (b) for the microflagellate *Monas* sp. Redrawn after Figures 3 and 4 of Sherr, Sherr, and Berman, 1983. Symbols represent different bacterial prey species.

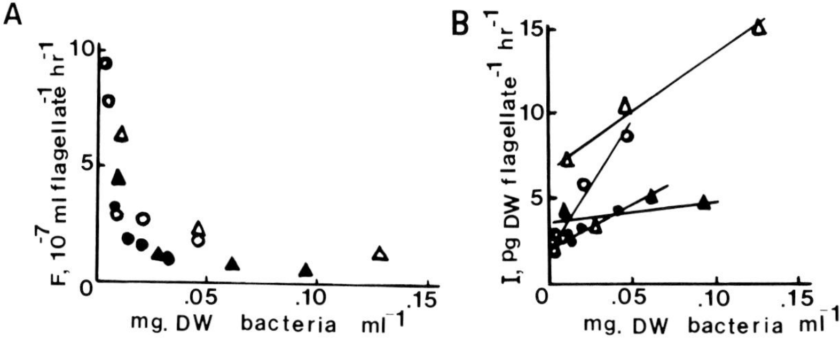

TABLE 4.3 Typical Flagellate Ingestion Rates, Clearance Rates, Growth Rates, and Gross Growth Efficiencies

Ingestion rate (bacteria/flagellate^{-1} d^{-1})	Clearance rate (μl/flagellate h^{-1})	Bacterial food concentration	Growth rate (doublings/d)	Gross growth efficiency (in percent)	Source
650–6,100	0.0014–0.08	10^6/ml	1.7	30	Fenchel (1982a–d)
400–900			5.0–9.0	43	Fenchel (1982a–d)
(18–41 chroococcoid cyanobacteria)	0.003–0.025	–	1.4–1.9	22	Landry et al. (1984)
0–7,200	0.001–0.014	10^5–10^7/ml	–	–	Davis and Sieburth (1984b)
240–1,800	0.0002–0.001	–	1.3–2.3	23.7–48.7	Sherr et al. (1983)
			2.7–8		Sherr and Sherr (1983b)
600	–	–	–	–	Daggett and Nerad (1982)
1,920–3,000	–	–	–	–	Kopylov et al. (1980)
450	0.002–.02	7–9 × 10^6/ml	0.4–1.2	35	Lucas et al. (1987)

to increased food concentration (Sherr et al., 1983; Fig. 4.2). Sherr et al. (1983) observed the highest ingestion rates for *Monas* sp. feeding on *Escherichia coli,* which, interestingly, supported the lowest flagellate yield. These higher rates may be the result of needing to eat more of a qualitatively less balanced food to survive and grow. The ingestion rates for *Monas* sp. feeding on all the other bacterial prey were similar below concentrations of 0.02 mg bacteria/ml dry weight. What effect bacterial prey motility has on ingestion rate, particularly for encounter feeders, remains to be demonstrated, although Fenchel (1982a,b) has suggested that increased prey motility should result in higher predation rates. An understanding of the effects of prey motility would provide for a better understanding of data such as is found in Sherr et al. (1983).

A detailed study of the feeding behavior of the omnivorous marine microflagellate *Paraphysomonas imperforata* has been carried out by Goldman and Caron (1985). This microflagellate becomes cannibalistic when food concentrations decrease to low levels. Many phytoplanktonic species were found in its food vacuoles, including *Dunaliella tertiolecta, Chlorella stigmatophora, C. capsulata, Isochrisis galbana, Porphyridium* sp, and *Phaeodactylum tricornutum,* as were many bacterial prey. *Nannochloris* sp., *Pavlova lutheri, Thalassiosira pseudonana, Chaetoceros simplex,* and *Thalassiosira weissflogii* were not eaten.

In most of the studies referred to earlier, the flagellates had diameters about 10 times their bacterial prey. It is of additional interest to note that during periods of starvation not only does cannibalism sometimes result (Goldman and Caron, 1985), but certain flagellates also become autophagous, digesting certain of their cellular components, including varius organelles, in particular the mitochondria (Fenchel, 1982c,e).

Abundance and Feeding Impact

Most of the quantitative estimates of heterotrophic microflagellate concentrations suggest a typical range of 10^2–10^4 flagellates/ml (see Table 4.2). Davis et al. (1985) found numbers ranging from 10^4 cells/ml in estuarine environments to 10^2 cells/ml in oceanic environments, with numbers generally decreasing with distance from shore and as a function of depth. For Limfjørden, Denmark, choanoflagellates were found to dominate the flagellate community, with nonpigmented chrysomonads, bicoecids, kinetoplastids, nonpigmented euglenids, dinoflagellates, and helioflagellates also found, as were amoebae and heliozoans (Fenchel, 1982d). In Fenchel's study no strong tendency for vertical zonation was found, although numbers were higher in the upper waters. Bacterial numbers increase in early August. This is followed about four days later by a flagellate numerical response that precipitates a decline in bacterial concentrations through the middle of the mouth. This decline in bacterial density is paralleled by concomitant flagellate decline. Following this August peak and decline in flagellate and bacterial biomass, little variability in bacterial abundance is observed. Fenchel calculated that about 20 percent of the Limfjørden water is filtered by the microflagellates.

For the Sea of Japan, it was noted that the spring diatom bloom seeds the thermocline with organic detritus, which initiates a heterotrophic phase of succession in which heterotrophic protozoan biomass exceeds phytoplanktonic biomass by four times (Sorokin, 1977). This leads to a bacteria–flagellate–ciliate-dominated food web, which is quickly superseded by a pulse in microcrustacea. These microcrustaceans appear to be heavily dependent on protozoan production.

In their studies of the waters around Sapelo Island, Georgia (USA), Sherr and Sherr (1983a) found no recognizable seasonal differences in heterotrophic nanoplankton abundances along near-shore transects, although differences were apparent, with decreasing numbers, from inshore to offshore sites. The majority of microflagellates encountered were less than 10 μm in diameter and consisted primarily of monads, particularly choanoflagellates and occasional bicoecids. Colorless cryptomonads were ubiquitous but never exceeded about 10 percent, by numbers, of the population. Positive correlations between heterotrophic flagellate and bacterial numbers have been observed by several workers (Fenchel, 1982d; Sherr and Sherr, 1983a; Sorokin, 1977).

Applying the flagellate numbers of Table 4.2 to the clearance and ingestion rates of Table 4.3 indicates that under varied sets of food and environmental conditions, between 10^4 and 10^8 bacteria/ml of seawater are removed by the flagellate community each day, which clear between 0.5 μl and 19 ml of water over that same 24-hour time interval and remove as much as 30–50 percent on average, of the bacterial production.

Other studies have also documented the important role played by the heterotrophic microflagellates in marine food webs (Sibbald and Albright, 1988; Barlow et al., 1988). Twenty-one to 60 percent of the total consumable carbon of the English Channel was found to be comprised of microheterotrophic protozoa and bacteria (Newell and Linley, 1984). In late summer as much as 60 percent of the total carbon flow, in this area, enters the bacteria, with 34–41 percent of that being respired by the bacteria and 9–12 percent by the microflagellates (Fig. 4.3). In general estimates of bacterivory range from about 20 percent to greater than 100 percent of daily bacterial production (Fenchel, 1982d; Wright and Coffin, 1984; Landry et al., 1984; Ducklow and Hill, 1985; Iben et al., 1986a; Wikner et al., 1986; Coffin and Sharp, 1988; McManus and Fuhrman, 1988a).

For Kaneohe Bay, Hawaii, Landry et al. (1984) found that 18 percent of the total community carbon (9 percent heterotrophic flagellate and 9 percent bacteria) was associated with the protozoa and bacteria. This value is similar to the 21 percent level observed in deeper, stratified waters by Newell and Linley (1984). In the Kaneohe Bay study the phagotrophic flagellates ingested 4.7 times their body carbon per day (28 pg C/d), which supported 1.4–1.9 doublings/d. Landry et al. concluded that the flagellates were food limited and did not control

FIGURE 4.3 Net carbon budget for the plankton community of the western English Channel estimated from biomass data and literature-derived estimates of balanced energy budgets for each group of organisms (consumption [C] = production [P] + respiration [R] + feces [F] + urine [U]). Values = percentage of total carbon flow through the consumer community. (Redrawn after Newell and Linley, 1984.)

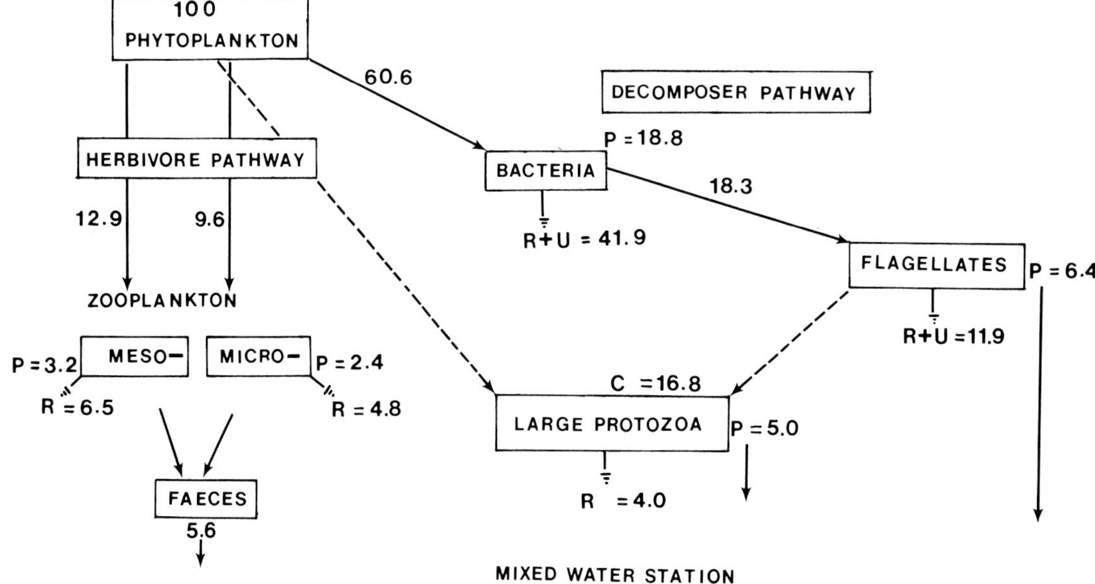

bacterial populations. They ingested 18–41 chroococcoid cyanobacteria and 400–900 bacteria/d/cell with corresponding average clearance rates of 0.0006 μl flagellate/h^{-1} (0.003–.025 μl flagellate^{-1} d^{-1} range). These values are lower but of the same order of magnitude as those found by Fenchel (1982b,d) for the Limfjørden system (650–6,100 bacteria flagellate^{-1} d^{-1}, equal to 14.4 times body volume ingested per day and clearance rates of 0.0014–0.08 μl flagellate^{-1} d^{-1}). Campbell and Carpenter (1986) estimated the grazing pressure of heterotrophic nanoplankton on *Synechococcus* spp., using the dilution technique of Landry and Hassett (1982) in conjunction with selective inhibitors. They noted that 37–52 percent of *Synechococcus* biomass (approximately 100 percent of its primary production) was consumed by these grazers at Northwest Atlantic coastal, oceanic, and warm core eddy sites. Tremaine and Mills (1987b) consider the validity of the critical assumptions of the dilution method for estimating grazing rates.

In addition to what might be considered typical bacterial, algal, and cyanobacterial prey, flagellates also feed on diatoms, including chain-forming types up to six times their own size (Suttle et al., 1986), and likely on picoplankton-sized (0.2–2 μm) algae such as those documented for the Atlantic Ocean by Johnson and Sieburth (1982). In addition, E. B. Sherr (1988) reported on the direct utilization of high molecular weight polysaccharides by heterotrophic flagellates.

It would seem logical to assume that bacterial abundance and biomass, which, as discussed earlier, are often significant in both coastal and pelagic water column environments, should in addition to the flagellates also support significant amebae and ciliate populations. However, in the water column, numbers of amebae appear to be too low to have a significant impact on bacterial populations, and predation by most ciliate species (but not all, see Sherr and Sherr, 1987, and later) at typical planktobacterial concentrations is just incidental (Fenchel, 1980, 1986a, 1987). This is not the case in benthic environments, where the higher concentrations of bacteria, coupled with the various evolutionary adaptations of ciliates to life in the sediments, make them extremely important bacterial predators.

Types of algal and bacterial prey affect growth rates (see Table 4.3) of flagellates as well as ciliates (Taylor and Berger, 1976a,b; Sherr et al., 1983; Stoecker et al., 1981; Rubin and Lee, 1976). Sherr et al. (1983) found that the population growth rates of *Monas* sp. varied with both type and concentration of food. All instantaneous growth rates measured in that study increased, as did food concentration. Similar results have been reported for bodonids (Gorjacheva et al., 1978). In the Sherr et al. study the highest growth rate for *Monas* sp. was r = 0.22/h. Growth rate maxima were achieved at about 0.01 mg dry-weight bacteria/ml, and maximum population abundance yields were linearly related to initial biomass of each, respective, species of bacteria. Sherr et al. found that the optimal temperature for growth of *Monas* sp. was 18–23°C, with 3°C providing the lowest yield and 30°C representing a severe stress condition.

In general, many microflagellates encyst when food conditions become unfavorable (Sherr et al., 1983). Also, in terms of bioenergetics, the dominant energy requirement is for growth with basal metabolic requirements and propulsion (flagella beating) energy requirements representing a very small percentage of the total (Sherr et al., 1983; Fenchel, 1982b; 1986a,b, 1987). A similar pattern of energy requirements seems to be true also for bacteria and yeast growth (Payne, 1970) and appears to be characteristic of small organisms whose populations exist in one of three alternate states: actively growing, declining, and encysted.

Doubling times observed by Fenchel (1982b,d) were about four times higher than those noted for Kaneohe Bay flagellates (Landry et al., 1984). For the open ocean side of Kaneohe Bay, flagellate and bacterial population sizes decreased by 25–30 percent, and primary production decreased by two orders of magnitude relative to the inner bay system. These findings from the Hawaii study suggest that certain ecosystems, particularly oligotrophic ones, are severely food limited and slow growing.

The Dinoflagellates

It is now recognized that many species of dinoflagellates, including many chloroplast-containing forms, are phagotrophic (Droop, 1953; Norris, 1969; Kimor, 1981; Spero and Montescue, 1981; Frey and Stoermer, 1980; Spero, 1982; Lessard and Swift, 1985; Morey-Gaines and Elbrachter, 1987). Lessard and Swift (1985) used a ^{3}H and ^{14}C dual-label radiotracer technique to measure grazing rates of nine species of dinoflagellates (including *Protoperidinium depressum, P. divergens, P. oceanicum, P. elegans, P. ovatum, Podolampas palmipes, Podolampas spinifer, Podolampas bipes,* and *Diplopsalis lenticulata*) isolated from natural populations and feeding on both heterotrophic (including bacteria) and autotrophic prey. Measured clearance rates varied between 0 and 28 μl per dinoflagellate per day, depending on species of predator, and showed no relationship to temperature. Most species took up both autotrophic and heterotrophic prey (*P. ovatum* and *P. palmipes,* which only removed heterotrophic prey, were the exceptions), with many showing a slight favoring of autotrophic prey. The clearance rates observed by Lessard and Swift are comparable to ciliate volume swept clear rates. Of particular importance is the observation by Lessard and Swift (1985) that the dinoflagellates appear to feed routinely on bacteria, even those less than 1 μm in size.

Although food selectivity has not been well documented for any dinoflagellates, bacteria, cyanobacteria, small dinoflagellates, ciliates, and diatoms have been cited as prey items (Lessard and Swift, 1985; Jacobson and Anderson, 1986; Gaines and Taylor, 1984; Kofoid and Swezy, 1921; Barker, 1935; Biecheler, 1952; Norris, 1969; Dodge and Crawford, 1970; Frey and Stoermer, 1980). Certain dinoflagellates, both thecate and naked forms, draw prey to their sulcal region by means of pseudopodial extensions or by flagella-generated water currents. It is at this sulcal region that ingestion takes place (Dodge and Crawford, 1970; Barker, 1935; Frey and Stoermer, 1980; Cachon and Cachon, 1974).

A novel prey capture technique in the dinoflagellates (Gaines and Taylor, 1984; Jacobson and Anderson, 1986) involves use of a large protoplasmic feeding veil/pallium to capture prey (Fig. 4.4). This veil/pallium is extruded between thecal plates of the feeding dinoflagellate, which separate to some degree, and is used to engulf prey (in somewhat of an amoeboid fashion). The prey are often as large as or larger than the pre veil/pallium-extruded dinoflagellate, and include diatoms (chain lengths up to 58 cells) as well as other prey. Digestion occurs extracellularly. Following digestion, the protoplasmic veil/pallium is "retracted." The whole process takes only 7–30 minutes to complete (Gaines and Taylor, 1984; Jacobson and Anderson, 1986). Spero (1982) reported that *Katodinium fungiforme* feeds by attaching to its prey and ingesting the prey cytoplasm through an extensible peduncle. Jacobson and Anderson (1986) report on the feeding of 18 species of thecate dinoflagellates from three genera (*Protoperidinum*, *Oblea* and *Zygobikodinium*).

FIGURE 4.4 Depiction of deployment of the feeding veil (fv), and capture of diatom prey (d), by the thecate dinoflagellate *Protoperidinium conicum*. (Based on the video micrographs of Gaines and Taylor, 1984.)

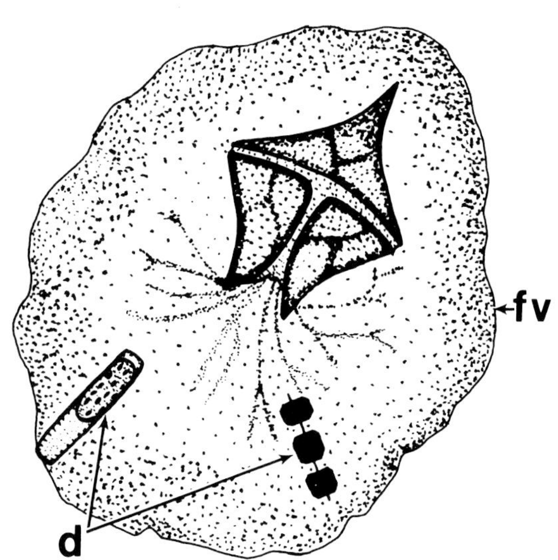

The significance of a voracious, obligate heterotrophic dinoflagellate, *Noctiluca scintillans*, on populations of the copepod *Acartia*, through predation on its eggs, has been well documented by Sekiguchi and Kato (1976). They reported that 33–39 percent of these dinoflagellates observed in Ise Bay, Japan, in May and June, contained *Acartia* sp. eggs and estimated that 74 percent of the eggs produced by this copepod were lost to *Noctiluca* during this May–June time period. Other dinoflagellates also affect copepod eggs. Cysts of *Dissodinium pseudolunula*, a dinoflagellate that parasitizes the eggs of *Pseudocalanus elongatus*, *Acartia* sp, and *Temora longicornis*, have been reported to be widely distributed in the eastern Atlantic Ocean.

Dinoflagellates also frequently appear associated with marine snow samples (Caron et al., 1982; Capriulo, Small, Sniezek, and Angel, unpublished data) and have recently been observed on the surfaces of colonial radiolarians (Angel, unpublished data). Despite this progress in understanding the feeding ecology of dinoflagellates, a quantitative understanding of the importance of this group of protozoa to the ecology of the seas is still far from realization. Such an accomplishment must await a burgeoning of research interest and data concerning this group of phagotrophic protozoa.

Marine Amebae

Foraminifera

An extensive amount of research on foraminiferan ecology has been carried out to date. Slightly over 4,000 described, extant species exist. (If extinct species are included, this number exceeds 40,000.) Of the 4,000 plus species, about 40 are planktonic, although their numbers can be quite high at times. The Foraminifera (henceforth referred to as forams), ubiquitous in the marine environment, are found in supralittoral to abyssal environments as both infaunal and epifaunal benthic, as well as planktonic dwellers, from tropical to arctic–antarctic latitudes. The life cycles of only about 20 species have been studied in detail (see Chap. 1). Many forams, because of the presence of endosymbionts, achieve large size (Lee et al., 1979) (nanometer to centimeter range) and have long life spans (ranging from months to years for some) that may be linked to lunar or annual cycles (Lee, in press; Lee et al., 1985; Lipps, 1982; Loeblich and Tappan, 1964; Lee, 1982a,b; Spindler et al., 1979).

Forams are characterized by the presence of granuloreticulopods and a test and are, along with the order Athalamida, the only rhizopods with anastomozing pseudopods that contain granular cytoplasm (Lipps, 1982). The pseudopods of certain forams have been found to terminate in sticky knobs (reviewed in Nisbet, 1984). In addition to the presence of granuloreticulopodia, the possession of a single nucleus at one life stage and numerous nuclei at another also distinguishes the forams from other single-celled organisms with which they share the typical cadre of cell structures and organelles (Grell, 1973; Lipps, 1982; Lee et al., 1985).

The forams use their unique pseudopods for food capture, digestion, locomotion, test construction, and possibly defense, although the actual mechanism of pseudopodial movement in these protozoa is still not well understood (Anderson and Bé, 1978; Lipps, 1982). Some benthic forams have been observed to leave and later return to their tests (reviewed in Lipps, 1982).

It was realized early on that the forams live both holozoically and osmotrophically (Myers, 1943a,b) and that symbionts play an important role in the nutrition of many benthic as well as planktonic forms. This presence in certain forams of algal endosymbionts, and their influence on various metabolic waste–gas exchange rates, may also explain why certain species achieve exceptionally large sizes (for single cells) that should otherwise be precluded by diffusion-related limitations.

Trophic mechanisms in forams include

1. Uptake of dissolved organic matter (DOM).
2. Herbivory.
3. Carnivory.
4. Omnivory.
5. Suspension feeding.
6. Scavenging, both epifaunally and infaunally.
7. Parasitism.
8. Cannibalism.
9. Symbiosis (endo- and ectocommensals).

These mechanisms have been reviewed in Lipps (1982). In general, forams appear to be opportunistic feeders who use a dense network of rhizopodia, some supported by skeletal $CaCO_3$ spines (spinose) and some without (nonspinose), with associated secretion of adhesive substances to capture prey (Bé, 1982). Feeding frequency and the natural environment have been shown to have a strong influence on the life span and final shell size of forams (Anderson et al., 1979; Bé et al., 1981). Hemleben and Spindler (1983) present a good overview of the state of current research on the living forams. The deep chlorophyll maximum of the western North Atlantic, which occurs between 0 and 95 m, has a strong influence on the abundance of planktonic forams, who appear to be able to track it (Fairbanks et al., 1980). This is especially true for the nonspinose forams, which rely heavily on algae for food (Hemleben et al., 1985). *Globigerinata truncatulinoides* and *G. hirsuita* collected from depths of up to 800 m were found to have no significant differences in identified food remains, when compared with surface-collected individuals (Hemleben et al., 1985; Hemleben and Spindler, 1983). Diatoms are the most easily recognized and most frequently reported food organism of forams, with small flagellates, filamentous algae, bacteria, fecal pellets, and micrometazoans also often reported (Hemleben et al., 1985; also see review of Bé, 1982; Lipps, 1982, 1983). Additionally, symbionts release proteins, peptides, polysaccharides, monosaccharides, amino acids, glycolic acid, and various vitamins for host uptake (Lee, 1974, 1980; Lee and McEnery, 1983). *Pyrocystis robusta,* a known commensal algae in certain planktonic forams (Bé et al., 1977) can, once captured by a foram, live totally unharmed for several days within the foram test (Hemleben et al., 1985).

Planktonic Forms

Planktonic forams have been known to exist at least since 1826 (Hemleben and Spindler, 1983). About 20 of the approximately 40 known planktonic species have been studied to date, and all show some evidence of carnivory (Caron and Bé, 1984; Anderson et al., 1979). In particular, the spinose forms are clearly carnivorous, whereas the nonspinore forms rely more heavily on herbivory (Caron and Bé, 1984; Bé et al., 1977; Anderson et al., 1979). At least one spinose species, *Hastigerina pelagica,* is exclusively carnivorous (Anderson and Bé, 1976a). Freshly collected specimens are typically found to have food between the spines, in various stages of digestion. Anderson and Bé (1976a) frequently found crustaceans in the bubble capsule of the spinose *Hastigerina pelagica* and among the spines of *Globigerinella aeguilateralis.* Foram color has been observed to change with food type, from green to red and orange-red.

The main food of nonspinose planktonic forams consists of diatoms, dinoflagellates, small flagellates, including coccolithophorids, with some multicellular animal tissue occasionally found (Bé et al., 1977; Bé, 1982; Hemleben et al., 1985; Anderson et al., 1979; Hemleben and Spindler, 1978;

TABLE 4.4 Some Known Foraminiferan Prey, Natural and from Culture Work

Foraminiferan species	Known species
Rosalina globularis	Diatoms and other algae
Rosalina leei	Bacteria, chlorophytes
Rosalina floridana	Diatoms, dinoflagellates, chlorophytes, chrysophytes, bacteria, yeast
Globigerinella aequilateralis	Diatoms, phytoflagellates, copepods, hyperiid amphipods
Hastigerina pelagica	Copepods, tunicates, hyperiid amphipods
Rubratella intermedia	Diatoms
Globigerinoides sacculifer	Doliolids, pteropods, acantharia, appendicularia, siphonophores, salps, radiolarians, copepods, algae, tintinnids, other ciliates, chaetognaths, eggs and larvae of invertebrates
Globigerinoides conglobatus	Copepods
Orbulina universa	Copepods
Globigerinoides ruber	Copepod, diatoms, phytoflagellates
Pulleniatina obliquiloculata	Diatoms, phytoflagellates, copepods
Globorotalia menardii	Tintinnids, diatoms, phytoflagellates, copepods
Globorotalia inflata	Diatoms, phytoflagellates
Globorotalia hirsuita	Diatoms, phytoflagellates
Globorotalia truncatulinoides	Diatoms, phytoflagellates
Globigerina bulloides	Diatoms, chlorophytes
Spiroloculina hyalina	Diatoms, chlorophytes, yeast, bacteria, chrysophytes
Elphidium poeyanum	Diatoms, chrysophytes, bacteria, yeast, chlorophytes
Anomalina sp.	Chlorophytes, bacteria, chrysophytes
Quinqueloculina spp.	Diatoms, chlorophytes, bacteria
Ammonia beccarii	Diatoms, dinoflagellates, chlorophytes, chrysophytes, bacteria, yeast, cyanobacteria
Spiroloculina hyalina	Bacteria
Notodendrodes antarctikos	Dissolved organics, diatoms, bacteria
Rotaliella heterocaryotica	Chlorophytes
Rotaliella roscoffensis	Chlorophytes
Nemogullmia longevariabilis	Bacteria
Metarotaliella parva	Diatoms
Haliphysema tumanowiczii	Seaweed pieces, diatoms, copepod parts
Glabratella sulcata	Pennate diatoms
Elphidium crispum	Diatoms
Carterina spiculotesta	Algae
Cibicides lobatulus	Algae
Spiculosiphon radiata	Other forams
Glabratella ornatissima	Cannibalism
Allogromia spp.	Diatoms, dinoflagellates, bacteria, yeast, cyanobacteria
Bolivina spp.	Diatoms, including *Nitzschia* sp.
Discorinopsis spp.	Diatoms, *chrysophaera* sp.
Cylindrogullmia alba	Bacteria
Calcituba polymorpha	Chrysophytes, diatoms
Discorbis spp., *Pyrgo* sp., *Triloculina* spp. *Bulimina* spp., *Patellina* spp., *Spirillina* spp. and *Robulus* spp.	*Nitzschia* spp. and *Navicula* spp.

Sources: Lee (1980); Lee et al. (1966); Anderson et al. (1979); Delaca et al. (1981); Lipps (1983); Caron and Bé (1984); Bé (1982); Hemleben et al. (1973).

Hemleben, 1982; Table 4.4). Species such as *Globigerina glutinata, G. hirsuita, G. inflata,* and *G. truncalulinoides* have been observed to contain 10s to 100s of diatom frustules in their cytoplasm, as well as mussel tissue.

Spinose species often have calanoid copepods as the main part of their diet, with cyclopoid and, rarely, harpacticoid copepods also sporadically taken. Digestion takes from 7 to 9 hours; and 9.5 to 26 hours are required for calanoid and cyclopoid prey, respectively. Other prey digested by spinose forams include ostracodes, siphonophores, acantharians, radiolarians, ciliates, flagellates, polychaete larvae, pteropods, heteropods, tunicates, chaetognaths, diatoms, and other algae (Bé, 1982; Hemleben and Spindler, 1983; Caron and Bé, 1984; Bé et al., 1977; Hemleben et al., 1977; Hemleben et al., 1985). Spinose forams catch and digest prey about one time per day.

Spinose species can catch copepods because of the support of their spines and the associated extended rhizopodial network (Bé, 1982). The spines serve the dual purpose of prey capture and channels for symbiont movement into and out of the foram test. Nonspinose forams cannot hold large live zooplankton, although they will accept immobilized or cut-up microcrustaceans and digest them (Anderson et al., 1979; Hemleben et al., 1985; Bé, 1982). Once prey are trapped in the rhizopodial network, escape attempts result in more entanglement and immobilization.

Secondary producers such as copepods are often the dominant zooplankters in oligotrophic central water masses, and this may be why the forams typically found in these regions are carnivorous spinose ones. Nonspinore forams are found more in eutrophic waters, where more algae production occurs, such as upwelling regions, at boundary current sites, and at times in central water masses (Bé, 1982). Table 4.5 is constructed after information in Bé (1982).

Nonspinose species are without symbionts, which are often found in spinose forams such as *Globigerinoides ruber, G. sacculifer, G. conglobatus, Orbulina universa,* which have dinoflagellate endosymbionts and *G. cristatus* and *G. aequilateralis,* which have haptophyte symbionts (Lee and McEnery, 1983; Lee, 1980, 1982; Bé, 1982; Hemleben and Spindler, 1983). Bé presented a film at the Villefranche sur mer 1981 NATO conference on marine pelagic protozoa that depicted the migration of algal symbionts into and out of the foram test along the pathway of the spines (described in Bé et al., 1977). The migration into and out of the test follows a diel cycle. From these observations and the preceding data, it can be concluded that the spines are an obligate requirement of the symbionts.

TABLE 4.5 Some Typical Oceanographic Regions and the Foraminiferan Species Found There

Oceanographic region	Foram species
1. Boundary current and upwelling regions	*Globoquadrina dutertrei*
2. Diatom rich subpolar and cold temperate regions	*Globorotalia inflata*
3. Winter–early spring North Sargasso Sea phytoplankton bloom	*Globorotalia hirsuita* *Globorotalia truncatulinoides*
4. Mixed zone Sargasso Sea and other nutrient poor oceanic regions	*Globigerinoides ruber, G. sacculifer, G. aequilateralis, G. conglobatus, Orbulina universa, Hastigerina pelagira*

Source: Bé (1982).

In addition to the previously mentioned foods, at least certain forams have an obligate requirement for bacteria in their diet (Muller and Lee, 1969) for sustained reproduction. Other foram species are capable of taking up dissolved organic matter (DOM) directly from water and sediment, through their cell surface (DeLaca et al., 1981). DeLaca (1982) reported the use of dissolved amino acids by the foram *Notodendrodes antarctikos*

Most of the foram species studied to date have been collected from shallow photic zone environments, and many types of food have been identified, including algae, protozoan, and metazoan forms (Loeblich and Tappan, 1964; Boltovskoy and Wright, 1976; Murray, 1973; Lee, 1974, 1980). Vigorously moving prey are not captured, nor are very large prey, because of the limitations of pseudopodial structures and function and the fact that forams do not appear to be capable of stunning or killing their prey. The upper limit of size for copepod prey seems to be about 1.5 mm (Caron and Bé, 1984). Caron and Bé found that copepods were the dominant prey of *G. sacculifer,* accounting for 44 percent of its prey intake. But they observed a wide variety of other prey captured and digested. This may be attributable to the paucity of food typical of oligotrophic waters studied by Caron and Bé and the probable opportunistic nature of these protozoans. This finding of a wide variety of prey contrasts markedly with the findings of Lee et al. (1966) and others that indicates very specific food selectivity in benthic forams (see later). It should be kept in mind, however, that the benthic environment, at least in shallow, eutrophic waters, where much of the benthic research has been conducted, is typically much richer in potential prey biomass than is the open ocean. Such a cornucopia may have allowed for the evolution of selectivity and specificity in foram diets as it has to "sloppy" feeding in estuarine copepods, which are rarely food limited. In addition to the copepods and algal prey listed earlier, tintinnid ciliates were also found to be common prey. Although appendicularians, radiolarians, and acantharians, were often abundant in the plankton, they only infrequently were found trapped in foram rhizopods. Low mobility of these planktonic forms may account for this, since the passive feeding forams would encounter relatively quiescent prey less frequently than they would more active prey (Caron and Bé, 1984). Organisms such as siphonophores, echinoderms, sipunculid larvae, polychaetes, and radiolarians were ensnared into the pseudopodial net at times but were later rejected, possibly because of some prey-associated disagreeable secretions. Caron and Bé calculated, assuming a 100 percent capture efficiency, that prey approximated spheres, as well as other simplifications, that *G. sacculifer* encounters copepods about once every 3.3 days and feeds continuously. Gametogenesis occurs faster with more feeding, regardless of the light regime, but a longer life span is possible if growth and maturity are retarded. Survival time is inversely related to feeding frequency because of the faster onset of gametogenesis with more food. In general, *G. sacculifer*'s life span varies on average between two and four weeks, depending on food availability. It appears that the species of zooplanktonic prey is not as critical as size and vigor of potential prey. The maximum prey size a foram is able to capture depends heavily on its morphology, swimming speed, vigor, and available area for subsequent rhizopodial attachment. Caron et al. (1987) studied the foram *Orbulina universa* and found that frequency of feeding influences rate of chamber formation as well as the frequency of gametogenesis and, therefore, its reproductive potential.

It has been shown that at least certain forams can attract both prey and potential symbionts by some chemical means. Lee et al. (1961, 1963) demonstrated that *Dunaliella sp.* changes from random to directional swimming toward the foram pseudopodial net in its presence. This has been termed the *Circean effect*. Also, *Dunaliella parva* stimulated rapid pseudopodial formation in starved cultures of *Globoratalia, Orbulina, Quinqueloculina,* and *Bolivina*. It is possible that other protists may also be able to control, by means of exuded chemicals, the activity of potential prey–symbionts (Belar, 1923, 1924).

Benthic Forms

Benthic forams appear to be much more selective in their feeding habits than planktonic forams. In one study, only five species of algae and bacteria out of a menu of some 28 species were consumed in significant quantities by *Rosalina leei, Allogromia laticollaris, Spiroloculina hyalina,* and *Ammonicum beccarii* (Muller, 1975). Two algal species, *Phaeodactylum tricornutum* and *Amphora* sp., were eaten in large quantities by all the species studied. The bacteria, however, when eaten, represented only a small contribution to the diet of these protozoans (0.1×10^6 to 8.3×10^6 mg foram^{-1} d^{-1}), when compared with algal food. *Rosalina leei*, one of the less selective forams, ate large amounts of only 12 of some 50 test foods in another study (Lee et al., 1966). *Globigerina bulloides* did not eat much of any of the test foods presented to it (Lee et al., 1966; Lee, 1982a). Lee and his coworkers observed threshold food concentrations of about 100 food cells/ml, with saturation reached at about 10^5 cells/ml. Additionally, differences between feeding rates of juvenile and mature forams also exist, with young cells eating as much as 200 percent more than adults grown on a comparable diet.

The food of benthic forams includes diatoms, algal swarm spores, filamentous algae, radiolarians, minute eggs, cysts, nauplii and other larvae, nematodes, small echinoderms, arthropods, and the like (Christiansen, 1971; Table 4.4), with cyanobacteria, dinoflagellates, chrysophytes, and most bacteria routinely rejected by many littoral zone species (Lee et al., 1966). Food age and concentration also affect feeding. Some forams, such as *Allogromia laticollaris*, grow better on algal mixtures than on single species (Lee et al., 1969). When food is scarce, certain forams may revert to digestion of symbionts (some of which are more resistant than others to digestion), which become reestablished later (Koestler et al., 1985).

In shallow photic-zone environments, forams attain high numbers in epiphytic communities of *Enteromorpha*, in early summer, and later spread to *Zostera, Ulva, Zanichellia, Polysiphonia,* and *Ceramium*, but typically not to *Fucus* or *Codium* communities (Lee et al., 1969). In environments where particulate food is scarce, such as in the oligotrophic deep sea, some forams use DOM. *Allogromia laticollaris*, for instance, takes up D-glucose (but not L-glucose) as well as Fructose 6-phosphate and ATP, when natural food is absent (Schwab and Hofer, 1979). At least certain agglutinated benthic forams have been shown to directly use dissolved organic carbon sources that are taken up through the cell surface (DeLaca et al., 1981). DeLaca (1982) also pointed out that *Notodendrodes antarctikos* uses dissolved amino acids. A large (up to 38-mm) agglutinated foram from an ice-covered embayment of the Ross Sea, Antarctica, captures bottom sediment, suspended by large benthic invertebrates, with cytoplasmic extensions. Feeding in this organism occurs only infrequently, with digestion occurring in the outside-test cytoplasm (DeLaca et al., 1980). Some benthic forams, such as *Elphidium crispum,* can distinguish between live and dead food, showing a preference for live (Murray, 1963).

Sessile forams, such as *Haliphysema tumanourczi* and *Gromia oviformis,* scavenge dead diatoms and other organic material found associated with algal substrates, as well as small pieces of seaweed and parts of decaying metazoans (Hedley, 1958; Christiansen, 1971). *Calcitula polymorpha* lives almost exclusively on filamentous algae up to 1–1.5 mm thick (Schaudinn, 1895).

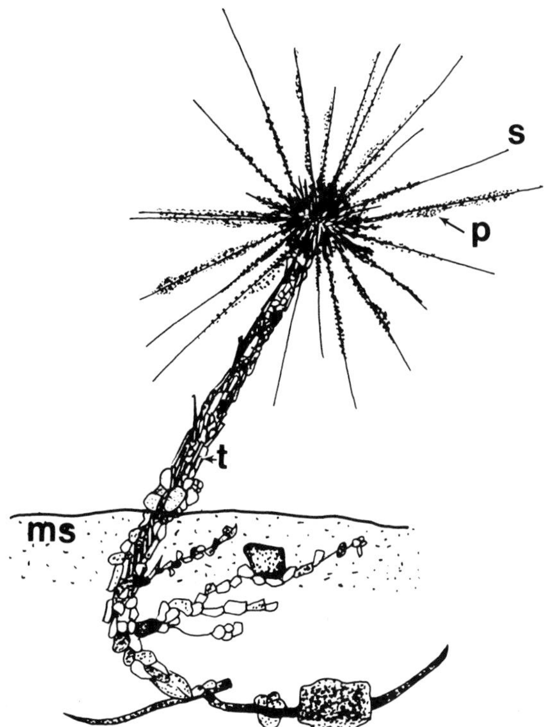

FIGURE 4.5 Testate (t) benthic foraminiferan *Marsipella arenaria,* which anchors itself in the mud and stands erect above the mud surface (ms), where spine-supported (s) pseudopodia (P) are used to suspension feed. (Redrawn after Christiansen, 1971.)

MUD BOTTOM FORAMS

A great number of benthic forams live in mud below the depth of algal growth. These forams often rely on dead organic particles for food. *Saccammina sphaerica* has a spherical chamber that lies half buried in the mud and extends its pseudopods into the sediment, where it ingests the organic remains of diatoms and other settled organics. In contrast, *Saccammina alba* has not been found to contain algal debris (reviewed in Christiansen, 1971). *Pilulina* shapes its test into a pit that food organisms fall into, and *Elphidum crispum* weaves itself into a feeding cyst formed from a pseudopodial network. *Streblus* sp. is found deeper in mud, with pseudopods reaching up to the mud surface. *Hypesammina* spp. have tube tests (one end closed) that they move about with the open end downward, thus contacting the mud surface, where they pick up debris as surface deposit feeders. *Nemogulmia longivariabilis* contains numerous nematocysts obtained from various hydroids such as *Boreohydra* (Nyholm, 1956). Some forams stand up from the bottom and take in suspended organic debris (e.g., *Marsipella arenaria, Dendrophrya erecta, Pelosina arborescens, Jaculella obtusata* (Fig. 4.5). *Bathysiphon filiformi* also uses organic substances in mud (LeCalvez, 1937). *Psammatodendrona arborescens* lives on amphipods, where it produces a highly branched, extensive pseudopodial net that it uses to filter-feed.

Another species, *Quinqueloculina* sp., feeds on large, dead organic remains, especially crustaceans, but not on dead algae or live food (Christiansen, 1971). *Spiculosiphon radiata* exclusively feeds on other forams. It has a hollow test (built of sponge spicules and prey tests; Fig. 4.6) from which extend many thin threads. *Astrorhiza limicola* lives flat on mud with an associated extensive pseudopodial network on the mud surface. This pseudopodial complex can hold and immobilize copepods, caprellids up to 30 mm in size, cumacea, nematodes, and slow echinoderms (Buchanan and Hedley, 1960). *Pilulina argentea* has a spherical test with one opening. It builds a cemented mineral particle–lined mud hollow from which it extends its pseudopods over the mud to capture copepods.

In most of the preceding cases, as is true for most forams, planktonic and benthic alike, prey death occurs from exhaustion.

SANDY AND ROCKY BOTTOM FORAMS

As average grain size increases, the concentration of dissolved organics and associated bacteria as well as algae decreases. This occurs mainly for two reasons: first, because of the decrease in surface area associated with larger sediment particles; sec-

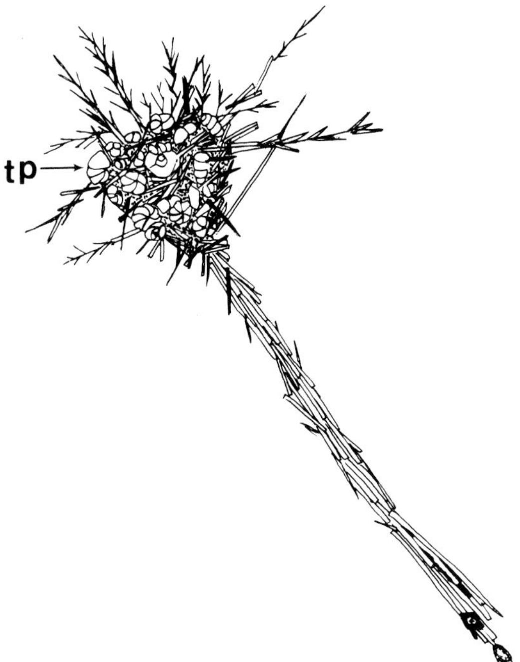

FIGURE 4.6 Testate benthic mud-dwelling foraminiferan *Spiculosiphon radiata*, a predator of other foraminifera. This foram builds its test out of the test remains of its prey. (Redrawn after Christiansen, 1964, as portrayed in Lipps, 1983.)

ond, because of the increased turbulence that not only prevents particles from settling, but also increases the likelihood of damage, particularly to more delicate life forms. With this decrease in organics and increased turbulence comes a decrease in abundance of forams, although they do occur in sandy habitats at times with high local diversity. *Astrorhiza limicola* is found on the surface of fine to medium sand bottoms, where it feeds by extending its pseudopods into sand interstitial spaces 2–3 mm deep (Lipps, 1983; Buchanan and Hedly, 1960).

Rocky bottoms have far less detritus and organics than sandy bottoms, and typically higher turbulence. Nonetheless, forams are found among the holdfasts of macroalgae and associated with algal mats where bacteria and diatoms are present. They also live on the thalli of larger algae and on algal slimes found on rock surfaces, as well as being associated with coralline algae. Forams are also found in the abyssal benthic environment (Lipps, 1983; Christiansen, 1971).

SESSILE FORAMINIFERA

Many forams (agglutinated and calcareous ones) are permanently attached by cement or organic membranes to substrates. This forces these species to feed in one place in a passive manner. Pseudopods of these forms are often extended many times the diameter of the test. To protect their pseudopods, some species, such as *Carterina spiculotesta* and *Cibicides lobatulus,* have coverings of agglutinated layers (made of spicules and sediment particles, respectively; DeLaca, 1982).

General Comments About Benthic Foraminifera

Certain nonattached herbivorous species will feed in one place when food is in sufficient supply but will move about when it is not (DeLaca, 1982; DeLaca et al., 1981). Carnivores do not stalk their prey but appear to catch them by random chance encounters brought about by currents, gravity, and prey movements. Suspension feeders are, by and large, also passive, not capable of generating feeding currents. *Elphidium crispum* suspends itself between stipes of coralline algae on sandy bottoms, where it suspension-feeds. The species *Spiculosiphon radiata* feeds on about 13 other species of foram and, as mentioned earlier, incorporates their tests (Fig. 4.6; Lipps, 1983; Christiansen, 1964). For benthic forms below the photic zone, detrital and bacterial scavenging is the main feeding mode. Most of these species live on the surface or burrow slightly to a few centimeters.

Some forams are parasitic (e.g., *Lagena* sp., Haynes, 1981; parasites are covered in this text only in occasional passing and briefly in Chapter 1). *Discorbis mediterraneansis* captures granules from other foram pseudopodial networks and digests them (LeCalvez, 1947; Lipps, 1983). Cannibalism in species such as *Globoratella ornatissima,* where juveniles eat each other, is also found (Myers, 1940). Some classic competition studies among forams have also been carried out. One such study (Muller, 1975) followed competition among three species: *Allogromia laticollaris, Rosalina leei,* and *Spiroloculina hyalina*. Results indicate that crowding limits reproduction in *A. laticollaris* but has little effect on the two other species. Interspecific competition for food exists between *R. leei* and *A. laticollaris,* but *S. hyalina* appears not to compete with the others for food.

Symbiosis

Symbiosis plays an important role in the nutrition of both planktonic and benthic forams (Taylor, this text and 1982; Lee, 1980) and makes it extremely difficult to quantify the trophodynamic significance of the forams to marine food webs. Symbiosis represents an adaptation to life in nutrient-poor environments and is likely the driving force in the evolution of larger size in the forams (Lee et al., 1979). At least seven species of planktonic forams possess symbiotic algae. Concentrations of symbionts in

Orbulina universa range from 130 to 4,400 cells per individual (Spero and Parker, 1985). *Globigerinoides sacculifer, G. ruber, G. conglobatus, Globigerinella aequilateralis, Globigerina cristata,* and *Globigerina falconensis* also contain endosymbionts, including dinoflagellates, diatoms, chlorophytes, rhodophytes, and isolated chloroplasts from diatoms or chrysomonads that persist as functioning units (Koestler et al., 1985; Hemleben and Spindler, 1983; Lipps, 1983; Taylor, 1982 and this text; Lee, 1980; Lopez, 1979), a condition that also has been observed in ciliates (McManus and Furhman, 1986; Laval-Peuto and Febvre, 1986; Stoecker et al., 1987). Algal symbiont biomass can, at times, equal as much as 75–90 percent of the total cytoplasmic volume of the foram host (Lee et al., 1965; Bé and Hutson, 1977).

The photosynthetic characteristics of the symbionts resemble those of sun-adapted algae, with saturation intensities (I_k value) of 386 μE m^{-2} s^{-1} and light-saturated photosynthetic rates of 1.72 $\times$ 10^6 μmol C $symbiont^{-1}$ h^{-1}, with no susceptibility to photoinhibition to 800 μE m^{-2} s;$^{-1}$ light intensities (Spero and Parker, 1985). Measurements suggest that a single large *Orbulina universa* is about 20,000 times more productive than an equivalent volume of seawater (Spero and Parker, 1985).

Forams may, at times, obtain all or a large part of their nutrition from endosymbionts by utilizing their extracellular metabolites. This obviously would relieve the ecological pressure to find and capture particulate food. *Heterostegina depressa* can obtain all its nutrient requirements from its symbionts (Röttger, 1972; Schmaljohann and Röttger, 1976, 1978), although at times it may take small amounts of particulate food (McEnery and Lee, 1981). Other species, such as *Amphisorus hemprichii, Amphistegina lobifera, Archaias angulatus,* and others, receive less than 10 percent of their carbon needs from symbionts and therefore must feed actively (Lee and Bock, 1976).

Algal symbionts not only are important to their foram hosts as a source of nutrition, but also contribute to test calcification (Duguay and Taylor, 1978). In some species, test wall modifications that enhance light transmission to the endosymbionts have been found (Ross and Ross, 1978; Haynes, 1965, 1981). Also, typically, symbiont frustules (diatoms) and/or plates (dinoflagellates) are not present while the symbiont is residing in the host (Lee, 1980). This loss of plates and frustules may be a further response to the need for endosymbionts to be exposed to specific light intensities and quality, for primary production. Lack of frustule–plates makes identification of endosymbionts very difficult, particularly since some symbionts occur in more than one species of foram. It has been found that foram host homogenate inhibits formation of new frustules of diatom endosymbionts, which, in the absence of the host, develop new frustules. This suggests a host control of frustule development. Additionally, host homogenate increases the level of photosynthate released by endosymbionts to as high as 76% for *Nitzschia voldenstriata*. Even free-living diatoms can be made to release photosynthate in this way (Lee et al., 1984). Symbiotic algae are routinely different species from those algae found in the waters surrounding the foram host, and the same foram species may contain different symbionts, depending on where it is being collected (Lee, 1980). When external food is limiting, many forams digest their endosymbionts in a way suggestive of food gardening (Lee, 1980).

Trophodynamic Significance of the Foraminifera

Because of the limited number of studies on the quantitative abundance of the forams, the complicating factors associated with symbiotic interactions and the uptake of dissolved organic matter (DOM), the paucity of quantitative studies of ingestion rates of different foram species or of the degree of diet specificity, a meaningful and accurate assessment of the importance of the forams to the trophodynamic ecology of the world ocean cannot yet be made. Many marine invertebrates and vertebrates eat forams. This aspect of their ecology is examined in detail later in this chapter, in the section on the fate of protozoan biomass.

One study of numerical abundance of planktonic forams in the western North Atlantic (Gulf Stream to Caribbean Sea) (Cifelli and Sacks, 1966) reported numbers ranging from 0.8 to 39/m^3, with highest concentrations observed in slope waters.

Kuile and Erez (1987) examined the uptake of inorganic carbon (as ^{14}C) and internal carbon cycling in certain benthic, endosymbiont-containing forams and found differences between perforate and imperforate species. Such differences are likely to extend to cycling of food carbon among different species, which, in turn, will further hinder our efforts to understand the global significance of these protists.

Testate and Naked Amebae, the Filosea, and the Acarpomyxea

The testate and naked amebae belong to the class Lobosea, subclasses Testacealobosia and Gymnamoebia, respectively (Lee et al., 1985). They are primarily bacterivores and herbivores and typically are found associated with surfaces, including the surfaces of seaweeds, where microalgae, bacterial, and detrital materials accumulate. There has been relatively little research of any ecological nature carried out on these marine amebae, with much of the current literature devoted to classification.

Naked amebae often round up upon contact or appear as wrinkled masses (personal observation). Some are active at dawn or dusk, whereas others are active at night. They typically glide along surfaces. Many of both the naked and testate amebae are freshwater types, especially fond of soil, bogs, and wet moss environments. Naked amebae are found in both nearshore and pelagic environments (Sieburth, 1979) and include many parasitic forms. Sawyer (1971, 1975) isolated and identified free-living forms from the upper Chesapeake Bay. Flagellated naked forms (ameboflagellates) are also found at times, and they populate bacterial films. They have a transient flagellated stage and form cysts (Sieburth, 1979).

In a study of oceanic amebae (Davis et al., 1978), 13 species of naked amebae were isolated from the North Atlantic between Rhode Island and Spain and from the Straights of Florida. The surface microlayer (at the sea–air interface, which typically exhibits higher concentrations of organics, bacteria, and protozoa; Sieburth et al., 1976) was found to contain an average of 34 amebae/L with a value of 1.4/L for subsurface waters down to 3,090 m. The overall range was from a low of 0.03 to a high of 149 amebae/L. These values compare with numbers typically in the 6×10^{-4} to 3.7×10^{-2}/L range for radiolarians and foraminiferans from similar locations. The most frequently encountered genera were *Clydonella, Acanthamoeba,* and *Platyamoeba*. The amebae were positively correlated with particulate ATP levels, which presumably, at least in part, represent their food source. A study carried out by Hopkins (1930) on *Flabellula mira* indicated that in this species, growth, reproduction, and locomotion depend on food concentration. In another study, Pussard and Rouelle (1986) observed that predation by *Acanthamoeba castellanii* and *Klebsiella aerogenes* stimulated bacterial growth and activity.

Testaceans have a single large aperture in their test, which is composed of either pseudochitinous material, reinforced cemented sand, or siliceous platelets. Most testaceans are freshwater, but some are associated with beach sand as well as plankton (Wailes, 1927, 1937). They are also found on bacterial-laden surfaces and textured surfaces of certain invertebrates, such as bryozoans (Sieburth, 1979). Haberey (1973a,b) described the feeding habits of some testaceans.

Filose amoebae (class Filosea; Lee et al., 1985) are mostly housed within monolocular tests. They have long, slender, clear to faintly granular, nonanastomosing pseudopods that they use for feeding. They can be found among the holdfasts of kelp and under rocks. Some naked forms exist, and these are often found associated with naked amebae (Lee et al., 1985).

The class Acarpomyxea represents a mixed group of much branched forms. Many species are found associated with coral reefs or associated with rich bottom sediments of deep fjords such as the Norwegian fjords (Lee et al., 1985). These forms are believed to be scavengers (Fig. 4.7).

In general, no assessment of the ecological importance of any of the preceding groups (naked and testate amebae) to the food web ecology of the seas can yet be made. Table 4.6 identifies some commonly observed naked and testate amebae, their average size, and preferred food and habitat.

Actinopods

The actinopods are marine amebae characterized by having radially stiffened axopods comprised of cross-linked bundles of microtubules, along with long, fine pseudopods (filopods). Most actinopods are marine coastal and pelagic, and some (e.g., certain heliozoans) are freshwater. This group of organisms includes four classes, namely, the Acantharia, Heliozoa, Polycystinea, and the Phaeodarea (Lee et al., 1985). The Polycystinea and Phaeodarea are commonly referred to as the radiolarians. Some acantharians and heliozoans are attached to substrates, particularly in various benthic habitats.

Febvre-Chevalier and Febvre (1986) have pub-

FIGURE 4.7 The naked amebae *Megamoebamyxa argillobia* with extended pseudopods (p). This species is a scavenger that lives associated with rich bottom sediments and is found at a depth of 40–70 m in the Norwegian fjords. (Redrawn after Lee et al., 1985.)

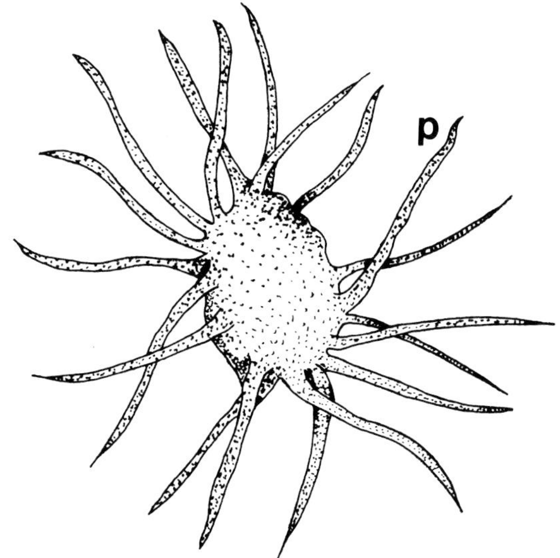

TABLE 4.6 Some Common Amebae with Their Size, Habitat, and Food

Species	Marine habitat	Size (μm)	Food type
Naked Amebae			
Trichamoeba schaefferi	Quiet bays and estuaries	150–175	Diatoms, bacteria, flagellates
Saccamoeba fulvum		50	Diatoms
Rhizamoeba polyura	Brackish water, mudflats	75	Bacteria
Striamoeba munda	Coastal waters		Herbivore
Clydonella vivax	Marine waters, salt springs	6	Bacteria
Clydonella sindermanni	Shallow bays		Bacteria
Unda schaefferi	Shallow bays	10–25	Bacteria
Platyamoeba flabellata	Shallow water		Bacteria
P. bursella	Shallow water		Bacteria
P. calcynucleolus	Shallow water		Bacteria
P. plurinucleolus	Shallow water		Bacteria
P. mainensis	Shallow water		Bacteria
P. douvresi	Shallow water		Bacteria
P. weinsteini	Shallow water		Bacteria
P. murchelanoi	Shallow water		Bacteria
Lingulamoeba leei	Shallow bays	20	Bacteria
Vannella mira	Coastal	15–25	Bacteria
Vannella crassa	Warm coastal	50–75	
Gibbodiscus gemma	Marine tide pools	30–40	Bacteria
Hyalodiscus angelovici	Shallow bays	24–38	Bacteria
Ovalopodium carrikeri	Shallow bays		Bacteria
Flabellula citata	Coastal and hot springs	15–75	Bacteria
Vexillifera telmathalassa	Shallow coastal	15–45	Bacteria
Vexillifera telma	Warm salt springs		
V. aurea	Warm coastal	60–80	Diatoms
V. minutissima	Shallow bays	5–8	Bacteria
Myorella gemmifera	Warm coastal	40–50	Herbivore
M. crystallus		25–30	Small flagellates
M. conipes	Coastal	100–170	Flagellates and diatoms
M. smalli	Similar to *M. conipes*		
M. corlissi	Similar to *conipes*		
M. bulla	Similar to *conipes*		
Triaenamoeba jackowskii	Shallow bays	13–22	Bacteria
Striolatus tardus	Shallow water	50–60	Herbivore
Dinamoeba acuum	Shallow water	50–60	Herbivore
Boveella obscura	Covered with debris and diatoms; shallow bays		Herbivore
Paramoeba eilhardi	Algal material, sand, coastal waters	70–90	Herbivore
P. pemiquidensis	Sand beaches	14–38	Bacteria
Pontifex maximus	Shallow water	100–500	Herbivore
Pseudovalkampfia emersoni	Commensal digestive tract and gills of crabs	18–30	Bacteria
Heteramoeba clara	Shallow water	20–70	Herbivore
Flabellula calkinsi			Cannibal
Testate Amebae and Filosea			
Pomoriella valkanovi	Sand beaches	40–50	

TABLE 4.6 *(continued)*

Species	Marine habitat	Size (μm)	Food type
Gromia spp.	Undersurfaces of stones, holdfasts of kelp		
Gromia oviformis	Tide pools	150–3,000	Herbivore
Lageniopsis valkanovi	Sandy beaches	26–30	
Rhumbleriella filosa	Sandy beaches	20–28	
Volutella hemispirales	Beach sand	25–30	
Microamphora pontica	Beach sand	15–18	
Amphorellopsis elegans	Beach sand	38–46	
Micropsammella retorta	Beach sand	34–40	
Chardezia caudata	Beach sand	40–50	
Alepiella tricornuta	Beach sand	90–110	
Psammonobioticus communis	Beach sand	38–48	
P. minutus, P. golemanskyi, P. balticus similar to *P. communis*			
Corythionella minima	Beach sand	30–40	

Source: Illustrated Guide, Lee et al. (1985); Sieburth Sea Microbes (1979).

lished an excellent review of axopodial movement in the actinopods. Axopods are both motile and contractile and serve in cytoplasmic transport functions. General cytoplasmic flow is probably caused, as it is in the rhizopods, by actomyosin interactions in close association with the cell membrane (Edds, 1981; Febvre-Chevalier and Febvre, 1986). Small particles and vesicles, mucocysts, dense granules, and kinetocysts are transported back and forth in axopodial cytoplasm at speeds of 0.5–2.5 μm/s in discontinuous and often erratic fashion (termed *saltatory transport*). In addition to this, cytoplasmic streaming occurs, and both are involved in food capture and ingestion, as well as in producing secretions and defecating.

In heliozoans, membrane stimulation often results in rapid axopodial contraction at velocities up to 5 mm/s (Haussmann and Patterson, 1982). Such rapid contraction is used for prey capture, which has been described by several workers (Suzaki et al., 1980; Haussmann and Patterson, 1982; Febvre-Chevalier and Febvre, 1986). In at least some heliozoans, such as *Actinocoryne,* rapid contraction is not accomplished by typical microtubular sliding but, rather, by a complete cataclysmic breakdown of axopodial microtubules, which is described in Febvre-Chevalier and Febvre (1986).

Although some data exist on preferred prey and feeding in heliozoans (Looper, 1928; Greissman, 1914; Bovee and Cordele, 1971), this group has, in general, not been well studied. Bovee and Cordele (1971) indicated that some populations of the heliozoan *Actinophrys sol* feed on gastrotrichs. Others prey only on ciliates or are generally described as omnivorous, feeding on yeasts, ciliates, flagellates, and metazoans (Sieburth, 1979). To feed on larger prey, some species of heliozoans may feed cooperatively, temporarily fusing to form a single large predator (Fig. 4.8). Similar behavior occurs in certain colonial radiolarians, which will be discussed later. Some heliozoans have endosymbiotic algae (zoochlorellae) associated with their cytoplasm (Sieburth, 1979). All are free-living and feed holozoically, as far as we know, and are cosmopolitan in marine, brackish, and primarily, fresh water. Most are benthic, living in the centimeter of water above the sediment, or adhere to hard substrates by axopods. Others are sessile, and still others are pelagic (as holoplankton or meroplankton). Most species require still, highly oxygenated water. Some are devoid of skeletal elements or have spicules arranged radially, and genera such as *Lithocolla* have agglutinated tests (Lee et al., 1985; Sieburth, 1979).

Heliozoans have three general types of extrusomes that are used in feeding: kinetocysts, dense bodies, and muciferous bodies. Heliozoans can, as single-cell predators capture prey as large as themselves. Table 4.7 lists some common marine species with approximate sizes.

Acantharians have skeletons composed of monoclinic crystaline strontium sulfate with 20 radial spines joined in the center. These spines emerge

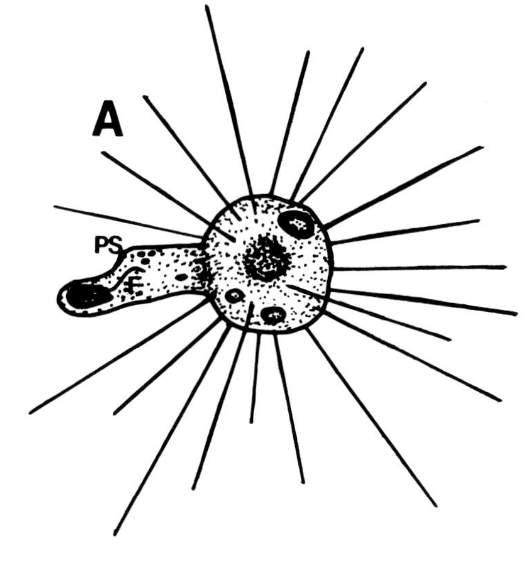

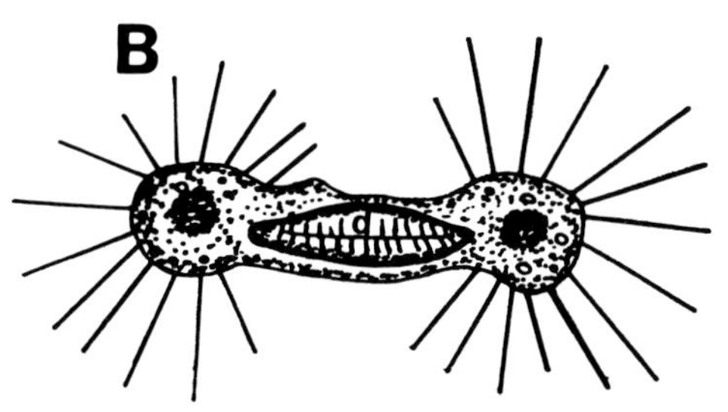

FIGURE 4.8 Feeding in the heliozoan *Cilophrys marina*: (a) solitary form capturing a flagellate (F) by pseudopodial (PS) engulfment and (b) a fused pair cooperatively ingesting a pennate diatom. (Redrawn after Griessmann, 1914, *Arch. Protistenk.* 32:1–78, as portrayed in Sieburth, 1979.)

from the cellular body along five latitudinal circles. Their endoplasm is contained by a thin fibrillar capsular membrane that is surrounded by a vacuolated ectoplasmic layer. Axopod numbers are constant and originate from a central axoplast. Acantharians often contain symbionts, sometimes called yellow cells, which are cryptomonads or dinoflagellate xanthellae that live in the cytoplasm (Lee et al., 1985). They strongly resemble the radiolarians, which also contain strontium sulfate crystals (Dror Angel, personal communication) during their swarmer stage. Since strontium sulfate readily dissolves after the death of these organisms, there is little information available from fossil records.

The significance of the heliozoans and acantharians to the marine food web has not been addressed in any significant way and is difficult to speculate on, because of a severe lack of information even on numerical abundance and distribution of these organisms. The heliozoans likely play a significant role as phagotrophs of at least benthic microbial biomass.

Radiolarians

Strictly holoplanktonic organisms, radiolarians are actually composed of the two orders Polycystina and Phaeodaria, which can be recognized as distinct from related sarcodines, such as the foraminiferans, testaceans, heliozoans, and acantharians, by the presence of a relatively thick, organic (consisting of mucoproteinaceous plates) capsular membrane or wall (central capsule), which divides the cell's cytoplasm into distinct endoplasm and ectoplasm (Lee et al., 1985; Anderson, 1983a; Sieburth, 1979; Fig. 4.9). Anderson's (1983a) textbook on the radiolarians is an excellent source of information on all aspects of this group of strictly marine pro-

TABLE 4.7 Some Common Heliozoans with Their Size and Habitat

Species	Habitat	Size (μm)
Actinophrys sol	Marine	50
Actinophrys vesiculata	Marine	—
Camptonema nutans	Marine	150
Orbulinella smaragdea	Salt ponds	30
Hedriaophrys hovassei	On algae	100–400
Cienkovskya mereschkowski	On bryozoans	45
Actinolophus pedunculatus	On bryozoans and other substrates	35
Actinolophus contractilis	Benthos, shallow water	50
Wagnerella borealis	Marine stalked, attached	175
Oxnerella maritima	Marine	20
Gymnosphaera albida	Benthic, shallow	70–100
Radiophrys annulifera	Marine	3

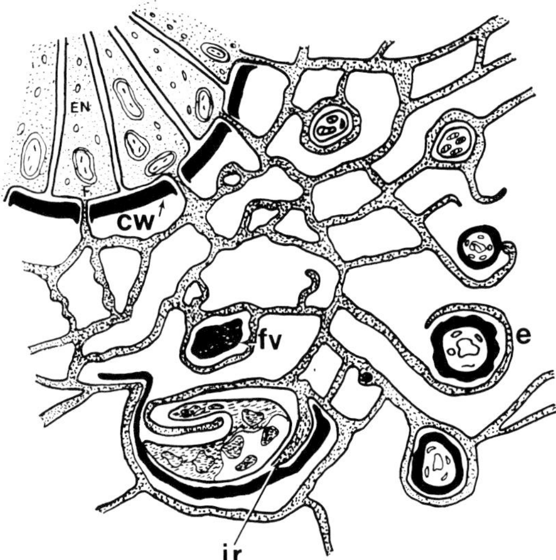

FIGURE 4.9 Schematic representation of prey capture by the carnivorous radiolarian. *Thalassicolla nucleata* depicting the intricate extracapsular network of rhizopodia surrounding the capsular wall (CW). The rhizopodia first ensnare and then engulf (e) prey, subsequently rupturing the exoskeleton, prying prey tissue away from the exoskeleton by means of invading rhizopodia (ir), and forming a food vacuole (fv). (Redrawn after Anderson, 1978b, *Tissue and Cell* 10:401–412.)

tozoa, including their cytology and ultrastructure, distribution and abundance, general ecology, and the like, and the *Illustrated Guide to the Protozoa* (Lee et al., 1985) is a good taxonomic reference. The preceding two references, along with Sieburth (1979), provide the base for much of the following general information.

The distinctive radiolarian central capsule has pores in it (with special endoplasm retaining structures called fusules) through which the axopods pass (Anderson, 1983; Hollande et al., 1970). The endoplasm contained within the central capsule is primarily responsible for carrying out reproductive and storage functions, whereas the ectoplasm is primarily responsible for controlling buoyancy and for food capture and digestion. The ectoplasm is vacuolated and in some species is highly alveolated, with air inclusions (termed the *calymma*) that aid in the control of buoyancy and vertical migration. It is believed that vertical position in the water column may be controlled in some species by expansion and contraction of alveoli in the extracapsulum (ectoplasm) as well as by controlling lipoidal composition of intracapsular vacuoles and by secreting buoyancy-enhancing fluids such as ammonium chloride (Anderson, 1980). The skeleton, when present, may act as a type of ballast control, allowing for a rapid descent when necessary.

Radiolarians rarely survive in near-shore water but for rare instances when they are brought there by near-shore ocean currents or when they are associated with deep, inland water masses, such as the Norwegian fjords (Anderson, 1983a).

The Polycystinea either have no skeletons or have one composed of amorphous silica (with or without spines) with various organic substances included (Lee et al., 1985). Axopods originate in an intracapsular organelle, the axoplast. The Phaeodaria also have a silica skeleton with associated organic substances as well as traces of magnesium, calcium, and copper. Spines are tubular and hollow, and reproduction in both orders is by binary fission or sporogenesis. At least one radiolarian species, *Sticholonche zanclea,* is motile, using axopodial appendages as oars (Hollande et al., 1967; Cachon et al., 1977). Anderson and Bennett (1985) examine the question of skeletal morphogenesis in solitary radiolarians.

The Polycystinea include the spumellarians, which have a spherical capsular wall, with uniform pores over its surface, and when present a spherical skeleton, and the nassellarians, solitary species, which have a single polar pore or aperture. By contrast, the Phaeodaria have three capsular openings, a principal one (the astropyle) and two additional, smaller ones (parapylae). Spumellarians and nassellarians are known to contain symbionts, whereas phaeodarians are not (Anderson, 1983a,b). In addition to the references given earlier, Petrushevskaya (1981) and Boltovsky (1981) provide excellent reviews on general radiolarian ecology.

SUBCELLULAR SPECIALIZATION

The degree to which radiolarians have compartmentalized and differentiated various functions at the subcellular level is truly amazing. As described earlier, digestion occurs principally in the extracapsulum. This has been demonstrated by the finding that three times more acid acryl phosphatase (a digestive vacuole marker enzyme) is found there than elsewhere in the cell, whereas 1.5 times more cytochrome oxidase (a respiratory marker enzyme) is found in the intracapsulum. The extracapsulum is the site of catabolism, of residence of the algal symbionts, as well as of prey capture and buoyancy control. In contrast, the intracapsulum carries on anabolism, lysosomal secretion, and food product storage. Additionally, the rhizopodial network itself is somewhat differentiated in function, with symbionts being held stationary or moving in one direction, in one type of vacuole, while prey are held in more reinforced vacuoles moving in alternate directions (Anderson, 1976a–c; 1980; 1981; 1983a,b; Anderson and Botfield, 1983).

FEEDING ECOLOGY OF THE RADIOLARIANS

Photic zone radiolarians feed on algae (especially diatoms), protozoa, copepods, and other invertebrates, and often contain zooxanthellae (Anderson, 1978a; Anderson, 1983a,b; Anderson et al., 1979; Table 4.8). The actual feeding habits in this group of protozoans have only been identified, with research evidence, for a few species. Most appear to be omnivorous and accept a mixed diet of diatoms, such colorless flagellates as the dinoflagellate *Crypthecodinium cohnii,* ciliates, and small crustaceans (Anderson, 1980; 1983a; Table 4.8). Colonial radiolarians are also found that consist of hundreds or more cells united by a network of rhizopodia and enclosed within a hollow gelatinous envelope (Anderson, 1980, 1983a; Swanberg and Harbison, 1980; Anderson et al., 1985). These colonial radiolarians may not accept food for several weeks, depending on the activity of the endosymbionts (Anderson, 1976a, 1978b).

PREY CAPTURE

Radiolarians snare prey on axopodia or within the peripheral network of rhizopodia not by narcotizing or disabling the prey, but by rhizopodial adherence to the prey (probably with mucuslike substances secreted by Golgi appratus–derived vesicles) with subsequent pseudopodial flow and engulfment, and food vacuole formation. No stinging cells have been observed. As prey struggle, further pseudopodial entrapment occurs; ultimately, the prey dies from exhaustion. Small prey, such as algae, are firmly enclosed in a vacuole surrounded by a large amount of cytoplasm. This contrasts dramatically with how symbionts are loosely held (Anderson, 1980). After capture, prey are carried to the pericapsular cytoplasmic envelope (sarcomatrix), where digestion occurs in a digestive vacuole. In large skeletal-bearing species with many axopods

TABLE 4.8 Radiolarian Prey, Natural and from Culture Work, Comprised of Data from the References Contained Within This Chapter

Radiolarian species	Prey
Spumellarians	
Thalassicolla nucleata and *T. spp.*	Diatoms (including *Thalassiosira pseudonana, T.* spp., *Skeletonema costatum, Amphora coffeaeformis*), colorless flagellates (e.g., *Crypthecodinium cohnii*), thecate algae and dinoflagellates (e.g., *Amphidinium carteri*), coccolithophores (e.g., *Coccolithus huxleyi*), copepods, mollusc larvae, *Artemia* nauplii, ostracods, larvaceans, crustacean larvae
Spongodrymus sp., Thalassicolla nucleata and sp. and *Physematium muelleri*	Tintinnids, copepods, mollusc larvae, larvaceans, ostracodes
Collodaria sp.	Crustacean larvae, calanoid and harpacticoid copepods
Collozoum inerme and *Sphaerozoum punctatum*	Copepods and ciliates, artemia nauplii, algal symbionts at times digested
Collozoum radiosum	*Nitzschia closterium*
Collozoum pelagicum	*Ceratium* sp.
Collozoum ameboides	Small jellyfish
Collozoum caudatum	Salp *Thalia democratica*
Collozoum colony	Siphonophore nematocysts
Sphaerozoum sp.	Chaetognaths and heteropods
Physematium mulleri	Copepods, tintinnids and other ciliates, Foraminifera, mollusc larvae, radiolarians, silicoflagellates, diatoms (e.g., *Chaetoceros*), acantharians, unidentified eggs, chaetognaths, dinoflagellates (e.g., *Ceratium*), other crustaceans, fish
Sphaerellaria sp., *Spongodrymus* sp., *Hexastylus* sp., and *Diplosphaera* sp.	Copepods, tintinnids plus nonloricate ciliates, copepod larvae, pteropods, trochophore larvae, larvaceans, acantharians, silicoflagellates, diatoms (e.g., *Thalassiosira* spp.), dinoflagellates (e.g., *Amphidinium carteri*), *Artemia* nauplii, colorless flagellates,coccolithophores
Nassellarian species	Predominantly bacteria

(e.g., *Spongodrymis* sp.) the contractile response of shorter axopodia is observed during prey capture (this is similar to events that occur in the heliozoans), which brings prey to the pericapsular cytoplasm.

Large prey, such as crustacean food items, are caught differently from smaller prey. This is due not only to size but also to motile strength, the presence of an exoskeleton, and a typical vigorous struggle. A radiolarian must first deal with the prey's exoskeleton before it can digest tissue. After contact and "acceptance as suitable," prey become progressively more tangled and encumbered by rhizopodial strands (Fig. 4.9; Anderson, 1980). The rhizopods ensnare the prey by flowing along the exoskeletal surface and engulfing the appendages. As this proceeds, the prey is drawn deeper into the extracapsular cytoplasm with associated increases of pressure exerted on the exoskeleton. This ultimately leads to a rupturing of the exoskeleton with subsequent rhizopodial penetration into the underlying tissue, which is slowly pryed from the exoskeleton and enclosed in a food vacuole, to be then moved to the sarcomatrix, where digestion proceeds (Figs. 4.9, 4.10; Anderson, 1980). The process of ingestion is faster when the predatory radiolarian is starving. The products of digestion are occasionally stored in extracapsulum, where, for example, oil droplets are sometimes found in well-fed individuals (Anderson, 1980).

Radiolarian pseudopods have small (Anderson, 1983a) osmiotrophic granules contained within small vacuoles that are distributed throughout the cytoplasm. These granules are released by exocytosis and are similar to the extrusomes of Heliozoa, which are used to capture prey (Bardele, 1976; Patterson, 1979; Hausmann and Patterson, 1982).

There is much differentiation in trophic patterns among radiolarians, which likely means that they occupy many ecological niches. This would explain their high diversity and, at times, abundance. Competition for food among some species may not be intense. It is also possible that algal, bacterial and zooplanktonic production, may be higher than what is needed by the radiolarians. However, this implies heavy predation on radiolarians, to keep their numbers down. The limited data on predators of these protozoans (see section in this chapter on the fate of the protozoan biomass) and rates of predation cannot settle this question at this time.

The long spines found in some radiolarians act as reinforcement support rods to which pseudopods attach and advance. This gives mechanical advantage to the cell, enabling it to send pseudopods out to greater distances from the axoplast and, with the extra structural support, in addition to greater adhesive surface area, to capture larger, more active prey (Anderson, 1980; 1983a). The evolution of spines then can be thought of as the probable result of competition for food with associated niche separation. The evolution of spines would allow some of these protozoan predators to take advantage of alternate prey availability and, therefore, would reduce competition for food among various species. Such a development would likely be favored by natural selection and could lead to the trophic diversification discussed earlier. Among large spumellarian species, there is an obvious lack of forms with skeletons, whereas smaller species have well-developed skeletons and associated radiating spines. This relationship of skeletons and spines to overall size may be a result of weight restrictions imposed on these planktonic protists. At some point, the cost of carrying around excess weight in a gravity-controlled pelagic environment in which it is profitable, from the standpoint of food intake at least, to remain in upper waters would outweigh the benefits of skeleton–spine development and would then be selected against. A similar cost–benefit question pertains to planktonic ciliates and their need for lorica (Margalef, 1982; Capriulo et al., 1982). Large radiolarians without skeletons catch prey by sheer mass of rhizopodia with its associated increased surface area and, therefore, adhesive strength. Radiolarian spines may also provide protection against predation and reduce sinking rates by increasing a cell's drag coefficient, in addition to adding structural support for prey capture.

Anderson (1978a, 1983a), working with the spumellarian *Thalassicolla nucleata*, found that prey were immobilized immediately upon contact with the rhizopodial surface and identified a complex prey selection process with at least two types of responses:

FIGURE 4.10 Representation of an invading rhizopodia (r and arrow) of *Thalassicolla nucleata*, prying prey tissue from the surface of a ruptured exoskeleton (ex). ff represents surrounding of prey tissue by other rhizopodia and the forming of a food vacuole. (Based on the transmission electron micrograph of Anderson, 1978b.)

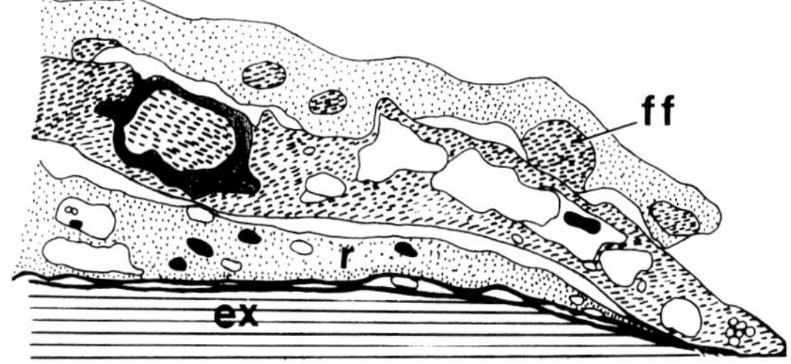

1. Rapid tactile response—prey apprehended or rejected and released (still swimming) instantly upon contact.
2. Delayed response—prey apprehended, enclosed in rhizopodial network, and later rejected (not swimming) or retained.

From this and other work with spumellarian species Anderson (1980, 1976b, 1978a) concluded that

a. radiolarians have a complex sensitivity to food and can discriminate during prey capture.
b. there is a differentiation of function of rhizopodial streaming to facilitate prey capture.
c. there is (as pointed out earlier) a specialization of extracapsular function within the ectoplasmic cytoplasm with associated mediation of food vacuole formation and lysosomal degradation of prey tissue.

Anderson (1977) also pointed out that small radiolarians, especially nassellarians and some spumellarians (of the 30- to 80-μm size), eat bacteria. Table 4.8 summarizes our current state of knowledge concerning known radiolarian prey. How important the radiolarians are at keeping oceanic bacterial concentrations near constant and what their impact on bacterial biomass may be relative to heterotrophic flagellates and ciliates is at this time unknown. That they may have a substantial impact on bacterial populations, however, cannot be ruled out.

Little detailed information exists on radiolarian natural prey or on how predation varies as a function of depth, latitude, living conditions, and the like. Swanberg (1983), Swanberg and Harbison (1980), Anderson (1978a, 1980), Swanberg and Anderson (1985), Anderson et al. (1983, 1984), Anderson and Botfield (1983) and Swanberg and Caron (in preparation) have begun to elucidate prey types and sources of nutrition. The preceding work has indicated that motile, large phytoplankton (such as the dinoflagellate *Amphidinium carteri* and the coccolithophore *Coccolithus huxleyi*) are eaten in greater amounts than smaller flagellates such as the haptophyte *Isochrisis galbana* and nonmotile diatoms such as *Thallasiosira fluviatiles*. Also, most radiolarians studied are omnivores eating more zooplanktonic than algal prey with phytoplankton to zooplankton (P/Z) ratios for larger spumellarians (0.5–0.8 mm) in the 6.7×10^{-2} range. This compares with values of 3×10^{-3} to 8×10^{-3} for planktonic forams (Anderson et al., 1984; Anderson, 1983a).

In a study carried out by Swanberg et al. (1986) on the biology of *Physematium muelleri* it was observed that a wide range of prey items, both microbial and somewhat larger prey, were taken. (Table 4.8 incorporates these data derived from Tables 1 and 2 of Swanberg et al., 1986). Symbiont average photosynthetic rates of 2.5 mg C mg^{-1} chl *a* h^{-1} were measured for an incorporated carbon value of 50 mg C h^{-1} radiolarian^{-1}. The carbon derived by the radiolarians from symbiont photosynthesis was insufficient to account for measured respiration, thus pointing out the obligate predation needs of these planktonic protozoans. In terms of numbers, dinoflagellates and ciliates were the most important prey items. However, when biomass is considered, the most significant prey are the copepods, which account for about 70 percent of the carbon consumed.

SYMBIONTS

When one considers the importance of radiolarians to the trophodynamic structure of the marine environment, the role of endosymbionts must be considered, as it was for the foraminifera. Anderson (1983a,b) provides an excellent review of our current state of knowledge concerning radiolarian symbiosis. Radiolarians with skeletons have been found to harbor three kinds of symbionts: dinoflagellates, prasinomonads, and probable prymmesiomonads (similar to those associated with acantharians). Only dinoflagellates have been found in colonial radiolarians and nassellarians (Swanberg, 1983; Anderson, 1983b; Anderson et al., 1985). Colonial radiolarians with symbionts can have a significant impact on the primary production of an area with as much as three times higher production associated with the colony than an equivalent volume of surrounding water (Swanberg, 1983; Khmeleva, 1967; Anderson et al., 1985). Anderson et al. (1985) have ^{14}C evidence for the translocation of photosynthetically derived carbon from algal symbionts to radiolarian hosts for both solitary and colonial forms. The amount of photosynthate assimilated by the host in colonial forms is a function of both light intensity and, therefore, depth, and varies from about 3.6–47.8 mg C per mg. Chl *a*/mg of protein at 35–239 μE m^{-2} s^{-1} light intensities. Label was later found in fatty acids, triglycerides, wax esters, and glycerol. Swanberg (1983) found that larger colonies photosynthesize at a higher chlorophyll-specific rate. Also, label appears in the radiolarian host only when it is kept in the light and not when kept in the dark.

In general, tropical water primary productivity is on average less than 100 mg C m^{-2} d^{-1} or about 50 g C m^{-2} yr^{-1} (Ryther, 1956). Mean radiolarian production probably equals about 1–2 percent of the primary production when integrated over the whole water column. Although this figure may seem small, it must be kept in mind that the radiolarians represent floating oases of production, highly condensed on microscales; as such, this may be an important component of the pelagic food web.

COLONIAL RADIOLARIANS

Colonial radiolarians are ubiquitous in temperate, subtropical, and tropical oceanic environments (Swanberg, 1983; Swanberg and Harbison, 1980). They have a dense individual cell-secreted gelatinous matrix with both endo- and ectocytoplasmic regions, which often contain symbionts such as dinoflagellates, as well as various vacuoles and digestive organelles (Anderson, 1983a,b; Anderson, 1976a–c; Brandt, 1885). They feed on small zooplankton, including appendicularians, copepods, larvaceans, tintinnids, ostracods, pteropod larvae, and others (Table 4.8) and may achieve meter-sized proportions. For *Collozoum inerme* there is evidence of moderate cropping of symbionts (Anderson, 1976a).

Some colonial radiolarians develop from solitary forms, such as *Thalassophyra sanguinolenta* and *T. pelagicum* (Brandt, 1902). Hollande and Enjumet (1953) documented this for *T. sanguinolenta* and *T. spiculosa*. The amount of food caught by a colonial radiolarian affects the photosynthetic rate of the algal symbionts, as is true for some foraminiferans (Swanberg and Harbison, 1980; Lee and Zucker, 1969). *Collozoum longiforme*, found in the Atlantic Ocean near the equator, has about 70 mg C/cell and a C:N ratio of 8.6, with about 14–28 symbionts/cell (Fig. 2b and Fig. 6 in Swanberg and Harbison, 1980). As with foram symbionts, the photosynthesis versus light intensity curves (*P* vs. *I* curves) are similar to those for certain dinoflagellates. *Collosphaera globularis* receives about 54×10^{-5} g C h^{-1} 100 $cells^{-1}$ from its symbionts (Anderson, 1978b).

Swanberg (1983) studied the role of five species of colonial radiolarians in the oligotrophic oceanic environment and found that carbon content was strongly correlated with cell numbers of each colony, with between 50 and 200 mg C/cell. The amount of chlorophyll *a* per radiolarian colony varied widely, however. Anderson (1980) reported, from culture work, about 20–30 algal cells per host cell in the light. These numbers were reduced about 15 percent after several days in the dark. The reduction in numbers of symbionts can likely be attributed to the fact that the host continually digests symbionts, which by cell division in the light reach a steady-state population size. In the dark, growth does not balance digestion of algae, thus resulting in a net loss of symbionts. The digestion of algal symbionts is suggestive of microbial gardening similar to that found in forams (see earlier discussion).

Colonial species of radiolarians such as *Sphaerozoum punctatum* contain about 36–50 symbionts $cell^{-1}$, so that a colony of 50 cells is home to about 3×10^3 symbionts at any given time. The colonial species *Collozoum longiforme* fixes about 378×10^{-9} μg C $cell^{-1}$ min^{-1} (Swanberg and Harbison, 1980). Large, solitary spumellarians hold about 5×10^3 algal cells/radiolarian and fix about 350×10^{-9} μg C $cell^{-1}$ min^{-1}, a value very similar to that observed in colonial species.

TROPHODYNAMIC SIGNIFICANCE

As discussed earlier, not enough data exist on ingestion rates and feeding preferences of the radiolarians to properly assess their quantitative importance to the ecology of the seas. Studies of their numerical abundance, however, coupled with the data on preferred prey (Tables 4.8 and 4.9) as well as productivity associated with their endosymbionts, suggest that they are likely to be important in at least certain oligotrophic habitats where they represent highly concentrated packets of biomass and primary productivity. Pacific Ocean radiolarian

TABLE 4.9 Some Representative Concentrations of Radiolaria at Various Locations

Abundance	Location	Depth	Source
242–18,730/m^3	Central Pacific Various latitudes	Upper 200 m	Renz (1976)
0.6–27/m^3 (highest in Caribbean)	Western North Atlantic Slope waters; Gulf Stream; Sargasso Sea; Antilles Current; Caribbean Sea		Cifelli and Sacks (1966)
3,000–4,000 colonial rads/m^3	Davis Straight	Top 50 m	Pavshtiks and Pan'Kova (1966)
16,000–20,000/m^3	Gulf of Aden		Khmeleva (1967)
0.04–14 colonial rads/m^3 to 540/m^3 in surface patches	—		Swanberg (1983)
10,000–16,000/m^3	Pacific Ocean	Surface to 150 m	Petrushevskaya (1971)
100–500/m^3	Pacific Ocean	150–5,000/m	Petrushevskaya (1971)
Little data	Indian and Antarctic waters		

concentrations are generally high relative to other areas, with overall world ocean concentrations ranging from around 1–20,000 radiolarians/m^3 in areas where they are found (Table 4.9). At least the upper range of these concentrations certainly suggest that at times they are an important component of the food web, particularly given the kinds of prey that serve as food for these protozoans. Their impact is likely to be greatest on algal, protozoan, and microcrustacean biomass. Anderson (1983) concludes that the polycystine radiolarians occupy at least five distinct to overlapping feeding niches: nanoherbivore, bacterivore, detritivore, omnivore, and endosymbiont derived nutrition. The fate of the resulting radiolarian biomass will be discussed in a later section of this chapter.

Ciliates

Ciliates are found in virtually all freshwater to marine environments, from lakes, rivers, ponds, bogs, saltmarshes, and estuaries to the deepest parts of the ocean, as members of both the benthos and water column, and often in extremely high concentrations (Lee et al., 1985). They have, for example, been documented, often in high diversity, in such extreme environments as the Pacific Ocean hydrothermal vent ecosystems (Small and Gross, 1985; Sniezek, 1987) and the ice-edge region of the Weddell Sea (Heinbokel and Coats, 1986), where concentrations as high as 12,000/L have been observed (Buck and Garrison, 1983). Many species appear to have the ability to tolerate changing physiochemical conditions, such as light, temperature, pH, salinity, and concentrations of dissolved gases, particularly oxygen and carbon dioxide (Lee, 1982a; Fauré-Fremiet, 1950). How changes in these conditions affect competitive hardiness among interacting species of ciliates exposed to fluctuating conditions, however, has been largely unstudied. It does appear, nonetheless, that food supply is at least one of the chief factors influencing ciliate community structure in aquatic ecosystems. Food can be taken in as dissolved nutrients of small molecular weight or directly as particulate matter (detrital or living) by endocytosis (see the following discussion).

Role of Dissolved Organics

With the exception of rare conditions in specific microhabitats, the concentrations of dissolved organics needed to support the growth of free-living protozoa, including ciliates, is higher than that which is typically present in nature (Fenchel, 1968a). Therefore, all free-living ciliates have mouths or some form of oral ingestatory apparatus (Fenchel, 1968a; Small and Lynn, 1985) and feed on particulate food. Dissolved organics absorbed onto particles could, at times, support the growth of ciliates, as well as other protozoa. However, for the most part it only represents an incidental uptake of nutrients (Fenchel, 1968a). The primary food of ciliates, then, are particulates composed of various combinations of bacteria, chroococcoid cyanobacteria, microalgae, diatoms, dinoflagellates, heterotrophic flagellates, and other protozoa including ciliates. In addition, the ciliates themselves represent an integral component of the marine food web both as a source and sink of carbon, phosphorus and nitrogen and as an important part of the decomposition cycle.

The Role of Ciliates in the Decomposition Cycle

A good account of the role of ciliates as bacterivores associated with the decomposition cycle can be found in Fenchel and Jørgensen (1977). In general, some research evidence (Legner, 1973) suggests that certain species of microphagous ciliates (e.g., *Colpidium campylum, Glaucoma chattoni, Cyclidium glaucoma*) have no specialization for a particular type of decomposing organic substrate. Other findings (Legner, 1964) point to a substrate dependence of, for example, *Glaucoma scintillans* for proteins, *Cyclidium citrullus* for carbohydrates, and *Spirostomum ambigiuum* and *Loxocephalus luridus* for bulk cellulose material. The discrepancies in results between these studies may be artifacts of experimental design or may, in fact, point out the high degree of individuality and specificity prevalent among certain groups of ciliates. The work by Legner (1973) demonstrated that *Colpidium campylum* was directly dependent only on the initial quantity of organic material present, and not on the quality of the substrate. These findings likely reflect the fact that requirements for bacterial biomass vary by species, with different ciliates exhibiting different preferred prey and half-saturation constants (Ks) for bacterial carbon. For example, Hamilton and Preslan (1969) found a Ks value of 0.486 mg C/L for the pelagic scuticociliate *Uronema* sp.

Regardless of the question of specificity, ciliates help catalyze decomposition of organic matter by

1. Stimulating bacterial growth by keeping, through predation, the bacteria growing below saturation levels (i.e., in log-phase growth) (Butterfield et al., 1931; Finlay, 1978; Johannes, 1964, 1965; Porter et al., 1979).
2. Prolonging high bacterial growth rates through

the stimulating effects of thermolabile substances derived from ciliates and released to the water (Straskrabova-Prokesova and Legner, 1966; Curds and Cockburn, 1968).

The preceding work suggests that the main interest of the ciliates associated with decomposing matter is not the organic substances themselves, but the microflora associated with the dead plant and animal material. This fact has been stressed by Fenchel (1970, 1972), who quantified the microbes associated with turtle grass detritus and found 3×10^9 bacteria, 5×10^7 flagellates, 5×10^4 ciliates, and 2×10^7 diatoms/g dry weight of detritus. Fenchel further pointed out that even the detritus-associated amphipods, which feed on both detrital-particle-complexes and their own fecal material, were primarily utilizing only the microbial fauna and flora.

Histophagous ciliates also exist that are important to the physical and chemical breakdown of dead tissue or dying or mechanically damaged animals (Fenchel, 1967; 1968a,b; 1969). The prostomatids are facultative histophages that also feed on algae, cyanophytes and other ciliates, while the scuticociliates specialize in feeding on bacteria associated with decaying animal tissue.

Some bacteria, small penate diatoms, cyanobacteria, and the like, attach to sand grains. In these instances, certain ciliates actually ingest the grains to obtain this food (e.g., *Strombidium* spp., *Discocephalus ehrenbergii*) (Fenchel, 1987). Detritus-associated ciliates also represent food for larger invertebrates, such as estuarine copepods (Heinle et al., 1977).

The Benthic Environment

Benthic ciliates feed on a wide variety of microbial food items including, for example in the case of the psalmnobiotic forms, other ciliates, cyanophytes, diatoms, and bacteria (Fenchel, 1987). Nonetheless, many of the ciliates found in benthic habitats appear to be primarily bacterivorous, and much of the quantitative research on ciliate feeding has been concerned with benthic ciliates. Fenchel (1980a–d) examined feeding rates and functional responses of 14 species of ciliates. He found that ingestion rates followed a saturation response, closely fitted by a hyperbolic function, with increasing food concentration, in the manner described by Holling (1959) and demonstrated for some planktonic ciliates (Heinbokel, 1978a,b; Capriulo, 1982; Verity, 1985). Saturation results from the limiting action of phagocytosis, which results in some clogging of the mouth region. Fenchel (1980a–d) also suggested that ciliates do not discriminate between different particles on the basis of properties other than size or shape and that food size is often very limiting to ingestion. This question of feeding selectivity among ciliates has not yet been resolved (Stoecker, 1988), particularly given the probable importance of chemosensory cueing among ciliates (Verity, 1988; Levandowsky et al., 1984; Levandowsky and Hauser, 1978; Hellung-Larsen et al., 1986). Nonetheless, Fenchel's (1987) mathematical interpretation of protozoan filter feeding, filter morphology, and cilia-generated flow patterns provided the groundwork for a serious demonstration of the highly influential, if not deterministic, role that size plays in the feeding of at least certain ciliate species. Somewhat contrary to the findings of Fenchel, others have noted that some ciliates are specific with respect to the bacteria and algae on which they feed and/or grow (Taylor and Berger, 1976a,b; Sawica et al., 1983; Repak, 1986, 1983; Albright et al., 1987; Jonsson, 1986; Skøgstad et al., 1987). Small (1973) found that only four of 26 strains of bacteria isolated along with the ciliate *Carchescin polypinum* supported its growth. Rapport et al. (1972) found that *Stentor coeruleus* selected among particles brought to its buccal cavity. Certain particles were preferentially rejected by localized ciliary reversal, although discriminatory abilities were better when organisms were satiated than when they were hungry. Also, this organism could discriminate between algal and nonalgal species in favor of nonalgal forms. Laybourn and Stewart (1975) found that the freshwater ciliate *Colpidium campylum* would not ingest dead bacteria. Turley et al. (1986), working with species of *Uronema* and *Euplotes*, found different food-related response times between ciliates, with *Uronema* being more opportunistic than *Euplotes* and *Euplotes* preferentially ingesting bacterial rods before cocci forms. Allbright et al. (1987) examined several ciliate species and found that all fed equally well or better on free as compared with attached bacteria, but for *Euplotes* sp., which showed a preference for attached forms. Similar findings of feeding selectivity have been reported for soil ciliates (Petz et al., 1985, 1986). Food specialization, then, may be part of a niche partitioning that accounts for the high species richness observed in benthic protozoan communities.

Some of Fenchel's observations suggest the existence in many ciliates, of a lower limit to the size of food that can be retained. For the spirotrichs he studied, the limit was about 1 μm. Other ciliates, such as the bacterivorous holotrichs studied by Fenchel, retained even the smallest prokaryote cells (i.e., 0.2–2 μm in size), although many ciliate species primarily feeding on bacterial-sized particles (0.2–1 μm) have strongly reduced clearance rates, because of the decreased rate of water flow through ciliary filters associated with smaller interciliary spacing and the resultant higher resistance to flow (Fenchel, 1980a–d). It has been thought that, because at their low water-processing capabilities, bacterivorous ciliates should rarely be found in the plankton, where bacterial concentrations are generally too low for the growth of ciliates with low clearance rates. Fenchel (1980a,b), Porter et al. (1979), and Berk et al. (1976) found that bacterivorous ciliates require concentrations of bacteria in the 10^6–10^8/ml range before positive growth occurs. Such concentrations are typical of benthic communities rich in organics and in the early successional stages of decomposition. Similar findings exist for other protozoa. For example, Danso and Alexander (1975) found that growth of amoebae populations depended on bacterial concentrations reaching the 10^6–10^7 cells/ml level. Concentrations of bacteria in the plankton (see Table 4.1) are typically below the preceding levels. Consequently, larger ciliates that feed on nanoplankton appear to be the dominant forms in the zooplankton. The recent work of Sherr and Sherr (1987 and unpublished data), which has documented extremely high specific clearance rates for certain small, aloricate oligotrich and scuticociliate planktonic ciliates feeding on bacteria (10–100 times greater than previous estimates), challenges this notion of the water column being a poor place for primarily bacterivorous ciliates. Albright et al. (1987) observed that certain spirotrich and free-swimming peritrichs can feed on planktobacteria at bacterial concentrations less than 10^7 cells/ml. Additionally, Rivier et al. (1986) reported success in growing the choreotrich *Strombidium sulcatum* (20–30 μm) on bacterial concentrations of 3–6 × 10^5 bacteria/ml. These findings should stimulate continued research in this area and a subsequent reinterpretation of this component of the planktonic microbial food web.

Taylor (1978a,b) found that thresholds exist in the numerical response of ciliates to their bacterial prey. Among the ciliates, considerable interspecific variation is present with respect to prey densities required for population growth and to the way the different species of ciliates are adapted to different prey. Although, because of low bacterial densities, life in the plankton appears to be difficult for most (but certainly, as discussed earlier, not all) ciliates that are primarily bacterivores, interstitial sediment and benthic surface sediment layers contain bacterial concentrations high enough for the growth of a wide variety of ciliates. These ciliates are likely to be from different taxonomic groups than the water column forms. Benthic ciliates typically ingest 80–120 percent of their body volume per hour, compared with 10–30 percent for larger individuals (Fenchel, 1980a–c). Benthic bacteria typically are associated with sediment surfaces, and benthic ciliates therefore tend to be surface grazers, or stalked forms feeding on nonmotile or attached prey as well as on motile prey (Fenchel, 1986a, 1987; Taylor, 1978). Feeding functional morphology of water column versus benthic ciliates exhibits differences that parallel the divergent needs for food collection in these different habitats. It should be stressed here that even pelagic ciliates may often be associated with suspended particles (i.e., floating seaweed, marine snow, remains of dead organisms, and the like) and that the concept of benthic versus planktonic is not always meaningful with respect to microorganisms. Not all "benthic" ciliates are found on the bottom. Some attach themselves either permanently or temporarily to, or are just loosely associated with, suspended living and nonliving particles. For example, the suctorian ciliate *Ephelota* sp. is found on the euphausiid *Meganyctiphanes norvegica* (Nicol, 1983) with as many as 82 percent of the euphausiids, at times hosting this epibiont, in the Bay of Fundy. Other typically "nonplanktonic" ciliates are routinely found associated with marine snow aggregates (Silver et al., 1984; Caron et al., 1982; Capriulo, unpublished data). This phenomenon will be discussed in this chapter's section on nontraditional habitats.

The maximum size of food ingested by a ciliate is determined to a large degree, but not entirely, by mouth diameter (Fenchel, 1980a–d; Capriulo, 1982; Heinbokel, 1978a,b; Spittler, 1973; Verity and Villareal, 1986). Mouth size, however, is not always constant and at times may vary (Hedin, 1975, 1976; Capriulo, 1982; Fenchel, 1980a–d), particularly with species that are polymorphic (Fenchel, 1967, 1968a,b; Kidder et al., 1940; Tuffrau, 1959). Pierce et al. (1978) noted four stages leading to cannibal giantism in *Blepharisma*. The transition to larger size is accompanied by a larger and more enhanced oral ciliature and increased complexity in associated architecture.

Such changes likely represent adaptations to exploit larger-sized food, such as ciliates, that occur during the later stages of succession (Sorokin, 1977), when decomposition-related organics and associated bacterial peaks, characteristic of the early stages of decomposition, are gone. Adaptations such as these enhance an organism's competitive capabilities and, therefore, niche space. Adaptations and events that reduce competition for a given species have been shown to increase ciliate density as well as distributional range (Stenson, 1984).

Muller and Lee (1977) studied competition of the ciliate *Euplotes vannus* with the foram *Allogromia laticollaris* and the nematode *Chromadorina germanica* in a benthic setting and found that *E. van-*

nus required high algal food concentrations for rapid growth (greater than 10^4 cells/ml), whereas the forams and nematodes required only greater than 10^2 and 10^3 cells/ml, respectively. This high food density requirement selected against *E. vannus* at low prey concentrations. When several competing ciliate species are considered under poor food conditions, the species with the lower individual energy requirements and greater percentages of stored reserves will prevail (Jackson and Berger, 1984, 1985a,b). Superimposed on such species-specific physiological advantages associated with food concentrations are specific requirements for food type. These interactions among benthic (as well as planktonic) ciliates and their prey, along with their high degree of specialization, results in definable patterns of habitat selection that can be somewhat delineated by taxonomic grouping (Table 4.10). An example of the extreme levels of specialization that can be realized was reported by Small (1973), who noted that some carnivorous free-swimming rhabdophorans had associated with them predatory carnivorous suctorian ciliates which attached to the peritrichs and fed both on the rhabdophorans and on the peritrichs. Further support of this notion of habitat specificity among aquatic ciliates is provided by Small and Gross (1985) and Sniezek (1987), who provide a detailed account of the diversity of ciliates (mostly bacterivorous ones) from two Pacific Ocean thermal vent communities. Their results indicate that all ciliates encountered in this ecosystem were probably new, endemic species.

TABLE 4.10 Ciliate Taxa that Predominate in Marine Habitats

I. Neritic environs
   - A. Shallow: still waters
      - 1. Salt marshes
         - Aufwuchs: Peritrichs: especially *Zoothamnium, Vorticella, Corthunia,* Suctoria; *Ephelota, Acineta, Stentor; Folliculina; Spathidiopsis*
         - Detritus and decay: Pleuronematine and philasterine scuticociliates; *Nassula* and *Frontonia* peniculines; prorodontid and metacystid prostomes
         - Diatoms: Phyllopharngians
      - 2. Sandy beaches with fine sands
         - Aerobic: Karyorelicteans and spirotricheans, strombidiid choreotrichs, nassophorean hypotrichs; pleuronematine scuticociliates
         - Anaerobic: metopid heterotrichs and geleiid karyorelicteans
      - 3. Mangroves: geleiidkaryorecteans
      - 4. Plankton communities
         - Surface, photic waters: choreotrichs, condylostomatid heterotrichs, didiniid litostomeans
         - Off bottom, aphotic waters: philasterine, scuticociliates, loxocephalines
      - 5. Oyster reefs—outside: See aufwuchs
         - inside: thigmotrichine and pleuronematine, scuticociliates and rhynchodine phyllopharyngids
   - B. Offshore subtidal
      - 1. Sea urchins: philasterine scuticociliates, euplotine hypotrichs
      - 2. Bivalve molluscs: pleuronematine and thigmotrichine scuticociliates; euplotine hypotrichs; rhynchodine phyllopharyngids
      - 3. Coelomic cavities: paranophryid and scuticociliates
      - 4. Kelps: ephelotid suctorians
   - C. Marine caves
      - 1. Animal decay: philasterine scuticociliates
      - 2. Plant decay: frontoniid nassophoreans and condylostomatid heterotrichs
      - 3. Sediments: *Kentrophorus* spp.

II. Oceanic—Pelagic Environs
   - A. Euphotic zone
      - 1. Plankton: choreotrichs, especially tintinnids and strobilidiids; *Zoothamnium pelagicum;* pleuronematid scuticociliates.
      - 2. Gyre kelp community: pseudocohnilembid and other philasterine scuticociliates
   - B. Decay-related marine "Snow"
      - 1. Upper layers: paralembid philasterine scuticociliates, spiroprorodontid phyllopharygeans
      - 2. Sediment traps: see Silver et al. (1984)
   - C. Hydrothermal vents (Juana deFuca Ridge; 21°N; Galapagos vents; Guaymas Basin; see special habitats section this chapter)
   - D. Upwelled, nutrient rich waters (Antartica Ice Water, off Peru, Chile, South America; see special habitats section this chapter)

Source: Unpublished notes of E. B. Small.

Several studies give detailed accounts of the ecology of benthic ciliates (Fenchel, 1967, 1968a,b, 1969, 1980a–d, 1987; Borror, 1963, 1968, 1980; Hartwig, 1980; Webb, 1965). All these studies dramatize the fact that ciliates are extremely important to the ecology of sediments, particularly fine sands and sulfureta as macro- and microphages (as defined by Dragesco, 1962). As mentioned earlier, many of them are polymorphic, with morphology and size dependent on the type of food offered (Fenchel, 1967, 1968a,b, 1969; Kidder et al., 1940; Tuffrau, 1959; Pierce et al., 1978). *Glaucoma vorax,* for instance, may feed as a saprozoan, or holozoically on bacteria, yeast, or ciliates. If it is fed large food, *G. vorax* becomes larger (as is true for *Blepharisma*), with an appropriately more size-accommodating mouth (Fenchel, 1968a). Similar reports exist for still other ciliates (Hewett, 1980). The ability of benthic ciliates to discriminate food based on size is also evident in *Diophrys scutum, Blepharisma clarissimum, Chlamydodon triquetrus,* and *Frontonia marina,* all of which are at times found together on the surface layer of bay sediments, all feeding on diatoms, yet with each species feeding on a differently sized species of diatom (Fenchel, 1968a). Differential growth responses to varying diet were found for the heterotrich ciliate *Fabrea salina* (Repak, 1983, 1986). Fenchel (1968a,b) observed that the division rate of *Colpidium colpoda* is dependent on the species of bacteria offered as food. Some food types actually support zero growth, as was also noted by Repak (1986). Such data stress the fact that high bacterial concentrations do not necessarily imply high standing stock or production of ciliates. Fenchel (1968a,b) reported habitat specificity in four species of *Remanella* from deeper layers of fine sand.

Some benthic (and water column) ciliate groups are dependent on sulfur bacteria for food. These include oligohymenophorans in the family Plagiophylidae and certain dysteriids, pleuronematines, and heterotrichs. Many of these ciliates actually discriminate between species of sulfur bacteria. In addition to the bacteria, diatoms, dinoflagellates, euglenoids, cryptomonads, and various phytomonads, colorless flagellates (e.g., *Peranema*) are also important to the diet of numerous benthic ciliates (Fenchel, 1968; Borror, 1963; Table 4.11), as are other protozoans, including ciliates and, at times, metazoans. Many ciliates feed on zooflagellates such as *Bodo* and *Rhynchomonas,* other ciliates, and rotiferes. The rhabdophoran groups are specialized ciliate hunters (Table 4.11) which possess proteolytic and paralytic toxicysts that are used to kill prey, often more than can be ingested.

Additional quantitative feeding work has been carried out on benthic ciliates (Fenchel, 1980a–d; Taylor, 1986). Fenchel found that the maximum rate of clearance, for the ciliates he studied, occurred at low food concentrations, whereas maximum ingestion rates were observed at high food concentrations. Taylor (1986) reported ingestion rates for *Colpidium colpoda* (freshwater ciliate) of 16,000 *Enterobacter aerogenes* bacteria h^{-1} ciliate^{-1} during exponential growth.

Planktonic Ciliates

Trophodynamic Significance

Zooplankton are ubiquitous in the world ocean and consume a major portion of its primary production. Many taxonomic groups are represented in the zooplankton, and the trophodynamic significance of each group differs both in intensity and as a function of time and space. Among the zooplankton, the copepods have long been considered the apparent dominant grazers of phytoplankton biomass (Harvey, 1937). Copepods have been extensively studied, and much is known about their feeding ecology (Jørgensen, 1966; Parsons et al., 1967; Mullin and Brooks, 1967; Mullin et al., 1975; Lehman, 1976; Richman et al., 1977; Mayzaud and Poulet, 1978; Poulet, 1978; Gifford et al., 1981; O'Connors et al., 1980; Gerritsen and Porter, 1982; Uye, 1986). In contrast, until recently, the microzooplankton (those zooplankters passing a 200-μm mesh net; Dussart, 1965) have received little attention, despite the fact that their potential importance has long been recognized (Lohmann, 1908; Steeman-Nielsen, 1962; Adams and Steele, 1966). In recent years, it has been suggested that the microzooplankton are responsible for removing a large percentage of phytoplankton primary production in certain marine ecosystems (Beers and Stewart, 1967, 1971; Capriulo and Carpenter, 1980, 1983; Landry and Hasset, 1982; Heinbokel and Beers, 1979; Verity, 1985; Burkill et al., 1987; Paranjape, 1987). The role of micro- and nanoplankton in marine food webs has recently been reviewed (Laval-Peuto et al., 1986; Sherr et al., 1986b).

Theoretical calculations, based on numerical abundance data and assumptions concerning feeding rates of microzooplankton in the eastern tropical Pacific, suggest that at times these organisms may consume as much as 70% of the daily phytoplankton organic carbon production (Beers and Stewart, 1971). Riley (1956) estimated that in Long Island Sound as much as 43 percent of the net carbon fixed annually by photosynthesis may be removed by the microzooplankton and bacteria in the water column. Using a direct measurement technique, Capriulo and Carpenter (1980) found that Long Island Sound microzooplankton removed up

TABLE 4.11 Diet of Ciliates Found Associated With Benthic Habitats

Ciliate group	Prey
Postciliodesmatophora	
Karyorelictea	
Trachelocercidae	ciliates, flagellates, diatoms in sand and detritus.
Loxodidae	interstitial in fine sands, diatoms, unicellular algae, flagellates, small ciliates, large bacteria.
Geleiidae	dinoflagellates, algae, diatoms, flagellates, ciliates.
Spirotrichea	
Heterotrichia	
Blepharismidae	small diatoms, unicellular algae, flagellates, sulfur bacteria, cyanobacteria, detritus diatoms, Rhodobacteria.
Condylostomatidae	ciliates, diatoms, flagellates, green algae, small metazoans.
Stentoridae	dinoflagellates, bacteria, diatoms, sulfur bacteria, some cannibalism, green flagellates.
Chattonidae	dinoflagellates, diatoms, bacteria, ciliates.
Peritromidae	algae, diatoms, sulfur bacteria, chlorophytes, cyanobacteria.
Metopidae	sulfur bacteria, purple sulfur bacteria, colorless bacteria.
Odontostomatidia	bacteria (bacilli and cocci).
Epaxellidae	colorless bacteria.
Mylestomatidae	sulfur bacteria, cyanobacteria.
Stichotrichida	
Oxytrichidae	diatoms, ciliates, sulfur bacteria, flagellates, other bacteria.
Rhabdophora	
Litostomatea	
Enchelyidae	
Enchelyodon	carnivore (eats *Strombidium, Cyclidium,* small ciliates, cryptomonads, diatoms, dinoflagellates).
Helicoprorodontidae	
Helicoprorodon	carnivore (eats ciliates; *Remanella brunnea; R. margaritifera* and *Kentrophorous*).
Spathidiidae	
Myriokaryon (formerly	flagellates, euglenoids, ciliates, histophagous.
Pseudoprorodon)	flagellates.
Trachelophylliidae	
Chaenea	carnivore (eats *Uronema; Cyclidium; Cohnilembus;* other scuticociliates).
Lacrymariidae	
Lacrymaria	ciliates (*Pleuronema, Cyclidium, Frontonia, Uronema*); flagellates; amoebae; cysts of *Nassula citrea;* dinoflagellates; diatoms.
Didiniidae	bacteria, dinoflagellates, green flagellates.
Amphileptidae	ciliates, flagellates, small metazoans.
Tracheliidae	carnivory on ciliates and small metazoans.
Trichostomatia	
Coelosomatidae	filamentous bacteria, algae, dinoflagellates, diatoms.
Cyrtophora	
Prostomatea	
Colepidae (e.g. *Coleps*)	dinoflagellates, bacteria, invertebrate tissue, diatoms, scuticociliates.
Plagiopogon	dinoflagellates, unicellular green algae including phytoflagellates; histophagous.
Holophryidae	
Holophyra	unicellular brown algae, diatoms, flagellates.
Placidae	
Placus	flagellates, euglenids, ciliates, diatoms, bacteria, *Mesodinium;* histophagus.
Plagiocampidae	
Plagiocampa	purple sulfur bacteria, small algae, peridineans; histophagous.
Prorodontidae	
Prorodon	algae, flagellates, rotifers, dioflagellates, cryptomonads, diatoms, dead nematodes; histophagous.
Nassophorea	
Nassulidae	diatoms, cyanobacteria.
Fontoniidae	diatoms, cyanobacteria, bacteria, animal prey.
(other peniculids)	purple sulfur bacteria, small cyanobacteria, flagellates.

TABLE 4.11 *(continued)*

Ciliate group	Prey
Microthoracidae	sand dwelling diatoms.
Hypotrichia	
Euplotidae	rhodobacteria, other bacteria, diatoms, detritus, sulfur bacteria, colorless flagellates, dinoflagellates, bacteria, cyanobacteria, yeast, ciliates.
Aspidiscidae	diatoms, bacteria.
Phyllopharyngea	
Chlamydodontidae	diatoms, green algae, cyanobacteria, purple sulfur bacteria, other bacteria.
Dysteriidae	diatoms, cyanobacteria.
Oligohymenophorea	
Cohnilembidae	bacteria, cellular debris, organic debris, metazoan tissue; parasitic.
Pleuronema	purple sulfur bacteria, dinoflagellates, diatoms, bacteria.
(other pleuronematines)	suspension feeders on bacteria, small flagellates, diatoms.
Plagiopylia	
Plagiopylidae	sulfur bacteria, purple and colorless bacteria, small diatoms, euglenids and other flagellates, cyanobacteria.

Based on data from Fenchel (1968), Borror (1963), and Small (unpublished data), and on the revised systematic scheme of Small and Lynn (1985).

to 41 percent of the chlorophyll *a* standing crop per day and, as a community, at times exhibited ingestion and filtration rates equal to those of Long Island Sound copepods. Landry and Hasset (1982), using a dilution technique, found that in the coastal water off of Washington, microzooplankton removed 6–24 percent of the phytoplankton standing crop per day, or 17–52 percent of the daily primary production. Burkill et al. (1987), combining HPLC pigment analysis with the Landry and Hasset dilution technique, found that for the coastal waters of the northeastern Atlantic the microzooplankton (consisting mostly of ciliates, choanoflagellates, and aloricate colorless flagellates) removed 13–65 percent of the algal standing crop per day. A study of the Canadian Arctic region, in summer, also employing the Landry and Hassett dilution technique, reported that 8–15 percent of the microalgal standing crop (>40 percent of the daily primary production) was removed by the microzooplankton (Paranjape, 1987).

Microzooplankton assemblages are often dominated by marine protozoa—termed *protozooplankton* by Sieburth et al. (1978). The heterotrophic flagellates and sarcodines have been discussesd. Among the protozooplankton, the tintinnid ciliates (Fig. 4.11a) (order choreotrichida, suborder tintinnina; Small and Lynn, 1985) are often the predominant group of larger protozoans (Vitiello, 1964; Sorokin, 1977; Heinbokel and Beers, 1979; Capriulo and Carpenter, 1983; Beers et al., 1980). Tintinnid ciliates have been found to remove up to 21 percent of the daily primary production in the Southern California Bight (Heinbokel and Beers, 1979); 27 percent of the yearly primary production in Long Island Sound (Capriulo and Carpenter, 1983); 16–26 percent of the yearly primary production in Narragansett Bay, Rhode Island (Verity, 1985, 1987); and 25 percent of the primary production in the waters of the open northern Baltic Sea (Leppanen and Bruun, 1986). Further emphasizing the importance of these ciliates to planktonic trophodynamics is the finding that Long Island Sound tintinnids

FIGURE 4.11 (a) Representation of a tintinnid ciliate of the genus *Tintinnopsis,* showing agglutinated lorica (L), oral membranelles (OM), and capsule containing striae (S, extrusomes likely used in prey capture).

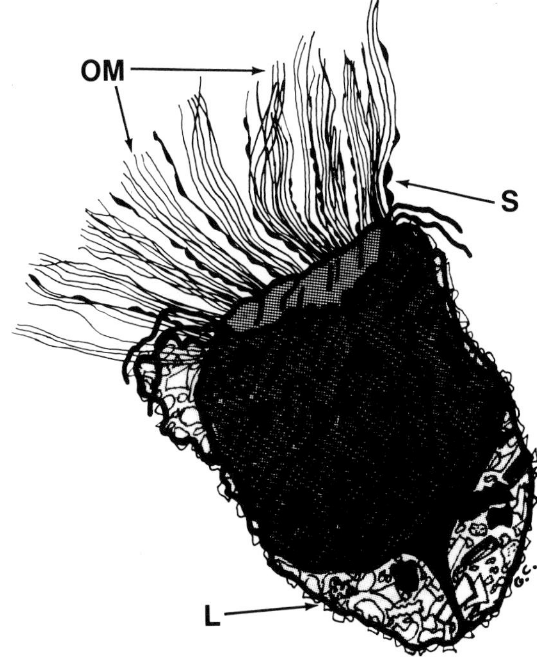

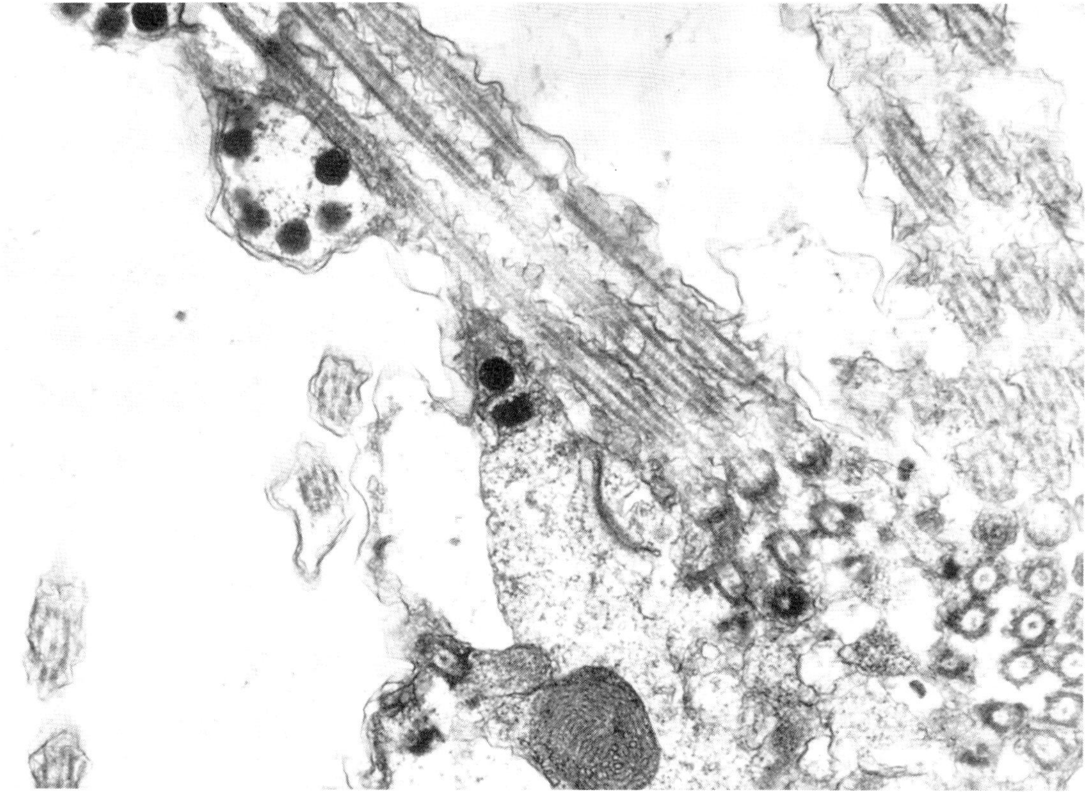

FIGURE 4.11 *(continued)* (b) Transmission micrograph, longitudinal section, through the oral membranelles of *T. parva*, showing the striae and capsules contained within. ([b] From Capriulo et al., 1986.)

exhibit annual community ingestion rates equal to those of the copepods (Capriulo and Carpenter, 1983).

Despite the strides made by the preceding studies, several problems, including those associated with death and/or growth of organisms over the duration of an experiment that result in measurement inaccuracies, still plague ciliate feeding research. In addition, feeding rates and related ecological data of aloricate ("naked") oligotrich ciliates, which often dominate the microzooplankton but are not retained well by plankton nets with meshes as small as 20 μm (Smetacek, 1981), have only rarely been measured (Rassoulzadegan, 1982; Sherr et al., 1986a; Rivier et al., 1981; Sherr and Sherr, 1987; Jonsson, 1986). Detailed studies of the feeding rates and distribution of these numerically dominant members of the protozooplankton represent one of the important next steps in evaluating the trophodynamic importance of the microzooplankton.

Investigations in both temperate and tropical waters have demonstrated that nanophytoplankton often represent a large proportion of algal biomass and can be responsible for as much as 80–99 percent of the observed phytoplankton productivity (Anderson, 1965; Yentsch and Ryther, 1959; Malone, 1971; Capriulo and Carpenter, 1983). Malone (1971), comparing the nanophytoplankton and netphytoplankton primary production and standing crop in tropical neritic and oceanic waters, concluded that the nanophytoplankton were the most important producers in all the environments studied. Recent evidence suggests that a major portion of the chlorophyll in many marine ecosystems consists of particles as small as 1 μm or less (Johnson and Sieburth, 1979; Morris and Glover, 1981) and that much of this material is composed of small eukaryotic phytoplankters (Johnson and Sieburth, 1982; Stockner and Antia, 1986).

Certain estuarine copepods have been shown to feed very inefficiently on small phytoplankton material when compared with such ciliates as tintinnids (Capriulo and Ninivaggi, 1982). Indeed, many copepods exhibit higher ingestion rates when fed larger algae than smaller algae (Mullin, 1963; Frost, 1972; O'Connors et al., 1976; Paffenhöfer and Knowles, 1978). O'Connors et al. (1980), working with the estuarine species *Temora longicornis* feeding on natural food, found that maximum ingestion rates increased linearly by a factor of 3.5 as food size increased from 5 to 30 μm in diameter (flagellate to diatom dominated). Capriulo and Ninivaggi (1982) noted that the ratio of copepod ingestion to tintinnid ingestion changed from 147:1 to 24:1 as the natural food biomass peak shifted from (16 to 4 μm; Fig. 4.12). In spite of this the ratio of maximum filtration rate remained at about 75:1 under both conditions. This highlights the inefficiencies with which at least certain estuarine copepods filter small-sized food particles. Therefore, although some overlap does occur in the sizes of natural food ingested by tintinnids and copepods (as suggested by Capriulo and Carpenter, 1980), each group has greatest feeding efficiency on different sizes of food. A logical consequence of the preceding argument is that tintinnids, and presumably other protozooplankton, should be increasingly more important, as secondary producers, as food size in an ecosystem decreases (because of either natural successional changes or man-induced changes). Since the nanoplankton are often dominant in many marine environments, as pointed out earlier, the importance of the protozooplankton is likely to be great throughout the world ocean.

Planktonic Ciliate Feeding

Few studies quantifying ingestion and filtration rates of planktonic ciliates can be found. Among those studies that have been carried out, tintinnids have been the primary experimental organisms (see Fig. 4.11a).

Qualitative analyses have shown the food of tintinnid ciliates to include bacteria (Campbell, 1926, 1927; Hollibaugh et al., 1980), phytoflagellates (Gold, 1968, 1969, 1973), diatoms (Campbell, 1926, 1927), dinoflagellates (Beers and Stewart, 1967; Stoecker et al., 1983; Stoecker et al., 1981), radiolarians and chrysomonads (Campbell, 1926, 1927), and smaller tintinnids (Blackbourn, 1974). Strong positive correlations between phytoplankton and tintinnid abundances have been found (Kimor and Golandsky, 1977; Sorokin, 1977), and the importance of detritus to the diet of certain protozoans has also been noted (Barnes et al., 1976).

Spittler (1973), performing direct counts on number of starch grains and yeast cells removed by several tintinnid species, reported filtration rates of 0.5–8.5 μl individual^{-1} h^{-1} (Table 4.12). Blackbourn (1974), using methods including direct observation, counts of accumulated food cells, and electronic particle counts of ingested latex beads, reported filtration rates of 3–38 μl individual^{-1} h^{-1} for several species. Heinbokel (1978a,b), using both a disappearance of cultured phytoplankton cells technique and a starch ingestion method, reported filtration rates of 1–10 μl individual^{-1} h^{-1} for both cultured and field-collected tintinnids and maximum ingestion rates on the order of 400 pg C tintinnid^{-1} h^{-1} (Table 4.12). For field-collected tintinnids feeding under very limited natural food conditions, maximum ingestion rates per individual of 5.3×10^5 μm^3/d for *Favella ehrenbergii* (Rassoulzadegan, 1978) and 4×10^5 μm^3 tintinnid^{-1} d^{-1} for *Stenosemmella ventricosa* (Rassoulzadegan and Etienne, 1981; Table 4.12) have been reported. Capriulo and Carpenter (1980), working with natural microzooplankton communities (consisting predominantly of tintinnids) feeding on natural food observed filtration rates of 1–85 μl individual^{-1} h^{-1} with corresponding ingestion rates of 0.001–0.017 ng chl *a* individual^{-1} h^{-1} and 0.1–87 cells removed individual^{-1} h^{-1}.

The preceding studies have been carried out with cultured tintinnids feeding on cultured algae (Heinbokel, 1978a) and field-collected tintinnids feeding on cornstarch (Heinbokel, 1978b) or under limited natural food conditions (Rassoulzadegan, 1978; Rassoulzadegan and Etienne, 1981; Capriulo and Carpenter, 1980). Cultured tintinnids have been shown by Gold (1971) and Gold and Pollingher (1971) to differ from freshly caught tintinnids in their growth and reproduction cycles. Heinbokel (1978a) pointed out that cultured tintinnids represent the result of adaptation or selection for success under artificial conditions, which may cause them to differ from field individuals in their feeding responses. Capriulo (1982) tested the validity of applying ingestion rates of cultured tintinnids measured under artificial food conditions to the field. In that study ingestion and filtration rates of seven species of field-collected tintinnids on natural particle suspensions were measured. Ingestion rates of 0.05×10^6 μm^3 to 1.3×10^6 μm^3 tintinnid^{-1} d^{-1} were observed with corresponding maximum filtration rates of 2–65 μl tintinnid^{-1} h^{-1} (Table 4.12). Filtration rates (among species) were independent of temperature, and highest rates were associated with larger food particles. The filtration rates observed in this study equaled rates reported for tintinnids feeding on cultured algae (Heinbokel, 1978a) when natural food sizes equal to those of the cultured algae were considered (i.e., 4 μm) but were higher when the entire spectrum of natural food sizes was considered. The discrepancy can be explained by carefully considering the assumptions made by both Heinbokel and Spittler that all food items of a size less than or equal to the maximum size a tintinnid species is capable of ingesting are eaten with equal efficiency. Examination of the size-specific filtration rate curves of Capriulo (1982) demonstrates that filtration efficiencies vary with food size. Also, Heinbokel and Spittler assume that the largest particle ingested in their experiments is the largest that a particular tintinnid

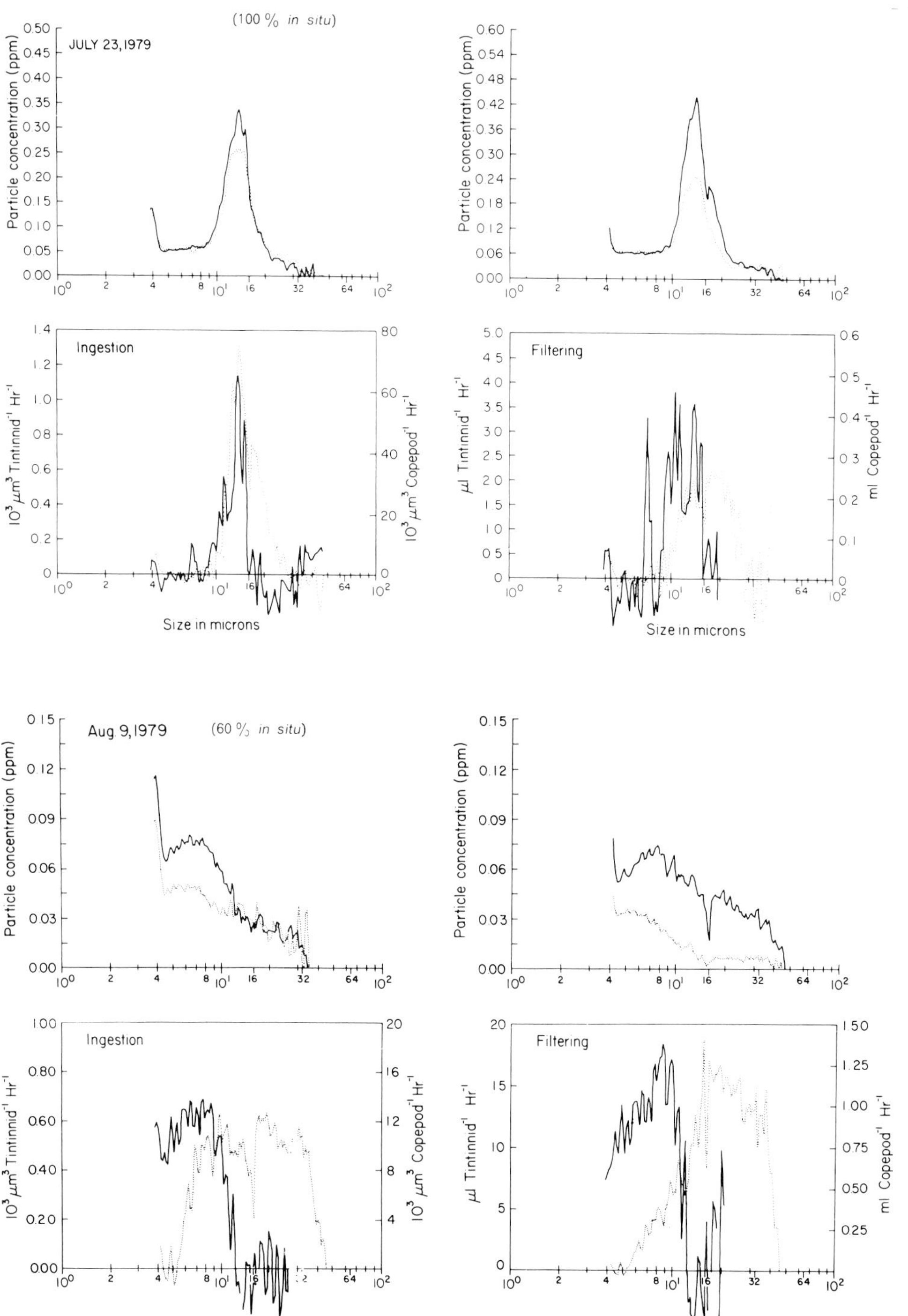

FIGURE 4.12 Comparison of ingestion and clearance rates of tintinnids and copepods (*Acartia tonsa*, adult females) feeding under two different natural particle regimes: (a) biomass peak at 16-µm equivalent spherical diameter (EDS) (dominant food was the dinoflagellate *Prorocentrum*) and (b) biomass peak at 4 µm ESD (dominant food items were small flagellates). Top curves are the size–biomass frequency distributions of the natural particle assemblages under the control (solid line) and experimental (dotted line) conditions, for the tintinnids (top left) and copepods (top right) at the termination of an experiment. Bottom curves represent the corresponding size-specific ingestion and filtration curves for the tintinnids (solid line) and copepods (dotted line), respectively. (After Capriulo and Ninivaggi, 1982.)

TABLE 4.12 Some Typical Ciliate Ingestion and Clearance Rates Measured by Different Techniques Employing Cultured and Natural Food and Inert Particles*

Ciliate	Ingestion rate	Clearance rate (μl cilliate^{-1} h^{-1})	Source
Lohmanniella spiralis	5.2–20.8 ng C ciliate^{-1} d^{-1}	2–9	Rassoulzadegan (1982)
	0.06 × 10^{6} μm^{3} ciliate^{-1} d^{-1}	1.6–26	Jonsson (1986)
Strombidium reticulatum	—	1.1–3.5	Jonsson (1986)
Strombidium vestitum	—	0.11–0.52	Jonsson (1986)
Small benthic ciliates	80–120% of body volume h^{-1}	—	Fenchel (1980b)
Large benthic ciliates	10–30% of body volume h^{-1}	—	Fenchel (1980b)
Colpidium colpoda	16,000 bacteria h^{-1}	—	Taylor (1986)
Colpidium campylum	—	0.04	Fenchel (1986)
Glaucoma scintillans	0.024–0.19 × 10^{6} μ^{3} d^{-1}	—	Fenchel (1986)
Uronema marinum	—	0.24	Fenchel (1986)
Vorticella	—	0.16	Fenchel (1986)
Halteria grandinella	—	0.67	Fenchel (1986)
Cyclidium glaucoma	—	0.003–0.02	Fenchel (1986)
Euplotes vannus	—	0.3–0.7	Fenchel (1986)
Favella ehrenbergii	20.8 ng C ciliate^{-1} d^{-1}	—	Rassoulzadegan (1978)
Stenosemella ventricosa	21.8 ng C ciliate^{-1} d^{-1}	—	Rassoulzadegan and Etienne (1981)
Stenosemella oliva	22 ng C ciliate^{-1} d^{-1}	—	Capriulo (1982)
Helicostomella subulata	8–19 ng C ciliate^{-1} d^{-1}	4.3–7.5	Verity (1985)
	18–42 ng C ciliate^{-1} d^{-1}	10–18	Capriulo (1982)
	0.4–8 ng C ciliate^{-1} d^{-1}	1–5	Heinbokel (1978a, b)
Tintinnidium fluviatile	8–48 ng C ciliate^{-1} d^{-1}	4–20	Capriulo (1982)
Eutintinnus pectinis	0.4–4 ng C ciliate^{-1} d^{-1}	1–5.5	Heinbokel (1978a, b)
Leprotintinnus bottnicus	—	0.5	Spittler (1973)
Tintinnopsis acuminata	0.4–8 ng C ciliate^{-1} d^{-1}	1–4	Heinbokel (1978a, b)
	2–6 ng C ciliate^{-1} d^{-1}	1.7–2.6	Verity (1985)
	3 ng C ciliate^{-1} d^{-1}	2	Capriulo (1982)
T. vasculum	4 ng C ciliate^{-1} d^{-1}	8	Capriulo (1982)
	8–19 ng C ciliate^{-1} d^{-1}	4.3–7.5	Verity (1985)
T. parva	68 ng C ciliate^{-1} d^{-1}	38	Capriulo (1982)
T. beroidea	33 ng C ciliate^{-1} d^{-1}	65	Capriulo (1982)
T. tubulosa	—	0.8	Spittler (1973)
T. parvula	—	1.7	Spittler (1973)
T. fimbriata	—	8.5	Spittler (1973)

*Natural food particles are converted to carbon units assuming a mg C/μm^{3} ratio of 5.2 × 10^{-11} (see Capriulo, 1982).

species could ingest. Capriulo (1982) has shown that when feeding on natural particles the absolute cutoff points of food sizes that can be ingested by particular tintinnid species are somewhat flexible. This is attributed (Capriulo, 1982) to membranelle–microtubule interconnections (Hedin, 1976) found in tintinnids. These connections likely allow regulation of the size of the lumen (cytostome/mouth) and suggest that the maximum food size that can be ingested may depend on the immediate past history of food that a tintinnid has been exposed to. The ability of aloricate ciliates to ingest food items even larger than themselves has been previously reported (Smetacek, 1981).

The size-specific ingestion rates of Capriulo (1982) indicate maximum food removal, by tintinnids, from biomass peaks, with filtration effort generally maximized on the largest food a species could handle. In addition, Capriulo (1982) noted a strong negative relationship between weight-specific ingestion and filtration rates and body weight in tintinnids. Tintinnids in this study exhibited saturation feeding responses when presented with several concentrations of natural food up to in situ levels.

The ingestion rates for tintinnids presented by Capriulo (1982), 0.05×10^6 to 1.3×10^6 μm^3 individual^{-1} d^{-1}) agree well with the maximum ingestion rates (0.4×10^6 and 0.57×10^6 μm^3 tintinnid^{-1} d^{-1}) reported for *Stenosemella ventricosa* (Rassoulzadegan and Etienne, 1981) and *Favella ehrenbergii* (Rassoulzadegan, 1978), respectively. Assuming a carbon to particle volume ratio of 5.2×10^{-11} mg C/μm^3 (Parsons et al., 1967) the ingestion rates reported by Capriulo (1982) ranged from 2.6 to 68 ng (Table 4.12) C ingested tintinnid^{-1} d^{-1}. These rates are similar to the values for *Petalotricha ampula* of 0.7–36 ng C tintinnid^{-1} d^{-1} reported in Table 2 of Rassoulzadegan and Etienne (1981), and although somewhat higher, they are also similar to the maximum rates of 10, 4, and 8 ng C tintinnid^{-1} d^{-1} reported by Heinbokel (1978a; Table 4.12).

Current food web models typically ignore potential pathways between chain-forming algae, such as diatoms, and the ciliates (both planktonic and benthic), predominantly because of their relative sizes, with many diatoms or their composite chains being equal to or larger in size than many ciliates. It is known, however, that protozoans, including ciliates, can ingest cells their own size and larger (Kahl, 1935; Makinnon and Hawes, 1961; Smetacek, 1981; Verity and Villareal, 1986; Suttle et al., 1986). Verity and Villareal (1986) also demonstrated that some diatoms provided sufficient nutrition for ciliate growth. Ciliates such as *Nassula* and *Pseudomicrothorax* also have been shown to ingest large sections of cyanobacterial filaments by breaking off the unusable portion through the crimping action of a pharyngeal basket with associated Capitula (Hausmann and Peck, 1978, 1979).

Loricate ciliates, such as the tintinnids, appear to have more rigid restrictions, imposed on them by lorica diameter (Spittler, 1973; Heinbokel and Beers, 1979; Capriulo, 1982; Verity, 1985), than nonloricate ciliates (Smetacek, 1981). However, for all ciliates, as well as other phagotrophic protozoa, it is likely that events that modify the size of food that would otherwise be too large for ingestion make that food available to various protozoan predators in a new, modified size. In this way an unrecognized food web pathway may exist by which large algal food items, such as chain-forming diatoms, are made available to small predators such as ciliates thanks to the size altering of particles during "sloppy" feeding by such organisms as marine microcrustaceans (e.g., copepods) and possibly other marine microinvertebrates. Modifications of food sizes of such things as chain-forming diatoms by estuarine copepods have been observed (O'Connors et al., 1976; Deason, 1980; Roman and Rublee, 1980; Capriulo and Ninivaggi, 1982; Donaghay, 1980; Dexter, 1984). Additionally, physical breakage of food by turbulence and water–sediment interactions also probably occurs. Along with others, I have conducted some recent research (Capriulo et al., 1988) that demonstrates enhanced ciliate growth in the presence of size-altered diatom chains. Turner et al. (1983) observed a significant negative relationship between length of diatom chains and numbers of smaller zooplankton over a year cycle in a temperate zone estuary, which suggests that such modifications may be occurring in the field.

Predator feeding-related modification of prey size is likely to be important at all levels of both planktonic and benthic food webs. Even ciliates have been shown to produce particles as a by-product of their feeding (Stoecker, 1984), although to what extent this is occurring and how it is related to thecate versus nonthecate prey as well as larger versus smaller ciliate predators is presently unknown. The particles produced by the ciliates represent sizes appropriate for ingestion by heterotrophic flagellates, whose growth may also be stimulated under these modified food circumstances and may be predominantly below the sizes that are routinely enumerated by particle counters.

Certain algal prey have been shown to have a negative effect on tintinnid ciliate growth. For instance, the red-tide flagellate *Olisthodiscus luteus* has been shown to depress growth in the tintinnids *Favella* sp. and *Tintinnopsis tubulosoides* when it is ingested or is in direct contact with those ciliates. This algae has also been shown to affect development of echinoderm eggs and plutei. In Narragansett Bay there is an inverse relationship between densities of *O. luteus* and tintinnid populations (Verity and Stoecker, 1982).

One of the larger tintinnids, *Favellas ehrenbergii,* has been shown to have an obligate food requirement for dinoflagellates (Stoecker et al., 1981) with *Gonyaulax tamarensis, G. polyedra, Heterocapsa* sp., *Scrippsiella trochoidea* all representing "good" food and *Prorocentrum mariaelebouriae* and *Amphidinum carterae* representing poor food. Little predation was observed on cryptophytes, haptophytes, diatoms, a chrysophyte, prasinophytes, or chlorophytes. In the field *Favella* sp. (as well as the ciliates *Balanion* sp. and *Strobilidium* sp.) were found to be positively correlated with dinoflagellate abundance more than with such physiochemical parameters as temperature or salinity (Stoecker et al.,

1984). In addition to its dinoflagellate prey, *Favella* sp. also feeds on other ciliates, such as *Balanion* (Stoecker et al., 1983).

Taniguchi and Kawakami (1983), working with the tintinnid ciliates *Eutintinnus lususundae* and *Favella taraikaensis*, found that these two ciliates did not survive in culture with *Chaetoceros gracilis*, *Chlorella* sp., *Dunaliella tertiolecta*, *Heterosigma* sp., and *Thalassiosira decipiens* (solitary forms), whereas *Prorocentrum minimum* provided a good diet, and *Isochrisis galbana* and *Rhodomonas* sp. supplied an excellent food regime. This study also provides one of the rare looks at competition and competitive exclusion between two ciliate species, with the successful species dependent on the intial starting conditions.

At least one tintinnid species, *Stenosemella ventricosa*, has been found to take up dissolved organic substances by both absorption and active transport (Pavillon and Rassoulzadegan, 1980), although what role this plays in its overall nutrition is unclear. Admiraal and Venekamp (1986) examined tintinnid grazing during blooms of *Phaeocystis*. Another study (Verity and Villareal, 1986) examined the relative value of various algae and cyanobacteria as food for ciliates. Their results demonstrated that *Tintinnopsis acuminata* and *Tintinnopsis vasculum* grew well when fed diatoms lacking threads or cetae and grew poorly on setae-possessing forms, although if the threads were reduced, growth improved. *T. vasculum* grew moderately well on dinoflagellates, and *T. acuminata* grew well on chlorophytes and prasinophytes. Both species experienced significant mortality when fed the chroococcoid cyanobacterium *Synechococcus*. As expected, *Isochrisis galbana* and *Dicraterra inornata* promoted good growth in both species. Both tintinnid species were able to feed and grow on food items greater in size than 50 percent of the tintinnid oral diameter, which supports my findings (Capriulo, 1982) that upper size limit of prey is somewhat flexible. In addition to this food-quality study, Verity (1986) has examined growth rates of natural tintinnid populations in Narragansett Bay. Another study (Hernroth, 1983) has demonstrated that tintinnids are important to the spring plankton community food web of the Gullmar Fjord in Sweden. A broad-spectrum examination of ciliate growth as a function of algal food (Skøgstad et al., 1987) indicated that flagellates of the classes Chrysophyceae, Cryptophyceae, and Dinophyceae were good to excellent foods, chlorococcal and flagellate Chlorophyceae were poor to good, Bacillariophycea (diatoms) were not acceptable for all but one species of ciliate studied, and cyanobacteria were poor food items. When food conditions (in general, any living conditions) are poor, many ciliates form cysts to survive the unfavorable condition (Reid, 1987; Paranjape, 1980).

Aloricate Planktonic Ciliates

As previously mentioned, aloricate planktonic ciliates have only rarely been studied, even though their densities at times greatly exceed that of loricate forms. The primary reason for their omission in many studies is their extremely delicate nature. They often do not survive fine-mesh net collections and are likely not to survive long experimental incarcerations. Manipulation of these forms many times results in cell lysing, an event that leaves no "dead bodies" to count at the termination of an experiment.

Despite the difficulties associated with working with these ciliates, Sherr et al. (1986a) examined the importance of small nanoplanktonic-sized (2–20 μm) aloricate ciliates to the plankton. It has been assumed that heterotrophic nanoplankton are comprised mostly of phagotrophic microflagellates (Fenchel, 1982a–e; Sieburth and Davis, 1982) and that pelagic ciliates were important microzooplankton mainly in the 20–200-μm range. Sherr and coworkers, however, point out that aloricate ciliates of less than 20 μm are found in both eutrophic and oligotrophic marine waters, at times equal to a very high percentage of heterotrophic nanoplanktonic biomass. Using a double-staining epifluorescence method, they have observed numbers in the 200–61,000/L range, and by examination of food vacuoles they found predominantly bacterial and coccoid cyanobacterial, rarely nanoplanktonic, food items. These findings and those of Sherr and Sherr (1987), as discussed earlier, challenge the notion (Fenchel, 1980a–d, 1984; Berk et al., 1976; Porter et al., 1979) that bacterivorous ciliates cannot maintain growth on suspensions of bacteria below 10^6–10^8 cells/ml. Sherr et al. (1986) and Sherr and Sherr (1987), therefore, illustrate that the production of suspended bacteria is sufficient to support the growth requirements, at least at times, of both apochlorotic microflagellates and ciliates.

Rassoulzadegan (1982), working with the oligotrichous naked ciliate *Lohmaniella spiralis*, measured ingestion rates for this ciliate of 4×10^3 to 16×10^3 μm^3 ciliate^{-1} h^{-1} with corresponding clearance rates of 2–9 μl ciliate^{-1} h^{-1} (Table 4.12). These rates are comparable with those reported for tintinnids (see preceding discussion and Table 4.12). This ciliate has a gross growth efficiency of 10–70 percent (Table 4.13) and feeds on flagellates (Sheldon et al., 1986). It is not uncommon to find strong correlations, in the field, between flagellate

TABLE 4.13 Ciliate Gross Growth Efficiencies

Ciliate	Gross growth efficiency (in percent)	Primary food	Reference
Uronema sp.	20–27	Bacteria	Lee et al. (1975) and Turley et al. (1986)
Euplotes sp.	10–19	Bacteria	Lee et al. (1975) and Turley et al. (1986)
Tintinnopsis acuminata	>50	*Isochrisis galbana*	Heinbokel (1978a,b)
Lohmanniella spiralis	10–70 (temperature dependent)	Natural food	Rassoulzadegan (1982)
Colpidium campylum	11	Bacteria	Laybourne and Stewart (1975)
Other ciliates	37–78	—	Referenced in Laybourne and Stewart (1975)
Favella ehrenbergi	67	Natural food	Rassoulzadegan (1976)
Assorted tintinnids	33	Phytoflagellates	Verity (1985)
Strombidium reticulatum	25–45	*Pyramemonas* sp.	Jonsson (1986)
Lohmanniella spiralis	32–42	*Pyramimonas* sp.	Jonsson (1986)

and ciliate abundance (Ibanez and Rassoulzadegan, 1977; Sorokin, 1977; Capriulo and Carpenter, 1983).

Smetacek (1981), examining the prozooplankton in the Kiel Bight, found that many nonloricate ciliates eat prey as large as or larger than themselves and are the major herbivore in the spring and autumn, with biomass levels comparable to those of summer metazoans (0.5 g C/m^2 levels). In another study, Endo et al. (1983) quantified naked ciliate densities and biomass in the western subtropical Pacific off of the Bonin Islands. Particle size selection and functional response based on latex beads experiments have been studied for three naked, marine, oligotrichous ciliates, including *Strombidium vestitum*, *S. reticulatum*, and *Lohmanniella spiralis* (Jonsson, 1986). Each species exhibited different particle size optima (2.1, 7.9, and 9.7 μm, respectively) and more could retain 1-μm particles. Additionally, no feeding thresholds were found. Clearance rates in these ciliates varied between 0.5 and 26 μl ciliate^{-1} h^{-1}. Certain small spirotrichs and free-swimming peritrichs have been shown to be selective with regard to bacterial prey, ingesting 12–36 times more unattached (free) bacteria than epibacteria (Albright et al., 1987).

There has been some suggestion (Banse, 1982a,b) that ciliates may not be as important to the marine pelagial as is now generally believed. One line of evidence proposed by Banse cites ciliate threshold feeding responses as a mitigating factor to consider, since below certain food concentrations no detectable feeding occurs. This inferred threshold, however, is likely as not to be an artifactual result of our inability to measure ciliate ingestion rates at extremely low food concentrations. Also, many ciliates ingest food both smaller and larger than the sizes suggested by Banse. In addition, some ciliate species may feast and famine, exhibiting high ingestion rates when food concentrations are adequate or good, followed by periods of low feeding rates or even encystment prior to another encounter with a high food concentration. Banse also cites low ciliate numbers in many ecosystems as a signpost of less feeding potential relative to metazoans. What this ignores is the possibility that the low numbers are a result of high metazoan predation. It is true that before we can fully document the trophodynamic importance of ciliates to marine food webs, we must obtain large amounts of field growth and production estimates as well as additional data quantitatively assessing the ciliates as prey (see the section in this chapter on the fate of protozoan biomass). Nonetheless, the data presently available are enough to generate a rough sketch that clearly highlights these organisms as a major component of the marine food web. Additionally, in recent years it has become more obvious that small ciliates are an important trophic component of marine food webs, particularly as grazers of bacteria (Gast, 1985; Rivier et al., 1986; Sherr et al., 1986a,b, 1987; Rassoulzadegan et al., 1988; Rassoulzadegan and Sheldon, 1986; Albright et al., 1987; Sherr and Sherr, 1987).

PHAGOCYTOSIS AND THE DIGESTION CYCLE

1. Endocytosis

All heterotrophic (and mixotrophic) protista require some form of preformed organic compounds for survival, growth, and reproduction. Preferred food can be in the particulate or dissolved phase and consists of any combination of bacteria, cyanobacteria, algae, detritus, other protozoa, metazoans, or dissolved organics. As discussed earlier, few, if any, protozoa can survive on dissolved organics at the concentrations that are typical of marine ecosystems. Some protozoa, such as the colonial radiolarians (see the preceding discussion) do routinely live on dissolved organics, but only after they secrete hydrolyzing enzymes into the external medium to degrade large nutritive macromolecules into smaller soluble units.

No matter which type of food is preferred by a particular heterotroph, it somehow must be brought by the feeding organism from the external medium to the cell's interior. Specific mechanisms by which food is located, collected, concentrated, and taken in vary as a function of major taxonomic grouping, individual species, and size of the food in question (see the following section on feeding biology). The essence of the uptake process, endocytosis, however, is the same for all heterotrophic protista. This process has been recently reviewed (Nisbet, 1984) and will only be briefly outlined, after Nisbet, here.

During endocytosis food material is enclosed in a sac formed from part of the plasma membrane (Fig. 4.13). If the material being taken in is in the dissolved phase, the process is typically referred to as pinocytosis; if it is in the particulate phase it is called phagocytosis. In reality, these two processes do not differ from each other; rather, they reflect our arbitrary size cutoff criteria for separating the particle size spectrum continuum (atomic to cosmological proportions) into dissolved versus particulate phase.

In some protozoa such as the sarcodines, all or most of the cell body surface may be involved in endocytosis. In other protists endocytosis may be restricted to a defined area, such as the base of the collar of tentacles in the choanoflagellates, the cytostome found in the wall of the flagella pocket of some trypanosomes, or a large unciliated surface of cell membrane (certain karyorelictean ciliates), or it may be even more highly restricted and structured as a definite mouth (cytostome; Small, 1983 and 1984), as is the case for some phagotrophic flagellates and most ciliates. The ciliate mouth, if present, may be simple with oral ciliature sometimes indistinguishable by light microscopy from somatic ciliature, or very elaborate, with the extensive development of oral polykineties. The oral cavity may be shallow, without noticeable ciliation, or may be deep and well developed, with an oral kinetid-lined infundibulum (e.g., any peritrich that swirls food into the cytostome and down into a cytopharynx (Small and Lynn, 1985; Fenchel and Small, 1980). Some ciliates may take up dissolved organics by "pinocytotic" endocytosis not at the cytostome but, rather, in parasomal sacs found

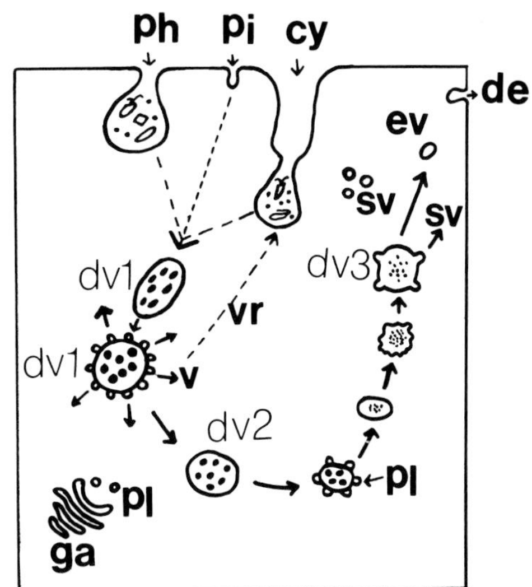

FIGURE 4.13 Conceptual representation of endocytosis. Dissolved to particulate-sized food is taken into a cell either through unspecialized areas of the cell surface, through an organized mouth or cytostome (cy), by pseudopodial phagocytosis (ph), or by pinocytosis (pi). After formation of a food vacuole the recently formed vacuole generally is quickly shuttled away from the site of formation, presumably to allow for the unencumbered production of additional vacuoles. At this point the vacuole enters the digestive vacuole 1 (dv 1) stage. The digestive vacuole now rounds up, releasing surplus membrane as small, coated vesicles (v) for recycling (vr) back to a new vacuole formation site(s). The vacuole continues to condense and shrink with accompanied water loss, to enter the dv 2 stage. At its fully condensed state, the vacuole becomes coated with enzyme containing primary lysosomes (pl). The enzymes are released into the vacuole, causing changes in pH to higher acidity levels. Vacuoles then enter dv 3 and begin to expand and take on an irregular shape as contents are digested. Secondary vacuoles (sv) that contain the digestion products are pinched off and transported to wherever they are needed. Undigested remains in egestion vacuoles (ev) are discarded (de) either at a fixed, organized site (cytoproct), at a nonspecific generalized area (cytopyge), or "randomly" anywhere along the cell surface. (ga = golgi apparatus.) (Modified after Nisbet, 1984.)

near the base of the cilia of cyrtophoran ciliates or by diffusion/osmosis through the general cell surface in other ciliates.

Once endocytosis is complete, a vacuole, now within the cell, breaks off from the plasma membrane. Smaller and even some larger vacuoles may then coalesce to form large food vacuoles.

2. Stimulation of the Endocytotic Response

Many substances have been identified as likely initiators of the endocytotic feeding response in various protozoa (reviewed in Nisbet, 1984). For example, hyaluronidase, trypsin and other amino acids, cytochrome *c*, peptones, lecithin, NaCl and other salts, and acridine orange have been found to induce feeding in free-living amebae. The carnivorous euglenid flagellate *Peranema* is stimulated by phospholipids such as lecithin and cephaline as well as by trypsin and some surface-active agents (e.g., tweens). Some ciliates (e.g., *Coleps*) respond to SH-bearing substances, including glutathione, cysteine, and phospholipids (Nisbet, 1984; Seravin and Orlovskaja, 1973, 1977).

In amebae, large numbers of ionized compounds that react with the glycocalyx (mucopolysaccharide coat) overlying the plasmalemma are necessary to initiate "pinocytotic" endocytosis. These inducers must have a positive charge and must bind to the cell surface, after which there is a cessation of movement (similar to the capping that results in amebae after contact with motile prey) and likely development of invagination channels and vacuoles. The presence in the external medium of Ca^{2+}, which likely stimulates contraction of microfibrils, is important (reviewed in Nisbet, 1984). The plasmalemma itself is mostly impenetrable to many carbohydrates and depends on inducer substances to initiate uptake. The actual quantities of solute imbibed varies with different inducers. Similar Ca^{2+}-mediated responses are probably involved in all protistan endocytotic events concerning live prey whose contact with cell surfaces and subsequent movements effect a surface charge separation and depolarization that results in prey engulfment. Ciliary reversals also result from changing external ionic concentrations or physical contact-mediated membrane depolarization which alter internal–external Ca^{2+}–K^{+} ion concentrations and cell membrane electrical potentials. Such ciliary reversals alter the random walk swimming patterns of ciliates (as well as flagellates) and may enable these organisms either to remain in the vicinity of high food concentrations (on a miscroscale) or to avoid predators. In addition to endocytotic uptake of food, various nutrients are taken up by diffusion, facilitated diffusion, and active transport.

3. Digestion

The digestive process in heterotrophic eukaryotes centers around the endocytotic vacuole (food vacuole) after it has broken away from the plasma membrane. Numbers and size of food vacuoles vary as a function of protist age and physiological state as well as with food quality, size, and motility. The process of digestion has been described for different protozoa by a number of authors (Nisbet, 1984; Elliott and Clemmons, 1966; Nilsson, 1977; Allen and Staehelin, 1981; Fok et al., 1982; Jurand, 1961; Fisher-Defoy and Hausmann, 1977; Rudzjinska, 1970, 1972, 1973; Allen, 1974, 1978; Roth, 1960; Stockem, 1973; Nillson, 1979; Martinez-Palomo, 1982; Chapman-Andresen, 1973; Hitchen and Butler, 1973; Sawicka et al., 1983; Berger and Pollock, 1981). The rates at which vacuoles are formed also vary as a function of the size and quality of the particles presented and the physiological state and age of the feeder. At least in some protozoa, size seems to be more of a controlling factor than quality, as evidenced by *Paramecium* that will take up bacteria or bacteria-sized latex beads at equal rates (Nisbet, 1984; Muller et al., 1965). The heterotrich ciliate *Climacostomum* can eat as many as 56 flagellates min^{-1}, first by creating a single large vacuole and then by producing smaller ones (Nisbet, 1984; Fischer-Defoy and Hausmann, 1977). *Epistylis* can produce about seven vacuoles/min (McKanna, 1973), whereas *Carchesium* produces about 1/min (Sugden, 1950). Although size and number of vacuoles formed per time vary to a large degree, for at least one ciliate, the heterotrich *Fabrea salina*, I have found (Capriulo, in preparation) that both vacuole passage time and digestion time (90 and 50 minutes, respectively, at 25°C) are constant and independent of initial food concentration. This may or may not be the case for all ciliates, but it is nonetheless interesting, because it suggests a fixed life expectancy for a vacuole, perhaps determined by factors such as enzymes controlling de novo membrane synthesis and internal membrane recycling. It is likely that membrane production and recycling events determine the upper limit of ingestion rates in protozoa, producing the observed Holling–Ivlev–Michaelis-Menton-like saturation functional responses. Nisbet (1984) has suggested that the total potential endocytotic vacuole volume may have an upper limit with regard to what percentage this represents of the total cell volume.

Release of a forming vacuole is elicited by contraction of various contractile proteins, which are stimulated by both Ca^{2+} ions and prey activity. Prey movement may cause a charge separation–depolarization of the vacuole membrane, thus facilitating separation. After separation a vacuole usually, but not always, moves posteriorly, to vacate the vicinity of a newly forming vacuole. Such vacuole movement is mediated by microtubule activity. At this point a vacuole and its contents are condensed (see Fig. 4.13) and surplus vacuole membrane is recycled to the site(s) of the new vacuole(s) formation (Nisbet, 1984). The food vacuole, now referred to as a stage 1 digestive vacuole (DVI), undergoes a pH shift (from basic or neutral to acidic), reaching at DVII a maximum acidity of perhaps 2 and a size about one quarter of the original. This pH shift is catalyzed by acid phosphatase released from primary lysosomes that have fused with the food vacuole and released into it their digestion lysozymes. As digestion proceeds to DVIII the vacuole begins to enlarge again, the pH shifts toward basic, and final food breakdown occurs. Digestive products are then pinched off as small vesicles or secondary vacuoles and are transported, as needed, around the cell. The remaining undigested or undigestible material, now contained within an egestion vacuole, is passed back to the outside through an organized discharge site (cytoproct; Small and Lynn, 1985), through a discharge area (cytopyge; Foissner, 1972), or somewhat more randomly through another area on the cell surface.

The entire endocytosis–digestion process is similar in sarcodines, ciliates, and flagellates (perhaps reflecting evolutionary kinships, with particular taxons and species exhibiting specific characteristics with regard to rates, and variations on a theme).

FEEDING BEHAVIOR AND PHYSIOLOGY

1. Flagellates

The heterotrophic flagellates include both phytomastigophoreans (some of which are chloroplast-bearing mixotrophs or autotrophs and others of which are colorless, obligate heterotrophs) and the zoomastigophoreans (all of which are obligate, nonpigmented heterotrophs, including many parasitic forms). The presence of a cytostome (mouth) is not a universal feature of the heterotrophic species, and ingestion may occur at unspecialized points on the cell surface by diffusion, active transport, and endocytosis. Some species are obligate photoautotrophs, whereas others are facultative autotrophs that supplement autotrophy by heterotrophic nutrition. Some flagellates, such as *Euglena gracilis*, can switch from autotrophy to heterotrophy in the absence of light, if a suitable carbon source is present.

Particle-feeding flagellates use flagella undulation to create feeding currents in small, semisessile flagellates (Sleigh, 1964). In the species studied by Sleigh, undulations were from flagella base to tip. In *Codonosiga* water currents follow the same direction as the undulation, whereas in *Actinomonas* and *Monas*, currents move in the opposite direction (i.e., from tip to base). Such coincidental or opposed water movement appears to depend on the absence or presence, respectively, of mastigonemes (hairlike naked microtubule projections arising from the flagella; Holwill and Sleigh, 1967; Fenchel, 1982a). Generated water currents move particles to a point on the flagellate cell surface, where phagocytosis takes place. Sleigh (1964) points out that in *Monas* food must be funnelled to contact the plasma membrane near the flagella base, whereas *Actinomonas* possesses food-catching pseudopods (filopodial type) for particle trapping. *Codonosiga* sports a projecting collar with internal rod supported microvilli, through which food particles pass as water is drawn in, after which they drop onto the base just outside the collar and are ingested.

The Phytomastigophorea

Included in the Phytomastigophorea (phytoflagellates) are the cryptomonads, dinoflagellates, euglenids, chrysomonads, heterochlorids, raphidomonads, prymnesiids (haptophytes), volvocids, prasinomonads and silicoflagellates (Leedale and Hebberd, 1985). The phytoflagellates exhibit many nutritional modes from complete autotrophy to obligate heterotrophy. A good general review of the ecology of phagotrophic phytoflagellates can be found in Sanders and Porter (1988). Estep et al. (1986) concentrate specifically on the significance of bacterivory by oceanic algal nanoflagellates.

The Euglenida exhibit a variety of nutritional characteristics, from obligate photoauxotrophy (*Euglena pisciformis*) and photoauxotrophy–facultative heterotrophy (e.g., *Euglena gracilis*) to obligate heterotrophy (e.g., *Astasia longa*). *Peranema, Heteronema,* and *Entosiphon* are colorless euglenoids that are particulate feeders possessing a permanent cytostome (Leedale, 1967). Members of the genus *Heteronema* feed on bacteria and coccoid algae (Lee et al., 1985; Loefer, 1931).

In addition to a permanent cytostome, *Peranema trichophorum* possesses a specialized feeding organelle, the rodorgan, with which it can eat other flagellates (Chen, 1950; Leedale, 1967; Mignot, 1966; Nisbet, 1974). During feeding the rodorgan is moved by an elaborate mechanism, accompanied by violent writhing and twisting of the organism. It appears that during feeding the rodorgan and the adjacent cytostome are pulled forward and the cytostome expands to allow large food items to be ingested. Withdrawal of these structures "sucks" food into the cytostomal sac, from which endocytosis produces food vacuoles (Nisbet, 1974). The ingestion organelle of *Entosiphon* is similar to *Peranema,* but also includes a siphon tube that is responsible for propeling food into the cell (Mignot, 1966). Triemer and Fritz (1987) closely examined the feeding apparatus of *Entosiphon sulcatum.* They found the elongate tube to extend the length of the cell and to be covered by a cap. The plasma membrane is continuous over both the siphon and the cap. During feeding, the siphon (Fig. 4.14) is extended forward until it is nearly flush with the cell apex. As the siphon moves forward, the cap is withdrawn. This spreads the siphon further, thus increasing the diameter of the opening. Upon retraction the cap slides back into place. The feeding apparatus has two types of vesicles associated with

FIGURE 4.14 Representation of the colorless phagotrophic euglenoid *Entosiphon sulcatum* with depiction of the feeding apparatus (fa), called a siphon, and cap (c) covering. During feeding the siphon protrudes from the cell, at which time the cap is withdrawn to the side, thus opening the siphon. Within the tube are four striated vanes that spread apart during feeding to create a large cavity into which food particles are drawn. (Drawing based on transmission micrographs of Triemer and Fritz, 1987.)

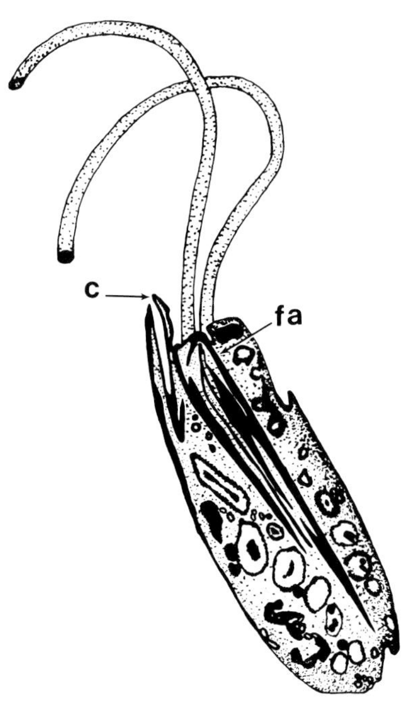

it that provide factors, such as calcium, needed for contraction, and possibly lytic enzymes needed for digestion, respectively. A similar apparatus has been observed in a recently described genus of phagotrophic euglenid, *Serpenomonas costata* (Triemer, 1986).

Peranema does not appear to be very selective in its choice of food, which includes bacteria, algae, yeast, flagellates, and the like (reviewed in Nisbet, 1984). The majority of colorless euglenoids are heterotrophic, with a nondistinct mouth.

The Volvocida and Cryptomonadida typically possess pigment, although some are colorless heterotrophs that include the acetate forms (Wise, 1959; Levandowsky and Hutner, 1979; Cosgrove and Swanson, 1952; Holz, 1954; Antia, 1980). A distinctive characteristic of the cryptomonads is the ejectosomes, which are extrusive organelles that are associated with defense (Klaveness, 1981). Phagocytosis has never been observed in the volvocids, silicoflagellates and prasinophytes. There is scant published evidence and more anecdotal evidence indicating phagocytosis in colorless cryptomonads (e.g., Mignot, 1965; Sherr and Sherr, unpublished data).

Phagotrophy in the obligate heterotroph *Ochromonas* has been discussed by Aaronson (1980). No phagocytosis has been found in any of the Volvocida, nor in individuals of the Silicoflagellida or Prasinomonadida. Both the prymnesiids (Haptophytes) and the chloromonads have colorless heterotrophic forms, with some haptophytes being phagotrophic. For example, *Chrysochromulina* can ingest particles up to several micrometers in diameter by extruding rhizopodia between scales and by means of muciferous bodies that eject long threads.

The Dinoflagellida are biflagellate, primarily planktonic phytomastigophoreans. Autotrophy, as discussed earlier, is important in the dinoflagellates but is often insufficient for their nutritional needs. Some have evolved a means of trapping particulate food and engulfing it at special regions of the sulcral groove. Particles are drawn in by flagella-generated currents and caught, in some species, by fine cytoplasmic nets, and in others, through the action of exploding trichocysts (e.g., *Gyrodinium*). Some forms, such as *Noctiluca,* are obligate heterotrophs with voracious appetites for prey, such as copepod eggs, crustacean larvae, and diatoms. *Noctiluca* possesses a permanent mouth at the base of a tentacle (Fig. 4.15), which it uses to capture food (Droop, 1954, 1959; Lee et al., 1985; Sieburth, 1979; Lessard et al., 1985). Some dinoflagellates, such as *Polykrikos,* produce nematocysts similar to those of coelenterates (Greuet, 1972; Greuet and Hovasse, 1977). *Gyrodinium pavillardi* has been observed to ingest algae, bacteria, and protozoa

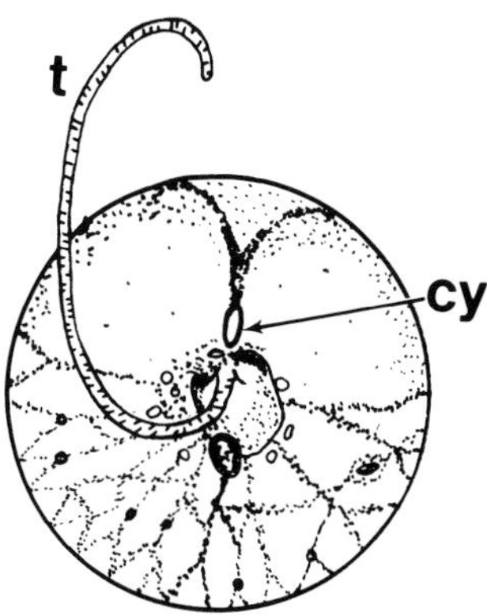

FIGURE 4.15 Representation of the colorless dinoflagellate *Noctiluca scintillans* depicting the tentacle (t) and cytostome (cy). (Redrawn after Mackinnon and Hawes, *An Introduction to the Study of Protozoa*. Oxford: Clarendon Press, 1961. As portrayed in Sieburth, 1979.)

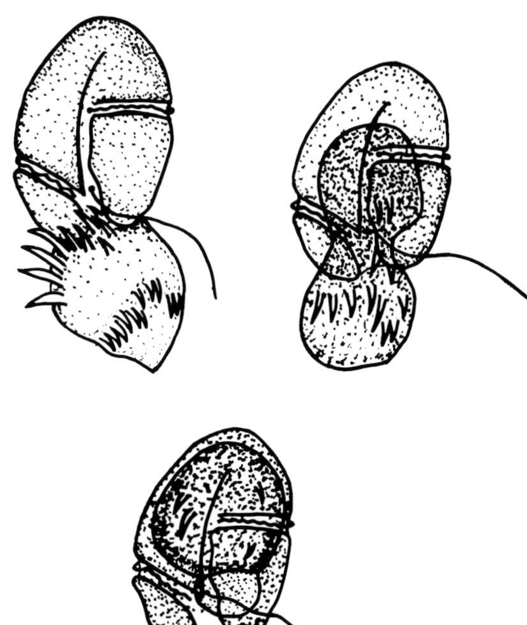

FIGURE 4.16 Representation of the capture and ingestion of the ciliate *Strombidium* by the dinoflagellate *Gyrodinium pavillardi*. Clockwise from top left: grasping, ingestion, and digestion of prey. (Redrawn after Biecheler, *Bull. Biol. Fr. Belgique* Suppl. 36, 1952. As portrayed in Sieburth, 1979).

(Biecheler, 1952; Sieburth, 1979), including the oligotrichous ciliate *Strombidium* (Fig. 4.16), which itself predates on other dinoflagellates, such as *Peridinium*. *Ceratium lunula* engulfs *Peridinium* (Norris, 1969). *Oxyrrhis marina* ingests bacteria, diatoms (including *Nitzschia*), and flagellates (Dodge and Crawford, 1974; Barker, 1935; Droop, 1954, 1959; Sieburth, 1979).

Most dinoflagellates are free-living, but some are symbionts (e.g., *Symbiodinium microadriaticum*) in marine invertebrates such as corals, jellyfish, sea anemones, foraminiferans, and giant clams; others are parasitic (e.g., the family Blasidinidae). *Dissodinium* parasitizes the eggs of copepods (Elbrachter and Drebes, 1978; Chatton, 1952; Drebes, 1969) or are ectoparasites (e.g., family Pyrocystidae) on marine invertebrates (Taylor and Seliger, 1979; Lee et al., 1985). Members of the genus *Duboscquella* parasitize tintinnids (Coats, 1988; Cachon, 1964; Cachon and Cachon, 1987), whereas *Mesodinium* parasitizes radiolarians, and *Syndinium* is found on the copepod *Paracalanus parvus*. Other heterotrophic dinoflagellates include members of the following genera: *Amphidinium* (salt marshes and sandy beaches), *Cochliodinium, Gymnodinium, Gyrodinium, Katodinium, Polykrikos,* and *Protoperidinium* (Lee et al., 1985).

Chrysomonodida are small, biflagellate phytoflagellates, some of which have a tendency to form colonies, that exhibit autotrophic, mixotrophic, and obligate heterotrophic life-styles. Heterotrophs capture food through the use of flagella-generated currents aided by pseudopodial ingestion and often a protoplasmic collar to concentrate the food. Some, such as *Ochromonas malhamensis,* capture food by flagella lashing. Dubowsky (1974) found that polystyrene latex particles stimulate feeding responses as readily as bacterial suspensions. Other species of *Ochromonas* exhibit similar behavior (Daley et al., 1973). *Ochromonas wyssotzke,* a freshwater form, attaches itself to a substrate by a threadlike extension from its posterior end, for feeding. It has two anterior flagella, one long and mastigoneme covered (hispid) and one short and smooth, which lies in a ventral furrow parallel to the cytostome. Water currents generated by the longer flagellum move toward the cell along the ventral furrow, where phagocytosis occurs. (The complete process takes about 20 seconds.) The attachment aids in the feeding process by improving efficiency (Fenchel, 1982a–e) as discussed later. The marine species *Paraphysomonas vestita* has an identical feeding pattern. Fenchel (1982b) has found evidence of cannibalism in *Ochromonas* sp. Some paraphysomonads feed on the purple sulfur bacteria *Chromatium okenii* (Korshikov, 1929; Sieburth, 1979).

Zoomastigophorea

All zoomastigophoreans (zooflagellates) are obligate heterotrophs, and many are parasitic. Many (e.g., *Hypernastiga*) have multiple flagella; others are uniflagellate. Free-living forms have a distinct cytostomal area for ingestion.

Choanoflagellates (collared flagellates) possess a delicate collar of tentacles enclosed and suspended in a lorica of silica rods or costae, surrounding the base of a single flagellum (Fig. 4.17). This collar acts as a funnel, trapping and concentrating particles (e.g., bacteria and detritus), which are drawn in by flagella undulation (Hibberd, 1975). Several studies have examined feeding-related structure in choanoflagellates (Laval, 1971; Leadbeater, 1972, 1973; Leadbeater and Morton, 1974; Leadbeater and Manton, 1974). In general, food material caught in the mucus covering the ring of tentacles is engulfed by advancing pseudopodia, which are structurally supported by the tentacles. Food ingested by pseudopodia pass into the body of the organism at a point within the collar between the protoplasm and the lorica (Leadbeater and Morton, 1974; Leadbeater and Manton, 1974; Manton et al., 1976). The actual feeding process can be divided into two phases: (1) food capture by the collar tentacles with the help of the flagella generated current; and (2) phagocytosis by pseudopodial engulfment.

The Kinetoplastida group includes the parasitic trypanosomads as well as the free-living, ubiquitous bodonids. The bodonids have a distinct cytostome that is bounded by a flap and a cytopharyngeal tube. Food particles such as bacteria are wafted into the cytostome by the action of a mastigoneme-covered flagellum. Brooker (1971) and Vickerman (1976) should be consulted for detailed information about the bodonids.

One particular bodonid, *Rhynchomonas*, has a cytostome at the end of a mobile scavenging, bilobed proboscis that is attached to the anterior flagellum. The cytostome is located at the tip of this proboscis, which is continuous with the cytopharynx (Nisbet, 1974; Burzell, 1973, 1975). The cytopharynx has a cytoskeleton, microtubular structure not unlike that found in certain euglenoid flagellates as well as suctorian and nassulid ciliates.

Pseudobodo tremulans (Fenchel, 1982a) has two flagella, the posterior one of which is smooth, passes through a ventral furrow, and is used for temporary attachment to particles or water films. The second anterior, long flagellum is hispid and drives water to the cytostome whose right side is bordered by a large lip (Fig. 4.18). Ingestion is similar to that found in *Ochromonas* and *Paraphysomonas*. Starving cells form swarmers and never attach (Fenchel, 1982a).

Pleuromonas jaculans, like *Bodo* and *Rhynchomonas*, is a somewhat typical representative of the free-living Kinetoplastida. It has two flagella, an

FIGURE 4.17 Representation of the choanoflagellate *Monosiga* sp. depicting the collar of tentacles (t) and single flagellum (f). Pseudopodial engulfment of prey occurs at the base of the tentacles. (Redrawn after Fenchel, 1982a.)

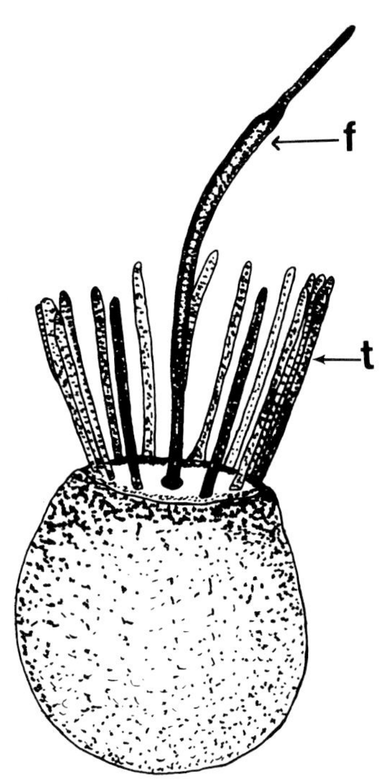

FIGURE 4.18 Representation of the nonloricate bicoecid flagellate *Pseudobodo tremulans* depicting the mastigoneme-covered (= hispid) anterior flagellum and a smooth posterior flagellum that passes through a ventral furrow and that is used for temporary attachment to particles or water films. (Redrawn after Fenchel, 1982a.)

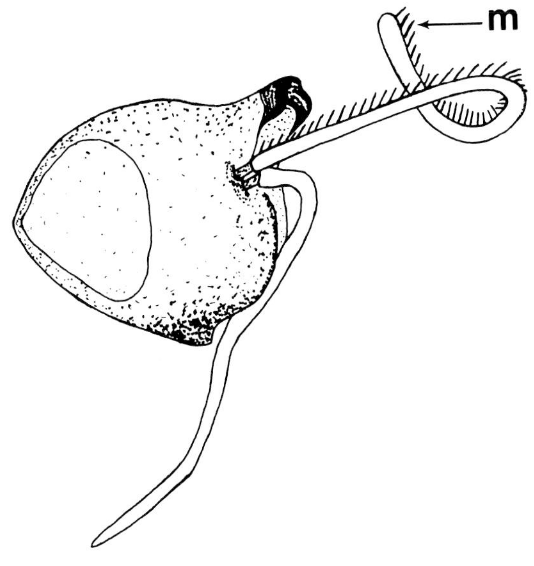

anterior hispid one and a long posterior one used for temporary attachment. The cytostome leads into a permanent pharynx, which is supported by microtubular bundles (Fig. 4.19). The anterior flagellum drives water past the cytostome, where phagocytosis occurs.

General Concepts Concerning Flagellate Feeding

Flagellates generally employ two trophic strategies (Fenchel, 1982a–e):

1. Predation by filtration through either a collar or a filter of some kind, where sieving occurs (e.g., Choanoflagellates, Chrysomonads).
2. Predation by direct encounter during flagellate movement, where food directly contacts a surface without projections (bodonids, monads, pseudobodonids, phagotrophic euglenids).

Fenchel (1982a) points out that food collection is a function of the velocity of the water current (produced by a single flagellum) and the area of the collecting surface. The water velocity through a filter is dependent on the filter porosity, so that as the interfilter filament distance decreases, water velocities also decrease with concomitant greater pressure drops over the filter (Fenchel, 1986a, 1987). For this reason water velocities through the filter of *Actinomonas* are greater than those of *Monosiga*. Also, Fenchel points out that only particles with center flow lines of less than one particle radius from the cell will be intercepted by direct impact (Fenchel, 1986a, 1987).

Free-swimming cells produce only small local velocity fields, whereas attached cells create much greater flow fields (Lighthill, 1976; Fenchel, 1986a, 1987). For example, a stationary *Ochromonas* has a three times greater volume swept clear than a free-swimming cell (Fenchel, 1982a). This suggests the likelihood that many, if not most, planktonic flagellates will attach during feeding, which may be why flagellate numbers are higher in nature when associated with suspended particles in the water.

Fenchel (1982a) considered the possibility that prey cell movement, in the form of either Brownian movement or self-directed motility, may enhance rates of capture by flagellates relative to capture of nonmotile prey. Since motile bacteria appear to be quantitatively unimportant in marine planktonic environments (Sieburth, 1979), such enhancement is unlikely. According to Fenchel, organisms that depend on prey motility, to any significant degree, for successful feeding should have a different morphology that would provide for uptake of food particles on the entire body surface, as opposed to a fixed cytostome. Another likely adaptation would be maximization of linear dimensions relative to cell volume similar to what is found in such planktonic testate amebae as the foraminiferans and radiolarians, which depend on random collisions between prey and pseudopodia for food. Such adaptations are not found in the flagellates.

Consumption rates in phagotrophic flagellates (in summary after Fenchel) are likely controlled by several factors, including

1. Time to phagocytize food, at which time another food item cannot be engulfed (= handling time).
2. Concentration of food particles in the environment.
3. Velocity of feeding currents.
4. Geometry of cells feeding organelle.
5. Size and shape of food.

Additionally, it is likely that rate of food vacuole production, which includes vesicle recycling time, and the availability and rate of synthesis of cellular

FIGURE 4.19 Representation of the kinetoplastid flagellate *Pleuromonas jaculans* depicting the two flagella that emerge from a cavity on the left side of the cell, the anterior one of which is mastigoneme (M) covered, and its distinct cytostome (c) structure, which leads into a permanent pharynx that is enforced by microtubule bundles. (Redrawn after Fenchel, 1982a.)

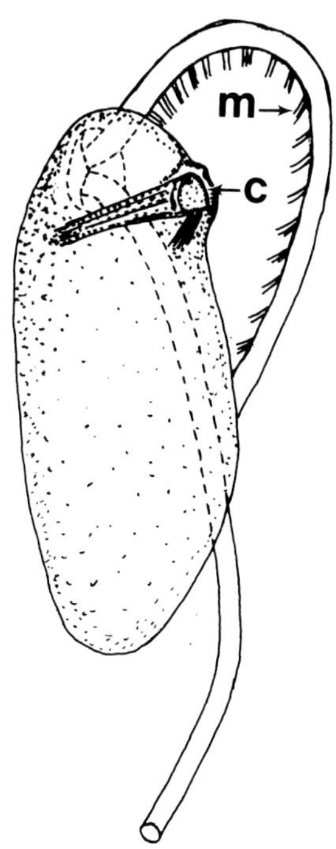

enzymes associated with such recycling, will also affect consumption rate.

2. Foraminiferans

Forams, as discussed earlier and reviewed by Lipps (1982, 1983), obtain their nutrition by uptake of dissolved organics, herbivory, carnivory (or a mixture of the two), suspension feeding, scavenging, parasitism, cannibalism, and/or symbiosis.

The uptake of dissolved organic matter in some benthic forams has been documented (DeLaca et al., 1980, 1981) and is carried out over the entire cell surface, as well as, in the case of rooted, agglutinated forams, through the root system (Fig. 4.20; Lipps, 1983). One study (DeLaca et al., 1980) documented the uptake of 11 amino acids and glucose. Many forams attach, either permanently or temporarily, to a substrate and passively harvest algae, bacteria, yeast, and the like, by means of extended pseudopodia. Some species, such as *Rosalina globularis,* move in search of food (Sliter, 1965). Pseudopods may be agglutinated or supported by spines to protect them from physical damage. Additionally, some forams move with pseudopods extended forward or back, gathering food as they move (Lipps, 1982, 1983).

FIGURE 4.20 Representation of the benthic foraminiferan *Notodenrodes antarctikos,* which lives on mud substrates (ms) at 33 m at New Harbor, McMurdo Sound, Antarctica. This organism is able to take up dissolved organic material through all parts of its body, especially the in-mud root system (rs). It also suspension feeds by means of extended, above-mud pseudopodia (P). (Redrawn after Lipps, 1983.)

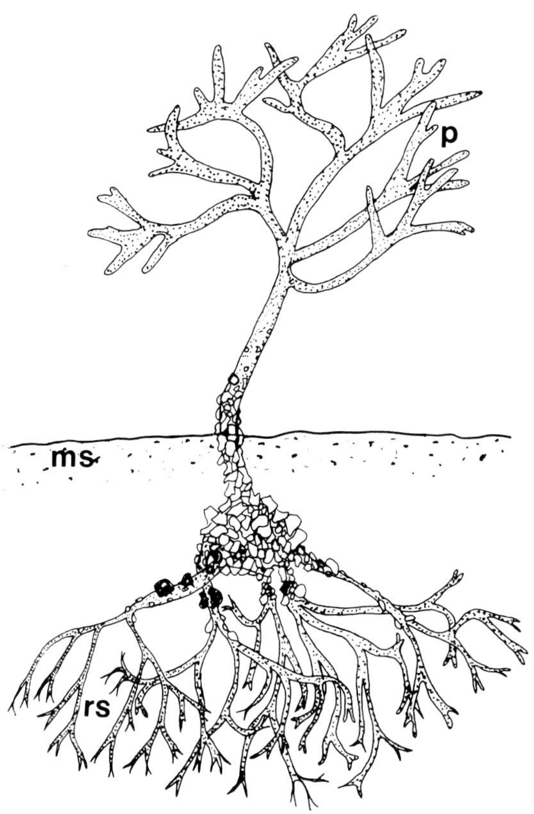

Carnivory has been widely reported for forams (Sandon, 1932; Boltovskoy and Wright, 1976; Murray, 1973) but has been studied in detail only for planktonic forams. Anderson and Bé (1976a,b) describe prey capture by the planktonic foram *Hastigerina pelagica* feeding on *Artemia* nauplii. *H. pelagica,* as well as *Globigerinella aequilateralis,* is often found with crustacean prey in its bubble capsule and spines. Anderson and Bé reported that prey contact with the foram spines initiates the feeding response, which begins with rhizopodial flow and ensnaring of the prey. As prey struggle continues, the rhizopodial entrapment strengthens and the prey gradually is drawn into the bubble capsule, where it typically dies of exhaustion. Adhesive substances are released from vacuoles in the rhizopodia, and this substance coats the prey. Rhizopods ultimately penetrate into crevices or cuticles of the prey, thereby invading underlying soft tissue and engulfing both lipid droplets and soft tissue particles. The engulfed particles are drawn into the foram cytoplasm, where the content of the vacuoles are digested. A similar process is found in the radiolarians (see the following discussion).

Suspension-feeding benthic foraminiferans at times extend pseudopods above the substratum into the overlying water column, where they may act as a sieve or may capture food by direct interception, inertial impaction, motile particle deposition, or gravitational deposition (Rubenstein and Koehl, 1977). Regardless of the precise method employed, all suspension feeding, from the perspective of the foraminiferan, is passive, with the foram unable to move water to facilitate prey capture.

Some benthic forams—such as *Astrorhiza limicola,* which uses its pseudopods within sand or silt sediment, where it captures interstitial organisms—act more like deposit feeders than suspension feeders. Scavenging benthic forams live either in the top 1–2 cm of sediment, where they tend to burrow for food, or on a hard substrate.

Parasitism of forams on other forams has been reported (LeCalvez, 1947; Todd, 1965; Banner, 1971; Boltovskoy and Wright, 1976). For example, *Fissurina marginata* captures granules from the pseudopodial network of *Discorbus mediterranensis* (LeCalvez, 1947). Cannibalism has also been reported in forams (e.g., *Glabratella ornatissima* feeding on its own gametes).

Forams attach to many living firm substrates, including marine plants, wood, other forams, hard skeletons of invertebrates, crabs, certain vertebrates, and the like. At least some of these attachments are beneficial from a feeding standpoint. For example, *Placopsilina cerromana* and *Bdelloidina vincentoronensis* attach to and encircle the pore opening region of the burrowing sponge *Entobia cretacea* (Bromley and Nordmann, 1971) or to the brachiopod *Tichosina floridensis,* each using the host's feeding current to obtain food (Zumwalt and DeLaca, 1980; Lipps, 1983). Some forams also attach to pelagic gulfweed and become passive riders (DeLaca and Lipps, 1972; Haward and Haynes, 1976; Bock, 1969; Spindler, 1980).

3. Actinopods

Bovee and Cordele (1971), Nisbet (1984), Patterson (1979), Hausmann and Patterson (1982), and Patterson and Hausmann (1981) examined the behavior of the heliozoan *Actinophrys sol* feeding on gastrotrichs and found that feeding involves a hyaline ectoplasm that forms special pseudopodia for different types of prey. For example, small, motionless food is captured by a straight, small pseudopod, which on contact with the prey forms a cup to engulf the food item. Large, motionless prey result in production of large, wide hyaline outgrowths that advance onto the food and expand in all directions to engulf it. Active prey cause production of a large, saclike pseudopod, with associated obstruction of prey escape effected by the axopods. When prey contact the adhesive protoplasmic surface of the axopod tips, they are pulled against the predator's body surface by retraction of the axopods, and adjacent axopods enclose the prey in a cytoplasmic pocket. For larger prey, several may feed cooperatively by temporarily fusing into one large predator (Fig. 4.8), digesting the prey and then separating. Patterson (1979) observed extension of five pseudopodial arms that coalesce around prey to form a deformable funnel (duration of about five minutes). This funnel engulfs the prey, which after 20 minutes are enclosed in an endocytotic digestive vacuole. Extrusomes lie below the plasmalemma and travel along the axopodial arms during feeding. Extrusomes may consist of adhesive material or may represent membrane replenishment material meant to subsidize that material used up in the process of phagocytosis. This whole feeding cycle takes about 24 hours to complete.

Radiolarian feeding shares characteristics found in both the forams and heliozoans and has already been described in the radiolarian section on general food web considerations.

4. Ciliates

A single cytostome with, in certain groups, associated specialized structures is routinely found in most ciliates. (Suctorian ciliates and parasitic forms such as those found in the astomatous and apostomatous ciliates are noted exceptions.) Small (1984) considers the evolution of ciliate oral structure. In ciliates the collection of food to be phagocytosed may be achieved by either active or passive searching or perhaps by a combination of both (Dragesco, 1962; Savoie, 1968; Bardele, 1974; Sleigh, 1973; Fenchel, 1980a–d; Small and Lynn, 1985).

Nassophorean, phyllopharyngean and certain colpodean ciliates possess a complex cytopharyngeal basket that is used to feed on filamentous algae or cyanobacteria and diatoms. This basket is composed of bundles of longitudinally oriented microtubules and lamellae held together by contractile, dense filaments that change shape by twisting, dilating, and/or constricting during feeding. Movement of this basket permits folding and breaking of prey filaments and cell chains (Tucker, 1968, 1972; Peck and Hausmann, 1980; Hausmann and Peck, 1978, 1979) and is found in such genera as *Phascolodon, Nassula, Pseudomicrothorax, Euigmostoma* and *Puytoraciella*. The ability of tintinnid ciliates to modify the particle size spectrum of their food has also been considered (Capriulo, 1982; Rassoulzadegan and Etienne, 1981; Hedin, 1976). Data of Capriulo et al. (1988) suggest that the hypotrich ciliate *Euplotes vannus* may possess the ability to alter chain length size of *Skeletonena costatum* by an as yet undetermined mechanism.

Feeding in *Didinium* involves the incidental contact of prey with its proboscis. This contact causes an instantaneous discharge of two kinds of extrusive organelles: (1) short pexicysts, which attach to the prey surface without penetration, and (2) long toxicysts, one end of which penetrates the prey discharing proteolytic and paralytic enzymes into the prey while the other end remains firmly attached to the ciliate. Cytoplasmic streaming pulls the discharged toxicyst inward, thereby drawing the prey slowly into the cytostome. As the prey is swallowed, it is compressed and its volume is reduced by elimination of fluid (Wessenberg and Antipa, 1970).

In suctorian ciliates (which are sessile and lack cilia and a permanent mouth as adults) each tentacle, which is composed of bundles of microtubules, may become a temporary mouth during feeding. The tentacles are composed of a shift with a terminal knob. Within the shaft is a series of microtubular lamellae that extend deep into the cytoplasm. The tentacle knob is not just a continuation of the shaft but has a unique, fine structure with associ-

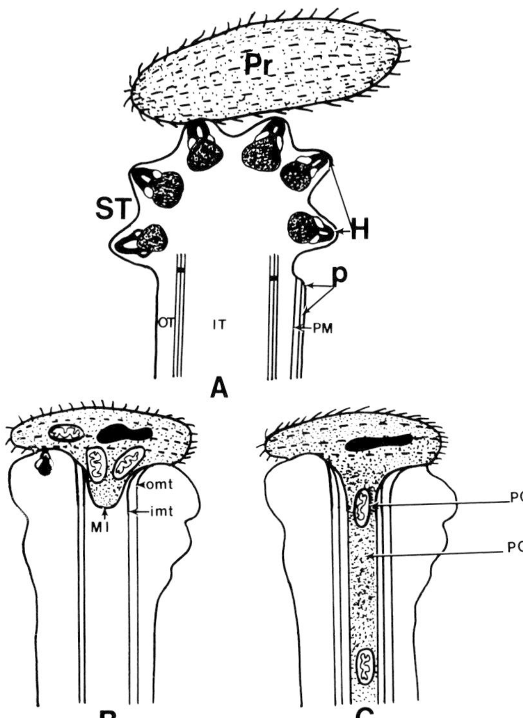

FIGURE 4.21 Prey capture in suctorian ciliates. Suctorians, which possess no permanent cytostome, produce temporary mouths as needed, at the tip of any tentacle (ST). When live prey (pr), usually other ciliates, come in contact with a tentacle(s), they are attacked by extrusomal bodies known as haptocysts (H) (a kind of toxicyst), which dissolves a small area of prey pellicle, thus allowing penetration of the prey by the suctorian tentacle, which is then transofrmed into a feeding tentacle (A). Predator and prey pellicles then fuse, which prevents leakage of prey contents. The tentacle microtubules, prior to feeding, form an outer (ot) and inner (it) tube (A). After prey capture and fusion with the tentacle, the outer (OMT) and inner (IMT) microtubules move into the tentacle head (or knob) to the site of prey attachment (B). The tentacle knob then invaginates (MI of B), moving down the inner tube to form a feeding tube within the microtubule ring. Prey organelles (PO), subcellular components, and cytostome (PC) flow down the feeding tube deep into the ciliate, where food vacuoles are formed by endocytosis. (PM = suctorian plasma membrane.) (Drawing based on transmission micrographs of Rudzinska, 1973.)

ated proteolytic and paralytic haptocysts. The haptocysts are used to capture prey by adhesion (Rudzinska, 1973; Fig. 4.21). During prey capture dense bodies and vessicles rush up from the cytoplasm through the outer tentacle tube and the prey is caught and ceases to move. Tentacles shorten and widen and microtubules from the tentacle shaft enter the knob, reach the tip, and bend outwardly, thereby pressing the knob against the prey cytoplasmic membrane, which fuses with the tentacle membrane. This fusion prevents leakage of prey cytoplasm. The expanded "knob" allows the proteolysed cytoplasm to be ingested and enclosed within food vacuoles formed by vessicles that had moved up into the knob. At the end of the tube a large food vacuole is thus formed; this ingestion process is completed when the prey cytoplasm is consumed and is followed by production of additional food vacuoles when other prey encounter the reformed tentacle tip. This process uses up much membrane, which is continually resupplied via vessicle recycling (Hackney and Butler, 1981; Nisbet, 1984). Suctorians also possess invaginated, plasma membrane–covered body pits through which additional macromolecular nutrients are taken in by pinocytosis (Rudzinska, 1980; Tucker, 1974; Bardele, 1974). Kent (1981) studied the life history response of a suctorian to varying food regime. Some suctorians are parasitic on other free-swimming ciliates. Others, such as *Choanophrya infundibulifera,* are detritus feeders (Hitchin and Butler, 1973). Detritus feeders, such as *C. infundibulifera,* are without haptocysts.

Dragesco (1962) gives a good account of feeding in ciliates. Salt (1967) examined general ecological questions concerning predator–prey relationships among ciliates, including hunger stimulus, time spent hunting, frequency of attack, search rate, rate of capture, growth, and the like. Questions concerning ciliate filter feeding have been expertly detailed in Fenchel (1986a). Fenchel (1980a–d) found that bacterivorous holotrichs retain food particles down to 0.1 μm in size, with highest efficiencies in the 0.3–1-μm size. Spirotrichs did not retain particles of less than 1–2 μm. Fenchel and Small (1980) examined the function of the oral cavity and associated organelles in the fresh water hymenostome ciliate *Glaucoma*. Feeding currents and prey retention in *Stylonychia, Stentor,* and *Vorticella* have also been described (Sleigh and Barlow, 1976; Sleigh and Aiello, 1972). Membranellar metachronal beat frequencies of 7–10 Hz were found for *Glaucoma* (Fenchel and Small, 1980), compared with 15 Hz for other hymenostomes and 40 Hz for peritrichs and polyhymenostomes (Fenchel, 1980a–d). In *Glaucoma* water moves anterior to posterior, creating a whirl on either side of the ciliate. Water movements exterior to the oral area are driven by those within the oral cavity by viscous forces. Water moving along the anterior to left anterior rim passes posteriorily between two pairs of membranelles, toward the mouth.

Foissner and Didier (1983) found that the colpodid ciliate *Pseudoplatyophyra nana* possesses a tubular structure (the oral trapeze) that is used to break open the cell wall of the prey and then suck out its content. This oral trapeze is surrounded by adoral and paroral membranelles.

In a study of *Chilodonella steini* Sawicka et al. (1983) found this ciliate to be a strict herbivore (diatom food only) that partitions ingested food among its daughter cells during division. .As is true for many ciliates, this ciliate stores membrane involved in food vacuole formation in the cytoplasm; from there it is recycled as necessary (Allen, 1974, 1978; Nilsson, 1979; Allen and Fok, 1980; Sawicka et al., 1983). Factors affecting the number of food vacuoles formed include the size of the ciliate cells, the phase of the cell cycle the ciliate is in, the number of vacuoles already present, and the rate of vacuole egestion. *C. steini* requires 100 food vacuoles to complete one cell cycle, as compared with 200 for *Paramecium* and *Tetrahymena* (Smith-Sonneborn and Rodermal, 1976; Rasmussen, 1976). Also, size of prey regulates, to some extent, size of food vacuole, as do such factors as level of starvation.

Actual mechanisms by which food is captured by large planktonic ciliates, and subsequently eaten, remain largely unknown. In particular, the significance of extrusomes in prey capture should be examined. At least for tintinnids it appears that capsule-containing striae may be involved in prey capture (Gold, 1979; Laval-Peuto, 1979; Capriulo et al., 1986; Fig. 4.11b). Additionally, the trophic position of ciliates in planktonic food webs needs to be reconsidered in light of recent work that indicates that many planktonic ciliates live as mixotrophs by retaining functional chloroplasts that are involved in intraciliate photosynthesis (Stoecker et al., 1987; Laval-Peuto et al., 1986; McManus and Fuhrman, 1986; Taylor, 1982, and this volume) or by harboring algal endosymbionts, as is the case for the ciliate *Mesodinium rubrum* (Lindholm, 1985).

FATE OF PROTOZOAN BIOMASS

Determination of the fate of protozoan biomass in marine ecosystems is of key importance to our understanding of marine food webs. Despite this fact, serious investigations concerning such trophic transfers have only recently begun. Finlay and Uhlig (1981) and Verity and Langdon (1984) have considered the food value of several marine and freshwater protozoa in various units of total calories, carbon, nitrogen, and ATP. In terms of elucidating trophic transfers, however, significant intraprotozoan carnivory suggests that repetitive biomass transfers routinely occur within the protozoan community. For example, nanoplanktonic heterotrophic flagellates, which feed primarily on bacteria and/or bacteria-sized eukaryotes, are preyed upon by microzooplanktonic protozoa (Azam et al., 1983; Sheldon et al., 1986). Dinoflagellates are an important part of the diet of at least certain ciliates (Stoecker et al., 1981; Capriulo, 1982; Capriulo and Carpenter, 1983; Smetacek, 1981; Verity and Vilareal, 1986). Various freshwater ciliates such as *Chilodonella, Colpidium, Cyclidium, Dexiotricha, Glaucoma, Leptopharynx, Paramecium, Stentor, Tetrahymena,* and *Trochilioides,* as well as certain suctorians and peritrichs, are food for other ciliates, including *Didinium, Euplotes, Tokophyra, Podophyra, Metacineta, Histriculus,* and *Acineta* (Small, 1973). *Blepharisma americanus* grows on *Glaucoma* and *Bursarus truncatella* feeds on *Colpidium campylum* (Fenchel, 1980a–d). Ciliates in such genera as *Cyclotrichium, Tiarina, Euplotes, Holophyra, Tertrahymena, Strombidium,* and *Woodruffia,* as well as certain tintinnids, feed on other ciliates (Faure-Fremiet, 1924; Kahl, 1930–1935; Sandon, 1983; Corliss, 1979; Kidder et al., 1940; Nilsson, 1979; Salt, 1967; Kloetzel, 1974; Fenchel, 1987). Rhabdophoran ciliates, such as *Prorodon, Coleps,* and *Lacrymaria,* feed on other ciliates and flagellates, whereas amphileptids feed on zooids of peritrich ciliates (Fenchel, 1987; Small, 1983). Species like *Didinium nasutum* feed exclusively on ciliates (Luckinbill, 1974). Predatory ciliates such as *Litonotus* and *Loxophyllum* slide along surfaces and feed on other ciliates, large diatoms, and filaments of *Oscillatoria* (Fenchel, 1987). Within-species cannibalism, as is found in the microflagellate *Paraphysomonas imperforata* (Goldman and Caron, 1985) and in the ciliates *Blepharisma* (Pierce et al., 1978), *Miamiensis avidis* and *Potomacus pottsi* (Small, personal communication) and intracellular autophagy of cellular organelles and components, as is found in certain flagellates (Fenchel, 1982c,e), are also known to occur in protozoa. Bacterial parasitism of protozoa and/or interprotistan parasitism such as is found between dinoflagellates and tintinnids (Coats, 1989; Cachon, 1964; Cachon and Cachon, 1987) also likely results in significant biomass transfers. Such internal (within and among trophic levels) cycling of carbon, with associated gross growth efficiencies and respiratory energy losses, could greatly affect secondary production levels and make estimation of that production difficult and controversial (Banse, 1982a,b). Food size considerations alone do not always help our understanding, since many protozoans, including flagellates (Suttle et al., 1986) and oligotrich ciliates (Smetacek, 1981), ingest food items larger than themselves. More detailed information on internal cycling of carbon within protozoan communities is needed before an accurate portrait of this component of the microbial food web can be completed.

Data are more available, although far from

complete, on metazoan trophic interactions with aquatic protozoa (Table 4.14). The red tide flagellate *Chattonella antigua* has been found to be fed upon by copepods, including *Acartia erythracea, Calanus sinicus, Centropages yamadai, Paracalanus parvus, Pseudodiaptomus marinus, Acartia pacifica, Calanopia thomponi, Labidocera rotunda,* and *Tortanus forcipatus* (Uye, 1986). Many crustaceans, including Cladocerans (Heinle et al., 1977; Porter et al., 1979), and copepods (Robertson, 1983; Turner and Anderson, 1983; Berk et al., 1977; Stoecker and Sanders, 1985; Sheldon et al., 1986; Sherr et al., 1986b; Pace and Orcutt, 1981; Laval-Peuto et al., 1986; Stoecker and Egloff, 1987; Elbourne, 1966; Klekowski and Shushkina, 1965; Korniyenko, 1976; Monakov, 1972; Shcherbina, 1970) eat ciliates. Ciliates have also been found to be part of the diet of snails and midges (Small, 1973), rotifers (Small, 1973; Gilbert, 1968, 1976, 1980; Guiset, 1977; Maly, 1969, 1975; Sorokin and Paveljevo, 1972; Beaver and Crisman, 1982; Archbold and Berger, 1985; Robertson and Salt, 1981), oligochaete worms (Green, 1954; Taylor, 1980), turbellarians (Kepner and Carter, 1931; Nuttycombe and Water, 1935; Sonneborn, 1930), fish (Korniyenko, 1970, 1971; Pinus, 1970; Stoecker and Govoni, 1984; Capriulo and Ninivaggi, 1982), and larval and postlarval ctenophores (Stoecker et al., 1987) and jellyfish (Stoecker et al., 1987). Smetacek (1981) suggests that predation by metazoans is the major factor regulating the size of protozoan populations in the Kiel Bight.

Foraminiferans are among the best understood and most studied of the marine protozoa. They are ingested by a myriad of marine organisms (see Table 4.14), including many deposit feeders, which often control foram densities and species diversity patterns not regulated by physiochemical conditions alone (Buzas, 1978, 1982; Buzas et al., 1977; Lankford and Phleger, 1973; Lipps and Ronan, 1974; Lipps and Valentine, 1970; Blegved, 1928; Dayton, 1971; Young et al., 1976; Virnstein, 1977). Selective and incidental consumers of forams are reviewed by Lipps (1983). The selective consumers are usually small and include the juvenile stages of larger predatory species (Brand and Lipps, 1982). Certain predators (e.g., scaphopods) can select among foram prey species (Bilyard, 1974; Shonman and Nybakhen, 1978), although there is more selection for forams in general than among foram species. Scaphopod predators have also been found to be capable of distinguishing between live and dead prey, with a preference for calcareous foram prey over agglutinated forams. The scaphopod *Dentalium entale stimpsoni* has an ingestion rate of 7,884 forams individual^{-1} y^{-1}. Some gastropod predators are also very selective, with *Cylichna attonis* feeding only on species of *Nonionella. Acteocina culcitella* has been found to predate on 13 species of forams (Lipps and Valentine, 1970; Lipps, 1983).

Adult predators of forams include opisthobranch gastropods, scaphopods (e.g., *Dentalium*), and other forams. Christiansen (1964) reported that the foram *Spiculosiphon radiata* attacks and consumes at least 13 other foram species. Forams are also predated upon by nematodes; shrimp, such as *Crangon vulgaris* (Myers, 1943); hermit crabs of the genus *Eupagurus* (Orton, 1927); the gastropod *Patella* (Pelseneer, 1935); the brittle star *Ophiothrix fragilis* (Mortensen, 1927); the crustacean *Gonoplax* (Mare, 1942); nine types of echinoderms; two molluscs and seven polychaetes (Yonge, 1954);

TABLE 4.14 Predators of Protozoan Biomass*

Prey	Predators
Heterotrophic microflagellates	Naked amebae; ciliates; dinoflagellates; microzooplankton, including copepods; cannibalism
Dinoflagellates	Ciliates; larger dinoflagellates; copepods
Radiolarians	
Solitary forms	Planktonic foraminiferans
Colonial forms	Amphipods (*Hyperietta* and *Oxycephalus clausi*) juvenile *Brachyscelus* sp., juvenile *Lycaea* sp., harpacticoid copepod *Miracia efferata*.
Foraminiferans	Selective consumers: other forams, nematods, polychaetes, gastropods, scaphopods; crustaceans
	Incidental consumers: flatworms, polychaetes, chitons, gastropods, nudibranchs, ammonites, bivalves, crustaceans, holothuroids, asteroids, ophiuroids, echinoids, crinoids, tunicates, fish
Ciliates	Cannibalism; other ciliates, crustaceans, including cladocerans, copepods; parasitism (e.g. by dinoflagellates) snails, midges, rotifers, turbellarians, oligochaete worms; fish, such as larval menhaden; ctenophores; jellyfish; dinoflagellates

*See text for references.

detritus-feeding gastropods (Morton, 1959); carnivorous prosobranchs (Marcus and Marcus, 1959; Graham, 1955); and polychaetes (Perkins, 1958). Gastropod borings have been found in the tests of *Quinqueloculina* and *Spiroplectammina* (Reyment, 1966), and *Rosalina globularis* was found with test borings and associated nematodes. Lipps (1973) considered predation on forams by reef fish. Also, the rock crab *Cancer irroratus,* the gastropod *Littorina littorea,* the starfish *Asterias vulgaris,* and the sea urchin *Strongylocentrotus drobachiensis* have been found with foram tests in their guts. Forams have been found to be parasitized by other forams, sporozoans, gregarinids, nematodes, amebae, and fungi (Boltovskoy and Wright, 1976).

Some predators from mud bottom environments, such as certain opisthobranchs that possess gizzardlike mills for crushing foram tests, are specialized to feed on forams (Morton, 1958; Moore, 1961). Small abyssal benthic predators of forams have been found, including various scaphopods, pectinids, and decapods (Lipps and Valentine, 1970; Knudsen, 1967; Sokolova, 1957). On sandy bottom shelf areas, shrimp, scaphopods, bulloid gastropods, and certain worms have been found to eat forams (Lipps and Valentine, 1970). On high-turbulence, rocky bottoms, forams are found living in algal mats, among algal holdfasts, or attached to coralline algae. Key predators in these environments have not yet been clearly identified.

Little data exist on the fate of planktonic foraminiferan and radiolarian biomass. Anderson (1983a) considers the possibility that predation on radiolarians is low because of a number of factors, including the following: (1) dinoflagellate symbionts associated with radiolarians give them a disagreeable taste (Withers et al., 1979); (2) radiolarians are large; (3) radiolarians are bioluminescent; and (4) radiolarians have spines and spicules. Planktonic forams have been found with solitary radiolarians snared in their rhizopodial net (Anderson and Bé, 1976a,b; Bé et al., 1977; Anderson et al., 1979). Swanberg and Harbison (1980) noted that juvenile stage hyperiid amphipods (*Hyperietta*) are obligate parasites of the colonial radiolarian *Collozoum longiforme*. The amphipods are found embedded in the center of the colony. The amphipod *Oxycephalus clausi,* which also predates on a wide variety of gelatinous zooplankton, including salps, ctenophores, medusae, siphonophores, and heteropods (Madin and Harbison, 1977; Harbison et al., 1977), is also found on *C. longiforme*. Additional species found associated with *C. longiforme* include juvenile Brachyscelus sp. and juvenile *Lycaea* sp. *Lycaea* sp. is a highly specific parasite of salps (Madin and Harbison, 1977). The harpacticord copepod *Miracia efferata* also lives on *C. longiforme* (as well as on *Rhizosolenia* mats; Carpenter et al., 1977).

SPECIAL HABITATS

1. Sea Surface Microlayer

The surface skin of ocean water has been recognized as a unique, highly enriched marine habitat for some time (MacIntyre, 1974a,b; Zaitsev, 1970; Naumann, 1917) and has been considered in a short review fashion by Sieburth (1979). Specific studies have considered organics, bacteria, algae, certain protozoa, including amebae and choanoflagellates, and the general ecology of this habitat (Baier, 1972; Baier et al., 1974; Dahlback et al., 1982; De Souza-Lima and Romano, 1983; Dietz et al., 1976; Garrett, 1965; Hardy, 1973; Hardy and Valett, 1981, 1982; Harvey, 1966; Harvey and Burzell, 1972; Harvey et al., 1983; Hermansson and Dahlback, 1983; Kjelleberg and Stenstrom, 1980; Kjelleberg et al., 1979; Norris, 1965; Larsson et al., 1974; Norkrans, 1980; Sieburth et al., 1976; Carlucci et al., 1985; Davis et al., 1978).

Concentrations of organics, as well as densities of autotrophic and heterotrophic microbes, have been found to be significantly enhanced in this microhabitat. For example, carbohydrates, dissolved organic carbon, and ATP are enriched in surface North Atlantic water by 1.6–3.1 times (Sieburth et al., 1976). Increases in organics, such as this, are accompanied by high bacteria and phagotrophic protozoa numbers (Sieburth et al., 1976; Davis et al., 1978). The microheterotrophs from oligotrophic waters utilize the organics at higher rates than similar heterotrophs from eutrophic waters (Carlucci et al., 1985). Davis et al. (1978) found sarcodine protozoan concentrations in the surface ocean microlayer to be 10^3 times higher than in the underlying water.

The surface microlayer of coral has similarly been shown to exhibit elevated levels of microbial activity relative to the overlying water, with up to 7.5 times more bacteria, 18–280-fold more chlorophyll *a* and 1.23–50 times higher rates of thymadine incorporation (Paul et al., 1986).

Enhanced heterotrophic activity such as is found at the air–sea interface surface layer is likely to occur elsewhere in the water column, wherever density discontinuity layers exist. Evidence for special conditions such as chlorophyll and heterotrophic maxima at the thermocline has long been in existence (Parsons, Takahashi, and Hargraves, 1984; Sorokin, 1977). Similar biological research on other density discontinuity layers, however, is lacking but should provide interesting data in the future. Townsend and Cammen (1985) did report a deep protozoan maximum in the Gulf of Maine, at 55–100 m, and considered the possibility that this protozoan spike was a response to settled organics that were derived from surface layers and that collected at a deep hydrographic anomaly.

2. Marine Snow

Marine snow represents a pelagic zone, water column microhabitat that supports elevated biological activity and possesses features of both pelagic and benthic systems (Caron et al., 1982; Davoll and Silver, 1986). Material collectively known as marine snow is composed of amorphous, fragile, particulate, macroscopic aggregates formed from flocculent, water column detrital material. This detrital material is composed of exudates and remains of both autotrophs and heterotrophs, including thecosomates, pteropods, discarded larvacean houses, assorted mucus nets, slough mucus from corals, ctenophores, phytoplankton, and bacterial remains and other detritus. Several studies detailing sinking rates, chemical composition, and autotrophic and heterotrophic components of marine snow have been carried out (Silver et al., 1978; Alldredge, 1972, 1976, 1979; Alldredge and Cox, 1982; Alldredge et al., 1986; Caron et al., 1982; Knauer et al., 1982; Pomeroy and Deibel, 1980; Silver et al., 1984; Shanks and Trent, 1980; Beers et al., 1986). Alldredge and Cohen (1987) found microscale chemical patches associated with marine snow and pelagic water column fecal pellets. They observed that the snow had persistent oxygen and pH gradients and high nutrient enrichment and metabolic activity. Oxygen levels associated with the snow were low but never anaerobic, and snow particles had evidence of denitrification, sulfate reduction, and methanogenesis. Alldredge and Cohen did find anaerobic fecal pellets in their study and attributed the anoxic state to poor oxygen penetration through the peritrophic membrane surrounding the pellets.

Snow particle concentrations of 1–10 particles/L, even in the deep sea, have been found, with each particle representing a microbial oasis (Beers et al., 1986; Alldredge and Cohen, 1987). Some typical concentrations of microbial organisms found associated with marine snow are presented in Table 4.15.

Aloricate and loricate ciliates are routinely observed associated with particles in even the most oligotrophic parts of the ocean (Caron et al., 1982; Silver et al., 1984; Davoll and Silver, 1986). Small et al. (1983) present a list of about 40 species of ciliates that have been found, to date, associated with marine snow. Silver et al. (1984) found ciliates accompanying sinking detritus to bathypelagic depths. These ciliates may control the rate of bacterial remineralization (through bacterivory) of sinking particles. This notion is supported by the fact that many of the ciliates found associated with snow particles are known bacterivores.

Beers et al. (1986) found algal enrichment of macroaggregates of the Southern California Bight of 6.2–1,300 times for numbers and 6.2–2,500 times for carbon. Large phytoplankton found associated with snow include *Trichodesmium* and *Rhizosolenia* (Caron et al., 1982). *Rhizosolenia* mats from the Sargasso Sea have also been found with elevated bacterial and other heterotrophic biomass (Caron et al., 1982; Carpenter et al., 1977). Davoll and Silver (1986) defined three phases of colonization of marine snow from larvacean houses, with the ratio of biomass of bacterivores to bacteria greater in phase 2, after the larvacean house has been abandoned and bacteria are in exponential growth.

In addition to their significance as free-floating

TABLE 4.15 Typical Concentrations of Microbiol Organisms Found Associated with Marine Snow

Organism	Concentration (per ml)	Enrichment factor	Source
Choanoflagellates and other heterotrophic flagellates	10^5	100–1,000 ×	Davoll and Silver (1986)
Ciliates (including oligotrichs and choreotrichs)	10^2–10^3	100–2,000 ×	Davoll and Silver (1986)
Autotrophs	10^4–10^6	75–1,000 ×	Davoll and Silver (1986)
Bacteria	10^8	80–300 ×	Davoll and Silver, (1986)
Bacterivorous flagellates (50% bodonids)	3×10^0 to 7×10^4		Caron et al. (1982)
Ciliates and amebae	3–23		Caron et al. (1982)
Total nanoplankton	10^3–10^5		Caron et al. (1982)

aggregates in pelagic waters, marine snow particles are important, inasmuch as they accumulate by passive, gravitational collection on bottom (benthic) sediments, as well as at density discontinuity layers in the water column, where they form a "false benthos" environment. All that is required for water column development of such a layer is that the rate of accumulation of particles at a given layer from above exceed that of deposition from that layer to layers below. This accumulated material represents a biomass enrichment for each affected discontinuity layer, as well as, ultimately, for the bottom sediments. Such layers attract mobile organisms and enhance growth of organisms associated with the aggregates (Silver et al., 1978). The aggregates and accumulated layer represent a locus for bacteria, microflagellates, ciliates, amoebae, dinoflagellates, and marine algal enrichment and for associated enhanced production of higher trophic levels.

3. Hydrothermal Vents

Hydrothermal vents have been discovered at both Pacific and Atlantic Ocean seafloor spreading centers (Ballard, 1977; Weisburd, 1984; Ballard, 1983; Pain, 1986). A plethora of prokaryotic and eukaryotic organisms have been found associated with individual vent sites. However, the microbiological research concerning individual vent sites has been primarily focused on bacteria. Vent bacteria include attached, free-living, and endosymbiotic forms, with a dominance of chemosynthetic types (Jannasch and Wirsen, 1979; Tuttle et al., 1983; Jannasch and Wirsen, 1985; Baross and Deming, 1985). Consideration of the protozoans has been lacking from most vent studies. Recently, Small and Gross (1985) reported preliminary results of the examination of protozoans from the 21-degrees-north vent site, listing sarcodines, chrysophytes, and six of eight systematic classes of ciliates (as recognized by Small and Lynn, 1985) as encountered organisms. Tunnicliffe et al. (1985) reported the occurrence of abundant folliculinid ciliates from the Juan de Fuca Ridge vent at 47 degrees north. Sniezek (1987) examined protargol-stained samples from the hydrothermal vent environment at 10 degrees 57 minutes north; 103 degrees, 40 minutes west (called 11 degrees north). There he found 48 taxonomically distinct species comprising all eight classes of the phylum Ciliophora as recognized by Small and Lynn (1985). Strong similarities were found between the ciliates encountered at this vent site and those found to inhabit sulfureta. Most of the ciliates found at vent sites are known bacterivores. Sniezek (1987) compared the ciliates from the 21 degrees north site to those from 11 degrees north.

4. Interstitial Spaces of Sediment

The interstitial spaces of sand and other sediment represent a special microhabitat with a unique chemistry, as compared with the overlying water. This environment is expertly discussed and reviewed in Fenchel (1987) from which most of the information presented below is derived. The 0.12–0.25-mm median grain size grouping of sediment produces the highest numbers of ciliates, whereas finer-grained material is dominated by nematodes and larger sizes are dominated by metazoans. Interstitial ciliates are typically long and narrow, oblong or leaf shaped, and ciliated on only one side, to ease sliding along sediment grains. Most ciliates possess the ability to adhere temporarily to surfaces and include various karyorelectids, heterotrichs, prostomatids, and haptorids. Smaller forms include hypotrichs, cyrtophorids, scuticociliates, prostomatids as well as some oligotrichs. Phagotrophic flagellates from these habitats include dinoflagellates, euglenids, and bodonids, and naked and testate amebae are also found. Because of the extreme chemical gradient present in sediments, much vertical zonation of species is found as a function of oxidation state, with differences most apparent between oxidized surface, redox discontinuity, and deep anoxic layers of sediment (Fenchel, 1987).

Deeper ocean sediment, beyond the continental shelf, is primarily composed of silt and clay-sized sediment. These types of sediment have little interstitial fauna and are dominated by surface layer protozoans such as scuticociliates and hypotrich ciliates, bodonid and nonpigmented chrysomonad flagellates, and foraminiferans (Fenchel, 1987). Giant sarcodines of the class Xenophyophorea have also been found on deep seamounts, where they alter hydrodynamic conditions and provide deep-sea metazoans with substrate, food, and refuge, while enhancing benthic diversity (Levin et al., 1986).

In the sulfureta, numerous sulfur bacteria are found along with many other chemolithotrophs. Here as many as 50 species of ciliates, including anaerobic forms, are found that feed on sulfur bacteria, cyanobacteria, diatoms, and flagellates (Fenchel, 1987).

5. Other Surfaces

Fenchel (1987) points out the importance of detrital (other than marine snow) and various solid surfaces to marine protozoa. Particulate material composed of the remains of macrophytes, seagrasses, mangrove debris, larger algae, dead phytoplankton, and metazoans are found in shallow and deeper sediments. Associated with this material are large amounts of bacteria, heterotrophic flagellates (bodonids, colorless euglenids, and chrysomonads), and ciliates (scuticociliates, hypotrichs). Inanimate solid surfaces (e.g., boat surfaces; Sieburth, 1979) are found to harbor bodonids; euglenids; amebae; foraminiferans; heliozoans; ciliates, including peritrichs (e.g., *Zoothamium* and *Vorticella*); heterotrichs; folliculinids; suctorians, dysterids (which feed on filamentous bacteria and cyanobacteria); hypotrichs, amphileptids; and others (reviewed in Sieburth, 1979; Fenchel, 1987).

6. Living Surfaces

Like most other surfaces found in the marine environment, the surfaces of living organisms support microbial life. These surfaces include aquatic vegetation, such as seaweeds (kelp, sea lettuce, etc.) seagrasses (eel and turtle grass), and Chordgrass (Sieburth, 1979; Sieburth and Thomas, 1973; Fenchel, 1970), as well as animal surfaces (Sieburth, 1975). Diatoms are known to be colonized by choanoflagellates (Taylor, 1982; Taylor, Chapter 9, this volume); vorticellids settle on bryozoans (Sieburth, 1975, 1979) and copepods (Herman and Mikwisky, 1964; Hirche, 1974); and suctorians attach to blue pontellid copepods (Sieburth et al., 1976) as well as to egg masses, larval forms, chitinous skeletons, shells, and the like (Sieburth, 1975). Large mammals, such as whales, also serve as a substrate for innumerable protozoans. Evans et al. (1986) recently found ciliates endemic to the baleen plates of whales.

7. Hydrodynamic Habitats

Ocean mixing processes, microscale to macroscale, such as upwelling and downwelling, Langmuir circulation cells, oceanic fronts (Bowman and Esaias, 1978), ocean currents (e.g., Gulf Stream and Kirushio), gyres, and the like, control to some extent microbial, including protistan, distributions. Such "habitats" are rarely considered with respect to protistan trophodynamics, yet they must be if we are to gain a true working understanding of the ecology of the seas. Hanson et al. (1986) and Roman et al. (1985) have begun to examine microbial activity and microzooplankton distribution patterns, respectively, associated with warm core and cold core Gulf Stream rings. These kinds of studies need to expand further into examination of protozoa. Additionally, microbes associated with ice shelves, edges, pack ice, and the like, should be studied with a design to quantify the significance of protozoan productivity. Some studies of protista from these habitats can be found (DeLaca et al., 1980; Lipps et al., 1977; Heinbokel and Coats, 1984, 1985, 1986; Buck and Garrison, 1983; Smith and Nelson, 1985; Nelson et al., 1984; Taylor and Lee, 1971; Silver et al., 1980; Takahashi, 1981; Marra et al., 1982; Hada, 1970; Buck, 1981; Brockel, 1981; Balech and El-Sayed, 1965) but represent only the beginning of the work necessary to understand the microbial food webs of these regions.

ACKNOWLEDGMENTS

This work was supported in part by grants from the Hudson River Foundation for Science and Environmental Research and New York State Sea Grant (National Oceans and Atmospheric Administration) and by a State University of New York at Purchase, President's Junior Faculty Development Award. I am indebted to my sister Clare for her painstaking and thorough typing and retyping of this manuscript, to Peter Catanese for his help with wordprocessing related matters, to John J. Lee for providing me access to his excellent reference collection, and to Art Repak for his help in the computer-aided alphabetizing of this chapter's references. I give special thanks to Gene Small for his exhaustive review of this chapter and for his expert assistance with regards to ciliate systematics.

REFERENCES

Aaronson, S. (1980) Descriptive biochemistry and physiology of the Chrysophyceae. In Levandowsky, M., and Hutner, S.H. (eds.), *Biochemistry and Physiology of Protozoa*, Vol. 2. Orlando, Fla.: Academic Press.

Adams, J.A., and Steele, J.H. (1966) Shipboard experiments on the feeding of *Calanus finmarchicus*. In Barnes, H. (ed.), *Some Contemporary Studies in Marine Science*. London: Allen and Unwin, pp. 19–36.

Admiraal, W., and Venekamp, L.A.H. (1986) Significance of tintinnid grazing during blooms of *Phaeocystis pouchetii* (Haptophyceae) in Dutch coastal waters. *Neth. J. Sea Res*. 20:61–66.

Allbright, L.J., Sherr, E.B., Sherr, B.F., and Fallon, R.D. (1987) Grazing of ciliated protozoa on free and particle-attached bacteria. *Mar. Ecol. Prog. Ser.* 38:125–129.

Alldredge, A.L. (1972) Abandoned larvacean houses: a unique food source in the pelagic environment. *Science* 177:885–887.

Alldredge, A.L. (1976) Discarded appendicularian houses as sources of food, surface habitats and particulate organic matter in planktonic environments. *Limnol. Oceanogr.* 21:14–23.

Alldredge, A.L. (1979) The chemical composition of macroscopic aggregates in two neritic seas. *Limnol. Oceanogr.* 24:855–866.

Alldredge, A.L., and Cohen, Y. (1987) Can microscale chemical patches persist in the sea? Microelectrode study of marine snow, fecal pellets. *Science* 235:689–691.

Alldredge, A.L., Cole, J.J., and Caron, D.A. (1986) Production of heterotrophic bacteria inhabiting macroscopic organic aggregates (marine snow) from surface waters. *Limnol. Oceanogr.* 31:68–78.

Alldredge, A.L., and Cox, J.L. (1982) Primary productivity and chemical composition of marine snow in surface waters of the Southern Calif. Bight. *J. Mar. Res.* 40:517–527.

Allen, R.D. (1974) Food vacuole membrane growth with microtubule-associated membrane transport in *Paramecium*. *J. Cell. Biol.* 63:904–922.

Allen, R.D. (1978) Membranes of ciliates: ultrastructure, biochemistry and fusion. In Poste, G., and Nicholson, G.L. (eds.), *Membrane Fusion*. Amsterdam: Elsevier/North Holland, pp. 657–763.

Allen, R.D., and Fok, A.K. (1980) Membrane recycling and endocytosis in *Paramecium* confirmed by horseradish peroxidase pulse-chase studies. *J. Cell. Sci.* 45:131–145.

Allen, R.D., and Staehelin, L.A. (1981) Digestive system membranes: freeze-fracture evidence for differentiation and flow in *Paramecium*. *J. Cell. Biol.* 89:9–20.

Andersen, P., and Fenchel, T. (1985) Bacterivory by microheterotrophic flagellates in seawater samples. *Limnol. Oceanogr.* 30:198–202.

Andersen, P., and Sorensen, H.M. (1986) Population dynamics and trophic coupling in pelagic microorganisms in eutrophic coastal waters. *Mar. Ecol. Prog. Ser.* 33:99–109.

Anderson, G.C. (1965) Fractionation of phytoplankton communities off the Washington and Oregon coasts. *Limnol. Oceanogr.* 10:477–480.

Anderson, O.R. (1976a) Ultrastructure of a colonial radiolarian *Collozoume inerme* and a cytochemical determination of the role of its zooxanthellae. *Tissue Cell,* 8:195–208.

Anderson, O.R. (1976b) A cytoplasmic fine-structure study of two spumellarian radiolaria and their symbionts. *Mar. Micropaleontol.* 1:81–99.

Anderson, O.R. (1976c) Fine structure of a collodarian radiolarian (*Sphaerozoum punctatum* Muller 1858) and cytoplasmic changes during reproduction. *Mar. Micropaleontol.* 1:287–297.

Anderson, O.R. (1977) Cytoplasmic fine structure of nassellarian radiolaria. *Mar. Micropaleontol.* 2:251–264.

Anderson, O.R. (1978a) Light and electron microscopic observations of feeding behavior, nutrition, and reproduction in laboratory cultures of *Thalassicolla nucleata*. *Tissue Cell* 10:401–412.

Anderson, O.R. (1978b) Fine structure of a symbiont-bearing colonial radiolarian, *Collosphaera globularis,* and ^{14}C isotopic evidence for assimilation of organic substances from its zooxanthellae. *J. Ultrastruct. Res.* 62:181–189.

Anderson, O.R. (1980) Radiolaria. In Levandowsky, M., and Hutner, S. (eds.), *Biochemistry and Physiology of Protozoa,* 2nd ed., Vol. 3. Orlando, Fla.: Academic.

Anderson, O.R. (1981) Radiolarian fine structure and silica deposition. In Simpson, T., and Volcani, B. (eds.), *Silicon and Siliceous Structures in Biological Systems*. Heidelberg: Springer-Verlag, pp. 347–379.

Anderson, O.R. (1983a) *Radiolaria*. New York: Springer-Verlag, 355 pp.

Anderson, O.R. (1983b) The radiolarian symbiosis. In Goff, L.J. (ed.), *Algal Symbiosis*. Cambridge University Press, pp. 69–89.

Anderson, O.R., and Bé, A.W.H. (1976a) A cytochemical fine structure study of phagotrophy in a planktonic foraminifer, *Hastigerina pelagica* (d'Orbigny). *Biol. Bull.* 151:437–449.

Anderson, O.R., and Bé, A.W.H. (1976b) The ultrastructure of a planktonic foraminifer *Globigerinoides sacculifer* (Brady) and its symbiotic dinoflagellates. *J. Foram. Res.* 6:1–21.

Anderson, O.R., and Bé, A.W.H. (1978) Recent advances in foraminiferan fine structure research. In Hedley, R.H., and Adams, C.G. (eds.), *Foraminifera,* Vol. 3. London: Academic Press, pp. 121–202.

Anderson, O.R., and Bennett, P. (1985) A conceptual and quantitative analysis of skeletal morphogenesis in living species of solitary radiolaria: *Euchitonia elegans* and *Spongaster tetras*. *Mar. Micropaleontol.* 9:441–454.

Anderson, O.R., and Botfield, M. (1983) Biochemical and fine structure evidence for cellular specialization in a large spumellarian Radiolarian *Thalassicolla nucleata*. *Mar. Biol.* 72:235–241.

Anderson, O.R., Spindler, M., Bé, A.W.H., and Hemleben, C. (1979) Trophic activity of planktonic foraminifera. *J. Mar. Biol. Assoc. U.K.* 59:791–799.

Anderson, O.R., Swanberg, N.R., and Bennett, P. (1983) Assimilation of symbiont-derived photosynthates in some solitary and colonial radiolaria. *Mar. Biol.* 77:265–269.

Anderson, O.R., Swanberg, N.R., and Bennett, P. (1984) An estimate of predation rate and relative preference for algal versus crustacean prey by a spongiose skeletal radiolarian. *Mar. Biol.* 78:205–207.

Anderson, O.R., Swanberg, N.R., and Bennett, P. (1985) Laboratory studies of the ecological significance of host–algal nutritional associations in solitary and colonial radiolaria. *J. Mar. Biol. Assoc. U.K.* 65:262–272.

Andersson, A., Larsson, U., and Hagstrom, A. (1986) Size-selective grazing by a microflagellate on pelagic bacteria. *Mar. Ecol. Prog. Ser.* 33:51–57.

Antia, N.J. (1980) Nutritional physiology and biochemistry of marine Cryptomonads and Chrysomonads. In Levandowsky, M., and Hutner, S.H. (eds.), *Biochemistry and Physiology of Protozoa,* Vol. 3. Orlando, Fla.: Academic Press, pp. 67–115.

Archbold, J.H.G., and J. Bérger (1985) A qualitative assessment of some metazoan predators of *Halteria grandinella*, a common freshwater ciliate. *Hydrobiologia* 126:97–102.

Azam, F., Fenchel, T., Field, J.G., Gray, J.S., Meyer-Reil, L.A., and Thingstad, F. (1983) The ecological role of water-column microbes in the sea. *Mar. Ecol. Prog. Ser.* 10:257–263.

Baier, R.E. (1972) Organic films on natural waters: their retrieval, identification and modes of elimination. *J. Geophys. Res.* 77:5062–5075.

Baier, R.E., Goupil, D.W., Perlmutter, S., and King, R. (1974) Dominant chemical composition of sea-surface films, natural slicks and foams. *J. Rech. Atmos.* 8:571–600.

Balech, E., and El-Sayed, S.Z. (1965) Microplankton of the Weddell Sea. *Antarctic Res. Ser.* 5:107–124.

Ballard, R.D. (1983) *Exploring our Living Planet.* Washington, D.C.: National Geographic Society, 366 pp.

Ballard, R.D. (1977) Notes on a major geographic find. *Oceanus* 20:35–44.

Banner, F.T. (1971) A new genus of the Planorbulinidae, an endoparasite of another foraminifer. *Rev. Esp. Micropaleontol.* 3:113–128.

Banse, K. (1982a) Mass-scaled rates of respiration and intrinsic growth in very small invertebrates. *Mar. Ecol. Prog. Ser.* 9:281–297.

Banse, K. (1982b) Cell volume, maximal growth rates of unicellular algae and ciliates, and the role of ciliates in the marine pelagial. *Limnol. Oceanogr.* 27:1059–1071.

Bardele, C. (1974) Transport of materials in the Suctorian tentacle. Symp. Soc. Exp. Biol. #28. Transport at the cellular level. Cambridge University Press, pp. 191–208.

Bardele, C. (1976) Particle movement in Heliozoan axopods associated with lateral displacement of highly ordered membrane domains. *Z. Naturforsch.* 31:189–194.

Barker, H.A. (1935) The culture and physiology of the marine dinoflagellates. *Arch. Microbiol.* 6:157–181.

Barlow, R.G., Burkill, P.H., and Mantoura, R.F.C. (1988) Grazing and degradation of algal pigments by the marine protozoan *Oxyrrhis marina*. *J. Exp. Mar. Biol. Ecol.* 119:119–129.

Barnes, R.S.K., Sattelle, D.B., Everton, I.J., Nicholas, W., and Scott, D.H. (1976) Intertidal sands and interstitial fauna associated with different stages of salt-marsh development. *Est. Coast. Mar. Sci.* 4:497–511.

Baross, J.A., and Demming, J.W. (1985) The role of bacteria in the ecology of black smoker environments. In Jones, M.L. (ed.), *The Hydrothermal Vents of the Eastern Pacific: An Overview. Bull. Biol. Soc. Washington* 6:355–371.

Bé, A.W.H. (1982) Biology of planktonic Foraminifera. In Broadhead, T.W. (ed.), *Foraminifera: Notes for a Short Course*. Univ. Tennessee Dept. Geol. Sci. Studies in Geology 6.

Bé, A.W.H., Caron, D.A., and Anderson, O.R. (1981) Effects of feeding frequency on life processes of the planktonic foraminifera *Globigerinoides sacculifer* (Brady) in laboratory culture. *J. Mar. Biol. Assoc. U.K.* 61:257–277.

Bé, A.W.H., Hemleben, C., Anderson, O.R., Spindler, M., Hacunda, J., and Tuntivate-Choy, S. (1977a) Laboratory and field observations of living planktonic foraminifera. *Micropaleontology* 23:155–179.

Bé, A.W.H., Hemleben, C., Anderson, O.R., Spindler, M., Hacunda, J., and Tuntivate-Choy, S. (1977b) Laboratory and field observations of living planktonic foraminifera. *Micropaleontology* 25:294–307.

Bé, A.W.H., and Hutson, W.H. (1977) Ecology of planktonic foraminifera and biogeographic patterns of life and fossil assemblages in the Indian Ocean. *Micropaleontology* 23:369–414.

Beaver, J.R., and Crisman, T.L. (1982) The trophic response of ciliated protozoans in freshwater lakes. *Limnol. Oceanogr.* 27:246–253.

Beers, J.R., and Stewart, G.L. (1967) Micro-zooplankton in the eutrophic zone at five locations across the California current. *J. Fish. Res. Bd. Can.* 24:2053–2068.

Beers, J.R., and Stewart, G.L. (1971) Microzooplankters in the plankton communities of the upper waters of the eastern tropical Pacific. *Deep-Sea Res.* 18:861–883.

Beers, J.R., Reid, F.M.H., and Stewart, G.L. (1980) Microplankton population structure in southern California nearshore waters in late spring. *Mar. Biol.* 60:209–226.

Beers, J.R., Trent, J.D., Reid, F.M.H., and Shanks, A.L. (1986) Macroaggregates and their phytoplanktonic components in the Southern California Bight. *J. Plankton Res.* 8:475–487.

Belar, K. (1923) Untersuchungen an *Actinophrys sol* Ehrenberg. I. Die morphologie des form wechsels. *Archiv. Protistenk* 46:1–96.

Belar, K. (1924) Untersuchungen an *Actinephrys sol* Ehrenberg. II. Béitrage zur Physiologie des Formwechsaels. *Archiv. Protistenk* 48:371–435.

Berger, J.D., and Pollock, C. (1981) Kinetics of food vacuole accumulation and loss in *Paramecium tetraurelia*. *Trans. Amer. Microsc. Soc.* 100:120–133.

Berk, S.G., Brownlee, D.C., Heinle, D.R., Kling, H.J., and Colwell, R.R. (1977) Ciliates as a food source for marine planktonic copepods. *Microbiol. Ecol.* 4:27–40.

Berk, S., Colwell, R.R., and Small, E.B. (1976) A study of feeding responses to bacterial prey by estuarine ciliates. *Trans. Amer. Microsc. Soc.* 95:514–520.

Berman, T., Nawrocki, N., Taylor, G.T., and Karl, D.M. (1987) Nutrient flux between bacteria, bacterivorous nanoplanktonic protists and algae. *Mar. Microb. Food Webs* 2:69–81.

Biecheler, B. (1952) Recherches sur les Peridiniens. *Bull. Biol. Fr. Belg.* 36(S):1–149.

Bilyard, G.R. (1974) The feeding habits and ecology of *Dentalium entale stimpsoni* Henderson. *Veliger* 17:126–138.

Blackbourn, D.J. (1974) The feeding biology of tintinnid protozoa and some other inshore microzooplankton. Ph.D. thesis. University of British Columbia (Vancouver), 224 pp.

Blegved, H. (1928) Quantitative investigations of bottom invertebrates in the Limfjord 1910–1927, with special reference to plaice food. *Rep. Danish Bio. Sta.* 34:33–52.

Bock, W.D. (1969) *Thalassia testudinum*, a habitat and means of dispersal for shallow water benthonic foraminifera. *Gulf Coast Assoc. Geol. Soc. Trans.* 19:337–340.

Boltovskoy, D. (1981) Radiolaria. In Boltovskoy, D. (ed.), *Atlas of the Zooplankton of the Southwestern Atlantic Ocean and Methods in Marine Zooplankton Research*. Instituto Nacional de Investigacion y Desarrolo Pesquero Mar del Plata, Argentina, pp. 261–316.

Boltovskoy, E., and Wright, R. (1976) *Recent Foraminifera*, The Hague: Junk.

Borass, M.E., Estep, K.W., Johnson, P.W., and Sieburth, J. McN. (1988) Phagotrophic phototrophs: the ecological significance of mixotrophy. *J. Protozool.* 35:249–252.

Borror, A.C. (1963) Morphology and ecology of the benthic ciliated protozoa of Alligator Harbor, Florida. *Arch. Protistenk.* 106:465–534.

Borror, A.C. (1968) Ecology of interstitial ciliates. *Trans. Amer. Microsc. Soc.* 87:233–243.

Borror, A.C. (1980) Spatial distribution of marine ciliates: micro-ecologic and biogeographic aspects of protozoan ecology. *J. Protozool.* 27:10–13.

Børsheim, K.Y. (1984) Clearance rates of bacteria-sized particles by freshwater ciliates, measured with nondisperse fluorescent latex beads. *Oecologia* (Berl.) 63:286–288.

Bovee, E.C., and Cordele, D.C. (1971) Feeding on gastrotrichs by the heliozoan *Actinophyrys sol. Trans. Amer. Microsc. Soc.* 90:365–369.

Bowman, M.J., and Esaias, W.E. (1978) *Oceanic Fronts and Coastal Processes*. Berlin: Springer-Verlag.

Brand, T.E., and Lipps, J.H. (1982) Foraminifera in the trophic structure of shallow-water antarctic marine communities. *J. Foram. Res.* 12:96–104.

Brandt, K. (1885) Die koloniebildenden Radiolarien (Sphaerozoeen) des Golfes von Neapel und der angrenzenden Meeresabschnitte. *Monogr. Fauna Flora Golfes Neapel* 13:1–272.

Brandt, K. (1902) Béitrage zur Kenntnis der Colliden. *Arch. Protistenk.* 1:59–88.

Brockel, K., von (1981) The importance of nanoplankton within the pelagic Antarctic ecosystem. *Kiel. Meeresforsch.* 5:61–67.

Bromley, R.G., and Nordmann, E. (1971) Maastrichtian adherent foraminifera encircling clionid pores. *Bull. Geol. Soc. Den.* 20:362–268.

Brooker, B.E. (1971) Fine structure of *Bodo saltans* and *Bodo caudatus* (Zoomastigophora: Protozoa) and their affinities with the Trypanosomatidae. *Bull. Br. Mus. Nat. Hist.* 22:81–102.

Buchanan, J.B., and Hedley, R.H. (1960) A contribution to the biology of *Astrorhiza limicola* (foraminifera). *J. Mar. Biol. Assoc. U.K.* 39:549–560.

Buck, K.R. (1981) A study of choanoflagellates (Acanthoecidae) from the Weddell Sea, including a description of *Diaphanoeca multiannulata* n. sp. *J. Protozool.* 28:47–54.

Buck, K.R., and Garrison, D.L. (1983) Protists from the ice-edge region of the Weddell Sea. *Deep-Sea Res.* 30:16–177.

Burkill, P.H., Mantoura, R.F.C., Llewellyn, C.A., and Owens, N.J.P. (1987) Microzooplankton grazing and selectivity of phytoplankton in coastal waters. *Mar. Biol.* 93:581–590.

Burney, C.M., Davis, P.G., Johnson, K.M., and Sieburth, J. McN. (1982) Diel relationships of microbiol trophic groups and *in situ* dissolved carbohydrate dynamics in the Caribbean Sea. *Mar. Biol.* 67:311–322.

Burzell, L.A. (1973) Observations on the proboscis-cyto-pharynx and flagella in *Rhynchomonas metabolita* Pshenen 1964 (Zoomastigophora Bodonidae). *J. Protozool.* 20:385–393.

Burzell, L.A. (1975) Fine structure of *Bodo curvifilus* Griessmann. *J. Protozool.* 22:35–39.

Butterfield, C.T., Purdy, W.C., and Theriault, E.J. (1931) Studies on natural purification in polluted waters. IV. The influence of plankton on the biochemical oxidation of organic matter. *Public Health Rep.* 46:393–426.

Buzas, M.A. (1978) Foraminifera as prey for benthic deposit feeders: results of predator exclusion experiments. *J. Mar. Res.* 36:617–625.

Buzas, M.A. (1982) Regulation of foraminiferal densities by predation in the Indian River, Florida. *J. Foram. Res.* 12:66–71.

Buzas, M.A., Smith, R.K., and Beem, K.A. (1977) Ecology and systematics of foraminifera in two *Thalassia* habitats, Jamaica, West Indies. *Smithsonian Contrib. Paleobiol.* 31:1–139.

Cachon, J. (1964) Contribution a l'etude des peridiniens parasites. Cytologie, cycles evolutifs. *Ann. Sci. Nat. Zool. Biol. Animal.* 6:1–158.

Cachon, P.J., and Cachon, M. (1974) Le systeme stomatopharyngien de *Kofoidinium* Pavillard. Comparisons avec celui divers Peridiniens libris et parasites. *Protistologica* 10:217–222.

Cachon, J., and Cachon M. (1987) Parasitic dinoflagellates. In: Taylor, F.J.R. (ed.) *The Biology of Dinoflagellates*. Blackwell Sci Pub., Oxford. pp. 571–610.

Cachon, P.J., Cachon, M., Tilney, L., and Tilney, M. (1977) Movement generated by interactions between the dense material at the ends of microtubules and non-actin-containing microfilaments in *Sticholonche zanclea*. *J. Cell. Biol.* 72:314–338.

Campbell, A.S. (1926) The cytology of *Tintinnopsis nucula* (FOL) Laack with an account of its neuromotor apparatus, division and a new intranuclear parasite. *Univ. Calif. Publ. Zool.* 29:179–236.

Campbell, A.S. (1927) Studies on the marine ciliate *Favella* (Jørgensen) with special regard to the neuromotor apparatus and its role in the formation of the lorica. *Univ. Calif. Publ. Zool.* 29:429–452.

Campell, L., and Carpenter, E.J. (1986) Estimating the grazing pressure of heterotrophic nanoplankton on *Synechococcus* spp. using the sea water dilution and selective inhibitor techniques. *Mar. Ecol. Prog. Ser.* 33:121–129.

Capriulo, G.M. (1982) Feeding of field collected tintinnid microzooplankton on natural food. *Mar. Biol.* 71:73–86.

Capriulo, G.M., and Carpenter, E.J. (1980) Grazing by

35 to 202 μm microzooplankton in Long Island Sound. *Mar. Biol.* 56:319–326.

Capriulo, G.M., and Carpenter, E.J. (1983) Abundance, species composition and feeding impact of tintinnid microzooplankton in central Long Island Sound. *Mar. Ecol. Prog. Ser.* 10:277–288.

Capriulo, G.M., Gold, K., and Okubo, A. (1982) Evolution of the lorica in tintinnids: a possible selective advantage. *Ann. Inst. Oceanogr.* 58S:319–324.

Capriulo, G.M., and Ninivaggi, D.V. (1982) A comparison of the feeding activities of field collected tintinnids and copepods fed identical natural particle assemblages. *Ann. Inst. Oceanogr.* 58(S):325–334.

Capriulo, G.M., Taveras, J., and Gold, K. (1986) Ciliate feeding: effect of food presence or absence on occurrence of striae in tintinnids. *Mar. Ecol. Prog. Ser.* 30:145–158.

Capriulo, G.M., Schreiner, R.A., and Dexter, B.L. (1988) Differential growth of *Euplotes vannus* fed fragmented versus unfragmented *Skeletonema costatum*. *Mar. Ecol. Prog. Ser.* 47:205–209.

Capriulo, G.M., Sherr, E.B., and Sherr, B.F. (1989) Trophic behavior and related community feeding activities of heterotrophic marine protists. NATO Advanced Study Institute series on The Role of Protozoa in Marine Ecological Processes. Springer-Verlag, Germany (in press).

Carlucci, A.F., Craven, D.B., and Henrichs, S.M. (1985) Surface-film microheterotrophs: amino acid metabolism and solar radiation effects on their activities. *Mar. Biol.* 85:13–22.

Caron, D.A. (1983) Technique for enumeration of heterotrophic and phototrophic nanoplankton using epifluorescence microscopy, and comparison with other procedures. *Appl. Environ. Microbiol.* 46:491–498.

Caron, D.A. (1987) Grazing of attached bacteria by heterotrophic microflagellates. *Microbiol. Ecol.* 13:203–218.

Caron, D.A., and Bé, A.W.H. (1984) Predicted and observed feeding rates of the spinose planktonic foraminifer *Globigerinoides sacculifer*. *Bull. Mar. Sci.* 35:1–10.

Caron, D.A., Davis, P.G., Madin, L.P., and Sieburth, J. McN. (1982) Heterotrophic bacteria and bacterivorous Protozoa in oceanic macroaggregates. *Science* 218:795–797.

Caron, D.A., Faber, Jr., W.W., and Bé, A.W.H. (1987a) Effects of temperature and salinity on the growth and survival of the planktonic foraminifer *Globigerinoides sacculifer*. *J. Mar. Biol. Assoc. U.K.* 67:323–341.

Caron, D.A., Faber, Jr., W.W., and Bé, A.W.H. (1987b) Growth of the spinose planktonic foraminifer *Orbulina universa* in laboratory culture and the effect of temperature on life processes. *J. Mar. Biol. Assoc. U.K.* 67:343–358.

Caron, D.A., Goldman, J.C., Andersen, O.K., and Dennett, M.R. (1985) Nutrient cycling in a microflagellate food chain: 2. Population dynamics and carbon cycling. *Mar. Ecol. Prog. Ser.* 24:243–254.

Caron, D.A., and Goldman, J.C. (1988) Experimental demonstration of the roles of bacteria and bacterivorous protozoa in plankton nutrient cycles. *Hydrobiologia* 159:27–40.

Carpenter, E.J., Harbison, G.R., Madin, L.P., Swanberg, N.R., Biggs, D.C., Hulburt, E.M., McAlister, V.L., and McCarthy, J.J. (1977) *Rhizosolenia* mats. *Limnol. Oceanogr.* 22:739–741.

Chapman-Andresen, C. (1973) Endocytotic processes. In Jeon, K.W. (ed.), *The Biology of Amoeba*. Orlando, Fla.: Academic Press.

Chatton, E. (1952) Classe des Dinoflagelles du Peridiniens. In Grasse, P.P. (ed.), *Zoologie, Anatomie Systematique, Biologie, Phylogenie Protozaires: Generalites Flagelles*, Vol. 1. Paris: Masson, pp. 309–390.

Chen, Y.T. (1950) The biology of *Peranema trichophorum*. *Quart. J. Microsc. Sci.* 91:279–308.

Chretiennot, M.J. (1974) Nanoplancton de flaques supralittorales de la region de Marseille. *Protistologica* 10:477–488.

Christiansen, B. (1964) *Spiculosiphon radiata*, a new foraminifera from northern Norway. *Astarte* 25:1–8.

Christiansen, B. (1971) Notes on the biology of foraminifera. Vie et Milieu. Troisième Sym. Européen de Biologie Marine S22:465–478.

Cifelli, R., and Sachs, Jr., K.N. (1966) Abundance relationships of planktonic Foraminifera and Radiolaria. *Deep-Sea Res.* 13:751–753.

Coats, D.W. (1988) *Duboscquella cachoni* n. sp., a parasitic dinoflagellate lethal to its tintinnine host *Eutintinnus pectinis*. *J. Protozool.* 35:607–617.

Coffin, R.B., and Sharp, J.H. (1988) Microbiol trophodynamics in the Delaware Estuary. *Mar. Ecol. Prog. Ser.* 41:253–266.

Corliss, J.O. (1979) *The Ciliated Protozoa*, 2nd ed. Elmsford, N.Y.: Pergamon.

Cosgrove, W.B., and Swanson, B.K. (1952) Growth of *Chilomonas paramecium* in single organic media. *Physiol. Zool.* 25:287–292.

Curds, C.R., and Cockburn, A. (1968) Studies on the growth and feeding of *Tetrahymena pyriformis* in axenic and monaxenic culture. *J. Gen. Microbiol.* 54:343–358.

Cynar, F.J., Estep, K.W., and Sieburth, J. McN. (1985) The detection and characterization of bacteria-sized protists in "protist free" filtrates and their potential impact on experimental marine ecology. *Microb. Ecol.* 11:281–288.

Daggett, P., and Nerad, T.A. (1982) Axenic cultivation of *Bodo edax* and *Bodo ancinatus* and some observations on feeding rate in monoxenic culture. *J. Protozool.* 29:290–291.

Dahlback, B., Gunnarsson, L.A.H., Hermansson, M., and Kjelleberg, S. (1982) Microbial investigations of surface microlayers, water column, ice and sediment in the Arctic Ocean. *Mar. Ecol. Prog. Ser.* 9:101–109.

Daley, R.J., Morris, G.P., and Brown, S. (1973) Phagotrophic ingestion of a blue green alga by *Ochromonas*. *J. Protozool.* 20:58–61.

Danso, S.K.A., and Alexander, M. (1975) Regulation of predation by prey density: the protozoan–*Rhizobium* relationship. *Appl. Microbiol.* 29:515–521.

Davis, P.G., Caron, D.A., and Sieburth, J. McN. (1978) Oceanic amoebae from the North Atlantic: culture, distribution and taxonomy. *Trans. Amer. Microsc. Soc.* 97:73–88.

Davis, P.G., Caron, D.A., Johnson, P.W., and Sieburth, J. McN. (1985) Phototrophic and apochlorotic components of picoplankton and nanoplankton in the North Atlantic: geographic, vertical seasonal and diel distributions. *Mar. Ecol. Prog. Ser.* 21:15–26.

Davis, P., and Sieburth, J. McN. (1984a) Differentiation and characterization of individual phototrophic and heterotrophic microflagellates by sequential epifluorescence and electron microscopy. *Trans. Amer. Microsc. Soc.* 103:221–227.

Davis, P.G., and Sieburth, J. McN. (1984b) Estuarine and oceanic microflagellate predation of actively growing bacteria: estimation by frequency of dividing-divided bacteria. *Mar. Ecol. Prog. Ser.* 19:237–246.

Davis, P.G., Sieburth, J. McN. (1982) Differentiation of the phototrophic and heterotrophic nanoplankton population in marine waters by epifluorescence microscopy. *Ann. Inst. Oceanogr.* (Paris) 58(S):249–260.

Davoll, P.J., and Silver, M.W. (1986) Marine snow aggregates: life history sequence and microbial community of abandoned larvacean houses from Monterey Bay, California. *Mar. Ecol. Prog. Ser.* 33:111–120.

Dayton, P.K. (1971) Competition, disturbance and community organization: the provision and subsequent utilization of space in a rocky intertidal community. *Ecol. Monogr.* 41:351–389.

Deason, E.E. (1980) Potential effect of phytoplankton colony breakage on the calculation of zooplankton filtration rates. *Mar. Biol.* 57:279–286.

DeLaca, T.E. (1982) Use of dissolved amino acids by the foraminifer *Notodendrodes antarctikos*. *Am. Zool.* 2:683–690.

DeLaca, T.E., Karl, D.M., and Lipps, J.H. (1981) Direct use of dissolved organic carbon by agglutinated benthic foraminifera. *Nature* 289:287–289.

DeLaca, T.E., and Lipps, J.H. (1972) The mechanism and adaptive significance of attachment and substrate pitting in the foraminiferan *Rosalina globularis* (d'Orbigny). *J Foramin. Res.* 2:68–72.

DeLaca, T.E., Lipps, J.H., and Hessler, R.R. (1980) The morphology and ecology of a new large agglutinated Antarctic foraminifer (Textulariina: Notodendrodidae nov.). *Zoo. J. Linn. Soc.* 69:205–224.

DeSouza-Lima, Y., and Romano, J.C. (1983) Ecological aspects of the surface microlayer. I. ATP, ADP and AMP contents, and energy charge ratios of microplanktonic communities. *J. Exp. Mar. Biol. Ecol.* 70:107–122.

Dexter, B.L. (1984) Developmental grazing capabilities of *Pseudocalnus* sp. and *Acartia clausi* (CI–adult). Ph.D. dissertation, Oregon State University, Corvallis, 166 pp.

Dietz, A.S., Albright, L.J., and Tuominen, T. (1976) Heterotrophic activities of bacterioneuston and bacterioplankton. *Can. J. Microbiol.* 22:1699–1709.

Dodge, J.D., and Crawford, R.M. (1970) The morphology and fine structure of *Ceratium hirundinella* (Dinophyceae). *J. Phycol.* 6:137–149.

Dodge, J.D., and Crawford, R.M. (1974) Fine structure of the dinoflagellate *Oxyrrhis marina* III. Phagotrophy. *Protistologica* 10:239–244.

Donaghay, P.L. (1980) Experimental and conceptual approaches to understanding algal-grazer interactions. Ph.D. dissertation, Oregon State University, Corvallis, 214 pp.

Dragesco, J. (1962) Capture et ingestion des proies chez les Infusoires cilies. *Bull. Biol. Fr. Belg.* 46:123–167.

Drebes, G. (1969) *Dissodinium pseudocalani* sp. nov. lin parasitischer Dinoflagellat auf Copepodeneiem. *Helg. Meer.* 19:58–67.

Droop, M.R. (1953) Phagotrophy in *Oxyrrhis marina*. *Nature* 172:250–252.

Droop, M.R. (1954) A note on the isolation of small marine algae and flagellates in pure culture. *J. Mar. Biol. Assoc. U.K.* 33:511–514.

Droop, M.R. (1959) Water soluble factors in nutrition of *Oxyrrhis marina*. *J. Mar. Biol. Assoc. U.K.* 38:605–620.

Dubowsky, N. (1974) Selectivity of ingestion and digestion in the chrysomonad flagellate *Ochromonas malhamensis*. *J. Protozool.* 21:295–298.

Ducklow, H.W. (1983) Production and fate of bacteria in the oceans. *BioScience* 33:494–501.

Ducklow, H.W., and Kirchman, D.L. (1983) Bacterial dynamics and distribution during a spring bloom in the Hudson River Plume, USA. *J. Plankton Res.* 5:333–355.

Ducklow, H., and Hill, S.M. (1985) The growth of heterotrophic bacteria in the surface waters of warm core rings. *Limnol. Oceanogr.* 30:239–259.

Duguay, L., and Taylor, D. (1978) Primary production and calcification by the sortid foraminifera *Archaias angulatus* (Fichtel and Moll). *J. Protozool.* 25:356–361.

Dussart, B.M. (1965) Les differentes categories de planction. *Hydrobiologia* 24:72–74.

Edds, K.T. (1981) Cytoplasmic streaming in a heliozoan. *Biosystems* 14:371–376.

Elbourne, C.A. (1966) Some observations on the food of *Cyclops strenuus* (Fischer). *Ann. Mag. Nat. Hist.* 9:227–231.

Elbrachter, M., and Drebes, G. (1978) Life cycles, phylogeny and taxonomy of *Dissodinium* and *Pyrocystis* (Dinophyta). *Helg. Meer.* 31:347–366.

Elliott, A.M., and Clemmons, G.L. (1966) An ultrastructural study of ingestion and digestion in *Tetrahymena pyriformis*. *J. Protozool.* 13:311–323.

Evans, D.A., Small, E.B., and Snyder, R. (1986) Investigation of ciliates collected from the baleen of fin and blue whales. Soc. of Protozool. (Abstract) p. 10.

Endo, Y., Hasumoto, H., and Taniguchi, A. (1983) Microzooplankton standing crop in the western subtropical Pacific off the Bonin Islands in winter, 1980. *J. Oceanogr. Jap.* 39:185–191.

Estep, K.W., Davis, P.G., Keller, M.D., and Sieburth, J. McN. (1986) How important are oceanic algal nanoflagellates in bacterivory? *Limnol. Oceanogr.* 31:646–650.

Fairbanks, R.G., Wiebe, P.H., and Bé, A.W.H. (1980) Vertical distribution and isotopic composition of living planktonic foraminifera in the western North Atlantic. *Science* 207:61–63.

Fauré-Fremiet, E. (1924) Contribution a la connaissance des infusoires planktoniques. *Bull. Biol. Fr. Belg.* (S) 6:1–171.

Fauré-Fremiet, E. (1950) Ecology of ciliate infusoria. *Endeavour* 9:183–187.

Febvre-Chevalier, C., and Febvre, J. (1986) Motility mechanisms in the actinopods (Protozoa). A review with particular attention to axopodial contraction/extension, and movement of nonaction filament systems. *Cell Motility and the Cytoskeleton* 6:198–200.

Fenchel, T. (1967) The ecology of marine microbenthos. I. The quantitative importance of ciliates as compared with metazoans in various types of sediments. *Ophelia* 4:121–137.

Fenchel, T. (1968a) The ecology of marine microbenthos. II. The food of marine benthic ciliates. *Ophelia* 5:73–121.

Fenchel, T. (1968b) The ecology of marine microbenthos. III. The reproductive potential of ciliates. *Ophelia* 5:123–136.

Fenchel, T. (1969) The ecology of marine microbenthos. IV. The structure and function of the benthic ecosystem, its chemical and physical factors and the microfauna communities with special reference to the ciliated protozoa. *Ophelia* 6:1–182.

Fenchel, T. (1970) Studies on the decomposition of organic detritus derived from the turtle grass *Thalassia testudinum*. *Limnol. Oceanogr.* 15:14–20.

Fenchel, T. (1972) Aspects of decomposer food chains in marine benthos. *Verh. Deut. Zool. Ges.* 65:13–22.

Fenchel, T. (1980a) Suspension feeding in ciliated protozoa: structure and function of feeding organelles. *Arch. Protistenk.* 123:239–260.

Fenchel, T. (1980b) Suspension feeding in ciliated protozoa: functional response and particle size selection. *Microbiol. Ecol.* 6:1–11.

Fenchel, T. (1980c) Suspension feeding in ciliated protozoa: feeding rates and their ecological significance. *Microbiol. Ecol.* 6:13–25.

Fenchel, T. (1980d) Relation between particle size selection and clearance in suspension-feeding ciliates. *Limnol. Oceanogr.* 25:733–738.

Fenchel, T. (1982a) Ecology of heterotrophic microflagellates. I. Some important forms and their functional morphology. *Mar. Ecol. Prog. Ser.* 8:211–223.

Fenchel, T. (1982b) Ecology of heterotrophic microflagellates. II. Bioenergetics and growth. *Mar. Ecol. Prog. Ser.* 8:225–231.

Fenchel, T. (1982c) Ecology of heterotrophic microflagellates. III. Adaptations to heterogeneous environments. *Mar. Ecol. Prog. Ser.* 9:25–33.

Fenchel, T. (1982d) Ecology of heterotrophic microflagellates. IV. Quantitative occurrence and importance as bacterial consumers. *Mar. Ecol. Prog. Ser.* 9:35–42.

Fenchel, T. (1982e) The bioenergetics of a heterotrophic microflagellate. *Ann. Inst. Oceanogr.* 58(S):55–60.

Fenchel, T. (1984) Suspended marine bacteria as a food source. In Fasham, M.J. (ed.), *Flows of Energy and Materials in Marine Ecosystems*. New York: Plenum, pp. 301–305.

Fenchel, T. (1986a) Protozoan filter feeding. In Corliss, J.O., and Patterson, D.J. (eds.), *Progress in Protistology*, Vol. 1. Bristol, England: Biopress, pp. 65–114.

Fenchel, T. (1986b) The ecology of heterotrophic microflagellates. In Marshall, K.C. (ed.), *Advances in Microbial Ecology*, Vol. 9. New York: Plenum, pp. 57–97.

Fenchel, T. (1987) *Ecology of Protozoa*. Berlin: Science Tech Pub./Springer-Verlag, 197 pp.

Fenchel, T., and Harrison, P. (1976) The significance of bacterial grazing and mineral cycling for the decomposition of particulate detritus. In Anderson, J.M., and Macfadyen, A. (eds.), *The Role of Terrestrial and Aquatic Organisms in Decomposition Processes*. Oxford: Blackwell, pp. 285–299.

Fenchel, T., and Jorgensen, B.B. (1977) Detritus food chains of aquatic ecosystems: the role of bacteria. In Alexander, M. (ed.), *Advances in Microbial Ecology*, Vol. 1. New York: Plenum, pp. 1–58.

Fenchel, T., and Patterson, D.J. (1986) *Percolomonas cosmopolitus* (Ruinen) N. gen., a new type of filter feeding flagellate from marine plankton. *J. Mar. Biol. Assoc. U.K.* 66:465–482.

Fenchel, T., and Small, E.B. (1980) Structure and function of the oral cavity and its organelles in the hymenostome ciliate *Glaucoma*. *Trans. Amer. Microsc. Soc.* 99:52–60.

Ferguson, R.L., and Rublee, P. (1976) Contribution of bacteria to the standing crop of coastal plankton. *Limnol. Oceanogr.* 21:141–145.

Finlay, B.J. (1978) Community production and respiration by ciliated protozoa in the benthos of a small eutrophic loch. *Freshwater Biol.* 8:327–341.

Finlay, B.J., and Uhlig, G. (1981) Calorific and carbon values of marine and freshwater protozoa. *Helg. Meer.* 34:401–412.

Fisher-Defoy, D., and Hausmann, K. (1977) Untersuchungen zun phagocytose bei Climacostomum *virens*. *Protistologica* 13:459–476.

Foissner, W. (1972) The cytopyge of ciliata. *Acta. Biol. Acad. Sci. Hung.* 23:161–174.

Foissner, W., and Didier, P. (1983) Nahrungsaufnahme, Lebenszyklus und Morphogenese von *Pseudoplatyophyra nana* (Kahl, 1926) (Ciliophora, Colpodida). *Protistologica* 19:103–109.

Fok, A., Lee, Y., and Allen, R.D. (1982) The correlation of digestive vacuole pH and size with the digestive cycle in *Paramecium caudatum*. *J. Protozool.* 29:409–414.

Frey, L.C., and Stoermer, E.F. (1980) Dinoflagellate phagotrophy in the upper Great Lakes. *Trans. Amer. Microsc. Soc.* 99:439–444.

Frost, B.W. (1972) Effects of size and concentration of food particles on the feeding behavior of the marine planktonic copepod *Calanus pacificus*. *Limnol. Oceanogr.* 17:805–815.

Fuhrman, J.A., Ammerman, J.W., and Azam, F. (1980) Bacterioplankton in the coastal euphotic zone: distribution, activity and possible relationships with phytoplankton. *Mar. Biol.* 60:201–207.

Fuhrman, J.A., and Azam, F. (1980) Bacterioplankton secondary production estimates for coastal waters of

British Columbia, Antarctica, and California. *Appl. Environ. Microbiol.* 39:1085–1095.

Fuhrman, J.A., and McManus, G.B. (1984) Do bacteria-sized marine eucaryotes consume significant bacterial production? *Science* 224:1257–1260.

Gaines, G., and Taylor, F.J.R. (1984) Extracellular digestion in marine dinoflagellates. *J. Plankton Res.* 6:1057–1061.

Garrett, W.D. (1965) Collection of slick-forming materials from the sea surface. *Limnol. Oceanogr.* 10:602–605.

Gast, V. (1985) Bacteria as a food source for microzooplankton in the Schlei Fjord and Baltic Sea with special references to ciliates. *Mar. Ecol. Prog. Ser.* 22:107–120.

Gerritsen, J., and Porter, K.G. (1982) The role of surface chemistry in filter feeding by zooplankton. *Science* 216:1225–1227.

Gifford, D.J., Bohrer, R.N., and Boyd, C.M. (1981) Spines on diatoms: do copepods care? *Limnol. Oceanogr.* 26:1057–1061.

Gilbert, J.J. (1968) Dietary control of sexuality in the rotifer *Asplanchna brightwelli* Gosse. *Physiol. Zool.* 41:14–43.

Gilbert, J.J. (1976) Selective cannibalism in the rotifer *Asplanchna sieboldi*: contact recognition of morphotype and clone. *Proc. Nat. Acad. Sci. U.S.A.* 73:3233–3237.

Gilbert, J.J. (1980) Observations on the susceptibility of some protists and rotifers to predation by *Asplanchma girodi. Hydrobiologia* 73:87–91.

Gold, K. (1968) Some observations on the biology of *Tintinnopsis* sp. *J. Protozool.* 15:193–194.

Gold, K. (1969) Tintinnida: feeding experiments and lorica development. *J. Protozool.* 16:507–509.

Gold, K. (1971) Growth characteristics of the mass reared tintinnid *Tintinnopsis beroidea Mar. Biol.* 8:105–108.

Gold, K. (1973) Methods for growing Tintinnida in continuous culture. *Am. Zool.* 13:203–208.

Gold, K. (1979) Scanning electron microscopy of *Tintinnopsis parva*: studies on particle accumulation and the striae. *J. Protozool.* 26:415–419.

Gold, K., Pfister, R.M., and Liguori, V.R. (1970) Axenic cultivation and electron microscopy of two species of choanoflagellida. *J. Protozool.* 17:210–212.

Gold, K., and Pollingher, U. (1971) Microgamete formation and the growth rate of *Tintinnopsis beroidea. Mar. Biol.* 11:324–329.

Goldman, J.C., and Caron, D.A. (1985) Experimental studies on an omnivorous microflagellate: implications for grazing and nutrient regeneration in the marine microbial food chain. *Deep-Sea Res.* 32:899–915.

Gorjacheva, N.V., Zukov, B.F., and Mylnikov, A.P. (1978) Biology of free living Bodonides. In *Biology and Systematics of Lower Organisms*. Transactions of the Institute of Biology of Inland Waters, U.S.S.R., Vol. 35, pp. 29–50.

Graham, A. (1955) Molluscan diets. *Proc. Malacol. Soc. Lond.* 31:144–159.

Green, J. (1954) A note on the food of *Chaetogaster diaphanus. Ann. Mag. Nat. Hist. Ser.* 12:842–844.

Greissman, K. (1914) Uber marine flagellaten. *Arch. Protistenk.* 32:1–78.

Grell, K. (1973) *Protozoology*. Berlin: Springer, 552 pp.

Greuet, C. (1972) La nature trichocystaire due cnidoplaste dans le complexe cnidoplaste nematocyste du *Polykrikos schwartzi* Butschli, *C.R. Hebd. Seances Acad. Sci. Paris*, Ser. D. 275:1239–1242.

Greuet, C., and Hovasse, R. (1977) A propose de la genese des Nematocystes de *Polykrikos schwartzi* Butschli. *Protistologica* 13:145–149.

Guiset, A. (1977) Stomach contents in *Asplanchna* and *Ploesoma Arch. Hydrobiol. Bech.* 8:126–129.

Haas, L. (1982) Improved epifluorescence microscopy for observing planktonic micro-organisms. *Ann. Inst. Oceanogr.* (Paris) 58(S):261–266.

Haas, L.W., and Webb, K.L. (1979) Nutritional mode of several non-pigmented micro-flagellates from the York River Estuary, Virginia. *J. Exp. Mar. Biol. Ecol.* 39:125–134.

Haberey M. (1973a) Die Phagocytose von Oscillatorien durch *Thecamoeba sphaeronucleolus*. I. Lichtoptische Untersuchung. *Arch. Protistenk.* 115:99–110.

Haberey, M. (1973b) Die Phagocytose von Oscillatorien durch *Thecamoeba sphaeronucleolus*. II. Electronenmikroscopische Untersuchung. *Arch. Protistenk.* 115:111–124.

Hackney, C.M., and Butler, R.D. (1981) Tentacle contraction in glycerinated *Discophyra collini* and the localization of HNN-Binding filaments. *J. Cell. Sci.* 47:65–75.

Hada, Y. (1970) The protozoan plankton of the Antarctic and *Subantarctic seas. JARE Sci. Rep. Ser. E.* 31:1–151.

Hamilton, R.D., and Preslan, J.E. (1969) Cultural characteristics of a pelagic marine hymenostome ciliate, *Uronema* sp. *J. Exp. Mar. Biol. Ecol.* 4:90–99.

Hanson, R.B., Pomeroy, L.R., and Murray, R.E. (1986) Microbial growth rates in a cold-core Gulf Stream eddy of the northwestern Sargasso Sea. *Deep-Sea Res.* 33:427–446.

Hanson, R.B., Shafer, D., Ryan, T., Pope, D.H., and Lowery, H.K. (1983) Bacterioplankton in Antarctic ocean waters during late austral winter: abundance, frequency of dividing cells, and estimates of production. *App. Environ. Microbiol.* 45:1622–1632.

Harbison, G.R., Biggs, D.C., and Madin, L.P. (1977) The associations of Amphipoda Hyperiidea with gelatinous zooplankton. II. Associations with Cnidaria, Ctenophora and Radiolaria. *Deep-Sea Res.* 24:465–468.

Hardy, J.T. (1973) Phytoneuston ecology of a temperate marine lagoon. *Limnol. Oceanogr.* 18:525–533.

Hardy, J.T. (1982) The sea surface microlayer: biology, chemistry, and anthropogenic enrichment. *Prog. Oceanogr.* 11:307–328.

Hardy, J.T., and Valett, M. (1981) Natural and microcosm phytoneuston communities of Sequim Bay, Washington. *Estuar. Coast. Shelf Sci.* 12:3–12.

Harris, R.H., and Mitchell, R. (1973) The role of polymers in microbial aggregations. *Ann. Rev. Microbiol.* 27:27–50.

Hartwig, E. (1980) A bibliography of the interstitial ciliates (Protozoa): 1926–1979. *Arch. Protistenk.* 123:422–438.

Harvey, G.W. (1966) Microlayer collection from the sea surface: a new method and initial results. *Limnol. Oceanogr.* 11:608–613.

Harvey, G.W., and Burzell, L.A. (1972) A simple microlayer method for small samples. *Limnol. Oceanogr.* 17:156–157.

Harvey, W., Lion, L.W., and Young, L.Y. (1983) Transport and distribution of bacteria and diatoms in the aqueous surface microlayer of a salt marsh. *Estuar. Coast. Shelf Sci.* 16:543–547.

Harvey, H.W. (1937) Note on selective feeding by *Calanus. J . Mar. Biol. Assoc. U.K.* 22:97–100.

Hausmann, K., and Patterson, D.J. (1982) Feeding in *Actinophrys* II. Pseudopod formation and membrane production during prey capture by a heliozoan. *Cell Motility* 2:9–24.

Hausmann, K., and Peck, R. (1978) Microtubules and microfilaments as major components of a phagocytic apparatus: the cytophrayngeal basket of the ciliate *Pseudomicrothorax dubius. Differentiation* 11:157–167.

Hausmann, K., and Peck, R. (1979) The mode of function of the pharyngeal basket of the ciliate *Pseudomicrothorax dubius. Differentiation* 14:147–158.

Haynes, J. (1965) Symbiosis, wall structure and habitat in foraminifera. *Cont. Cushman Found. Foram. Res.* 16:40–43.

Haynes, J.R. (1981) *Foraminifera*. New York: Wiley.

Hayward, N.J.B., and Haynes, J.R. (1976) *Chlamys opercularis* (Linnaeus) as a mobile substrate for foraminifera. *J. Foram. Res.* 6:30–38.

Hedin, H. (1975) On the ecology of tintinnids on the Swedish west coast. *ZOON* 3:125–140.

Hedin, H. (1976) Microtubules and microfilaments in the tintinnid ciliate *Ptychocylis minor* Jørgensen. *ZOON* 14:3–10.

Hedley, R.H. (1958) A contribution to the biology and cytology of *Haliphysena* (Foraminifera). *Proc. Zool. Soc. Lond.* 130:569–576.

Heinbokel, J.F. (1978a) Studies on the functional role of tintinnids in the Southern California Bight. I. Grazing and growth rates in laboratory cultures. *Mar. Biol.* 47:177–189.

Heinbokel, J.F. (1978b) Studies on the functional role of tintinnids in the Southern California Bight. II. Grazing rates of field populations. *Mar. Biol.* 47:191–197.

Heinbokel, J.F., and Béers, J.R. (1979) Studies on the functional role of tintinnids in the Southern California Bight. III. Grazing and impact of natural assemblages. *Mar. Biol.* 52:23–32.

Heinbokel, J.F., and Coats, D.W. (1984) Reproductive dynamics of ciliates in the ice-edge zone. *Antarct. J. U.S.* 19:111–113.

Heinbokel, J.F., and Coats, D.W. (1985) Ciliates and nanoplankton in Arthur Harbor, Dec. 1984–Jan. 1985. *Antarct. J. U.S.* 19:135–136.

Heinbokel, J.F., and Coats, D.W. (1986) Patterns of tintinnine abundance and reproduction near the edge of seasonal pack-ice in the Weddell Sea, November, 1983. *Mar. Ecol. Prog. Ser.* 33:71–80.

Heinle, D.R., Harris, R.P., Ustach, J.F., and Flemer, D.A. (1977) Detritus as food for estuarine copepods. *Mar. Biol.* 40:341–353.

Hellung-Larsen, P., Lerck, V., and Tommerup, N. (1986) Chemoattraction in *Tetrahymena*: on the role of chemokinesis. *Biol. Bull.* 170:357–367.

Hemleben, C. (1982) Cytologic and ecologic aspects of morphogenesis and structure of recent and fossil protistan and pteropod skeletons. *N. Jb. Geol. Palaont.* (Abh.) 164:83–95.

Hemleben, C., and Spindler, M. (1978) Cytological and ecological aspects of morphogenesis in recent and fossil protistan skeletons. *N. Jb. Geol. Palaontol.* (Abh.) 157:77–85.

Hemleben, C., and Spindler, M. (1983) Recent advances in research on living planktonic foraminifera. In Meulenkamp, J.E. (ed.), *Utretch Micropaliontological Bulletin No. 30. Reconstruction of Marine Paleoenvironments.*

Hemleben, C., Spindler, M., Breitinger, I., and Deusen, W.G. (1985) Field and laboratory studies on the ontogeny and ecology of some globorotalid species from the Sargasso Sea off Bermuda. *J. Foram Res.* 15:254–272.

Herman, S.S., and Mikwisky, J.A. (1964) Infestation of the copepod *Acartia tonsa* with the stalked ciliate *Zoothamium. Science* 146:543–544.

Hermansson, M., and Dahlback, B. (1983) Bacterial activity at the air/water interface. *Microbiol. Ecol.* 9:317–328.

Hernroth, L. (1983) Marine pelagic rotifers and tintinnids—important trophic links in the spring plankton community of the Gullmar Fjord, Sweden. *J. Plankton Res.* S:835–846.

Hewett, S.W. (1980) Prey-dependent cell size in a protozoan predator. *J. Protozool.* 27:311–313.

Hibberd, D.J. (1975) Observations on the ultrastructure of the chaonoflagellate *Codosiga botrytis* (Ehr) Saville-Kent with special reference to the flagellar apparatus. *J. Cell Sci.* 17:191–219.

Hirche, H.-J. (1974) Die Copepoden *Eurytemora affinis* POPPE and *Acartia tonsa* DANA und ihre Bésiedlung durch *Myoschiston centropagidarum* PRECHT (Peritricha) in der Schlei. *Kieler Meeresforsch.* 30:43–64.

Hitchen, E.T., and Butler, R.D. (1973) Ultrastructural studies of the commensal suctorian, *Choanophrya infundibulifera* Hartog. I. Tentacle structure, movement and feeding. *Z. Zellorsch.* 144:37–57.

Hobbie, J.E., Daley, R.J., and Jasper, S. (1977) Use of Nucleopore filters for counting bacteria by fluorescence microscopy. *Appl. Environ. Microbiol.* 33:1225–1228.

Hollande, A., and Enjumet, M. (1953) Contribution à l'etude biologique des Sphaerocollides (Radiolaires Collodaires et Radiolaires polycyttaires) et de leurs parasites. *Ann. Sci. Nat. Zool.* 15:99–183.

Hollande, A., Cachon, M., and Valentin, J. (1967) Infrastructure des axopodes et organization generale de *Sticholonche zanclea* Hertwig (Radiolaire Sticholonchidea). *Protistologica* 3:155–166.

Hollande, A., Cachon, J., and Cachon, M. (1970) La significance de la membrane capsulaire des Radiolarians et ses rapports avec le plasmalemme et les membranes due reticulum endoplasmique. Affinites entre

Radiolaires, Heliozaires et Peridiniens. *Protistologica* 6:311–318.

Hollibaugh, J.T., Fuhrman, J.A., and Azam, F. (1980) Radioactive labeling of natural assemblages of bacterioplankton for use in trophic studies. *Limnol. Oceanogr.* 25:172–181.

Holling, C.S. (1959) Some characteristics of simple types of predation and parasitism. *Can. Entomol.* 91:385–398.

Holwill, M.E.J., and Sleigh, M.A. (1967) Propulsion by hispid flagella. *J. Exp. Biol.* 42:267–276.

Holz, G.G., Jr. (1954) The oxidative metabolism of a cryptomonad flagellate *Chilomonas paramecium. J. Protozool.* 1:114–120.

Hopkins, D.L. (1930) The relation between food, the rate of locomotion and reproduction in the marine amoeba *Flabellula mira. Biol. Bull.* 58:334–343.

Ibanez, F., and Rassoulzadegan, F. (1977) A study of the relationships between pelagic ciliates (oligotrichina) and planktonic nanoflagellates of the Bay of Villefranche sur Mer. Analysis of Chronological series. *Ann. Inst. Oceanogr.* 53:17–30.

Jackson, K.M., and Berger, J. (1984) Survival of ciliate protozoa under starvation conditions and at low bacterial levels. *Microbiol. Ecol.* 10:47–59.

Jackson, K.M., and Berger, J. (1985a) Survivorship curves of ciliated protozoa under starvation conditions and at low bacterial levels. *Protistologica* 21:17–24.

Jackson, K.M., and Berger, J. (1985b) Life history attributes of some ciliated protozoa. *Trans. Amer. Microsc. Soc.* 104:52–63.

Jacobsen, D.J., and Anderson, D.M. (1986) Thecate heterotrophic dinoflagellates: feeding behavior and mechanisms. *J. Phycol.* 22:249–258.

Jannasch, H.W., and Wirsen, C.O. (1979) Chemosynthetic primary production at east Pacific sea floor spreading centers. *BioScience* 29:592–598.

Jannasch, H.W., and Wirsen, C.O. (1985) The biochemical versatility of chemosynthetic bacteria at deep-sea hydrothermal vents. In Jones, M.L. (ed.), *The Hydrothermal Vents of the Eastern Pacific: an Overview. Bull. Biol. Soc. Wash.* 6:325–334.

Johannes, R.E. (1964) Phosphorous excretion and body size in marine animals: microzooplankton and nutrient regeneration. *Science* 146:923–924.

Johannes, R.E. (1965) Influence of marine protozoa on nutrient regeneration. *Limnol. Oceanogr.* 10:434–442.

Johnson, P.W., Huai-Shu, X., and Sieburth, J. McN. (1982) The utilization of chroococcoid cyanobacteria by marine protozooplankters but not by calanoid copepods. *Ann. Inst. Oceanogr.* 58(S):297–308.

Johnson, P.W., and Sieburth, J. McN. (1979) Chroococcoid cyanobacteria in the sea; a ubiquitous and diverse phototrophic biomass. *Limnol. Oceanogr.* 24:928–935.

Johnson, P.W., and Sieburth, J. McN. (1982) *In situ* morphology and occurrence of eucaryotic phototrophs of bacterial size in the picoplankton of estuarine and oceanic waters. *J. Phycol.* 18:318–327.

Jonsson, P.R. (1986) Particle size selection, feeding rates and growth dynamics of marine planktonic oligotrichous ciliates (Ciliophora: Oligotrichina) *Mar. Ecol. Prog. Ser.* 33:265–277.

Jørgensen, C.B. (1966) Biology of suspension feeding. London: Pergamon, 357 pp.

Jurand, A. (1961) An electron microscope study of food vacuoles in *Paramecium aurelia. J. Protozool.* 8:125–130.

Kahl, A. (1930–1935) Urtiere oder Protozoa. I. Wimpertiere oder Ciliata (Infusoria). In Dahl, F. (ed.), *Die Tierwelt Deutschlands,* Teil 18, 21, 25, 30. Jena: G. Fisher.

Kent, E.B. (1981) Life history responses to resource variation in a sessile predator, the ciliate protozoan *Tokophrya lemnarum* Stein. *Ecology* 62:296–302.

Kepner, W.A., and Carter, J.S. (1931) Ten well-defined new species of *Stenostomum. Zool. Any.* 93:108–123.

Khmeleva, N.N. (1967) Rol'radiolyarii pri otsenke pervichnoi produktsii v krasnom more i adenskom zalive. *Dok. Akad. Nauk. SSSR* 172:1430–1433.

Kidder, G.W., Lilly, D.M., and Claff, C.L. (1940) Growth studies on ciliates. IV. The influence of food on the structure and growth of *Glaucoma vorax* sp. nov. *Biol. Bull.* 78:9–23.

Kimor, B. (1981) The role of phagotrophic dinoflagellates in marine ecosystems. *Kieler Meersforsch.* (Sondbd) 5:164–173.

Kimor, B., and Golandsky, B. (1977) Microplankton of the gulf of Elat: aspects of seasonal and bathymetric distribution. *Mar. Biol.* 42:55–67.

Kirchman, D.L., Ducklow, H.W., and Mitchell, R. (1982) Estimates of bacterial growth from changes in uptake rates and biomass. *App. Environ. Microbiol.* 44:1296–1307.

Kjelleberg, S., and Stenstrom, T.A. (1980) Lipid surface films: interaction of bacteria with free fatty acids and phospholipids at the air/water interface. *J. Gen. Microbiol.* 116:417–423.

Kjelleberg, S., Stenstrom, T.A., and Odham, G. (1979) Comparative study of different hydrophobic devices for sampling lipid surface films and adherent microorganisms. *Mar. Biol.* 53:21–25.

Klaveness, D. (1981) *Rhodomonas lacustris*: ultrastructure of the vegetative cell. *J. Protozool.* 28:83–90.

Klekowski, R.Z., and Shushkina, E.A. (1965) Ernahrung. Atmung, Wachstum und Energie Umformung in *Macrocyclops albidus* (Jurine). *Verh. Int. Ver. Limnol.* 16:399–418.

Kloetzel, J.A. (1974) Feeding in ciliated protozoa. I. Pharyngeal disks in *Euplotes*: a source of membrane for food vacuole formation? *J. Cell Sci.* 15:379–401.

Knauer, G.A., Heber, D., and Cipriano, F. (1982) Marine snow: major site of primary production in coastal waters. *Nature* (Lond.) 300:630–631.

Knudsen, J. (1967) The deep-sea Bivalvia. *Sci. Rep. John Murray Exped. 1933–1934* 11:237–243.

Koestler, R.J., Lee, J.J., Reidy, J., Sheryll, R.P., and Xenophontos, X. (1985) Cytological investigation of digestion and re-establishment of symbiosis in the larger benthic foraminifera *Amphistegina lessonii Endocyt. Cell Res.* 2:21–54.

Kofoid, C.A., and Swezy, O. (1921) The free-living unarmoured dinoflagellates. *Mem. Univ. Calif.* 5:1–538.

Kopylov, A.I., Mamayevua, R.I., and Batsanin, S.F.

(1980) Energy balance of the colorless flagellate *Parabodo attenuatus*. *Oceanology* 20:705–708.

Korniyenko, G.S. (1970) The role of protozoa in pond plankton. *Hydrobiol. J.* 6:27–32.

Korniyenko, G.S. (1971) The role of infusoria in the feeding of larvae of herbivorous fish. *J. Ichthyol.* 11:241–246.

Korniyenko, G.S. (1976) Contribution of infusoria to the nutrition of *Acanthocyclops vernalis* and *Cyclops vicinius*. *Hydrobiol. J.* 12:62–65.

Korshikov, A.A. (1929) Studies on the chrysomonads. I. *Physomonas vestita* Stokes. *Arch. Protistenk.* 67:253–290.

Kudo, R.R. (1966) *Protozoology*, 5th ed. Springfield, Ill.: Charles C. Thomas.

Kuile, B., and Erez, J. (1987) Uptake of inorganic carbon and internal carbon cycling in symbiont-bearing benthonic foraminifera. *Mar. Biol.* 94:499–509.

Landry, M.R., Haas, L.W., and Fagerness, V.L. (1984) Dynamics of microbial plankton communities: experiments in Kaneohe Bay, Hawaii. *Mar. Ecol. Prog. Ser.* 16:127–133.

Landry, M.R., and Hassett, R. P. (1982) Estimating the grazing impact of marine microzooplankton. *Mar. Biol.* 67:283–288.

Lankford, R.R., and Phleger, F.B. (1973) Foraminifera from the near-shore turbulent zone, western North America. *J. Foram. Res.* 3:101–132.

Larsson, K., Odham, G., and Sodergren, A. (1974) On lipid surface films on the sea. I. A simple method for sampling and studies of composition. *Mar. Chem.* 2:49–57.

Laval, M. (1971) Ultrastructure et mode de nutrition du Choanoflagellate *Salpingoeca pelagica*, sp. nov.—comparison avec les choanocytes des Spongiaires. *Protistologica* 7:325–336.

Laval-Peuto, M., and Febvre, M. (1986) On plastid symbiosis in *Tontonia appendiculariformis* (Ciliophora, Oligotrichina). *Biosystems* 19:137–158.

Laval-Peuto, M., Gold, K., and Storm, E.R. (1979) The ultrastructure of *Tintinnopsis parva*. *Trans. Amer. Microsc. Soc.* 98:204–212.

Laval-Peuto, M., Heinbokel, J.F., Anderson, O.R., Rassoulzadegan, F., and Sherr, B.F. (1986) Role of micro- and nanozooplankton in marine food webs. *Insect. Sci. Appl.* 7:387–395.

Laybourn, J.E., and Stewart, J.M. (1975) Studies on consumption and growth in the ciliate *Colpidium campylum* Stokes. *J. Animal Ecol.* 44:165–174.

Leadbeater, B.S.C. (1972) Fine structural observations on some choanoflagellates from the coast of Norway. *J. Mar. Biol. Assoc. U.K.* 52:67–79.

Leadbeater, B.S.C. (1973) External morphology of some marine choanoflagellates from the coast of Yugoslavia. *Archiv. Protistenk.* 115:234–252.

Leadbeater, B.S.C., and Manton, I. (1974) Preliminary observations on the chemistry and biology of the lorica in a collared flagellate (*Stephanoeca diplocostata* Ellis). *J. Mar. Biol. Assoc. U.K.* 54:269–276.

Leadbeater, B.S.C., and Morton, C. (1974) A microscopical study of a marine species of *Codosiga* (James-Clark) (Choanoflagellata) with special reference to the ingestion of bacteria. *Biol. J. Linn. Soc.* 6:337–347.

Leadbeater, B.S.C., and Morton, C. (1975) A microscopical study of a marine species of *Codosiga* James-Clark (Choanoflagellata) with special reference to the ingestion of bacteria. *Biol. J. Linn. Soc.* (Lond.) 6:337–348.

LeCalvez, J. (1937) Un foraminifère géant Bathysiphon filiformis G.O. Sars. *Arch. Zool. Exper. Gen.* 79:82–88.

LeCalvez, J. (1947) *Entosolenia marginata*, foraminifère apogamique ectoparasite d'un autre foraminifère *Discorbis vilardeboanus*. *CR Acad. Sci.* 224:1448–1450.

Lee, J.J. (1974) Towards understanding the niche of foraminifera. In Hedley, R.H., and Adams, C.G. (eds.), *Foraminifera*, Vol. 1. London: Academic Press, pp. 208–257.

Lee, J.J. (1980) Nutrition and physiology of the foraminifera. In *Biochemistry and Physiology of Protozoa*, 2nd ed., Levandowsky, M., and Hutner, S. (eds.), Vol. 3. New York: Academic Press.

Lee, J.J. (1982a) Perspective on algal endosymbionts in larger *Foraminifera*. *Int. Rev. Cytol.* S14:49–77.

Lee, J.J. (1982b) Physical, chemical and biological quality related food-web interactions as factors in the realized niches of microzooplankton. *Ann. Inst. Oceanogr.* 58(S):19–29.

Lee, J.J. Foraminifera. In Margulis, L., Chapman, D., and Corliss, J.O. (eds.), *Protoctista* (in press).

Lee, J.J., and Bock, W. (1976) The importance of feeding in two species of sortitid foraminifera with algal symbionts. *Bull. Mar. Sci.* 26:530–537.

Lee, J.J., Freudenthal, H.D., Muller, W.A., Kossoy, V., Pierce, S., and Grossman, R. (1963) Growth and physiology of foraminifera in the laboratory. III. Initial studies of *Rosalina floridana* (Cushman). *Micropaleontology* 9:449–466.

Lee, J.J., Freudenthal, H.D., Kossoy, V., and Bé, A.W.H. (1965) Cytological observations on two planktonic foraminifera, *Globigerina bulloides* d'Orbigny, 1826, and *Globigerinoides ruber* (d'Orbigny, 1839) Cushman, 1927. *J. Protozool.* 12:531–542.

Lee, J.J., Hutner, S.H., and Bovee, E.C. (1985) An *Illustrated Guide to the Protozoa*. Lawrence, Kansas: Society of Protozoologists and Allen Press, 629 pp.

Lee, J.J., McEnery, M., Pierce, S., Freudenthal, H.D., and Muller, W.A. (1966) Tracer experiments in feeding littoral foraminifera. *J. Protozool.* 13:659–670.

Lee, J.J., McEnery, M.E., and Rubin, H. (1969b) Quantitative studies on the growth of *Allogromia laticollaris* (Foraminifera). *J. Protozool.* 16:377–395.

Lee, J.J., McEnery, M.E., Kahn, E., and Shuster, F. (1979) Symbosis and the evolution of larger foraminifera. *Micropaleontology* 25:118–140.

Lee, J.J., and McEnery, M.E. (1983) Symbiosis in foraminifera. In Goff, L.J. (ed.), *Algal Symbiosis*. Cambridge, University Press.

Lee, J.J., Muller, W.A., Stone, R.J., McEnery, M.E., and Zucher, W. (1969a) Standing crop of foraminifera in sublittoral epiphyte communities of a Long Island Salt marsh. *Mar. Biol.* 4:44–61.

Lee, J.J., Pierce, S., Tentchoff, M., and McLaughlin, J.A. (1961) Growth and physiology of foraminifera in the laboratory. I. Collection and maintenance. *Micropaleontology* 7:461–466.

Lee, J.J., Saks, N.M., Kapiotou, F., Wilen, S.H., and Shilo, M. (1984) Effects of host cell extracts on cultures of endosymbiotic diatoms from larger foraminifera. *Mar. Biol.* 82:113–120.

Lee, J.J., and Zucker, W. (1969) Algal flagellate symbiosis in the foraminifer. *Arch. J. Protozool.* 16:71–81.

Leedale, G.F. (1967) *Euglenid Flagellates*. Englewood Cliffs, N.J.: Prentice-Hall.

Leedale, G.F., and Hibberd, D.J. (1985) Class I. Phytorrastigophorea Calkins, 1909. In Lee, J.J., Hutner, S.H., and Bovee, E.C. *An Illustrated Guide to the Protozoa*. Lawrence, Kansas: Soc. of Protozool. and Allen Press.

Legner, M. (1964) Annual observations on ciliates inhabiting the natant vegetation of two naturally polluted pools. *Vest. Csl. Zool. Spol.* 33:193–213.

Legner, M. (1973) Experimental approach to the role of protozoa in aquatic ecosystems. *Amer. Zool.* 13:177–192.

Lehman, J.T. (1976) The filter-feeder as an optimal forager and the predicted shapes of feeding curves. *Limnol. Oceanogr.* 21:501–516.

Leppanen, J-M., and Bruun, J.E. (1986) The role of pelagic ciliates including the autotrophic *Mesodinium rubrum* during the spring bloom of 1982 in the open northern Baltic proper. *Ophelia* (S)4:147–157.

Lessard, E.J., and Swift, E. (1985) Species-specific grazing rates of heterotrophic dinoflagellates in oceanic waters, measured with a dual-label radioisotope technique. *Mar. Biol.* 87:289–296.

Levandowsky, M., Cheng, T,. Kehr, A., Kim, J., Gardner, L., Silvern, L., Tsang, L., Lai, G., Chung, C., and Prakash, E. (1984) Chemosensory responses to amino acids and certain amines by the ciliate *Tetrahymena*: a flat capillary assay. *Biol. Bull.* 167:322–330.

Levandowsky, M., and Hauser, D.C.R. (1978) Chemosensory responses of swimming algae and protozoa. *Int. Rev. Cytol.* 53:145–210.

Levandowsky, M., and Hutner, S. (1979) Metabolism of the acetate flagellates. Subcellular structure and function in the acetate flagellates. In Levandowsky, M., and Hutner, S. (eds.), *Biochemistry and Physiology of Protozoa*, Vol. 2. Orlando, Fla.: Academic Press, pp. 9–65.

Levin, L.A., DeMaster, D.J., McCann, L.D., and Thomas, C.L. (1986) Effects of giant protozoans (class: xenophyophorea) on deep-seamount benthos. *Mar. Ecol. Prog. Ser.* 29:99–104.

Lighthill, J. (1976) Flagellar hydrodynamics. *SIAM Rev.* 18:161–230.

Lindholm, T. (1985) *Mesodinium rubrum*—a unique photosynthetic ciliate. In Jannasch, H.W., and Williams, P.J. LeB. (eds.), *Advances in Aquatic Microbiology*. London: Academic Press, pp. 1–48.

Linley, E.A.S., Newell, R.C., and Lucas, M.I. (1983) Quantitative relationships between phytoplankton, bacteria and heterotrophic microflagellates in shelf waters. *Mar. Ecol. Prog. Ser.* 12:77–89.

Lipps, J.H. (1973) Predation by reef fish on foraminifera. *Geol. Soc. Am. Abstr. Progr.* 5:73.

Lipps, J.H. (1982) Biology/Paleobiology of Foraminifera. In Broadhead, T.W. (ed.), *Foraminifera: Notes for a Short Course*. Univ. of Tennessee Dept. of Geol. Sci. Studies in Geology 6, pp. 1–21.

Lipps, J.H. (1983) Biotic interactions in benthic foraminifera. In Tevesy, M.J.S., and McCall, P.L. (eds.), *Biotic Interactions in Recent and Fossil Bénthic Communities*. New York: Plenum, pp. 331–376.

Lipps, J.H., and Ronan, T.E. (1974) Predation on foraminifera by the polychaete worm *Diopatra*. *J. Foramin. Res.* 4:139–143.

Lipps, J.H., and Valentine, J.W. (1970) The role of foraminifera in the trophic structure of marine communities. *Lethaia* 3:279–286.

Loeblich, Jr., A.R., and Tappan, H. (1964) Sarcodina, chiefly "thecamoebians" and Foraminiferida. In Moore, R.C. (ed.), *Treatise on Invertebrate Paleontology,* Part C. Protista 2,Vols. 1–2. New York: Geol Society of America.

Loefer, J.B. (1931) Morphology and binary fission in *Heteronema acus* (Ehrb.) Stein. *Arch. Protistenk.* 74:449–470.

Lohmann, H. (1908) Untersuchungen zur Feststellung des vollstandigen Gehaltes des Meeres an Plankton. *Wiss. Meeresunters* (Abt. Kiel) 10:131–370.

Looper, J.B. (1928) Observations on the food reactions of *Actinophrys sol*. *Biol. Bull.* 54:485–502.

Lopez, E. (1979) Algal chloroplasts in the protoplasm of three species of benthic foraminifera: taxonomic affinity, viability and persistence. *Mar. Biol.* 53:201–211.

Lucas, M.I., Probyn, T.A., and Painting, S.J. (1987) An experimental study of microflagellate bacterivory: further evidence for the importance and complexity of microplanktonic interactions. *S. Afr. J. Mar. Sci.* 5:791–808.

Luckinbill, L.S. (1974) The effects of space and enrichment on a predator–prey system. *Ecology* 55:1142–1147.

MacIntyre, F. (1974a) Chemical fractionation and sea-surface microlayer processes. In Goldberg, E.D. (ed.), *The Sea,* Vol. 5. *Marine Chemistry*. New York: Wiley, pp. 245–299.

MacIntyre, F. (1974b) The top millimeter of the ocean. *Sci. Amer.* 230:62–77.

Madin, L.P., and Harbison, G.R. (1977) The associations of amphipods Hyperiidea with gelatinous zooplankton. I. Associations with Salpidae. *Deep-Sea Res.* 24:449–463.

Makinnon, D.L., and Hawes, R.S. (1961) *An Introduction to the Study of Protozoa*. New York: Oxford University Press, 506 pp.

Malone, T.C. (1971) The relative importance of nanoplankton and netplankton as primary producers in tropical oceanic and neritic phytoplankton communities. *Limnol. Oceanogr.* 16:633–639.

Maly, E.J. (1969) A laboratory study of the interaction

between the predatory rotifer *Asplanchna* and *Paramecium*. *Ecology* 50:59–73.

Maly, E.J. (1975) Interactions among the predatory rotifer *Asplanchna* and two prey, *Paramecium* and *Euglena*. *Ecology* 56:346–358.

Manton, I., Sutherland, J., and Leadbeater, B.S.C. (1976) Further observations on the fine structure of marine collared flagellates (Choanoflagellata) from arctic Canada and west Greenland: species of *Parvicorbicula* and *Pleurasiga*. *Can. J. Bot.* 54:1932–1955.

Marcus, E., and Marcus, E. (1959) Studies on Olividae. Biol. Fac. Fil. Cien. Letr. Univ. Sao Paulo 232. *ZOOL* 22:99–188.

Mare, M.F. (1942) A study of marine benthic community with special reference to the micro-organisms. *J. Mar. Biol. Assoc. U.K.* 25:517–554.

Margalef, R. (1982) Some thoughts on the dynamics of populations of ciliates. *Ann. Inst. Oceanogr.* 58(S):15–18.

Marra, J.L., Burckle, H., and Ducklow, H.W. (1982) Sea ice and water column plankton distributions in the Weddell Sea in late winter. *Antarct. J. U.S.* 27:111–112.

Martinez-Palomo, A. (1982) The Biology of *Entamoeba histolytica*. New York: Research Studies Press/Wiley.

Mayzaud, P., and Poulet, S.A. (1978) The importance of the time factor in the response of zooplankton to varying concentrations of naturally occurring particulate matter. *Limnol. Oceanogr.* 23:1144–1154.

McEnery, M., and Lee, J.J. (1981) Cytological and fine structural studies of three species of symbiont-bearing large foraminifera from the Red Sea. *Micropaleontology* 27:71–83.

McKanna, J. (1973) Cyclic membrane flow in the ingestive-digestive system of peritrich protozoans 2. Cup-shaped coated vesicles. *J. Cell Sci.* 13:677–686.

McManus, G.B., and Fuhrman, J.A. (1986a) Bacterivory in seawater studied with the use of inert fluorescent particles. *Limnol. Oceanogr.* 31:420–426.

McManus, G.B., and Fuhrman, J.A. (1986b) Photosynthetic pigments in the ciliate *Laboea strobila* from Long Island Sound, USA. *J. Plankton Res.* 8:317–327.

McManus, G.B., and Fuhrman, J.A. (1988a) Control of marine bacterioplankton populations: measurement and significance of grazing. *Hydrobiologia* 159:51–62.

McManus, G.B., and Fuhrman, J.A. (1988b) Clearance of bacteria-sized particles by natural populations of nanoplankton in the Chesapeake Bay outflow plume. *Mar. Ecol. Prog. Ser.* 42:199–206.

Meyer-Reil, L. (1977) Bacterial growth rates and biomass production. In Rheinheimer, G. (ed.), *Microbial Ecology of a Brackish Water Environment*. Heidelburg: Springer Verlag, pp. 223–236.

Meyer-Reil, L.-A., Bolter, M., Liebezert, G., and Schramm, W. (1979) Short term variations in microbiological and chemical parameters. *Mar. Ecol. Prog. Ser.* 1:1–6.

Mignot, J.P. (1965) Etude ultrastructure de *Cyathomonas truncata* From. (Flagellé Cryptomonadine). *J. Microscopie* 4:239–252.

Mignot, J.P. (1966) Structure et ultrastructure de quelques Euglenomonadines. *Protistologica* 2:51–117.

Monakov, A.V. (1972) Review of studies on feeding of aquatic invertebrates conducted at the Institute of Biology of Inland Waters, Acad. Sci. USSR. *J. Fish. Res. Bd. Can.* 29:363–383.

Moore, D.R. (1961) The marine and brackish water mollusca of the state of Mississippi. *Gulf Res. Rep.* 1:1–58.

Morey-Gaines, G., and Elbrachter, M. (1987) Heterotrophic nutrition. In Taylor, F.J.R. (ed.), *The Biology of Dinoflagellates*. Oxford: Blackwell.

Morris, I., and Glover, H. (1981) Physiology of photosynthesis by marine coccoid cyanobacteria—some ecological implications. *Limnol. Oceanogr.* 26:957–961.

Mortensen, T. (1927) *Handbook of the echinoderms of the British Isles*. London: Oxford University Press, 471 pp.

Morton, J.E. (1958) *Molluscs*. London: Hutchinson Library, 232 pp.

Morton, J.E. (1959) The adaptations and relationships of the Xenophoridae (Mesogastropoda). *Proc. Malacol. Soc. Lond.* 33:89–101.

Müller, M., Rohlich, P., and Toro, I. (1965) Studies on feeding and digestion in protozoa. 7. Ingestion of polystyrene latex particles and its early effect on acid phosphatase in *Paramecium micronucleatum* and *Tetrahymena pyriformis*. *J. Protozool.* 12:27–34.

Muller, W.A. (1975) Competition for food and other niche-related studies of three species of salt marsh foraminifera. *Mar. Biol.* 31:339–351.

Muller, W.A., and Lee, J.J. (1969) Apparent indispensability of bacteria in foraminiferan nutrition. *J. Protozool.* 16:471–478.

Muller, W.A., and Lee, J.J. (1977) Biological interactions and the realized niche of *Euplotes vannus* from the salt marsh Aufwuchs. *J. Protozool.* 24:523–527.

Mullin, M.M. (1963) Some factors affecting the feeding of marine copepods of the genus *Calanus*. *Limnol. Oceanogr.* 8:239–250.

Mullin, M.M., and Brooks, E.R. (1967) Laboratory culture, growth rate and feeding behavior of a planktonic marine copepod. *Limnol. Oceanogr.* 12:657–666.

Mullin, M.M., Stewart, E.F., and Fuglister, F.J. (1975) Ingestion by planktonic grazers as a function of concentration of food. *Limnol. Oceanogr.* 20:259–262.

Murray, J.W. (1963) Ecological experiments on Foraminiferida. *J. Mar. Biol. Assoc. U.K.* 43:621–642.

Murray, J.W. (1973) *Distribution and Ecology of Living Benthic Foraminiferids*. London: Heinemann.

Myers, E.H. (1940) Observations of the origin and fate of flagellated gametes in multiple tests of *Discorbis* (foraminifera). *J. Mar. Biol. Assoc. U.K.* 24:201–226.

Myers, E.H. (1943a) Life activities of foraminifera in relation to marine ecology. *Proc. Amer. Phil. Soc.* 86:439–459.

Myers, E.H. (1943b) Biology, ecology and morphogenesis of a pelagic foraminifer. *Stanford Univ. Publ. Univ. Ser. Biol. Sci.* 9:1–40.

Naumann, E. (1917) Béitrag Zur Kenntnis des Teichnannoplankton. II. Uber das Neuston des Susserwassers. *Biol. Zentralbl.* 37:98–106.

Nelson, D.M., Gordon, L.I., and Smith, W.O. (1984) Phytoplankton dynamics of the marginal ice zone of the Weddell Sea, November and December, 1983. *Antarct. J. U.S.* 19:105–107.

Newell, R.C., and Linley, E.A.S. (1984) Significance of microheterotrophs in the decomposition of phytoplankton: estimates of carbon and nitrogen flow based on the biomass of plankton communities. *Mar. Ecol. Prog. Ser.* 16:105–119.

Newell, S.Y., and Fallon, R.D. (1982) Bacterial productivity in the water column and sediments of the Georgia (USA) coastal zone: estimates via direct counting and parallel measurement of thymidine incorporation. *Microbiol. Ecol.* 8:33–46.

Newell, S.Y., Sherr, B.F., Sherr, E.B., and Fallon, R.D. (1983) Bacterial response to presence of eukaryotic inhibitors in water from a coastal marine environment. *Mar. Environ. Res.* 10:147–157.

Nicol, S. (1983) *Ephelota sp.* a suctorian found on the euphausiid *Meganyctiphanes norvegica*. *Can. J. Zool.* 62:744–745.

Nilsson, J.R. (1979) Phagotrophy in *Tetrahymena*. In Levandowsky, M., and Hutner, S.H. (eds.), *Biochemistry and Physiology of Protozoa*, 2nd ed. New York: Academic Press, pp. 339–379.

Nilsson, J.R. (1977) On food vacuoles in *Tetrahymena pyriformis* GL. *J. Protozool.* 24:502–507.

Nisbet, B. (1974) An ultrastructural study of the feeding apparatus of *Peranema trichophorum*. *J. Protozool.* 21:39–48.

Nisbet, B. (1984) *Nutrition and Feeding Strategies in Protozoa*. London: Croom Helm, 280 pp.

Norkrans, B. (1980) Surface microlayers in aquatic environments. In Alexander, M. (ed.), *Advances in Microbial Ecology*. New York: Plenum, pp. 51–85.

Norris, D.R. (1969) Possible phagotrophic feeding in *Ceratium lunula* Schimper. *Limnol. Oceanogr.* 14:448–449.

Norris, R.E. (1965) Neustonic marine Craspedomonadales (choanoflagellates) from Washington and California. *J. Protozool.* 12:589–602.

Nuttycombe, J.W., and Waters, A.J. (1935) Feeding habits and pharyngeal structure in *Stenostomum*. *Biol. Bull.* 69:439–446.

Nygaard, K., Børsheim, K.Y., and Thingstad, T.F. (1988) Grazing rates on bacteria by marine heterotrophic microflagellates compared to uptake rates of bacterial-sized monodisperse fluorescent latex beads. *Mar. Ecol. Prog. Ser.* 44:159–165.

Nyholm, K.-G. (1956) On the life cycle of the foraminiferan *Nemogullmia longivariabilis*. *Zool. Biol. Upps.* 31:483–496.

O'Connors, H.B., Biggs, D.C., and Ninivaggi, D.V. (1980) Particle-size dependent maximum grazing rates for *Temora longicornis* fed natural particle assemblages. *Mar. Biol.* 56:65–70.

O'Connors, H.B., Small, L.F., and Donaghay, P.L. (1976) Particle size modification by two size classes of the estuarine copepod *Acartia clausi*. *Limnol. Oceanogr.* 21:300–308.

Orton, J.H. (1927) On the mode of feeding of the hermit crab *Eupagurus bernhardus* and some other Decapoda. *J. Mar. Biol. Assoc. U.K.* 14:909–921.

Pace, M.L. (1988) Bacterial mortality and the fate of bacterial production. *Hydrobiologia* 159:41–49.

Pace, M.L., and Bailiff, (1987) An evaluation of a fluorescent microsphaere technique for measuring grazing rates of phagotrophic microorganisms. *Mar. Ecol. Prog. Ser.* 40:185–193.

Pace, M.L., and Orcutt, Jr., J.D. (1981) The relative importance of protozoans, rotifers and crustaceans in a freshwater zooplankton community. *Limnol. Oceanogr.* 26:822–830.

Paffenhöfer, G.A., and Knowles, S.C. (1978) Feeding of marine planktonic copepods on mixed phytoplankton. *Mar. Biol.* 48:143–152.

Pain, S. (1986) Hot spots in the Atlantic Ocean. *New Scientist* 112:29.

Paranjape, M.A. (1980) Occurrence and significance of resting cysts in a hyaline tintinnid *Helicostomella subulata* (Ehr.) Jørgensen. *J. Exp. Mar. Biol. Ecol.* 48:23–33.

Paranjape, M.A. (1987) Grazing by microzooplankton in the eastern Canadian arctic in summer 1983. *Mar. Ecol. Prog. Ser.* 40:239–246.

Parslow, J.S., Doucette, G.J., Taylor, F.J.R., and Harrison, P.J. (1986) Feeding by the zooflagellate *Pseudobodo* sp. on the picoplanktonic prasinomonad *Micromonas pusilla*. *Mar. Ecol. Prog. Ser.* 29:237–246.

Parsons, T.R., Le-Brasseur, J.R., and Fulton, J.D. (1967) Some observations on the dependence of zooplankton grazing on the cell size and concentration of phytoplanktonic blooms. *J. Oceanogr. Soc. Jap.* 23:10–17.

Parsons, T.R., Takahashi, M., and Hargraves, B. (1984) *Biol. Oceanogr. Proc.* Oxford: Pergamon.

Patterson, D.J. (1979) On the organization and classification of the protozoan *Actinophrys sol* Ehrenberg, 1830'. *Microbios*. 26:165–208.

Patterson, D.J., and Hausmann, K. (1981) Feeding by *Actinophrys sol* (Protista, Heliozoa): 1. Light microscopy. *Microbios*. 31:39–55.

Paul, J.H., DeFlaun, M.F., and Jeffrey, W.H. (1986) Elevated levels of microbial activity in the coral surface microlayer. *Mar. Ecol. Prog. Ser.* 33:29–40.

Pavillon, J.F., and Rassoulzadegan, F. (1980) Un aspect de la nutrition chez *Stenosemella ventricosa* (Tintinnide): absorption de quelque substances organique dissoutes marquees au ^{14}C. *J. Rech. Oceanogr.* 4:53–61.

Pavshtiks, E.A., and Pan'Kova, L.A. (1966) O pitanii pelagicheskoi molodi morskikh okunei roda *Sebastes* planktonan v Devisovom prolive. Materialy nauchnoi sessii Polyarnogo nauchnoissledovatel'skogo Instituta morskogo rybnogo khozyaistva i okeanografii 6:87.

Payne, W.J. (1970) Energy yields and growth of heterotrophs. *Ann. Rev. Microbiol.* 24:17–52.

Peck, R., and Hausmann, K. (1980) Primary lysosomes of the ciliate *Pseudomicrothorax dubius*: cytochemical identification and role in phagocytosis. *J. Protozool.* 27:401–409.

Pelseneer, P. (1935) Essai d'ethologie zoologique d'apres l'etude des mollusques: *Acad. R. Bélg. Cl. Sci. Pub. Fond, Agnathon de Potter* 1:1–662.

Perkins, E.J. (1958) The food relationships of the microbenthos, with particular reference to that found at Whitstable, Kent. *Ann. Mag. Nat. Hist.* 13:84–87.

Petrushevskaya, M.G. (1971) Radiolaria in the plankton and recent sediments from the Indian Ocean and Antarctic. In Funnell, B.M., and Riedel, W.R. (eds.), *The Micropaleontology of Oceans*. Cambridge: Cambridge University Press, pp. 319–329.

Petrushevskaya, M.G. (1981) Radiolaria, Order Nassellaria. Acad. Sci of the Soviet Union Zoological Institute. Leningrad: Leningrad Science Publishers.

Petz, W., Foissner, W., and Adam, H. (1985) Culture, food selection and growth rate in the mycophagous ciliate *Grossglockneria acuta* Foissner, 1980: First evidence of autochthonous soil ciliates. *Soil Biol. Biochem.* 17:871–875.

Petz, W., Foissner, W., Wirnsberger, E., Krautgartner, W.D., and Adam, H. (1986) Mycophagy, a new feeding strategy in autochthonous soil ciliates. *Naturwissen. Schaften* 73: (S):560–561.

Pierce, E., Isquith, I.R., and Repak, A.J. (1978) Quantitative study of cannibal-giantism in *Blepharisma*. *Acta Protozool.* 17:493–501.

Pinus, G.N. (1970) Feeding of the Azov tyul'ka larvae and the effect of conditions on the nutrition and abundance of the brood. *J. Ichthyol.* 10:519–527.

Pomeroy, L.R. (1974) The ocean's food web, a changing paradigm. *Bioscience* 24:499–504.

Pomeroy, L.R. (1980) Microbial roles in aquatic food webs. In Colwell, R.R. (ed.), *Proc. of Am. Soc. Microb. Conference on Aquatic Microbial Ecology*. National Oceans and Atmospheric Administration, Washington, D.C.

Pomeroy, L.R., and Deibel, D. (1980) Aggregation of organic matter by pelagic tunicates. *Limnol. Oceanogr.* 25:643–652.

Porter, K.G., and Feig, Y.S. (1980) The use of DAPI for identifying and counting aquatic microflora. *Limnol. Oceanogr.* 25:943–948.

Porter, K.G., Pace, M.L., and Battey, J.F. (1979) Ciliate protozoans as links in freshwater planktonic food chains. *Nature* 277:563–565.

Poulet, S.A. (1978) Comparison between five coexisting species of marine copepods feeding on naturally occurring particulate matter. *Limnol. Oceanogr.* 23:1126–1143.

Pussard, M., and Rouelle, J. (1986) Predation de la microflore effet des protozoaires sur la dynamique de population bacterienne. *Protistologica* 22:105–110.

Rapport, D.J., Bérger, J., and Reid, D.B.W. (1972) Determination of food preference of *Stentor coeruleus*. *Biol. Bull.* 142:103–109.

Rasmussen, L. (1976) Nutrient uptake in *Tetrahymena pyriformis* Carlsberg. *Res. Com.* 41:143–167.

Rassoulzadegan, F. (1978) Dimensions et taux d'ingestion des particules consommees par un tintinnide: *Favella ehrenbergii* (Clap. et Lachm.) Jorg. Cilie pelagique. *Ann. Inst. Oceanogr.* (Paris) 54:17–24.

Rassoulzadegan, F. (1982) Dependence of grazing rate, gross growth efficiency and food size range on temperature in a pelagic oligotrichous ciliate *Lohmanniella spiralis* Leeg. fed on naturally occurring particulate matter. *Ann. Inst. Oceanogr.* 58:177–184.

Rassoulzadegan, F., and Etienne, M. (1981) Grazing rate of the tintinnid *Stenosemella ventricosa* (Clap. and Lachm.) Jørg. on the spectrum of the naturally occurring particulate matter from the Mediterranean neritic area. *Limnol. Oceanogr.* 26:258–270.

Rassoulzadegan, F., Laval-Peuto, M., and Sheldon, R.W. (1988) Partition of the food ration of marine ciliates between pico- and nanoplankton. *Hydrobiologia* 159:75–88.

Rassoulzadegan, F., and Sheldon, R.W. (1986) Predator–prey interactions of nanozooplankton and bacteria in an oligotrophic marine environment. *Limnol. Oceanogr.* 31:1010–1021.

Reid, P.C. (1987) Mass encystment of a planktonic oligotrich ciliate. *Mar. Biol.* 95:221–230.

Renz, G.W. (1976) The distribution and ecology of radiolaria in the central Pacific plankton and surface sediments. *Bull. Scripps Inst. Oceanogr.* 22:1–267.

Repak, A.J. (1983) Suitability of selected marine algae for growing the marine heterotrich ciliate *Fabrea salina*. *J. Protozool.* 30:52–54.

Repak, A.J. (1986) Suitability of selected bacteria and yeasts for growing the estuarine heterotrich ciliate *Fabrea salina* (Henneguy). *J. Protozool.* 33:219–222.

Reyment, R.A. (1966) Preliminary observations on gastropod predation in the western Niger Delta, *Palaeogeogr. Palaeoclimatol. Palaeoecol.* 2:81–102.

Richman, S., Heinle, D.R., and Huff, R. (1977) Grazing by adult estuarine calanoid copepods of the Chesapeake Bay. *Mar. Biol.* 42:69–84.

Riley, G.A. (1956) Oceanography of Long Island Sound, 1952–1954. IX. Production and utilization of organic matter. *Bull. Bingham Oceanogr. Coll.* 15:324–341.

Rivier, A., Brownlee, D.C., Sheldon, R.W., Rassoulzadegan, F. (1986) Growth of microzooplankton. A comparative study of bactivorous zooflagellates and ciliates. *Mar. Microbiol. Food Webs* 1:51–60.

Robertson, J.R. (1983) Predation by estuarine zooplankton on tintinnid ciliates. *Estuar. Coast Shelf Sci.* 16:27–36.

Robertson, J.R., and Salt, G.W. (1981) Responses in growth, mortality, and reproduction to variable food levels by the rotifer *Asplanchna girodi*. *Ecology* 62:1585–1596.

Roman, M.R. (1978) Ingestion of the blue-green alga *Trichodesminum* by the harpactacoid copepod, *Macrosetella gracilis*. *Limnol. Oceanogr.* 23:1245–1247.

Roman, M.R., and Rublee, P.A. (1980) Containment effects in copepod grazing experiments: a plea to end the black box approach. *Limnol. Oceanogr.* 25:982–990.

Roman, M.R., Gauzens, A.L., and Cowles, T.J. (1985). Temporal and spatial changes in epipelagic microzooplankton and mesozooplankton biomass in warmcore Gulf Stream ring 82-B. *Deep-Sea Rs.* 32:1007–1022.

Ross, C.A., and Ross, J.R.P. (1978) Adaptive evolution in the soritids *Marginopora* and *Amphisorus* (Foraminiferida). *Scan. Electro. Microsc.* 2:53–60.

Roth, L.E. (1960) Electron microscopy of pinocytosis and food vacuoles in *Pelomyxa*. *J. Protozool.* 7:176–185.

Röttger, R. (1972) Analyse von Wachstumskurwen von *Heterostegina depressa* (foraminifera: Nummulitidae). *Mar. Biol.* 17:228–242.

Rubenstein, D.I., and Koehl, M.A.R. (1977) The mechanisms of filter feeding: some theoretical considerations. *Amer. Nat.* 111:981–994.

Rubin, H.A., and Lee, J.J. (1976) Informational energy flow as an aspect of the ecological efficiency of marine ciliates: I. *Theor. Biol.* 62:69–91.

Rudzinska, M.A. (1970) The mechanism of food intake in *Topkophrya infusionum* and ultrastructural changes in food vacuoles during digestion. *J. Protozool.* 17:626–641.

Rudzinska, M.A. (1972) Ultrastructural localization of acid phosphatase in feeding *Tokophrya infusionum*. *J. Protozool.* 19:618–629.

Ruszinska, M.A. (1973) Do Suctoria really feed by suction? *BioScience* 23:87–94.

Rudzinska, M.A. (1980) Internalization of macromolecules from the medium in Suctoria. *J. Cell. Biol.* 84:172–183.

Ryther, J.H. (1956) Photosynthesis in the ocean as a function of light intensity. *Limnol. Oceanogr.* 1:61–70.

Salt, G.W. (1967) Predation in an experimental protozoan population (*Woodruffia-Paramecium*) *Ecol. Monogr.* 37:113–144.

Sanders, R.W., and Porter, K.G. (1986) Use of metabolic inhibitors to estimate protozooplankton grazing and bacterial production in a monomictic eutrophic lake with an anaerobic hypolimnion. *Appl. Environ. Microbiol.* 52:101–107.

Sanders, R.W., and Porter, K.G. (1988) Phagotrophic phytoflagellates. In Marshall, K.C. (ed.), *Advances in Microbial Ecology*, Vol. 10. New York: Plenum.

Sandon, H. (1932) *The Food of Protozoa*. Cairo: Misr-Sokkar Press.

Savoie, A. (1968) Les cilies histophages en Biologie cellulaire. Annali Dell'Universita di Ferrara Sezione III. *Biol. Anim.* 3:65–71.

Sawicka, F., Kaczanowski, A., and Kaczanowski, J. (1983) Kinetics of ingestion of food vacuoles during the cell cycle of *Chilodonella steini*. *Acta Protozoologica* 22:157–167.

Sawyer, T.K. (1971) Isolation and identification of free living marine amoebae from the upper Chesapeake Bay, Maryland. *Trans. Amer. Microsc. Soc.* 90:43–51.

Sawyer, T.K. (1975) Marine amoebae from surface waters of Chincoteague Bay, Virginia: two new genera and nine new species within the families Mayorellidae, Falbellulidae and Stereomyxidae. *Trans. Amer. Microsc. Soc.* 94:71–92.

Schaudinn, F. (1895) Untersuchurgen an Foraminiferen. I. Calcituba polymorpha Roboz. *Z. Wiss. Zool.* 59:191–232.

Schmaljohann, R., and Röttger, R. (1976) Die Symbioten der Grossforaminifere *Heterostegina depressa* sind Diatomeen. Naturwissenschaften 63:486–487.

Schmaljohann, R., and Röttger, R. (1978) The ultrastructive and taxonomic identity of the symbiotic algae of *Heterostegina depressa* (Foraminifera, Nummulitidae). *J. Mar. Biol. Assoc. U.K.* 58:227–237.

Schwab, A.D., and Hofer, H.W. (1979) Metabolism in the protozoan *Allogromia laticollaris*. *Z. Mikrosk-anat. Forsch. Leipzig* 93:715–727.

Sekiguchi, H., and Kato, T. (1976) Influence of Noctiluca's predation on the *Acartia* population in Ise Bay, Central Japan. *J. Oceanogr. Soc. Jap.* 32:195–198.

Seravin, L.N., and Orlovskaja, E.E. (1973) Factors responsible for food selection in protozoa. In dePuytorac, P., and Grain, J. (eds.), *Progress in Protozoology*. 4th Int. Congress Univ. deClermont-Ferrand, p. 371.

Seravin, L.N., and Orlovskaja, E.E. (1977) Feeding behavior of unicellular animals. I. The main role of chemoreception in the food choice of carnivorous protozoa. *Acta Protozool.* 16:309–332.

Servais, P., Billen, G., and Vives-Rego, J. (1985) Rate of bacterial mortality in aquatic environments. *Appl. Envir. Microbiol.* 49:1448–1454.

Shanks, A.L., and Trent, J.D. (1980) Marine snow: sinking rates and potential role in vertical flux. *Deep-Sea Res.* 27A:137–143.

Shcherbina, T.V. (1970) New data on the nutrition of *Eucyclops* (s. str.) *serrulatus* (Fisch.). *Hydrobiol. J.* 6:36–39.

Sheldon, R.W., Nival, P., and Rassoulzadegan, F. (1986) An experimental investigation of a flagellate-ciliate-copepod food chain with some observations relevant to the linear biomass hypothesis. *Limnol. Oceanogr.* 31:184–188.

Sherr, E.B. (1988) Direct utilization of high molecular weight polysaccharide by heterotrophic flagellates. *Nature* 335:348–351.

Sherr, B.F., and Sherr, E.B. (1983a) Enumeration of heterotrophic microprotozoa by epifluorescence microscopy. *Estuar. Coast. Shelf Sci.* 16:1–7.

Sherr, E., and Sherr, B. (1983b) Double staining epifluorescence technique to assess frequency of dividing cells and bacterivory in natural populations of heterotrophic microprotozoa. *Appl. Environ. Microbiol.* 46:1388–1393.

Sherr, E.B., and Sherr, B.F. (1987) High rates of consumption of bacteria by pelagic ciliates. *Nature* 325:710–711.

Sherr, B.F., Sherr, E.B., Andrew, T.L., Fallon, R.D., and Newell, S.Y. (1986a) Trophic interactions between heterotrophic protozoa and bacterioplankton in estuarine water analyzed with selective metabolic inhibitors. *Mar. Ecol. Prog. Ser.* 32:169–179.

Sherr, B.F., Sherr, E.B., and Berman, T. (1982) Decomposition of organic detritus: a selective role for microflagellate protozoa. *Limnol. Oceanogr.* 27:765–769.

Sherr, B.F., Sherr, E.B., and Berman, T. (1983) Growth, grazing and ammonia excretion rates of a heterotrophic microflagellate fed with four species of bacteria. *Appl. Environ. Microbiol.* 45:1196–1201.

Sherr, E.B., Sherr, B.F., Fallon, R.D., and Newell, S.Y. (1986b) Small, aloricate ciliates as a major component of the marine heterotrophic nanoplankton. *Limnol. Oceanogr.* 31:177–183.

Sherr, B.F., Sherr, E.B., and Fallon, R.D. (1987) Use of monodispersed, fluorescently labeled bacteria to estimate *in situ* protozoan bactivory. *App. Environ. Microbiol.* 53:00–00.

Sherr, B.F., Sherr, E.B., and Newell, S.Y. (1984) Abundance and productivity of heterotrophic nanoplankton in Georgia coastal waters. *J. Plankton Res.* 6:195–202.

Sherr, E.B., Sherr, B.F., and Paffenhöfer, G.-A. (1986c) Phagotrophic protozoa as food for metazoans: a "missing" trophic link in marine pelagic food webs. *Mar. Micrdobiol. Food Webs* 1:61–80.

Shonman, D., and Nybakhen, J.W. (1978) Food preferences, food availability and food resource partitioning in two sympatric species of cephalaspidean opisthobranchs. *Veliger* 21:120–126.

Sibbald, M.J., Albright, L.J., and Sibbald, P.R. (1987) Chemosensory responses of a heterotrophic microflagellate to bacteria and several nitrogen compounds. *Mar. Ecol. Prog. Ser.* 36:201–204.

Sibbald, M.J., and Albright, L.J. (1988) Aggregated and free bacteria as food sources for heterotrophic microflagellates. *Appl. Environ. Microbiol.* 54:613–616.

Sieburth, J. McN. (1975) *Microbial Seascapes.* Baltimore: University Park Press, 248 pp.

Sieburth, J. McN. (1979) *Sea Microbes.* New York: Oxford University Press, 491 pp.

Sieburth, J. McN., and Davis, P.G. (1982) The role of heterotrophic nanoplankton in the grazing and nurturing of planktonic bacteria in the Sargasso and Caribbean Sea. *Ann. Inst. Oceanogr.* 58(S):285–296.

Sieburth, J. McN., Smetacek, V., and Lenz, J. (1978) Pelagic ecosystem structure: heterotrophic compartments of the plankton and their relationships to plankton size fractions. *Limnol. Oceanogr.* 23:1256–1263.

Sieburth, J. McN., and Thomas, C.D. (1973) Fouling on eelgrass (*Zostera marina* L.). *J. Phycol.* 9:46–50.

Sieburth, J. McN., Willis, P.-J., Johnson, K.M., Burney, C.M., Lavoie, D.M., Hinga, K.R., Caron, D.A., French, III, F.W., Johnson, P.W., and Davis, P.G. (1976) Dissolved organic matter and heterotrophic microneuston in the surface microlayer of the North Atlantic. *Science* 194:1415–1418.

Sieracki, M.E., Haas, L.W., Caron, D.A., and Lessard, E.J. (1987) Effect of fixation on particle retention by microflagellates: underestimation of grazing rates. *Mar. Ecol. Prog. Ser.* 38:251–258.

Silver, M.W., Gowing, M.M., Brownlee, D.C., and Corliss, J.O. (1984) Ciliated Protozoa associated with oceanic sinking detritus. *Nature* (Lond.) 309:246–248.

Silver, M.W., Mitchell, J.G., and Ringo, D. (1980) Siliceous nanoplankton. II. Newly discovered cysts and abundant choanoflagellates from the Weddell Sea, Antarctica. *Mar. Biol.* 58:211–217.

Silver, M.W., Shanks, A.L., and Trent, J.D. (1978) Marine snow; microplankton habitat and source of small-scale patchiness in pelagic populations. *Science* 201:371–373.

Skøgstad, A.L., Granskog, L., and Klaveness, D. (1987) Growth of freshwater ciliates offered planktonic algae as food. *J. Plankton Res.* 9:503–512.

Sleigh, M.A. (1964) Flagella movement of the sessile flagellates *Actinomonas, Condonosiga, Monas,* and *Poteriodendron. Quart. J. Microsc. Sci.* 105:405–414.

Sleigh, M.A. (1973) *The Biology of Protozoa.* London: Edward Arnold.

Sleigh, M.A., and Aiello, E. (1972) The movement of water by cilia. *Acta Protozool.* 11:265–277.

Sleigh, M.A., and Barlow, D. (1976) Collection of food by *Vorticella. Trans. Amer. Microsc. Soc.* 95:482–486.

Sliter, W.V. (1965) Laboratory experiments on the life cycle and ecologic controls of *Rosalina globularis* d'Orbigny. *J. Protozool.* 12:210–215.

Small, E.B. (1984) An essay on the evolution of ciliophoran oral cytoarchitecture based on descent from within a karyorelictean ancestry. *Origins of Life* 13:217–228.

Small, E.B. (1973) A study of ciliate protozoa from a small polluted stream in east-central Illinois. *Amer. Zool.* 13:225–230.

Small, E.B., and Gross, M.E. (1985) Preliminary observations of protistan organisms, especially ciliates, from the 21°N hydrothermal vents of the eastern Pacific: an overview. *Bull. Biol. Soc. Wash.* 453–464.

Small, E.B., and Lynn, D. (1985) Phylum Ciliophoria. In Lee, J.J., Hutner, S.H., and Bovee, E.C. (eds.), *An Illustrated Guide to the Protozoa.* Lawrence, Kansas: Society of Protozoology/Allen Press, pp. 393–575.

Small, E.B., Neum, B., Caron, D., and Davis, P. (1983) Ciliates associated with Gulf Stream "snow." *J. Protozool.* (Abst.) 15A.

Smetacek, V. (1981) The annual cycle of protozooplankton in the Kiel Bight. *Mar. Biol.* 63:1–11.

Smith, W.O., and D.M. Nelson (1985) Phytoplankton bloom produced by a receding ice edge in the Ross Sea: spatial coherence with the density field. *Science* 227:163–166.

Smith-Sonneborn, J., and Rodermal, S.R. (1976) Loss of endocytic capacity in aging *Paramacium.* The importance of cytoplasmic organelles. *J. Cell Biol.* 71:575–588.

Sniezek, Jr., J. (1987) An examination of the ciliate fauna at the black smoker hydrothermal vents of 10° 57′N, 103° 46′W. M.S. thesis, University of Maryland at College Park, 161 pp.

Sokolova, M.N. (1957) The feeding of some carnivorous deep-sea benthic invertebrates on the far eastern seas and the northwest Pacific Ocean. *Trans. Inst. Okeanol. Akad. Nauk. SSSR* 20:279–301.

Sonneborn, T.M. (1930) Genetic studies on *Stenostomum incaudatum* (nov. sp.), 1. The nature and origin of differences among individuals formed during vegetative reproduction. *J. Exp. Zool.* 57:57–108.

Sorokin, Y.I. (1977) The heterotrophic phase of plankton succession in the Japan Sea. *Mar. Biol.* 41:107–117.

Sorokin, Y.I. (1978) Decomposition of organic matter and nutrient regeneration. In Kinne, O. (ed.), *Marine Ecology,* Vol. 4. Wiley-Interscience, pp. 501–616.

Sorokin, Y.I., and Paveljeva, E.B. (1972) On the quantitative characteristics of the pelagic ecosystems of Dalnee Lake (Kamchatka). *Hydrobiologia* 40:519–552.

Spero, A.J. (1982) Phagotrophy in *Gymnmodinium fungiforme* (Pyrrophyta): the peduncle as an organelle of ingestion. *J. Phycol.* 18:356–360.

Spero, H.J., and Montescue, M.D. (1981) Phagotrophic feeding and its importance to the life cycle of the holozoic dinoflagellate, *Gymnodinium fungiforme. J. Phycol.* 17:43–51.

Spero, H.J., and Parker, S.L. (1985) Photosynthesis in the symbiotic planktonic foraminifer *Orbulina universa* and its potential contribution to oceanic primary productivity. *J. Foram. Res.* 15:273–281.

Spindler, M. (1980) The pelagic gulfweed *Sargassum natans* as a habitat for the benthic Foraminifera *Planorbulina acervalis* and *Rosalina globularis. N. Jb. Geol. Paleontol. Mh.* 9:569–580.

Spindler, M., Hemleben, C., Bayer, U., Bé, A.W.H., and Anderson, O.R. (1979) Lunar periodicity of reproduction in the planktonic foraminifer *Hastigerina Pelagica. Mar. Ecol. Prog. Ser.* 1:61–64.

Spittler, P. (1973) Feeding experiments with tintinnids. *Oikos* (S) 15:128–132.

Steeman-Nielsen, E. (1962) The relationship between phytoplankton and zooplankton in the sea. Rapp. P.-V. Neun. *Cons. Perm. Int. Explor. Mer.* 153:178–182.

Stenson, J.A.E. (1984) Interactions between pelagic metazoans and protozoan zooplankton, an experimental study. *Hydrobiologia* 111:107–112.

Stockem, W. (1973) Morphological and cytochemical aspects of endocytosis and intracellular digestion in amoebae. In de Puytorac, P., and Grain, J. (eds.), *Progress in Protistology* (4th Int. Congress Univ. de Clermont-Ferrand), p. 402.

Stockner, J.G., and Antia, N.J. (1986) Algal picoplankton from marine and freshwater ecosystems: a multidisciplinary perspective. *Can. J. Fish. Aquatic Sci.* 43:2472–2503.

Stoecker, D.K. (1984) Particle production by planktonic ciliates. *Limnol. Oceanogr.* 29:930–940.

Stoecker, D.K. (1988) Are marine planktonic ciliates suspension-feeders? *J. Protozool.* 35:252–254.

Stoecker, D.K., Davis, L.H., and Anderson, D.M. (1984) Fine scale spatial correlations between planktonic ciliates and dinoflagellates. *J. Plankton Res.* 6:829–841.

Stoecker, D., Davis, L.H., and Provan, A. (1983) Growth of *Favella* sp. (Ciliata: Tintinnina) and other microzooplankters in cages incubated *in situ* and comparison to growth *in vitro*. *Mar. Biol.* 75:293–302.

Stoecker, D.K., and Egloff, D.A. (1987) Predation by *Acartia tonsa* Dana on planktonic ciliates and rotifers. *J. Exp. Mar. Biol. Ecol.* 110:53–68.

Stoecker, D.K., and Govoni, J.J. (1984) Food selection by young larval gulf menhaden. (*Brevoortia partronus*). *Mar. Biol.* 80:299–306.

Stoecker, D.K., Michaels, A.E., and Davis, L.H. (1987a) Grazing by the jellyfish *Aurelia aurita* on microzooplankton. *J. Plankton Res.* 9:901–915.

Stoecker, D.K., Michaels, A.E., and Davis, L.H. (1987b) Large proportion of marine planktonic ciliates found to contain functional chloroplasts. *Nature* 326:790–792.

Stoecker, D.K., and Sanders, N.K. (1985) Differential grazing by *Acartia tonsa* on a dinoflagellate and a tintinnid. *J. Plankton Res.* 7:85–100.

Straskrabova-Prokesova, V., and Legner, M. (1966) Interrelations between bacteria and protozoa during glucose oxidation in water. *Int. Rev. Gesamten Hydrobiol.* 51:279–293.

Sugden, B. (1950) A study of the feeding and excretion of the ciliate *Carchesium,* in relation to the clarification of sewage effluent. PhD. thesis, University of Leeds.

Suttle, C.A., Chan, A.M., Taylor, W.D., and Harrison, P.J. (1986) Grazing of planktonic diatoms by microflagellates. *J. Plankton Res.* 8:393–398.

Suzaki, T., Shigenoka, Y., Watanobe, S., and Toyohara, A. (1980) Food capture and ingestion in the large heliozoan. *Echinosphaerium nucleofilum. J. Cell Sci.* 42:61–75.

Swanberg, N.R. (1983) The trophic role of colonial radiolaria in oligotrophic oceanic environments. *Limnol. Oceanogr.* 28:655–666.

Swanberg, N.R., and Anderson, O.R. (1985) The nutrition of radiolarians: trophic activity of some solitary Spumellaria. *Limnol. Oceanogr.* 30:646–652.

Swanberg, N.R., Anderson, O.R., Lindsey, J.L., and Bennett, P. (1986) The biology of *Physematium muelleri:* trophic activity. *Deep-Sea Res.* 33:913–922.

Swanberg, N.R., and Harbison, G.R. (1980) The ecology of *Collozoum longiforme,* sp. nov., a new colonial radiolarian from the equatorial Atlantic Ocean. *Deep-Sea Res.* 27A:715–732.

Takahashi, E. (1981) Loricate and scale bearing protists from Lutzow-Holm Bay, Antarctica. I. Species of the Acanthoecidae and the Centrohelida found at a site selected in the fast ice. *Antarct. Rec.* 73:1–22.

Taniguchi, A., and Kawakami, R. (1983) Growth rates of ciliate *Eutintinnus lususundae* and *Favella taraikaensis* observed in the laboratory culture experiments. *Bull. Plankton Soc. Jap.* 30:33–40.

Taylor, D.L., and Lee, C.C. (1971) A new Cryptomonad from Antarctica: *Cryptomonas cryophila* sp. Nov. *Archivs fur Mikrobiologie* 75:269–280.

Taylor, D.L., and Seliger, H.H., eds. (1979) Toxic dinoflagellate blooms. *Dev. Mar. Biol.* 1. New York: Elsevier/North Holland.

Taylor, F.J.R. (1982) Symbioses in marine microplankton. *Ann. Inst. Oceanogr. Paris* 58(S):61–90.

Taylor, G.T., and Pace, M.L. (1987) Validity of eucaryote inhibitors for assessing production and grazing mortality of marine bacterioplankton. *Appl. Environ. Microbiol.* 53:119–128.

Taylor, W.D. (1978a) Growth responses of ciliate protozoa to the abundance of their bacterial prey. *Microbiol. Ecol.* 4:207–214.

Taylor, W.D. (1978b) Maximum growth rate, size and commonness in a community of bactivorous ciliates. *Oecologia* 36:263–272.

Taylor, W.D. (1980) Observations on the feeding and growth of the predaceous oligochaete *Chaetogaster langi* on ciliated protozoa. *Trans. Amer. Microsc. Soc.* 99:360–367.

Taylor, W.D. (1986) The effect of grazing by a ciliated protozoan on phosphorus limitation of heterotrophic bacteria in batch culture. *J. Protozool.* 33:47–52.

Taylor, W.D., and Berger, J. (1976a) Growth of *Colpidium campylum* in monaxenic batch culture. *Can. J. Zool.* 54:392–398.

Taylor, W.D., and Berger, J. (1976b) Growth responses of cohabitating ciliate protozoa to various prey bacteria. *Can. J. Zool.* 54:1111–1114.

Todd, R. (1965) A new *Rosalina* (Foraminifera) parasitic on a bivalve. *Deep-Sea Res.* 12:831–837.

Townsend, D.W., and Cammen, L.M. (1985) A deep protozoan maximum in the Gulf of Maine. *Mar. Ecol. Prog. Ser.* 24:177–182.

Tremaine, S.C., and Mills, A.L. (1987a) Inadequacy of the eucaryote inhibitor cycloheximide in studies of protozoan grazing on bacteria at the freshwater-sediment interface. *App. Environ. Microbiol.* 53:1969–1972.

Tremaine, S.C., and Mills, A.L. (1987b) Tests of the critical assumptions of the dilution method for estimating bacterivory by microeucaryotes. *Appl. Environ. Microbiol.* 53:2914–2921.

Triemer, R.E. (1986) Observations on a recently described genus of phagotrophic euglenoid *Serpenomonas costata*. *J. Protozool.* 33:412–415.

Triemer, R.E., and Fritz, L. (1987) Structure and operation of the feeding apparatus in a colorless Euglenoid, *Entosiphon sulcatum*. *J. Protozool.* 34:39–47.

Tucker, J.B. (1968) Fine structure and function of the cytopharyngeal basket in the ciliate *Nassula*. *J. Cell Sci.* 3:493–514.

Tucker, J.B. (1972) Microtubule arms and propulsion of food particles inside a large feeding organelle in the ciliate *Phascolodon vorticella*. *J. Cell Sci.* 10:883–903.

Tucker, J.B. (1974) Microtubule arms and cytoplasmic streaming and microtubule bending and stretching of intertubule links in the feeding tentacle of the suctorian ciliate *Tokophrya*. *J. Cell Biol.* 62:424–437.

Tuffrau, M. (1959) Polymorphisme par anisotomie chez le cilié *Euplotes balteatus* Dujardin. *C.R. Acad. Sci.* 248:3055–3057.

Tunnicliffe, V., Juniper, S.K., and deBurgh, M.E. (1985) The hydrothermal vent community on axial seamount, Juan de Fuca Ridge. In Jones, M.L. (ed.), *The Hydrothermal Vents of the Eastern Pacific: An Overview*. *Bull. Biol. Soc. Wash.* 6:453–464.

Turley, C.M., Newell, R.C., and Robins, D.B. (1986) Survival strategies of two small marine ciliates and their role in regulating bacterial community structure under experimental conditions. *Mar. Ecol. Prog. Ser.* 33:59–70.

Turner, J.T., and Anderson, D.M. (1983) Zooplankton grazing during dinoflagellate blooms in a Cape Cod embayment with observations of predation upon tintinnids by copepods. *P.S.Z.N.I. Mar. Ecol.* 4:359–374.

Turner, J.T., Bruno, S.F., Larson, R.J., Staker, R.D., and Sharma, G.M. (1983) Seasonality of plankton assemblages in a temperate estuary. *P.S.Z.N.I. Mar. Ecol.* 4:81–99.

Tuttle, J.H., Wirsen, C.O., and Jannasch, H.W. (1983) Microbial activities in the emitted hydrothermal waters of the Galapagos Rift vents. *Mar. Biol.* 73:293–299.

Uye, S. (1986) Impact of copepod grazing on the red-tide flagellate *Chattonella antigua*. *Mar. Biol.* 92:35–43.

Verity, P.G. (1985) Grazing, respiration, excretion and growth rates of tintinnids. *Limnol. Oceanogr.* 30:1268–1282.

Verity, P.G. (1986) Growth rates of natural tintinnid populations in Narragansett Bay. *Mar. Ecol. Prog. Ser.* 29:117–126.

Verity, P.G. (1987) Abundance, community composition, size distribution and production rates of tintinnids in Narragansett Bay, Rhode Island. *Estuar. Coast Shelf Sci.* 24:671–690.

Verity, P.G. (1988) Chemosensory behavior in marine planktonic ciliates. *Bull. Mar. Sci.* 43:772–782.

Verity, P.G., and Langdon, C. (1984) Relationships between lorica volume, carbon, nitrogen and ATP content of tintinnids in Narragansett Bay. *J. Plankton Res.* 6:859–868.

Verity, P.G., and Stoecker, D. (1982) Effects of *Olisthodiscus luteus* on the growth and abundance of tintinnids. *Mar. Biol.* 72:79–87.

Verity, P.G., and Villareal, T.A. (1986) The relative food value of diatoms, dinoflagellates, flagellates and cyanobacteria for tintinnid ciliates. *Arch. Protistenkd.* 131:71–84.

Vickerman, K. (1976) The diversity of kinetoplastid flagellates. In Lumsden, W.H.R., and Evans, D.A. (eds.), *Biology of the Kinetoplastida V.I.* London: Academic Press, pp. 5–8.

Virnstein, R.W. (1977) The importance of predation by crabs and fishes on benthic infauna in Chesapeake Bay. *Ecology* 58:1199–1217.

Vitiello, P. (1964) Contribution a l'etude des tintinnids de la Baie d'Alger. *Pelagos* 2:5–42.

Wailes, G.H. (1927) Rhizopodia and Heliozoa from British Columbia. *Ann. Mag. Nat. Hist.* 9th Ser. 20:153–156.

Wailes, G.H. (1937) Canadian Pacific fauna. 1. Protozoa (a, Lobosa; b, Reticulosa; c, Heliozoa; d, Radiolaria) Biol. Bd. Canada, Toronto, 14 p.

Webb, M.G. (1965) An ecological study of brackish water ciliates. *J. Anim. Ecol.* 25:148–175.

Weisburd, S. (1984) Seafloor vents found in the Atlantic. *Sci. News* 126:246.

Wessenberg, H., and Antipa, G.A. (1970) Capture and ingestion of *Paramecium* by *Didinium nasutum*. *J. Protozool.* 17:250–270.

Wikner, J., Andersson, A., Normark, S., and Hagstrom, A. (1986) Use of genetically marked minicells as a probe in measurement of predation on bacteria in aquatic environments. *Appl. Environ. Microbiol.* 52:4–8.

Wise, D.L. (1959) Carbon nutrition and metabolism in *Polytomella caeca*. *J. Protozool.* 6:19–23.

Withers, N.W., Kokke, W.C.M.C., Rohmer, M., Fenical, W.H., and Djerassi, C. (1979) Isolation of sterols with cyclopropyl-containing side chains from the cultured marine alga *Peridinium foliaceum*. *Tetrahedron Lett.* 38:3605–3608.

Wright, R.T., and Coffin, R.B. (1984) Measuring microzooplankton grazing on planktonic marine bacteria by its impact on bacterial production. *Microbiol. Ecol.* 10:137–149.

Yentsch, C.S., and Ryther, J.H. (1959) Relative significance of the net phytoplankton and nanoplankton in the waters on Vineyard Sound. *J. Cons. Perm. Int. Exp. Mer.* 24:231–238.

Yonge, C.M. (1954) Food of invertebrates. *Tabulae Biol.* 21:25–45.

Young, D.K., Buzas, M.A., and Young, M.W. (1976) Species densities of macrobenthos associated with seagrass: a field experimental study of predation. *J. Mar. Res.* 34:577–592.

Zaitsev, Yu. P. (1970) In Vinogradow, K.A. (ed.), *Marine Neustonology*. Israel Prog. Sci. Transl. Jerusalem, 207 pp.

Zumwalt, G.S., and Delaca, T.E. (1980) Utilization of brachiopod feeding currents by epizoic foraminifera. *J. Paleontol.* 54:477–484.

5

Nutrition and Growth of Marine Protozoa

JOHN J. LEE

INTRODUCTION AND PERSPECTIVE

I will never forget the excitement and joy of our late colleague E. Fauré-Fremiet as he mused over his six decades of interest in the complex nature of natural assemblages containing protozoa. As he turned to leave my lab (he was an octogenarian then), he paused for a second and returned for a final comment. He reflected, "Ah, but the best is yet to come." How far have we really come in the 20 years since Fauré-Fremiet (1967) summarized his impressions of the chemical aspects of ecology (his terms). Be your own judge. Most questions on the complexities of the interactions of protozoan populations with the microbial assemblages upon which they feed remain unresolved. However, we have begun to understand the syntax of community relationships. By building conceptual frameworks we have gained additional perspective to ask much more sophisticated questions of ecosystem dynamics at scales meaningful to protozoan ecology (e.g., Lee, 1980a; Lee, 1982; Fenchel, 1982b).

In analyzing feeding mechanisms and nutritional requirements we must constantly remind ourselves to avoid static concepts of habitats and niches. Studies have shown that growth rate constants for the same organism can vary over an order of magnitude according to environmental and nutritional factors (e.g., Rubin and Lee, 1976; Curds and Cockburn, 1971; Fenchel, 1982). By extension we interpret these ranges as adaptations to life in patchy environments (e.g., Wiens, 1976; Rubin and Lee, Chap. 6 of this volume).

As one attempts to relate nutrition and growth of protozoa in meaningful ("real-life") ecological terms, one is torn between the extremes of minute details of the nutrition of individual species and the context of those species in complex and heterogeneous communities dynamically changing in time and space. Disregarding, for the present, other biological aspects of niche, one of the most difficult factors to evaluate contextually and energetically is that of food quality. It has been argued on theoretical grounds that one of the keys to the success of protozoa throughout evolutionary history is their unicellularity (Lee, 1980a). Although unicellularity usually limits their size to within a cubic centimeter or less, and to some degree their ranges of response to environmental signals, there is good reason to believe that during evolution they have refined and honed their genetic and metabolic machinery to be energetically optimized and responsive to changes in heterogeneous habitats. The optimization principles are logical extensions of those that have long been recognized as operating in bacteria. In a habitat with heterogeneous distributions of prey a protozoan could be quite successful if it were able to synthesize adaptive enzymes to handle food it is ingesting. The energetic gain by not having complete batteries of constitutive enzymes to digest myriads of bacterial and algal cellular constituents that end up in food vacuoles has to be offset by a favorable cost–benefit ratio when the enzymes are synthesized. To do this a protozoan must graze on a food source patch long enough for it to tool up and pay the cost of metabolic adaptations. It follows logically that organisms adapted for this type of feeding strategy could be poorly adapted energetically to survive in communities in which mixtures of different types of food organisms are more homogeneously distributed.

We can look at food quality in a slightly different way. The value of an organism as food is both an intrinsic property of its molecular constitution and the molecular needs and metabolic capabilities of its protozoan consumer. The closer the consumed is matched to the energetic and metabolic needs of the consumer, the less energy is lost in degradation and new synthesis. I have suggested previously (Lee, 1980a) that we can look at the diet of a protozoan as being divisible into three categories: bulk metabolites and nutrients, stimulating metabolites, essential metabolites. Bulk nutrients are those that provide sources of energy and supply general metabolites needed for synthesis, growth, and maintenance. Stimulating nutrients are those that the host is able to synthesize itself but that, if brought into the metabolic pool, advance the cell cycle by reducing the time and energy that would otherwise be expended on their synthesis. Essential metabolites are those that the animal is unable to synthesize and that must be included in the diet.

The value of food in webs is customarily estimated in caloric terms. These energy units are easy to measure under standardized conditions, easily repeated and verified, and easy to conceptualize. However, we have argued (Rubin and Lee, 1976; Chap. 6, this volume) that such conceptualizations are oversimplifications and lack insight into the uniqueness of biological organization. It costs energy to synthesize the metabolites, macromolecules, and biological structure that we recognize as the essence of a particular type (species) of living organism. This uniqueness, or molecular improbability, cannot be easily recognized by present physical or chemical methods of analysis (e.g., calorimetry). By its nature it represents stored energy in the form of molecular information. Most of this informational energy is lost when prey is ingested and structure is digested and degraded. It is rare when hosts conserve the organelles of prey in their entirety and co-opt them for their own use (e.g., Lopez 1979; Taylor, Chap. 9 of this volume). Food quality factors (informational energy flow) enter into consideration at the molecular scavenging level, which is when digestive processes have released metabolites that can be channeled into synthetic pathways. This aspect underlies the great range in ecological growth efficiencies observed in protozoa (reviewed by Calow, 1977). High efficiencies (96–99 percent) are found in the relatively sedentary ciliates *Stentor coreleus* and *Podophrya fixa* (Laybourn, 1976, a,b). High net production efficiencies were also found in *A. proteus* (50–82 percent) at 15° and 20°C (Rogerson, 1981; Griffin, 1960), *Acanthamoeba* (58 percent; Heal, 1967), and *Chaos Chaos* (Griffin, 1960). The ecological efficiencies of *Euplotes vannus* and *Uronema marinum,* two benthic marine ciliates from the same benthic habitat, were tested in monoxenic culture. Their ecological growth efficiencies varied with their diet (Rubin and Lee, 1976). *Uronema* was the more efficient of the two ciliates. Its ecological growth efficiency approached 21 percent on diets of three chlorophyte species but dropped to only 6 percent on other species. *Euplotes* highest growth efficiency was only 10–12 percent on the best of the food organisms tested and dropped to 2 percent on three other species. Rubin and Lee suggested a way to evaluate informational energy flow by calculating the energetic gain in processing particular food species as a ratio of the calorific content of the food to the consumer's energy expended as follows:

$$\mathrm{EG} = \frac{F_i - C_i}{R_i - R_i k}$$

where

EG is energetic gain.
R_i is the total respiration of the consumer growing on a diet of food i per unit time.
k is the percentage of the total respiration used for maintenance (Stebbing, 1974).
F_i is the number of food organisms of type i ingested by the consumer per unit time.
C_i is the caloric value of a single food organism of type i.

From this expression we can see that the energetic gain from processing high informational molecules is highest as R_i approaches maintenance respiration ($R_i k$), or in other terms, the metabolic cost of processing the food is minimal. The food organism with the lowest EG value that would still support continuous growth and reproduction of the ciliate species was chosen as a base to which all other food species could be compared. The difference between the energetic cost of growth on the lowest-EG food and other foods was then expressed in terms of energy saved per food organism processed per consumer generation. The units of energy saved were expressed as cyberons. This unit of information gain in food web transformation can be translated to other energy units through fundamental respiration relations following the reasoning of Forrest and Walker (1971). To interconvert ATP saved per generation readily to cyberons, the cyberon value is set equal to the hydrolysis of high-energy phosphate bonds such as phosphoenolpyruvate (~ 15 kcal). Cyberons are practically calculated as follows:

$$\text{Cyberon content} = \mathrm{Cyb_{ase}} = \frac{(1 - k) \times R_i \times 36.0}{F_i C_i \times 22.4 \times 10^6}$$

where

k, R_i, F_i, C_i are defined as before,
$\mathrm{Cyb_{ase}}$ is the EG value of the food chosen as a base converted to ATP.

The informational content of food can then be related to ecological efficiency (E_e). Traditionally, E_e has been stated as

$$E_e = \frac{P}{I}$$

where

P is production.
I is ingestion.

This can be restated to include an informational value factor IV, by looking at both *P* and *I* in terms of this additional component:

$$E_e = \frac{P}{I} = \frac{\text{AB}}{N_i\,(C_i + p\text{IV}_i)} = \frac{\text{RC}_i,\ \text{IV}_i)}{N_i\,(C_i + pIV_i)}$$

where

P and *I* are defined as before.
AB is the change in biomass.
N_i is the number of food organisms of type *i* ingested to advance the consumer to its next generation.
*p*IV is the potential informational value of food source *i*, C_i is the caloric value of food type *i*.
$f(C_i,\ \text{IV}_i)$ is a function that relates production (*P*) to the caloric value of the food source and the informational content usable by the consumer.

Returning to niche and ecosystem considerations, informational energy processing is a mechanism for lowering competition and increasing the diversity of habitat to provide many more niches for highly specialized and rapidly reproductively responsive (*r* selected) organisms. For example, when Muller and Lee (1977) found that one benthic marine ciliate, *Euplotes vannus,* was a poor competitor for food in gnotobiotic laboratory cultures but increased in numbers five times faster in cultures of certain species of algae in which it was the only herbivore, they concluded it was best adapted to being a migrating initial colonizer of fresh blooms of particular species of algae. This opportunistic species is probably able to maintain itself between blooms by consuming algae and bacteria less energetically valuable to it ($E_e \approx 2$ percent; Rubin and Lee, 1976).

FLAGELLATES

Since the most recent review containing information on the nutrition and growth of marine protozoa in gnotobiotic culture by Provasoli (1977) there has been comparatively very little new research in the area. It is to be hoped that the lack of interest is only cyclical. Most of the work on flagellates was done by Provasoli's Haskins Lab group and his students. Four colorless flagellates—*Crypthecodinium cohnii, Noctiluca miliaris, Diaphancoca grandis,* and *Acanthoecopsis* sp.—received their attention. Although *C. cohnii* has been observed to be holozoic, it seems quite well adapted for osmotrophy. Initially, two strains of *C. cohnii,* GC and PR, were isolated in axenic culture on a medium containing yeast digest and acetate (Provasoli and Gold, 1962; Gold and Baren, 1966). The GC strain initially required histidine as a nitrogen source (other amino acids tested were not utilized), but subcultures were weened to grow on only ammonium salts. Best growth, however, was obtained in media containing $(NH_4)_2SO_4$, histidine, and betaine. The PR strain utilized glutamic acid and alanine as well as ammonium salts at optimum temperatures (20–28°C). Both strains could use PO_4^{-3}, sodium glycerol phosphate, and nucleic acids as P sources. A variety of low-molecular-weight organic compounds (e.g., glucose, glycerol, ethanol, acetic, succinic, and fumaric acids) served as carbon and energy sources. Biotin was required by both strains, and thiamin is stimulatory. Provasoli and Gold (1969) suggested that *C. cohnii* could be quite a useful bioassay organism for biotin in natural seawaters. *C. cohnii* has one unusual physiological characteristic for a marine organisms. Its optimum pH is quite low (pH 5.7–7.2). Overall, its nutritional requirements seem quite in consonance with what it probably encounters in the decaying seaweed habitats in which it is normally found. Interest in dinoflagellate genetics has resulted in the isolation of more than 200 strains, all of which grow well in Gold and Baren's (1966) medium (Beam and Himes, 1984).

Noctiluca miliaris was not easily isolated in axenic culture (McGinn and Gold, 1969). It grew monoxenically on *Platymonas tetrathele* and was eventually isolated axenically on a medium containing diatomaceous earth particles. Presumably, the particles stimulated formation of food vacules that were necessary for the uptake of metabolites. Axenic cultures could be grown in a medium containing 12 amino acids, sodium glycerol phosphate, biotin, thiamin, and vitamin B_{12}, a mixture of other vitamins, glucose, a small amount (10 mg/L) of RNA, DNA, and various salts.

The two bacterial feeding choanoflagellates, *Diaphaneca grandis* and *Acanthoecopsis* sp., isolated in axenic culture by Gold and his co-workers (1970) apparently have complex nutritional requirements. *D. grandis* grew osmotrophically in a medium containing salts, acetate, several vitamins, and liver concentrate. Using the same medium as a base, *Acanthoecopsis* needed the addition of proteose peptone, glucose, and pyruvate to grow.

The nutritional requirements of the voracious phagotrophic dinoflagellate *Oxyrrhis marina* were wrested from this beast after almost 20 years of toil by M. Droop (1953) and his co-workers (Droop and Pennock, 1971). Acetate or ethanol serve equally well as the major carbon and energy source. Although NO_3^{-1}, NH_4^{+1}, urea, and many amino acids did not serve as nitrogen sources, valine, alanine, and proline satisfied this requirement (Droop, 1959a,b). Vitamin B_{12} and thiamine (or its thiazole moiety) were essential. Biotin was stimulatory. *Oxyrrhis* has one, perhaps unique, growth factor: ubiquinone or plastoquinone (Droop and Doyle, 1966). The requirement is quite specific, since modifications of the nucleus or side chain of the molecule (e.g., vitamins K and E and chromaol) were inactive. It also requires a sterol. It can use dehydrocholesterol, cholestanol, cholestenone, cholestane, sito, stigma, and ergosterol but not vitamins D_2 or D_3 (Droop and Pennock, 1971). The need for the water-insoluble and light-labile ubiquinone and plastoquinone suggests that even in its nutrient-rich natural habitat, the brackish supralittoral zone, *Oxyrrhis* could not survive as an osmotroph. Since both are normal components of chloroplasts, feeding on algae in the habitat presents little problem in meeting this nutritional need. Along these lines, Droop (1966) found that *Oxyrrhis* grows well on a variety of chlorophytes, eustigmaphytes, prasinophytes, diatoms, cryptophytes, and rhodophytes.

The heterotrophic abilities and B vitamin requirements of many dinoflagellates have been known for quite some time (reviewed by Provasoli and McLaughlin, 1963; Provasoli and Carlucci, 1974; Loeblich III, 1984). Many photosynthetic dinoflagellates require biotin, thiamin, or B_{12}; some require combinations of these three vitamins. The growth in the light of three *Gyrodinium* spp. was stimulated by the addition of asparagine, alanine, glycine, and glutamate (Provasoli and McLaughlin, 1963). More recently, Baden and Mende (1978), Wright and Hobbie (1966), and Morrill and Loeblich (1979) have tested other marine species. None of the photosynthetic species tested grow heterotrophically in the dark. However, five species were stimulated by 50 m*M* glycerol in dim light (Cheng and Antia, 1970; Morrill and Loeblich, 1979). A recent good source for the methodology and media for isolating and culturing dinoflagellates is the review by Guillard and Keller (1984).

Recently, there has been considerable interest in the nano- and picoplankton and the roles that microflagellates may play in the trophic dynamics and nutrient regeneration in pelagic waters (e.g., Pomeroy, 1970; Hass and Webb, 1979; Sieberth et al., 1977; Haas, 1982; Goldman and Caron 1985, and Chap. 7 of this volume; Fenchel, 1982d, Capriulo, Chap. 4 of this volume). As correctly pointed out by Fenchel (1982a), these small flagellates are quite diverse morphologically and belong to a number of unrelated groups (choanoflagellates, bicoecids, kinetoplastids, euglenoids, dinoflagellates, and chrysomonads). In his bioenergetic and feeding experiments with four marine microflagellates (*Monosiga* sp., *Paraphysomonas vestita, Pseudobodo tremulans, Actinomonas mirabilis*), and two limnic species (*Ochromonas* sp. and *Pleuromonas jaculans*), Fenchel (1982b) found that growth of all the species in batch cultures was a function of bacterial concentrations within a range of 1–5 × 10^7 bacteria/ml. Flagellate cell volume and morphology varied greatly, depending upon their nutritional state (Fenchel, 1982b,c). For some species (e.g., *Monosiga* sp. and *Pseudobodo tremulans*) the volume of starved cells fell to only 20–25 percent of the volume of actively growing cells; for the others it fell to 50 percent of growing cells. Highest growth rates (μ_m) were in the range of 0.15–0.25 h corresponding to generation times between 2.8 and 4.6 hours. The data obtained by Fenchel (1982a–c) in his studies seem to mesh well with his field observations (Fenchel, 1982d) and support the thesis that microflagellates are the main consumers of pelagic bacteria.

A strain of *Paraphsomonas imperforata* isolated from Vineyard Sound, Massachusetts, grazes on phytoplankton as well as on bacteria. Its growth rate on the six species of algae upon which it fed (*Dunaliella tertiolecta, Chlorella stigmataphora, Chlorella capsulata, Isochrysis galbana, Porphyridium* sp., and *Phaeodactylum tricornutum*) was fairly uniform (2.4/d at 24°C and 1.4/d at 20°C) but was lower than when it was feeding on bacteria (3.5/d, Goldman and Caron, 1985). An important conclusion from this study was that conceptualizations of marine microbial food webs involving protozoa should be flexible enough to cover greater size range ratios (2–10) between predators and food species and also flexible enough to include omnivorous feeding behavior. They also concluded that microflagellates play significant roles in nutrient (nitrogen) regeneration.

SARCODINIDS

The nutrition of foraminifera has received considerable attention (reviewed by Lee, 1974, 1980b). Although many of the studies have been agnotobiotic assessments aimed at improving culture yields, there have also been some that detailed requirements in gnotobiotic cultures. Several generalizations emerge. Feeding by foraminifera is erratic when food is scarce (below 1 × 10^3 ml^{-1}) and proportional to concentration within a range of 10^3–10^6 cells (e.g., Bradshaw, 1961; Lee et al., 1966). Evidence seems to indicate that feeding and or assimilation is quite selective in the benthic foraminifera that have been studied. For example, one species, *Allogromia* sp. (NF), the only species to be isolated in monoxenic culture, seems to be a bacterivore (Lee and Pierce, 1963). Another species of *Allogromia, A. laticollaris,* was much more fecund on diets of particular mixtures of algae than it was on a diet of single algal species (Lee et al., 1969). Comparative data on the fecundity of gnotobiotic cultures of *Rosalina leei, Quinqueloculina lata, Spiroloculina hyalina,* and *A. laticollaris* on different diets of bacteria, single species of algae, and mixtures of algae clearly indicated some niche separation on the basis of diet (Muller and Lee, 1969). In general, a number of small, weakly silicified pennate diatoms (*Amphora* spp. [*Halamphora*], *Nitzschia hungarica, brevirostris, N. acicularis, N. ovalis, Phaeodactylum tricornutum, Cylindrotheca closterium*) and some small unicellular chlorophytes (*Dunaliella* spp., *Nannochloris* sp., *Chlorococcum* sp., *Chlorella* spp.) seem good nutritional sources for many shallow benthic and littoral foraminifera (Muller, 1975; Lee et al., 1969; Lee and Muller, 1973, Muller and Lee, 1969). The foraminifera did not reproduce continuously on bacteria-free diets, which led to the inference that bacteria contain a nutritional factor(s) necessary for the foraminifera but unavailable, or not available in sufficient quantities, in an exclusively algal diet (Muller and Lee, 1969).

Although planktonic foraminifera have not yet been cultured in the laboratory, feeding and survival have been the subject of a number of studies (Bé et al., 1977; Spindler et al., 1984; Anderson and Bé, 1976; Anderson et al., 1979). *Globorotalia truncatulinoides* survived best (21 days) when offered a diet of the coccolithophorid *Emiliania huxeyi,* compared with only 16 days for controls (Anderson et al., 1979). Survival times on unialgal diets of *Thalassiosira pseudonana* (centric diatom) or *Gymnodinium* sp. (dinoflagellate) were less (14 days) than those of controls. *G. truncatulinoides* survived even longer (35 days) when it was fed *Artemia* nauplii once every three days. Mean survival was reduced to only 22 days when the feeding interval was extended to 12 days. *Hastigerina pelagica* also survived longest (27 days) on a diet of *Artemia* nauplii. Controls survived, on the average, only 16 days. Mean survival of this species was about the same if the animals were fed daily or every six days but declined (17 days) if the interval between feeding was stretched to 12 days. The authors concluded that these two planktonic species are omniverous and capable of surviving for long periods of time in the fluctuating patches of prey they are likely to encounter in the seas where they flourish. Feeding experiments with another planktonic foraminifer, *Globigerinoides sacculifer,* suggested that growth rates were increased in proportion to frequency of feeding (one to seven days). Starved controls did not grow (Bé et al., 1981). In fact, some of the starved animals shrank in size by resorbing chambers. Increased feeding frequency correlated directly with frequency to build a saclike final chamber, a prelude to gametogenesis. Almost all the fed animals (~87 percent) underwent gametogenesis.

Spindler and co-workers (1984) studied feeding behavior in a variety of planktonic foraminifera in laboratory cultures in order to gain more comparative insight into the food requirements and digestive abilities of the group. Spinose species fed mainly on zooplankton, whereas the nonspinose species tested (*Globorotalia truncatulinoides, G. hirsuta, G. inflata, Pulleniatina obliquiloculata*) fed chiefly on diatoms. The latter group seemed less able to capture and hold active copepods. Their digestion times were also longer. Most of the calanoid copepods tested (*Calocalanus parvo, Euchaeta marina, Clausocalanus* sp., *Arcatia spinata, Undinula vulgaris*) were accepted by the spinose foraminifera tested (*Globigerinoides sacculifer, G. ruber, Globigerinella aequilateralis, Orbulina universa,* and *Hasterigena pelagica*) and digested within seven to nine hours. *E. marina,* however, always escaped after capture by *Globigerinoides sacculifer.* The other three were held and digested by *G. sacculifer,* which also was not able to digest the harpacticoid and only one of the cyclopoid species tested. *G. sacculifer* was able to completely digest the nauplii of *Artemia salina* in 3.5 hours. Besides copepods, most other smaller zooplankton (e.g., pteropods, chetognates, tunicates, tintinnids, radiolarians, acantharians, polychaete and gastropod larvae, ostacods) were captured and digested by *G. sacculifera.* Of the other spinose species tested, *G. ruber* seemed to be least adapted to feed on copepods because its spines tend to shed relatively easily. Spindler et al., using the data they obtained on digestion times and the percentage of freshly collected specimens with food remains, deduced that spinose planktonic foraminifera catch and digest at least one calanoid copepod equivalent per day. This seems in agreement with the Anderson (1983) estimates of the zooplankton–phytoplankton protein consumption ratio by *G. sacculifera* (185X) and *G. ruber* (117X), but it is greater than the earlier deduction of Bé and co-workers (1981) that *G. sacculifer* gets food less than every seventh day.

The contributions of endosymbiotic algae to the nutrition of the various foraminiferan hosts that bear them is not yet fully understood (Lee, 1980b, 1983; Lee and McEnery, 1983; Kuile and Erez, 1984).

For example, the data obtained by microelectrode studies of the photosynthetic abilities of the dinoflagellate-bearing planktonic foraminifer *Globigernoides sacculifer* suggested that the respiration of the host–symbiont system was only 17 percent of the gross photosynthesis at light saturation (Jorgensen et al., 1985). With such an excess of organic production the foraminiferan could theoretically cover its carbon requirements from algal photosynthates. However, as Jörgensen et al. (1985) and others (Bé et al., 1981; Caron et al., 1981) note, the animals require additional food to grow in the laboratory. It has been suggested that feeding might be the source of micronutrients (N and P) and vitamins for these protozoa, which abound in oligotrophic waters (Lee et al., 1979; Lee 1980, 1984; Hallock, 1981). Experiments with symbiont-bearing larger foraminifera, with one exception, have usually suggested that feeding is the major carbon pathway in nutrition. Estimates of carbon flow based on radionuclide tracer feeding and respirometric measurements of four species of larger foraminifera (*Archaias angulatus, Sorites marginalis, Amphisorus hemprichii,* and *Amphistegina lobifera*) give feeding–photosynthesis ratios of approximately 10:1 (Lee and Bock, 1976; Lee et al., 1980). *Heterostegina depressa* seems to be an exception. It seems to grow well at its optimal irradiance (600 lux) in the absence of obvious quantities of food organisms (Röttger, 1972, 1976; Röttger et al., 1980). It is clear that in all the larger foraminifera, light-mediated processes, possibly through the photosynthetic activities of symbionts or their byproducts (Kremer et al., 1980), are required to maintain the health of host–symbiont systems and to stimulate growth (Duguay, 1983; Hallock, 1981; Duguay and Taylor, 1978; Kuile and Erez, 1984). Growth rates of two species (*Amphistegina lobifera* and *Amphisorus hemprichii* incubated in the Gulf of Elat were an order of magnitude higher in the light than in the dark (Kuile and Erez, 1984). Feeding in the dark increased growth rates three times the rate of starved controls incubated in the dark. In the Caribbean the calcification rate of three species (*Archaias angulatus, Sorites marginalis,* and *Cyclorbiculina compressa*) was approximately two to three times that of dark rates (Duguay, 1983).

In recent years there has been considerable interest in the trophic dynamics and nutrition of radiolaria (reviewed by Anderson, 1983a,b). Laboratory studies of prey selectivity in freshly collected specimens of *Thalassiocolla* sp., *Collozoum inerme, Sphaerozoum punctatum, Spongodrymus* sp., *Hexastylus* sp., and *Diplosphaera* sp., indicate that many species are omnivorous (Anderson, 1976, 1978, 1980, 1983b, 1985; Swanberg et al., 1986a–c). They accept in culture calanoid and harpacticoid copepods, nauplii, dinoflagellates, coccolithophorids, ciliates, and so on. Fine structural studies suggested that some Nassellaria consume bacteria (Anderson, 1977). Examination of living radiolarians collected in particle traps in the North Pacific central gyre suggests that the deeper-dwelling phaeodarians were feeding on marine snow (Gowing, 1986). Phaeodarian food vacuoles contained mainly amorphous materials, diatom fragments, bacteria, and strips of crustacean cuticles, mostly the same types of debris and organisms found in the traps. The content of the vacuoles of two species (*Conchidium caudatum* and *Euphysetta elegans*), from 700 and 900 meters respectively, were an exception. The content of the former seemed to contain only diatom fragments, and that of the latter had darkly stained particles and nematocyst remains.

Radiolaria with algal symbionts have adapted at least two strategies for obtaining nourishment from their symbionts. Radionuclide tracer studies (^{14}C) have shown passage of symbiont fixed photosynthetates into the central capsules of hosts, and fine structural studies have shown symbionts in the process of digestion by their hosts. In this regard colonial radiolaria are almost ideal experimental organisms. The central capsule of *Collosphaera globularis* has such fine pores that symbionts cannot pass through them. Anderson (1978) incubated colonies in $H^{14}CO_3^-$ seawater in the light and dark (controls) and then gently disrupted the colonies and washed the central capsules free of symbionts. The central capsules of this radiolarian were much more heavily labeled in the light than in the dark. The longer the radiolaria were incubated in the light, the greater was the labeling. Analysis of the labeled product in the central capsules of *Collosphaera huxleyi* showed that 38 percent was water soluble, 20 percent was lipid soluble, and the remainder was insoluble. Within the lipid-soluble fraction a substantial percentage of the label was associated with the triglyceride and wax ester fractions (Anderson et al., 1985). In *Thalassiocola nucleata* approximately 80 percent of the total carbon incorporated in the extract from the central capsules was in the lipoidal fraction; in *Collosphaera* sp. it was approximately 90 percent (Anderson, 1985). Recent studies of *Physematium muelleri* suggest that it obtains over half its organic carbon through the photosynthetic activities of its symbionts. The remainder comes from predation on copepods and small prey (Swanberg et al., 1986a,b).

CILIATES

Only a half dozen ciliates have been isolated in axenic culture: *Miamiensis avidus, Miamiensis* sp., *Paramecium calkinsi, Uronema marinum, Uronema nigricans, Parauronema virginiatum* (reviewed by Provasoli, 1977). Holidic (chemically defined) media that support the ciliates in axenic culture are quite complex mixtures containing 18 amino acids, nucleotides, lipids, and B vitamins (folic acid, biotin, nicotinamide, Ca pantothenate, pyridoxal HCl, riboflavin, thiamine, DL-thioctic acid). The studies on holidic media (Lee et al., 1971; Hanna and Lilly, 1974; Soldo and Merlin, 1972) suggest that the ciliates isolated so far have nutritional requirements similar to those of *Paramecium aurelia* (Soldo and Van Wagtendonk, 1969) and *Tetrahymena pyriformis* (Holz, 1973).

Tangentially related to both nutrition and ecophysiology is work done on the effects of copper and zinc on the growth of two planktonic ciliates, *Balanion* sp. and *Favella* sp. (Stoecker et al., 1986). The growth of *Favella* was optimal at $1 \times 10^{-13} M$ Cu^{2+} and between 1×10^{-11} and 1×10^{-12} M Zn^{+}. Above these values both ions inhibited growth. In contrast, *Balanion* grew optimally at a Zn ion concentration of 1×10^{-10} M and at a Cu^{2+} concentration between 1×10^{-12} and 1×10^{-13} M. The ecological implications of such findings are that trace metal ion activities could be important niche parameters for marine ciliates and could influence the productivity and species composition of the assemblages of which they are a part. This aspect of marine protozoan nutrition–ecophysiology begs further study.

Gold's (1970) suggestion that the nutritional needs of tintinnids might be met osmotrophically has found support in a study by Pavillon and Rassoulzadegan (1980). Using ^{14}C radionuclide tracer tags, they found that *Stenosemella ventricosa* very rapidly takes up dissolved free amino acids (V_{max} 5 μg at C g^{-1} ciliate h^{-1}). This could be a significant carbon pathway during blooms of algae or during their crashes.

Myxotrophy also seems to be a nutritional mode in a number of pelagic ciliates. One of the most interesting is the cosmopolitan coastal species *Myrionecta* (formerly *Mesodinium*) *rubrum*. Fine structural studies by Hibberd (1977), Oakley and Taylor (1978), and Grain and co-workers (1982) have shown that this ciliate is the host for incomplete endosymbiotic cryptophytes. All the populations of *M. rubrum* have chloroplasts, their associated pyrenoids, and mitochondrial complexes in membrane-bounded compartments (CML) that are separate from the one that surrounds the nucleus. Since the ciliate lacks a cytostome, has never been shown to feed or have food vacuoles, and has often been shown to fix carbon at high rates in its blooms ($0.5–2.2 \times 10^3$ mg C m^{-3} h^{-1}), there is a presumption that it is a photolithotrophic host–symbiont system with some abilities to take up soluble organic compounds from seawater (reviewed by Lindholm, 1985). Recently, it has been shown that a number of ciliates (e.g., *Tontonia appendiculariformis, Laboea strobila,* and *Strombidium* sp.) temporarily husband chloroplasts of their prey (Leval-Peuto et al., 1986; Stoecker and Silver, 1987). In *T. appendiculariformis,* well-preserved sequestered chloroplasts constitute a 10μm-thick layer at the periphery in the median and posterior of the ciliate. Digestive vacuoles are separate and are linked together by a continuous network of dense endoplasmic reticulum, which is a distinctive organizational pattern for a ciliate (Laval-Peuto et al., 1986). Fine structural study suggests that the sequestered plastids are morphologically diverse, originating from dinoflagellates, pyrmnesians, and diatoms (Laval-Peuto et al., 1986). *Laboea strobila* contains as much as 200 pg chl *a*/cell and some *Strombidium* spp. have as much as 100 pg chl *a*/cell (Stoecker and Silver, 1987). The chloroplasts are functional, and the survival and growth of the ciliates is higher in the light than in the dark. The sequestered chloroplasts in *L. strobilia* and *Strombidium* spp. have relatively long life spans (more than two weeks) (Stoecker and Silver, 1987) and are able to fix ^{14}C-labeled carbon (Stoecker et al., 1987).

Selective feeding and growth have been the focus of much recent research (Stoecker et al. 1981, 1983, 1984, 1986; Stoecker and Evans, 1985; Rivier et al., 1986; Scott, 1985; Verity and Villareal, 1986; Gifford, 1985). *Favella* sp., a tintinnid, is apparently a selective dinoflagellate predator (Stoecker et al., 1981). *Balanion* sp. also feeds preferentially on dinoflagellates in mixtures of algae (Stoecker, 1986). In these experiments the growth of *Balanion* was compared on rations of three dinoflagellates (*Heterocapsa triquetra, H. pygmaea,* and *Provocentrum minimum*), a cryptophyte (*Chroomonas salina*), a *prymnesiophyte* (*Isochrysis galbana*), a prasinophyte (*Pyraminimonas* sp.), a chlorophyte (*Dunaliella tertiolecta*), and a cyanobacterium (*Synechococcus* sp.). In the first experiment *Balanion* were given each alga separately or in mixtures of two or three kinds of algae. *Balanion* preferentially fed on the two *Heterocapsa* species in mixtures containing cryptophytes and green flagellates even when the other algae were more abundant. Feeding behavior was also reflected in the growth rates of *Balanion* on diets of different algae or mixtures of algae. The growth of *Balanion* was 3.2 divisions a day on a diet of *Heterocapsa triquetra* but only one division a day on a diet of *Pyramimonas* sp. On diets of *Synechococcus, Dunaliella tertiolecta,* and *Isochrysis galbana,* division rates were even lower. After nine months in culture the maximum division rate slowed to only 1.1 per day on a diet of *Heterocapsa triquetra,* and it would not grow at all with *Isochrysis galbana, Pyrami-*

monas or *Synechococcus* as sole sources of food (Stoecker, 1986).

Since both *Balanion* and *Favella* feed on *Heterocapsa triquetra*, Stoecker and Evans (1985) investigated interactions of populations in multispecies cultures. At the outset one might expect that the effects of two grazers on a single prey species might be additive. Each ciliate imposes a different grazing pressure on the dinoflagellate: *Favella* ingests at a higher rate, but *Balanion* has a shorter generation time. Furthermore, *Favella* ate *Balanion* when populations of the dinoflagellate were reduced. Thus, one of the competitors also becomes a predator when joint prey becomes reduced after exploitation. Adding the extra link (*Heterocapsa–Balanion–Favella*) lowered the relative gross production efficiency of the system algae and allowed more algae to survive because of the reduced grazing pressure by *Balanion*.

It is sometimes difficult to separate the effects of feeding parameters from nutrition. The presence of B-chitin threads in *Thalassiosira weissflogii* has been suggested as a way to increase the effective size of a diatom and reduce its grazability by microzooplankton (Gifford et al., 1981). Two marine oligotrichs (*Strombidium* and *Strombidinopsis*) grew well in enriched seawater containing the dinoflagelates *Heterocapsa* and *Scrippsiella trochoidea* but not when fed *Thalassiosira* (Gifford, 1985).

The effects of feeding of centric diatoms with thread extrusions through their strutted processes (e.g., *Thalassiosira* and *Cyclotella*) was more thoroughly studied by Verity and Villareal (1986). They compared the feeding and growth rates of two tintinnids (*Tintinnopsis vasculum* and *I. acuminata*) on cultures of different algae and on centric diatoms grown in stationary or shaken cultures. In the latter, the threads tend to be shorter. Both tintinnids grew rapidly (1.5 divisions d^{-1}) when fed diatoms lacking threads or small threads or setae (*Chaetoceros*) and did not grow on diatoms with setae. They both also grew well (1.3 divisions d^{-1}) on the prymnesiophytes *Dicrateria, Isochrysis,* and *Pavlova. I. vasculum* also grew well on the dinoflagellates tested (*Gyrodinium estuariale, Heterocapsa pigmaea, Prorocentrum triestinum*). *T. acuminata* did not grow on dinoflagellates but grew equally well on (1.6–1.9 doublings d^{-1}) *Pavlovia lutheri, Nannochloris oculata,* and *Micromonas* sp. Neither tintinnid grew on a diet of the cyanobacterium *Synechococcus*. The chrysophyte *Olisthodiscus luteus* inhibits tintinnid growth at low concentrations and is toxic to them at bloom concentrations (Verity and Stoecker, 1982). The "take home" lesson of these recent studies will have a chilling effect on ecologists with leanings toward simplistic descriptions of food webs with microzooplankton. Both size and food quality are important factors influencing the growth of marine protozoa. It would appear at this juncture that much more detailed work on feeding and nutrition will be needed before the key molecular aspects become more obvious.

REFERENCES

Anderson, O.R. (1976) A cytoplasmic fine-structure study of two spumellarian radiolaria and their symbionts. *Mar. Micropaleontol.* 1:81–99.

Anderson, O.R. (1977) Cytoplasmic fine structure of nasselarian radiolaria *Mar. Micropaleotol.* 2:251–264.

Anderson, O.R. (1978) Fine structure of a symbiont-bearing colonial radiolarian, *Collosphaera globularis*, and ^{14}C isotopic evidence for assimilation of organic substances from its zooxanthellae. *J. Ultrastructure Res.* 62:181–189.

Anderson, O.R. (1983a) *Radiolaria*. New York: Springer-Verlag, 355 pp.

Anderson, O.R. (1983b) Radiolarian symbiosis. In Goff L.J. (ed.), *Algal Symbiosis*. Cambridge, England: Cambridge University Press.

Anderson, O.R., and Bé, A.W.H. (1976) A cytochemical fine structure study of phagotrophy in a planktonic foraminifer, *Hastigerina pelagica* (d'Orbigny). *Biol. Bull.* (Woods Hole) 151:437–449.

Anderson, O.R., Spindler, M., Bé, A.W.H., and Hemleben, C. (1979) Trophic activity of planktonic foraminifera. *J. Mar. Biol. Assoc. U.K.* 59:791–799.

Anderson, O.R., Swanberg, N.R., and Bennett, P. (1983) Assimilation of symbiont photosynthates in some solitary and colonial Radiolaria. *Mar. Biol.* 77:265–269.

Anderson, O. R., Swanberg, N.R., and Bennett, P. (1984) An estimate of predation rate and relative preference for algal versus crustacean prey by a spongiose skeletal radiolarian. *Mar. Biol.* 78:205–207.

Anderson, O.R., Swanberg, N.R., and Bennett, P. (1985) Laboratory studies of the ecological significance of host–algal nutritional associations in solitary and colonial radiolaria. *J. Mar. Biol. Assoc. U.K.* 65:263–272.

Baden, D.G., and Mende, T.J. (1978) Glucose transport and metabolism in *Gymnodinium* breve. *Phytochemistry* 17:1553–1558.

Bé, A.W.H., Caron, D.A., and Anderson, O.R. (1981) Effects of feeding frequency on life processes of the planktonic foraminifer *Globerginoides sacculifera* (Brady) in laboratory culture. *J. Mar. Biol. Assoc. U.K.* 61:257–277.

Bé, A.W.H., Hemleben, C., Anderson, O.R., Spindler, M., Hacunda, J., and Tuntivate-Choy, S. (1977) Laboratory and field observations of living planktonic foraminifera. *Micropaleontology* 23:155–179.

Beam, C.A., and Himes, M. (1984) Dinoflagellate genetics. In Spector, D.L. (ed.), *Dinoflagellates*. Orlando, Fla.: Academic Press, pp. 263–299.

Bradshaw, J.S. (1961) Laboratory experiments on the

ecology of foraminifera. *Cushman Found. Foram. Res. Contr.* 12:87–106.

Calow, P. (1977) Conversion efficiencies in heterotrophic organisms. *Biol. Rev.* 52:385–409.

Caron, D.A., Bé, A.W., and Anderson, O.R. (1981) Effects of variations in light intensity on life processes of the planktonic foraminifer, *Globigerinoides sacculifer,* in laboratory culture. *J. Mar. Biol. Assoc. U.K.* 62:435–451.

Cheng, J.Y., and Antia, N.J. (1970) Enhancement by glycerol of phototrophic growth of marine planktonic algae and its significance to the ecology of glycerol pollution. *J. Fish. Res. Board Can.* 27:335–346.

Curds, C.R., and Cockburn, A. (1971) Continuous monoxenic culture of *Tetrahymena pyriformis. J. Gen. Microbiol.* 66:95–108.

Droop, M.R. (1953) Phagotrophy in *Oxyrrhis marina. Nature* 172:250.

Droop, M.R. (1959) Water soluble factors in the nutrition of *Oxyrrhis marina. J. Mar. Biol. Assoc. U.K.* 38:605–620.

Droop, M.R. (1966) The role of algae in the nutrition of *Heteramoeba clara* droop with notes on *Oxyrrhis marina* Dujardin and *Philodina roseola* Ehrenberg. In Barns, H. (ed.), *Some Contemporary Studies in Marine Science*. London: George Allen and Urwin.

Droop, M.R., and Pennock, J.F. (1971) Terpenoid quinones and steroids in the nutrition of *Oxyrrhis marina. J. Mar. Biol. Assoc. U.K.* 51:455–470.

Droop, M.R., and Doyle, J. (1966) Ubiquinone as a protozoan growth factor. *Nature* 212:1474–1475.

Duguay, L.E. (1983) Comparative laboratory and field studies on calcification and carbon fixation in foraminiferal-algal associations. *J. Foram. Res.* 13:252–261.

Duguay, L.E., and Taylor, D.L. (1978) Primary production and calcification by the soritid foraminifer *Archais angulatus* (Fichtel & Moll). *J. Protozool.* 25:356–361.

Fauré-Frémiet, E. (1967) Chemical aspects of ecology. In Flovkin, M., and Scheer, B.T., eds., *Chemical Zoology*. Orlando, Fla.: Academic Press, pp. 21–54.

Fenchel, T. (1982a) Ecology of heterotrophic microflagellates. II. Bioenergetics and growth. *Mar. Ecol. Prog. Ser.* 8:225–231.

Fenchel, T. (1982b) Ecology of heterotrophic microflagellates. III. Adaptations to heterogeneous environments. *Mar. Ecol. Prog. Ser.* 9:25–33.

Fenchel, T. (1982c) Ecology of heterotrophic microflagellates. IV. Quantitative occurrence and importance as consumers of bacteria. *Mar. Ecol. Prog. Ser.* 9:35–42.

Fenchel, T. (1982d) The bioenergetics of a heterotrophic microflagellate. *Ann. Inst. Oceanogr.*, Paris. 58:55–60.

Forrest, W.W. and Walker, D.J. (1971) The generation and utilization of energy during growth. In Rosen, A.H. (ed.), *Microbial Physiology,* Vol. 5. Orlando, Fla.: Academic Press, pp. 213–214.

Gifford, D.J. (1985) Laboratory culture of marine planktonic oligotrichs (Ciliophora, Oligotrichida). *Mar. Ecol. Prog. Ser.* 23:257–267.

Gifford, D.J., Bohrer, R.N., Boyd, C.M. (1981) Spines on diatoms: Do copepods care? *Limnol. Oceanogr.* 26:1057–1061.

Gold, K. (1970) Cultivation of marine ciliates (Tintinnida) and heterotrophic flagellates. *Helgol. Wiss. Meeresunters* 20:264–271.

Gold, K. and Baren, C.F. (1966) Growth requirements of *Gyrodinium cohnii. J. Protozool.* 13:255–257.

Gold, K., Pfister, R.N., and Liguori, V.R. (1970) Axenic cultivation and electron microscopy of two species of choanoflagellida. *J. Protozool.* 17:210–212.

Goldman, J.C., and Caron, D.A. (1985) Experimental studies on an omnivorous microflagellate: Implications for grazing and nutrient regeneration in the marine microbial food chain. *Deep-Sea Res.* 32:899–915.

Gowing, M.M. (1986) Trophic biology of phaeodarian radiolarians and the flux of living radiolarians in the upper 2000 M of the North Pacific Central Gyre. *Deep-Sea Res.* 33:655–674.

Grain, J., dePuytorac, P., and Grolière, C.A. (1982) Quelques précisions sur l'ultrastructure et la position systématique du cilié *Mesodinium rubrum* et sur la constitution de ses symbiontes chloroplastiques. *Protistologica* 18:7–21.

Griffin, J.L. (1960) Improved mass culture of amoebae. Exp. Cell. Res. 21:170–178.

Guillard, R.L., and Keller, M.D. (1984) Culturing dinoflagellates. In Spector, D.L. (ed.), *Dinoflagellates*. Orlando, Fla.: Academic Press, pp. 391–443.

Haas, L.W., and Webb, K.L. (1979) Nutritional mode of several non-pigmented microflagellates from the York River Estuary, Virginia. *J. Exp. Mar. Biol. Ecol.* 39:125–134.

Hallock, P. (1981) Light dependence in *Amphistegina. J. Foram. Res.* 11:40–46.

Hanna, B.A., and Lilly, D.M. (1974) Growth of *Uronema marinium* in chemically defined medium. *Mar. Biol.* 26:153–160.

Heal, O.W. (1967) Quantitative studies on soil amoebae. In Graff, O., and Satchell, J.E. (eds.), *Progress in Soil Biology*. Amsterdam: pp. 120–126.

Hibberd, D.J. (1977) Observations on the ultrastructure of the cryptomonad endosymbiont of the red-water ciliate *Mesodinium rubrum. J. Mar. Biol. Assoc. U.K.* 57:45–61.

Holz, G.G. (1973) The nutrition of Tetrahymena: essential nutrients, feeding and digestion. In Elliot, A.M. (ed.), *Biology of Tetrahymena*. Stroudsbourg, Pa.: Dowden, Hutchinson and Ross, pp. 89–98.

Jorgensen, B.B., Erez, J., Revsbech, N.P. and Cohen Y. (1985) Symbiotic photosynthesis in a planktonic foraminiferan, *Globigerinoides sacculifer* (Brady), studied with microelectrodes. Limnol. Oceanogr. 30:1253–1267.

Kremer, B.P., Schmaljohann, R., and Rottger, R. (1980) Features and nutritional significance of photosynthates produced by unicellular algae symbiotic with larger foraminifera. *Mar. Ecol. Prog. Ser.* 2:225–228.

Kuile, B., and Erez, J. (1984) *In situ* growth rate experiments on the symbiont-bearing foraminifera *Amphistegina lobifera* and *Amphisorus hemprichii. J. Foram. Res.* 14:262–276.

Laybourn, J.E.M. (1976a) Energy budgets for *Stentor* Coeruleus Ehrenberg (Ciliophora). *Oecololgia* (Berlin) 22:431–437.

Laybourn, J.E.M. (1976b) Respiratory energy losses in *Podophrya fixa* Muller in relation to temperature and nutritional status. *J. Gen. Microbiol.* 96:203–208.

Lee, J.J. (1974) Towards understanding the niche of foraminifera. In Hedley, R.H., and Adams, C.G. (eds.), *Foraminifera,* Vol. 1. Orlando, Fla.: Academic Press, pp. 208–260.

Lee, J.J. (1980a) Informational energy flow as an aspect of protozoan nutrition. *J. Protozool.* 27:5–9.

Lee, J.J. (1980b) Nutrition and physiology of the foraminifera. In Levandowsky, M., and Hutner, S.H. (eds.), *Biochemistry and Physiology of Protozoa,* Vol. 3. Orlando, Fla.: Academic Press, pp. 43–66.

Lee, J.J. (1982) Physical, chemical and biological quality related food-web interactions as factors in the realized niches of microzooplankton. *Ann. Inst. Oceanogr.* (Paris) 58:19–30

Lee, J.J. (1983) Perspective on algal symbionts in larger foraminifera. *Int. Rev. Cytol.* 14 (suppl.): 49–77.

Lee, J.J., and Bock, W.D. (1976) The importance of feeding in two species of soritid foraminifera with algal symbionts. *Bull. Mar. Sci.* 26:530–537.

Lee, J.J., and McEnery, M.E. (1983) Symbiosis in foraminifera. In Goff, L.J. (ed.), *Algal Symbiosis.* Cambridge: Cambridge University Press, pp. 37–68.

Lee, J.J., McEnery, M.E., and Garrison, J.R. (1980) Experimental studies of larger foraminifera and their symbionts from the Gulf of Elat on the Red Sea. *J. Foram. Res.* 10:31–47.

Lee, J.J., McEnery, M.E., and Rubin, H. (1969) Quantitative studies on the growth of *Allogromia laticollaris* (Foraminifera). *J. Protozool.* 16:377–395.

Lee, J.J., McEnery, M.E., Kahn, E.G., and Schuster, F.L. (1979) Symbiosis and the evolution of larger foraminifera. *Micropaleontology,* 25:118–140.

Lee, J.J., McEnery, M.E., Pierce, S., Freudenthal, H.D., and Muller, W.A. (1966) Tracer experiments in feeding littoral foraminifera. *J. Protozool.* 13:659–670.

Lee, J.J., and Muller, W.A. (1973) Trophic dynamics and niches of salt marsh foraminifera. *Amer. Zool.* 13:215–223.

Lee, J.J., and Pierce, S. (1963). Growth and physiology of foraminifera in the laboratory; Part 4—Monoxenic culture of an allogromiid with notes on its morphology. *J. Protozool.* 10:404–411.

Lee, J.J., Tietjen, J.H., and Mastropaolo, C.A. (1971) Axenic culture of the marine hymenostome ciliate *Uronema marinum* in a chemically defined medium. *J. Protozool.* 18 (suppl.): 10.

Leual-Peuto, M., Salvano, P., Gayol, P., and Greuet, C. (1986) Mixotrophy in marine planktonic ciliates: ultrastructural study of *Tontonia appendiculariformis* (Ciliophora), *Oligotrichina. Mar. Microbiol. Food Webs* 1:81–104.

Lindholm, T. (1985) *Mesodinium rubrum*—a unique photosynthetic ciliate. *Adv. Aquatic Microbiol.* 3:1–48.

Loeblich, A.R. III. (1984) Dinoflagellate physiology and biochemistry. In Spector, D.L. (ed.) *Dinoflagellates.* Orlando, Fla.: Academic Press, pp. 300–342.

Lopez, E. (1979) Algal chloroplasts in the protoplasm of three species of benthic foraminifera: taxonomic affinity, viability and persistance. *Mar. Biol.* 53:201–211.

McGinn, M.P., and Gold, K. (1969) Axenic cultivation of *Noctiluca scintillans. J. Protozool.* 16 (suppl.): 13.

Morrill, L.C., and Loeblich, A.R., III. (1979). An investigation of heterotrophic and photoheterotrophic capabilities in marine pyrrophyta. *Phycologia* 18:394–404.

Muller, W.A. (1975) Competition for food and other niche-related studies of three species of Salt Marsh foraminifera. *Mar. Biol.* 31:339–351.

Muller, W.A., and Lee, J.J. (1969) Apparent indispensability of bacteria in foraminiferan nutrition. *J. Protozool.* 16:471–478.

Muller, W.A., and Lee, J.J. (1977) Biological interactions and the realized niche of *Euplotes vanus* from the Salt Marsh Aufwuchs. *J. Protozool.* 24:523–527.

Oakley, B.R., and Taylor, F.J.R. (1978) Evidence for a new type of endosymbiotic organization in a population of the ciliate *Mesodinium rubrum* from British Columbia. *Biosystematics* 10:361–369.

Pavillon, J.F., and Rassoulzadegan, F. (1980) Un aspect de la nutrition chez *Stenosemella ventricosa* (Tintinnide): absorption de quelque substances organique dissoutes marquees ou ^{14}C. *J. Rech. Oceanogr.* 4:53–61.

Pomeroy, L.R. (1970) The strategy of mineral cycling. *Ann. Rev. Ecol. Syst.* 1:171–190.

Provasoli, L. (1977) Cultivation of animals: anexic cultivation. In Kinne, O. (ed.), *Marine Ecology,* Vol. III. New York: Wiley, pp. 1295–1320.

Provasoli, L., and Carlucci, A.F. (1974) Vitamins and growth regulators. In Stewart, W.D.P. (ed.), *Algal Physiology and Biochemistry.* Berkeley: University of California Press, pp. 241–787.

Provasoli, L., and Gold, K. (1962) Nutrition of the American strain of *Gyrodinium cohnii. Arch. Microbiol.* 42:196–203.

Provasoli, L., and McLaughlin, J.J.A. (1963). Limited heterotrophy of some photosynthetic dinoflagellates. In Oppenheimer, C.H. (ed.), *Symposium on Marine Microbiology.* Springfield, Ill.: Charles C. Thomas, pp. 105–113.

Rivier, A., Brownlee, D.C., Sheldon, R.W. and Rassoulzadegan, F. (1985) Growth of microzooplankton: a comparative study of bactivorous zooflagellates and ciliates. *Mar. Microbiol. Food Webs* 1:51–60.

Rogerson, A. (1981) The ecological energetics of *Amoeba proteus* (protozoa). *Hydrobiologia* 85:117–128.

Röttger, R. (1972) Die kultur von Heterostegina depressa (Foraminifera: Nummulitidae *Mar. Biol.* 15:150–159.

Röttger, R. (1976) Ecological observations of *Heterostegina depressa* in the laboratory and in its natural habitat. *Maritime sediments.* Special publ. 1:75–79.

Röttger, R., Irwan, A., Schmaljohann, R., and Francisket, L. (1980) Growth of the symbiont-bearing foraminifera *Amphistegina lessonii* and *Heterostegina depressa*. In Schwemmler, W., and Schenk, H.E.A. (eds.), *Endocytobiology, Endosymbiosis and Cell Biology*, Vol. 1. Berlin: Walter deGruyter, pp. 125–132.

Rubin, H.A., and Lee, J.J. (1976). Informational energy flow as an aspect of the ecological efficiency of marine ciliates. *J. Theoret. Biol.* 62:65–91.

Scott, J.M. (1985). The feeding rates and efficiencies of a marine ciliate, *Strombidium* sp., grown under chemostat steady-state conditions. *J. Exp. Mar. Biol. Ecol.* 90:81–95.

Sieburth, J. McN., Johnson, K.M., Burney, C.M., and Lavoie, D.M. (1977) Estimation of *in situ* rates of heterotrophy using diurnal changes in dissolved organic matter and growth rates of picoplankton in diffusion culture. *Helgolander Wiss. Meeresunters.* 30:565–574.

Soldo, A.T., and Merlin, E.J. (1972) The cultivation of symbiont-free marine ciliates in axenic medium. *J. Protozool.* 19:519–524.

Soldo, A.T., and Wagtendonk, W.J. Van (1969) The nutrition of *Paramecium aurelia* Stock 299. *J. Protozool.* 16:500–506.

Spindler, M., Hemleben, C., Salomons, J.B., Smit, L.P. (1984) Feeding behavior of some planktonic foraminifers in laboratory cultures. *J. Foram. Res.* 14:237–249.

Stoecker, D.K., Cucci, T.L., Hulburt, E.M., and Yentsch, C.M. (1986) Selective feeding by *Balanion* sp. (Ciliata: Balanionidae) on phytoplankton that best support its growth. *J. Exp. Mar. Biol. Ecol.* 95:113–130.

Stoecker, D.K., and Evans, G.T. (1985) Effects of protozoan herbivory and carnivory in a microplankton food web. *Mar. Ecol. Prog. Ser.* 25:159–167.

Stoecker, D., Davis, L.M., and Anderson, D.M. (1984) Fine scale spatial correlations between planktonic ciliates and dinoflagellates. *J. Plankton Res.* 6:829–842.

Stoecker, D., Guillard, R.R.L., and Kavee, R.M. (1981) Selective predation by *Favella ehrenbergii* (Tintinnia) on and among dinoflagellates. *Biol. Bull.* 160:136–145.

Stoecker, D., Davis, L., and Provan, A. (1983) Growth of Favella sp. (Ciliata, Suborder Tintinnia) and other microzooplankters in cages incubated in situ and comparison to growth in vitro. *Mar. Biol.* 75:293–302.

Stoecker, D.K., Michaels, A.E., and Davis, L.H. (1987) Large proportion of marine planktonic ciliates found to contain functional chloroplasts. *Nature* 326:790–792.

Stoecker, D.K., and Silver, M.W. (1987) Chloroplast retention by marine planktonic ciliates. *N.Y. Acad. Sci. Ann.* 503:562–565.

Stoecker, D.K., Sunda, W.G., and Davis, L.H. (1986) Effects of copper and zinc on two planktonic ciliates. *Mar. Biol.* 92:21–29.

Swanberg, N.R., Anderson, O.R., Lindsey, J.L., and Bennett, P. (1986) The biology of *Physematium muellerii:* trophic activity. *Deep-Sea Res.* 33:913–922.

Swanberg, N., Bennett, P., Lindsey, J.L., and Anderson, O.R. (1986a) The biology of coelodendrid: a mesopelagic phacodarian radiolarian. *Deep-Sea Res.* 33:15–25.

Swanberg, N., Bennett, P., Lindsey, J.L., and Anderson, O.R. (1986b) A comparative study of predation in Caribbean radiolarian populations. *Mar. Microbiol. Food Webs* 1:105–118.

Verity, P.G., and Villareal, T.A. (1986) The relative food value of diatoms, dinoflagellates, flagellates, and cyanobacteria for tintinnid ciliates. *Arch. Protistenk.* 131:71–84.

Verity, P.G., and Stoecker, D. (1982) Effects of *Olisthodiscus luteus* on the growth and abundance of tintinnids. *Mar. Biol.* 72:79–87.

Wiens, J.A. (1976) Population responses to patchy environments. *Ann. Rev. Ecol. Syst.* 7:81–120.

Wright, R.T., and Hobbie, J.E. (1966) Use of glucose and acetate by bacteria and algae in aquatic ecosystems. *Ecology* 47:447–464.

6

Food Quality and Microzooplankton Patchiness: An Automata Model

HOWARD A. RUBIN
JOHN J. LEE

INTRODUCTION

At almost every scale in which they have been examined "real-world" environments have heterogeneous qualities that exert powerful influences on the distributions of organisms, their interactions, and their adaptations (Levin, 1976; Heller, 1980; Oaten, 1977; Wiens, 1976; Pyke, 1984). In pelagic marine systems it is easier to understand and model how physical processes (e.g., wind shear, mixing, upwelling) contribute to the discontinuous environmental patchwork in which we find marine protozoa than it is to grasp many of the biological interactions (e.g., Levin and Segel, 1976). Optimal foraging theory predicts that foraging behavior becomes refined in time (revolutionary scale), so that it is characterized by traits that enhance individual fitness (e.g., Maynard Smith, 1978; Corio, 1983; Pyke, 1984). Models of foraging must include a number of random variables, such as the rates of encountering patches; the gain of energy obtained from various food types; relative abundances of food types of various ranks of quality; differential energetic costs of handling and searching for food; degree of satiation at each encounter; perhaps other, yet to be defined characters; and some rules for departure from patches (McNair, 1979; Murdoch, 1969; Pulliam, 1974, 1975; Pyke et al., 1977; Pyke, 1984; Estabrook and Dunham, 1976). The assumptions are, of course, that we can characterize the relationships between foraging behavior and fitness.

In many of our models of the lower levels of marine food webs very simplistic approaches seem quite attractive. Track the phytoplankton and you track the zooplankton dependent on them. Effective size and abundance of the phytoplankton will affect and determine the zooplankton species that can harvest them. If such a reductionist "black box" approach is inherently true at the microzooplankton level, then we must ask ourselves why we have such species diversity at this level of organization. Are there any inherent biological characteristics that might bring more structure to planktonic communities than ephemeral populations linked in time and space by casual size and abundance relationships?

As we contemplated many questions on linkage in microzooplanktonic food webs we could not help being drawn into the contemporary fabric of evolutionary theories, optimum design, and molecular genetics. Feedback pathways seem to have played important roles in the evolution, or fine tuning, of biological systems at every level from the molecular to the ecosystem. Margalef (1968, 1973, 1982) has drawn the attention of protozoologists to the cybernetic qualities of ecosystems. In his scheme, ecosystems are channels of information with three different layers or subchannels. One channel, the genetic, is organized into individual structures. The "truly ecological channel" is based on interactions between different cohabiting species and is expressed in the relative constancy and regular changes of their numbers. In the third channel, the cultural, information is expressed in the learning and experience of animals and is transmitted to future generations outside the genetic channel.

The nature of the second information channel is intriguing to contemplate. How is the information in the channel stored? Here Margalef's analysis may be somewhat incomplete or misleading. Some information is stored in the structure, diversity, and interrelationships of the populations, assemblages, and communities within the ecosystem, but there is also much ecological information stored in the genetic repertoire of the individuals of each species. The latter idea prompted us to initiate an experimental program at the interorganism level (Rubin and Lee, 1976; discussed in Chap. 5). Very briefly, we were able to calculate and measure both informational and calorific transfer in food webs with species of marine ciliates. A pivotal aspect of the experimental design was our ability to feed the animals nonlimiting amounts of different species of food organisms with equal calorific value but potentially different informational content for the consumers. Differences in ecological efficiency on different diets were thus a reflection of molecular information transfer and utilization. In other words, although food organism has a specific total energetic value, only some percentage of it is available to the ingesting organism. This percentage is a function of the ingesting organism's ability to recognize and utilize molecules from the food organism. This is what we consider to be the informational content of a food organism. It impacts ecological efficiency because actual energy transfer is a function of this informational value, and the ingesting organism may have to incur an energetic cost to break ingested molecules down into recognizable and therefore usable components.

We also found that lags in growth and reproduction occurred when the animals were switched from some algal diets to others. The parallel to bacterial induction lags was striking. Four simple types of operon control circuits are well recognized (Stebbing, 1974), which are, in principle, examples of either positive or negative control at the molecular level. When genes for enzyme synthesis are under positive control, expression is possible only when an active inducer protein is present, whereas when the genes are under negative control, they are expressed until they are switched off by interaction with active repressor proteins. At the time of our earlier experiments (Rubin and Lee, 1976) we felt that the effects of molecular control circuits observable in the ciliates tested should be applicable to population dynamics at the community level, but we lacked a suitable modeling technique to explain the mechanics. Time has given us new tools and ideas to do so. We here examine the subset of energy flow we call food quality and, consider how it might contribute to the patchiness of microzooplankton.

METHODOLOGY AND RATIONALE FOR MODELING APPROACHES

The early elegant applications of automata theory to developing biological and social systems (Arbib, 1967, 1969; Pask, 1969) gave us insight into the characteristics and parameters of such models. Our first step was to construct a "macro" model in which we conceptualized the types of internal molecular information processing machinery needed by our automatons. Accepting the premise that optimum design and energy cost–benefit considerations are focal points for the evolutionary process (Rosen, 1967; Maynard Smith, 1978; Pyke, 1984; Curio, 1983), we determined that our automatons must have mechanisms by which information could be recognized, processed, and utilized. A preliminary model was then constructed to illustrate the linkages between these components (Fig. 6.1). After capture and possible mechanical disruption, food is first ingested and digested to some degree. This is the entry point to our model information-processing network. A recognition device is needed

FIGURE 6.1 An automaton conceptual representation of food processing and related metabolic steps in the marine ciliate *Euplotes vanus*.

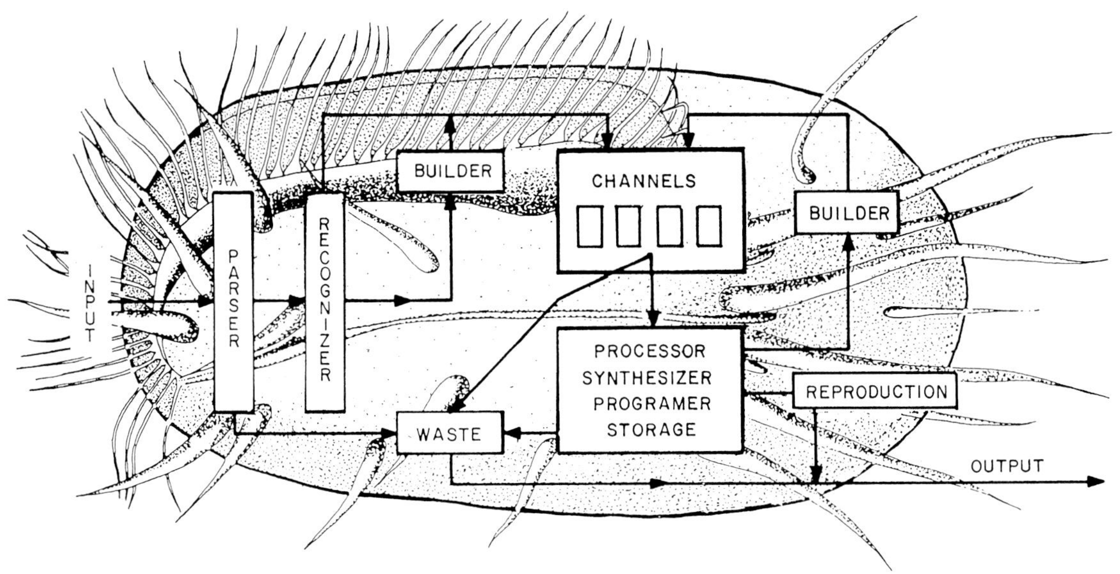

to determine whether a particular food component is usable by the organism. If the partially digested food is not in usable form, a digestive mechanism iteratively breaks it into smaller units before forwarding it to the recognizer. Food components are repeatedly broken down until either their subunits are recognized or they are brought down to a level below which no further digestion is possible. If they are not recognized at this point, they are discarded as waste. If they are recognized, they are then sent to the proper metabolic pathway. From this stage further refinement was done based on our experimental work (Rubin and Lee, 1976). In our food-switching experiments we were able to show that there were lags in growth when certain species of algae were fed after other species. In light of our present knowledge of adaptive and constitutive enzymes (Stebbing, 1974), it seemed reasonable to include in the model a mechanism that simulated the time delays that occur in adaptively constructing metabolic pathways. Hence, after recognizing a component that the organism has the potential to process, it must determine if it has the proper processing pathway available or if it must generate it—analogous to the process of dynamically loading a subprogram in a computer system. Once a pathway is available, the food component can be processed.

Our model organism has the ability to perform many operations concurrently. Materials and energy are constantly being drawn from pools or shunted for use from various life processes. Our model also has the ability to reproduce when temporal and energetic conditions reach specific values. All activities can be thought to be coordinated by a central control module that operates as an organismic system supervisor at all times.

In practice, our model can be used to visualize the digestive and intermediary metabolic processes of ciliates as a composite of many functioning feedback systems. Ecological growth efficiency is realized, therefore, as a manifestation of the metabolic and digestive systems' ability to react to random fluctuations in the quality of the food species available for capture in the environment. A generalist organism that can eat, process, and grow on a broad spectrum of food sources would be at a distinct energetic advantage if its digestive and metabolic mechinery could maximize recovery of distinctive and needed molecules in its processing of food while minimizing processing cost. In both experimental animals, *Uronema marinum* and *Euplotes vanus*, ecological growth efficiency varied greatly on different diets, from 20 percent for *U. marinum* feeding on certain chlorophytes to only 6 percent for *U. marinum* when feeding on other algae. Ecological growth efficiency for *E. vanus* ranged from 2 to 12 percent depending on diet. More important, time lags were found when each species was switched from one diet to another. The data obtained clearly indicated that nutritional quality factors (what we have deemed molecular information) were important determinants for energy flow, since reproduction was neither strictly a function of the feeding rate nor strictly coupled with the calorific values of the food organisms themselves.

It appeared to us that our ciliate food utilization model could be used to simulate interspecific interactions at the community level by distributing models of various types of herbivores to be considered in our model, within a community matrix. It seemed to us that an automaton representation could be used for the simulation and analyses of community behavior under the influence of both informational and calorific transfers.

Formally, automata models can realize populations as nets of finite state machines (Apter, 1967). We can define a finite automata as quintuple:

$$M = (X, Y, Q, \delta, \lambda)$$

where

X is the input set
Y is the output set
Q is the state set
δ is the next output function
λ is the next state function

(Arbib, 1969). In our organismic model we can treat the organism as such a multicomponent entity. Food, temporal, and other environmental factors are represented by the set X. X may be considered by some to be infinite. However, the organism as modeled is not affected by every possible member of X, that is, it has limited ranges of temperature, salinity, pH, and so on, and cannot eat or digest every possible food. We consider a subset of X, X_r, to be the subset of usable input that, for purposes of our simulation, is finite. Growth, reproduction, energy storage, waste energy, respiration, and excretion are included in the output set. When reproduction is an output, a complete copy of the automaton is produced and allowed to function independently within the system (Apter, 1967). All new automatons start in an initial state denoted by q_0. At any instant, the animal is at some stage of its life cycle processing and metabolizing (q_k). Q, the set of states, is an n-tuple of states denoted by a single metastate q_k. The states q_1 through q_n are processing states for utilization of food species f_1 through f_n (in X). To reach the reproductive processing state, the automata must proceed through a series of state transitions corresponding to advances along metabolic pathways and through their life cycle. Life cycle and metabolic transitions may be occurring simultaneously. The number of substates, in particular metabolic processing pathways, and the number of active pathways determine food processing time. The reproduction time of the organism therefore may vary as a function of the input characteristics (i.e., particular food species, mixtures of food species, external conditions, etc.). The model also includes a mechanism to account for processing delays that are a consequence of switching input foods. Some metabolic pathways are constitutive (i.e., they are always present); other pathways are adaptive (i.e., they are present when needed but have a finite existence and must be reconstructed by the automaton from memory, when necessary, resulting in some measurable time delay [switching time]). The model required some means of input recognition to ensure orderly and efficient processing. This is modeled by a finite state acceptor, a binary output construct that parses input to substrings of decreasing levels of complexity until the input components can be recognized and processed along some pathway. A finite state acceptor of strings with a vocabulary represented by T is a machine whose input set 1 is from T and whose output set has two members, 0 and 1. If the output is 1, the string is accepted (recognized). A state of the recognizer that produces an output of 1 is an accepting state. In our model, T corresponds to X_r, which is repeatedly parsed. After each parse the substrings are each processed by the recognizer. Any string not recognized is parsed again and reprocessed. When a component is recognized by the acceptor, it is forwarded to the proper processing pathway.

A flowchart for the operation of an individual in our simulated population of automatons is shown in Figure 6.2.

The simulation of this mode used a program written in FORTRAN IV and executed by an IBM 360/50 computer. The environment was represented as a matrix containing 4,096 lattice points. Food sources in the environment were represented by associating a code describing each food type (A,B,C,D,E). The characteristics of each food type were determined experimentally (Rubin and Lee, 1976). The characteristics of each food type were encoded in a table that was referenced during processing by each automaton. Each automaton was given a "memory" in which parameters describing food being processed were stored.

When population size was not an initial condition, 100 ciliates of either or both species were the inoculum. Abiotic conditions were not varied in these simulations. The simulation program used actual experimental data (Rubin and Lee, 1967), that is,

1. Feeding rate on each.
2. Generation time.
3. Switching time lags.
4. Reproduction rates of the various species of algae at 25°C.

Variables were population sizes, distribution of food, and food quality. The following simulated microcosms were investigated:

I. *A single ciliate species with*
   a. A randomly distributed food of low informational value and patches of a food of high value.
   b. A randomly distributed food of high informational value and patches of a food of low informational value.
   c. Both food types randomly distributed.

II. *Two ciliate species*
   a. Patches of two food species, each having a high informational value for only one of the ciliates.
   b. Patches of a single food of high informational value to both.
   c. A randomly distributed single food species of high informational value to both.
   d. In the same environment as in item c but with a delayed inoculum of the second species of ciliate.
   e. Patches of five different foods–three foods of high informational value to only one ciliate, one food of high informational value to the other ciliate, one food of value to both.
   f. With the same foods as in item e but randomly distributed.

The results of the simulation were displayed at 10-day intervals for 365 days. Spatial distributions of all organisms, population sizes, and biomass were recorded. Each simulation was replicated 10 times to take into account the stochastic nature of the model. Representative results of each class of simulation are presented.

Results

Modeling was more successful in mimicking many field observations than might have been predicted at the onset of this work. If thermodynamic considerations govern the system, as in the simulation, patchy distribution of ciliates and their food was more probable and more productive. Random community organizations eventually became patchy and cyclical. Results were as follows:

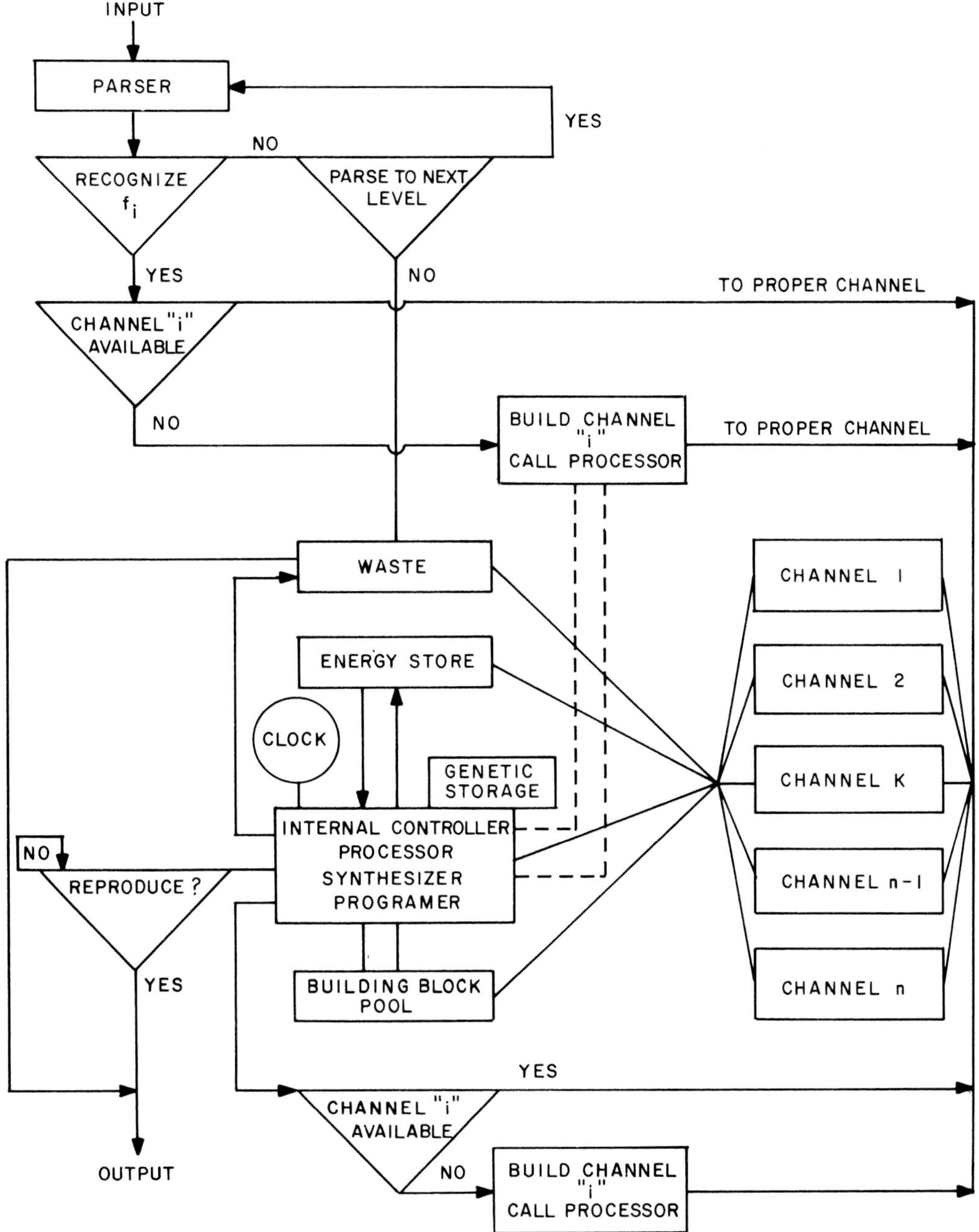

FIGURE 6.2 A more detailed conceptual model of how informational energy in food could be recognized and processed in the marine ciliates *Euplotes vanus* and *Uronema*.

Ia In the simulations in which a single ciliate species was placed in an environment containing a food of high informational value in patches and another food, randomly distributed, of low informational content, the model ciliates fed selectively on the better food. Selection was realized by a rapid increase in ciliate population size when a patch of preferred food was encountered. Organisms that encountered the poorer food continued reproduction and growth at lower rates. The ciliates virtually ignored the poorer food at boundaries of the preferred food patches. Patches of ciliates reached their maximum density after 20 days (Fig. 6.3b). At this time, ciliate food uptake exceeded the production rate of the food population. Feeding on the patches continued until the food density was

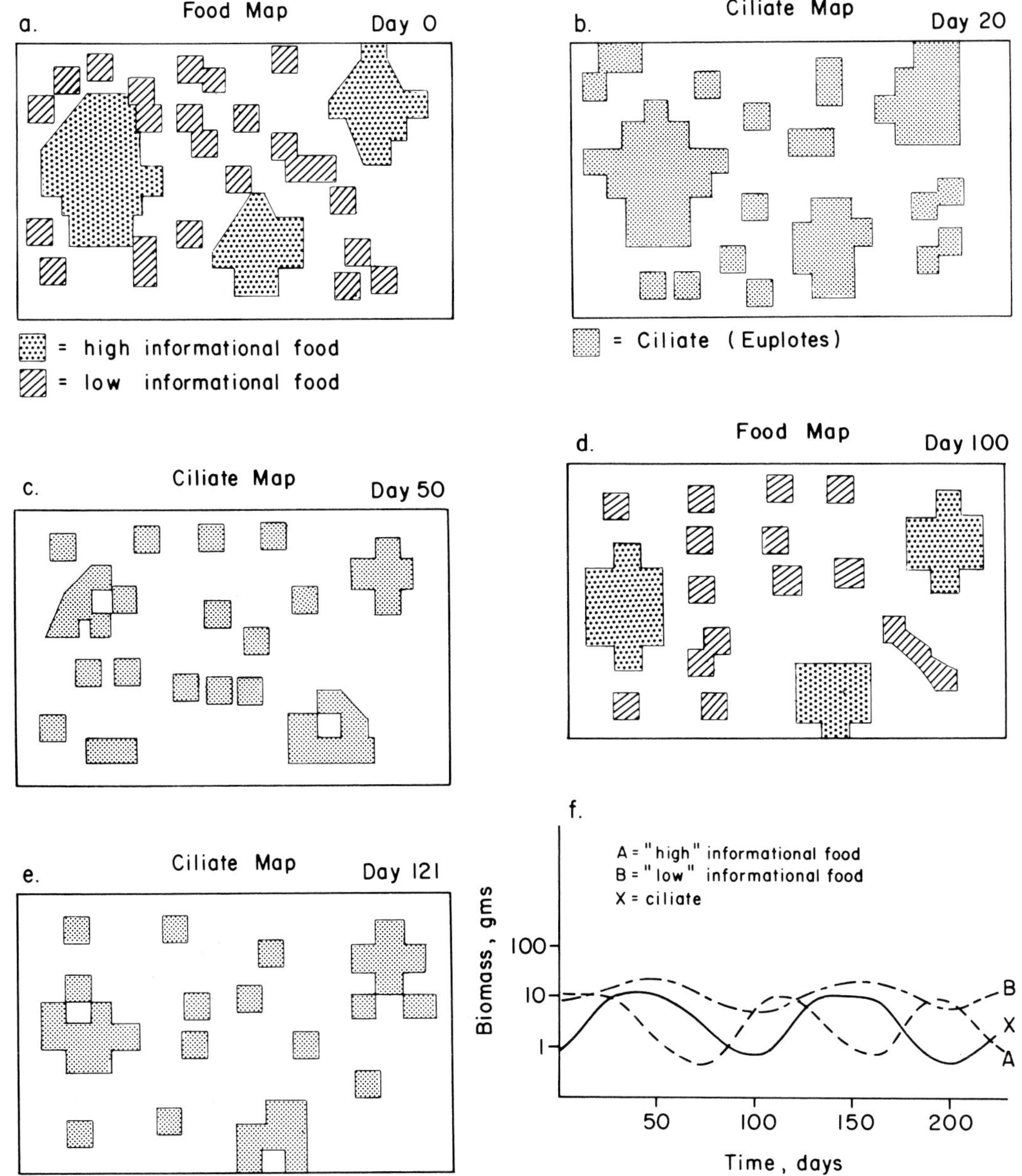

FIGURE 6.3 An automaton simulation of the growth of a population of a marine ciliate, *Euplotes vanus*, in a "sea" that contains randomly distributed food species with low information value and patches of food species with high informational value. (a) Initial distribution of food. (b) Distribution of ciliates after 20 days. (c) Distribution of ciliates after 50 days. (d) Distribution of food after 100 days. (e) Distribution of ciliates after 121 days. (f) Fluxes of the biomass of food organisms and ciliates over the course of 200 days of simulation.

below threshold level for the ciliate, at which point the population started to decrease (day 50, Fig. 6.3c). The ciliates feeding on the poorer food and reproducing much slower were not affected by this density decrease. With the absence of an abundant ciliate population in the depleted patch areas, the food organism population was able to regenerate to its previous level (day 100, Fig. 6.3d). When sufficient concentrations were reached and the new patches were located by the randomly moving ciliates, ciliate patches formed again and the whole cycle repeated (day 121, Fig 6.3e). As environmental abiotic factors were not considered in this model, the cycle repeated (Fig. 6.3f).

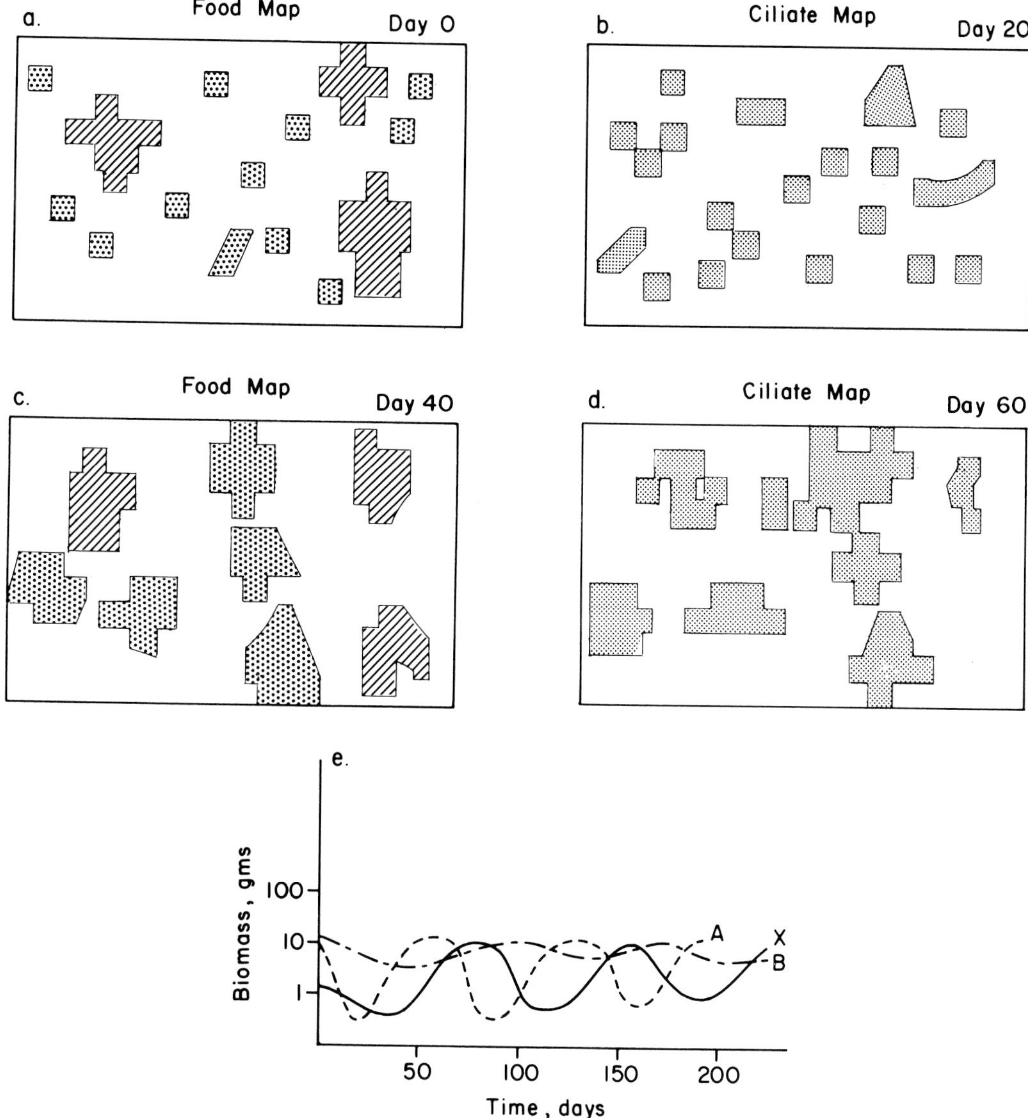

FIGURE 6.4 An automaton simulation of the growth of a population of a marine ciliate, *Euplotes vanus*, in a "sea" that contains randomly distributed food with high informational content and patches of food with low informational content for the ciliate. Symbols the same as in Figure 6.3: (a) Initial distribution of food. (b) Distribution of ciliates after 20 days. (c) Distribution of food after 40 days. (d) Distribution of ciliates after 60 days. (e) Fluxes of the food organisms and ciliates over the course of 200 days of simulation.

Ib When the reverse food distribution was simulated, patches of a low informational value food and a randomly distributed better food, results were similar (Fig. 6.4a–d). Initially, the randomly distributed better food was grazed down to below the threshold of detection by the ciliates. The patches of poor food were then the only feeding sites available to maintain the population. As the poorer food was not efficiently processed, its patches remained fairly stable with a small population of ciliates feeding on them. The absence of an abundant ciliate population allowed the better food to grow without any grazing decreases. Patches were formed, at which point the ciliates located them and exploited them by blooming (day 60). The cycle noted in the previous simulation were then initiated (Fig. 6.4e).

Ic When both of the foods were available in the environment in a random distribution, the ciliate population remained unproductive bio-

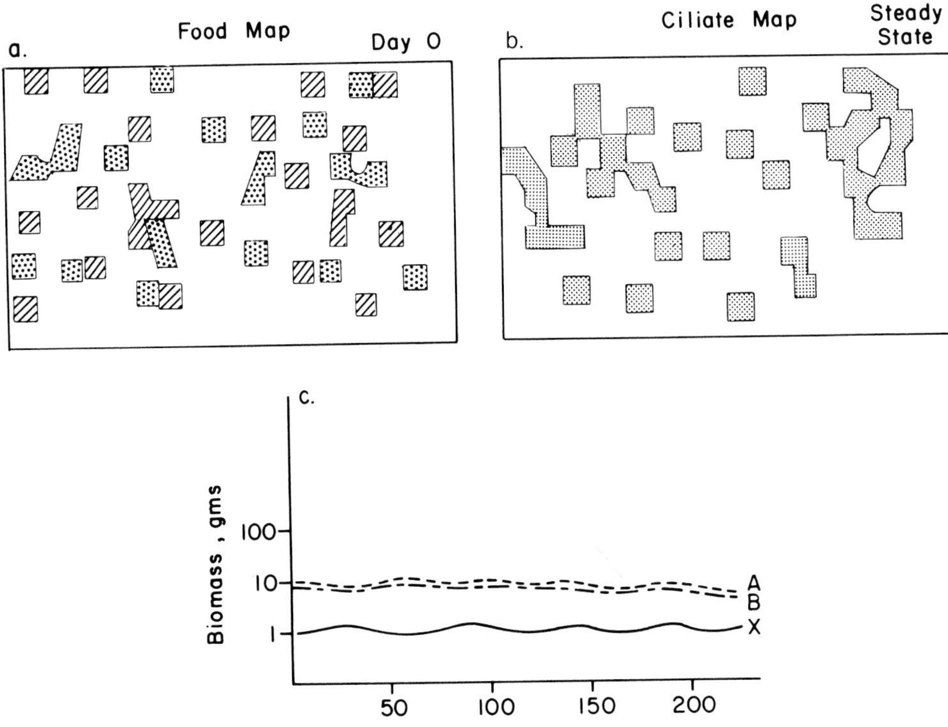

FIGURE 6.5 An automaton simulation of the growth of a population of a marine ciliate, *Euplotes vanus* in a sea of randomly distributed food organisms belonging to two different species. One species has high informational value to the ciliate, whereas the other has low quality. Symbols the same as in Figure 6.3: (a) Initial distribution of the food. (b) Steady-state population distribution of the ciliate. (c) Fluxes of the food organisms and ciliates over the course of 200 days of simulation.

mass 2 g) when compared with the patchy environment (200 g at maximum (Fig. 6.5a–c). This can be accounted for by the continual switching between foods necessitated by such an organization. Equal probabilities of encountering either food prevented the ciliates from exploiting either one.

IIa When two species of ciliates with different information-processing abilities were placed in a patchy environment of two foods, each of high value to only one ciliate, (and low to the other), only one ciliate was able to achieve full biotic potential. Each ciliate was most abundant on the particular food it could process most efficiently with a low level of other ciliate present on the same patch. The food patch depletion rate was dependent upon the generation time of the dominant consumer. The patches of food on which the ciliate with the shorter generation time grew were depleted first. The population was then forced to migrate to the patches of food where the other, slower-reproducing ciliate was dominant. The competition for food on these patches prevented the second ciliate from achieving its full potential. However, as the depleted food patches regenerated, patches of the more efficient ciliate again developed. The system continued to cycle in this fashion (Fig. 6.6).

IIb Competition for a single food of high informational value distributed in patches resulted in population distributions and population sizes that were a function of inoculum size, generation time, and a random variable. Random movement simulated in the environment determined which ciliate located a patch first and therefore had an expoitation advantage. Two such cases were documented (Fig. 6.7a,b). In most cases the ciliate with shorter generation time was dominant after a given period of time in which it was able to counter the effects of initial random collision with a food source. The single food was quickly overgrazed by both species and the entire population collapsed (time 70).

IIc Competition for a single randomly distributed food of high informational value to both ciliates resulted in a rather rapid extinction of both populations. As population centers were

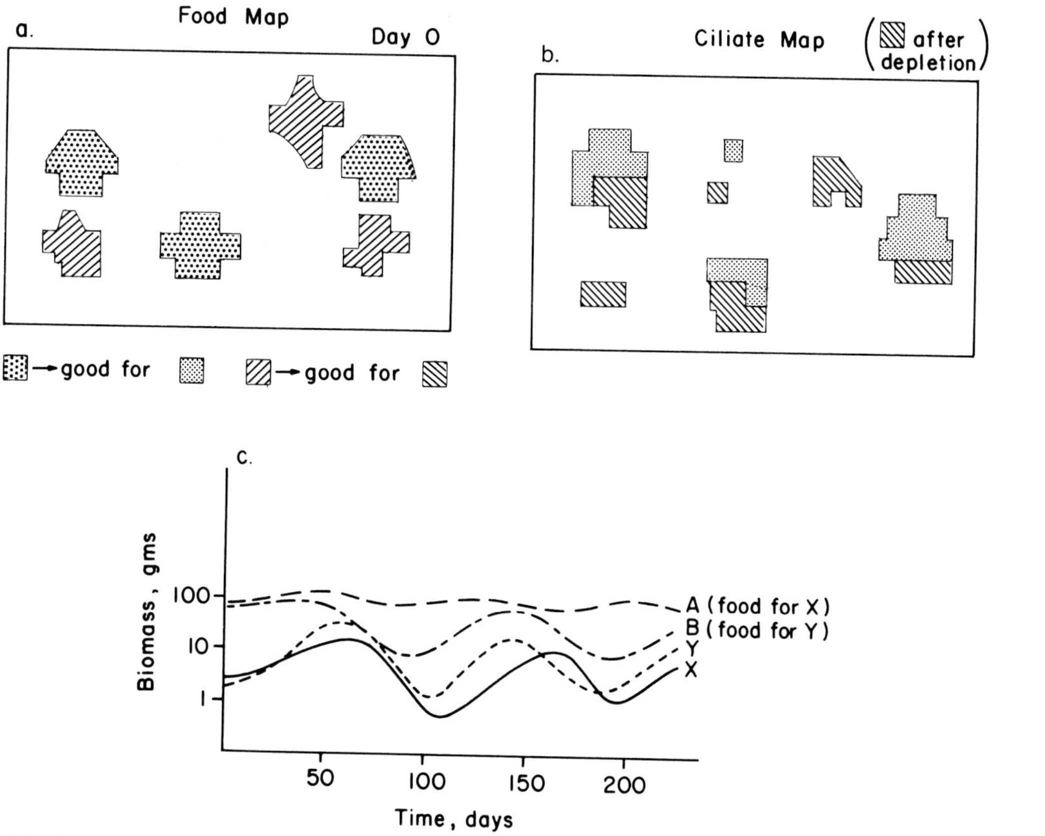

FIGURE 6.6 An automaton simulation of the growth of the populations of two species of marine ciliates, *Euplotes vanus* and *Uronema marinum,* in a "sea" that contains patches of two different species of alga food, each of which has high informational value for only one of the ciliate species. (a) Initial distribution of the food. (b) Ciliate distribution after the patches of algae have been depleted. (c) Fluxes of food organisms and ciliates during 200 days of simulation.

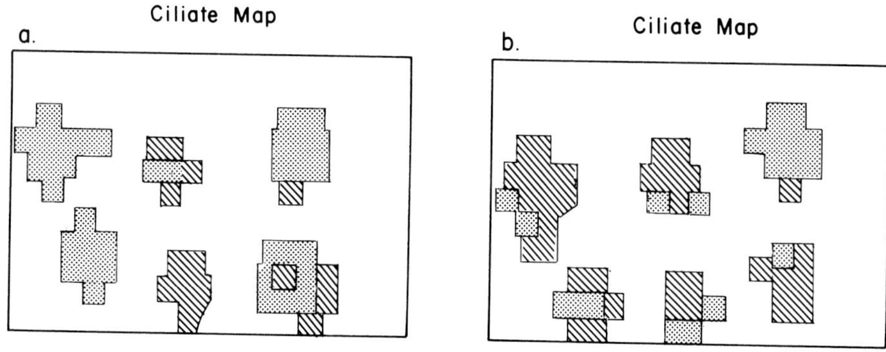

FIGURE 6.7 An automaton model of populations of the same two species of ciliates as Figure 6.6 with only a single species of food with high informational value for both species of ciliates. Symbols the same as in Figure 6.6: (a) Initial distribution of ciliates. (b) Distribution of ciliates after 50 days. (c) Fluxes of ciliates and food over 150 days of simulation (Figure 6.8).

randomly distributed about the system and were not as spatially distinct as in the patchy systems, competition was more fierce (Fig. 6.8). The system was only able to support one half the biomass (50 g) at its maximum of the similar patchy system (100 g). The species with the shorter generation time was greatest in abundance, as it had a reproductive advantage.

IId The advantage of the species with the shorter generation time could be counteracted by decreasing its inoculum size in comparison with

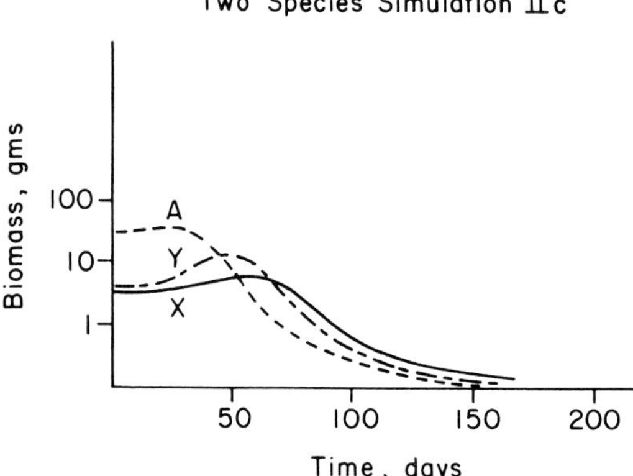

FIGURE 6.8

that of the other ciliate. If the inoculum proportions of the two species were in the ratio of the generation times, the dominance of the population could be shifted for a period of time. In all cases, however, the less abundant species could catch up to the slower.

IIe A natural environment for the two organisms was simulated by having five food species of different informational value present: three foods preferred by one ciliate, one preferred by the other, and one of high value to both. The same cyclic behavior of ciliate patch formation was observed (Fig. 6.9). The generalist ciliate with four possible foods was greatest in abundance; however, its patches were not as dense as those of the specialist, with only two available foods. The depletion time of the food patches was a function of the individual food uptake rate of the ciliate feeding on the patch and its generation time. The shared food species acted as a safety valve for the growth of both food organisms in the event that their primary patch feeding areas were depleted. It supported populations of both organisms at all times, but the combined population remained at a low level (50 g) compared with the biomass that could be supported when only a single species was growing on the patch (70 g).

IIf When the same foods were randomly distributed, the specialist was not able to exploit its food source because of the interference in switching caused by coincidence with other foods (Fig. 6.10). It eventually was extinct and generalist species were able to remain, using the shared food as a safety valve. Total system biomass was less than before (40 g), because of inefficiencies generated by generalist switching lags.

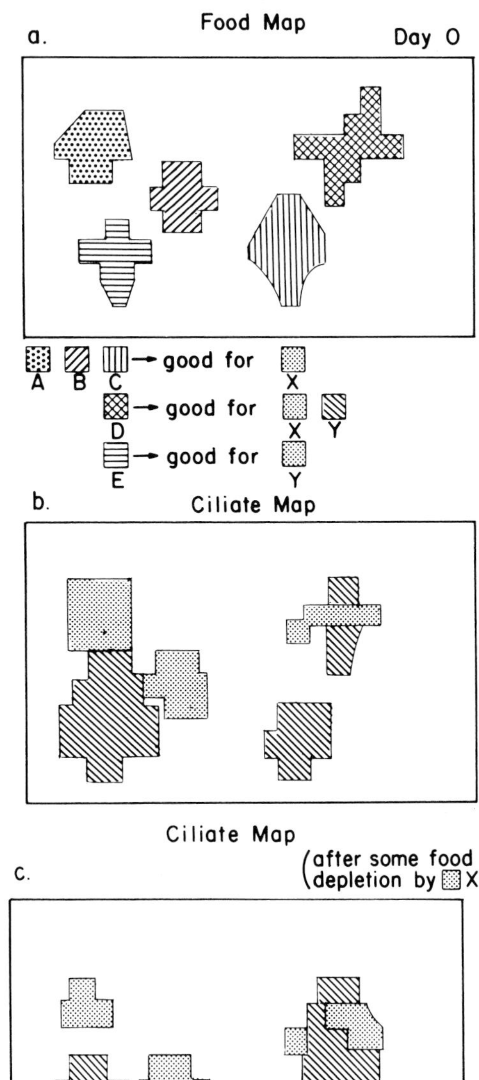

FIGURE 6.9 An automaton model of the growth of populations of two species of marine ciliates in a complex environment of patchy food species of which have some high informational value for one or the other species of ciliate or both. (a) initial distribution of the food. (b) initial distribution of the ciliates on the randomly encountered food. (c) distribution of the ciliates after some depletion of the ▩ X.

Discussion

Secondary productivity in our model was directly linked to temporal and spatial distribution of various food organisms. Simulated environments with patchy food distribution were as much as 60 times

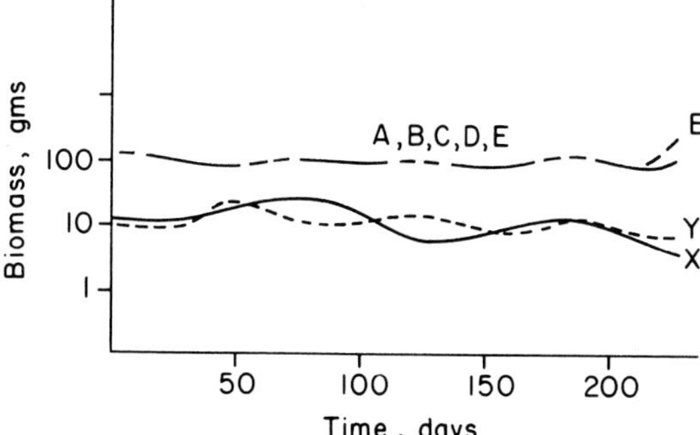

FIGURE 6.10 The same automaton simulation as in Figure 6.9 except that the food is randomly distributed. Biomass flux over the course of 200 days.

more productive than environments that were randomly distributed even though food inocula were equal. Since patchiness is a quality of most environments, particularly aquatic ones (Pyke, 1984; Wiens, 1976; Weibe, 1970), one wonders if it is not a mechanism for lowering competition and increasing diversity of the habitat to provide more niches for highly specialized and very productive organisms. As such, the phenomenon favors exploitation of high-informational-content foods. This is logical if optimization principles (Schoener, 1971; Rosen, 1967) govern ecosystem development. This is not meant to deemphasize the importance of the converse situation, which has some of the same optimal characteristics. Uniform, almost pure monocultures of some plants and their consumers are as highly productive as patchy ones.

In our simulations, interpatch distance (within critical bounds) directly affected secondary productivity through switching and inter- and intraspecific competition. When patches of different species of food are too close, switching between foods can become important. Switching, in a general sense, is a spatial phenomenon but is usually not considered in this construct. Usually, switching is considered only as a function of abundance relations (Murdock, 1969; Pyke, 1984). The energetic costs of metabolic adaptation for the ciliates as a result of switching food sources are not trivial (Rubin and Lee, 1976). At the extreme, switching time can equal generation time. Other aspects of our simulation behavior have strong biological analogues. Patterns of growth of automata in mixed food environments and their varying degrees of success in terms of total biomass are indicative of "lock and key" relationships between specialized environmental conditions and automaton characteristics. The fact that the hypothetical populations of automata reached stable but oscillatory levels indicates that such populations have implicit homeostatic mechanisms.

How widely applicable is an automaton model to protozoan distribution and productivity in the "real" world? We must constantly remind ourselves that we should never be satisfied with reductionist approaches that look at particular subsets in order to make the analysis easier. Many other factors come into play in nature. For example, Stoecker et al. (1984) found that the horizontal distributions of *Favella* sp., *Balanion* sp., and *Strombilidium* were positively correlated with the patchy distribution of dinoflagellates but not with temperature or food. The ciliates *Favella* and *Balanion* tracked with their dinoflagellate food throughout the diel cycle even though they are not phototactic in the laboratory. Stoecker and co-workers note in their analysis and discussion that chemotaxis (e.g., Antipa et al., 1983) could lead to aggregation within ciliate patches. Superimposed on this might be differential grazing impact on the phytoplanktion if food quality factors are brought into play.

It would be presumptuous for us to conclude that we have done any more than plant an idea, or look into the "black box" we could call ecological growth efficiency. An automaton model has qualities that seem to bridge the molecular and ecosystem aspects of information transfer in protozoan nutrition. It is a model that could easily be subsumed by or built into larger models that consider additional aspects of protozoan populations in ecosystems. Like any other model or theory, the automaton model challenges us to test its validity or breath of application. We all look forward optimistically to studies that will clarify this aspect of protozoan ecology.

REFERENCES

Antipa, G.A., Martin, K., and Rintz, M.T. (1983) A note on the possible ecological significance of chemotaxis in certain ciliated protozoa. *J. Protozool.* 30:55–57.

Apter, M.J. (1967) *Cybernetics and Development.* Elmsford, N.Y.: Pergamon.

Arbib, M.A. (1967) Automata theory and development. *J. Theor. Biol.* 14:131–156.

Arbib, M.A. (1969) *Theories of Abstract Automata.* Englewood Cliffs, N.J.: Prentice-Hall.

Curio, E. (1983) Time-energy budgets and optimization. *Experientia* 39:25–34.

Estabrook, G.F., and Dunham, A.E. (1976) Optimal diet as a function of absolute abundance, relative abundance, and relative value of available prey. *Amer. Nat.* 110:401–413.

Heller, R. (1980) On optimal diet in a patchy environment. *Theor. Popul. Biol.* 17:201–214.

Levin, S.A. (1976) Population dynamic models in heterogeneous environments. Ann. Rev. Ecol. Syst. 7:287–310.

Levin, S.A., and Setgel, L.A. (1976) An hypothesis to explain the origin of planktonic patchiness. *Nature*, 259:659.

Margalef, R. (1968) *Perspectives in Ecological Theory*. Chicago: University of Chicago Press.

Margalef, R. (1973) Some critical remarks on the usual approaches to ecological modeling. *Inv. Pesq.* 37:621–640.

Margalef, R. (1982) Some thoughts on the dynamics of populations of ciliates. *Ann. Inst. Oceanogr.* (Paris) 58:15–18.

Maynard Smith, J. (1978) Optimization theory in evolution. *Ann. Rev. Ecol. Syst.* 9:31–56.

McNair, J.N. (1979) A generalized model of optimal diets. *Theor. Popul. Biol.* 15:159–170.

Murdoch, W.W. (1969) Switching in general predators: experiments on predator specificity and stability of prey populations. *Ecol. Monogr.* 39:335–354.

Oaten, A. (1977) Optimal foraging in patches: a case for stochasticity. *Theor. Popul. Biol.* 12:263–285.

Pask, G. (1969) Computer simulated development of populations of automata. *Math Biosci.* 4:101–127.

Pulliam, H.R. (1974) On the theory of optimal diets. *Amer. Nat.* 108:59–75.

Pulliam, H.R. (1975) Diet optimization with nutrient constraints. *Amer. Nat.* 109:765–768.

Pyke G.H. (1984) Optimal foraging theory: a critical review. *Ann. Rev. Ecol. Syst.* 15:523–575.

Pyke, G.H., Pulliam, H.R., and Charnov, E.L. (1977) Optimal foraging: a selective review of theory and tests. *Quart. Rev. Biol.* 52:137–154.

Rosen, R. (1967) Optimality principles in Biology. New York: Plenum.

Rubin, H.A., and Lee, J.J. (1976) Informational energy flow as an aspect of the ecological efficiency of marine ciliates. *J. Theor. Biol.* 62:69–91.

Schoener, T.W. (1971) Theory of feeding strategies. *Ann. Rev. Ecol. Syst.* 11:369–404.

Stebbing, N. (1974) Precursor pools and endogenous control of enzyme synthesis and activity in biosynthetic pathways. *Bact. Rev.* 38:1–28.

Stoecker, D.K., Davis, L.H., and Anderson, D.M. (1984) Fine scale spatial correlations between planktonic ciliates and dinoflagellates. J. Plankton Res. 6:829–842.

Wiebe, R.W. (1970) Small scale distribution in oceanic zooplankton. *Limn. Ocean.* 15:205–218.

Wiens, J.A. (1976) Population responses to patchy environments. *Ann. Rev. Ecol. Syst.* 7:81–120.

7

Protozoan Nutrient Regeneration

DAVID A. CARON
JOEL C. GOLDMAN

The utilization of energy by marine organisms is a unidirectional process. Energy is "fixed" by living organisms to form organic material during photosynthesis and chemoautotrophy. As the organic material passes through the food web, this energy is slowly released back to the environment (primarily as heat) when the chemical bonds within these compounds are broken and reformed. In contrast, utilization of the elements involved in these processes is largely cyclic. The same molecules released to the environment during the metabolism of living organisms or the decomposition of nonliving organic matter are reassimilated by primary producers and used to make new organic material. Because some of these elements are relatively rare in the ocean, the rate at which they become available to primary producers can have a controlling influence on the rate of primary productivity. Therefore, to understand the processes controlling primary productivity in the ocean, it is necessary to understand the processes by which these essential nutrients are remineralized from living and nonliving organic material.

There have been a number of reviews on nutrient regeneration by protozoa (Johannes, 1968; Fenchel and Jørgenson, 1977; Stout, 1980; G. Taylor, 1982). However, the importance of protozoa in the remineralization of nutrients in the ocean is not a settled issue. The marine environment encompasses a diverse set of microbial ecosystems that differ in their chemical and physical composition (Sieburth, 1979). These differences result in a protozoan fauna that varies spatially and temporally according to the conditions imposed by the surrounding environment. To date, there is little information on the relative contribution of protozoa to nutrient regeneration in many of these environments. There is also little or no information on specific groups of protozoa. For example, there are no reports of nutrient excretion rates by sarcodines, and very few studies of phosphorus excretion rates by any group of protozoa. Therefore, at present there is not sufficient information to evaluate properly their impact on nutrient regeneration in the ocean. Despite this paucity of data, there is a growing consensus that protozoa are important in the regeneration of nutrients in the marine environment. This conclusion has been drawn largely from indirect or circumstantial evidence, but as we will discuss later, results from recent studies support the original hypothesis by Johannes (1965) that protozoa are important in the regeneration of nutrients in the ocean.

Much of the circumstantial evidence for an important role of protozoa in nutrient regeneration has been generated by plankton and microbial ecologists. Planktonic protozoa have been implicated as major regenerators of nutrients for a number of reasons. One reason is the abundance of ciliates and heterotrophic microflagellates. Because of the development of new counting methodologies (Dale and Burkill, 1982; Davis and Sieburth, 1982; Haas, 1982; Caron, 1983; Sherr and Sherr, 1983), it has been demonstrated that the densities of protozoa in the plankton are much higher than were previously believed (Beers et al., 1982; Davis et al., 1985). The trophic role that protozoa play in plankton communities also makes them likely candidates for the regeneration of significant amounts of nutrients in the ocean. Nanoplanktonic protozoa (2–20 μm in size) apparently are the major consumers of bacterioplankton (Fenchel, 1982d; Sieburth and Davis, 1982; Sherr et al., 1983; Davis and Sieburth, 1984; Sieburth, 1984). As such they consume a significant portion of the energy fixed by primary productivity and then transformed into bacterial biomass (Azam et al., 1983). In addition, numerous protozoan species can prey directly on phytoplankton or other protozoa (Heinbokel and Beers, 1979; Capriulo, 1982; Capriulo and Carpenter, 1980, 1983; Goldman and Caron, 1985). Collectively, this group of microorganisms, by ingesting a significant part of the particulate organic material present in plankton communities, may play an important role as remineralizers in the marine environment.

Additional circumstantial support for the importance of remineralization by protozoa comes from evidence that phytoplankton growth rates in pelagic waters may be higher than previously believed (Goldman et al., 1979; Laws et al., 1984). Most of the pool of nutrients supporting the growth of oceanic phytoplankton apparently comes from nutrient regeneration taking place in situ, since it has been estimated that only about 10 percent of the nutrients supporting oceanic primary production are a result of advective inputs (Eppley and Peterson, 1979). In addition, measured rates of nutrient release by marine plankton communities often are in balance with rates of nutrient uptake (Caperon et al., 1979; Glibert, 1982; Harrison, 1983).

A large role in nutrient regeneration processes has been documented in recent years for the microzooplankton size class largely through size fractionation studies. In such studies, plankton organisms are divided into various size groups by size differential filtration, and the remineralization rate of each size group is determined separately. Harrison (1980; see his Table 3) recently presented a summary of the available data for nitrogen regeneration in marine waters. Nitrogen regeneration by organisms in the "microheterotroph" size category (<100 μm) accounted for 13–100 percent of the calculated nitrogen requirement of the phytoplankton assemblage over a range of environmental conditions. Similar results have been obtained in other studies for nitrogen (Glibert, 1982; Paasche and Kristiansen, 1982; Harrison et al., 1983) and phosphorus (Harrison, 1983; Herbland, 1984). Although the microzooplankton size class is usually dominated numerically by protozoa, it is not always possible to attribute nutrient regeneration by the microzooplankton solely to the protozoa in this size category. The presence of micrometazoa such as copepod nauplii and the lack of consistency among investigators in the filters used to collect samples in these studies complicate the interpretation of these data. Nevertheless, as will be discussed in subsequent sections, protozoan excretion products appear to be important sources of nutrients for many plankton communities.

Regeneration of nutrients in the sediments of estuarine and coastal waters and their release to the overlying waters also may be an important source of nitrogen and phosphorus. The release of remineralized nitrogen from the sediments of a variety of coastal environments is an important source of NH_4^+ for the phytoplankton in the overlying waters (Davis, 1975; Rowe et al., 1975; Rowe et al., 1977; Conway and Whitledge, 1979; Raine and Patching, 1980; Pomroy et al., 1983; Boynton and Kemp, 1985). Although it has been demonstrated that protozoa occur in large densities in the benthos (Fenchel, 1969; Lee et al., 1969), it is difficult to estimate their relative importance as producers of regenerated nutrients in these environments because of the presence of a dense bacterial and metazoan fauna.

Despite our inability in the past to quantify the amount of nutrient regeneration in the ocean attributable directly to protozoa, it has been concluded, largely from the circumstantial evidence described earlier, that protozoa play an important role in this process. In this chapter we present a summary of much of the existing data on the regeneration of nutrients by marine protozoa. Pertinent information also is included from investigations in fresh water. We review the types of nitrogen- and phosphorus-containing excretion products of protozoa and the rates at which these compounds are excreted, and we discuss the data in light of our current understanding of nutrient regeneration by marine metazoa. In addition, we review some of the controversy surrounding theories and existing data on nutrient excretion by protozoa and discuss possible resolutions to these problems. We deal exclusively with nitrogen and phosphorus cycling, since these elements are considered to be of primary importance in aquatic ecosystems.

PROTOZOAN EXCRETION PRODUCTS

Protozoa excrete a variety of phosphorus- and nitrogen-containing compounds. Existing information on various groups of protozoa is summarized in Table 7.1, but most of these investigations were conducted to examine the production of one or a few specific substances. Most likely a variety of other compounds were excreted but not measured.

Phosphorus Excretion Products

Phosphorus-containing excretion products of protozoa have not been well characterized, but most of the phosphorus excreted by protozoa apparently is released as PO_4^{-3} (Johannes, 1964; Taylor and Lean, 1981; Güde, 1985; Andersen et al., 1986). Taylor and Lean (1981), using the radiotracer $^{32}PO_4^{-3}$, found that most of the ^{32}P released by the freshwater ciliate *Strombidium viride* was in the form of PO_4^{-3}. Release of the label occurred by two processes. The initial loss of phosphorus from the ciliate took place rapidly in large steps, and proceeded only for a short time following transfer of ciliates to unlabeled water. They hypothesized that this process reflected defecation (emptying of the digestive vacuoles) by the ciliate. In addition to these large, discontinuous losses of phosphorus, the ciliate also released phosphorus in a slow, continuous process. Taylor and Lean (1981) noted that small amounts of dissolved organic phosphorus (DOP) compounds also were released by the ciliate.

TABLE 7.1 Summary of the Nitrogen- and Phosphorus-Containing Excretion Products of Protozoa

Species	Excreted compound: Nitrogen	Excreted compound: Phosphorus	Reference
Ciliates			
Paramecium sp.	Mostly urea*		Weatherby (1929)
Spirostomum sp.	Mostly urea		Weatherby (1929)
Didinium sp.	Mostly NH_4^+		Weatherby (1929)
Glaucoma sp.	Mostly NH_4^+		Doyle and Harding (1937)
Paramecium caudatum	NH_4^+		Cunningham and Kirk (1941)
Colpidium campylum	Mostly NH_4^+, some urea		Nardonne (1949)
Colpidium campylum	Urea		Nardonne and Wilbur (1950)
Paramecium aurelia	NH_4^+, hypoxanthine, dihydrouracil, small amounts of adenine and guanine		Soldo and Wagtendonk (1961)
Euplotes crassus		PO_4^{-3}	Johannes (1964)
Euplotes trisulcatus		PO_4^{-3}	Johannes (1964)
Euplotes vannus		PO_4^{-3}	Johannes (1964)
Uronema sp.		PO_4^{-3}	Johannes (1964)
Tetrahymena pyriformis	Hypoxanthine		Leboy et al. (1964)
Parauronema acutum	Hypoxanthine†		Soldo et al. (1978)
Strombidium viride		PO_4^{-3}	Taylor and Lean (1981)
Euplotes vannus	NH_4^+		Gast and Horstmann (1983)
Tintinnopsis vasculum	NH_4^+		Verity (1985b)
Tintinnopsis acuminata	NH_4^+		Verity (1985b)
Flagellates			
Bodo caudatus	NH_4^+		Lawrie (1935)
Ochromonas malhamensis	Urea		Lui and Roels (1970)
Monas sp.	NH_4^+		Sherr et al. (1983)
Ochromonas sp.	NH_4^+, dissolved free amino acids	PO_4^{-3}	Andersson et al. (1985)
Natural population of unidentified flagellates	NH_4^+		Wambeke and Bianchi (1985)
Natural population of unidentified flagellates	NH_4^+	PO_4^{-3}	Güde (1985)
Paraphysomonas imperforata	NH_4^+, urea		Goldman et al. (1985)
Paraphysomonas imperforata		PO_4^{-3}	Andersen et al. (1986)
Amebae			
Amoeba proteus	Triuret (carbonyldiurea)†		Griffin (1960)
Amoeba dubia	Triuret†		Griffin (1960)
Chaos choas	Triuret†		Griffin (1960)
Amoeba proteus	Triuret†		Carlstrom and Møller (1961)
Amoeba proteus	Mostly NH_4^+; some glutamic acid; small amounts of uracil, adenine and thymidine		DeVincentiis et al. (1964)

*This product was not found in subsequent investigation.

†This product is released as crystalline "refractile bodies."

In agreement with these results, Andersen et al. (1986) also observed that PO_4^{-3} was the principal phosphorus excretory product of the omnivorous microflagellate *Paraphysomonas imperforata*. For example, the release of phosphorus as PO_4^{-3} and DOP during an eight-day period averaged 82 percent and 18 percent, respectively, of the total amount of dissolved phosphorus released by the microflagellate in four experimental trials where *P. imperforata* was fed the diatom *Phaeodactylum tricornutum*.

Little work has been done to identify the DOP compounds released by protozoa, and their potential importance as a phosphorus source for primary producers or bacteria is largely unknown. However, significant concentrations of DOP are known to occur in the ocean. Jackson and Williams (1985), and Smith et al. (1985) have shown that these compounds cycle rapidly in planktonic microbial assemblages. Therefore, it is likely that in addition to PO_4^{-3}, DOP compounds released by protozoa play a significant role in the cycling of phosphorus in the ocean.

Nitrogen Excretion Products

It was concluded from early reports on the nitrogen-containing excretion products of protozoa that urea, and perhaps uric acid, might be the most important nitrogen-containing products released by some protozoa (Weatherby, 1929; Nardonne, 1949). These conclusions have not been supported by subsequent studies, and it now seems clear that NH_4^+ is the main nitrogen-containing excretion product for most protozoa (see Table 7.1). Ammonium is excreted by protozoa primarily by the deamination of amino acids (Soldo and Wagtendonk, 1961). Although this compound accounts for most of the nitrogen released by protozoa, a variety of other nitrogen-containing compounds are released in much smaller amounts, including amino acids themselves. Andersson et al. (1985), for example, found that *Ochromonas* sp. released dissolved free amino acids (primarily serine, aspartic acid, glutamic acid, and ornithine) but that these compounds accounted for only 0.02 percent of the total amount of nitrogen ingested by this microflagellate when preying on bacteria.

Urea also is produced in small quantities by a number of protozoa. Goldman et al. (1985) observed that urea excretion by the microflagellate *Paraphysomonas imperforata* was approximately 15 percent of the total excreted nitrogen. They did find, however, that urea excretion was curtailed when the phytoplankton or bacterial prey were nitrogen limited. In addition, Seaman (1954) demonstrated that urease activity in the ciliate *Tetrahymena pyriformis* was pH sensitive. Therefore, although the ability of a protozoan species to produce urea is determined ultimately by the metabolic pathways present or absent in the species, the biochemical nature of its food and/or changes in urease activity may control urea excretion. Thus, conflicting reports on urea excretion by some protozoa may be reconciled by comparing differences in experimental protocols between studies (see Soldo and Wagtendonk, 1961, for discussion).

In addition to NH_4^+ and urea, excretion of hypoxanthine, dihydrouracil, and small amounts of adenine and guanine (all end products of purine and pyrimidine metabolism), have been observed in a few species of ciliates (Soldo and Wagtendonk, 1961; Leboy et al., 1964; Soldo et al., 1978). Soldo and Wagtendonk (1961) found that dihydrouracil is excreted by *Paramecium aurelia* because of the ciliate's inability to degrade the purine nucleus. Soldo et al. (1978) similarly demonstrated that the marine ciliate *Parauronema acutum* lacks the enzyme xanthine oxidase and therefore is not able to degrade the purine nucleus. They concluded that the "refractile bodies" observed in this species are crystals of hypoxanthine, the end product of purine metabolism. Hypoxanthine has also been demonstrated in the freshwater ciliates *Tetrahymena pyriformis* and *Paramecium aurelia* (Soldo and Wagtendonk, 1961; Leboy et al., 1964), but crystalline structures were not produced in these freshwater species. The production of hypoxanthine appears to be a unique feature of ciliates (Soldo et al., 1978).

Crystalline inclusions also have been observed as a common feature in many amebae (see references in Griffin, 1960). These crystals have been identified as triuret (carbonyldiurea), and it has been suggested that, analogous to the situation in some ciliates, they are the end product of purine metabolism in amebae (Griffin, 1960; Carlstrom and Møller, 1961). Nonetheless, NH_4^+ appears to be the main nitrogen-containing excretory product of amebae (DeVincentiis et al., 1964).

As is the case with phosphorus-containing organic compounds, there is little information on the fate of nitrogen-containing organic compounds released by protozoa. It has been demonstrated that several organic nitrogen compounds excreted by protozoa, including urea (Harrison et al., 1985) and hypoxanthine (Antia et al., 1980), are assimilable directly by primary producers. Other organic nitrogen compounds may not be assimilated directly by phytoplankton but may be taken up and incorporated by bacteria (Jackson and Williams, 1985). Data are not available on the dissolution rates or susceptibility to microbial degradation of crystalline nitrogen-containing end products of protozoan metabolism. This pathway represents a potential sink for nitrogen in the ocean. However, because of the small number of species that are known to use this pathway, the influence of crystal formation on nitrogen cycling probably is unimportant.

Although the data on protozoan nutrient regeneration products are limited, there is little doubt that NH_4^+ and PO_4^{-3} are the major excretion products. Therefore, the importance of nutrient regeneration by protozoa relative to regeneration by bacteria and metazoa (to be discussed in subsequent sections) is largely dependent on the rates of nitrogen and phosphorus release by protozoa, and not on the chemical nature of their excretion products.

PROTOZOAN NUTRIENT REGENERATION RATES

There is now considerable evidence that protozoa are important nutrient remineralizers. For example, from microcosm studies of terrestrial and aquatic ecosystems we know that protozoa are the main nutrient regenerators in detrital communities, whereas the bacteria are primarily a sink, not a source, for regenerated nutrients. Cole et al. (1978), Sinclair et al. (1981), and Woods et al. (1982) have shown in their terrestrial microcosm studies that phosphorus and nitrogen were immobilized by incorporation into bacterial biomass in the absence of protozoa, and only when protozoa were present were nutrients released from the bacterial biomass. Comparable results have been obtained in studies of detritus decomposition in aquatic ecosystems (Johannes, 1965; Sherr et al., 1982).

The most direct evidence that protozoa are important in the nutrient regeneration process comes from a limited number of controlled laboratory studies (Table 7.2). Protozoan excretion rates have been particularly difficult to measure because of the potential artifact of nutrient excretion or uptake by the prey (usually bacteria) during the course of an experiment. For this reason, and because of the lack of sophisticated techniques for quantifying nutrient excretion, most of the early measurements of protozoan excretion rates (Weatherby, 1929; Lawrie, 1935) must be held suspect. Likewise, by measuring excretion rates of protozoa in axenic cultures (Nardonne, 1949; Nardonne and Wilbur, 1950; Soldo and Wagtendonk, 1961), it is possible to avoid the potential complication of nutrient excretion or uptake by living prey, but it is doubtful that excretion rates of protozoa growing on defined me-

TABLE 7.2 Weight-Specific Ammonium Regeneration Rates Estimated for Protozoa*

Species	Weight-specific excretion rate (μg N mg^{-1} DW h^{-1})	Comments	Reference
Ciliates			
Didinium sp.	1.5	Assumed vol. = 10^6 μm^3	Weatherby (1929)
Paramecium aurelia	1.0, 4.0	For growing and resting cells, respectively; assumed vol. = 10^5 μm^3; axenic	Soldo and Wagtendonk (1961)
Tintinnopsis vasculum	1.5–3.0	5–15°C; algal food	Verity (1985b)
Euplotes vannus	3.0	25°C; algal or bacterial food	Gast and Horstmann (1983)
Colpidium campylum	0.3	Assumed vol. = 5.4×10^4 μm^3; axenic	Nardonne (1949), Nardonne and Wilbur (1950)
Tintinnopsis acuminata	4.3–10	15–25°C; algal food	Verity (1985b)
Flagellates			
Bodo caudatus	3.0	Assumed vol. = 250 μm^3	Lawrie (1935)
Paraphysomonas imperforata	3–20	20–24°C; bacterial or diatom food	Goldman et al. (1985)
Monas sp.	6.4–51	3–30°C; bacterial food	Sherr et al. (1983)
Ochromonas sp.	28	Room temperature; bacterial food	Andersson et al. (1985)
8-μm^3 unidentified flagellate	140	10°C; mixed natural bacterial flora food	Wambeke and Bianchi (1985)

*For those studies in which information on cell volume of the protozoan was not included, the cell volume was assumed in calculating weight-specific excretion rate. Cell volume was converted to dry weight assuming specific gravity = 1.0, and (0.2) (wet weight) = (dry weight).

dia will be the same as when the food source is natural prey. Nevertheless, these excretion values have been included in Table 7.2 for completeness.

The range of reported NH_4^+ excretion rates of microflagellates and ciliates varies tremendously, from 0.3 to 140 μg N mg^{-1} dry weight h^{-1}. However, the value of 140 μg N mg^{-1} dry weight h^{-1} from the field study of Wambeke and Bianchi (1985) must be considered an approximation, since this rate was estimated from a microcosm containing other organisms that may have contributed to excretion. Despite the caveats that must be placed on some of the earlier measurements, they are still in reasonably good agreement with excretion rates that have been measured during the last few years. For example, in one recent study (Goldman et al., 1985), NH_4^+ excretion rates of a microflagellate growing on living phytoplankton prey in the absence of bacteria averaged approximately 10 μg N mg^{-1} dry weight h^{-1} between 20° and 24°C. Comparison of nitrogen excretion in bacterized cultures of microflagellates and phytoplankton, and in axenic cultures of phytoplankton, confirmed that the microflagellates themselves, not the algae or attendant bacterial population, were important nutrient regenerators.

In general, on a weight-specific basis, NH_4^+ excretion rates of microflagellates are larger than those of ciliates (see Table 7.2). This difference appears to be related to protozoan cell size. There is still a considerable debate among researchers concerning the validity of the proposed relationship between body weight and weight-specific metabolism of protozoa. Relationships describing decreasing weight-specific nutrient excretion rate with increasing body weight have been well documented for zooplankton (Figs. 7.1 and 7.2). Because protozooplankters are much smaller than most metazooplankters, they are thought to have weight-specific nutrient regeneration rates that are high relative to the metazooplankton (Johannes, 1968). Therefore, given equivalent biomasses of protozoa or metazoa, more nutrient should be regenerated by the protozoa because of their higher weight-specific regeneration rates.

Johannes (1964, 1965) and Buechler and Dillon (1974) suggested that protozoa may have extremely high rates of nutrient turnover per unit biomass. Johannes (1964) calculated the amount of time required for four species of ciliates to excrete an amount of phosphorus equivalent to their own phosphorus content (body-equivalent excretion time = BEET). From his results (BEETs = a few minutes to a few hours) he concluded that protozoa must excrete a large percentage of their ingested phosphorus and that they must be important nutrient remineralizers in aquatic ecosystems. Similar results obtained by Buecher and Dillon (1974)

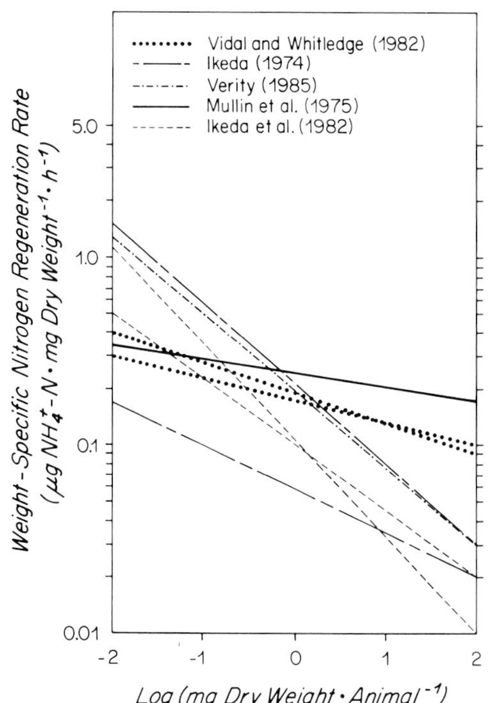

FIGURE 7.1 Linear regressions of weight-specific excretion rate of NH_4^+ as a function of body weight for mixed assemblages of marine zooplankton. The regressions were calculated from the data of Ikeda (1974), Mullin et al. (1975), Ikeda et al. (1982), Vidal and Whitledge (1982), and Verity (1985a).

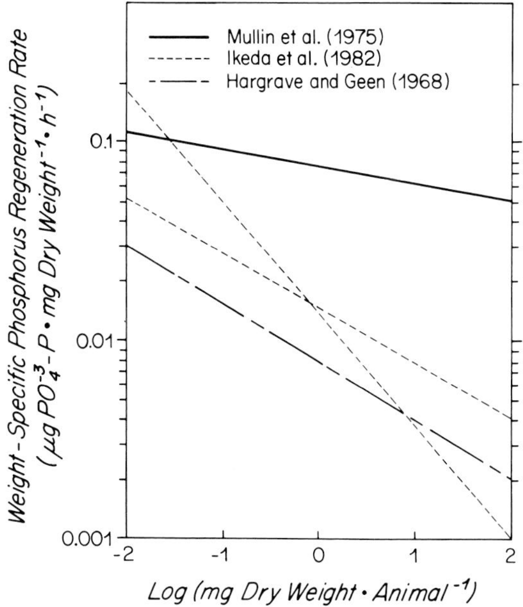

FIGURE 7.2 Linear regressions of weight-specific excretion rates of PO_4^{-3} as a function of body weight for mixed assemblages of marine metazoa (primarily zooplankton). The regressions were calculated from the data of Hargrave and Geen (1968), Mullin et al. (1975), and Ikeda et al. (1982).

(BEETs <1 h for ciliates) appeared to support this hypothesis. However, other investigators have not been able to find such rapid turnover rates in protozoa, and the methodologies employed in the studies of Johannes (1964) and Buechler and Dillon (1974) have been challenged. Taylor and Lean (1981), for example, suggested that the filtration procedure used by Johannes (1964) and Buechler and Dillon (1974) may have been too harsh for delicate protozoa. In addition, they argued that the BEET values of both studies were artificial because the protozoa were not uniformly labeled with ^{32}P. Taylor and Lean (1981), in contrast, obtained much lower turnover rates for phosphorus in the freshwater ciliate *Strombidium viride* (2.7–30 h). These latter results may have been biased by their use of a symbiont-containing ciliate (which undoubtedly regenerates nutrients at rates dissimilar to those of aposymbiotic species), but a subsequent study using an aposymbiotic ciliate (Taylor, 1986) also gave rates much slower than those of Johannes (1964) and Buechler and Dillon (1974).

Several inconsistencies exist with the BEET data of Johannes (1964) and Buechler and Dillon (1974). For example, experimentally derived BEET values of about one day for nitrogen excretion by the microflagellate *Paraphysomonas imperforata* (Goldman et al., 1985) are orders of magnitude greater than what would be predicted from the weight-specific relationship of Johannes (1964). Similarly, if the BEET values obtained by Buechler and Dillon (1974) for *Paramecium* spp. (4.3–38 min) are converted to weight-specific excretion rates of NH_4^+ (assuming cell volume = 10^6 μm^3, specific gravity = 1, [0.08] [wet weight] = organic carbon, and C:N ratio = 7), rates between 100 and 660 μg N mg^{-1} dry weight h^{-1} are obtained. These rates are greater than the highest excretion rates reported for other ciliates and for most flagellates of much smaller size. Additionally, if ciliates growing exponentially have a gross growth efficiency (GGE) of 50 percent, and nutrient concentration in the prey is comparable to that in the protozoan, a BEET time of 20 minutes would mean that the protozoan would have to ingest 600 percent of its own body weight in food each hour. This is a conservative estimate because it is assumed that all egested nutrient is released in remineralized form. Reported ingestion rates are far lower both for ciliates (Heinbokel, 1978; Fenchel, 1980) and flagellates (Fenchel, 1982b; Sherr et al., 1983; Davis and Sieburth, 1984).

Despite the inconsistencies present in the results of Johannes (1964) and Buechler and Dillon (1974), the rates of NH_4^+ excretion measured for protozoa (see Table 7.2) are still considerably higher than those for most metazoa (see summary in Table 7.3). Most of the values for NH_4^+ excretion by protozoa in Table 7.2 exceed the ranges of observed excretion rates by marine metazoa. Unfortunately, most studies of PO_4^{-3} excretion by protozoa do not provide enough information to allow conversion to weight-specific excretion rates. However, Andersen et al. (1986) calculated that the maximum phosphorus regeneration rate for *Paraphysomonas imperforata* was 2.2 μg P mg^{-1} dry weight h^{-1} when fed bacteria. These rates are greater than the weight-specific regeneration rates observed for marine metazoa (see Table 7.3).

The protozoan excretion rates in Table 7.2 have been presented in order of decreasing protozoan cell volume. Higher weight-specific excretion rates generally are associated with smaller protozoa. (However, a number of the reported excretion rates for larger protozoa are from older studies.) When these weight-specific NH_4^+ excretion rates are plotted according to protozoan cell volume (Fig. 7.3), a relationship between cell size and weight-specific excretion rate is apparent with a trend similar to the relationship that has been described for zooplankton (see Figs. 7.1 and 7.2). To investigate the applicability of metazoan weight-specific excretion rate regressions to protozoa, several of these regressions that are based on substantial data sets have been extrapolated to predict weight-specific excretion rate in protozoan-sized animals (Table 7.4). Dry weights corresponding to the approximate upper and lower size limits for zooplankton (10^2 and 10^{-2} mg), a "ciliate-sized" organism (2×10^{-4} mg = 10^6 μm^3), and a "flagellate-sized" organism (2×10^{-8} mg = 10^2 μm^3) were investigated. Weight-specific excretion rates are presented as a range for studies where more than one regression line was reported. The overall ranges for nitrogen excretion rate predicted from these regressions were 0.38–7.48 μg N mg^{-1} dry weight h^{-1} for a "ciliate-sized" animal and 0.87–562 μg N mg^{-1} dry weight h^{-1} for a "flagellate-sized" animal. The weight-specific excretion rates given in Table 7.2 are essentially all within the range of excretion rates predicted by the zooplankton regressions.

The potential applicability of metazoan weight-specific excretion rate regressions to protozoa is further supported by the results of some field investigations. Taylor (1984), for example, compared the uptake and release rates of ^{32}P by freshwater protozoa and small metazoa in Lake Ontario and concluded that the weight-specific rate of phosphorus turnover (uptake and release) was inversely related to body size. Harrison (1978) drew a similar conclusion from his studies on coastal waters when he observed that marine plankton <183 μm in size accounted for 90 percent of the nitrogen remineralization in the plankton even though they constituted only 40 percent of the plankton biomass. Clearly, more measurements of protozoan excre-

TABLE 7.3 A Summary of the Ranges of Calculated Weight-Specific Nitrogen and Phosphorus Regeneration Rates of Marine Metazoa

Organism(s)	Observed regeneration rates (μg N or P mg^{-1} DW h^{-1})	Reference
Nitrogen		
Mixed zooplankton, mostly copepods; summary of 13 studies	0.01–1.80	Corner and Davies (1971)
Mixed zooplankton, including amphipods, jellyfish, and ctenophores	0.01–0.35	Jawed (1973)
Zooplankton, including 14 major taxa	0.009–1.1	Ikeda (1974)
Zooplankton, including seven major taxa	0.01–4.5	Ikeda et al. (1982)
Zooplankton, summary of 10 studies, including eight major taxa	0.001–2.1	George and Fields (1984)
Overall range	0.001–4.5	
Phosphorus		
Mixed zooplankton, Doboy Sound, Georgia	0.21–0.28	Satomi and Pomeroy (1965)
Zooplankton, summary of 13 studies	0.02–0.46	Corner and Davies (1971)
Meganyctiphanes norvegica (13°C)	0.01–0.32	Roger (1978)
Zooplankton, including seven major taxa	0.005–0.24	Ikeda et al. (1982)
Euphausia superba (1°C)	0.03	Hirche (1983)
Overall range	0.005–0.46	

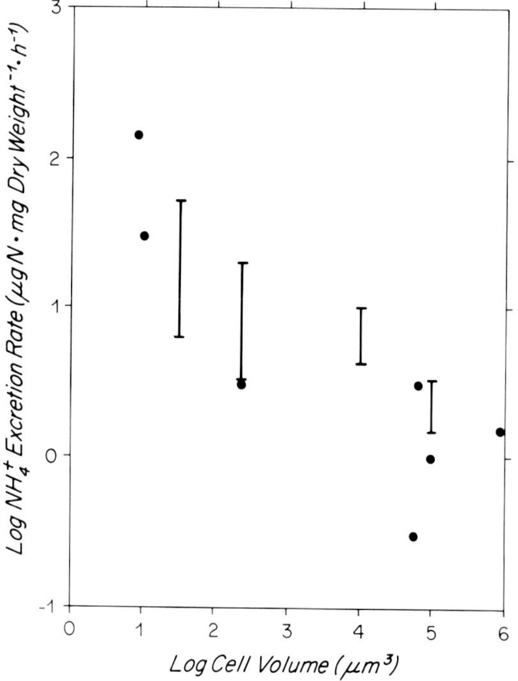

FIGURE 7.3 Ammonium weight-specific excretion rates of protozoa as a function of cell volume. (Data taken from Table 7.2.) Where necessary, cell volumes were converted to dry weight assuming a specific gravity of 1.0 and (0.2)(wet weight) = dry weight. Excretion rates given as a range in Table 7.2 are expressed as vertical lines.

tion rate are required before such a relationship can be confirmed, but the data available so far support the possibility that a relationship exists between weight-specific excretion rate and body weight in protozoa as in metazoan zooplankton.

NUTRIENT REGENERATION VS. INCORPORATION

The degree to which protozoa regenerate ingested nutrients is intimately related to growth efficiency. Nutrients that are contained in the ingested prey clearly are not available for recycling if they are used to form protozoan biomass. Therefore, growth efficiencies place an upper limit on the amount of nutrients that will be regenerated directly by protozoa. High gross growth efficiencies (GGE), often >50 percent, have been determined experimentally for a variety of protozoan species (Table 7.5). Their efficiencies generally are comparable to the highest efficiencies observed for heterotrophic organisms (Calow, 1977). High growth efficiencies of protozoa, based on parameters such as volume, carbon, or calories, are in agreement with estimates of nutrient regeneration efficiency. For example, Andersson et al. (1985) found that approximately 13 percent (NH_4^+) of the total ingested nitrogen and 30 percent (PO_4^{-3}) of the ingested phosphorus was

TABLE 7.4 Predicted Weight-Specific Nitrogen and Phosphorus Regeneration Rates of Marine "Zooplankton" and "Protozoan-Sized" Organisms Based on Linear Regressions of Observed Nutrient Regeneration Rates of Marine Metazoa.*

Organism(s)	Predicted excretion rate				Regression employed
	Metazooplankton		Protozooplankton		
	(10^2 mg DW)†	(10^{-2} mg DW)‡	(2×10^{-4} mg DW)§	(2×10^{-8} mg DW)#	
Nitrogen					
Mixed zooplankton (range of tropical to boreal species)	0.02–0.03	0.17–1.51	0.38–7.48	2.62–324	Ikeda (1974)
Mixed zooplankton (20°C)	0.17	0.34	0.45	0.87	Mullin et al. (1975)
Mixed zooplankton from Great Barrier Reef (warm and cold seasons)	0.01–0.02	0.51–1.10	3.33–7.08	282–562	Ikeda et al. (1982)
Mixed planktonic crustaceans (subtropical and boreal species)	0.09–0.10	0.29–0.39	0.45–0.74	1.26–3.30	Vidal and Whitledge (1982)
Oceanic copepods	0.03	1.29	6.16	251	Verity (1985a)
Overall range =	0.01–0.17	0.17–1.51	0.38–7.48	0.87–562	
Phosphorus					
Marine planktonic and benthic crustaceans	0.002	0.03	0.10	1.61	Hargrave and Geen (1968)
Mixed zooplankton (20°C)	0.05	0.11	0.16	0.34	Mullin et al. (1975)
Mixed zooplankton from Great Barrier Reef (warm and cold seasons)	0.001–0.004	0.05–0.18	0.15–1.68	1.66–322	Ikeda et al. (1982)
Overall range =	0.001–0.05	0.03–0.18	0.10–1.68	0.34–322	

*Upper† and lower‡ size limits for zooplankton were estimated from the literature. Upper§ and lower# size limits for protozoa correspond to cell volumes of 10^6 μm³ and 10^2 μm³, respectively, assuming a specific gravity = 1.0, and (0.2) (wet weight) = (dry weight). All values are μg N or μg P mg^{-1} dry weight h^{-1}.

released by the microflagellate *Ochromonas* sp. when feeding on bacteria. Similarly, Goldman et al. (1985) observed that approximately 50 percent of the nitrogen present in bacteria prey or diatom prey was regenerated as NH_4^+ + urea by the microflagellate *Paraphysomonas imperforata* over the entire growth cycle (early exponential growth to late stationary growth) of the protozoan. This latter result is in agreement with the high GGE based on carbon (44 percent) that has been observed for this microflagellate (Caron et al., 1985). From these results we conclude that although protozoa are important remineralizers of organic material, a substantial fraction of the ingested nutrients is retained as protozoan biomass, where it is presumably available to higher trophic levels (Azam et al., 1983).

Although protozoa do regenerate and release some nutrients during starvation, it is unlikely that they release substantial amounts of their cellular pool of nutrients under starvation conditions. Goldman et al. (1985) observed that nitrogen remineralization by *P. imperforata* slackened considerably as the microflagellate entered the stationary growth phase. Similarly, resting cells of *Paramecium aurelia* had lower NH_4^+ excretion rates than growing cells (Soldo and Wagtendonk, 1961). These results are in agreement with published accounts of reduced nutrient regeneration by metazoan zooplank-

TABLE 7.5 Summary of Reported Gross Growth Efficiencies of Protozoa

Species	Gross growth efficiency (%)	Comments	Reference
Ciliates			
Entodinium caudatum	50	39°C; bacterial food	Coleman (1964)
Colpoda steinii	78	30°C; bacterial food	Proper and Garver (1966)
Tetrahymena pyriformis	9 and 50	25°C; axenic and bacterial food, respectively	Curds and Cockburn (1968)
Uronema sp.	5–16	20°C; continuous culture bacterial food	Hamilton and Preslan (1970)
Tetrahymena pyriformis	54	Continuous culture, bacterial food	Curds and Cockburn (1971)
Didinium nasutum	8–25	*Paramecium caudatum* food	Sleigh (1973)
Dileptus cygnus	7–32	*Colpidium colpoda* food	referenced in Klekowski and Fischer (1975)
Spirostomum ambiguum	15	*Colpidium colpoda* food	referenced in Klekowski and Fischer (1975)
Colpidium campylum	3–11	10°–20°C; bacterial food	Laybourn and Stewart (1975)
Stentor coeruleus	64–82	15°–20°C; *Tetrahymena* food	Laybourn (1976a)
Podophrya fixa	50–66	15°C; *Colpidium campylum* food	Laybourn (1976b)
Uronema marinum	6–20	25°C; algal food	Rubin and Lee (1976)
Euplotes vannus	2–12	25°C; algal food	Rubin and Lee (1976)
Tintinnopsis acuminata	50	18°C; algal food	Heinbokel (1978)
Strombidium sp.	5–11	21°C; continuous culture, algal food	Scott (1985)
Favella sp.	21–69	15°C; *Heterocapsa triquetra* food	Stoecker and Evans (1985)
Balanion sp.	25–61	15°C; *Heterocapsa triquetra* food	Stoecker and Evans (1985)
Tintinnopsis acuminata	17–72 ($\bar{x}$ = 33)	15°–25°C; algal food	Verity (1985b)
Tintinnopsis vasculum	30–76 ($\bar{x}$ = 33)	5°–15°C; algal food	Verity (1985b)
Flagellates			
Parabodo attenuatus	18–20	bacterial food	Kopylov et al. (1980)
Mixed nanoflagellate assemblage	14–44	23°–25°C; bacterial food	Kopylov and Moiseev (1980)
Ochromonas	34	20°C; bacterial food	Fenchel (1982b)
Pleuromonas jaculans	43	20°C; bacterial food	Fenchel (1982b)
Monas sp.	24–45	20°C; bacterial food	Sherr et al. (1983)
Monas sp.	12–49	3°–30°C; bacterial food	Sherr et al. (1983)
Unidentified 8 μm^3 flagellate	23	8°–10°C; bacterial food	Wambeke and Bianchi (1985)
Paraphysomonas imperforata	44	20°–24°C; bacterial or diatom food	Caron et al. (1985)
Amebae			
Acanthamoeba sp.	37	25°C; yeast food	Heal (1967)
Amoeba proteus	4–47	10°–20°C; *Tetrahymena pyriformis* food	Rogerson (1981)

ton during starvation (Mayzaud, 1976). Nutrient regeneration by protozoa during starvation must take place at the expense of cellular structures that are autophagocytized (Fenchel, 1982c). Therefore, once nutrients are incorporated into protozoan biomass, it is probable that they are regenerated primarily by the consumption of protozoa by other organisms or through death and bacterial decomposition, and not by direct remineralization by protozoa.

The rate (and perhaps the magnitude) of nutrient regeneration by protozoa is undoubtedly affected by the chemical composition of the prey, but there are still few data on this subject. In a recent study, Goldman et al. (1985) found that nitrogen regeneration efficiency of the microflagellate *Paraphysomonas imperforata* when growing exponentially on phosphorus-limited diatoms was approximately 3 × as great as when the prey was nitrogen limited (Goldman et al., 1985). However, as seen in the summary plot in Figure 7.4, there were no clear differences in the total amount of nitrogen remineralization when summed over the whole experimental period. Differences in the amount of phosphorus regenerated in the two treatments were much greater than those for nitrogen. Virtually no regenerated phosphorus was released by the microflagellates during exponential growth on P-limited diatoms (20 percent was released when they were fed N-limited diatoms), and only 12 percent of the initial particulate organic phosphorus was released as PO_4^{-3} over the entire experimental period when P-limited diatoms were the prey (as opposed to 57 percent when the food source was N-limited diatoms).

FIGURE 7.4 Total regenerated nitrogen (NH_4^+ + urea) and phosphorus (PO_4^{-3}) in cultures of the microflagellate *Paraphysomonas imperforata* feeding on nonnutrient-limited (A), nitrogen-limited (B), and phosphorus-limited (C) diatoms (*Phaeodactylum tricornutum*) at different growth stages of the microflagellate. Open bars are regenerated phosphorus, and closed bars are regenerated nitrogen. (Data taken from Goldman et al. [1985] and Andersen et al. [1986].)

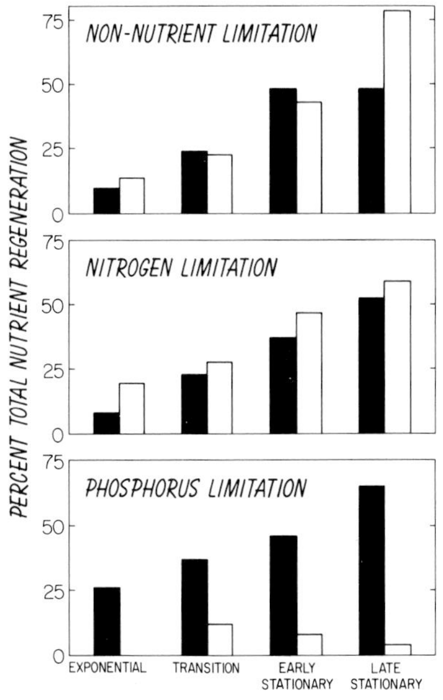

In addition to the effect of the prey C:N or C:P ratios, a probable explanation for the differences observed in nitrogen and phosphorus regeneration in the latter study is the fact that regenerated nitrogen and phosphorus are the result of distinct physiological processes in the protozoan cell. As stated previously, nitrogen is released primarily as NH_4^+ resulting from the deamination of amino acids. However, amino acid decomposition would have little effect on phosphorus regeneration, since most of this latter element is contained in nucleic acids, cellular lipids, and soluble compounds of low molecular weight. Therefore, it is not unexpected that nitrogen and phosphorus remineralization might differ in magnitude and rate within a single population.

Sherr et al. (1983) found differences in the nitrogen remineralization rate of the microflagellate *Monas* sp. when it was fed different species of bacteria. These differences also appear to be related to food quality, since growth of the microflagellate was not limited by prey density. Food quality is known to affect growth rates of protozoa (Rubin and Lee, 1976; Curds and Bazin, 1977), and it has been shown to affect nutrient excretion rates of planktonic copepods (Ikeda, 1977; Debs, 1984). However, the use of different prey species does not always affect protozoan growth and nutrient regeneration rates. Goldman and Caron (1985) observed that both the growth rate and NH_4^+ regeneration efficiency of *P. imperforata* were the same when it was fed several types of phytoplankton. Clearly, the effect of prey chemical composition on nutrient regeneration by protozoa requires more investigation.

Finally, growth conditions may affect the amount or rate of nutrient regeneration by protozoa. GGEs of some protozoa are affected by temperature (the variability of the GGE for each species in Table 7.5 is primarily an effect of temperature; e.g., see Rassoulzadegan, 1982), and nutrients not incorporated by protozoa as a result of suboptimal growth conditions may be released in remineralized form. Sherr et al. (1983) measured higher NH_4^+ excretion rates for *Monas* sp. at 3° and 30°C than at two intermediate temperatures. This result presumably was caused by reduced GGE of the microflagellate at temperatures near the lower and upper temperature limits for this protozoan.

NUTRIENT REGENERATION: BACTERIA VERSUS PROTOZA

Much of the controversy surrounding the role of marine microbes in nutrient regeneration has focused on the relative contributions of bacteria and protozoa. Most of the early work centered on nutrient regeneration in detrital food chains, and primary emphasis was given to the role that protozoa play as bacterial consumers. Two general views concerning nutrient regeneration arose from this work. According to one view, bacteria are the primary nutrient remineralizers in detrital food chains, and the main role of protozoa is to graze the bacterial population, thereby maintaining bacteria in a state of "physiological youth" (e.g., Fenchel and Harrison, 1976). In contrast, the other view considers bacteria to be primarily "sinks," not sources, of regenerated nutrients, and the protozoa, by virtue of very high weight-specific regeneration rates, are the main nutrient regenerators in these ecosystems (Johannes, 1968). Although there is little question that rates of decomposition and nutrient regeneration are enhanced in the presence of protozoa, there is still considerable disagreement on which microorganisms are responsible for the bulk of nutrient regeneration. These conflicting views have not easily been reconciled, and ample data in the literature appear to support either claim.

BACTERIA AND NUTRIENT REGENERATION

Bacteria are primarily responsible for the decomposition of nonliving organic material in the ocean. It has been estimated that between 10 and 50 percent of the organic carbon resulting from primary production in the ocean is assimilated or decomposed by bacteria (Fuhrman and Azam, 1982). It has therefore been assumed that marine bacteria play a key role in the nutrient regeneration process. In support of this hypothesis, it has been shown that bacteria may remineralize significant quantities of the nitrogen and phosphorus contained in detritus. Barsdate et al. (1974), for example, examined the decomposition of leaves of the macrophyte *Carex aquatilis* in bacterized microcosms with and without protozoa feeding on the bacteria. They observed that decomposition proceeded faster in the presence of protozoan grazers but that this increase in decomposition rate was apparently not due to enhanced nutrient regeneration by the protozoa. Based on their calculations of the phosphorus content of bacteria ingested by the protozoa, they concluded that protozoan nutrient regeneration was not sufficient to explain phosphorus turnover in the microcosms. However, the turnover rate of this nutrient by bacteria was substantially greater in grazed systems. These results have been interpreted as evidence for an indirect role of protozoa in the regeneration of nutrients. According to Fenchel and Harrison (1976), protozoan grazing maintains the bacterial population in a state of "physiological youth," and this active bacterial population is primarily responsible for nutrient cycling in the system.

Other investigators also have suggested that protozoa play a minimal role as remineralizers, based on the results of microcosm studies. Fenchel (1977) investigated the regeneration of nutrients from detritus in microcosms with and without protozoa. His conclusion that protozoa are not the primary decomposers in these microcosms was based on the observation that when inorganic nutrient (phosphorus) was added to the microcosms, the decomposition of the organic material was enhanced even in those microcosms that contained protozoa. The decomposition of organic material in microcosms that received little or no inorganic nutrient enrichment was only slightly enhanced by the presence of protozoa. Fenchel reasoned that protozoa were not efficient mineralizers because they were unable to maintain the decomposition process by regenerating nutrients contained in the bacterial biomass. However, it must be recognized that if there was only a small amount of nutrient present in the microcosms initially, then mineralization would proceed slowly even in the case of efficient phosphorus remineralization by the protozoa.

FACTORS CONTROLLING BACTERIAL NUTRIENT REGENERATION

In reviewing the question of nutrient regeneration in aquatic ecosystems, Johannes (1968) pointed out that most investigations dealing with this topic up to that time were biased to some degree by the preconceived notion that bacteria were the primary mineralizers of nitrogen and phosphorus. However, Johannes noted that when nutrients are in short supply in the bacterial substrate, inorganic nutrients will be assimilated from the environment. In soil ecology this process is referred to as "nutrient immobilization" because the amount of nutrients available to plants for growth is reduced (Alexander, 1961). More recently, it has been shown that aquatic bacteria are equipped physiologically to compete with phytoplankon for dissolved inorganic nutrients present at very low concentrations (Currie and Kalff, 1984a). In fact, it is now accepted that bacteria may be a significant "sink" for inorganic nutrients in a number of terrestrial (Coleman et al., 1978, 1983; Woods et al., 1982) and aquatic environments (Hargrave and Geen, 1968; Faust and Correll, 1976; Eppley et al., 1977; Harrison et al., 1977; Horstmann and Hoppe, 1981; Krempin et al., 1981; Currie and Kalff, 1984b; Güde, 1985).

The degree to which bacteria and algae compete for dissolved nutrients is affected by the nitrogen and phosphorus content of the organic material assimilated by the bacteria. Currie and Kalff (1984c) demonstrated that freshwater bacteria could outcompete algae for PO_4^{-3}-phosphorus in chemostats run under a variety of conditions. They observed that when organic material containing no phosphorus was added to the chemostats, the bacteria took up a larger portion of the available PO_4^{-3} (i.e., bacterial biomass increased and algal biomass decreased) relative to unenriched chemostats. Rhee (1972) also observed that bacteria interfered with the growth of the alga *Scenedesmus* when cultures of the alga were enriched with glucose. The author concluded that this phenomenon was due to competition for PO_4^{-3} between the bacteria and the alga. Similarly, Parsons et al. (1981) observed that the addition of 1–5 mg/L of glucose to enclosed seawater columns resulted in lower algal productivity because of enhanced competition for nitrogen between the bacteria and phytoplankton. The ecological implications of bacteria–phytoplankton nutrient competition have been discussed recently (Bratbak and Thingstad, 1985).

The C:N and C:P ratios of the bacterial substrate influence directly the degree to which bacteria will excrete or take up nutrients. Net regeneration (or uptake) of a nutrient is the difference between the amount of nutrient in the substrate taken up by the bacteria and the amount needed to produce bacterial biomass. The amount of nutrient in the substrate can be expressed as the amount of substrate ingested (I, expressed as carbon) divided by the carbon:nutrient (i.e., nitrogen or phosphorus) ratio in the substrate. Likewise, the amount of nutrient required for growth can be expressed as the amount of substrate ingested (I) multiplied by the gross growth efficiency (GGE) and divided by the carbon:nutrient ratio of the bacteria. Therefore, excretion rate (E) of the nutrient (N or P) can be expressed as

$$E = \frac{I}{\text{C:Nu}_{\text{sub}}} - \frac{I \times \text{GGE}}{\text{C:Nu}_{\text{bac}}} \tag{1}$$

where C:Nu_{sub} and C:Nu_{bac} are the carbon:nutrient ratios of the substrate and bacteria, respectively. It is assumed in this equation that egested nutrients are released by the bacteria as regenerated inorganic compounds and that DOC released by the bacteria is an insignificant fraction of the carbon budget (i.e., the loss of organic carbon is due primarily to respiration). A negative excretion rate (E) in equation (1) indicates the rate of nutrient *uptake*.

An important implication of equation (1) for bacterial nutrient regeneration is the controlling influence of C:Nu_{sub}, C:Nu_{bac}, and GGE on E. When $(\text{C:Nu}_{\text{sub}})^{-1} > (\text{GGE}) \times (\text{C:Nu}_{\text{bac}})^{-1}$, nutrients will be regenerated (barring luxury uptake of nutrients by the bacteria). On the other hand, when $(\text{C:Nu}_{\text{sub}})^{-1} < (\text{GGE}) \times (\text{C:Nu}_{\text{bac}})^{-1}$, nutrients will be taken up from the environment. Therefore, although the magnitude of E is affected by the ingestion rate, whether or not nutrients are regenerated or taken up (positive E or negative E) is dictated solely by the relative magnitudes of GGE and the C:Nu ratios.

Because gross growth efficiency is defined as $\text{GGE} = (I - R) \times I^{-1}$ (where R is respiration rate expressed as carbon cell $^{-1}$ time $^{-1}$), equation (1) can be rewritten as

$$E = \left(\frac{R}{1 - \text{GGE}}\right) \times \left(\frac{1}{\text{C:Nu}_{\text{sub}}} - \frac{\text{GGE}}{\text{C:Nu}_{\text{bac}}}\right) \tag{2}$$

Therefore, the excretion rate (or uptake rate) of the bacteria also can be expressed as a function of the respiration rate, GGE of the bacteria and the carbon:nutrient ratios of the bacteria and substrate. Where the carbon:nutrient ratios are the same for substrate and bacteria, equation (2) simplifies to

$$E = \frac{R}{\text{C:Nu}} \tag{3}$$

Various forms of these equations have been derived previously (Fenchel and Blackburn, 1979; Blackburn, 1983; Lancelot and Billen, 1985).

It is clear from these equations that nutrient regeneration is closely coupled to growth and the carbon:nutrient ratios of the substrate and bacteria. These equations also can be applied to predator–prey relationships, provided the carbon:nutrient ratios of egested POC and DOC are the same as the ratios of the predators, that they can be quantified, or that they constitute a small fraction of the total carbon budget. An example of the relationships between nitrogen excretion rate, GGE, and carbon:nitrogen ratios is given in Figure 7.5 based on ingestion rate information for the microflagellate *Paraphysomonas imperforata* (Caron et al., 1985). Nitrogen regeneration is indicated by positive values in this figure. Negative values indicate that nitrogen in the consumed prey is not sufficient to maintain the desired C:N ratio in the predator.

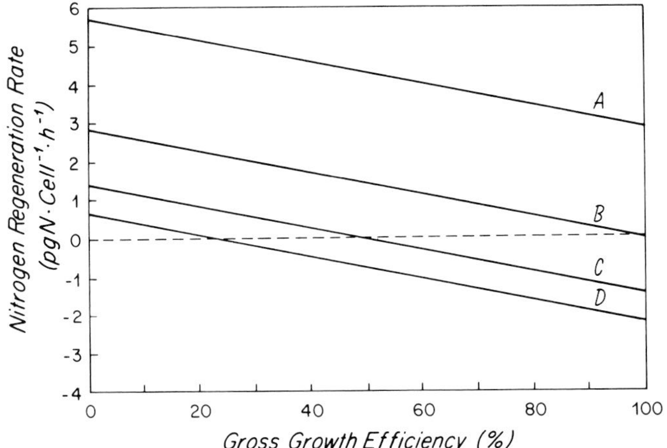

FIGURE 7.5 Examples of changes in the nitrogen regeneration rate (E) based on equation (1) as a function of gross growth efficiency of a predator and the carbon:nitrogen ratios of the predator and prey. The ingestion rate (20 pg C $cell^{-1}$ h^{-1}) is based on experiments with the microflagellate *Paraphysomonas imperforata* (Caron et al., 1985). Positive values are nutrient excretion; negative values are nutrient uptake. Carbon:nitrogen ratios (by weight) for prey and predator are 3.5:7 (A), 7:7 (B), 14:7 (C), and 28:7 (D).

These latter conditions would result in nutrient uptake by organisms that are capable of doing so (e.g., bacteria).

Two important points are evident from Figure 7.5. First, increasing GGE of a predator results in lower nutrient excretion rates (indicated by negative slopes). This relationship (that nutrient excretion is the complement of growth efficiency) is intuitively obvious (see "Nutrient Regeneration vs. Incorporation" in this chapter), but has been overlooked in the literature. Second, nutrient excretion rates at a single GGE are affected significantly by differences in the $(C{:}N_{prey}) \times (C{:}N_{pred})^{-1}$ ratio (indicated by the amount to which the lines are vertically offset from each other). This vertical offset has an effect on the x-intercept and, therefore, on the amount of nutrient regeneration/uptake. For example, at a GGE of 50 percent, a $(C{:}N_{prey}) \times (C{:}N_{pred})^{-1}$ ratio of 1 results in nutrient regeneration (positive E), whereas a ratio of 4 results in nutrient uptake (negative E).

It is important to note that the absolute values of the ingestion rate or respiration rate will change the slopes of the lines in Figure 7.5 but will not affect the x-intercepts. Whether or not nutrients will be released or taken up is dictated solely by GGE and the carbon:nutrient ratios. However, the absolute rates of nutrient uptake or release (slopes of the lines) are affected by ingestion and respiration rates, and therefore, the nitrogen excretion rates in Figure 7.5 do not necessarily depict absolute rates for *P. imperforata*. Nevertheless, they are useful for demonstrating the general effects of gross growth efficiency and nutrient composition on nutrient regeneration/uptake.

In the context of bacterial nutrient regeneration, it is clear from equations (1)–(3) that enriching the growth medium with organic material that does not contain phosphorus or nitrogen increases the overall C:N and C:P ratios of the bacterial substrate, which decreases the $(C{:}Nu_{sub})^{-1}$ term in equation (2), thus enhancing bacterial uptake of these nutrients from the pool of dissolved inorganic nutrients (or retention of a greater percentage of the intracellular nutrients). These effects have been demonstrated experimentally. When amino acids (low C:N ratio) are added to seawater, a significant fraction of the nitrogen remineralization can be attributed to the bacteria (Hollibaugh, 1978; Wambeke and Bianchi, 1985). However, Hollibaugh (1978) observed that adding glucose to seawater samples already enriched with the amino acid L-arginine reduced the amount of NH_4^+ released by the bacteria. Ammonium release in these samples was inversely proportional to the overall C:N ratio of the organic substrates added to the samples.

Results from these studies provide valuable insight into the mechanisms controlling the role of bacteria in nutrient regeneration and uptake. From these experiments we conclude that the potential exists for bacteria to be either a source or a sink for remineralized nutrients in aquatic ecosystems. The potential for this "dual" role by bacteria is best exemplified in the "microbial food loop" of Azam et al. (1983). However, despite this mechanistic understanding of nutrient regeneration by bacteria, it is difficult to measure directly the overall importance of bacteria as nutrient regenerators in the ocean. The quality and relative proportions of organic compounds that serve as bacterial substrates in aquatic ecosystems are too poorly known to estimate the average C:N or C:P ratios of these materials. For example, nitrogen-rich compounds (e.g., amino acids) and nitrogen-poor compounds (e.g., combination of carbohydrates) are considered to be important bacterial substrates in the ocean (Hollibaugh et al., 1980; Burney et al., 1981, 1982). The relative importance of these materials as bacterial substrates could have a pronounced effect on nutrient assimilation or regeneration by marine bacteria (Fasham, 1985). Until methods become available to characterize and quantify bacterial substrates in nature, other methods (such as size fractionation) will have to be employed to measure bacterial regeneration.

MODELING THE RESULTS OF EXPERIMENTAL STUDIES

It is obvious from the discussion in the preceding pages that there is an apparent contradiction in the literature concerning the relative importance of nutrient regeneration by protozoa. Although remineralization by protozoa has been shown in some studies to be unimportant relative to remineralization by bacteria, high weight-specific excretion rates for protozoa have been observed in other studies. Some of these discrepancies may be due to differences in the activity (i.e., uptake vs. regeneration) of bacteria in different environments. In addition, some of these discrepancies may arise from the use of inappropriate methodologies for measuring protozoan regeneration rates. More difficult to reconcile are studies that show that nutrient regeneration rates of protozoa may be very low. We have already recounted the results of Fenchel's (1977) studies on detritus decomposition. He concluded that protozoa were not efficient mineralizers in his experiments because they were unable to maintain the decomposition process by regenerating nutrients tied up in bacterial biomass. These results are quite different from the results of Sherr et al. (1982), who showed that the rate of decomposition of *Peridinium* cells by bacteria was accelerated when protozoa were present. Sherr et al. (1982) concluded that the accelerated rate of decomposition was due to regeneration of phosphorus by the protozoa because the addition of PO_4^{-3} to protozoan-free cultures resulted in a decomposition rate comparable to that in unenriched cultures containing protozoa.

Results from recent studies in organic geochemistry may hold the key to reconciling the results of these studies and still allow for the existence of rapid nutrient regeneration rates by protozoa. A significant correlation has been observed between nitrogen accumulation in decomposing detritus and the appearance of humic substances (Rice, 1982). Reactive phenolic and carbohydrate compounds, formed by bacterial exoenzymes, condense with nitrogen to form amino-phenols and amino-sugars. Rice (1982) hypothesized that nitrogen in these compounds is relatively recalcitrant (i.e., difficult for bacteria to assimilate) and that this nonlabile nitrogen may be one explanation for nitrogen immobilization in marine detritus. In support of this hypothesis, Melillo et al. (1984) have shown that relatively little of the nitrogen in detritus is present as microbial biomass. The formation of unassimilable condensation products, even in a system where protozoa regenerate most of the nutrients contained in their food, would eventually result in the depletion of limiting nutrient available to the bacteria. If more nutrients are added to such a system, one would expect renewed bacterial uptake (and subsequent protozoan consumption of bacteria) and reestablishment of active microbial decomposition until the nutrients being recycled are slowly but inextricably immobilized in relatively recalcitrant compounds.

Viewed in this context, we conclude that protozoa were not necessarily inactive as nutrient regenerators in the microcosm studies of Fenchel (1977). The fact that the presence of protozoa resulted in more decomposition (relative to ungrazed microcosms) when nutrients were added suggests that protozoa kept these nutrients "active" (i.e., cycling back and forth between bacteria and protozoa) for a longer period of time than in ungrazed microcosms. Therefore, it may be possible to explain some of the apparent contradiction presented by these microcosm studies.

A conceptual model for the decomposition of organic detritus that incorporates the processes described earlier is presented in Figure 7.6. In this idealized view of decomposition, it is assumed that the C:N and C:P ratios in the detritus are greater than these ratios in bacterial biomass and that dissolved inorganic nutrients as well as nutrients contained in the organic detritus are incorporated into bacterial biomass. Protozoa feeding on the bacteria regenerate some portion of the nutrients contained in the bacteria. Most of these regenerated nutrients are taken up by the bacteria and are again incorporated into bacterial biomass during the decomposition of more organic detritus. However, some small portion of the regenerated nutrients are condensed with reactive phenols and carbohydrates and are lost from the biological cycle. As the nutrients cycle from the bacteria to the protozoa and back again, more nutrients are lost to these unassimilable condensation products. Eventually, the bulk of the nutrients are removed from the biological cycle and the decomposition rate of the detrital material decreases.

This model is meant to describe nutrient cycling in a detrital food web, and we have not attempted

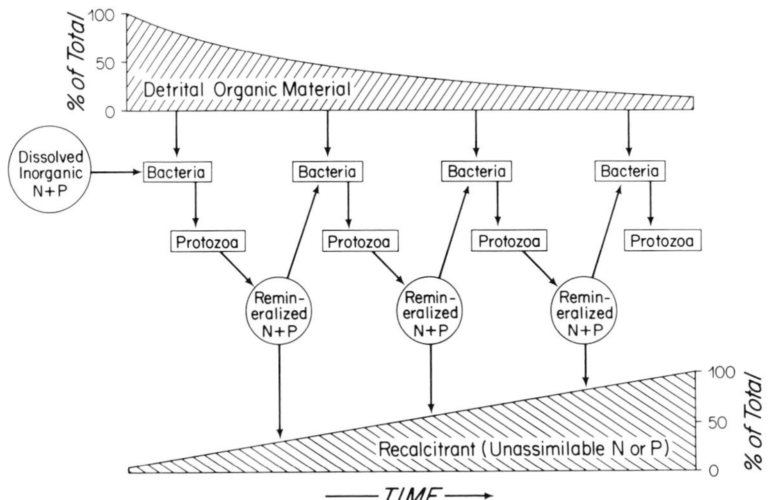

FIGURE 7.6 Conceptual model for the role of protozoa in the cycling of nitrogen and phosphorus in detrital food chains. (See text for details.)

to incorporate the role of photosynthetic organisms into it. In addition, we have assumed that the carbon:nutrient ratios of the detritus are high and that bacteria primarily assimilate, not regenerate, nutrients. Although this assumption appears to fit most of the existing data on detritus decomposition, it is possible that bacteria may function as nutrient regenerators when the carbon:nutrient ratios of the detritus are lower than those required for balanced growth of bacteria (see Fig. 7.5). Also, it is important to recognize that, according to this model, nutrients can be lost from the detrital food chain not only through the formation of unassimilable condensation products, but also by their incorporation into protozoan biomass. Thus, a number of processes commonly observed in detritus decomposition studies are incorporated into the model: (1) enhanced bacterial growth rates in the presence of protozoan grazers without concomitant bacterial nutrient regeneration (bacteria assimilate nutrients in this model, not regenerate them); (2) an important role for protozoa as nutrient remineralizers; (3) nitrogen accumulation in the detritus, but not in the microbial biomass; (4) a reduction in the percentage of organic material remaining with time; (5) a reduction in decomposition *rate* with time, because of the loss of limiting nutrient to condensation products.

NUTRIENT CYCLING AT THE "MICRO" LEVEL

It is well known that the distribution of free-living protozoa and other microorganisms often is heterogeneous at the microscopic level (Wimpenny, 1981). This heterogeneity can be caused by a variety of trophic and/or biochemical interactions between protozoa and other microorganisms. The most tightly coupled relationships, both spatially and biochemically, are those in which species exist in a symbiotic relationship. The diversity of symbiotic relationships that exist between protozoa and other microorganisms is the subject of another chapter in this book (see Chapter 9).

Other factors also give rise to heterogeneous distributions of microorganisms. Microscale concentrations of prey organisms can provide a favorable microenvironment for the growth of protozoa, leading to faster rates of division and thereby larger protozoan biomasses in those microenvironments. For example, it has been demonstrated that macroscopic detrital aggregates (so-called marine snow) in marine plankton communities are highly enriched with protozoa and other microorganisms relative to the surrounding seawater (Silver et al., 1978, 1984; Caron et al., 1982b, 1986). These elevated densities of protozoa may be due to enhanced growth rates of protozoa feeding on the large populations of microorganisms associated with the aggregate environment. Elevated densities of protozoa in these detrital aggregates may also result from the migration of protozoa to these microenvironments, since chemosensory abilities are known to exist among protozoa (Hauser et al., 1975; Levandowsky and Hauser, 1978; Antipa et al., 1983; Fitt, 1985; Spero, 1985). Heterogeneous distributions may also arise in physically homogeneous environments, however (Cairns and Yongue, 1968). These associations are apparently a result of behavioral aggregation of protozoa and may indicate the tendency for trophic or nutrient relationships to be established among microorganisms at the microorganismal level.

Because of the physical proximity of microorganisms in many of these associations, there is a strong probability that nutrient regeneration by protozoa often is closely coupled with uptake by other microorganisms. This probability is particularly strong in the case of symbiotic relationships among protozoa. A wide variety of microorganisms have been described as the symbionts of protozoa, including bacteria (Gill and Vogel, 1963; Fenchel et al., 1977; Lee at al., 1985), cyanobacteria (see review by F. Taylor, 1982) and numerous eucaryotic algae (see reviews by Lee, 1980, F. Taylor, 1982, Lee et al., 1985). The majority of these associations can be characterized as a heterotrophic component (protozoan) and an autotrophic component (cyanobacterium or alga). Although energy and nutrient relationships within many of these associations are not yet clear, it is presumed that the symbionts provide a source of energy for their host either directly, through the digestion of some symbionts, or indirectly, by the translocation of organic compounds from symbionts to host. In turn, nutrients regenerated by the protozoan host's metabolic processes are utilized during photosynthesis by the symbionts, creating a "closed loop" for the transfer of nutrients between host and symbionts.

Confirmation of the physiological relationships that have been proposed to take place between host and symbiont has been hampered by methodological limitations. Separation of the host and symbiont in order to identify the transfer of materials between the two components poses a formidable task, and direct evidence of symbiont-derived nutrition has been provided mostly through the use of the radioisotope ^{14}C. It has been demonstrated from these studies that the host–symbiont complex can incorporate $^{14}CO_2$ and that this photosynthetically derived organic carbon is translocated from symbiont to protozoan host in ciliates (Barber et al., 1969; Brown and Nielson, 1974), foraminifera (Lee, 1980) and radiolaria (Anderson, 1978; Anderson et al., 1983, 1985). Host–symbiont relationships in benthic and planktonic foraminifera (Bé et al., 1977; Lee, 1980) also have been studied by observing the effect of symbiont activity on growth and survivorship of the foraminiferal host. From these studies it has been concluded that symbiont-derived nutrition in these associations is important for the normal growth and development of many marine sarcodines (Lee, 1980; Bé et al., 1981, 1982; Caron et al., 1982a).

Direct confirmation of the transfer of regenerated nitrogen from host to photosynthetic symbionts has not been clearly demonstrated and has been hindered largely by the lack of a suitable isotope for this element. However, from studies with marine coelenterates it has been shown that symbiont-containing organisms released less NH_4^+ (or incorporated more NH_4^+) than aposymbiotic organisms or symbiont-containing organisms kept in the dark (Muscatine and D'Elia, 1978; Muscatine et al., 1979; Wilkerson and Muscatine, 1984). It is clear from these results that the NH_4^+ produced by the host was utilized by the symbiotic algae. A tight coupling between nutrient regeneration and uptake undoubtedly also exists between protozoan hosts and their photosynthetic symbionts.

Nutrient release by protozoa may, in fact, be a mechanism for bringing potential hosts and free-swimming algal symbionts together. Fitt (1984, 1985) has demonstrated that the symbiotic dinoflagellate *Symbiodinium microadriaticum* displayed positive chemotaxis toward nitrogen-containing compounds, including NH_4^+, urea, and some amino acids. Concentrations of NH_4^+ of $<1\ \mu M$ elicited a positive response. Release of nitrogen-containing waste products by protozoa may attract algae that are acceptable as symbionts into close proximity of the protozoa and thereby enable the establishment of the symbiotic association.

Symbiosis is a highly structured evolutionary outcome in response to efficient nutrient and energy cycling between heterotrophs and autotrophs. Other less evolutionarily advanced associations exist among marine microorganisms in which nutrient regeneration by protozoa and nutrient uptake by primary producers are spatially structured at the microscopic level. A number of planktonic algae from marine and freshwater environments form macroscopic colonies that contain large populations of protozoa and other microorganisms. For example, colonies of the marine diatoms *Rhizosolenia* (Carpenter et al., 1977; Alldredge and Silver, 1982; Caron et al., 1982b) and *Thalassiosira partheneia* (Elbrächter and Boje, 1978; Caron, unpublished data) have diverse protozoan assemblages associated with them. Similarly, colonies of the planktonic cyanobacterium *Oscillatoria* (*Trichodesmium*) often harbor populations of amebae (Anderson, 1977) and ciliates (Caron, unpublished data). These protozoan associates may contribute significantly to the pool of regenerated nutrients required by the diatoms and cyanobacteria.

Marine snow particles are also potentially important microenvironments for spatial and temporal coupling of nutrient regeneration and uptake. Macroscopic aggregates are centers of intense nutrient regeneration in the plankton (Shanks and Trent, 1979; Glibert et al., 1988). The numbers of protozoan populations associated with these aggregates are often several orders of magnitude greater than those in the surrounding seawater, and they are undoubtedly responsible for a significant fraction of the remineralization of nutrients in these microenvironments. Photosynthetic microorganisms are abundant in these aggregates (Caron et al., 1986) and measured rates of primary productivity sometimes are high compared with the surrounding seawater (Alldredge and Cox, 1982; Knauer et al., 1982). Therefore, it appears that marine snow in surface waters of the ocean affords a mechanism whereby photosynthetic microorganisms can be in close physical proximity to the microorganisms responsible for nutrient regeneration, thereby avoiding the potential for nutrient limitation of growth. A similar role has been proposed for microscopic aggregates (Goldman, 1984), many of which also have large microbial populations associated with them (Pomeroy, 1983).

PERSPECTIVES FOR FUTURE RESEARCH

There are, at present, relatively few available data on excretion rates of remineralized nutrients by individual species of protozoa. Although the results obtained from studies of nutrient cycling in natural microbial assemblages (such as in experiments with enclosed water columns or studies of the decomposition of detritus) are valuable for a thorough understanding of protozoan activity, the biological and biochemical interactions in these microcosms are often too complex to differentiate with certainty the role of protozoa. There is a need for simpler experimental protocols that will yield straightforward answers on the ability of protozoa to remineralize nutrients contained in ingested food. In addition, it is necessary to consider the potential problems associated with nutrient regeneration or uptake by prey organisms when designing these experiments and to account for this activity in the determination of protozoan regeneration rates. Only with baseline data on the ability of individual species to regenerate nutrients can we hope to interpret protozoan interactions in more complex microbial communities.

Determination of the weight-specific excretion rates for individual species of protozoa will provide the information required to define and quantify better the relationship between weight-specific excretion rate and body weight (see Fig. 7.3), as well as provide information on the effect that food quality and trophic mode has on nutrient regeneration by protozoa. Although consumption of bacterial biomass is often considered to be the main role of protozoa in the ocean, protozoa consume a remarkable diversity of food types, ranging from bacteria (Fenchel, 1982b) and phytoplankton (Heinbokel and Beers, 1979; Capriulo and Carpenter, 1980; Goldman and Caron, 1985) to relatively large zooplankton (Caron and Bé, 1984). It is likely that the magnitude and/or rate of nutrient regeneration is related to the chemical composition of the prey. Ikeda (1977), for example, observed that differences in the NH_4^+ excretion rates for herbivorous and carnivorous copepods were caused by differences in the nitrogen content of the food organisms and the catabolic pathways of the copepod species.

Along with the need for more measurements of protozoan regeneration *rates* is the need to know the total *amount* of nutrients regenerated by various protozoa (overall regeneration efficiency). This measurement is an upper limit of the fraction of ingested nutrients that can be released directly by protozoa, as well as an estimate of the amount that must be regenerated by other means (i.e., consumption by other organisms or death and decomposition). It is, therefore, necessary to characterize nutrient regeneration throughout the whole growth cycle of a species and not solely during exponential growth. Undoubtedly, not all protozoa are growing exponentially at all times in the ocean. Natural assemblages probably are composed of individuals in all stages of growth. Determination of the total amount of ingested nutrients that can be regenerated by protozoa is vital for placing an upper limit on the potential contribution of these organisms in oceanic nutrient cycles. For example, it is estimated that 90 percent of the nutrients used by phytoplankton in the open ocean are regenerated in situ by microorganisms <10 μm in size (Eppley and Peterson, 1979; Glibert, 1982). If we assume that this regeneration is carried out mainly by heterotrophic microflagellates and small ciliates with regeneration efficiencies no greater than 50 percent, then at least three trophic steps would be necessary to account for the observed regeneration (Goldman and Caron, 1985). Therefore, by an examination of the regeneration efficiencies of individual protozoan species, we may gain important insights into how nutrients move through aquatic food webs. In summary, the major nitrogen and phosphorus excretion products of most protozoa appear to be NH_4^+ and PO_4^{-3}, but baseline data on the rates and magnitude of nutrient regeneration by protozoa are still rare. Based on the available data, particularly the high weight-specific excretion rates of protozoa coupled with size-fractionation studies showing the importance of the <100-μm size class of heterotrophs in nutrient regeneration, we conclude that these microbes play a pivotal role in the regeneration of inorganic nitrogen and phosphorus in marine ecosystems.

ACKNOWLEDGMENTS

We are grateful to Drs. T. Fenchel, L.R. Pomeroy, J. McN. Sieburth, D.K. Stoecker, W.D. Taylor, P.G. Verity, and one anonymous reviewer for their comments on the manuscript. This work was supported by NSF Grant OCE-8117715 and a Woods Hole Oceanographic Institution Mellon Study Award (D.A.C.) and NSF Grants OCE-838578 and OCE-8511283 (J.C.G.). Contribution No. 6114 from Woods Hole Oceanographic Institution.

REFERENCES

Alexander, M. (1961) *Introduction to Soil Microbiology*. New York: Wiley.

Alldredge, A.L., and Cox, J.L. (1982) Primary productivity and chemical composition of marine snow in surface waters of the Southern California Bight. *J. Mar. Res.* 40:517–527.

Alldredge, A.L., and Silver, M.W. (1982) Abundance and production rates of floating diatom mats (*Rhizosolenia castracanei* and *R. imbricata* var. *shrubsolei*) in the Eastern Pacific Ocean. *Mar. Biol.* 66:83–88.

Andersen, O.K., Goldman, J.C., Caron, D.A. and Dennett, M.R. (1986) Nutrient cycling in a microflagellate food chain: III. Phosphorus dynamics. *Mar. Ecol. Prog. Ser.* 31:47–55.

Anderson, O.R. (1977) The fine structure of a marine amoeba associated with a blue-green alga in the Sargasso Sea. *J. Protozool.* 24:370–376.

Anderson, O.R. (1978) Fine structure of a symbiont-bearing colonial radiolarian, *Collosphaera globularis*, and ^{14}C isotopic evidence for assimilation of organic substances from its zooxanthellae. *J. Ultrastruct. Res.* 62:181–189.

Anderson, O.R., Swanberg, N.R., and Bennett, P. (1983) Assimilation of symbiont-derived photosynthesis in some solitary and colonial radiolaria. *Mar. Biol.* 77:265–269.

Anderson, O.R., Swanberg, N.R. and Bennett, P. (1985) Laboratory studies of the ecological significance of host-algal nutritional associations in solitary and colonial radiolaria. *J. Mar. Biol. Assoc. U.K.* 65:263–272.

Andersson, A., Lee, C., Azam, F., and Hagstrom, A. (1985) Release of amino acids and inorganic nutrients by heterotrophic marine microflagellates. *Mar. Ecol. Prog. Ser.* 23:99–106.

Antia, N.J., Berland, B.R., and Bonin, D.J. (1980) Proposal for an abridged nitrogen turnover cycle in certain marine planktonic systems involving hypoxanthine-guanine excretion by ciliates and their reutilization by phytoplankton. *Mar. Ecol. Prog. Ser.* 2:97–103.

Antipa, G.A., Martin, K. and Rintz, M.T. (1983) A note on the possible ecological significance of chemotaxis in certain ciliated protozoa. *J. Protozool.* 30:55–57.

Azam, F., Fenchel, T., Field, J.G., Gray, J.S., Meyer-Reil, L.A., and Thingstad, F. (1983) The ecological role of water-column microbes in the sea. *Mar. Ecol. Prog. Ser.* 10:257–263.

Barber, R.T., White, A.W., and Siegelman, H.W. (1969) Evidence for a cryptomonad symbiont in the ciliate *Cyclotrichium meunieri*. *J. Phycol.* 5:86–88.

Barsdate, R.J., Fenchel, T., and Prentki, R.T. (1974) Phosphorus cycle of model ecosystems: significance for decomposer food chains and effect of bacterial grazers. *Oikos* 25:239–251.

Bé, A.W.H., Hemleben, C., Anderson, O.R., Spindler, M., Hacunda J., and Tuntivate-Choy, S. (1977) Laboratory and field observations of living planktonic foraminifera. *Micropaleontology* 23:155–179.

Bé, A.W.H., Caron, D.A., and Anderson, O.R. (1981) Effects of feeding frequency on life processes of the planktonic foraminifer *Globigerinoides sacculifer* in laboratory culture. *J. Mar. Biol. Assoc. U.K.* 61:257–277.

Bé, A.W.H., Spero, H.J., and Anderson, O.R. (1982) Effects of symbiont elimination and reinfection on the life processes of the planktonic foraminifer *Globigerinoides sacculifer*. *Mar. Biol.* 70:73–86.

Beers, J.R., Reid, F.M.H., and Stewart, G.L. (1982) Seasonal abundance of the microplankton in the North Pacific central gyre. *Deep-Sea Res.* 29:227–245.

Blackburn, T.H. (1983) The microbial nitrogen cycle. In: Krumbein, W.E. (ed.), *Microbial Geochemistry*. Oxford, England: Blackwell, pp. 63–89.

Boynton, W.R., and Kemp, W.M. (1985) Nutrient regeneration and oxygen consumption by sediments along an estuarine salinity gradient. *Mar. Ecol. Prog. Ser.* 23:45–55.

Bratbak, G., and Thingstad, T.F. (1985) Phytoplankton-bacteria interactions: an apparent paradox? Analysis of a model system with both competition and commensalism. *Mar. Ecol. Prog. Ser.* 25:23–30.

Brown, J.A., and Nielson, P.J. (1974) Transfer of photosynthetically produced carbohydrate from endosymbiotic Chlorellae to *Paramecium bursaria*. *J. Protozool.* 21:569–570.

Buechler, D.G., and Dillon, R.D. (1974) Phosphorus regeneration in freshwater *Paramecia*. *J. Protozool.* 21:339–343.

Burney, C.M., Davis, P.G., Johnson, K.M., and Sieburth, J.McN. (1981) Dependence of dissolved carbohydrate concentrations upon small scale nanoplankton and bacterioplankton distributions in the Western Sargasso Sea. *Mar. Biol.* 65:289–296.

Burney, C.M., Davis, P.G., Johnson, K.M., and Sieburth, J.McN. (1982) Diel relationships of microbial trophic groups and in situ dissolved carbohydrate dynamics in the Caribbean Sea. *Mar. Biol.* 67:311–322.

Cairns, J., Jr., and Yongue, W.H., Jr. (1968) The distribution of protozoa on a relatively homogeneous substrate. *Hydrobiologia* 31:65–72.

Calow, P. (1977) Conversion efficiencies in heterotrophic organisms. *Biol. Rev.* 52:385–409.

Caperon, J., Schell, D., Hirota, J. and Laws, E. (1979) Ammonium excretion rates in Kaneohe Bay, Hawaii, measured by a ^{15}N isotope dilution technique. *Mar. Biol.* 54:33–40.

Capriulo, G.M. (1982) Feeding of field collected tintinnid microzooplankton on natural food. *Mar. Biol.* 71:73–86.

Capriulo, G.M., and Carpenter, E.J. (1980) Grazing by 35 to 202 μm micro-zooplankton in Long Island Sound. *Mar. Biol.* 56:319–326.

Capriulo, G.M., and Carpenter, E.J. (1983) Abundance, species composition and feeding impact of tintinnid micro-zooplankton in Central Long Island Sound. *Mar. Ecol. Prog. Ser.* 10:277–288.

Carlstrom, D., and Møller, K.M. (1961) Further observations on the native and recrystallized crystals of the amoeba *Amoeba proteus*. *Exp. Cell. Res.* 24:393–404.

Caron, D.A. (1983) Technique for enumeration of heterotrophic and phototrophic nanoplankton, using epifluorescence microscopy, and comparison with other procedures. *Appl. Environ. Microbiol.* 46:491–498.

Caron, D.A., and Bé, A.W.H. (1984) Predicted and observed feeding rates of the spinose planktonic foraminifer *Globigerinoides sacculifer*. *Bull. Mar. Sci.* 35:1–10.

Caron, D.A., Bé, A.W.H. and Anderson, O.R. (1982a) Effects of variations in light intensity on life processes of the planktonic foraminifer *Globigerinoides sacculifer* in laboratory culture. *J. Mar. Biol. Assoc. U.K.* 62:435–451.

Caron, D.A., Davis, P.G., Madin, L.P. and Sieburth, J.McN. (1986) Enrichment of microbial populations in macroaggregates (marine snow) from surface waters of the North Atlantic. *J. Mar. Res.* 44:543–565.

Caron, D.A., Davis, P.G., Madin, L.P., and Sieburth, J.McN. (1982b) Heterotrophic bacteria and bacterivorous protozoa in oceanic macroaggregates. *Science* 218:795–797.

Caron, D.A., Goldman, J.C., Andersen, O.K. and Dennett, M.R. (1985) Nitrogen cycling in a microflagellate food chain: II. Population dynamics and carbon cycling. *Mar. Ecol. Prog. Ser.* 24:243–254.

Carpenter, E.J., Harbison, G.R., Madin, L.P., Swanberg, N.R., Biggs, D.C., Hulbert, E.M., McAllister, V.L. and McCarthy, J.J. (1977) *Rhizosolenia* mats. *Limnol. Oceanogr.* 22:739–741.

Cole, C.V., Elliott, E.T., Hunt, H.W. and Coleman, D.C. (1978) Trophic interactions in soils as they affect energy and nutrient dynamics. V. Phosphorus transformations. *Microbiol. Ecol.* 4:381–387.

Coleman, D.C., Anderson, R.V., Cole, C.V., Elliott, E.T., Woods, L., and Campion, M.K. (1978) Trophic interactions in soil as they affect energy and nutrient dynamics. IV. Flows of metabolic and biomass carbon. *Microbiol. Ecol.* 4:373–380.

Coleman, D.C., Reid, C.P.P., and Cole, C.V. (1983) Biological strategies of nutrient cycling in soil systems. *Adv. Ecol. Res.* 13:1–55.

Coleman, G.S. (1964) The metabolism of *Escherichia coli* and other bacteria by *Entodinium caudatum*. *J. Gen. Microbiol.* 37:209–223.

Conway, H.L., and Whitledge, T.E. (1979) Distribution, fluxes and biological utilization of inorganic nitrogen during a spring bloom in the New York Bight. *J. Mar. Res.* 37:657–668.

Corner, E.D.S., and Davies, A.G. (1971) Plankton as a factor in the nitrogen and phosphorus cycles in the sea. *Adv. Mar. Biol.* 9:101–204.

Cunningham, B., and Kirk, P.L. (1941) The chemical metabolism of *Paramecium caudatum*. *J. Cell. Comp. Physiol.* 18:299–316.

Curds, C.R., and Bazin, M.J. (1977) Protozoan predation in batch and continuous culture. *Adv. Aquatic Microbiol.* 1:115–176.

Curds, C.R., and Cockburn, A. (1968) Studies on the growth and feeding of *Tetrahymena pyriformis* in axenic and monoxenic culture. *J. Gen. Microbiol.* 54:343–358.

Curds, C.R., and Cockburn, A. (1971) Continuous monoxenic culture of *Tetrahymena pyriformis*. *J. Gen. Microbiol.* 66:95–108.

Currie, D.J., and Kalff, J. (1984a) A comparison of the abilities of freshwater algae and bacteria to acquire and retain phosphorus. *Limnol. Oceanogr.* 29:298–310.

Currie, D.J., and Kalff, J. (1984b) The relative importance of phytoplankton and bacterioplankton in phosphorus uptake in freshwater. *Limnol. Oceanogr.* 29:311–321.

Currie, D.J., and Kalff, J. (1984c) Can bacteria outcompete phytoplankton for phosphorus? A chemostat test. *Microbiol. Ecol.* 10:205–216.

Dale, T., and Burkill, P.H. (1982) "Live counting"—a quick and simple technique for enumerating pelagic ciliates. *Ann. Inst. Oceanogr.* (Paris) 58(S):267–276.

Davies, J.M. (1975) Energy flow through benthos in a Scottish sea loch. *Mar. Biol.* 31:353–362.

Davis, P.G., Caron, D.A., Johnson, P.W. and Sieburth, J.McN. (1985) Phototrophic and apochlorotic components of picoplankton and nanoplankton in the North Atlantic: Geographic, vertical, seasonal and diel distributions. *Mar. Ecol. Prog. Ser.* 21:15–26.

Davis, P.G., and Sieburth, J.McN. (1982) Differentiation of the photosynthetic and heterotrophic populations of nanoplankters by epifluorescence microscopy. *Ann. Inst. Oceanogr.* (Paris) 58(S):249-259.

Davis, P.G., and Sieburth, J.McN. (1984) Predation of actively growing bacterial populations by estuarine and oceanic microflagellates. *Mar. Ecol. Prog. Ser.* 19:237–246.

Debs, C.A. (1984) Carbon and nitrogen budget of the calanoid copepod *Temora stylifera:* Effect of concentration and composition of food. *Mar. Ecol. Prog. Ser.* 15:213–223.

DeVincentiis, M., Salvatore, F., and Zappia, V. (1964) Nitrogen catabolism in *Amoeba proteus*. *Exp. Cell. Res.* 35:204–206.

Doyle, W.L., and Harding, J.P. (1937) Quantitative studies on the ciliate *Glaucoma*. Excretion of ammonia. *J. Exp. Biol.* 14:462–469.

Elbrächter, M., and Boje, R. (1978) On the ecological significance of *Thalassiosira partheneia* in the Northwest African upwelling area. In Boje, R., and Tomezak, M. (eds.), *Upwelling Ecosystems*. Berlin: Springer-Verlag, pp. 24–31.

Eppley, R.W., and Peterson, B.J. (1979) Particulate organic matter flux and planktonic new production in the deep ocean. *Nature* 282:677–680.

Eppley, R.W., Sharp, J.H., Renger, E.H., Perry, M.J., and Harrison, W.G. (1977) Nitrogen assimilation by phytoplankton and other microorganisms in the surface waters of the Central North Pacific Ocean. *Mar. Biol.* 39:111–120.

Fasham, M.J.R. (1985) Flow analysis of materials in the marine euphotic zone. In Ulanowicz, R.E., and Platt, T. (eds.), *Ecosystem Theory for Biological Oceanography. Can. Bull. Fish. Aquat. Sci.* 213:139–162.

Faust, M.A., and Correll, D.L. (1976) Comparison of bacteria and algal utilization of orthophosphate in an estuarine environment. *Mar. Biol.* 34:151–162.

Fenchel, T. (1969) The ecology of marine microbenthos. IV. Structure and function of the benthic ecosystem, its chemical and physical factors and the microfauna communities with special reference to the ciliated protozoa. *Ophelia* 6:1–182.

Fenchel, T. (1977) The significance of bactivorous protozoa in the microbial community of detrital particles. In Cairns, J., Jr., (ed.), *Aquatic Microbial Communities*. New York: Garland, pp. 529–544.

Fenchel T. (1980) Suspension feeding in ciliated protozoa: Feeding rates and their ecological significance. *Microb. Ecol.* 6:13–25.

Fenchel, T. (1982b) Ecology of heterotrophic microflagellates. II. Bioenergetics and growth. *Mar. Ecol. Prog. Ser.* 8:225–231.

Fenchel, T. (1982c) Ecology of heterotrophic microflagellates. III. Adaptations to heterogeneous environments. *Mar. Ecol. Prog. Ser.* 9:25–33.

Fenchel, T. (1982d) Ecology of heterotrophic microflagellates. IV. Quantitative occurrence and importance as bacterial consumers. *Mar. Ecol. Prog. Ser.* 9:35–42.

Fenchel, T., and Blackburn, T.H. (1979) Bacteria and mineral cycling. London: Academic Press, 225 pp.

Fenchel, T., and Harrison, P. (1976) The significance of bacterial grazing and mineral cycling for the decomposition of particulate detritus. In Anderson, J.M. (ed.), *The Role of Terrestrial and Aquatic Organisms in Decomposition Processes*. Oxford, England: Blackwell Scientific, pp. 285–299.

Fenchel, T.M., and Jørgensen, B.B. (1977) Detritus food chains of aquatic ecosystems: the role of bacteria. *Adv. Microbiol. Ecol.* 1:1–58.

Fenchel, T., Perry, T., and Thane, A. (1977) Anaerobiosis and symbiosis with bacteria in free-living ciliates. *J. Protozool.* 24:154–163.

Fitt, W.K. (1984) The role of chemosensory behavior of *Symbiodinium microadriaticum*, intermediate hosts, and host behavior in the infection of coelenterates and molluscs with zooxanthellae. *Mar. Biol.* 81:9–17.

Fitt, W.K. (1985) Chemosensory responses of the symbiotic dinoflagellate *Symbiodinium microadriaticum* (Dinophyceae). *J. Phycol.* 21:62–67.

Fuhrman, J.A., and Azam, F. (1982) Thymidine incorporation as a measure of heterotrophic bacterioplankton production in marine surface waters: evaluation and field results. *Mar. Biol.* 66:109–120.

Gast, V., and Horstmann, U. (1983) N-remineralization of phyto- and bacterioplankton by the marine ciliate *Euplotes vannus*. *Mar. Ecol. Prog. Ser.* 13:55–60.

George, R.Y., and Fields, J.R. (1984) Ammonia excretion in the antarctic krill *Euphausia superba* in relation to starvation and ontogenetic stages. *J. Crust. Biol.* 4:263–272.

Gill, J.W., and Vogel, H.J. (1963) A bacterial endosymbiote in *Crithidia (Stigomonas) oncopelti:* biochemical and morphological aspects. *J. Protozool.* 10:148–152.

Glibert, P.M. (1982) Regional studies of daily, seasonal and size fraction variability in ammonium remineralization. *Mar. Biol.* 70:209–222.

Glibert, P.M., Dennett, M.R., and Caron, D.A. (1988) Nitrogen uptake and NH_4^+ regeneration by pelagic microplankton and marine snow from the North Atlantic. *J. Mar. Res.* 46:837–852.

Goldman, J.C. (1984) Conceptual role for microaggregates in pelagic waters. *Bull. Mar. Sci.* 35:462–476.

Goldman, J.C., and Caron, D.A. (1985) Experimental studies on an omnivorous microflagellate: implications for grazing and nutrient regeneration in the marine microbial food chain. *Deep-Sea Res.* 32: 899–915.

Goldman, J.C., Caron, D.A., Andersen, O.K., and Dennett, M.R. (1985) Nutrient cycling in a microflagellate food chain: I. nitrogen dynamics. *Mar. Ecol. Prog. Ser.* 24:231–242.

Goldman, J.C., McCarthy, J.J., and Peavey, D.G. (1979) Growth rate influence on the chemical composition of phytoplankton in oceanic waters. *Nature* 279:210–215.

Griffin, J. (1960) The isolation, characterization, and identification of the crystalline inclusions of the large free-living amebae. *J. Biophys. Biochem. Cytol.* 7:227–234.

Güde, H. (1985) Influence of phagotrophic processes on the regeneration of nutrients in two-stage continuous culture systems. *Microbiol. Ecol.* 11:193–204.

Haas, L.W. (1982) Improved epifluorescence microscopy for observing planktonic microorganisms. *Ann. Inst. Oceanogr.* (Paris) 58(S):261–266.

Hamilton, R.D., and Preslan, J.E. (1970) Observations on the continuous culture of a planktonic phagotrophic protozoan. *J. Exp. Mar. Biol. Ecol.* 5:94–104.

Hargrave, B.T., and Geen, G.H. (1968) Phosphorus excretion by zooplankton. *Limnol. Oceanogr.* 13:332–342.

Harrison, W.G. (1978) Experimental measurement of nitrogen remineralization in coastal waters. *Limnol. Oceanogr.* 23:684–694.

Harrison, W.G. (1980). Nutrient regeneration and primary production in the sea. In Falkowski, P.G. (ed.), *Primary Productivity in the Sea*. New York: Plenum, pp. 433–460.

Harrison, W.G. (1983) Uptake and recycling of soluble reactive phosphorus by marine microplankton. *Mar. Ecol. Prog. Ser.* 10:127–135.

Harrison, W.G., Azam, F., Renger, E.H., and Eppley, R.W. (1977) Some experiments on phosphate assimilation by coastal marine plankton. *Mar. Biol.* 40:9–18.

Harrison, W.G., Douglas, D., Falkowski, P., Rowe, G., and Vidal, J. (1983) Summer nutrient dynamics of the Middle Atlantic Bight: nitrogen uptake and remineralization. *J. Plankton Res.* 5:539–556.

Harrison, W.G., Head, E.J.H., Conover, R.J., Longhurst, A.R., and Sameoto, D.D. (1985) The distribution and metabolism of urea in the eastern Canadian Arctic. *Deep-Sea Res.* 32:23–42.

Hauser, D.C.R., Levandowsky, M., Hutner, S.H., Chunosoff, L., and Hollowitz, J.S. (1975) Chemosensory responses by the heterotrophic marine dinoflagellate *Crypthecodinium cohnii*. *Microbiol. Ecol.* 1:246–254.

Heal, O.W. (1967) Quantitative feeding studies on the soil amoebae. In Graff, O., and Satchell, J.E. (eds.), *Progress in Soil Biology*. Amsterdam: North Holland, pp. 120–126.

Heinbokel, J.F. (1978) Studies on the functional role of tintinnids in the Southern California Bight. I. grazing and growth rates in laboratory cultures. *Mar. Biol.* 47:177–189.

Heinbokel, J.F., and Beers, J.R. (1979) Studies on the functional role of tintinnids in the Southern California Bight. III. Grazing impact of natural assemblages. *Mar. Biol.* 52:23–32.

Herbland, A. (1984) Phosphorus uptake in the euphotic layer of the Equatorial Atlantic Ocean. Methodological observations and ecological significance. *Océanogr. Trop.* 19:25–40.

Hirche, H.-J. (1983) Excretion and respiration of the antarctic krill *Euphausia superba*. *Polar Biol.* 1:205–209.

Hollibaugh, J.T. (1978) Nitrogen regeneration during the degradation of several amino acids by plankton communities collected near Halifax, Nova Scotia, Canada. *Mar. Biol.* 45:191–201.

Hollibaugh, J.T., Carruthers, A.B., Fuhrman, J.A., and Azam, F. (1980) Cycling of organic nitrogen in marine plankton communities studied in enclosed water columns. *Mar. Biol.* 59:15–21.

Horstmann, U., and Hoppe, U.G. (1981) Competition in the uptake of methylamine/ammonium by phytoplankton and bacteria. *Kieler Meeresforsch.* 5:110–116.

Ikeda, T. (1974) Nutritional ecology of marine zooplankton. *Mem. Fac. Fish. Hokkaido Univ.* 22:1–97.

Ikeda, T. (1977) The effect of laboratory conditions on the extrapolation of experimental measurements to the ecology of marine zooplankton. IV. Changes in respiration and excretion rates of boreal zooplankton species maintained under fed and starved conditions. *Mar. Biol.* 41:241–252.

Ikeda, T., Fay, E. Hing, Hutchinson, S.A., and Boto, G.M. (1982) Ammonia and inorganic phosphorus excretion by zooplankton from inshore waters of the Great Barrier Reef, Queensland. I. Relationship between excretion rates and body size. *Aust. J. Mar. Freshwater Res.* 33:55–70.

Jackson, G.A., and Williams, P.M. (1985) Importance of dissolved organic nitrogen and phosphorus to biological nutrient cycling. *Deep-Sea Res.* 32:223–235.

Jawed, M. (1973) Ammonia excretion by zooplankton and its significance to primary productivity during summer. *Mar. Biol.* 23:115–120.

Johannes, R.E. (1964) Phosphorus excretion and body size in marine animals: microzooplankton and nutrient regeneration. *Science* 146:923–924.

Johannes, R.E. (1965) Influence of marine protozoa on nutrient regeneration. *Limnol. Oceanogr.* 10:434–442.

Johannes, R.E. (1968) Nutrient regeneration in lakes and oceans. In Droop, M.R., and Wood, E.J.F. (eds.) *Advances in Microbiology of the Sea*. Orlando, Fla.: Academic Press, pp. 203–213.

Klekowski, R.Z., and Fischer, Z. (1975) Review of studies on ecological bioenergetics of aquatic animals. *Pol. Arch. Hydrobiol.* 22:345–373.

Knauer, G.A., Hebel, D., and Cipriano, F. (1982) Marine snow: Major site of primary production in coastal waters. *Nature* 300:630–631.

Kopylov, A.I., Mamayeva, T.I., and Batsanin, S.F. (1980) Energy balance of the colorless flagellate *Parabodo attenuatus* (Zoomastigophora, Protozoa). *Oceanology* 20:705–708.

Kopylov, A.I., and Moiseev, E.S. (1980) Effect of colorless flagellates on the determination of bacterial production in seawater. *Dokl. Acad. Sci. U.S.S.R., Biol. Sci. Sec.* 252:272–274.

Krempin, D.W., McGrath, S.M., Soohoo, J.B., and Sullivan, C.W. (1981) Orthophosphate uptake by phytoplankton and bacterioplankton from the Los Angeles Harbor and Southern California Coastal Waters. *Mar. Biol.* 64:23–33.

Lancelot, C., and Billen, G. (1985) Carbon-nitrogen relationships in nutrient metabolism of coastal marine ecosystems. *Adv. Aquatic Microbiol.* 3:263–321.

Lawrie, N.R. (1935) Studies in the metabolism of protozoa. I. The nitrogeneous metabolism and respiration of *Bodo caudatus*. *Biochem. J.* 29:588–598.

Laws, E.A., Redalje, D.G., Haas, L.W., Bienfang, P.K., Eppley, R.W., Harrison, W.G., Karl, D.M., and Marra, J. (1984) High phytoplankton growth and production rates in oligotrophic Hawaiian coastal waters. *Limnol. Oceanogr.* 29:1161–1169.

Laybourn, J. (1976a) Energy budgets for *Stentor coerulus* Ehrenberg (Ciliophora). *Oecologia* 22:431–437.

Laybourn, J.E.M. (1976b) Energy consumption and growth in the suctorian *Podophrya fixa* (Protozoa: Suctoria). *J. Zool. Lond.* 180:85–91.

Laybourn, J.E.M. and Stewart, J.M. (1975) Studies on consumption and growth in the ciliate *Colpidium campylum* Stokes. *J. Anim. Ecol.* 44:165–174.

Leboy, P.S., Cline, S.G., and Conner, R.L.. (1964) Phosphate, purines and pyrimidines as excretory products of *Tetrahymena*. *J. Protozool.* 11:217–222.

Lee, J.J. (1980) Nutrition and physiology of the foraminifera. In Levandowsky, M., and Hutner, S.H.

(eds.), *Biochemistry and Physiology of Protozoa,* 2nd ed. Orlando, Fla.: Academic Press, Vol. 3. pp. 43–66.

Lee, J.J., Muller, W.A., Stone, R.J., McEnery, M., and Zucker, W. (1969) Standing crop of foraminifera in sublittoral epiphytic communities of a Long Island salt marsh. *J. Mar. Biol.* 4:44–61.

Lee, J.J., Soldo, A.T., Reisser, W., Lee, M.J., Jeon, K.W., and Görtz, H.-D. (1985) The extent of algal and bacterial endosymbioses in protozoa. *J. Protozool.* 32:391–403.

Levandowsky, M., and Hauser, D.C.R. (1978) Chemosensory responses of swimming algae and protozoa. *Int. Rev. Cytol.* 53:145–210.

Lui, N.S.T., and Roels, O.A. (1970) Nitrogen metabolism of aquatic organisms. I. The assimilation and formation of urea in *Ochromonas malhamensis*. *Arch. Biochem. Biophys.* 139:269–277.

Mayzaud, P. (1976) Respiration and nitrogen excretion of zooplankton. IV. The influence of starvation on the metabolism and biochemical composition of some species. *Mar. Biol.* 37:47–58.

Mellilo, J.M., Naiman, R.J., Aber, J.D., and Linkins, A.E. (1984) Factors controlling mass loss and nitrogen dynamics of plant litter decaying in northern streams. *Bull. Mar. Sci.* 35:341–356.

Mullin, M.M., Perry, M.J., Renger, E.H., and Evans, P.M. (1975) Nutrient regeneration by oceanic zooplankton: A comparison of methods. *Mar. Sci. Comm.* 1:1–13.

Muscatine, L., and D'Elia, C.F. (1978) The uptake, retention, and release of ammonium by reef corals. *Limnol. Oceanogr.* 23:725–734.

Muscatine, L., Masuda, H., and Burnap, R. (1979) Ammonium uptake by symbiotic and aposymbiotic reef corals. *Bull. Mar. Sci.* 29:572–575.

Nardonne, R.M. (1949) Nitrogenous excretion in *Colpidium campylum*. *Fed. Proc.* 8:117.

Nardonne, R.M., and Wilber, C.G. (1950) Nitrogenous excretion in *Colpidium campylum*. *Proc. Soc. Exper. Biol. Med.* 75:559–561.

Paasche, E., and Kristiansen, S. (1982) Ammonium regeneration by microzooplankton in the Oslofjord. *Mar. Biol.* 69:55–63.

Parsons, T.R., Albright, L.J., Whitney, F., Wong, C.S., and Williams, P.J. le B. (1981) The effect of glucose on the productivity of seawater: an experimental approach using controlled aquatic ecosystems. *Mar. Environ. Res.* 4:229–242.

Pomeroy, L.R. (1983) Origin and distribution of flocculent aggregates. EOS, *Trans. Amer. Geophys. Union* (Abstr.) 64:1020.

Pomroy, A.J., Joint, I.R., and Clarke, K.R. (1983) Benthic nutrient flux in a shallow coastal environment. *Oecologia.* 60:306–312.

Proper, G., and Garver, J.C. (1966) Mass culture of the protozoa *Colpoda steini*. *Biotechnol. Bioeng.* 8:287–296.

Raine, R.C.J., and Patching, J.W. (1980) Aspects of carbon and nitrogen cycling in a shallow marine environment. *J. Exp. Mar. Biol. Ecol.* 47:127–139.

Rassoulzadegan, F. (1982) Dependence of grazing rate, gross growth efficiency and food size range on temperature in a pelagic oligotrichous ciliate *Lohmanniella spiralis* Leeg., fed on naturally occurring particulate matter. *Ann. Inst. Oceanogr.* (Paris) 58:177–184.

Rhee, G.-Y. (1972) Competition between an alga and an aquatic bacterium for phosphate. *Limnol. Oceanogr.* 17:505–514.

Rice, D.L. (1982) The detritus nitrogen problem: new observations and perspectives from organic geochemistry. *Mar. Ecol. Prog. Ser.* 9:153–162.

Roger, C. (1978) Axote et phosphore chez un crustace macroplanktonique, *Meganyctiphanes norvegica* (M. Sars) (Euphausiacea): excretion minérale et constitution. *J. Exp. Mar. Biol Ecol.* 33:57–83.

Rogerson, A. (1981) The ecological energetics of *Amoeba proteus* (Protozoa). *Hydrobiologia* 85:117–128.

Rowe, G.T., Clifford, C.H., and Smith, K.L. (1977) Nutrient regeneration in sediments off Cape Blanc, Spanish Sahara. *Deep-Sea Res.* 24:57–63.

Rowe, G.T., Clifford, C.H., Smith, K.L., and Hamilton, P.L. (1975) Benthic nutrient regeneration and its coupling to primary productivity in coastal waters. *Nature* 255:215–217.

Rubin, H.A., and Lee, J.J. (1976) Informational energy flow as an aspect of the ecological efficiency of marine ciliates. *J. Theor. Biol.* 62:69–91.

Satomi, M., and Pomeroy, L.R. (1965) Respiration and phosphorus excretion in some marine populations. *Ecology* 46:877–881.

Scott, J.M. (1985) The feeding rates and efficiencies of a marine ciliate, *Strombidium* sp., grown under chemostat steady-state conditions. *J. Exp. Mar. Biol. Ecol.* 90:81–95.

Seaman, G.R. (1954) Enzyme systems in *Tetrahymena pyriformis* S. VI. Urea formation and breakdown. *J. Protozool.* 1:207–210.

Shanks, A.L., and Trent, J.D. (1979) Marine snow: microscale nutrient patches. *Limnol. Oceanogr.* 24:850–854.

Sherr, B., and Sherr, E. (1983) Enumeration of heterotrophic microprotozoa by epifluorescence microscopy. *Estuar. Coast. Shelf Sci.* 16:1–7.

Sherr, B.F., Sherr, E.B., and Berman, T. (1982) Decomposition of organic detritus: a selective role for microflagellate protozoa. *Limnol. Oceanogr.* 27:765–769.

Sherr, B.F., Sherr, E.B. and Berman, T. (1983) Grazing, growth, and ammonium excretion rates of a heterotrophic microflagellate fed with four species of bacteria. *Appl. Environ. Microbiol.* 45:1196–1201.

Sieburth, J.McN. 1979 *Sea Microbes*. New York: Oxford University Press, 491 pp.

Sieburth, J.McN. (1984) Protozoan bacterivory in pelagic marine waters. In Hobbie, J.E., and Williams, P.J. leB. (eds.) *Heterotrophic Activity in the Sea*. New York: Plenum, pp. 405–444.

Sieburth, J.McN., and Davis, P.G. (1982) The role of heterotrophic nanoplankton in the grazing and nurturing of planktonic bacteria in the Sargasso and Caribbean Sea. *Ann. Inst. Oceanogr.* 58(S):285–296.

Silver, M.W., Gowing, M.M., Brownlee, D.C., and

Corliss, J.O. (1984) Ciliated protozoa associated with oceanic sinking detritus. *Nature* 309:246–248.
Silver, M.W., Shanks, A.L., and Trent, J.D. (1978) Marine snow: microplankton habitat and source of small-scale patchiness in pelagic populations. *Science* 201:371–373.
Sinclair, J.L., McClellan, J.F., and Coleman, D.C. (1981) Nitrogen mineralization by *Acanthamoeba polyphaga* in grazed *Pseudomonas paucimobilis* populations. *Appl. Environ. Microbiol.* 42:667–671.
Sleigh, M. (1973) *The biology of protozoa*. London: William Clowes, 315 pp.
Smith, R.E.H., Harrison, W.G., and Harris, L. (1985) Phosphorus exchange in marine microplankton communities near Hawaii. *Mar. Biol.* 86:75–84.
Soldo, A.T., Godoy, G.A. and Larin, F. (1978) Purine-excretory nature of refractile bodies in the marine ciliate *Parauronema acutum*. *J. Protozool.* 25:416–418.
Soldo, A.T., and Wagtendonk, W.J. (1961) Nitrogen metabolism in *Paramecium aurelia*. *J. Protozool.* 8:41–55.
Spero, H.J. (1985) Chemosensory capabilities in the phagotrophic dinoflagellate *Gymnodinium fungiforme*. *J. Phycol.* 21:181–184.
Stoecker, D.K., and Evans, G.T. (1985) Effects of protozoan herbivory and carnivory in a microplankton food web. *Mar. Ecol. Prog. Ser.* 25:159–167.
Stout, J.D. (1980) The role of protozoa in nutrient cycling and energy flow. *Adv. Microbiol. Ecol.* 4:1–50.
Taylor, F.J.R. (1982) Symbioses in marine microplankton. *Ann. Inst. Oceanogr.* (Paris) 58(S):61–90.
Taylor, G.T. (1982) The role of pelagic heterotrophic protozoa in nutrient cycling: A review. *Ann. Inst. Oceanogr.* (Paris) 58(S):227–241.
Taylor, W.D. (1984) Phosphorus flux through epilimnetic zooplankton from Lake Ontario: relationship with body size and significance to phytoplankton. *Can. J. Fish. Aquatic Sci.* 41:1702–1712.
Taylor, W.D. (1986) The effect of grazing by a ciliated protozoan on phosphorus limitation of heterotrophic bacteria in batch culture. *J. Protozool.* 33:47–51.
Taylor, W.D., and Lean, D.R.S. (1981) Radiotracer experiments on phosphorus uptake and release by limnetic microzooplankton. *Can. J. Fish. Aquatic Sci.* 38:1316–1321.
Verity, P.G. (1985a) Ammonia excretion rates of oceanic copepods and implications for estimates of primary production in the Sargasso Sea. *Biol. Oceanogr.* 3:249–283.
Verity, P.G. (1985b) Grazing, respiration, excretion and growth rates of tintinnids. *Limnol. Oceanogr.* 30:1268–1282.
Vidal, J., and Whitledge, T.E. (1982) Rates of metabolism of planktonic crustaceans as related to body weight and temperature of habitat. *J. Plankton Res.* 4:77–84.
Wambeke, F. van, and Bianchi, M.A. (1985) Bacterial biomass production and ammonium regeneration in Mediterranean seawater supplemented with amino acids. 2. Nitrogen flux through heterotrophic microplankton food chain. *Mar. Ecol. Prog. Ser.* 23:117–128.
Weatherby, J.H. (1929) Excretion of nitrogenous substances in protozoa. *Physiol. Zool.* 2:375–394.
Wilkerson, F.P., and Muscatine, L. (1984) Uptake and assimilation of dissolved inorganic nitrogen by a symbiotic sea anemone. *Proc. Roy. Soc. Lond.* B221:71–86.
Wimpenny, J.W.T. (1981) Spatial order in microbial ecosystems. *Biol. Rev.* 56:295–342.
Woods, L.E., Cole, C.V., Elliott, E.T., Anderson, R.V., and Coleman, D.C. (1982) Nitrogen transformations in soil as affected by bacterial-microfaunal interactions. *Soil Biol. Biochem.* 14:93–98.

8

Protozoan Respiration and Metabolism

DAVID A. CARON
JOEL C. GOLDMAN
TOM FENCHEL

Protozoan respiration represents the sum of the dissimilatory processes that take place during metabolism in order to meet the energy demands of motility, food capture–absorption, osmotic regulation, growth, reproduction, and so on. In most protozoa these processes take place aerobically and the respiration rate can be measured conveniently as O_2 uptake or CO_2 generation. In the former case, carbon utilization can be estimated from the rate of oxygen utilization via the respiratory quotient (RQ).

All major eukaryote metabolic pathways of energy generation have been demonstrated in the protozoa. Indeed, all enzymes of the glycolytic and Krebs cycle are present, and oxidative phosphorylation is carried out as effectively in these unicellular organisms as in mammalian mitochondria (Danforth, 1967; Hill, 1972; Wichterman, 1986). These enzymatic pathways are present in species from all major protozoan taxa, and therefore, the degree to which specific pathways function in particular species appears to be related more to ecological strategy than to phylogeny.

From an energetics point of view, respiration represents that portion of the energy contained in the ingested food that is not assimilated, egested, or excreted. Assuming that the latter two quantities are small (which may not always hold true), respiration can be expressed as the fraction of the ingested food that is not incorporated into biomass, or

$$R = I \times (1 - \text{GGE}) \qquad (1)$$

where R is the amount of carbon respired, I is the amount of carbon ingested, and GGE is the gross growth efficiency of the protozoan (fraction of prey carbon converted to protozoan carbon). The gross growth efficiencies that have been reported for protozoa are relatively high, and values of 40–60 percent appear to be typical (see summary in Table 7.5, Caron and Goldman, Chap. 7). Based on these efficiencies, it is clear that protozoa transform a substantial portion of the carbon contained in their food into protozoan biomass, thus placing an upper limit on the amount of organic carbon that will be respired.

From an ecological perspective, it has been recognized that protozoan respiration in aquatic environments may constitute an important portion of the total system respiration. The bases for this realization are much the same as those outlined in the previous chapter on nutrient regeneration (Caron and Goldman, Chap. 7). Through the use of improved counting techniques it has been shown that the densities of protozoa in the ocean are greater than previously believed (Davis et al., 1978; Dale and Burkill, 1982; Davis and Sieburth, 1982; Haas, 1982; Caron, 1983; Sherr and Sherr, 1983; Davis et al., 1985; Gifford, 1985; Sherr et al., 1986). Also, protozoa are now considered to be the primary consumers of bacterioplankton (Haas and Webb, 1979; Fenchel, 1982d; Linley et al., 1983; Sherr et al., 1984; Sieburth, 1984; Andersen and Fenchel, 1985) and of chroococcoid cyanobacteria (Johnson et al., 1982; Iturriaga and Mitchell, 1986), and they may be important consumers of phytoplankton (Heinbokel, 1978; Heinbokel and Beers, 1979; Capriulo, 1982; Azam et al., 1983; Goldman and Caron, 1985; Suttle et al., 1986; Parslow et al., 1986). Finally, the weight-specific metabolic rates of protozoa generally are greater than those of larger zooplankton (see the next section). Therefore, given equivalent amounts of growing protozoan and zooplankton biomass, more carbon typically will be processed (respired or incorporated) by the protozoa.

Strong support for the importance of protozoa in the respiration of plankton communities has come from the results of size fractionation studies, which show that most of the total respiration is carried out by microorganisms <10 μm in size (Pomeroy and Johannes, 1968; Williams, 1981). However, a significant contribution of bacteria to community respiration has been indicated in these studies. Thus, like the situation with nutrient regeneration, the relative importance of bacterial and protozoan respiration is presently unclear.

The analogy between protozoan respiration rate

and nutrient excretion rate is, of course, not fortuitous. Nutrient excretion is the result of dissimilatory processes in the cell that release nutrients associated with food constituents that are respired by the protozoan. The rate of release of these nutrients is related mathematically to the respiration rate of the protozoan through the equation . . .

$$E = \left(\frac{R}{1 - \mathrm{GGE}}\right) \times \left(\frac{1}{\mathrm{C{:}Nu_{prey}}} - \frac{\mathrm{GGE}}{\mathrm{C{:}Nu_{pro}}}\right) \quad (2)$$

where E and R are the nutrient excretion rate (g nutrient time^{-1} cell^{-1}) and respiration rate (g C time^{-1} cell^{-1}), GGE is the gross growth efficiency (based on carbon), and $\mathrm{C{:}Nu_{pro}}$ and $\mathrm{C{:}Nu_{prey}}$ are the carbon:nutrient ratios of the protozoan and prey, respectively (Fenchel and Blackburn, 1979; Billen, 1984; Caron and Goldman, Chap. 7).

WEIGHT-SPECIFIC RESPIRATION RATES

It has been demonstrated that the metabolic rates (i.e., respiration rates) and growth rates of protozoa are covariant (Fenchel and Finlay, 1983). Another way of expressing this relationship is that, during balanced growth, the gross growth efficiency of a species (and therefore the percentage of ingested food that is respired) does not change with different growth rates (Fenchel, 1982b). Therefore, the concept of "basal metabolism" has little applicability to protozoa, and growth (macromolecular synthesis) constitutes the predominant energy-requiring process of the cell. In fact, from theoretical calculations, it can be demonstrated that the energy requirements of processes other than growth (e.g., motility) are an insignificant portion of the energy budget (Sleigh, 1973).

A significant positive correlation also exists between protozoan respiration rate (nl O_2 cell^{-1} h^{-1}) and cell size as demonstrated previously by Fenchel and Finlay (1983). This correlation is expected because a larger biomass of organisms will consume more oxygen per unit of time. However, an interesting feature of this relationship is that respiration rate does not increase as rapidly as protozoan cell size (slope < 1.0). Therefore, as body weight increases, the respiration rate per unit of protozoan biomass (the "weight-specific respiration rate") decreases. This relationship is evident as a negative slope for a linear regression of weight-specific respiration rate when plotted as a function of protozoan cell size (Fig. 8.1; Table 8.1).

Because of this inverse correlation between body size and respiration rate, weight-specific respiration rates that have been determined for protozoa (Table 8.1) are generally greater than rates measured for poikilothermic marine metazoa (summarized by Ikeda, 1985). However, these weight-specific respiration rates of protozoa are not a result of lower growth efficiencies, which in fact are relatively high for protozoa (see Table 7.5 in Caron and Goldman, Chap. 7). Rather, they reflect the ability of these microorganisms to process an amount of food equivalent to their own weight in a relatively short time.

FIGURE 8.1 Weight-specific respiration rates of protozoa as a function of cell volume. (Data taken from Table 8.1.) For those data expressed as a range, the two extremes were used in the calculation of the regression line.

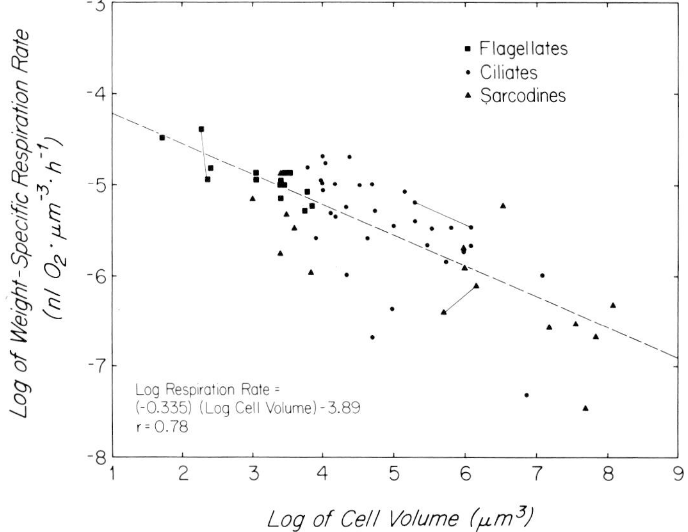

TABLE 8.1 Relationship of Weight-Specific Respiration Rates of Protozoa and Cell Volumes*

Species	Cell volume (μm^3)	Weight-specific respiration rate (nl O_2 μm^{-3} h^{-1})	Reference
Amebae			
Acanthamoeba sp.	3.00×10^3	4.5×10^{-6}	Hamburger (1975)
A. castellanii	3.88×10^3	3.3×10^{-6}	Byers et al. (1969)
A. castellanii	2.49×10^3	1.3×10^{-5}	Baldock et al. (1980)
A. castellanii	2.49×10^3	1.7×10^{-6}	Baldock et al. (1982)
A. castellanii	1.00×10^3	6.7×10^{-6}	D. Lloyd (pers. comm.) to Fenchel and Finlay (1983)
A. palestinensis	6.76×10^3	1.1×10^{-6}	Reich (1948)
Allogromia laticollaris	7.10×10^7	2.1×10^{-7}	Lee and Muller (1973)
Amoeba proteus	9.36×10^5	1.9×10^{-6}	Korohoda and Kalisz (1970)
A. proteus	$0.50–1.40 \times 10^6$	$3.8–7.7 \times 10^{-7}$	Rogerson (1981)
Chaos carolinense	3.68×10^7	2.9×10^{-7}	Pace and Belda (1944)
C. carolinense	5.00×10^7	3.4×10^{-8}	Holter and Zeuthen (1949)
C. carolinense	1.50×10^7	2.6×10^{-7}	Scholander et al. (1952)
Difflugia sp.	9.75×10^5	1.2×10^{-6}	Zeuthen (1943)
Rosalina leei	1.25×10^8	4.7×10^{-7}	Lee and Muller (1973)
Spiroloculina hyalini	3.40×10^6	5.7×10^{-6}	Lee and Muller (1973)
Flagellates			
Astasia klebsii	6.00×10^3	8.1×10^{-6}	von Dach (1942)
A. longa	2.46×10^3	9.2×10^{-6}	Padilla and James (1960)
A. longa	2.47×10^3	6.9×10^{-6}	Hunter and Lee (1962)
A. longa	2.47×10^3	1.1×10^{-5}	Wilson (1963)
A. longa	5.57×10^3	5.0×10^{-6}	Wilson and James (1963)
Chilomonas paramecium	1.12×10^3	1.1×10^{-5}	Mast et al. (1936)
C. paramecium	2.75×10^3	9.6×10^{-6}	Hutchens (1941)
C. paramecium	2.75×10^3	1.3×10^{-5}	Hutchens et al. (1948)
C. paramecium	3.0×10^3	1.3×10^{-5}	Holz (1954)
C. paramecium	1.12×10^3	1.3×10^{-5}	Johnson (1962)
Euglena gracilis var. bacillaris	7.07×10^3	5.7×10^{-6}	Buetow (1961)
Ochromonas sp.	2.50×10^2	1.5×10^{-5}	Fenchel and Finlay (1983)
Paraphysomonas imperforata	$1.80–2.30 \times 10^2$	$1.1–4.0 \times 10^{-5}$	Caron et al. (1985)
Pleuromonas jaculans	5.00×10^1	3.2×10^{-5}	Fenchel and Finlay (1983)
Ciliates			
Bresslaua insidiatrix	3.30×10^4	9.7×10^{-6}	Scholander et al. (1952)
Colpidium campylum	4.30×10^4	2.6×10^{-6}	Hall (1938)
C. steinii	1.00×10^4	8.4×10^{-6}	Proper and Garver (1966)
Didinium nasutum	1.18×10^6	2.3×10^{-6}	Laybourn (1977)
D. nasutum	2.94×10^5	2.3×10^{-6}	Laybourn (1977)
Paramecium aurelia	1.00×10^5	3.5×10^{-6}	Pace and Kimura (1944)
P. aurelia	2.00×10^5	3.9×10^{-6}	Stewart (1964)
P. caudatum	6.40×10^5	3.3×10^{-6}	Pace and Kimura (1944)
P. caudatum	$2.00–12.0 \times 10^5$	$6.2–3.7 \times 10^{-6}$	Cunningham and Kirk (1942)
P. caudatum	9.50×10^5	1.9×10^{-6}	Scholander et al. (1952)
P. caudatum	5.30×10^5	1.4×10^{-6}	Khlebovich (1974)

TABLE 8.1 *(continued)*

Species	Cell volume (μm^3)	Weight-specific respiration rate (nl O_2 μm^{-3} h^{-1})	Reference
Podophrya fixa	4.95×10^4	4.2×10^{-7}	Laybourn (1976)
Spirostomum ambiguum	1.20×10^7	1.0×10^{-6}	Khlebovich (1974)
Stentor coeruleus	7.30×10^6	4.8×10^{-8}	Laybourn (1975)
Tetrahymena geleii	5.40×10^4	5.0×10^{-6}	Ormsbee (1942)
T. geleii	2.36×10^4	2.0×10^{-5}	Pace and Lyman (1947)
T. pyriformis	5.00×10^4	1.0×10^{-5}	Hamburger and Zeuthen (1957)
T. pyriformis	1.50×10^4	4.2×10^{-6}	McCashland and Kronschnabel (1962)
T. pyriformis	1.00×10^4	2.0×10^{-5}	Lovlie (1963)
T. pyriformis	1.00×10^4	1.0×10^{-5}	Chen (1970)
T. pyriformis	8.00×10^3	2.6×10^{-6}	Laybourn (1976)
T. pyriformis	1.10×10^4	1.7×10^{-5}	Khlebovich (1974)
T. pyriformis	2.15×10^4	5.5×10^{-6}	Baldock et al. (1982)
T. pyriformis	2.15×10^4	1.0×10^{-6}	Baldock et al. (1982)
T. pyriformis	6.00×10^3	1.1×10^{-5}	Lloyd et al. (1978)
T. pyriformis	1.30×10^4	4.9×10^{-6}	A. Cowling (pers. comm.) to Fenchel and Finlay (1983)
T. pyriformis	1.48×10^4	9.9×10^{-6}	Fenchel and Finlay (1983)
Tintinnopsis acuminata	9.20×10^3	1.03×10^{-5}	Verity and Langdon (1984), Verity (1985)
T. vasculum	1.42×10^5	8.2×10^{-6}	Verity and Langdon (1984)
Tracheloraphis sp.	3.40×10^5	3.1×10^{-6}	Vernberg and Coull (1974)

*Respiration rates (normalized to 20°C assuming a Q_{10} of 2) and cell volumes as summarized by Fenchel and Finlay (1983) were used to calculate weight-specific respiration rates for all values except Caron et al. (1985) and Verity (1985). Data for growing cells only were employed.

Fenchel and Finlay (1983) found that extrapolation of the regression line between respiration rate and body size for metazoan poikilotherms as reported by Hemmingsen (1960) closely approximated the regression obtained for protozoa in their study. Similarly, Caron and Goldman (see Chap. 7) pointed out that the regression lines describing the weight-specific nitrogen and phosphorus excretion rates versus body size for metazoan zooplankton also may be applicable to protozoan nutrient excretion if they are extrapolated to the appropriate cell size. The latter speculation was based on an examination of the relatively few measurements of protozoan nutrient excretion rates that are presently available. Considerably more information is available on the respiration rates of protozoa (Fig. 8.1; Table 8.1; also summarized by Fenchel and Finlay, 1983). However, these more extensive respiration data are in general agreement with the speculation concerning the remineralization of major nutrients by protozoa. The analogy of the relationships between protozoan cell size and respiration rate or nutrient excretion rate is not surprising, given that these processes are physiologically related (equation [2]).

It is clear from a close inspection of Figure 8.1 that there is considerable variability in the relationship between protozoan respiration rate and cell size. This variability is further increased if respiration data for senescent or encysted protozoa are included (Fig. 1 in Fenchel and Finlay, 1983). However, although the physiological state of the cells and food type have profound effects on the observed respiration rate (see subsequent sections), independent measurements performed on actively growing cells of a single species also may differ by an order of magnitude (see Table 8.1). This variability may be due to a variety of artifacts, including the accuracy and sensitivity of the different instruments and methods used to measure respiration, the accuracy of the estimation of protozoan cell volume, the Q_{10} value chosen to normalize the data to 20°C, and the contribution of the prey to respiration in the experimental system. More work is required to differentiate the contribution of these artifacts from the effects of environmental and nutritional variables.

ANAEROBIC METABOLISM

Although most protozoa respire aerobically, anaerobiosis is well known in a variety of taxa, including intestinal flagellates (Danforth, 1967), rumen ciliates (Hungate, 1966; Coleman, 1979), and a number of free-living species (Webb, 1961; Fenchel, 1969; Fenchel et al., 1977; Finlay, 1985). Among free-living forms, there appear to be both facultatively anaerobic (or microaerophilic) species and truly anaerobic species, the latter characterized by a lack of mitochondria and cytochrome oxidase activity (Fenchel et al., 1977).

The pathways of energy generation by protozoa in anaerobic environments are still poorly characterized, but presumably energy is derived from several fermentative processes. There are several consequences of this form of metabolism for protozoa. Organic compounds often are not respired to CO_2, and a variety of organic acids have been identified as the end products of protozoan fermentation (Coleman, 1979). The metabolic rate of these species therefore cannot be accurately measured solely by the appearance of carbon dioxide. In addition, 0_2 is not available as a terminal electron acceptor and reducing equivalents must be removed by other means. In some species this occurs by the production of molecular hydrogen (Müller, 1980). In one novel case, the facultatively anaerobic ciliate *Loxodes* may employ dissimilatory nitrate respiration to accomplish this task (Finlay, 1985; Finlay et al., 1983).

The removal of organic acids and hydrogen (the latter can be inhibiting to protozoan metabolism) produced by anaerobic respiration is believed to be accomplished by the bacteria occurring in these environments. For example, there is now quite convincing evidence that methanogenic bacteria that are symbiotic in some anaerobic ciliates use H_2 produced by the ciliates to reduce carbon dioxide to methane (Vogels et al., 1980; van Bruggen et al., 1983; Lee et al., 1985a). The maintenance of a low partial pressure of hydrogen enhances the efficiency of the formulative pathways. These metabolic relationships, or consortia, may be one explanation for the high occurrence of ecto- and endosymbiotic bacteria with protozoa from anaerobic environments (Fenchel et al., 1977).

A final consequence of anaerobiosis, which is clear from thermodynamic considerations, is that energy yield from fermentative respiration is low relative to aerobic respiration. This difference is reflected in a low gross growth efficiency for these species.

STARVATION/ENCYSTMENT AND METABOLISM

The concept of balanced growth is useful for mathematical analysis of protozoan metabolism at the population level but probably occurs rarely among natural assemblages. Rather, the results of recent field studies indicate that a "boom and bust" situation may be more common for microbial populations, as evidenced by cyclic and temporally offset increases and decreases in the abundance of protozoa and their prey (Fenchel, 1982d; Laake et al., 1983; Davis et al., 1985). The overall metabolism of an assemblage will therefore be an integration of the physiological states of the individuals within the assemblage. This lack of predictive value of the respiratory rates obtained in laboratory studies has been noted previously (Fenchel and Finlay, 1983).

One of the most common factors affecting the metabolic rates of protozoa is the temporary absence of food. Protozoa typically respond to starvation in one of two ways. Some species encyst during adverse conditions such as starvation to form relatively recalcitrant cells characterized by low metabolic activity and long survival capabilities (Corliss and Esser, 1974). Other species do not encyst and rely on stored energy reserves, reductions in metabolic activity, autophagocytosis, or a combination of these alternatives to survive the starvation period (Fenchel, 1982c). The benefit of this latter strategy presumably is to allow a rapid response when conditions again become favorable for growth of the species.

Although encystment undoubtedly results in reduced metabolic activity and thus a conservation of energy, few data are available on the respiration rates of encysted protozoa, and it is therefore difficult to estimate the energetic "savings" of encystment as a mechanism for avoiding starvation. It has been shown that the respiration rates of single cells of the ciliate *Bresslaua insidiatrix* decreased during encystment, were below detectability for encysted cells, and rose sharply at the time of excystment (Scholander et al., 1952). In addition, weight-specific respiration rates have been shown to be less for encysted cells than for nonencysted cells of the same species (Table 8.2). However, too few investigations have been conducted to consider these values representative of all species.

TABLE 8.2 Comparative Data on the Weight-Specific Respiration Rates (WSRR) of Encysted and Nonencysted Protozoa

Species	Growth rate	Weight-specific respiration rate (nl O_2 μm^{-3} h^{-1})	$\frac{WSRR_{encysted}}{WSRR_{nonencysted}}$ (%)	Reference
Acanthamoeba palestinensis	Actively growing	1.07×10^{-6}	36	Reich (1948)
	Encysted	3.80×10^{-7}		
Colpoda cucullus	Actively growing	1.90×10^{-6}	32	Pigon (1959)
	Encysted	6.02×10^{-7}		
Urostyla grandis	Starved	1.02×10^{-5}	9	Pigon (1954)
	Encysted	9.04×10^{-7}		

This scarcity of experimental data on the metabolic activity of protozoan cysts has been recognized previously (Corliss and Esser, 1974). In addition to prolonging survival under starvation conditions, cysts may serve other purposes (e.g., sexual recombination); therefore, energy conservation may not be the primary goal of encystment in some cases. That is, cysts of two similarly sized species of protozoa may differ significantly in their weight-specific respiration rates. Clearly, if cyst formation represents a significant long-term decrease in the respiration rate, then encystment will be an important mechanism for reducing energy consumption for species whose food is unavailable for extended periods of time (e.g., seasonally). This hypothesis, although not a new one, must be verified experimentally with more species, and the "residual" metabolic rates of encysted cells must be accurately determined.

The observation that many nonencysting protozoa are capable of surviving for significant periods of time in the absence of food is strong indirect support for the previous conclusion that growth is by far the most energy-requiring function of these organisms. However, even in the absence of cell growth, protozoa expend some energy for motility, osmotic regulation, and so on, resulting in the respiration of energy storage products (if present) or cellular structural components. The chemical composition of these components will affect the amount of carbon respired per unit of O_2 consumed (i.e., the respiratory quotient, RQ). However, as with other aspects of starvation, there is little information as to which cellular components are respired by protozoa during starvation. A range of materials are undoubtedly utilized because the large reductions in cell volume that are observed during the starvation of these microorganisms (Holter and Zeuthen, 1949; Fenchel, 1982c; Caron et al., 1985) could not be realized by respiring any single component in the cell. Fenchel (1982c) observed a reduction in the size and number of mitochondria, a reduction in the volume of the nucleus and nucleolus, and the appearance of autophagous vacuoles in starving microflagellates. Based on these observations and on the amount of NH_4^+ regenerated by microflagellates in the stationary phase of growth (Fig. 7.4 in Caron and Goldman, Chap. 7), it is probable that cellular proteins are an important source of energy for starving protozoa. However, despite our lack of understanding concerning the types and amounts of compounds that are respired, it is clear that respiration during starvation may result in the loss of up to 80 percent of the protozoan biomass before death occurs.

WEIGHT-SPECIFIC RESPIRATION RATE DURING STARVATION

Jackson and Berger (1984) have hypothesized that weight-specific respiration rate is an important consideration for nonencysting protozoa under starvation conditions. These investigators found a weak inverse correlation between 50 percent survivorship times of nonencysting protozoa and their weight-specific respiration rates but not with cell size. That is, for two species of protozoa with similar cell volumes, the species with the lower weight-specific respiration rate had the longer survivorship time. For two species with similar weight-specific respiration rates but different cell volumes, the two species had similar 50 percent survivorship times.

This hypothesis is intuitively appealing because it is logical that the slower an organism respires its energy reserves, the longer it should be able to survive without food. Also, since weight-specific respiration rates increase with decreasing cell size (see Fig. 8.1), we would expect a direct correlation between survivorship and protozoan cell size. However, Jackson and Berger (1984) were unable to find a general relationship between survivorship and cell size, perhaps because of the small number of species examined. Another possible explanation is that many species of protozoa when starved, do not maintain the same weight-specific respiration rate as when they are actively growing. It is clear from a comparison of the weight-specific respiration rates for growing or starved (but not encysted) populations of protozoa (Table 8.3) that most species reduce these rates in response to starvation. Furthermore, the three largest decreases for the species listed in Table 8.3 were for three small flagellated protozoa.

If the values in Table 8.3 are averaged by taxonomic groups, amebae possess a limited ability to reduce their weight-specific respiration rates ($\bar{x}$ = 22 percent), ciliates are intermediate ($\bar{x}$ = 51 percent), and flagellates on average have the greatest

TABLE 8.3 Weight-Specific Respiration Rates (WSRR) of Actively Growing and Starved Protozoa*

Species	Weight-specific respiration rate (nl O_2 μm^{-3} h^{-1}) Growing cells	Starved cells	$\frac{WSRR_{starved}}{WSRR_{growing}}$ (%)	Cell volume of growing cells (μm^3)	References
Amebae					
Amoeba proteus	1.25×10^{-6}	3.03×10^{-7}	24	1.6×10^{6}	Brachet (1955); Kalisz (1973); Korohoda and Kalisz (1970); Rogerson (1981)
Chaos carolinense	1.94×10^{-7}	2.57×10^{-7}	132	5.0×10^{7}	Holter and Zeuthen (1949); Pace and Belda (1944); Pace and McCashland (1951); Scholander et al. (1952)
Flagellates					
Astasia klebsii	8.08×10^{-6}	8.80×10^{-7}	11	6.0×10^{3}	von Dach (1942)
Astasia longa	8.09×10^{-6}	3.25×10^{-6}	40	5.6×10^{3}	Hunter and Lee (1962); Padilla and James (1960); Wilson (1963); Wilson and James (1963)
Chilomonas paramecium	1.19×10^{-5}	4.40×10^{-6}	37	3.0×10^{3}	Holz (1954); Hutchens (1941); Hutchens et al. (1948); Johnson (1962); Mast et al. (1936)
Ochromonas sp.	1.54×10^{-5}	1.00×10^{-6}	6.5	2.5×10^{2}	Fenchel and Finlay (1983)
Paraphysomonas imperforata	1.50×10^{-5}	7.35×10^{-5}	49	2.3×10^{2}	Caron et al. (1985)
Pleuromonas jaculans	3.22×10^{-5}	2.10×10^{-6}	6.5	5.0×10^{1}	Fenchel and Finlay (1983)
Ciliates					
Podophrya fixa	4.24×10^{-7}	5.17×10^{-7}	122	5.0×10^{4}	Laybourn (1976)
Spirostomum ambiguum	1.06×10^{-6}	2.21×10^{-7}	21	8.0×10^{6}	Khlebovich (1974); Specht (1935)
Stentor coeruleus	$4.9–7.5 \times 10^{-8}$	1.13×10^{-6}	1800	7.3×10^{6}	Laybourn (1975); Whitely (1960)
Tetrahymena geleii	1.25×10^{-5}	2.69×10^{-6}	22	5.4×10^{4}	Baker and Baumberger (1941); Ormsbee (1942); Pace and Lyman (1947)
Tetrahymena pyriformis	8.86×10^{-6}	2.63×10^{-6}	30	5.0×10^{4}	Baldock et al. (1982); Chen (1970); Conner and Cline (1967); Hamburger and Zeuthen (1957); Khlebovich (1974); Laybourn (1976); Lovlie (1963); McCashland and Kronschnabel (1962); Ryley (1952); van de Vijver (1966)

*Respiration rates are averages for several studies in most cases.

capacity for reducing their weight-specific respiration rates ($\bar{x}$ = 75 percent). Interestingly, the order by taxa of average cell volumes are amebae > ciliates > flagellates, indicating a possible negative correlation between protozoan cell volume and the ability to reduce the weight-specific respiration rate. This relationship has been described previously (Fenchel and Finlay, 1983) and is evident in much earlier studies. For example, Scholander et al. (1952; see their Fig. 1) plotted oxygen consumption versus cell volume for individual cells of three species of protozoa. The slopes generated for each species were not identical and increased with decreasing cell size. That is, as the cell volume of a species decreased during starvation, the weight-specific respiration rate decreased faster for small species of protozoa than for larger ones. This same relationship can be seen in a similar study of Fenchel and Finlay (1983; see their Fig. 4), in which they pointed out that although the ameba *Chaos carolinensis* (cell volume = 5×10^7 μm^3) did not reduce its weight-specific respiration rate in the starvation experiments of Holter and Zeuthen (1949), the respiration rate of the microflagellate *Ochromonas* sp. (cell volume = 240 μm^3) decreased by an order of magnitude in the study by Fenchel (1982c).

Two examples of the ability of protozoa to reduce their weight-specific respiration rates are given in Figure 8.2, which shows the change in their respiration rates as a percentage of their respiration rates at the end of the exponential growth phase. In these examples the two extremes in physiological responses of protozoa during starvation are described. The ciliate *Tetrahymena pyriformis* showed little change in its weight-specific respiration rate during seven days of starvation, whereas the microflagellate *Ochromonas* sp. showed an immediate and dramatic reduction in its rate.

FIGURE 8.2 Percent change in the weight-specific respiration rate of two species of protozoa during starvation. Changes are given as a percentage of the respiration rate at the end of the exponential growth phase (time = 0). Data points were calculated from Fenchel (1982c) for *Ochromonas* sp. and from Finlay et al. (1983) for *Tetrahymena pyriformis*.

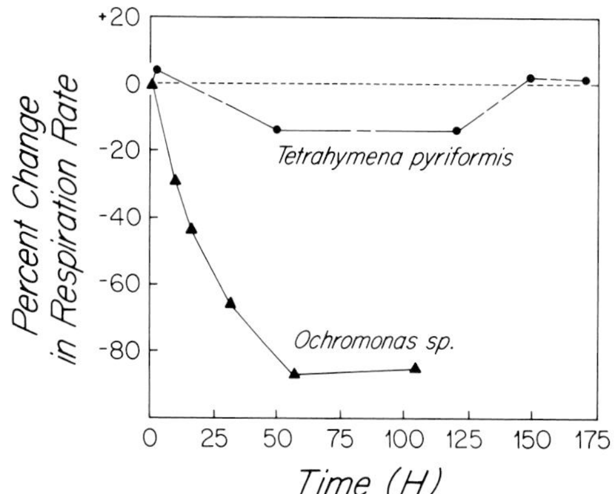

Therefore, it would appear that, although larger protozoa have an enhanced survival ability (relative to small protozoa) because of their intrinsically lower weight-specific respiration rates (see Fig. 8.1), they may be unable to reduce their weight-specific respiration rates to the degree to which the smaller protozoa are able. These offsetting characteristics may be the reason for the lack of a significant correlation between survivorship and cell volume as recognized by Jackson and Berger (1984).

Unfortunately, there are still too few data to evaluate the relationship properly between protozoan cell size and the ability to reduce weight-specific respiration rate during starvation. Large inconsistencies are present in the available data. For example, unlike the microflagellates *Ochromonas* sp. and *Pleuromonas jaculans* studied by Fenchel (1982c), the microflagellate *Paraphysomonas imperforata* did not show an immediate reduction in weight-specific respiration rate during starvation, and the reduction that eventually was observed was much less than that for the former two flagellates (see Table 8.3; Caron et al., 1985). Similarly, although the large ameba *Chaos carolinense* apparently has no ability to reduce its weight-specific respiration rate during starvation, *Amoeba proteus* is capable of a significant reduction (see Table 8.3). These "inconsistencies" make it difficult to draw firm conclusions concerning the relationship between cell size and the ability to reduce respiration rate and undoubtedly reflect alternative strategies for coping with temporal heterogeneity in the biological and physical variables of specific environments.

FOOD QUALITY

Food quality, like food availability, has a pronounced effect on protozoan metabolism. As shown by Caron and Goldman (Chap. 7), food quality can affect the nutrient regeneration rate and/or the total amount of nutrients remineralized from ingested food. Similarly, the amount of carbon respired for energy or incorporated into protozoan biomass may be affected by the type of food ingested.

The theoretical basis for this argument is that if the composition of the food is similar to that of the consumer, then comparatively little energy is required to convert the food biomass into consumer biomass. On the other hand, if the composition of the food differs substantially from that of the consumer, then a large amount of the energy contained in the food must be utilized to reduce the food components to simple units in order to enable synthesis of consumer biomass from these metabolites. In the context of this discussion, then, food quality affects respiration through its effect on the relative proportions of food that are respired for energy or incorporated into the protozoan cell (i.e., gross growth efficiency).

The energetic and ecological importance of this aspect of protozoan nutrition has been recognized recently (Rubin and Lee, 1976; Lee, 1980). Rubin and Lee (1976) expressed the net energy yield or "energetic gain" of consuming a food item as a ratio of the energy content of the item to the energy expended for processing the item (the latter term expressed in units of respiration). This ratio reaches a maximum when the amount of respiration associated with processing the food is minimal.

Conceptually, equations such as the one described by Rubin and Lee (1976) are an informative way of assessing the importance of food quality for protozoan growth because they provide a means of estimating the importance of a parameter that is difficult to measure (food quality) by using parameters that are directly measurable (calorie content, protozoan respiration). However, determining energetic gain for a suite of protozoa and food species and using this information to predict ecological advantage is not yet straightforward. For example, Scott (1980) noted that the ecological efficiency of the rotifer *Brachionus plicatilis* was not correlated with the caloric content of the food but was highest when major biochemical constituents (carbohydrate, protein, lipid) were approximately equal. The author argued that compatibility between the chemical composition of food and consumer was most important for efficient use of ingested food. Based on these results, it is probable that caloric content will not always be useful for estimating food quality.

Similarly, the usefulness of respiration as a measure of energetic gain is based on the assumption that the fraction of the ingested organic material which is egested by a protozoan is insignificant or constant for all food types. That is, it is assumed that a constant amount of food is used for growth *or* respiration for all food types. However, the amount of egested material may vary for different food species (e.g., particularly recalcitrant cell structures may be egested rather than degraded). Therefore, it is not always clear how much food has been incorporated into protozoan biomass for the amount of food respired.

In addition to these potential problems, there are other questions that have not yet been answered concerning food quality. We still have very little information on the extent to which a protozoan must reduce its food before the constituents can be "recognized" as biosynthetic units and incorporated into protozoan biomass. That is, is it necessary to reduce all proteins into amino acids and all polysaccharides into monosaccharides, or can larger units be used in biosynthesis? The use of large food molecules in biosynthesis is a basic question in the study of food quality because this process represents a significant energy "savings" by eliminating several catabolic and anabolic reactions.

Finally, there is little available information concerning the effects of physical or chemical factors (e.g., temperature) on food quality. These effects may be in the form of changes in enzyme activity that might eliminate or activate specific metabolic pathways for food processing. Thus, a particular species may have a greater energetic gain than a second species for a particular food at one temperature but a lower energetic gain compared with the second species for the same food at a different temperature. The observation that gross growth efficiency of protozoa often changes with temperature is indirect evidence that this "switching" may be important (Laybourn and Stewart, 1975; Rassoulzadegan, 1982; Sherr et al., 1983; Verity, 1985). However, these data must be interpreted with caution because it is often overlooked that the cell volumes of some protozoa (and algae) increase with decreasing temperature during balanced growth (James and Read, 1957; Lee and Fenchel, 1972; Gold and Morales, 1975; Goldman, 1977; Rogerson, 1981).

The complications described earlier do not negate the usefulness of measurements of food quality and its relationship to protozoan metabolism, but they limit its present usefulness for estimation of energetic gain. The accurate determination of energetic gain as it relates to food quality undoubtedly will provide important information on the competitive advantages and disadvantages for protozoan species in natural communities.

TEMPERATURE

Temperature has a direct effect on the metabolic rate of protozoa. The overall importance of this effect on the respiration rate of a species is due, in part, to the effect that temperature has on other growth processes of the species. For example, if the amount of energy required for anabolic processes does not change as a result of an increase in temperature, then the relative proportions of food respired for energy or incorporated into protozoan biomass will not change. However, if anabolic processes require a greater amount of energy at the higher temperature, then proportionately more carbon will be respired at the higher temperature and a lower gross growth efficiency will result. This latter condition seems particularly plausible near the upper temperature limit when thermolysis of enzymes plays a significant role.

A summary of how protozoan respiration rates change with respect to temperature is given in Figure 8.3. Respiration rate has been expressed as the log weight-specific respiration rate in this figure because of the effect that temperature may have on protozoan cell volume (James and Read, 1957; Lee and Fenchel, 1972; Gold and Morales, 1975; Rogerson, 1981). Cell volume often, but not always, decreases with increasing temperature, so changes in a respiration rate expressed on a per cell basis may be misleading.

The respiration rates in Figure 8.3 span several orders of magnitude and are dependent on a number of factors, including cell volume of the species and the physiological state of the individuals. However, with a few exceptions, the slopes of these lines (the relative change in respiration rate with a change in temperature) are similar. That is, changes in the weight-specific respiration rate of protozoa due to a change in temperature do not show any clear relationship with the magnitude of the respiration rate or with protozoan cell volume, but they appear to be a relatively constant proportion of the absolute rate.

FIGURE 8.3 Changes in the weight-specific respiration rate of protozoa as a function of temperature. Data points resulting in negative slopes were omitted from the graph. (See text for discussion.) Species identifications are (A) *Chaos carolinensis* (*Pelomyxa carolinensis*), Pace and Belda (1944); (B) *Paramecium aurelia* and (C) *Paramecium caudatum*, Pace and Kimura (1944); (D) *Tetrahymena geleii*, Pace and Lyman (1947); (E) *Spirostomum ambiguum* (assumed volume = 8 × 10^6 μm³), Sarojini and Nagabhushanam (1966); (F) *Frontonia leucas* (G) *Spirostomum teres*, (H) *Paramecium aurelia*, (I) *Vorticella microstoma* and (J) *Tetrahymena pyriformis*, Laybourn and Finlay (1976); (K) *Saccamoeba limax* and (L) *Vannella* sp., Laybourn-Parry et al. (1980); (M) *Tintinnopsis acuminata* and (N) *Tintinnopsis vasculum*, Verity (1985); (O) *Paraphysomonas imperforata*, Caron et al. (1986).

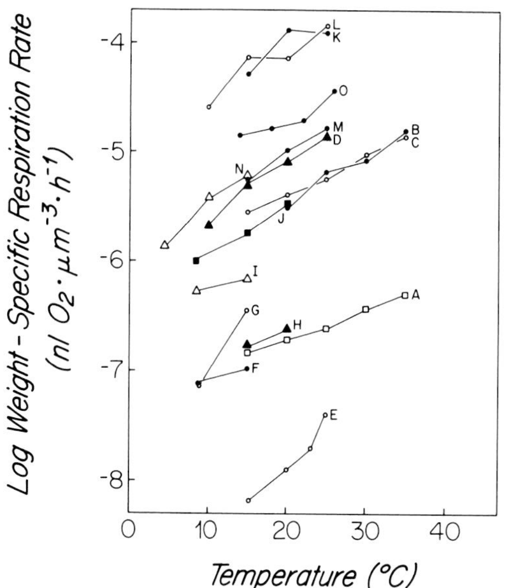

A common method for expressing this relationship is the Q_{10} value, which indicates the relative change in a physiological rate with a 10°C change in temperature according to the following equation by Prosser (1961).

$$\log \frac{R_1}{R_2} = \log Q_{10} \frac{(t_1 - t_2)}{10} \quad (3)$$

Q_{10} values were calculated for the species in Figure 8.3 for each temperature increment that resulted in a positive slope (see the following for a discussion of negative slopes). The overall range of these Q_{10} values was 1.39–10.5 (excluding one extremely high value of 37.3 for *Spirostomum ambiguum*), with a median value of 2.42. Therefore, for a 10°C temperature increase, the weight-specific respiration rate of protozoa on average will increase by a factor of approximately 2.5. This Q_{10} value is larger than the average values that have been obtained for oxygen uptake rates by metazoan zooplankton (Ikeda, 1985) but is in agreement with Q_{10} values calculated for other rate processes of protozoa (Finlay, 1977; Fenchel, 1968).

Data points that resulted in negative slopes in Figure 8.3 ($Q_{10} < 1.0$) were not plotted. These values represent a decrease in respiration rate with an increase in temperature. These decreases commonly are observed at higher temperature as the upper temperature limit for growth of a species is approached (species D, E, G, and I in Fig. 8.3 displayed this behavior). Values of $Q_{10} < 1.0$ also occur infrequently at the lowest temperatures tested. For example, the weight-specific respiration rate of *Saccamoeba limax* decreased by 43 percent when the temperature was increased from 10° to 15°C, but increased with a subsequent temperature increase (Laybourn-Parry et al., 1980). These decreases are more difficult to explain, but may be related to an increase in the amount of energy required to maintain cell function at low temperature. Sherr et al. (1983) observed an increase in NH_4^+ regeneration rate and a decrease in gross growth efficiency (i.e., a relative increase in catabolic processes) of the bacterivorous microflagellate *Monas* sp. at 3°C relative to these processes at two higher temperatures.

From these results, we hypothesize that a greater fraction of the food ingested by this microflagellate was respired at 3°C than at the warmer temperatures. This effect of low temperature on the survival, growth, and respiration of protozoa requires more investigation.

Although relative changes in protozoan respiration rate are similar throughout the optimal temperature range of many species (slopes in Fig. 8.3), the upper and lower temperature limits for growth are species-specific and will affect the success of a species in a community. Given two species of similar cell size, weight-specific respiration rate, and Q_{10} values (e.g., species B and C in Fig. 8.3), as the temperature approaches the upper limit of one species but not the other, the latter species would have a competitive advantage because it is able to put more energy into growth. Therefore, temperature is undoubtedly a major determinant of the geographical range of many protozoan species, as well as the relative abundance of species at a particular time and place (Borror, 1980; Bamforth, 1981). A knowledge of the effects that temperature has on protozoan metabolism is essential for a thorough understanding of species competition and success in aquatic ecosystems.

PROTOZOAN–ALGAL SYMBIOSES

The diversity of relationships between protozoa and symbiotic microorganisms has recently been reviewed by Taylor (1982) and Lee et al. (1985a, 1985b) and is the subject of another chapter of this book (see Chapter 9). Many of these associations involve photosynthetic protists that are maintained as endosymbionts in the protozoan cytoplasm. These associations are well developed in many shallow-dwelling species of protozoa but are particularly abundant among the large sarcodines. Algal symbionts have been documented and studied from numerous species of foraminifera (Bé et al., 1977; Lee et al., 1979), radiolaria (Anderson, 1983a 1983b; Anderson et al., 1985), and acantharia (Febvre and Febvre-Chevalier, 1979). In addition, intact photosynthesizing chloroplasts have been documented in several species of pelagic ciliates (Barber et al., 1969; Smith and Barber, 1979; McManus and Fuhrman, 1986) and foraminifera (Lopez, 1979).

The presence of algal endosymbionts in many protozoa complicates the determination of respiration rate for these species. In the light, the utilization of dissolved O_2 and the production of CO_2 by the protozoan host is partially or wholly compensated by CO_2 uptake and O_2 production by the endosymbiotic algae during photosynthesis. In contrast, symbiont respiration in the dark is additive to that of the host, resulting in a respiration rate that cannot be attributed solely to the protozoan. The changing contribution of symbiont productivity and respiration results in large shifts in the observed respiration rate of the host–symbiont association over a diel cycle and makes it difficult to determine the protozoan respiration rate.

The photosynthetic abilities of the endosymbiotic algae of some protozoa result in a net production of O_2 by the host–algal association in the light. For example, large monospecific blooms of the ciliate *Mesodinium rubrum* (*Cyclotrichium meunieri*) show a net increase in dissolved O_2 in the light (Barber et al., 1969). Similarly, using microelectrodes to measure oxygen gradients at the surface of a planktonic foraminifer, Jorgensen et al. (1985) showed that respiration of a single *Globigerinoides sacculifer* and its dinoflagellate symbionts in the dark was 3 nmol $O_2\ h^{-1}$, whereas symbiont photosynthesis resulted in the production of ~18 nmol $O_2\ h^{-1}$ at light saturation (a net increase of 15 nmol $O_2\ h^{-1}$). Therefore, O_2 production by the symbionts in the light may reduce or negate the respiration rate observed for the protozoan host. These observations have led to the use of terms such as *functional autotroph* to describe the activity of protozoa with algal endosymbionts (Taylor, 1982).

The potential implications of symbiosis for nutrient cycling between protozoa and their symbionts were discussed in Chapter 7. A similar situation undoubtedly exists for O_2 and CO_2 cycling in these associations. The overall oxygen utilization or production reflects the *net* consumption or production of the host–algal complex. Although it may eventually be possible to determine respiration due to the protozoan component, it is probably more ecologically pertinent to consider the host and algae as a single complex system. These associations form an efficient pathway of energy flow among populations of microorganisms and probably should be excluded from data sets used to describe the physiological processes of protozoa (see Fig. 8.1).

Symbionts of protozoan hosts also may contribute to the metabolism of the host in much more subtle ways than by affecting the overall respiration rate of the host–symbiont complex. These "contributions" may take several forms, including the production of specific growth factors for the nutrition of the host, removal of growth-inhibiting metabolites, providing energy or materials for skeletogenesis, or supplying (still poorly characterized) biochemical or physiological cues to initiate or complete developmental and reproductive processes. The range of symbioses involving protozoa and the potential adaptive significance of these associations have been reviewed recently (Lee et al., 1985a, 1985b). Although many of these facultative or obligatory relationships are not yet well characterized, it is probable that the dependence of the protozoan host on the establishment and maintenance of these symbioses can be explained as "metabolic consortia" between the partners. The importance of specific symbiont metabolites for the survival and growth of a protozoan may far outweigh the energy contained in these metabolites and can have a controlling influence on the metabolism of the host.

ECOLOGICAL APPLICATION OF DATA ON METABOLISM

The role of free-living protozoa is presently a topic of great interest and intensive research in marine ecology (Porter et al., 1985). This interest has developed as a result of the growing realization that a very significant fraction of the primary production in many aquatic ecosystems flows through the protozoan trophic link either directly via the ingestion of cyanobacteria and microalgae or via the ingestion of bacteria in the detrital food loop (Azam et al., 1983). Because of the pivotal role played by these microorganisms in aquatic communities, it is imperative that we understand the rates at which protozoa feed and the fate (incorporation–remineralization) of the ingested biomass.

These two aspects of protozoan ecology (in situ feeding rates and protozoan energetics) have traditionally been addressed in separate manners: either in laboratory (energetics) or field-oriented (feeding rates) programs. Numerous techniques for estimating protozoan grazing rates have been developed and applied in recent years (see Chap. 4). The accuracy of these methods depends on (among other things) the ability to measure grazing of mixed assemblages of protozoa in natural systems with as few perturbations or methodological artifacts as possible. In contrast, studies of protozoan energetics have typically been conducted on single species of protozoa under carefully controlled and often highly artificial laboratory conditions. These conditions are necessary in order to be able to assess accurately the amount of organic carbon and nutrients in prey organisms that is transformed into protozoan biomass (e.g., Caron et al., 1985; Goldman and Caron, 1985; Goldman et al., 1985; Andersen et al., 1986).

These two directions of research in protozoan ecology provide highly complementary types of information. To rationalize the grazing rates of protozoa observed in field studies, data on metabolism must be available to determine if the protozoan trophic step represents an energy "link" or "sink" in marine food webs. For example, high growth efficiencies would indicate a considerable "link" in the transfer of materials and energy to higher trophic levels, whereas low efficiencies would imply an important role for protozoa as remineralizers (i.e., a "sink"). This information can be obtained most appropriately under exacting laboratory conditions. Although laboratory-based studies are not natural situations, unlike field experiments they provide an environment where variables can be examined in a systematic manner. Thus, they are a way of obtaining essential baseline data on the limits of protozoan metabolism (i.e., growth and regeneration efficiencies, maximum growth rates) and the effects of environmental variables, food quality, and starvation on the survival and metabolism of individual species.

Protozoa constitute a significant fraction of the living biomass in a number of marine ecosystems and are important consumers of autotrophic and heterotrophic biomass. A major goal of marine ecology in recent years has been to model accurately energy and nutrient flow through this component of the microbial assemblage. Based on the preceding discussion, this goal will not be realized without continued research on the growth and metabolism of ecologically important species.

ACKNOWLEDGMENTS

This work was supported by NSF Grant OCE-8511283 (Joel C. Goldman) and a Woods Hole Oceanographic Institution Mellon Study Award (David A. Caron). Contribution No. 6385 from the Woods Hole Oceanographic Institution.

REFERENCES

Andersen, O.K., Goldman, J.C., Caron, D.A., and Dennett, M.R. (1986) Nutrient cycling in a microflagellate food chain: III. Phosphorus dynamics. *Mar. Ecol. Prog. Ser.* 31:47–55.

Andersen, P., and Fenchel, T. (1985) Bacterivory by microheterotrophic flagellates in seawater samples. *Limnol. Oceanogr.* 30:198–202.

Anderson, O.R. (1983a) *Radiolaria*. New York: Springer-Verlag, 355 pp.

Anderson, O.R. (1983b) The radiolarian symbiosis. In Goff, L.J. (ed.) *Algal Symbiosis*. Cambridge, England: Cambridge University Press, pp. 69–89.

Anderson, O.R., Swanberg, N.R. and Bennett, P. (1985)

Laboratory studies of the ecological significance of host-algal nutritional associations in solitary and colonial radiolaria. *J. Mar. Biol. Assoc. U.K.* 65:263–272.

Azam, F., Fenchel, T., Field, J.G., Gray, J.S., Meyer-Reil, L.A., and Thingstad, F. (1983) The ecological role of water-column microbes in the sea. *Mar. Ecol. Prog. Ser.* 10:257–263.

Baker, E.G.S., and Baumberger, J.P. (1941) The respiratory rate and the cytochrome content of a ciliate protozoan (*Tetrahymena gelei*). *J. Cell. Comp. Physiol.* 17:285–303.

Baldock, B., Baker, J.H., and Sleigh, M.A. (1980) Laboratory growth rates of six species of freshwater Gymnamoebia. *Oecologia* (Berlin). 47:156–159.

Baldock, B.M., Rogerson, A. and Berger, J. (1982) Further studies on respiratory rates of freshwater amoebae (Rhizopoda, Gymnamoebia). *Microb. Ecol.* 8:55–60.

Bamforth, S.S. (1981) Protist biogeography. *J. Protozool.* 28:2–9.

Barber, R.T., White, A.W., and Siegelman, H.W. (1969) Evidence for a cryptomonad symbiont in the ciliate *Cyclotrichium meunieri*. *J. Phycol.* 5:86–88.

Bé, A.W.H., Hemleben, C., Anderson, O.R., Spindler, M., Hacunda, J., and Tuntivate-Choy, S. (1977) Laboratory and field observations of living planktonic foraminifera. *Micropaleontology* 23:155–179.

Billen, G. (1984) Heterotrophic utilization and regeneration of nitrogen. In Hobbie, J.E., and Williams, P.J. leB. (eds.), *Heterotrophic Activity in the Sea*. New York: Plenum, pp. 313–355.

Borror, A.C. (1980) Spatial distribution of marine ciliates: micro-ecologic and biogeographic aspects of protozoan ecology. *J. Protozool.* 27:10–13.

Brachet, J. (1955) Recherches sur les interactions biochimiques entre le noyau et le cytoplasme chez les organismes unicellulaires. I. *Amoeba proteus*. *Biochim. Biophys. Acta* 18:247–268.

Bruggen, J.J.A. van, Stumm, C.K., and Vogels, G.D. (1983) Symbiosis of methanogenic bacteria and sapropelic protozoa. *Arch. Microbiol.* 136:89–95.

Buetow, D.E. (1961) Variation of the respiration of protozoan cells with length of centrifuging. *Anal. Biochem.* 2:242–247.

Byers, T.J., Rudick, V.L., and Rudick, M.J. (1969) Cell size, macromolecule composition, nuclear number, oxygen consumption and cyst formation during two growth phases in unagitated cultures of *Acanthamoebae castellanii*. *J. Protozool.* 16:693–699.

Capriulo, G.M. (1982) Feeding of field collected tintinnid microzooplankton on natural food. *Mar. Biol.* 71:73–86.

Caron, D.A. (1983) Technique for enumeration of heterotrophic and phototrophic nanoplankton, using epifluorescence microscopy, and comparison with other procedures. *Appl. Environ. Microbiol.* 46:491–498.

Caron, D.A., Goldman, J.C., Andersen, O.K., and Dennett, M.R. (1985) Nitrogen cycling in a microflagellate food chain: II. Population dynamics and carbon cycling. *Mar. Ecol. Prog. Ser.* 24:243–254.

Caron, D.A., Goldman, J.C., and Dennett, M.R. (1986) Effect of temperature on growth, respiration and nutrient regeneration by an omnivorous microflagellate. *Appl. Environ. Microbiol.* 52:1340–1347.

Chen, M-L. (1970) Effect of energy source on growth and respiration of *Tetrahymena pyriformis*. *Bull. Inst. Zool., Academia Sinica*. 9:1–5.

Coleman, G.S. (1979) Rumen ciliate protozoa. In Levandowsky, M., and Hutner, S.H. (eds.), *Biochemistry and Physiology of Protozoa*, 2nd ed., Vol. 2. Orlando, Fla.: Academic Press, pp. 381–408.

Conner, R.L., and Cline, S.G. (1967) Some factors governing respiration, glucose metabolism and iodoacetate sensitivity in *Tetrahymena pyriformis*. *J. Protozool.* 14:22–26.

Corliss, J.O., and Esser, S.C. (1974) Comments on the role of the cyst in the life cycle and survival of free-living protozoa. *Trans. Amer. Micros. Soc.* 93:578–593.

Cunningham, B., and Kirk, P.L. (1942) The oxygen consumption of single cells in *Paramecium caudatum* as measured by a capillary respirometer. *J. Cell. Comp. Physiol.* 20:119–134.

Dach, H. von (1942) Respiration of a colorless flagellate *Astasia klebsii*. *Biol. Bull.* 82:356–371.

Dale, T., and Burkill, P.H. (1982) "Live counting"—a quick and simple technique for enumerating pelagic ciliates. *Ann. Inst. Oceanogr.* (Paris) 58(S):267–276.

Danforth, W.F. (1967) Respiratory metabolism. In Chen, T.T. (ed.), *Research in Protozoology*, Vol. 1. Oxford, England: Pergamon, pp. 201–306.

Davis, P.G., Caron, D.A., Johnson, P.W., and Sieburth, J.McN. (1985) Phototrophic and apochlorotic components of picoplankton and nanoplankton in the North Atlantic: geographic, vertical, seasonal and diel distributions. *Mar. Ecol. Prog. Ser.* 21:15–26.

Davis, P.G., Caron, D.A., and Sieburth, J.McN. (1978) Oceanic amoebae from the North Atlantic: culture, distribution, and taxonomy. *Trans. Amer. Micros. Soc.* 96:73–88.

Davis, P.G., and Sieburth, J.McN. (1982) Differentiation of the photosynthetic and heterotrophic populations of nanoplankters by epifluorescence microscopy. *Ann. Inst. Oceanogr.* (Paris) 58(S):249–259.

Febvre, J., and Febvre-Chevalier, C. (1979). Ultrastructural study of zooxanthellae of three species of acantharia (Protozoa: Actinopoda), with details of their taxonomic position in the Prymnesiales (Prymensiophyceae, Hibberd, 1976). *J. Mar. Biol. Assoc. U.K.* 59:215–226.

Fenchel, T. (1968) The ecology of marine microbenthos. III. The reproductive potential of ciliates. *Ophelia* 5:123–136.

Fenchel, T. (1969) The ecology of marine microbenthos. IV. Structure and function of the benthic ecosystem, its chemical and physical factors and the microfauna communities with special reference to the ciliated protozoa. *Ophelia* 6:1–182.

Fenchel, T. (1982b) Ecology of heterotrophic microflagellates. II. Bioenergetics and growth. *Mar. Ecol. Prog. Ser.* 8:225–231.

Fenchel, T. (1982c) Ecology of heterotrophic microflagellates. III. Adaptations to heterogeneous environments. *Mar. Ecol. Prog. Ser.* 9:25–33.

Fenchel, T. (1982d) Ecology of heterotrophic microflagellates. IV. Quantitative occurrence and importance as bacterial consumers. *Mar. Ecol. Prog. Ser.* 9:35–42.

Fenchel, T., and Blackburn, T.H. (1979) *Bacteria and mineral cycling*. London: Academic Press, pp. 225.

Fenchel, T., and Finlay, B.J. (1983) Respiration rates in heterotrophic, free-living protozoa. *Microb. Ecol.* 9:99–122.

Fenchel, T., Perry, T., and Thane, A. (1977) Anaerobiosis and symbiosis with bacteria in free-living ciliates. *J. Protozool.* 24:154–163.

Finlay, B.J. (1977) The dependence of reproductive rate on cell size and temperature in freshwater ciliated protozoa. *Oecologia.* 30:75–81.

Finlay, B.J. (1985) Nitrate respiration by protozoa (*Loxodes* spp.) in the hypolimnetic nitrite maximum of a productive freshwater pond. *Freshwater Biol.* 15:333–346.

Finlay, B.J., Span, A.S.W., and Harman, J.M.P. (1983) Nitrate respiration in primitive eukaryotes. *Nature* 303:333–336.

Gifford, D.J. (1985) Laboratory culture of marine planktonic oligotrichs (Ciliophora, Oligotrichida). *Mar. Ecol. Prog. Ser.* 23:257–267.

Gold, K., and Morales, E.A. (1975) Seasonal changes in lorica sizes and the species of Tintinnida in the New York Bight. *J. Protozool.* 22:520–528.

Goldman, J.C. (1977) Temperature effects on phytoplankton growth in continuous culture. *Limnol. Oceanogr.* 22:932–936.

Goldman, J.C., and Caron, D.A. (1985) Experimental studies on an omnivorous microflagellate: Implications for grazing and nutrient regeneration in the marine microbial food chain. *Deep-Sea Res.* 32:899–915.

Goldman, J.C., Caron, D.A., Andersen, O.K., and Dennett, M.R. (1985) Nutrient cycling in a microflagellate food chain: I. Nitrogen dynamics. *Mar. Ecol. Prog. Ser.* 24:231–242.

Haas, L.W. (1982) Improved epifluorescence microscopy for observing planktonic microorganisms. *Ann. Inst. Oceanogr.* (Paris) 58(S):261–266.

Haas, L.W., and Webb, K.L. (1979) Nutritional mode of several non-pigmented microflagellates from the York River Estuary, Virginia. *J. Exp. Mar. Biol. Ecol.* 39:125–134.

Hall, R.L. (1938) The oxygen consumption of *Colpidium campylum*. *Biol. Bull.* 75:395–408.

Hamburger, K. (1975) Respiratory rate through the growth-division cycle of *Acanthamoebae* sp. *C.R. Trav. Lab., Carlsberg.* 40:175–185.

Hamburger, K., and Zeuthen, E. (1957) Synchonous divisions in *Tetrahymena pyriformis* as studied in an organic medium. *Exp. Cell. Res.* 13:443–453.

Heinbokel, J.F. (1978) Studies on the functional role of tintinnids in the Southern California Bight. I. Grazing and growth rates in laboratory cultures. *Mar. Biol.* 47:177–189.

Heinbokel, J.F., and Beers, J.R. (1979) Studies on the functional role of tintinnids in the Southern California Bight. III. Grazing impact of natural assemblages. *Mar. Biol.* 52:23–32.

Hemmingsen, A.M. (1960) Energy metabolism as related to body size and respiratory surfaces and its evolution. *Rep. Steno. Mem. Hosp.* (Copenhagen) 9:1–110.

Hill, D.L. (1972) *The Biochemistry and Physiology of Tetrahymena*. Orlando, Fla.: Academic Press, 230 pp.

Holter, H., and Zeuthen, E. (1949) Metabolism and reduced weight in starving *Chaos chaos*. *Compt.-rend. Lab. Carlsberg, Ser. Chim.* 26:277–296.

Holz, G.G. (1954) The oxidative metabolism of a cryptomonad flagellate *Chilomonas paramecium*. *J. Protozool.* 1:114–120.

Hungate, R.E. (1966) *The Rumen and Its Microbes*. Orlando, Fla.: Academic Press, 533 pp.

Hunter, F.R., and Lee, J.W. (1962) On the metabolism of *Astasia longa* (Jahn). *J. Protozool.* 9:74–78.

Hutchens, J.O. (1941) The effect of the age of the culture on the rate of oxygen consumption and the respiratory quotient of *Chilomonas paramecium*. *J. Cell. Comp. Physiol.* 17:321–332.

Hutchens, J.O., Podolsky, B., and Morales, M.F. (1948) Studies on the kinetics and energetics of carbon and nitrogen metabolism of *Chilomonas paramecium*. *J. Cell. Comp. Physiol.* 32:117–141.

Ikeda, T. (1985) Metabolic rates of epipelagic marine zooplankton as a function of body mass and temperature. *Mar. Biol.* 85:1–11.

Iturriaga, R., and Mitchell, B.G. (1986) Chroococcoid cyanobacteria: a significant component in the food web dynamics of the open ocean. *Mar. Ecol. Prog. Ser.* 28:291–297.

Jackson, K.M., and Berger, J. (1984) Survival of ciliate protozoa under starvation conditions and at low bacterial levels. *Microb. Ecol.* 10:47–59.

James, T.W., and Read, C.P. (1957) The effect of incubation temperature on the cell size of *Tetrahymena pyriformis*. *Exp. Cell. Res.* 13:510–516.

Johnson, B.F. (1962) Influence of temperature on the respiration and metabolic effectiveness of *Chilomonas*. *Exp. Cell Res.* 28:419–423.

Johnson, P.W., Xu, H., and Sieburth, J.McN. (1982) The utilization of chroococcoid cyanobacteria by marine protozooplankters but not by calanoid copepods. *Ann. Inst. Oceanogr.* (Paris) 58(S):297–308.

Jorgensen, B.B., Erez, J., Revsbech, N.P., and Cohen, Y. (1985) Symbiotic photosynthesis in a plankton foraminiferan, *Globigerinoides sacculifer* (Brady), studied with microelectrodes. *Limnol. Oceanogr.* 30:1253–1267.

Kalisz, B. (1973) Stimulation of respiration in *Amoeba proteus* by inducers of pinocytosis. *Folia Biol.* 21:169–172.

Khlebovich, T.V. (1974) Rate of respiration in ciliates of different sizes. *Tsitologiya.* 16:103–110.

Korohoda, W., and Kalisz, B. (1970) Correlation of respiratory and motile activities in *Amoeba proteus*. *Folia Biol.* (Krakow) 18:137–143.

Laake, M., Dahle, A.B., Eberlein, K., and Rein, K. (1983) A modelling approach to the interplay of carbohydrate, bacteria and non-pigmented flagellates in a controlled ecosystem experiment with *Skeletonema costatum*. *Mar. Ecol. Prog. Ser.* 14:71–79.

Laybourn, J. (1975) Respiratory energy losses in

Stentor coeruleus Ehrenberg (Ciliophora). *Oecologia* (Berlin) 21:273–278.

Laybourn, J. (1977) Respiratory energy losses in the protozoan predator *Didinium nasutum* Müller (Ciliophora). *Oecologia* (Berlin) 27:305–309.

Laybourn, J.E.M. (1976) Respiratory energy loss in *Podophrya fixa* Müller in relation to temperature and nutritional status. *J. Gen. Microbiol.* 96:203–208.

Laybourn, J.E.M., and Stewart, J.M. (1975) Studies on consumption and growth in the ciliate *Colpidium campylum* Stokes. *J. Anim. Ecol.* 44:165–174.

Laybourn, J., and Finlay, B.J. (1976) Respiratory energy losses related to cell weight and temperature in ciliated protozoa. *Oecologia* (Berlin) 24:349–355.

Laybourn-Parry, J., Baldock, B., and Kingsmill-Robinson, J.C. (1980) Respiratory studies on two small freshwater amoebae. *Microb. Ecol.* 6:209–216.

Lee, C.C., and Fenchel, T. (1972) Studies on ciliates associated with sea ice from Antarctica. II. Temperature responses and tolerances in ciliates from Antarctic, temperate and tropical habitats. *Archiv. Protistenk.* 114:237–244.

Lee, J.J. (1980) Informational energy flow as an aspect of protozoan nutrition. *J. Protozool.* 27:5–9.

Lee, J.J., Lee, M.J., and Weis, D.S. (1985a) Possible adaptive value of endosymbionts to their protozoan hosts. *J. Protozool.* 32:380–382.

Lee, J.J., McEnery, M.E., Kahn, E.G., and Schuster, Fl. (1979) Symbiosis and the evolution of larger foraminifera. *Micropaleontology.* 25:118–140.

Lee, J.L., and Muller, W.A. (1973) Trophic dynamics and niches of salt marsh foraminifera. *Amer. Zool.* 13:215–223.

Lee, J.J., Soldo, A.T., Reisser, W., Lee, M.J., Jeon, K.W., and Görtz, H.-D. (1985b) The extent of algal and bacterial endosymbiosis in protozoa. *J. Protozool.* 32:391–403.

Linley, E.A.S., Newell, R.C., and Lucas. M.I. (1983) Quantitative relationships between phytoplankton, bacteria and heterotrophic microflagellates in shelf waters. *Mar. Ecol. Prog. Ser.* 12:77–89.

Lloyd, D., Phillips, C.A., and Statham, M. (1978) Oscillations of respiration, adenine nucleotid levels and heat evolution in synchronous cultures of *Tetrahymena pyriformis* ST prepared by continuous-flow selection. *J. Gen. Microbiol.* 106:19–26.

Lopez, E. (1979) Algal chloroplasts in the protoplasm of three species of benthic foraminifera: Taxonomic affinity, viability and persistence. *Mar. Biol.* 53:201–211.

Lovlie, A. (1963) Growth in mass and respiration rate during the cell cycle of *Tetrahymena pyriformis. C.R. Trav. Lab.,* (Carlsberg) 33:377–413.

Mast, S.O., Pace, D.M., and Mast, L.M. (1936) The effect of sulfur on the rate of respiration and on the respiratory quotient in *Chilomonas paramecium. J. Cell. Comp. Physiol.* 8:125–140.

McCashland, B.W., and Kronschnabel, J.M. (1962) Exogenous factors affecting respiration in *Tetrahymena pyriformis. J. Protozool.* 9:276–279.

McManus, G.B., and Fuhrman, J.A. (1986) Photosynthetic pigments in the ciliate *Laboea strobila* from Long Island Sound, U.S.A. *J. Plankton Res.* 8:317–327.

Müller, M. (1980). The hydrogenosome. In Gooday, G.W., Lloyd, D. and Trine, A.P.J. (eds.), *The Eukaryotic Microbiol Cell.* Soc. Gen. Microbiol. Symp. 30. Cambridge, England: Cambridge University Press, pp. 127–142.

Ormsbee, R.A. (1942) The normal growth and respiration of *Tetrahymena geleii. Biol. Bull.* 82:423–437.

Pace, D.M., and Belda, W.H. (1944) The effect of food content and temperature on respiration in *P. carolinensis* Wilson. *Biol. Bull.* 86:146–153.

Pace, D.M., and Kimura, K.K. (1944) The effect of temperature on the respiration of *P. caudatum* and *P. aurelia. J. Cell. Comp. Physiol.* 24:173–183.

Pace, D.M., and Lyman, E.D. (1947) Oxygen consumption and carbon dioxide elimination in *Tetrahymena gelii. Biol. Bull.* 92:210–216.

Pace, D.M., and McCashland, B.W. (1951) Effects of low concentration of cyanide on growth and respiration in *Pelomyxa carolinensis*—Wilson (18424). *Proc. Soc. Exp. Biol. Med.* 76:165–168.

Padilla, G.M., and James, T.W. (1960) Sychronization of cell division in *Astasia longa* on a chemically defined medium. *Exp. Cell Res.* 20:401–415.

Parslow, J.S., Doucette, G.J., Taylor, F.J.R., and Harrison, P.J. (1986) Feeding by the zooflagellate *Pseudobodo* sp. on the picoplanktonic prasinomonad *Micromonas pusilla. Mar. Ecol. Prog. Ser.* 29:237–246.

Pigon, A. (1954) Respiration and cytochrome oxidase content in certain Infusoria. *Bull. Acad. Pol. Sci. Cl. II. Ser.* (Biol.). 11:131–134.

Pigon, A. (1959) Respiration of *Colpoda cucullus* during active life and encystment. *J. Protozool.* 6:303–308.

Pomeroy, L.R., and Johannes, R.E. (1968) Occurrence and respiration of ultraplankton in the upper 500 meters of the ocean. *Deep-Sea Res.* 15:381–391.

Porter, K.G., Sherr, E.B., Sherr, B.F., Pace, M., and Sanders, R.W. (1985) Protozoa in planktonic food webs. *J. Protozool.* 32:409–415.

Proper, G. and Garver, J.C. (1966) Mass culture of the protozoa *Colpoda steini. Biotechnol. Bioeng.* 8:287–296.

Prosser, C.L. (1961) Oxygen: respiration and metabolism. In Prosser, C.L., and Brown, F.A., Jr. *Comparative Animal Physiology.* Philadelphia: W.B. Saunders, pp. 153–197.

Rassoulzadegan, F. (1982) Dependence of grazing rate, gross growth efficiency and food size range on temperature in a pelagic oligotrichous ciliate *Lohmanniella spiralis* Leeg., fed on naturally occurring particulate matter. *Ann. Inst. Oceanogr.* (Paris) 58:177–184.

Reich, K. (1948) Studies on the respiration of an Amoeba, *Mayorella palestinensis. Physiol. Zool.* 21:390–412.

Rogerson, A. (1981) The ecological energetics of *Amoeba proteus* (Protozoa). *Hydrobiologia* 85:117–128.

Rubin, H.A., and Lee, J.J. (1976) Informational energy flow as an aspect of the ecological efficiency of marine ciliates. *J. Theor. Biol.* 62:69–91.

Ryley, J.F. (1952) Studies on the metabolism of Protozoa. 3. Metabolism of the ciliate *Tetrahymena pyriformis* (*Glaucoma piriformis*). *Biochem. J.* 52:483–492.

Sarojini, R., and Nagabhushanam, R. (1966) Exogenous factors affecting respiration in the ciliate, *Spirostomum ambiguum*. *J. Anim. Morph. Physiol.* 13:95–102.

Scholander, P.F., Claff, C.L., and Sveinsson, S.L. (1952) Respiratory studies of single cells. II. Observations on the oxygen consumption in single protozoans. *Biol. Bull.* 102:178–184.

Scott, J.M. (1980) Effect of growth rate of the food alga on the growth/ingestion efficiency of a marine herbivore. *J. Mar. Biol. Assoc. U.K.* 60:680–702.

Sherr, B., and Sherr, E. (1983) Enumeration of heterotrophic microprotozoa by epifluorescence microscopy. *Estuar. Coast Shelf Sci.* 16:1–7.

Sherr, B.F., Sherr, E.B., and Berman, T. (1983) Grazing, growth, and ammonium excretion rates of a heterotrophic microflagellate fed with four species of bacteria. *Appl. Environ. Microbiol.* 45:1196–1201.

Sherr, E.B., Sherr, B.F., Fallon, R.D., and Newell, S.Y. (1986) Small, aloricate ciliates as a major component of the marine heterotrophic nanoplankton. *Limnol. Oceanogr.* 31:177–183.

Sherr, B.F., Sherr, E.B., and Newell, S.Y. (1984) Abundance and productivity of heterotrophic nanoplankton in Georgia coastal waters. *J. Plankton Res.* 6:195–202.

Sieburth, J.McN. (1984) Protozoan bacterivory in pelagic marine waters. In Hobbie, J.E., and Williams, P.J. leB. (eds.), *Heterotrophic Activity in the Sea*. New York: Plenum, pp. 405–444.

Sleigh, M. (1973) *The Biology of Protozoa*. London: William Clowes, 315 pp.

Smith, W.O., Jr., and Barber, R.T. (1979) A carbon budget for the autotrophic ciliate *Mesodinium rubrum*. *J. Phycol.* 15:27–33.

Specht, H. (1935) Aerobic respiration in *Spirostomum ambiguum* and the production of ammonia. *J. Cell. Comp. Physiol.* 5:319–333.

Stewart, J.M. (1964) The measurements of oxygen consumption in paramecia of different ages. *J. Protozool. Suppl. Abs.* 119:39–40.

Suttle, C.A., Chan, A.M., Taylor, W.D., and Harrison, P.J. (1986) Grazing of planktonic diatoms by microflagellates. *J. Plankton Res.* 8:393–398.

Taylor, F.J.R. (1982) Symbioses in marine microplankton. *Ann. Inst. Oceanogr.* (Paris) 58(S):61–90.

Verity, P.G. (1985) Grazing, respiration, excretion and growth rates of tintinnids. *Limnol. Oceanogr.* 30:1208–1282.

Verity, P.G., and Langdon, C. (1984) Relationships between lorica volume, carbon, nitrogen, and ATP content of tintinnids in Narragansett Bay. *J. Plankton Res.* 6:859–868.

Vernberg, W.B., and Coull, B.C. (1974) Respiration of an interstitial ciliate and benthic energy relationships. *Oecologia* (Berlin). 16:259–264.

van de Vijver, G. (1966) Studies on the metabolism of *Tetrahymena pyriformis* GL I. Influence of substrates on the respiratory rate. *Enzymologia* 31:363–381.

Vogels, G.D., Hoppe, W.F., and Stumm, C.K. (1980) Association of methanogenic bacteria with rumen ciliates. *Appl. Environ. Microbiol.* 40:608–612.

Webb, M.G. (1961) The effects of thermal stratification on the distribution of benthic protozoa in Esthwaite water. *J. Anim. Ecol.* 30:137–151.

Whitely, A. (1960) Interactions of nucleus and cytoplasm in controlling respiratory patterns in regenerating *Stentor coeruleus*. *C.R. Trav. Lab.* (Carlsberg) 32:49–62.

Wichterman, R. (1986) *The Biology of Paramecium*, 2nd ed. New York: Plenum, 599 pp.

Williams, P.J. leB. (1981) Microbial contribution to overall marine plankton metabolism: direct measurements of respiration. *Oceanol. Acta* 4:359–364.

Wilson, B.W. (1963) The oxidative assimilation of acetate by *Astasia longa* and the regulation of cell respiration. *J. Cell. Comp. Physiol.* 62:49–56.

Wilson, B.W. and James, T.W. (1963) The respiration and growth of synchronized populations of the cell *Astasia longa*. *Exp. Cell Res.* 32:305–319.

Zeuthen, E. (1943) A cartesian diver micro-respirometer with a gas volume of 0.1 μl. Respiration measurements with an experimental error of 2×10^{-5} μl. *C.R. Trav. Lab.* (Carlsberg) 24:479–518.

9

Symbiosis in Marine Protozoa

F. J. R. TAYLOR

This chapter will summarize physically close, nonparasitic associations involving nonphotosynthetic marine protists. In the traditional, broad context, *Protozoa* also includes the photosynthetic flagellates. However, in keeping with the ecological focus of this volume, only associations in which at least one partner is nonphotosynthetic will be included. Although many such associations are known and some, particularly those involving foraminiferans and radiolarians, have been reviewed in detail relatively recently (Anderson, 1983a,b; Lee, 1983; Lee and McEnery, 1983; Lee and Hallock, 1987), there has been only one previous comprehensive review on symbioses in marine microorganisms (Taylor, 1982). Many examples have been given by Sieburth (1979), and very brief reviews were incorporated in the general protozoological texts by Anderson (1988) and Fenchel (1987).

One way in which these associations can be approached is to distinguish between those with photosynthetic, versus the nonphotosynthetic associants, the former seeming to provide the greatest opportunity for cycling of metabolites and, therefore, a more stable association with a nonphotosynthetic host. As a variation on this, one can distinguish between obligate and facultative associations. Unfortunately, many of the symbioses are known only from microscopic observations, with little or no experimental information regarding the nature of the associations. Some may be ephemeral (delayed digestion or rejection?), and others may involve subtle damage to the host.

There are also significantly different ecological requirements of planktonic and benthic associations, both of which are considered here.

Finally, one may distinguish between *ectosymbioses* (epibionts: external) and *endosymbioses* (endobionts: internal), these representing fundamental differences in intimacy with the host. Endosymbiosis, when used in a broad context, may encompass radically different habitats, ranging from body cavities to intracellular environments. In this chapter a distinction will be made between such *endozoic* inhabitants and those that are intracellular. For the latter the term *cytobionts* has been coined (Taylor, 1983; endocytobionts: Schwemmler, 1984). In this text there has been a frequent need to refer to photosynthetic symbionts–cytobionts residing in association with nonphotosynthetic hosts. For the sake of brevity these are referred to as *photobionts* throughout.

The ecto- versus endosymbiotic distinction will provide the framework for this chapter. After brief outlines of the known associations involving marine, nonphotosynthetic protist hosts, their ecological significance will be discussed. Further information on symbioses can be obtained from the volumes by Jeon (1983—intracellular), Goff (1983—algal), a symposium on protozoa as hosts (*J. Protozool.*, 1985) and two recent general texts by Ahmadjanian and Paracer (1986) and Smith and Douglas (1986).

ECTOSYMBIOSES

As a general rule, ectosymbioses, although sometimes highly specialized, are usually less intimate and less modified associations than endosymbioses. Typically, they are separated into *epiphytic* (on plants) versus *epizoic* (on animals) associations. The term *epibiotic,* simply signifying that the organism occurs on a living surface, is also useful. Marine protozoa may be either the active colonizers or the relatively passive hosts in such epibiotic associations.

Marine Protozoa as Hosts for Epibionts

Unlike the walls of some phytoplankters, such as diatoms or coccolithophorids, the walls of marine protozoans, with the exception of ciliate loricae, are covered, or coverable, by cytoplasm. One might expect that a living membrane surface, even with a "coat" of glycoprotein, is less readily colonizable by epizoic organisms than a passive mineral or organic one. However, surface bacteria are commonly present on the surface of both flagellates and ciliates in bacteria-rich environments, such as the guts of herbivores.

Bacteria on the Surface of Ciliates

In the marine environment bacteria can be abundant on the surfaces of benthic, sand-dwelling ciliates. A particularly interesting association is found in the ribbonlike genus *Kentrophoros* (Raikov, 1974), which "farms" sulfur bacteria. The ciliate's body is highly modified, with the cilia restricted to the ventral surface and a total reduction of the mouth. The upper surface is covered by a "lawn" of obligately associated, elongate sulfur bacteria. The ciliate feeds off these by phagocytosing them anywhere on the upper surface and they can be seen in digestive vacuoles within the host.

Indeed, many of the ciliates living in oxygen-poor or anoxic sediments have surface bacteria (Fenchel et al., 1977). The patterns they form on the surface depend on the species involved. For example, in *Sonderia* they lie parallel with the surface, whereas in *Parablepharisma* they are packed vertically into a dense layer through which the cilia penetrate. The bacterial abundance can be great, estimated to reach 1–20 percent of the ciliate biomass (Fenchel et al., 1977; Fenchel, 1987). The functional significance of most of these bacteria is not known. However, these ciliates lack mitochondria, which would be of little or no use in such low-oxygen conditions, and the bacteria may provide some supplementary respiratory capacity (particularly those found inside the ciliate cells; see the following discussion). Some may be sulfur oxidizers, converting sulfide to sulfur or sulfate, similar to those found symbiotically in metazoans occurring in sulfide-rich submarine hot springs.

Diatoms on Ciliates

Tintinnid ciliates construct external vaselike or tubular loricae within which the animals can retract, presumably for defense. The walls may be entirely organic or foreign particles may be added to it during construction (agglutinated). Diatom frustules and coccoliths are commonly observed on the loricae of many pelagic tintinnids with agglutinated bowls, such as *Codonellopsis* (Taylor, 1973), but these are apparently the remains of feeding. However, there are several cases of attachment of living diatoms to the surface of tintinnid loricae that involve predictable partners (Pavillard, 1916; Taylor, 1982). Several species of *Eutintinnus* (*E. apertus, E. pinguis, E. lusus-undae*), in which the lorica is open at both ends, can be found in tropical waters with two closely related *Chaetoceros* species: *C. tetrastichon* (Fig. 9.1) and *C. dadayi* (Fig. 9.2), attached to the anterior outer surface. Both diatoms have hollow, spinulated setae, several of which seem to "embrace" the lorica, and consist of chains that are invariably three cells long. The tintinnids are able to swim rapidly, despite the diatoms adhering to them (personal observation) and so these associations might be conventionally classified as *phoresy* (Pavillard, 1916), that is, transport, in which the sessile organism derives benefit from the transporting host, but probably not *vice versa*. Although the reproduction of these associations has not been studied, it is clear that there must be some synchronization of division between their reproductive rates; otherwise chains of greater than three diatom cells would be seen. Nothing is known of any nutritional interaction.

A similar association can be found between *E. medius* and the chain-forming pennate diatom *Pseudoeunotia doliolus*, although the latter is far more frequently found free in the plankton.

Cyanobacteria (Blue-Green Algae) on Dinoflagellates and Ciliates

The more morphologically elaborate, nonphotosynthetic, dinophysoid dinoflagellates usually have coccoid cyanobacteria attached to their outer surfaces in the area between the upper and lower girdle lists (Norris, 1967; Taylor, 1980). Before their true identity was known these cells were referred to as "phaeosomes" (brown bodies). They are now considered to members of the genera *Synechococcus* and *Synechocystis* (Norris, 1967). In *Ornithocercus* and *Parahistioneis* they occur as a belt occupying the inner region of the cingulum between the wide, flaring girdle lists (Fig. 9.4). In *Histioneis* the transparent girdle lists are strongly modified, the upper girdle list being long, narrow, and trumpetlike, and the lower list forming the walls of a chamberlike cingular space in which the phaeosomes occur (Fig. 9.5). The dinoflagellates appear to have become morphologically adapted to the symbiosis, creating a "greenhouse" in which to contain the cyanobacteria. This process is taken to extreme by *Citharistes* (Fig. 9.6), in which the "phaeosome chamber" is created by the lower girdle list and the distorted cell body, with a small opening barely big enough for the cyanobacteria to pass through. Norris (1967) also reported intracellular examples of the same cyanobacterial species (see the following discussion). *Synechococcus carcerius*, 8–10 × 3–5 μm, is pink to purplish-red in color, presumably because of a predominance of phycoerythrin over phycocyanin. In fact, the majority of the free-living marine *Synechococcus* are red, rather than blue. (Their characteristics and ecological roles are discussed by Glover, 1985, who, however, omits mention of the symbiotic species.) But the freely occurring species are much smaller, typically around 2 μm in diameter. *Synechocystis consortia* cells are spherical, 6–

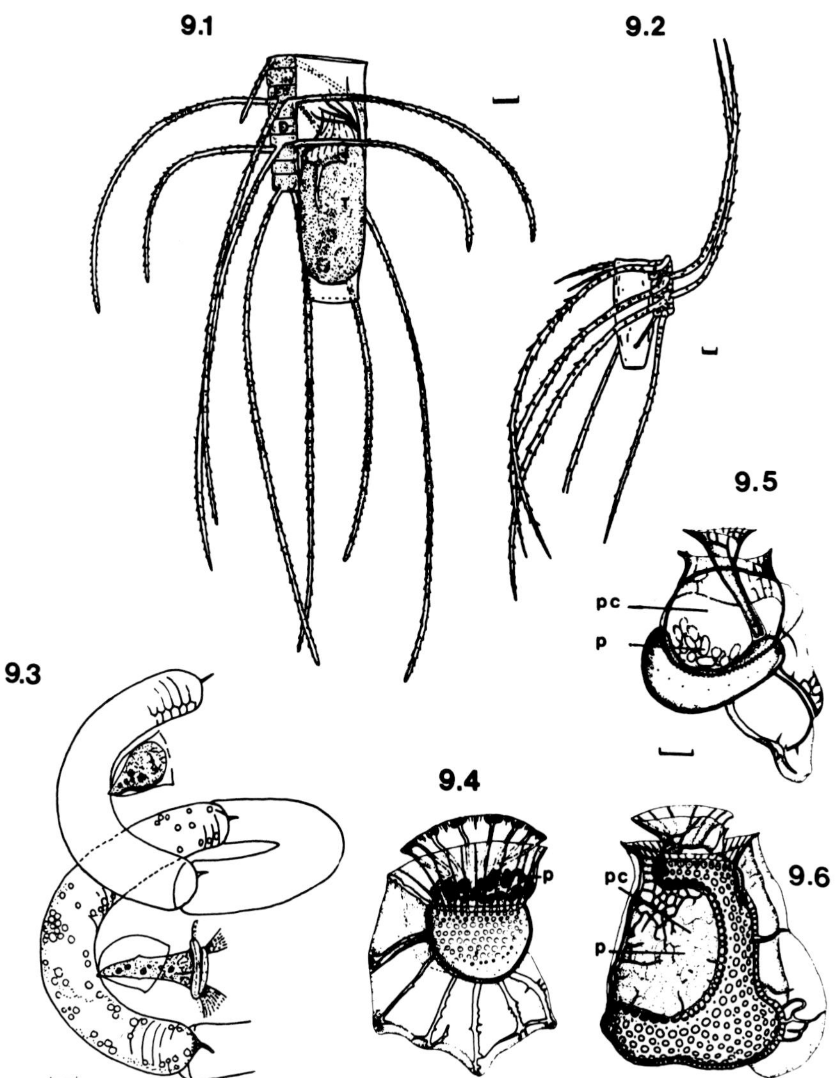

FIGURE 9.1–9.6 Scale indicates 10 μm. 9.1 A typically three-celled chain of the diatom *Chaetoceros tetrastichon* attached to the lorica of the tintinnid *Eutintinnus pinguis*. (From Taylor, 1982.) 9.2 The diatom *Chaetoceros dadayi* on the lorica of *Eutintinnus apertus*. (From Cupp, 1943.) 9.3 Two individuals of *Vaginicola* sp. attached to a chain of the diatom *Rhizosolenia phuketensis*. (From Taylor, 1982.) 9.4 The dinoflagellate *Ornothocerus thumii* with cells of the cyanobacterium *Synechococcus carcerius* attached within the girdle. (From Taylor, 1976.) 9.5 *Histioneis panda*, the lower girdle list forming a "phaeosome Chamber" to house its *Synechococcus*. (From Taylor, 1976.) 9.6 *Citharistes apsteinii*, in which the phaesome chamber is almost an enclosed cavity. p = *Synechococcus carcerius* ("phaesomes"); pc = phaesome chamber. (From Taylor, 1976.)

8 μm, and are usually blue-gray in color. If their larger size is a response to symbiotic conditions, it is possible that these species are not unique to these associations and may simply be sequestered from the surrounding water, not undergoing digestion for some reason. The author has seen a delicate protoplasmic cone resembling a "feeding veil" (Gaines and Taylor, 1984) protruding from the upper sulcal region of *Ornithocercus magnificus* (unpublished observations on live material, Barbados), and this could act as a means of entrapping other cells. Cells of the size of these symbiotic species are not found freely in the water.

On several occasions the planktonic, colonial peritrich *Zoothamnium pelagicum* has been observed to have large numbers of coccoid cyanobacteria attached to the cuticle of its zooids (Dragesco, 1948; Laval, 1968), but they are not constantly present, being more common in summer in the Mediterranean. Again, the nature of any interaction is unknown.

Epibionts on the Surfaces of Larger Foraminifera

The surfaces of benthic foraminiferans, particularly the soritids, are covered with colonies of bacteria, diatoms, and unicellular green algae. Small species of the pennate diatom *Amphora* spp. (*Halamphora* group) can fit in the crevices where the calcareous plates covering the chamberlets of *Amphisorus hemprichii* join together (Lee, 1983). The same foraminiferan is often colonized by a unicellular green alga that creates pockets in the lateral walls of the test. A smaller foraminiferan, heavily colonized by a species of *Fragilaria*, is shown in Figure 9.7.

Marine Protozoa as Active Epibionts

Ciliates and Flagellates on Diatoms

Marine planktonic diatoms may serve as surfaces for epiphytic peritrich ciliates. In tropical waters it is commonplace to see vorticellid ciliates, primarily of the genera *Vaginicola, Vorticella,* and *Zoothamnium* on diatoms (Taylor, 1982). *Vorticella oceanica* is commonly found on chains of two closely related species of *Chaetoceros: C. coarctatum* and *C. densum. Zoothamnium pelagicum,* which usually occurs free in the plankton, can also attach to *Rhizosolenia alata*. The location of the ciliates is sometimes quite specific. In *R. phuketensis,* for example, *Vaginicola* occurs only on the inner curve of the spiral chain (Taylor, 1982; see Fig. 9.3). The ciliary motion of the epibionts moves the chains through the water, but not in a directed fashion like the tintinnid associations described earlier. Another common association involving *Vaginicola* is its occurrence on the valves of large single cells of *Coscinodiscus centralis*. Presumably all these examples of epiphytism by habitually benthic ciliates (other than *Zoothamnium*) are essentially adventitious, the inert diatom walls substituting for inanimate surfaces. As with the tintinnid associations, however, the diatoms may benefit by the flushing of their surfaces with new water and may be maintained in the euphotic zone if their motive partners are phototactic.

Dinoflagellates on the Surface of Planktonic Foraminifera

The planktonic foraminiferan *Hastigerina pelagica* produces a frothy outer "bubble capsule," to which the large, coccoid dinoflagellates *Pyrocystis noctiluca, P. fusiformis, P. robusta,* and *P. lunula* often adhere (Fig. 9.8) (Alldredge and Jones, 1973; Bé et al., 1977; Hemleben and Spindler, 1983). These dinoflagellates, enclosed in a tough, continuous wall (the pellicle, homologous to a cyst wall), are also found on the surfaces of the rhizopodial network of *Globigerinoides sacculifer* and *G. ruber.* As these dinoflagellates are common in the surrounding water, it is evident that they are trapped from it. It is not known why they are not drawn further into

FIGURE 9.7 An unidentified benthic foraminiferan with chains of a diatom, *Fragilaria* sp., growing on its test (original).

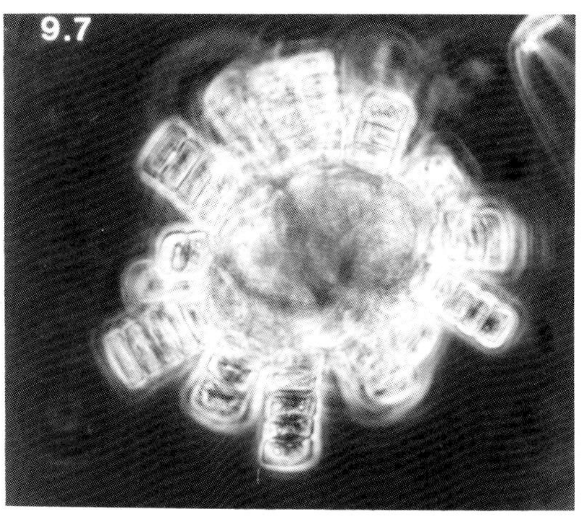

FIGURE 9.8 The planktonic foraminiferan *Hastigerina pelagica* with many individuals of the dinoflagellate *Pyrocystis fusiformis* attached to its bubble capsule. (From Bé et al., 1977.) ×51.

the cytoplasm. The tough nature of the pellicle may be the reason they are not digested, but they are not rejected either. Being photosynthetic, they offer their hosts the possibility of leaked photosynthates, but nothing is known of any physiological interactions, so far.

In addition to those in contact with the surface of the Foraminifera, there are also dinoflagellates which occur regularly swimming in between the spines, possibly feeding on the particles trapped on them. One species of *Gyrodinium* is such an associant (unpublished observation, Barbados; and H. Spero, personal communication).

Ectocommensals with Marine Metazoans

A casual inspection of the feeding and breathing cavities of larger marine animals reveals a variety of marine protozoa taking advantage of the shelter, increased flow, and food in such environments. Since these have relatively free contact with the exterior environment, although possibly interrupted, as in the mantle cavities of intertidal bivalve molluscs, they are not considered as endozoic here. Some are undoubtedly parasitic. Despite their potential importance to their animal hosts, there has been very little study of these communities in marine environments, other than studies on specific parasites.

Many maintain themselves by swimming, but others may have sophisticated attachment organelles. An example of the latter is the fascinating ciliate *Ellobiophrya donacis* (Fig. 9.9), which attaches itself to the gills of the bivalve lamellibranch *Donax vittatus* by wrapping twin posterior stalks around the gill bars in a clamplike manner (Chatton and Lwoff, 1929). Reproduction is longitudinal, and it has to avoid "letting go" while this process continues. In this position it undoubtedly robs its host of some of the food trapped on the gills but does not appear to be significantly harmful to it. This type of feeding and shelter in a larger animal host has been termed *inquilinism*.

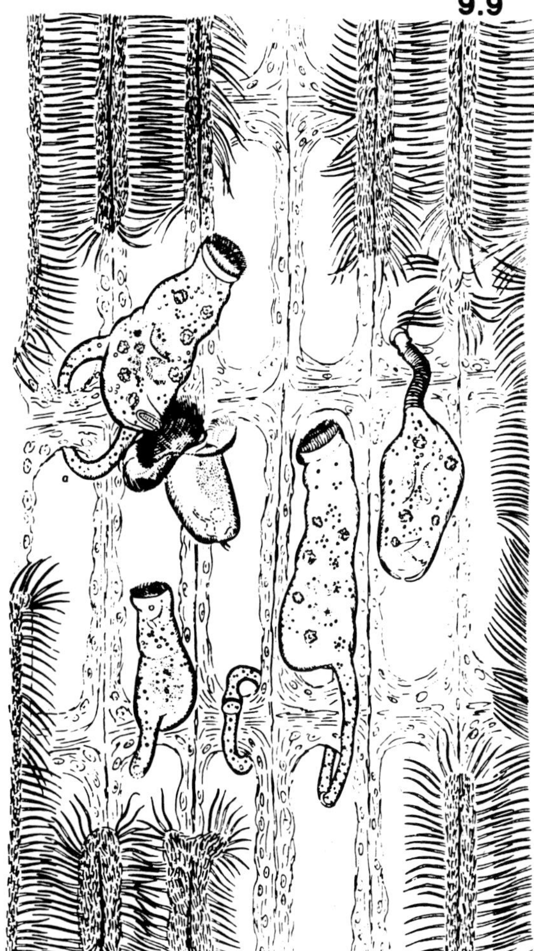

FIGURE 9.9 Cells of the ciliate *Ellobiophrya donacis* attached to the gill bars of the lamellibranch *Donax vitatus* by means of their bifurcated stalks. (From Chatton and Lwoff, 1929.)

ENDOSYMBIOSES

Endozoic Marine Protozoa

Surprisingly little systematic study has been carried out on the gut protozoa of marine animals compared with those in land animals, particularly herbivorous insects and mammals (for reviews see Ahmadjanian and Paracer, 1986, and Smith and Douglas, 1986). However, a start has been made recently with the study of gut protozoa in a herbivorous surgeon fish, *Acanthurus nigrofuscus*, by Fischelson et al. (1985). They found that, in addition to abundant bacteria, particularly spirilla, there were trichomonad flagellates and (in all fish examined) very high densities of an unusual, large (up to 500 μm) cigar-shaped protozoan, capable of locomotion without flagella. No mitochondria could be observed, and the symbionts were not identified to group. It was present in one other species of the host genus, but not in other related genera. The presence of trichomonads accords with their well-known presence in the guts of terrestrial herbivores, particularly termites, where they are thought to contribute to the break-down of cellulose. The rod-shaped organisms could be parasitic, vanishing from the gut with starvation of the host.

Cytobioses

Within marine nonphotosynthetic protists one can find many examples of cytobioses, ranging from very minimally integrated, facultative associations in which the cytobionts literally swim freely in host vacuoles (such as those in *Noctiluca;* see the following discussion), to highly integrated, obligate associations in which it is difficult to tell the partners apart.

Inhabitants of the Vacuole of Noctiluca

The unusual, phagotrophic dinoflagellate *Noctiluca scintillans* (= *N. miliaris*) is swollen into a subspherical shape by a large, low-density, low-pH, fluid-filled vacuole whose function is primarily to assist in flotation. In most temperate coastal waters *Noctiluca* is renowned for its production of bioluminescence (not in the North Pacific) and for its common production of spectacularly orange "red water." However, in the northern Indian Ocean and southeast Asian waters it is a cause of "green water" instead, because of the presence of as many as 12,000 minute (2–6 μm long), photosynthetic prasinomonad flagellates (Fig. 9.10) swimming within its buoyancy vacuole. At first they were assigned to the euglenoids and named *Protoeuglena noctilucae* (Subrahmanyan, 1954), but a later ultrastructural study (Sweeney, 1976) established their prasinomonad identity and led to the species being renamed *Pedinomonas noctilucae*. In these green blooms virtually all cells are similarly infected and are presumably transmitted during division. However, *Noctiluca* periodically undergoes "sporulation" (possibly gametogenesis), and it is not known if the infection can be transmitted through this phase.

Sweeney (1971) was able to demonstrate that the prasinomonads could satisfy the feeding requirements of *Noctiluca* if kept in the light but not in the dark, although the mechanism(s) involved are not known. Feeding the *Noctiluca* with another green flagellate, *Dunaliella,* led to loss of the prasinomonads.

The full geographic extent of green *Noctiluca* has not been clearly established. They extend into the Andaman Sea, Bay of Bengal, Arabian Sea (west coast of India), the South China Sea, and the waters of the East Indies.

In the more typical examples of cytobiosis described later, the cytobionts are located within the cytoplasm in small vacuoles or, more rarely, embedded within the cytoplasm itself. Marine rhizopods are particularly prone to cytobiosis, photosynthetic and otherwise.

FIGURE 9.10 Two prasinomonad flagellate cells (f), *Pedinomonas noctilucae,* within the vacuole of the dinoflagellate *Noctiluca scintillans.* (From Sweeney, 1976.) ×15,000

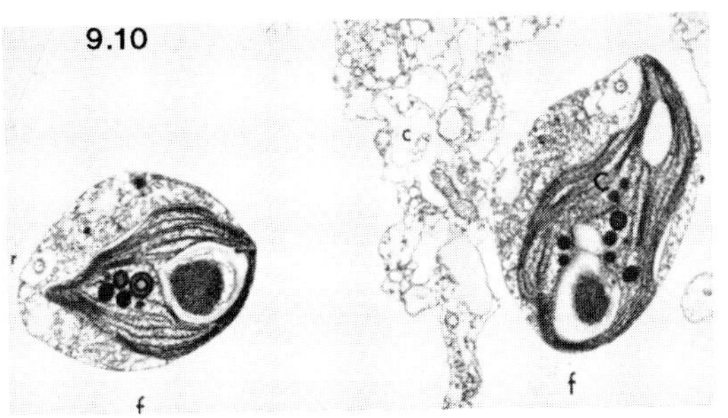

Acantharian Photobioses

Acantharians frequently can be seen to contain brown photosynthetic bodies in their ectoplasm or, in advanced forms, in the endoplasm, but these have not been systematically studied. According to Tregouboff (1953), the only acantharians to lack photobionts are deep-water, primitive forms (holocanthids), giving figures of 6 to 300 per acantharian for the rest. I have observed (unpublished fluorescence obs., Sargasso Sea and eastern Caribbean) that photobionts may be absent from some individuals of a species possessing them. Within the group as a whole, there is great variety in the morphology of the photobionts. Some are of similar size and appearance to dinoflagellate "zooxanthellae", i.e., around 8-10μm, with dinoflagellate ultrastructure (Hollande & Carré, 1974), but many are much smaller, 3-6μm in diameter. Hollande & Carré (1974) showed that the ultrastructure of those occurring in *Acanthometra* differed in having a conventional nucleus and plastids with immersed pyrenoids. Electron microscopy has revealed that the photobionts of *Lithoptera muelleri, Acanthometra pellucida,* and *Amphilonche elongata* are prymnesiomonads (Febvre and Febvre-Chevalier, 1979), the group that includes chrysochromulinids and coccolithophorids. Similar cells have been seen in some solitary radiolaria and foraminifera (Fig. 9.11). Coccoliths are not present on these cytobionts, but external skeletal structures are often lost in cytobiosis. In other instances the brown bodies may be large and irregularly oval in shape (Taylor, unpublished observation).

Unfortunately, because the group is so little studied ecologically (partly because of the difficulty in preserving them), it is difficult to assess the impact of these symbioses. However, as noted previously (Taylor, 1982), qualitative observations in the Sargasso Sea suggest that much of the chlorophyll above the subsurface chlorophyll maximum may be

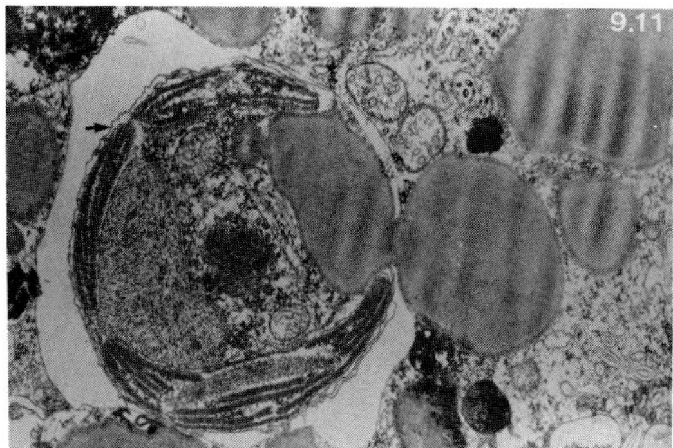

FIGURE 9.11 A prymnesiomonad zooxanthella from the planktonic foraminiferan *Globigerinella aequilateralis*. (From C. Hemleben.) ×24,600

FIGURE 9.12 An outer portion of the colonial radiolarian *Spaerozoum neapolitanum* showing the dinoflagellate zooxanthellae (arrows) embedded in the common mucilage. (From Brandt, 1881.)

present in acantharians, with the exception of picoplankton and occasional blooms of filamentous cyanobacteria.

Radiolarians (Spumellaria and Nasselaria)

In solitary and colonial spumellarian and nasselarian radiolarians most of the photobionts are dinoflagellate zooxanthellae and are also ectoplasmic, occupying vacuoles in the extracapsulum (see Fig. 9.13). They are not located within the central capsule, the apertures of which are too small for their passage (Anderson, 1983a,b). Phaeodarians, which usually live much deeper in the water column, appear to lack photosynthetic cytobionts, although the greenish-brown, pigmented mass known as "phaeodium" was erroneously interpreted by some early authors as being of zooxanthellar origin.

The colloquial term *zooxanthellae,* used loosely to refer to yellowish or brown photobionts in animal hosts, is derived from the genus *Zooxanthella,* whose type species, *Z. nutricola,* was described first from the colonial radiolarian *Collozoum inerme* (Brandt, 1881; Fig. 9.12 is of a similar colonial species, *Sphaerozoum neapolitanum*). Its ultrastructure (Fig. 9.13) has been described by Anderson (1976), and similar organization, with multiple stalked pyrenoids penetrated by thylakoids, has been seen in other radiolaria and the siphonophore *Velella* (Hollande and Carré, 1974). D. Taylor (1974) reported that the ultrastructure of this zooxanthellar type resembles the usually free-living genus *Amphidinium,* and it could be argued that *Zooxanthella* spp. should be "sunk" into the latter, older genus (Blank and Trench, 1986). However, it is evident that the symbiont of *Collozoum inerme,* like most photobionts, divides in the nonmotile, coccoid life-cycle phase (Anderson, 1976a) and not in the motile phase, thus differing fundamentally from *Amphidinium*. This same distinction is usually

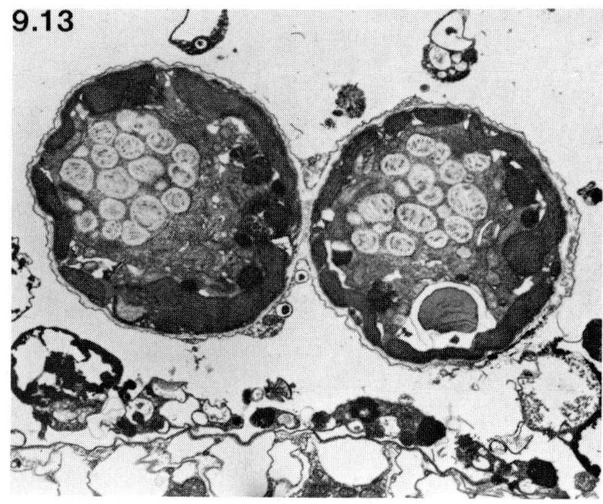

FIGURE 9.13 Two cells of the dinoflagellate *Zooxanthella nutricola* in the colonial radiolarian *Collozoum inerme*. (From Anderson, 1976.) ×8,300

used to separate symbiotic *Symbiodinium microadriaticum,* from the genera *Gymnodinium* and *Woloszynskia,* whose form its motile phase resembles (Taylor, 1982).

Not all the brownish bodies of radiolaria are dinoflagellates. As noted earlier, small cells whose ultrastructure resembles the prymnesiomonads, possibly coccolithophorids, found in some acantharians, have also been seen in some solitary and colonial radiolarians (Anderson et al., 1983a).

"Zoochlorellae" (i.e., green coccoid photobionts) have also been observed in the periphery of some spumellarians (e.g., by Norris [1967], who suggested that, unlike the "zooxanthellae," they may be digested by the host). Electron microscopy of two solitary astrosphaerid spumellarians revealed that their green photobionts were prasinomonads, the flagella of which were reduced to their bases, with some vestigial scales and a well-developed osmiophilic wall. In the zoochlorellae from *Spongodrymus* four flagellar bases were seen, consistent with several prasinomonad genera. These features have been interpreted by Anderson (1983b) as indicating less integration with their hosts than the dinoflagellate photobionts, whose walls are usually greatly reduced. *Thallasolampe margarodes* has only prasinomonad symbionts, whereas *Spongodrymus* sp. can have prasinomonads, prymnesiids, and dinoflagellate photobionts.

An interesting behavioral control of the symbionts by their hosts consists of moving them to dispersed, peripheral positions during the day and drawing them in, close to the capsule, at night. This occurs in both the solitary and colonial forms and is presumably an adaption to increase plastid illumination, mimicking the plastid movements observed in many planktonic, centric diatoms. A similar phenomenon is also known in some planktonic foraminifera (see the following discussion and Figs. 9.14 and 9.15).

Anderson et al. (1983, 1984) have summarized the data on photosynthetic activity and evidence of translocation of photobiont photosynthate to the hosts. Using two- to eight-hour incubations and light intensities of 35–230 $\mu E\ m^{-2}\ s^{-1}$, they found that 9–14 percent of the inorganic carbon fixed by the photobionts in *Thallasicola nucleata* and *Collosphaera huxleyi* was translocated to the central capsule of their hosts. The photosynthetic rates of the photobionts are roughly equivalent to those of free-living dinoflagellates. An interesting observation is the enhanced photosynthesis that occurs when the hosts are well fed with other food, suggesting that there may be increased raw material availability (nitrogen?) under the latter circumstances. There is evidence, both enzymatic and ultrastructural, for digestion of the photobionts, at least in *Collozoum inerme*.

Photobionts are not transmitted during sexual reproduction and so they must be acquired from the surrounding water in each sexual generation.

Planktonic Foraminifera

At the time of writing, 10 species of planktonic foraminifera (i.e., roughly one quarter) have been reported to contain photobionts (Table 9.1). Not all have been checked yet, but it has been observed that it is usually the spinose forms that possess photobionts, the nonspinose forms examined so far lacking them (Hemleban and Spindler, 1983). A report of their presence in *Globigerina bulloides* (Spindler and Hemleben, 1980) was subsequently stated to be erroneous (Hemleben and Spindler, 1983).

A good example of one of the spinose species is *Globigerinoides sacculifer* (Brady), (see Figs. 9.14 and 9.15), which usually makes up 20 percent or more of tropical planktonic foraminiferan populations (Bé and Tolderlund, 1971). Typically, it contains approximately 500 dinoflagellate photobionts per cell, which are moved by it in a diel rhythm. During the day most of the 5–8 μm photobionts are

FIGURES 9.14–9.16. The symbiosis in *Globigerinoides ruber*. Figures 9.14, 9.15 The expanded (daytime) and contracted (nighttime) positions of the zooxanthellae in the pseudopodia.

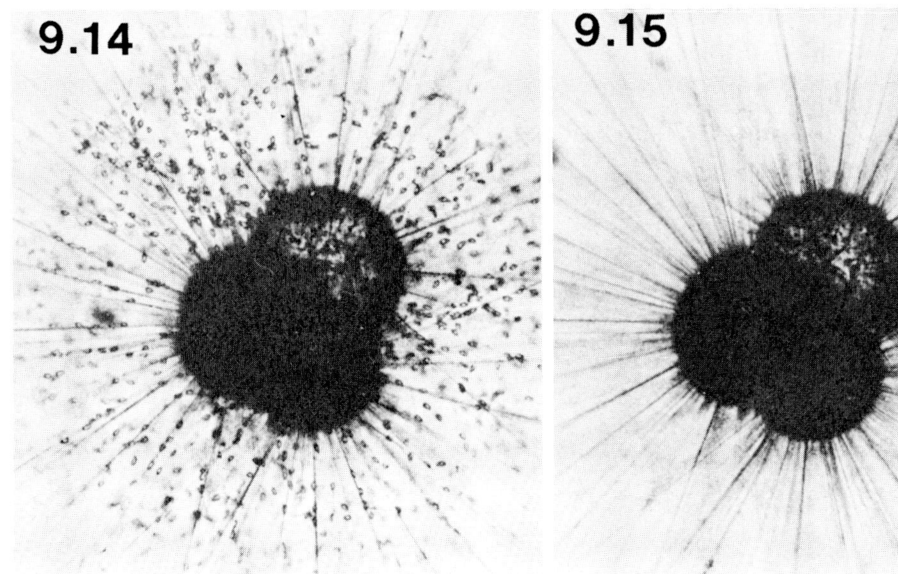

TABLE 9.1 Planktonic Foraminiferans with Intracellular Photobionts

Host	Cytobiont(s)	Source(s)
Orbulina universa	*Gymnodinium* sp.	1,5,6
Globigerinoides conglobatus	*Gymnodinium* sp.	5
Globigerinoides ruber	*Gymnodinium* sp.	1,3,5
Globigerinoides sacculifer	*Gymnodinium* sp.	2,5
Globigerionoides glutinata	Prymnesiomonad?	4*
Globigerinella aequilateralis	Prymnesiomonad	5
Globorotaliella menardii	Prymnesiomonad	1*
Pulleniatina obliqueloculata	Prymnesiomonad	1*
Globigerina cristata	Prymnesiomonad	5
Globigerina falconensis	?	6

(1) Bé et al. (1977); (2) Anderson and Bé (1976); (3) Lee et al. (1965); (4) Lee and McEnery (1983); Hemleben and Spindler (1983); (6) Spero and Parker (1985).

*These were first recorded as chlorophytes, but the ultrastructure illustrated was inconsistent with this. *Hastigerina pelagica* has ectosymbiotic dinoflagellates (see earlier text).

situated in rhizopods extended outside the shell, forming a brown halo around the cell. At night they are withdrawn deep into the shell, where they are gathered in the earliest-formed chambers (Anderson and Bé, 1976). The photobionts are unusual in that they appear to remain in a state equivalent to the motile phase (a similar cell from *Globigerinoides ruber* is shown in Fig. 9.16), although lacking flagella, rather than the cystlike, coccoid state that dinoflagellate "zooxanthellae" usually assume within their hosts. The presence of extremely thin platelike structures within the peripheral vesicles, while within the host, resembles the genera *Gymnodinium* Stein (into which *Aureodinium* Dodge has been sunk; see Taylor, 1980) and *Zooxanthella* Brandt, but the pattern on the surface has not been described yet. When cultured free from the host the cells remain continuously motile, rather than alternating between a coccoid and motile state.

Using a microelectrode technique, Jørgensen et al. (1985) studied individual cells of *G. sacculifer*, recording a photosynthetic activity of 18 nmol O_2 h^{-1} and a respiration of 3 nmol O_2 h^{-1} foraminiferan^{-1}. Thus, there is a high net photosynthesis per host during the day. The compensation light intensity was found to be 26–30 μE m^{-2} s^{-1}, corresponding to a depth of 60 m (45 m on a diel basis) in the Gulf of Eilat (Aqaba) at the time the study was done. Saturation intensities of 160–170 μE m^{-2} s^{-1} occurred at more than 20 m.

Despite this more-than-ample excess of production, both Bé et al. (1981) and Jørgensen et al. (1985) found that it was necessary to feed *G. sacculifer* with nauplii to keep the organism healthy. The latter authors concluded that this may be essential to satisfy nitrogen or phosphorus demand by the consortium.

Orbulina universa, which has been recently subjected to a similar detailed study (Spero and Parker, 1985), is very similar in its activities and appears to house the same species of photobiont. It has been named *Gymnodinium bei* by Spero. Spero and Parker counted numbers per host ranging from 130 to 3,300, although it was estimated that large, mature, spherical specimens may contain as many as 23,000 photobionts per host cell.

Although the beneficial effects on the growth of the host by the symbionts have been established experimentally (and also survival—63 percent less in dark or DCMU treated starved cells: Bé et al., 1983), the role of the photobionts in skeletogenesis is less clear. With the use of the photosynthetic inhibitor DCMU, Bé et al. (1983) showed that shell growth rate and shell size were reduced the same as in dark controls in *G. sacculifer*. However, Erez (1983) obtained evidence that suggested that there was a positive light effect but not due to the presence of the photobiont.

Photobionts of Benthic Foraminifera

It appears that all the larger, shallow-water, benthic foraminifera, belonging to 11 families, can harbor photosynthetic cytobionts (Fig. 9.17) (for reviews see Lee, 1980; Lee and McEnery, 1983; Lee et al., 1985a,b). They are most common in the tropics, occurring as large discs, stars, cylinders, oblate sphaeroids (reaching nearly half a centimeter in diameter) on the bottom, attached to such surfaces as old coral rubble, sands, or sea grasses (*Halophila, Thalassia*). Many of the symbiont-containing species show apparent structural adaptations to enhance illumination in these associations, including their disclike forms, pores, or windowlike thinnings of the wall, which may also enhance the passage of nutrients and carbon dioxide to them (Leutenegger and Hansen, 1979).

In contrast to the planktonic forms and other cases of cytobiosis described here, the benthic foraminifera exhibit a great variety of photobionts, including the presence of more than one type within the same individual, although that is uncommon (Lee, 1980, 1983; Leutenegger, 1984). There are three principal groups of photobionts: dinoflagellates (*Symbiodinium, Amphidinium*?), pennate dia-

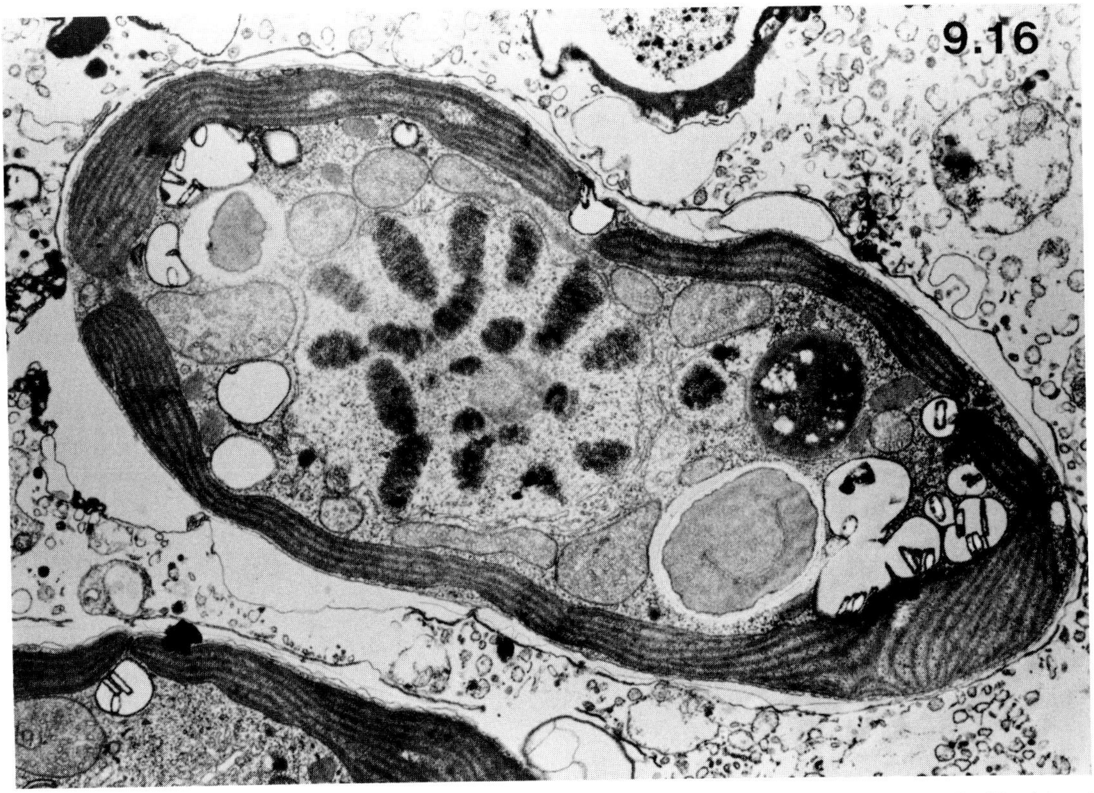

FIGURE 9.16 Electron micrograph section through the endobiont, *Gymnodinium (bei?)*. (From C. Hemleben.) ×14,500

FIGURE 9.17 The benthic foraminiferan *Heterostegina depressa*. ×50.

toms (*Fragilaria, Nitzschia*), and chlorophycean green algae (*Chlamydomonas hedleyi, C. provasolii, Chlorella*), but a unicellular red alga, *Porphyridium*, has also been found within some (Table 9.2). The diatoms are very small (<10 μm) and some are known only as symbionts.

As might be expected, the presence of photobionts is linked with illumination, although some photobiont-containing species, such as *Borelis schlumbergeri*, can occur as deep as 70 m and prefer shaded locations at shallower depths (10 m) (Leutenegger, 1984a). Phototaxis is exhibited by the hosts, depending on light intensity (Lee et al., 1980). Those harboring diatoms tend to occur deeper than the others (exceptions are *Calcarina* and *Baculogypsina;* Leutenegger, 1984a), and there is experimental evidence (summarized by Lee et al., 1985a) that suggests that the diatoms are adapted to low light levels, their compensation depth occurring roughly between 40 and 50 m in clear tropical waters such as those in the Gulf of Eilat (Aqaba) or the Red Sea. Growth of the hosts can be shown to be related to light levels without the presence of external food, and some photoinhibition can occur at high intensities, depending on the partners. Those with green algae are commonly found in high-light-intensity, shallow turtle grass (*Thalassia*) beds, and the dinoflagellate-bearing species are also relatively shallow.

Within their hosts the diatoms lose their distinctive silica walls (Fig. 9.18) and can only be identified when cultured separately from their hosts, whereupon they resume normal frustule (wall)

TABLE 9.2 Photobionts of Benthic Foraminifera

Soritids: dinoflagellates—probably *Symbiodinium* (*Amphisorus, Broekina, Peneroplis*); greens (*Archais, Sorites*); reds (*Peneroplis*)

Alveolinids: diatoms (*Borelis schlumbergeri*)

Amphisteginids: diatoms—*Fragilaria shiloi, Nitschia frustulum symbiotica, N. panduriformis continua, Nitzschia laevis,* are the most abundant in *Amphistegina lobifera, A. Lessonii* and *A. papillosa.* Cells of the green alga *Chlorella* sp. or other pennate diatoms may also be present.

Nummulitids—(*Operculina ammonoides, Heterocyclina tuberculata, Heterostegina depressa*): diatoms—*Nitzschia panduriformis, N. valdestriata, Achnanthes* sp.; unidentified green alga (*Heterostegina depressa*)

Calcarinids—(*Calacarina*): diatoms—*Fragilaria shiloi, Achnanthes* sp. *Cocconeis* sp., *Navicula hanseniana, Navicula* sp., *Nitzschia frustulum symbiotica, Nitzschia panduriformis continua, Amphora roettgeri* (Lee and co-workers, in progress)

Spirillinids—(*Spirillina, Dendritina peneroplis*): red alga (*Porphyridium purpureum*; Lee and Hawkins, in press)

Others—*Baculogypsina sphaerulata:* diatoms in shallow water

Cycloorbiculina: green alga—*Chlamydomonas hedleyi*

Marginopore: dinoflagellate—*Gymnodinium vertebralis*

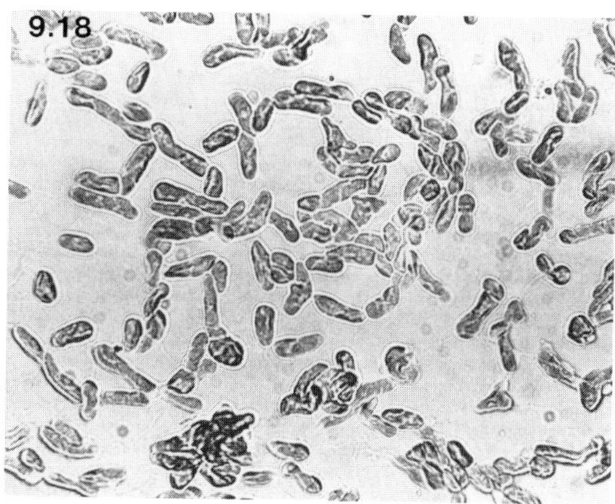

FIGURE 9.18 The diatom endobionts in the nonwalled state in which they occur in the host. ×950 (Both from Röttger, 1972.)

growth. In other respects the photobionts are essentially unaltered, growing much as they would outside their hosts. This, together with the continued feeding by the hosts, seems to suggest that the associations are only slightly adapted. However, the range of photobionts found is small compared with the potential range of organisms that can be taken up from their surroundings. The host wall structures and peripheral location of the photobionts do appear to be adaptations to photobiosis.

Lee and co-workers (review Lee, 1983) have isolated different endosymbionts within different natural populations of the same host species, suggesting that the associations are not species specific, whereas Leutenegger's (1984a,b) fine structural data have led her to conclude that there is strong species specificity. It has been suggested that benthic foraminifera are highly adapted because of the wide range of symbionts they can support (Lee et al., 1985b), although it could also be argued that they support a wide range because they have poorly developed recognition–digestion–rejection mechanisms and are thus more subject to fortuitous infections.

Cryptic Photosynthetic Cytobionts in Dinoflagellates

Dinoflagellates not only may be photosynthetic cytobionts of uni- and multicellular hosts, but also may harbor photosynthetic cytobionts of their own. Two well-known cases are *Kryptoperidinium* (= *Glenodinium*) *foliaceum* and *Durinskia* (= *Peridinium*) *baltica,* both of which were assumed to produce very anomalous pigments until it was discovered that their photosynthetic apparati were not originally theirs: they belong instead to foreign cytobionts (Dodge, 1971; Tomas, Cox and Steidinger, 1973; Withers and Haxo, 1975; Jeffrey and Vesk, 1976). The brown-pigmented cytobionts, possessing chlorophylls $a + c_1$ and c_2 and with fucoxanthin as a major xanthophyll, are so ramified within their hosts that they are hard to recognize except for the presence of two nuclei in each cell. The cytobiont nuclei have the conventional eukaryotic appearance, whereas the hosts have the typical dinokaryotic nuclear organization. The cytobionts are separated from their hosts by a single membrane. The hosts produce the usual dinoflagellate storage product (starch) in their cytoplasm. The associations are permanent and can be cultured indefinitely as if they were photosynthetic dinoflagellates. However, pigmented wild populations of *D. baltica* with only a dinokaryon have been observed in Danish coastal waters (Taylor, 1979), and it remains to be determined whether these have dinoflagellate pigments. In both associations there is an "eyespot" in the host cytoplasm, surrounded by a double membrane, which has been interpreted as a vestige of a former dinoflagellate photosynthetic system (Taylor, 1979).

The observation of blue plastids in some dinoflagellates is a further pigment puzzle that appears to have a similar explanation, at least in some species of *Gymnodinium*, such as *G. acidotum* (Wilcox and Wedemayer, 1984) and *G. eucyaneum* (= *G. cyaneum* Hu; Li et al., 1979; Zhang et al., 1982), both of which are freshwater species containing cytobionts of cryptomonad origin. One cannot readily extrapolate from these cases to other blue dinoflagellates or be sure of their permanence, however. *Amphidinium wigrense* has been found to harbor cryptomonad plastids only, each surrounded by three membranes (Wilcox and Wedemayer, 1985). This condition can be interpreted as a greater degree of incorporation of a cryptomonad cytobiont with the host so that only the plastids remain (Wilcox and Wedemayer, 1985), or the plastids may be temporarily maintained after the partial digestion of ingested cryptomonads (see the following discussion of plastid maintenance) J. Larsen (personal communication) has observed what appears to be the latter in marine, nonphotosynthetic sand dinoflagellates. Clearly, the answer to this can be obtained from the culture of the organisms in question. This has not been accomplished so far.

Mesodinium rubrum—A Permanently Photosynthetic Ciliate

In fresh water there are several examples of ciliates that maintain photosynthetic symbionts, usually "zoochlorellae" (Chlorophyceae, mostly the genus *Chlorella*), of which the most studied is *Paramecium bursaria* (Smith and Douglas, 1986).

In marine waters the best-known example of a permanently photosynthetic ciliate, caused by symbiosis, is the gymnostome *Mesodinium rubrum* (also known as *Cyclotrichium meunieri* and *Myrionecta rubra*); it has been reviewed in detail by Lindholm (1985). It is extremely widespread in coastal areas, from polar waters to upwelling zones of the tropics (Taylor et al., 1971) and is a common cause of harmless "red water" because of its periodic abundance in numbers (millions per liter) sufficient to color the water, during which times the chlorophyll concentrations may exceed 1,000 mg/m^3. Its redness is due to the presence of phycoerythrin-rich, functionally photosynthetic organelles (plastids) of cryptomonad origin, situated in islands of foreign cytoplasm (Fig. 9.19) that also contain cryptomonad mitochondria (Taylor et al., 1969; Oakley and Taylor, 1978). One island contains a cryptomonad nucleus. Thus, the association constitutes a unique example of a photosynthetic cytobiont that has proliferated its plastids and fragmented within the host, the whole remaining fully functional and reproduced by mechanisms as yet unknown. In one English population the cytobiont components were gathered in a common cytoplasm, or in two groups (Hibberd, 1977), but this

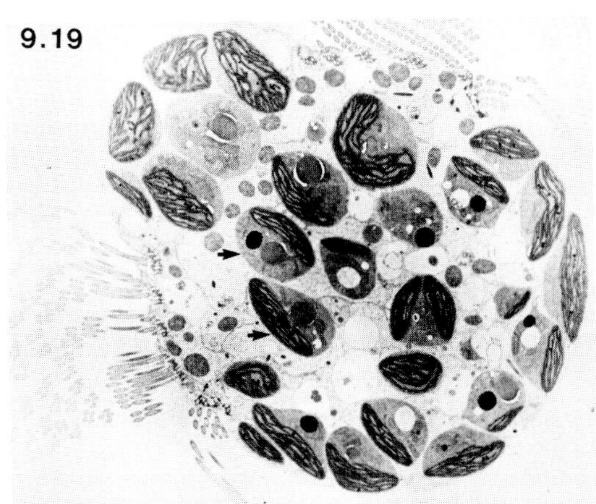

FIGURE 9.19 A low-power electron microscope section through the ciliate *Mesodinium rubrum*, showing the cryptomonad symbiont "islands" (arrows) containing plastids and mitochondria. (Original from material prepared by Oakley and Taylor, 1978.) ×4,500

has been observed only once. The host exhibits modifications apparently in response to its photosynthetic condition, including the loss of pellicular alveoli (improved illumination?) and loss of its mouth, swimming backward relative to its nonsymbiotic relatives. It is capable of very rapid darting movements, because of sweeps of its extended cirri, interspersed with motionless pauses or moderate turning accomplished by its coronal cilia, and may carry out vertical migrations to as deep as 30 m at night, returning to the surface during the day.

It is an active photosynthesizer and, as its peak activity, has achieved the highest levels of primary production yet recorded in seawater (1-2 g · C · m^{-3} · h^{-1}; Smith and Barber, 1979), although levels closer to 10s of milligrams over the same period may be more usual. This photosynthesis is not all used up by the host and it usually accumulates starch in the vicinity of its plastids. Its high levels of productivity are probably due to the opportunity for rapid cycling offered by cytobiosis (Taylor and Harrison, 1983).

Plastid Retention

Plastid retention, in which plastids are retained in an apparently functional state from ingested food organisms, is really a pseudosymbiosis, but it has some ecological similarities and appears to be ecologically significant (Stoecker et al., 1987). It superficially resembles the situation in *Mesodinium rubrum* (discussed earlier) and has a similar consequence (i.e., the nonphotosynthetic cell harboring the foreign bodies becomes photosynthetic). However, it differs in a very fundamental sense in that the associations are not permanent and do not involve whole cytobionts.

It is well known that some sacoglossan molluscs (seahares) retain the plastids from the siphonous green algae that they feed upon and that the plastids remain functional within the animal tissues for periods of weeks. The animals thus become "leaves that crawl" (Trench, 1975). Less well known is the fact that several planktonic ciliates and benthic foraminiferans (and perhaps some dinoflagellates) also retain plastids from their food in apparently functional condition. There are two early reports on pigmented ciliate blooms, associated with photosynthetic activity and not due to *M. rubrum*. Burkholder et al. (1967) found that a brown oligotrich ciliate off Puerto Rico fixed roughly 290 μg C L^{-1} h^{-1}, with 1.0–1.4 mg C · mg chl a^{-1}. A gymnostome, *Prorodon* sp., formed dense reddish blooms off Point Barrow, Alaska (Holm-Hansen et al., 1970); its physiology *en masse* resembled a phytoplankton bloom. McManus and Fuhrman (1986) showed that *Laboea strobila* contained regular photosynthetic pigments that were not breakdown products, although they were not sure of the reason for this.

The first unequivocal demonstration of plastids in a unicellular host was the ultrastructural study of the same (unnamed) *Prorodon* species as that seen in Alaska, by Blackbourn et al. (1973). They showed that it, and a species of *Strombidium* (Fig. 9.20), contained plastids that could have come from any member of the "Chrysophyte Complex," although most probably from diatoms, since these are the primary chrysophytan items in the diet of the ciliates. The plastids were in remarkably good condition, although in some ciliate cells other plastids were obviously undergoing digestion. In the latter cases it was the most peripheral plastids that were in good shape, those nearer the center (cytostome?) breaking down.

FIGURE 9.20 An electron micrograph showing plastids (arrows) retained in the cytoplasm of the ciliate *Strombidium* sp. (Original from material prepared by Blackbourn et al., 1973.) ×8,500

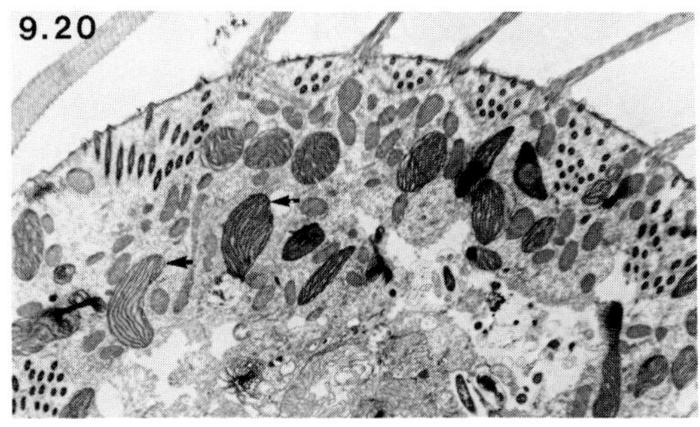

In benthic foraminifera clusters of undigested diatom plastids have been observed in *Metarotaliella parva* (E. Lanners in Taylor, 1974; illustrated in Lee, 1983, who identified the donor to be a species of *Entomoneis*). Plastid retention has also been observed in *Elphidium williamsoni* and *Nonion germanicum* in a brackish Danish fjord (Lopez, 1979). Using ^{14}C tracer methodology Lopez demonstrated that at light saturation (10 K lux) the plastids in *E. williamsoni* assimilated 2.3 × 10^{-3} mg C · mg^{-1} · h^{-1}. Lee and colleagues (in press) found that *Elphidium crispum* from Mombasa and Eilat also retained plastids as well as other organellar remnants (pyrenoids, mitochondria, and occasionally nuclei). Uptake and incorporation of ^{14}C in *E. crispum* was comparable to the rates found in *Heterostegina depressa* and *Operculina ammonoides* collected from the same depth.

Although not confirmed as such, Norris (1967) had suggested earlier that the golden, spindle-shaped bodies in a marine *Euplotes*like ciliate might have been plastids, the bodies rounding up on release. Plastid retention has also been clearly demonstrated by electron microscopy in the pelagic ciliate *Tontonia appendiculariformis* by Laval-Peuto and Febvre (1986; see also Laval-Peuto et al., 1986). No dividing forms of the plastids could be seen. Each was found to be surrounded by three membranes, the outer of which seemed to originate from the ciliate.

Proof of the functionality of the plastids within the ciliates and their ecological significance has come from the work of Stoecker et al. (1987). Isolated ciliates (*Laboea strobila* and two unidentified species of *Strombidium*) with plastids still required an external food source to survive. The plastids are not digested on starvation, however. At 150–200 μE m^{-2} s^{-1} the ciliates were capable of fixing 3.1pg C pg chlorophyll a^{-1} h^{-1} (to 586 pg C cell^{-1} h^{-1}). However, values can be highly variable, presumably depending on the number of plastids present per cell. Since an average of roughly 45 percent of the oligotrich ciliates in the vicinity of Woods Hole over the year was found to contain plastids (rising to over 90 percent in the summer), this is clearly a factor that must be considered in estimates of primary productivity from coastal localities.

Nonphotosynthetic Cytobionts

Ultrastructural studies have revealed incidentally that bacterialike cells are commonly found in the cytoplasm of some organelles of both freshwater and marine phagotrophic protozoans. In foraminifera it is evident that phagocytosed bacteria may be an essential part of their diet, being rapidly digested (Muller and Lee, 1969). Anderson (1977) observed numerous bacterialike bodies that he thought might be symbiotic, contained within individual vacuoles in the cytoplasm of the marine amoeba *Hartmanella* sp. Gram-negative bacteria have been observed free in the cytoplasm or even within the permanently closed nuclei of some photosynthetic dinoflagellates, such as *Gyrodinium instriatum* and *Kryptoperidinium* (*Glenodinium*) *foliaceum* (Silva and Franca, 1985), even though these are photosynthetic species. Bacteria are so commonly found in the cytoplasm of phagotrophs that the phenomenon must be looked upon as an infectious hazard of the feeding method. For example, the phagotrophic dinoflagellates *Noctiluca scintillans* and *Polykrikos herdmanni* have been found to contain numerous bacteria free in their cytoplasm, not surrounded by vacuoles (Lucas, 1982). A clear zone around the bacterial walls has been interpreted as being possibly due to the presence of electron-transparent bacterial capsules. Bacteria have also been commonly seen in the cytoplasm of marine ciliates, such as the tintinnids *Petalotricha ampulla* (Laval, 1972) and *Cyttarocylis brandti* (Laval-Peuto, 1975). In the mitochondrialess ciliates, which also lack cytochrome oxidase, found in anoxic marine sediments, there are rich intracellular bacterial communities that may be of distinctive species composition (Fenchel et al., 1977). They may play a respiratory function, similar to bacteria in other mitochondrialess protists.

One well-studied marine example of bacterial endocytobiosis is that of the gram-negative bacteria, originally termed *xenosomes*, which occur in the cytoplasm of the benthic ciliate *Parauronema acutum* (reviewed by Soldo, 1983). The cytobionts are most similar to Rickettsiae but are motile. Infected ciliates can be made cytobiont-free by X-irradiation and then reinfected, apparently by escaping from newly formed food vacuoles. Succinate is the most stimulating substrate for cytobiont respiration. No nutrients could be omitted from cultures of infected, relative to noninfected, cells. The presence of multicopy DNA in the cytobionts is similar to that found in several other cytobiotic bacteria and may be of adaptive significance to cytobionts (Soldo, 1983). The bacteria are lethal to the closely related *Uronema nigricans* when it is exposed in a suspension of as little as 100 cells per ciliate.

In *Euplotes minuta* another bacterialike cytobiont, termed *epsilon* (reviewed by Heckman, 1983), causes the death of noninfected "sensitive" strains of the same marine ciliate species as well as E. *crassus*. This "killer" property of the infected cells is very similar to the more well-known killer strains of the freshwater species *Paramecium tetraurelia*, which owe their toxicity to a bacterial cytobiont *Caedibacter* (formerly termed *kappa*). *E. crassus* harbors approximately 500–1,000 bacterialike particles, termed *eta* (Rosati et al., 1976) or *B2* (Rosati and Verni, 1977), that also kill sensitives of the same species. In addition, several other particle types are morphologically distinguishable in the cytoplasm of *E. crassus*. Strains of both *Euplotes* species may also kill sensitives during conjugation, referred to as mate killing (Heckman, 1983), although the particles responsible may be different from those responsible for remote killing. This is also similar to the *P. aurelia*-complex cytobionts. The adaptive significance of these killing traits is not known.

In addition to those bacteria that inhabit the cytoplasm, some have been found to enter the nuclei of their hosts. Rod-shaped cytobionts occur in the macronucleus of *E. crassus* (Rosati and Verni, 1976) and in *Zoothamnium pelagicum* (Laval, 1970). In the latter they can move out into the cytoplasm. Highly specialized intranuclear bacteria have been studied in freshwater ciliates (see review by Goertz, 1983). As noted earlier, intranuclear bacteria are also known from marine dinoflagellates (Silva and Franca, 1985), although they seem to have little effect on their hosts.

DISCUSSION

Ecological Significance

The most obvious ecological significance of symbioses (and the pseudosymbiosis of plastid maintenance) is the conversion of heterotrophs into partial or fully functional autotrophs by the acquisition of photobionts (Taylor, 1982). Many, if not all, of the foraminiferans and radiolarians continue to feed, somewhat like the well-known coral symbiosis, even though they show adaptive features that have evidently evolved in response to the symbiosis, such as those structural or behavioral features that enhance illumination. However, in the most highly evolved, such as some of the dinoflagellates and the ciliate *Mesodinium rubrum*, the hosts appear to be nutritionally satisfied by their photobionts. Photosynthetic rates adequate or more than adequate to satisfy the host's requirements have also been measured for single foraminifera (*Orbulina universa*, Spero and Parker, 1985) and colonial radiolarians (*Collozoum longiforme*, Swanberg and Harbison, 1980) and the continued feeding may be necessary to supplement nitrogen supply to the photobionts.

In essence the foraminiferal and radiolarian symbioses in tropical waters represent tightly cycling mini-ecosystems ("cytocosms," Taylor, 1983), packages of high productivity in an otherwise oligotrophic environment. It appears that the high photosynthetic performances recorded in these systems (see Table 9.2) are at least partly due to the same key factor as that found in the ecologicaly similar coral reefs: the rapid cycling permitted by close proximity of the associants. Normalized to cell volume dinoflagellates symbiotic in other protists outperform their free-living counterparts (those from *Orbulina universa* fixing more than five times the carbon per cubic micrometer than *Ceratium furca* or *Gonyaulax polyedra*), although normalization to chlorophyll content (figures not available) might produce a different picture. They are also higher than those symbiotic in coelenterates. Although the latter are also intracellular there is presumably a greater shading effect from the host tissue.

It would clearly result in severe underestimates if such "functional autotrophs" were excluded from primary productivity estimates based on community analyses (Taylor, 1982; Elbrächter, 1984), particularly if ciliate blooms were interpreted solely as micrograzers (a view perpetuated as recently as 1985 by Porter et al.). Furthermore, such concentrated sources of productivity can seriously affect estimates of oligotrophic phytoplankton productivity if even single cells are present in the incubation bottles (Spero and Parker, 1985).

Although the foraminiferans and radiolarians are such concentrated centers of productivity, their low concentrations in the tropical water column (1–5 m^3 for larger foraminiferans with maxima of 130 m^3 in patches [Almogi-Laban, 1984]; 10–30 m^3, with patches reaching 540 m^3 for colonial radiolaria [Swanberg, 1983]) mean that they are not usually the principal contributors to primary productivity except in the most concentrated patches. Spero and Parker (1985) estimated that the contribution from foraminiferal symbiont production would amount to approximately 1 percent of the total, although reaching 25 percent in patches. Khmeleva (1967) claimed that colonial radiolaria were producing three times as much as the phytoplankton in the Gulf of Aden. Blooms of *Mesodinium rubrum* have resulted in extremely high productivity measurements (2,187 mg C m^{-3} h^{-1} off Peru [Smith and Barber, 1979]) with 1,000 mg C commonly obtained.

Mode of Establishment

In many of the marine sarcodinian examples the associations are "open," that is, they are perpetuated only through asexual reproduction and the cytobionts must be reacquired from the surrounding water in each sexual generation. It appears that in some instances this may not be entirely due to accidental contact, the potential cytobionts being chemically attracted to their hosts. Lee (1983) has dubbed this the "Circean Effect" (but spelled it "Cercian") in benthic foraminifera, but it probably applies to planktonic species as well. There is a temptation to look for some highly evolved pheromonelike substance, but it may be simply chemotaxis towards excreted ammonia or another algal nutrient.

Establishment also requires avoidance of digestion, the mechanism(s) of which has not been determined for these marine examples. Similarly, they are not rejected and do not overgrow their hosts. The latter could be due simply to limitations imposed by nutrient supply, although the possibility of synchronizing factors in cell division or active repression are possible. These problems are still central to the study of many symbioses (Smith & Douglas, 1986).

In some of the associations described earlier, such as *Mesodinium rubrum* and some of the dinoflagellates harboring cytobionts, the symbionts are continuously transmitted, a condition referred to as "closed," and the symbionts have effectively become surrogate organelles of their hosts. The evolutionary implications of this have been discussed by many authors (see Taylor, 1983; Ahmadjanian and Paracer 1986).

Future Research Needs

It is evident that in some cases there has not been even a systematic descriptive analysis of the associated communities. This is particularly true of the nonparasitic protozoa associated with marine invertebrates and vertebrates, although a start has been made to investigate herbivorous fish gut microorganisms by Fischelson et al. (1985).

The best-studied cases are the foraminiferal associations. A major difficulty in the study of *Mesodinium rubrum*, the most sophisticated of the associations described here, is the lack of success in culturing it, despite its common and often abundant occurrence. Its ecological importance makes this a high-priority goal.

REFERENCES

Ahmadjanian, V., and Paracer, S. (1986) Symbiosis. An Introduction to Biological Associations. London: Clark, 212 pp.

Alldredge, A.L., and Jones, B.M. (1973) *Hastigerina*

pelagica: foraminiferal habitat for planktonic dinoflagellates. *Mar. Biol.* 22:131–135.

Almogi-Laban, A. (1984) Population dynamics of planktic Foraminifera and Pteropoda—Gulf of Aqaba, Red Sea. *Proc. Kon. Nederl. Akad. Wetensch,* Ser. B., 87:481–511.

Anderson, O.R. (1976) Ultrastructure of a colonial radiolarian *Collozoum inerme* and a cytochemical determination of the role of its zooxanthellae. *Tissue Cell* 8:195–208.

Anderson, O.R. (1977) Fine structure of a marine amoeba associated with a blue-green alga in the Sargasso Sea. *J. Protozool.* 24:370–376.

Anderson, O.R. (1983a) The radiolarian symbiosis. In Goff, L.J. (ed.), *Algal Symbiosis.* Cambridge, England: Cambridge University Press, pp. 69–89.

Anderson, O.R. (1983b) *Radiolaria.* New York: Springer-Verlag, 355pp.

Anderson, O.R. (1988) *Comparative Protozoology, Ecology, Physiology, Life History.* New York: Springer-Verlag, 482 pp.

Anderson, O.R., and Bé, A.W.H. (1976) The ultrastructure of a planktonic foraminifer *Globigerinoides sacculifer* (Brady) and its symbiotic dinoflagellates. *J. Foraminiferal Res.* 6:1–21.

Anderson, O.R., Swanberg, N.R., and Bennett, P. (1983a) Fine structure of yellow-brown symbionts (Prymnesiida) in solitary radiolaria and their comparison with similar acantharian symbionts. *J. Protozool.* 30:718–722.

Anderson, O.R., Swanberg, N.R., and Bennett P. (1983b) Assimilation of symbiont-derived photosynthates in some solitary and colonial radiolaria. *Mar. Biol.* 77:265–269.

Anderson, O.R., Swanberg, N.R., and Bennett, P. (1984) Laboratory studies of the ecological significance of host–algal nutritional associations in solitary and colonial radiolaria. *J. Mar. Biol. Assoc. U.K.* 65:263–272.

Bé, A.W.H., Anderson, O.R., Faber, W.W.,Jr., and Caron, D.A. (1983) Morphological and fine structure sequence of events during gametogenesis in planktonic foraminifer *Globigerinoides sacculifer* (Brady). *Micropaleontology* 29:310–325.

Bé A.W.H., Caron, D.A., and Anderson, O.R. (1981) Effects of feeding frequency on life processes of the planktonic foraminifer *Globigerinoides sacculifer* (Brady) in laboratory culture. *J. Mar. Biol. Assoc. U.K.* 61:257–277.

Bé, A.W.H., Hemleben, C., Anderson, O.R., Spindler, M., Hacunda, J., and Tuntivate-Choy, S. (1977) Laboratory and field observations of living planktonic foraminifera. *Micropaleontology* 29:155–179.

Bé, A.W.H., and Tolderlund, D.S. (1971) Distribution and ecology of living planktonic foraminifera in surface waters of the Atlantic and Indian Oceans, in Funnell, B.M. and Riedel, W.R. (eds.), *The Micropaleontology of Oceans.* Cambridge, England: Cambridge University Press, pp. 105–149.

Blackbourn, D.J., Taylor, F.J.R., and Blackbourn, J. (1973) Foreign organelle retention by ciliates. *J. Protozool.* 20:286–288.

Blank, R.J., and Trench, R.K. (1986) Nomenclature of endosymbiotic dinoflagellates. *Taxonomy* 35:286–294.

Brandt, K. (1881) Ueber das Zusammenleben von Thiere und Algen. *Ver. Physiol. Ges.* Berlin 1881/82, 4/5:570–574.

Burkholder, P.R., Burkholder, L.M., and Almodovar, L.R. (1967) Carbon assimilation of marine flagellate blooms in neritic waters of southern Puerto Rico. *Bull. Mar. Sci. Gulf Carib.* 17:1–15.

Chatton E., and Lwoff, A. (1929) Contribution à l'étude de l'adaption. *Ellobiophyra donacis* Ch. et Lw., péritriche vivant sur les branchies de l'Acéphale *Donax vittatus* da Costa. *Bull. Biol. Belg.* 63:321–349 + pls. 8–10.

Dodge, J.D. (1971) A dinoflagellate with both a mesokaryotic and eukaryotic nucleus. 1. Fine structure of the nuclei. *Protoplasma* 73:145–157.

Dragesco, J. (1948) Sur la biologie du *Zoothamnium pelagicum* (Du Plessis). *Bull. Soc. Zool. Fr.* 73:130–134.

Elbrächter, M. (1984) Functional types of marine planktonic primary producers and their relative significance in the food web. In Fasham, M.J.R. (ed.), *Flows of Energy and Materials in Marine Ecosystems.* New York: Plenum, pp. 199–122.

Erez, J. (1983) Calcification rates, photosynthesis and light in planktonic foraminifera, in Westbroek, P., and De Jongh, E.W. (eds.), *Biomineralization and Biological Metal Accumulation.* Dordrecht, Holland: Reidel, pp. 307–312.

Febvre, J., and Febvre-Chevallier, C. (1979) Ultrastructural study of zooxanthellae of three species of Acantharia (Protozoa: Actinopoda), with details of their taxonomic position in the Prymnesiales (Prymnesiophyceae Hibberd, 1976). *J. Mar. Biol. Assoc. U.K.* 59:215–226 + 7 pls.

Fenchel, T., Perry, T., and Thane, A. (1977) Anaerobiosis and symbiosis with bacteria in free-living ciliates. *J. Protozool.* 24:154–163.

Fenchel, T. (1987) *Ecology of Progozoa.* Berlin: Science Tech Pub./Springer-Verlag, 197 pp.

Fischelson, L., Montgomery, W.L., and Myrberg, A.A., Jr. (1985) A unique symbiosis in the gut of tropical herbivorous surgeonfish (Acanthuridae: Teleosti) from the Red Sea. *Science* 229:49–51.

Gaines, G., and Taylor, F.J.R. (1984) Extracellular digestion in marine dinoflagellates. *J. Plankt. Res.* 6:1057–1061.

Goertz, H.D. (1983). Endonuclear symbionts in ciliates, in Jeon, K.W. (ed.), *Intracellular Symbiosis, Int. Rev. Cytol.,* Suppl. 14:146–176.

Glover, H.E. (1985) The physiology and ecology of the marine cyanobacterial genus *Synechococcus. Adv. Aquat. Microbiol.* 3:49–107.

Goff, L. (1983) *Algal Symbiosis: A Continuum of Interaction Strategies.* Cambridge, England: Cambridge University Press, 216 pp.

Heckman, K. (1983) Endosymbionts of *Euplotes,* in Jeon, K.W. (ed.), *Intracellular Symbiosis, Int. Rev. Cytol.,* Suppl. 14:111–144.

Hemleben, C., and Spindler, M. (1983) Recent advances in research on living foraminifera. *Utrecht Micropaleontol. Bull.* 30:141–170.

Hibberd, D.J. (1977) Observations on the ultrastructure of the cryptomonad endosymbiont of the red water ciliate *Mesodinium rubrum*. *J. Mar. Biol. Assoc. U.K.* 57:45–61.

Hollande, A., and Carré (1974) Les xanthelles des radiolaires sphaerocollides, des acanthaires et de *Vellela vellela*: infrastructure–cytochimie–taxonomie. Protistologica 10:573–601.

Holm-Hansen, O., Taylor, F.J.R., and Barsdate, R.J. (1970) A ciliate red tide at Barrow, Alaska. *Mar. Biol.* 7:37–46.

Jeffrey, S.W., and Vesk, M. (1976) Further evidence for a membrane-bound endosymbiont within the dinoflagellate *Peridinium foliaceum*. *J. Phycol.* 12:450–455.

Jeon K.W. (ed.) (1983) *Intracellular Symbiosis*. *Int. Rev. Cytol.*, Suppl. 14.

Jøergensen, B.B., Erez, J., Revsbech, N.P., and Cohen Y. (1985) Symbiotic photosynthesis in a planktonic foraminiferan, *Globigerinoides sacculifer* (Brady), studied with microelectrodes. *Limnol. Oceanogr.* 30:1253–1267.

Journal of Protozoology (1985) Symposium "Symbiosis in Protozoa" (organisers Lee, J.J., and Corliss, J.O.). *J. Protozool.* 32:371–403.

Khmeleva, N.N. (1967) Rol' radiolayarii pri otsenke pervichnoi produktsii v Krasnon More i Adenskom Zalive. *Dokl. Akad. Nauk SSSR* 172:1430–1433.

Laval, M. (1968) *Zoothamnium pelagicum* Du Plessis, cilié péritrich planctonique: morphologie, croissance et comportement. Protistologica 4:333–363 + 4pls.

Laval, M. (1970) Présence de bactéries intranucléaires chez *Zoothamniuim pelagicum*. Abstr. 7th. Int. Congr. Micr. Electr. Grenoble 10970:403–404.

Laval, M. (1972) Ultrastructure de *Petalotricha ampulla* (Fol). Comparaison avec d'autres Tintinnides et avec les autres orders de Ciliés. Protistologica 8:369–386.

Laval-Peuto, M. (1975) Cortex, perilemme et reticulum vesiculeux de *Cyttarocylis brandti* (Cilié Tintinnide). Les Ciliés à perilemme. Protistologica 11:83–98.

Laval-Peuto, M., and Febvre, M. (1986) On plastid symbiosis in *Tontonia appendiculariformis* (Ciliophora Oligotrichina). Biosystems 19:137–158.

Laval-Peuto, M., Salvano, P., Gayol, P., and Greuet, C. (1986) Mixotrophy in marine planktonic ciliates: ultrastructural study of *Tontonia appendiculariformis* (Ciliophora, Oligotrichina). Mar. Microb. Food Webs 1:81–104.

Lee, J.J. (1980) Nutrition and physiology of the Foraminifera. Chapter 2 in Levandowsky M., and Hutner, S.H. (eds.), *Biochemistry and Physiology of Protozoa,* Vol. 3. Orlando, Fla.: Academic Press, pp. 43–66.

Lee, J.J. (1983) Perspective on algal symbionts in larger foraminifera. *Int. Rev. Cytol.* Suppl. 14:49–77.

Lee, J.J., and Hallock, P. (1987) Algal symbiosis as the driving force in the evolution of larger Foraminifera. *Ann. N.Y. Acad. Sci.* 503:330–347.

Lee, J.J., Lee, M.J., and Weis D.S. (1985a) Possible adaptive value of endosymbionts to their protozoan hosts. *J. Protozool.* 32:380–382.

Lee, J.J., and McEnery, M.E. (1983) Symbiosis in foraminifera. In Goff, L.J. (ed.) *Algal Symbiosis—A Continuum of Interaction Strategies*. Cambridge, England: Cambridge University Press, pp. 37–68.

Lee, J.J., Soldo, A.T., Reisser, W., Lee, M.J., Jeon, K.W., and Goertz, H.-D. (1985b) The extent of algal and bacterial endosymbioses in protozoa. *J. Protozool.* 32:391–403.

Li, J.-Y. Qiao, Y.-J., Chen, X.-C., and Zhang, H.-G. (1979) On the two nuclei of *Gymnodinium cyaneum*. *Kexue Tongbao* 24:461–462 (in Chinese).

Lindholm, T. (1985). *Mesodinium rubrum*—a unique photosynthetic ciliate. *Adv. Aquat. Microbiol.* 3:1–48.

Lopez, E. (1979) Algal chloroplasts in the protoplasm of three species of benthic foraminifera: traxonomic affinity, viability and persistance. *Mar. Biol.* 53:201–211.

Lucas, I.A.N. (1982). Observations on *Noctiluca scintillans* Macartney (Ehrenb.) (Dinophyceae) with notes on an intracellular bacterium. *J. Plankton Res.* 4:401–409.

McManus, G.B., and Fuhrman, J.A. (1986) Photosynthetic pigments in the ciliate *Laboea strobilia* from Long Island Sound. *J. Plankton Res.* 8, 317–327.

Muller, W.A., and Lee, J.J. (1969) Apparent indispensability of bacteria in Foraminifera nutrition. *J. Protozool.* 16:471–478.

Norris, R.E. (1967) Algal consortisms in marine plankton. In Krishnamurthy, V. (ed.), *Proceedings Seminar on Sea, Salt and Plants,* 1965, Central Salt and Mar. Chemicals Res. Inst., Bhavnagaer, India, pp. 178–189 + 1 pl.

Oakley, B.R., and Taylor, F.J.R. (1978) Evidence for a new type of endosymbiotic organization in a population of the ciliate *Mesodinium rubrum* from British Columbia. *BioSystems* 10:361–369.

Pavillard, J. (1916) Flagellés nouveaux, épiphytes des Diatomées pélagiques. *C.R. Acad. Sci.* (Paris) 163:65–68.

Porter, K.R., Sherr, E.B., Sherr, B.W., Pace, M., and Sanders, R.W. (1985). Protozoa in planktonic food webs. *J. Protozool.* 32:409–415.

Raikov, I.B. (1974) Étude ultrastructurale des bacteriés épizoiques de *Kentrophoros latum* Raikov, cilié holotriche mesopsammique. *Cah. Biol. Mar.* 15:379–393.

Rosati, G., and Verni, F. (1977) Bacterial-like endosymbionts in *Euplotes crassus*. Abstr. Vth. Int. Congr. Protozool, New York, p. 443.

Rosati, G., Verni, F., and Luporini, P. (1976) Cytoplasmic bacteria-like symbionts in *Euplotes crassus* (Ciliate Hypotrichida). *Monit. Zool. Ital.* (NS) 10:449–460.

Schwemmler, W. (1984) *Reconstruction of Cell Evolution. A Periodic System*. Boca Raton, Fla.: CRC Press.

Sieburth, J.M. (1979) *Sea Microbes*. New York: Oxford University Press, 491 pp.

Silva, E. de S., and Franca S. (1985) The association dinoflagellate-bacteria: their ultrastructural relationship in two species of dinoflagellates. *Protistologia* 21:429–446.

Silver, M. (1983) Symbiotic protists and complex photosynthetic associations of the oligotrophic North Pacific Ocean. Abstr. AGU/ASLO Ocean Sciences meeting, New Orleans, Jan. 1984.

Smith, D.C., and Douglas, A.E. (1986) *The Biology of Symbiosis*. London: Arnold, 302 pp.

Smith, J.W., and Barber, R.T. (1979) A carbon budget for the autotrophic ciliate *Mesodinium rubrum*. *J. Phycol*. 15:27–33.

Soldo, A.T. (1983) The biology of the Xenosome, an intracellular symbiont. In Jeon, K.W. (ed.), *Intracellular Symbiosis, Int. Rev. Cytol.*, Suppl. 14:79–109.

Spero, H.J., and Parker, S.L. (1985) Photosynthesis in the symbiotic planktonic foraminifer *Orbulina universa*, and its potential contribution to oceanic primary productivity. *J. Foram. Res*. 15:273–281.

Spindler, M., and Hemleben, C. (1980) Symbionts in planktonic foraminifera (Protozoa). In Schwemmler, W., and Schenk, H.E.A. (eds.), *Endocytobiology*, Berlin: De Gruyter, pp. 133–140.

Stoecker, D.K., Michaels, A.E., and Davis, L.H. (1987) Large proportion of marine planktonic ciliates found to contain functional chloroplasts. *Nature* (Lond.) 326:790–792.

Stoecker, D.K., and Silver, M.W. (1987) Chloroplast retention by marine planktonic ciliates. *Ann. N.Y. Acad. Sci*. (in press).

Swanberg, N.R. (183) The trophic role of colonial radiolaria in oligotrophic oceanic environments. *Limnol. Oceanogr.* 28:655–666.

Swanberg, N.R., and Harbison, G.R. (1980) The ecology of *Collozoum longiforme* sp. nov., a new colonial radiolarian from the equatorial Atlantic Ocean. *Deep-Sea Res*. 27:715–732.

Taylor, D.L. (1974) Symbiotic marine algae: taxonomy and biological fitness. In Vernberg, W. (ed.), *Symbiosis in the Sea*. Columbia: University of South Carolina Press, pp. 245–262.

Taylor, F.J.R. (1973) Application of the scanning electron microscope to the study of tropical microplankton. *J. Mar. Biol. Assoc. India*, 14:55–60.

Taylor, F.J.R. (1974) Implications and extensions of the Serial Endosymbiosis Theory of the origin of eukaryotes. *Taxonomy* 23:229–258.

Taylor, F.J.R. (1979) Symbionticism revisited: a discussion of the evolutionary impact of intracellular symbiosis. *Proc. Roy. Soc. B*. 204:267–286.

Taylor, F.J.R. (1980) On dinoflagellate evolution. BioSystems 13:65–108.

Taylor, F.J.R. (1982) Symbioses in marine microplankton. *Ann. Inst. Oceanogr.* (Paris), N.S., Suppl. 58:61–90.

Taylor, F.J.R. (1983) Some eco-evolutionary aspects of intracellular symbiosis. In Jeon, K.W. (ed.), *Intracellular Symbiosis, Int. Rev. Cytol.* Suppl. 14:1–28.

Taylor, F.J.R., Blackbourn, D.J., and Blackbourn, J. (1969)

Taylor, F.J.R., Blackbourn, D.J., and Blackbourn, J. (1971) The red water ciliate *Mesodinium rubrum* and its "incomplete symbionts": a review including new ultrastructural observations. *J. Fish. Res. Bd. Canada* 28:391–407.

Taylor, F.J.R., and Harrison, P.G. (1983) . . . . . In Schenk, H.E.A., and Schwemmler, W. (eds.) *Endocytobiology* vol. II, Berlin: De Gruyter, pp. 827–842.

Tomas, R.N., Cox, E.R., and Steidinger, K.A. (1973) *Peridinium balticum* (Levander) Lemmermann, an unusual dinoflagellate with a mesokaryotic and an eukaryotic nucleus. *J. Phycol*. 9:91–98.

Tregouboff, G. (1953) Classe des Acanthaires. In Grassé, P.-P. (ed.), *Traité de Zoologie,* Vol. 1(2). Paris: Masson, pp. 271–320.

Trench, R.K. (1975) Of "leaves that crawl": functional chloroplasts in animal cells. Symp. Soc. Exper. Biol. (Symbiosis) 29:229–265.

Wilcox, L.W., and Wedemayer, G.J. (1984). *Gymnodinium acidotum* Nygaard (Pyrrhophyta), a dinoflagellate with an endosymbiotic cryptomonad. *J. Phycol.* 20:236–242.

Wilcox, L.W., and Wedemayer, G.J. (1985). Dinoflagellate with blue-green chloroplasts derived from an endosymbiotic eukaryote. *Science* 227:192–194.

Withers, N., and Haxo, F.T. (1975). Chlorophyll c^1 and c^2 and extraplastidic carotenoids in the dinoflagellate, *Peridinium foliaceum* Stein. *Plant Sci. Letters* 5:7–15.

Zhang, X., Liu, Q., Wang, H., and Li, S. (1982) Preliminary separation and characteristics of phycocyanin in blue-green *Gymnodinium*. Kexue Tongbao 27:1000–1003.

A Glossary of Terms Pertaining to the Ecology of Marine Protozoa

ARTHUR J. REPAK

AABW: Antarctic Bottom Water.

abdomen: segment of radiolarian skeleton usually deposited last following the thorax.

abyssal: pertaining to deep water (>1,000 m).

acantharian: an actinopod ameba with skeletons composed of strontium sulfate, some species with 20 radial spines joined at the center.

acanthodesmiids: nassellarian radiolaria with D-shaped ring or latticed, bilobed chamber with internal D-shaped sagittal ring.

acarpomyxea: a class of ameba.

acathodesmins: D-shaped ringed nassellarian radiolaria with symbiotic algae; subdivision of acanthodesmiids.

accreted terranes: a geologic formation produced by the addition of material onto the surfaces, as in the growth of crystals and other solid bodies.

acetate flagellate: phytoflagellates capable of utilizing acetate, fatty acids, and alcohols as a carbon source; including euglenids, cryptomonads, and volvocids.

actinommids: ray-spined spumellarian radiolaria.

actinomonads: a heliozoan of the taxonomic order Ciliophryida.

actinopod: marine amebae characterized by radially stiffened axopods containing an internal shaft composed of cross-linked bundles of microtubules and filopodia. Includes radiolaria, acantharia, and heliozoa.

adoral zone of membranelles: found in heterotrichous ciliates; consists of numerous adoral membranelles arranged in parallel series at the left side of buccal apparatus.

advection: large-scale water transport processes in the sea.

agamont: diploid product of asexual generation in foraminifera and other protozoa.

agglutinated test: walls entirely composed of foreign particles added during construction.

agnotobiotic: not free of other fauna or flora in a particular host.

albaillellids: RIS subdivision; skeleton consisting of three spines in a plane forming a triangle; riblike elements may protrude from the spines, giving the skeleton a jointed appearance.

algalivorous: eater of algae; does not pertain to nutritional value of food organisms.

allogromiid: an order of foraminifera that is characterized by a smooth, flexible, membranous test with and without agglutinate material.

allopatric speciation: speciation by branching and by geographic and reproductive isolation.

aloricate: "naked," without a lorica.

alveolinid: benthic foraminifera.

amoeboflagellate: lobopodial amoeba capable of forming a flagellate stage in its life cycle (e.g., *Naegleria*).

amebulae: cysts of amebae or sponges.

amino acid racemization: process of forming a mixture of stereoisomers from a pure amino acid stereoisomer.

amphileptids: ciliates characterized by an subapical oral area and a laterally compressed body that is ciliated on both sides (e.g., order Pleurostomatida).

amphipod: a member of the order Amphipoda, class Crustacea.

amphipyndacids: nassellarian radiolaria with small, spherical, poreless cephali with several postcephalic joints.

amphisteginids: benthic foraminifera with chambers trochoid to asymmetrically lenticular. Dominant in neritic waters.

anadromous salmonids: salmon that spawn in fresh water and migrate to open marine waters to live as adults.

anisogamous: producing flagellated gametes that look identical but differ in overall size.

antarctisins: organisms found in the antarctic.

anticlinal: geologic beds inclined in opposite directions.

anticyclonic gyres: clockwise flow of water; may refer to the ocean's gyral systems (e.g., the Sargasso Sea).

aperture: opening larger than pores, as in radiolaria and foraminifera.

apochlorotic microflagellates: 20–200-μm-sized colorless flagellates.

apogamic: production of successive asexual generations.

appendicular houses: a gelatinous, often barrel-shaped housing created by appendicularian tunicates.

aragonite: calcium carbonate differing from calcite in crystallization and with a higher specific gravity.

archaeocyathid: pleosponges; extinct marine organisms that lack any known anatomic equivalents, either fossil or living.

areal: denoting unoccupied ground.

arenaceous: sandy; also pertains to type of protist shell composed of nonbiogenic (primarily mineral) particles (e.g., tintinnid lorica).

artiscins: loaf-shaped spumellarian radiolaria with an outer latticed shell, often with equilateral constriction and caps or spongy columns.

artostrobid: nassellarian radiolarian with cephalis consisting of a eucephalic lobe with other lobes and tubes.

astropyle: a principal opening in the capsule of phaeodarian radiolaria.

astrosphaerid: spumellarian radiolarian with latticed shells containing numerous spines.

aufwachs: periphyton community; organisms living on submerged substrates.

autocyclic: self-cycling.

autoecology: the study of the interrelations of individual organisms and their environment; often experimental and inductive.

autogamous: self-fertilizing.

autophagous: self-digestion of cellular components.

axenic: denoting growth in "pure" culture.

axoneme: an arrangement of microtubules in a nine doublets and two central singlets found in cilia and flagella.

axoplast: intracapsular organelle from which axopodia originate in radiolaria.

axopodia: straight, slender pseudopodia extending radially from the surface of the body (e.g., heliozoan, radiolaria, etc.) and possessing an axial rod.

bacterioplankton: suspended bacteria nurtured by soluble organic matter released by metabolizing and decaying organisms throughout the water column.

bacterivore: ingester of bacteria from a given environment; no indication of nutrient value.

bar (beam): rodlike cross connection between skeletal structures of a radiolarian.

bathyal: area from the edge of the continental shelf to 5,000 m in depth.

bathypelagic zone: area of the pelagic zone of the ocean below zone of light penetration between 200 and 4,000 m.

BEETS: body-equivalent excretion time.

beloids: spumellarian radiolaria with skeletons of scattered spicules.

bicoecid: colorless flagellates usually with lorica and heterkont flagellation; similar to choanoflagellates.

biochronology: study of the development of species over large amounts of time.

bioenergetics: study of the flow and transformation of energy that occurs in living forms.

biofacies: the general appearance or character of a geologic formation in a given region that is distinguished by its fossil deposits.

biogeography: study of the abundance and geographic distribution of species of organisms.

biological species concept: According to Ernst Mayr, biological species are "groups of actually or potentially interbreeding natural populations which are reproductively isolated from other groups" or "groups of populations the gene exchange between which is limited or prevented by Nature by one, or by a combination of several, reproductive isolating mechanisms" (Theodsius Dobzhansky).

biostratigraphy: stratigraphic correlations with fossils.

biserial: two columns of foraminiferan chambers added on alternate sides of a test.

bodonids: colorless flagellates similar to *Bodo*.

bolide: a large, brilliant meteor.

bolide-impact hypothesis: asteroidal theory put forth by Luis and Walter Alvarez to explain extinction during the Cretaceous period.

boreal: pertaining to the north wind.

boundary currents: currents that flow in a north–south direction from areas of accumulation to areas of removal.

bryozoa: colonial aquatic invertebrates resembling the hydroid forms of coelenterates (now known as Entoprocta).

bubble capsule: a frothy outer area on certain planktonic foraminifera to which large coccoid dinoflagellates adhere (e.g., *Hastigerina pelagica*).

buliminids: foraminiferal members of the family Buliminidae with chambers of test in a high trochospiral, three chambers per whorl; aperture loop-shaped.

bulk metabolite: provide sources of energy and supply general metabolites needed for synthesis, growth, and maintenance.

bulloid gastropod: mollusc member of the subclass Opisthobranchia, order Cephaiaspidea; shell reduced or absent.

by-spines: external projections not supported by bars in radiolaria.

calanoid copepod: a near microscopic crustacean member of the order Calanoidea.

calcareous hyaline perforate wall type: foraminiferan test walls composed of calcite arranged either in radial pattern or with the *c*-axis oblique to the wall surface and with pores.

calcarinids: benthic foraminifera.

calcium compensation depth (CCD): deepest limit of accumulation of sedimentary carbonate as determined by solubility factors.

calymma: the highly alveolated ectoplasm of a radiolarian; responsible for buoyancy and for food capture and digestion.

cannobotryids: nassellarian radiolaria with cephalis of eucephalic lobe and other lobes that may extend as tubes.

capping: a fluorescent labeling technique to detect antibody-binding, plasma membrane intrinsic proteins of the plasma membrane. After disruption of microtubules associated with the cytoplasmic surface of the membrane, the intrinsic proteins of the plasma membrane form a cluster or cap on the cell surface.

caprellid: a benthic worm.

carbonate lysocline (CL): level at which the percentage of carbonate of sediments decreases abruptly due to dissolution of the carbonate by pressure effects.

carnivory: (meat-eating) removal of multicellular life forms from an environment. This term has been adopted by microbial ecologists to describe the feeding behavior among protozoa.

carpocanids: nassellarian radiolaria with small cephali recessed into the thorax; thorax usually elongate and possessing longitudinally arranged pores.

cenodiscids: spumellarian radiolaria with a single latticed shell without spines, less interpore area, thick skeletons, and no tubes.

CENOP: cenozoic paleoceanography project.

central capsule: a central cytoplasmic mass surrounded by a porous wall containing endoplasm of a radiolarian responsible for reproductive and storage functions.

central waters: a water mass in the center of a gyre.

centrohelid: a taxonomic order of heliozoans usually with one central bulky microtubular organizing center.

cephalis: first radiolarian skeletal segment.

chaetognath: arrow worm belonging to the order Chaetognatha.

chamberlets: the smaller embryonic chambers found in forams.

chaunacanthid: a taxonomic order of acantharean actinopods with 20 radial spines quadricarenated and connected at their bases.

choanoflagellates: heterotrophic collared flagellates.

chromatophores: a colored plastid.

chroococcoid cyanobacteria: spherical, marine planktonic cyanobacteria, either single or in microcolonies.

chrysochromulinid: autotrophic members of the order Prymnesiida characterized by the possession of a coiled haptonema 18 × cell length and flagella one and a half to three times as long as the cell.

chrysomonid: an autotrophic biflagellated phytomastigophorid, member of the order Chyrsomonadida.

ciliature: refers to the characteristic pattern of somatic and/or buccal cilia in ciliates.

cingulum: transverse groove of a dinoflagellate.

Circean effect: the chemical attraction of potential cytobionts or prey to their hosts.

circling disease: whirling disease caused by a myxosporid, *Myxosoma cerebralis* in salmonid fish.

circumpolar deep water: water circulating in the vicinity of the polar regions at deep water depths.

cirrus: a conical "foot" structure composed of numerous compound, fused cilia and found on the ventral surface of hypotrichous ciliates.

cladistics: a method of classification based on shared characteristics indicating an evolution stemming from a branching of a common ancestral group.

cladoceran: pertaining to crustaceans belonging to the order Cladocera, commonly called water fleas.

cladogram: a branching diagram generated from a cladistic approach to systematics.

clastic: breaking up into fragments.

clastic sediments: sediments composed of fragments of rock; types are conglomerate, sandstone, and shale.

CLIMAP: long-range climate investigation mapping and planning program: a program to determine paleoclimates based on sedimentary fossil evidence.

cline: continuous graduation of form differences in a population of a species, correlated with geographic or ecological distribution.

cluster analysis: methods of grouping variables according to magnitudes and interrelationships among their correlation coefficients.

coccidia: a subclass of parasitic protozoa characterized by triphasic life histories with asexual reproduction occurring with epithelial or circulatory cells.

coccodiscids: radiolaria with lens-shaped shells containing a latticed center and spongy chambered girdle or arms.

coccolithophorid: phytomastigophorids covered with coccoliths, microscopic-sized organic scales on which calcium carbonate is crystalized as calcite or aragonite.

coiling ratios: a phenotypical comparison of the number of tests of individuals in a given population of planktonic foraminifera coiling in one direction (e.g., right-hand coil) as opposed to another.

collosphaerids: spumellarian radiolaria with spherical central capsules, including colonial radiolaria with symbiotic algae.

colpodid ciliate: a holotrichous ciliate with two nearly equal-sized oral polykinetids; body often reniform. Taxonomically placed in the class Colpodea, order Colpodidae.

commensal: members of different species living symbiotically, where one species receives benefit and the other is unaffected.

compensation light intensity: refers to the light intensity at the compensation depth where light level equals approximately 1 percent of the surface light intensity.

conoid: cone-shaped structure found on the anterior portion of a apicomplexan merozoite, formed by several spirally coiled electron-dense microtubules inside the polar ring.

continental shelf: the seabed underwater extension adjacent to the coast to the continental slope (≈200 m isobath).

continental slope: boundary between continental mass and true ocean basin, ranging from about 200 to 3,000–4,000 m.

contour currents: an ocean current following along isopycnal lines approximately parallel to the bathymetric contours.

coralline: calcareous forms (e.g., red algae).

coronal cilia: cytostome-bearing cilia, as found in haptorid gymnostome ciliates.

cortical shell: outer portion of a multishelled or single-shelled spumellarian radiolarian.

corythoecid: filosid amoeba housed in a vaselike test composed of scattered scales.

costae: silica rods found in the lorica of choanoflagellates; ribs.

crinoids: a class of echinoderms commonly called sea-lilies or feather stars.

cryptomonad: biflagellate asymmetric cells flattened dorsoventrally and bounded by a periplast; cells may be red, blue, olive-yellow, brown, or green. Color instability is a taxonomic trait.

cryptophorine gymnostome: a ciliate that possesses a cytopharyngeal basket composed of longitudinal rods composed of microtubular fibrils and filaments.

cryptophyte: a member of the Cryptophyceae.

cubosphaerids: spumellarian radiolaria with latticed shells with polar and equatorial spines.

cumacea: an order of crustacea that are marine sand or mud, littoral dwelling peracaridans of approximately the same size as mysids.

current-regime: currently in existence.

cyanobactericae: formerly called blue-green algae.

cyberons: units of energy saved in EG formulas; see also EG.

cyclopoid: members of the crustacean order Cyclopoidea (e.g., *Cyclops*).

cyrtentactins: nassellarian radiolaria which looks like a cross between entactinids and lophophaenins.

cyrtophorid: a ciliate classified in the subphylum Cyrtophora, order Cyrtophorida; characteristics include distinct ciliature that cover most of the ventrum; at least one kinetal fragment displaced to dorsum; oral region usually behind preoral suture.

cyrtostype: having a cytopharyngeal apparatus similar to those of a hypostome ciliate.

cytobiont: intracellular symbionts.

cytocosms: tightly cycling mini-ecosystems.

cytokinesis: division of the cytoplasm following nuclear division in mitosis or meiosis.

cytopharyngeal apparatus: a nonciliated tubular passageway in ciliates leading from the cytostome into the cytoplasm. The structure is particularly strengthened.

cytopharynx: a nonciliated tubular passageway in ciliates leading from the cytostome into the cytoplasm, without strengthened walls; found in some flagellates and ciliates.

cytoproct: a permanent slitlike opening in the posterior of most ciliates for discharge of egesta.

cytopyge: a site of discharge of egesta in ciliates which is much less organized than the cytoproct.

cytoskeleton: any secreted inorganic or organic material, or various microfibrillar or microtubular structures found in, on, or below the surface of a cell.

cytostome: a permanent oral opening through which food material may pass into a food vacuole in the endoplasm of a cell.

cytotomy: asexual reproduction as in foraminifera.

datums: first or last occurrences.

DCMU 3-[3,4-dichlorophenyl]-1, 1-dimethylurea): a herbicide that acts to stop oxygen evolution and photophosphorylation; also called Diuron.

death assemblage: faunal assemblages of extinct protista useful in paleoenvironmental analyses.

decapod: a member of the largest order of Crustacea, order Decopoda, which includes shrimp, crabs, lobsters, and crayfish.

deme: a group of organisms in the same taxon.

detrivore: eaters of detritus.

deuteroconch: second chamber of a foraminiferan.

diachronous: (time-transgressive) referring to a rock unit that is of varying age in different areas or that cuts across some planes or biostratigraphic zones.

dictyocorins: spongodiscid spumellarian radiolaria with spongy siliceous discs containing three arms.

diel thermal layer convection: a thermal layer cycle that follows a 24-hour process.

diel vertical migration patterns: vertical migration patterns linked to the light–dark cycles.

dinoflagellate: a biflagellated unicellular protist characterized by being mesokaryotic, and often photoautotrophic. The two flagella are typically each located in a transverse groove encircling the cell and in a longitudinally oriented groove, the sulcus.

dinokaryon: cytobiont of cell nucleus of a dinoflagellate.

dinophysoid dinoflagellate: a member of the family Dinophysidae with a plated theca, body laterally compressed; a sawtooth suture divides the body; and plates bordering the cingulum and sulcus are expanded as wings.

discoidal: pertaining to a disc.

disequilibrium: refers to equilibration after one generation of random mating of the dissimilarity of autosomal allelic frequencies in the sexes, but genotypic frequencies will not become equilibrated until the second generation of random mating.

doliolid: member of the class Thaliacea, subphylum Urochordata, that includes the solitary North Atlantic genus *Doliolum*.

DOM: dissolved organic matter.

DOP: dissolved organic phosphorus.

downwelling: the downward advective transport of water; inverse of upwelling.

drift currents: surface current.

DSDP: Deep Sea Drilling Project.

dysteriid: a member of the Phyllopharyngian ciliate family Dysteriidae characterized by the possession of a flexible podite; body may be laterally compressed. The left ventral body kineties are broken into preoral and midventral postoral fields.

eastern boundary currents: flow from temperate areas to equatorial zones along the western coast of continents or eastern ocean basin boundaries.

echiuroids: marine worms similar to sipunculids in size and general habitat.

ecophenotypic polymorphism: external variations in form of a given population of organisms adapted to a given environment.

ecophenotypic variation: variation in the external features of a given group of organisms adapted to a given habitat.

ecophysiology: a study of the physiology of a given group of organisms adapted to living in a given environment.

ectobiont: symbiont living attached to or associated with external surface of a host.

ectocommensals: a commensal on the outer surface of another organism.

ectosymbiont: ectobiont

ectosymbiosis: symbiosis occuring on or outside of host's body.

Ee: ecological efficiency; P/I; where P = production and I = ingestion.

EG (energy gain): $F_i - C_i/R_i - R_i\ (k)$; where R_i = total respiration growing on food i/unit time; k = percent of R_i used for maintenance; F_i = number of food organisms of type i ingested/unit time; C_i = caloric value of single type i food organism.

ejectosome: extrusive organelles associated with defense.

El Niño: in northern winter, the easterly flowing countercurrent, to the Peru Current, that turns southward and close along the coast of Equador.

elasmobranchs: synonymous with chondrichthyes.

endobiont: symbiont living within a host.

endocytosis: process of taking in food material by engulfment within the food vacuoles formed by invaginations of the plasma membrane or through an existing buccal apparatus.

endomitosis: division of the nucleus without division of the cell, resulting in polyploidy.

endosome: single conspicuous RNA-containing body not directly connected to any chromosome that persists and divides during cell division.

endozoic: residing inside a host.

energy gain (EG): see EG.

entactiniids: spumellarian radiolaria with a spherical skeleton with latticed wall structure; bars running from center of skeleton.

epiphytes: plants living attached to plants.

epibacteria: attached to particles suspended in water, or living on inert particles settled in the bottom sediments.

epibenthic: living attached to the ocean floor.

epibiont: external, surface dwelling organism.

epibiotic: referring to an organism on a living surface.

epicontinental: found or located in or on a continent.

epifaunal: referring to animals living on the surface of sand or sedimentary bottoms.

epifluorescence: an ultraviolet fluorescence technique useful in counting living microorganisms; blue light excitation inducing pigment or specific stain fluorescence.

epimerite: anterior-most portion of a separate gregarine separated from protomerite by a septum; serves as a holdfast organelle.

epipelagic: upper region of the pelagic zone.

epiphytic: living as an epiphyte.

episodic deposition: incidental deposition.

epizoic: ectosymbiont on animals.

epsilon: bacterialike cytobiont that causes death of noninfected sensitive strains of *Euplotes minuta* and *E. crassus*.

erathem: time-stratification unit.

Erdschreiber medium: a nonspecific culture medium, for growing marine, algae containing soil extract.

essential metabolite: metabolites that the organism cannot synthesize and that must be included in the diet.

eta: a bacterialike cytobiont killer found in *Euplotes crassus*.

eucrytidin: a nassellarian radiolarian with more than one postcephalic chamber and nonquadrangular pores. Subdivision of theoperids.

euglenid: an anteriobiflagellate, green or colorless, plastic phytoflagellate with stigma; helically symmetrical; no cell wall but with a proteinaceous pellicle.

euhaline: describing waters containing between 30 and 40 parts per thousand of dissolved salts (i.e., in most cases, normal seawater).

euhedal: an organism found at various parts in deep sea trenches.

eurybathyal: referring to an organism found in a wide range of depths of water.

eurytopic: referring to an organism found in a wide variety of environments.

eustigmaphyte: a member of a new algal class separated from the Xanthophyceae; characterized by motile cells with a stigma at the anterior end independent of the single chloroplast; one emergent anterior flagellum with a bilateral array of mastigonemes.

eutrophic: describing a body of water with an abundant supply of nutrients and a high rate of formation of organic matter by photosynthesis.

exocytosis: the process of expelling undigested contents from a processed food vacuole to the outside environment.

extracapsulum: a layer of cytoplasm outside of the central capsule of radiolaria.

extrusome: membrane-bound extrusible bodies of diverse functions (e.g., feeding, defense, etc.) found in numerous protozoa.

false benthos: a hypothesized layer of accumulated particulate matter—living, detrital, and inert—that gravitates to density discontinuity layers, dependent on the respective rate of flux of particles from one layer to the next.

faunal turnover: changes in animal population numbers and species composition.

feeding selectivity: the ability of a given protist either to actively select food organisms in the act of feeding or to devour a greater percentage of one group as opposed to others.

feeding veil: a delicate protoplasmic cone protruding from the upper sulcal region and acting as a means of prey entrapment in some dinoflagellates.

feeding cyst: temporary state found in forams resulting from an excessive food in the granuloreticulopodia. The organism weaves, accumulating the pseudopodial network about itself, which is filled with trapped food.

feeding net: pseudopodial networking for food capture in various ameboid forms.

feet: radiolarian structures that may or may not be

latticed; possess a rib and may be terminal or subterminal.

fileose amoeba: member of the class Filosea possessing long, slender, clear to faintly granular nonanastomosing pseudopods used for feeding.

flexostyle: the tubular enrolled chamber of a foraminiferan test immediately following the proloculus.

flysch: an association of certain types of marine sedimentary rocks characteristic of deposition in a foredeep.

folliculinid ciliates: loricate, epipzoic or epiphytic, large ciliates with conspicuous winglike extensions bearing left serial oral polykinetids and holotrichous in body ciliature.

foraminifera: amoeboid marine protozoa members of the class Granuloreticulosea, order Foraminiferida; conspicuous by virtue of their distinctive tests and granuloreticulose pseudopodia; characteristically, the life cycle of benthic species involves alternation of generations.

forma tepida: warm-water form.

forma typica: typical form.

frustule: outer siliceous shell of a diatom.

fusiform: spindle-shaped.

fusules: specialized pores found in radiolarian central capsule pores through which axopods pass peripherally into the external environment.

fusulinids: an extinct group of forams containing microgranular calcite in their test.

fusulinine: relating to fusulinids.

gametocytotomont: a multinucleated stage in the nonclassical gamogony portion of the life cycle of the foraminiferan *Allogromia laticollaris,* resulting in internal gametes that fuse to form zygotes leading to G1 agamont cell.

gametogamy: union of gametes (syngamy) in phytomonads and forams.

gamont: sexual aspect of the life cycle of forams in which the organisms are haploid and uninucleate.

gamontogamy: union of foraminiferan gamonts and dinoflagellate "gametes."

gastrotrich: a microscopic, bilateral, pseudocoelomatous metazoan with an unsegmented cuticle covered with spines or scales.

geosyncline: a linear belt of subsidence on a continent, in which great thicknesses of sediments have accumulated.

geronts: an "old world" organism; referring to "old age."

GGE: gross growth efficiency.

giantism: (gigantism); pertaining to the formation of giants sometimes two to three times the original protozoan trophozoite in size and organelle number.

girdle: latticed structure in radiolaria; the cingulum of a thecated dinoflagellate; also, pertaining to certain wreaths of ciliature in certain ciliates.

girdle list: a high membranous ridge perpendicular to the wall and bordering the girdle of a dinoflagellate.

glacioeustatic: pertaining to the worldwide change in sea level produced by the successive withdrawal and melting of ice sheets.

glycocalyx: mucopolysaccaride coat external to the cell membrane.

gnotobiotic: a germ-free animal into which a defined microflora is introduced.

Gondwanaland: according to the theory of drifting continents, one of the two segments of the original protocontinent.

granulorecticulopodia: delicate, finely granular or hyaline pseudopodia, usually characteristic of forams.

green water: a green-colored marine bloom occurring in the northern Indian Ocean and southeast Asian waters, associated with *Noctiluca scintillans* infected with the prasinomonad flagellate *Pedinomonas noctilucae,* located in the buoyancy vacuole.

gregarines: parasitic protozoa classified among the Apicomplexa; some aseptate, others septate; complex life cycles.

gregarinid: usually referring to a member of the largest gregarine family (Gregarinidae).

gross growth efficiency: (GGE) growth–ingestion as a measure of energy transfer between a respective producer and consumer.

guild: a group of species having similar niches and performing similar ecological roles.

gymnospore: (heliospore): rosettelike formation of gamonts in the gregarine family Porosporidae.

gymnostome: (naked mouth): an older classification of holotrichous ciliates.

gyre: closed circles oceanic circulation of water masses under the influence of zonal wind patterns.

hadal: denoting the benthic region that occurs in the great oceanic trenches with depths up to 10,000 m.

hagiastrids: spumellarian radiolaria with a flat disk with spongy rectangular mesh and two to four radial arms with spines.

haplosporidan: an older classification of protozoan parasites of invertebrates and fish characterized by the production of "simple spores" and ameboid sporonts.

haptocysts: extrusive organelles used to capture prey by adhesion; found in tentacles of suctorian ciliates.

haptophyte: an alga with all the characteristics of members classified among the Prymnesiophycea. Motile cells bearing two equal, subequal, or unequal acronematic flagella. In some cells, haptonema arise close to the pair of flagella.

haptorid: a ciliate with cytopharynx not permanently inverted (order Haptorida); oral dome supported by microtubular sets along which toxicysts occur.

harpacticoid: a member of an order of largely benthic copepods; some parasitic.

helioflagellate: a protozoan that possesses long, thin, immotile or slowly motile, filamentous pseudopodia as well as a flagellum.

heliozoa: planktonic, primarily freshwater actinopodial "sun-shaped" protists, characterized by the possession of axopodia.

herbivory: eating of metaphytan life forms; has been used to describe those protozoa that derived nourishment from ingesting algae.

heterochronous evolution: pertaining to a genetic shift in timing of the development of a tissue or anatomical part, on in the onset of a physiological process, relative to an ancestor.

heterokaryosis: having more than one kind of nucleus (e.g., ciliates and foraminiferans).

heterokont: a biflagellate phytoflagellate with one smooth flagellum and the other covered with mastigonemes.

heterophasic: pertaining to the foraminiferan life cycle of alternation of generations.

heteropod: a class of pelagic sea snails.

heterotrichs: a group of ciliates belonging to the order Heterotrichida.

hispid: an anterior long flagellum covered with mastigonemes.

histophagous: relating to an organism that ingests dead tissue of dying or mechanically damaged animals.

holes: larcoid with gates in radiolaria.

holidic: chemically defined.

holocanthid: a member of an order of Acantharea characterized by possession of 10 diametric spines simply crossed.

holophyrid: rhadophorean ciliates with apical oral areas and cytostomes not proceeded by invagination; lined by body kineties.

holoplankton: zooplankton living their entire life cycle as plankton in the pelagic environment.

holothurians: an order of Echinodermata commonly called sea cucumbers.

holotrichs: ciliates whose entire body surface is usually uniformly ciliated.

holozoic: ingestion of whole organisms.

hydroid: coelenterates with tentacles and mouth at the apical end; sometimes sessile.

hydrothermal vent: submarine vents where volcanic outgassing and related geologic activity occurs; heated by volcanic activity found in both the Pacific and Atlantic Ocean sea floor spreading centers.

hymenostome: ciliates with a subterminal, ventrally located buccal cavity composed of an undulating membrane to the right of the buccal cavity and an AZM to the left.

hyperiid: an amphipodal crustacean with the characteristics of the suborder Hyperiidea, which are entirely marine and pelagic.

hypersaline: denoting very salty water, with salinity values exceeding 32 o/oo.

hyponeustonic: microneustonic organisms.

hyposaline: not very salty water with salinity less than 18 o/oo.

hypotrich: ciliates that possess ventrally and anteriorly located cirri used in locomotion and feeding.

index fossil: specific fossils useful in identifying a given rock formation by virtue of their being characteristic of and only present in that formation.

infaunal: denoting animals that burrow or are buried in marine sediments.

infaunal standing crop: total amount of living matter in the bottom at a given moment of time.

informational energy flow: food quality factors.

infundibulum: a deep tubular buccal cavity as found among peritrichous ciliates.

inter alia: "among other things."

intracapsulum: endoplasm of a radiolarian.

intraprotozoan carnivory: eating of protozoa by other protozoa (e.g., *Didinium* and *Paramecium*).

isomorphous: referring to alternation of generations in which the generations are structurally identical.

isospores: spores of only one kind.

isotope disequilibrium: lack of balance in the frequency of breakdown of a given isotope; also, an isotope composition that is not in equilibrium with the surrounding environment.

iterative evolution: repeated development of new life forms from a limited stock; repeated independent evolution.

kappa: a cytobiont bacterium (*Caedibacter*) causing death in *Paramecium tetraurelia*.

karyokinesis: division of the nucleus, as in mitosis or meiosis.

karyology: study of the structure and function of the cell nucleus.

karyorelictid: a member of a primitive ciliate class of the same name; typically long, vermiform, flattened species; extremely contractile; many with one barren surface; two to many macronuclei.

kinetid: basic repeating organizational unit of the ciliate surface, consisting of one or more kinetosomes and certain intimately associated structures or organelles.

kinetocyst: an extrusome.

kinetoplast: a large, mitochondrionlike body with a localized deposit of DNA in a swollen part, occurring most often near the flagellar bases, as in *Trypanosoma* and other members of the order Kinetoplastida.

Ks value: half-saturation constant (mg/L) related to nutrient uptake kinetics; nutrient concentration for which uptake is half maximal.

kummerform chamber: in planktonic forams, an abnormally shaped, smaller than normal chamber in the shell—one of the last-formed chambers—exhibiting a flattened or distorted configuration rather than the more spherical shape.

labyrinthulids: a funguslike protozoan belonging to the order Labyrinthulida.

lamellar wall growth: walls laid down in sheets or layers, as in the formation of some foraminiferan shells.

lamellibranch: a bivalve mollusc (e.g., clams)

Langmuir circulation cells: parallel pairs of counterrotating convection cells driven by surface winds with associated alternate zones of upwelling and downwelling.

larvacean: a member of a class of tunicates with neotenic adults found in the marine plankton.

lateral movement: large-scale transport of water in currents.

lentelliptical: elliptical, lens shaped.

light intensity curves: *P* vs. *I*, that is, photosynthesis versus light intensity curves.

limnic: pertaining to ponds, lakes, or streams.

liospherids: spumellarian radiolaria with nested concentric shells without obvious spines.

lithelids: hagiastrid radiolaria with coiled latticed shells.

lithology: study of the physical characteristics of rocks or stratigraphic units.

littoral: pertaining to that marginal shallow-water region between high and low tides.

log-linear survivorship curves: an ecological mathematical analysis of proportion of organisms surviving up to a point in time.

lophophaenins: nassellarian radiolaria with cephalis and thorax about equally developed, usually with two lobes in cephalis; apical, dorsal, and primary spines obvious and extended beyond latticed shell.

lorica: a test, envelope, case, basket, or shell secreted and assembled by certain protozoa, usually open at one end.

lotic: pertaining to rapidly flowing waters.

lysocline: the depth of the ocean at which shells of calcareous species dissolve.

macroconjugant: the larger member of a pair of conjugating ciliates.

macrofossiliferous: pertaining to those strata containing large fossils.

macrogamont (macrogametocyte): a gamont that will become a female gamete.

macrophage (histiocyte): a large cell of the immune system that engulfs foreign particles.

macrophytic: feeding on large particles of food.

macroplankton: plankton falling within the 0.2–2-mm range size.

magnetostratigraphy: branch of geology dealing with the magnetic content of stratified rock.

main spines: polar or principal spines in radiolaria.

marine snow: composed of amorphous, fragile, particulate, macroscopic aggregates formed from flocculent, water column detrital material in the pelagic zone.

mass extinction: widespread extinction of large populations (or communities) of organisms in a given range of time (e.g., dinosaurs).

mastigoneme: hairlike naked microtubular projections arising perpendicular to the major axis of a flagellum.

medullary: internal shells of spumellarian radiolaria.

megalospheric: denoting a stage in the life cycle of certain forams characterized by a large proloculum and uninucleate condition.

megaplankton: plankton of sizes >2 mm.

meiofauna: the smaller invertebrate fauna of the sea bottom; intermediate in size between microfauna and macrofauna.

membrane bound compartments: membranes separating chloroplasts, associated pyrenoids, and mitochondrial complexes from one that surrounds the nucleus.

merogony: production of merozoites from a meront.

meroplankton: life forms that spend only part of their life as plankton (e.g., larvae of benthic invertebrates and benthic and pelagic fish).

merozoite: stage of the life cycle of an intracellular protozoan parasite produced by multiple division.

mesopelagic: relating to organisms living at depths ranging from 60 to 100 fathoms in the ocean.

metacercariae: a juvenile fluke produced by cercaria that forms after encystment and metamorphosis.

microbial loop: the aspect of aquatic food web dealing with protista, fungi, and bacteria.

microbial gardening: differential enhancement of microbial growth and subsequent feeding on microbes by other organisms; includes digestion of algal symbionts harbored by certain pelagic foraminifera and perhaps some colonial radiolaria.

microbiocöenose: all microorganisms in a given ecosystem; microbiotic component of an ecosystem.

microconjugant: smaller member of a pair of conjugating ciliates, usually absorbed into the macroconjugant.

microcrustacea: crustacea ranging in size from 200 μm to 2 mm.

microflagellate: flagellates as small as 200 μm.

microgamete: smaller member of a pair of anisogametes.

microgametocyte: microgamont; produces microgametes by either binary or multiple fission.

microgranular wall type (porcellaneous wall type): A wall made up of tiny calcite crystals in a randomly arranged fashion. More pores present and wall appears translucent or opaque and white or buff colored.

microheterotroph: usually, a heterotroph of <200 μm in size.

micrometazoa: usually, zooplankton ranging in size from 100 μm to 2 mm.

micronemes: numerous convoluted and elongate, electron-dense organelles found in the conoid and extending anteriorly in the body of sporozoites of most sporozoa.

microphagous: denoting ingestion of particulate matter in the pico- to nano- size range.

microplankton: plankton ranging in size from 20 to 200 μm.

microspheric: relating to a stage in the life cycle of some forams that is characterized by a small proloculum and a multinucleated condition.

microsporida: intracellular parasitic protozoa forming spores of a unicellular origin and containing one sporoplasm.

microsporidoses: diseases caused by microsporozoa.

microzooplankton: zooplankton ranging in size from 20 to 200 μm.

miliolinid: a foraminiferan with a smooth, porcellaneous test made of calcite crystals.

milky barracuda: a muscular disease condition in barracuda caused by a myxosporidan, *Kudoa thyrsites*.

mixotrophic: feeding both holotrophically and autotrophically; capable of holotrophic or saprotrophic nutrition.

monophyletic: descending from a single stock or individual.

monad: a biflagellated, free-living flagellate.

monolocular test: a foraminiferan test with a single opening.

monosporoblast: a micro- and myxosporidan cell that produces a single spore internally.

monothalamic: a foraminiferan with a single-chambered test.

monoxenic: a parasite that is host specific.

muciferous bodies: extrusomes that eject long, usually sticky threads.

mucocyst: a subpellicular, membrane-bound, saccular, or rod-shaped extrusome; usually involved in cyst formation.

mucron: an anterior attachment organelle found in aseptate gregarines that becomes embedded in the host.

multiform: a foraminiferan test composed of numerous chambers.

multivariate statistics: branch of statistics dealing with the simultaneous variation of two or more variables.

myxosporida: obligate parasitic protozoa of poikilothermic vertebrates, mainly fish and annelids; form spores of a multicellular origin with one or more polar capsules and sporoplasms.

NADW: North Atlantic Deep Water.

nannofossil: a fossil 2–20 μm in size smaller than a microfossil.

nanoherbivorous: devouring algae ranging in size from 2 to 20 μm.

nanoplankton: plankton ranging in size from 2 to 20 μm.

nassellaria: polycystine radiolaria with cone-shaped shells and a single polar pore.

nauplii: one-eyed larvae with three pairs of appendages, hatched from the egg of most crustacea.

nebenkorper (associated bodies): structures (intracellular symbionts) so integrated with a host as to appear to be an organelle (e.g., attached to the nucleus of *Paramecium eilhardi*).

nematocyst: the thread capsules of coelenterates produced by cnidoblasts.

nemertines: members of the phylum Rhynchocoela (Nemertina); elongated, flattened worms; mostly marine and benthic.

neogene: denoting the geologic epochs Pleistocene, Pliocene, and Miocene of the Tertiary period.

neosympatry: character displacement in size of two species.

nepiont: proloculus.

neritic: denoting areas over the continental shelf bordered by a 200-m depth line.

nummulitids: benthic foraminifera.

nutrient regeneration: the coastal upwellings, in temperate seas that replenish nutrients during the summer, when they would otherwise be depleted; also, production of nutrients by metabolic activities of bacteria, protists, and various metazoans.

O_2-minimum zone: the portion of a water column in which oxygen is used up by respiration and decomposition as fast as or faster than it is produced.

oceanic front: a sharp boundary between water masses of different properties; a special manifestation of mixing in the ocean.

oligotrich: a ciliate belonging to the order Oligotrichida; with greatly reduced somatic ciliature and a conspicuous AZM with an anterior ring of tactile cilia.

oligotrophic: denoting nutrient-poor waters with poor productivity.

ontogeny: the development of an individual, from conception to death.

oocyst: encysted sporozoan zygote.

opisthobranch gastropod: molluscan subgroup displaying stages of distortion, many with bilateral symmetry; shell reduced or absent; includes nudibranchs and sea hares.

oral trapeze: an organelle found in *Pseudoplatyphyra nana* used to break open the cell wall of its prey and suck out its contents.

ornamental ribbing: nonciliated surface of the right side of the buccal cavity of many oligohymenophoran ciliates.

orogeny: mountain making.

orosphaerids: spumellarian radiolaria with large spherical or cup-shaped skeletons with a coarse polygonal lattice and club-shaped spines.

orthogenesis: the evolution of a species in definite lines that are predetermined by the constitution of the germ plasm.

orthorhombic: pertaining to a system of crystallization characterized by three unequal axes intersecting at right angles.

osmotroph: saprozoic.

ostracod: mussel or seed shrimp belonging to the class Crustacea.

paleoscenids: RIS subdivision; skeleton consisting of four diverging basal spines with riblike elements protruding and topped by two to four "apical" spines.

paleoactinommids: all Paleozoic spumellaria with single latticed shells or two or more concentric shells lacking internal spicular system of entactinids and lattice structure of rotasphaerids.

Paleogene: a combined name for Oligocene, Eocene, and Paleocene.

paleoceanography: the study of the ancient seas.

palynomorphs: fossil pollen forms.

Pangaea: the ancient single supercontinent.

panmixis: random mating.

pansporoblast: a multinucleate body, found in life cycle of myxosporidians, that gives rise to one or more sporoblasts.

Panthallassia: the ancestral Pacific Ocean.

paramembranelles: each of several adoral buccal membranelles characteristic of free-living heterotrich ciliates; unique in having its transverse microtubules limited to kinetosomes of the left row of its infraciliary base.

parapatric speciation: speciation between contiguous populations in a continuous cline.

paraphysomonad: a colorless monad with siliceous body scales.

parapylae: small capsular openings in the capsules of phaeodarian radiolaria.

parapyle: accessory tubular apertures found near the astropyle of phaeodarian actinopods.

paroral: near the buccal structure.

patagium: skeletal, often spongiosa despositions connecting arms in radiolaria.

pectinid: scallop.

pelagic: pertaining to open waters of the ocean.

penaeid: a kind of shrimp.

pennate diatom: a diatom whose valve is arranged along a central line; also member of order Pennales.

penultimate: next to last.

peristome: unported base in some radiolaria; mouth of certain protozoa.

peritrichous: denoting oral ciliature that is arranged in a circular fashion, as in *Vorticella*.

pexicyst: extrusive organelle that attaches to prey without penetration; found in *Didinium*.

phacodiscids: radiolaria with lens-shaped outer shells.

phaeodaria: radiolarians with organic silica-rich skeletons and three capsular openings.

phaeodium: greenish-brown, pigmented mass found in phaeodarian radiolaria.

phaeosome chamber: area created by the lower girdle list and the distorted body of *Citharistes*, with a small opening barely big enough for the cyanobacterium to pass through.

phaeosome (brown body): coccoid cyanobacterium attached to the outer surface of a donophysoid dinoflagellate in the area between the upper and lower girdle lists.

phagocytosis: an ameboid feeding process involving the formation of engulfing pseudopodia to surround potential food material.

phagotrophic: eating in an ameboid fashion; ingestion of solid food.

phaseliformins: spongodiscids radiolaria with subellipsoidal spongy, pouch-shaped shell.

phoresy: organism deriving benefit from the transporting host (but probably not vice versa).

photoinnıbition: inhibition of photosynthesis due to the effect of strong light intensities, especially near the ocean surface or in the lab.

photobiont: photosynthetic symbiont residing in the cytoplasm of nonphotosynthetic hosts.

phycoerythrin: a red biliprotein pigment found in Cyanobacteriacea and red algae.

phyletic gradualism: gradual evolutionary development of a given species.

phytomastigophorean: photoautotrophic flagellates.

picoplankton: plankton <2 μm in size.

pinacoid: a heliozoan with a skeleton of scales.

pinocytosis: the taking of surrounding fluids into small vesicles within the cytoplasm of a cell.

piscine flagellates: flagellates found in a fish pond.

plagiacanthins: nassellarian radiolaria thorax and sometimes walls of cepharlis reduced with faceted spines.

plagonids: nassellarian radiolaria with simple forms, composed of radial spines united in a common center and supporting the central capsule but lacking a wicker-work lattice.

planispiral: denoting certain foraminifera that are coiled in a single plane.

planktobacteria: bacteria free-living in water.

plasmalemma: peripheral cell membrane usually applied to amoebae.

plasmodium: multinucleated mass of protoplasm enclosed by a plasma membrane; usually used in reference to slime molds.

plasmotomy: form of division, binary or multiple, of a multinucleated, single-celled organism.

plastogomy: cytoplasmic fusion, as in certain slime molds or during conjugation in certain ciliates.

plate tectonics: study of sea floor spreading.

plectopyramins: Subdivision of theoperids. These organisms have one postcephalic cone-shaped joint with longitudinally arranged quadrangular pores.

pleuronemid: a ciliate member of the family Pleuronematidae, characterized by a prominent postequatorial velum; with long, stiff, caudal cilia.

polar capsules: a thick-walled vesicle found only in myxosporidian spores containing the polar filament.

polar filament: an anchoring nonmembrane-bound filament found in the polar capsules of myxosporidia.

polaroplast: an organelle found in the spore of microsporids involved in the ejection of tubular filament; consists of a stack of smooth membranes continuous with the membranes of the tubular filament; explosive ejection of filament involves increased internal hydrostatic pressure in the polaroplast because of physical contact and chemical interaction, resulting in increased permeability of the polar cap.

polar ring: apical ring of thickened electron-dense material in sporozoites, merozoites, and certain other stages of sporozoa.

polar tube: an organelle found in microsporidia that is analogous to the polar filament.

polychaete: a member of the oldest and largest group of segmented worms, the Annelida.

polycystine: radiolaria with either no skeletons or those composed of opaline silica.

polyhymenostomes: spirotrichs.

polymorphism: the existence of three or more distinctly different body forms within a given species.

polyphyletic: multiple evolutionary origins for a given taxonomic group.

polythalamic: many-chambered, as in the foraminifera.

pontellid copepods: members of the family Pontellidae, class Crustacea.

porcellaneous wall type: a kind of foraminifera test appearing opaque white in reflected light.

postabdominal: denoting chambers following the radiolarian abdomen.

prasinomonads: green algae with scales on body and flagella; taxonomic order Prasinomonadida.

prasinophyte: prasinomonad.

primary lysosome: a vesicle containing digestive enzymes that is the immediate product of the golgi apparatus used in intracellular digestion.

principal components analysis: factor analysis (a branch of statistics) principal axes are calculated through hyperellipsoids representing large amounts of data.

prograding delta: a seaward-extending delta produced by progressive deposition of sediments from rivers or shoreline processes.

proloculum: initial embryonic chamber of a foraminiferan.

protostomatid: a member of the ciliate class Protostomatea, order Protostomatida; oral area and cytostome apical.

protomerite: anucleate segment of the trophont stage of a septate gregarine.

protozooplankton: protozoan component of the plankton.

prymaescophycean: a haptophyte flagellate.

prymnesiomonad: prymaescophycean.

pseudoaulophacids: hagiastrid radiolaria with triangular lens either spongy or mesh with prominent marginal spines.

pseudobodonid: a flagellate similar to bodonids (div. Phaeophycophyta).

pseudochitinous: denoting the composition of the outer membrane of the central capsule of a radiolarian.

pseudoepibacteria: bacteria in tight zones around suspended and/or settled particles where attachment and detachment may routinely occur.

pseudosymbiosis: the process by which plastids are retained in an apparently functional state from ingested food organisms.

pterocorythids (pterocorids): nassellarian radiolaria with large elongate cephalis with one or more postcephalic chambers.

pteropod (see butterfly): an unshelled mollusc with fluttering, winglike feet living in the pelagic zone.

punctuated equilibrium: a state in which species may exist for a long time more or less unchanged. Then some environmental change or major genetic mutation results in rapid changes in the gene pool, resulting in the emergence of a new species with its own genetic equilibrium.

punctuated evolution: see punctuated equilibrium.

pylentonemids: radiolaria with external features of Cyrtellaria and internal structures of Periaxoplastida.

pylome: a large pore or passageway into the radiolarian skeleton.

pylonids: hagiastrid radiolaria with elliptical shells with gates.

pyrolysis: the decomposition of organic substances by heat.

Q-study: a study of the relationships between samples in terms of their species composition, involving the calculation of a coefficient of association between all pairs of samples.

radial bar: internal bar.

radiolaria: strictly holoplanktonic protista consisting of two orders, Polycystina and Phaeodaria; characteristically possessing thick, organic capsular membranes that divide the cytoplasm into ectoplasmic and endoplasmic regions. Many species produce siliceous skeletons.

radiolaria incentrae sedis: (RIS): nassellarianlike radiolaria.

radiolaria incertae sedis: spumellarianlike radiolaria.

radiolarian oozes: siliceous oozes composed of skeletons of radiolarian and frustules of diatoms.

radiolarian skeleton: a siliceous, often ovuate, deposition supporting the extracapsular cytoplasm believed to be a possible ballast control allowing for rapid descent and for help in capturing prey.

radiometric dating: the procedure of calculating the absolute ages of rocks and minerals that contain radioactive isotopes.

rectilinear: uniserial.

Red Queen hypothesis: an explanation given by Van Valen (1965) for an evolutionary "law" regarding the extinction of subtaxa at a stochastically constant rate within a relatively homogeneous higher taxon. According to Hallam (1976), "the hypothesis states that increase in the momentary fitness of any set of species within a specific adaptive zone acts to reduce the fitness of all other species within that zone, in such a way as to cause the effective environment of all species to deteriorate in a stochastically constant manner through time."

red water: a spectacular orange colored nontoxic marine phenomenon in temperate coastal waters; associated with *Noctiluca scintillans* blooms. Other dinoflagellate blooms can be associated with similar kinds of water discoloration (e.g., *Gonyaulax tamarensis* [toxic and red tide]).

respiratory quotient (RQ): ratio of the carbon dioxide produced to the oxygen consumed.

rhabdophorines: lower gymnostome ciliates (i.e., the protostomatea and litostomatea).

rhabdos: a tubular cytopharyngeal apparatus found among certain kinds of ciliates.

rhizopodia: narrow lobopodia.

rhodophyte: red algae.

rhoptries: electron-dense, tubular or saccular organelles found among the Apicomplexa extending back from the anterior region inside the conoid.

rib: raised radiolarian structure within the terminal segment.

rod organ: specialized feeding organelle found in the flagellate *Peranema trichophorum*.

rotaformids: subdivision of acanthodesmiids—nassellarian radiolaria with lens-shaped central area enclosing a nassellarian cephalis connected to an outer ring by radial bars.

rotaliinid: foraminifera with a rotalinne wall and members of the suborder Rotalinna.

rotaliid wall: a chamber found among foraminifera composed primarily of a single layer continuous as a thin lamella over the entire previously formed exterior surface of the foraminiferan test.

rotasphaerids: paleozoic spumellarian radiolaria with single, spherical latticed shell with angular meshes lacking the internal spicular system of entactinids and displaying one or more points on shell with five radiating strong, straight lattice bars.

sacoglossan molluscs: sea hares of the molluscan order Sacoglossa.

saggital ring: in certain radiolaria, a loop or band of variable size and shape reinforcing the latticed wall in a medial vertical plane and from which processes may extend.

salmonid: bony fish that are classified among the Salmonidae.

salps: thaliacean urochordates.

saltation: the action of jumping.

saltational evolution: a type of evolution that proceeds by leaps and bounds through the production of mutants that differ grossly from the parent.

saltatory transport: organelles and particles transported back and forth in cytoplasm in a discontinuous, often erratic fashion.

sarcomatrix: pericapsular cytoplasmic envelope of a radiolarian used for digestion of the contents of a food vacuole.

saturation light intensity: the light intensity that maximizes the photosynthetic rate.

saturnalins: actinommid radiolaria with a spherical latticed skeleton possessing a circular outer ring connected to a centrally located latticed shell.

scaphopod: a class of small molluscs with tusk-shaped shells from the mouth of which a lobed foot and cephalic tentacles emerge.

schizogony: formation of daughter cells among sporozoa by multiple fission.

schizont: the cell that undergoes schizogony.

schizozoite: the product of schizogony.

schwagerinid wall: a wall found in a fusuline foraminifera genus *Schwagerina* in which the spiral curves for shell diameter decline in slope in later stages of growth.

scuticociliate: a member of a major taxonomic group of hymenostome ciliates that possess a transient multikinetosomal structure, a scutica, in each presumptive filial cell during stomatogenesis.

sea-floor spreading: the process by which plate tectonics and the movement of continents over time result in major rifts in the ocean floor, where separation occurs.

secondary lysosome: the product of the fusion of primary lysosomes with a phagosome (or food vacuole) to produce a digestive vacuole.

selective feeding: feeding preferentially on a given food organism.

semelperous: suicidal.

series: time-stratification unit.

sethoperins: nasselarian radiolaria with cephalis and six latticed plates extending from the cephalis or modified as a "basketlike" thorax.

sethophoromins: nassellarian radiolaria with large, hemispherical cephalis and with a flattened, umbrella-shaped thorax.

shelfal: pertaining to the continental shelf.

silicoflagellate: marine autotrophic flagellates with internal, ornate skeletons composed of silica and one or two flagella.

silled basin: tabular masses that are concordant underlying a circular downfolded geologic structure.

siphonophore: a large floating colonial cnidarian composed of polypoid and medusoid elements (e.g., *Physalia physalia*).

siphon tube: ingestion organelle of *Entosiphon* responsible for propelling food into the flagellate cell.

sipunculid: marine unsegmented wormlike animals; characteristically possessing a crown of short, hollow tentacles on the anterior end.

skeletogenesis: cytoplasmic desposition of the skeleton of radiolaria.

skeleton: radiolarian test.

sori: a cluster of reproductive organs in plants.

soritid: benthic foraminifera.

spinae: spines.

spinose: with spines.

spirillinids: benthic foraminifera.

spirotrichs: a class of ciliates characterized by the possession of conspicuous right and left oral and/or preoral ciliature.

spongasterins: spongodiscid radiolaria possessing a spongy disk with margins faceted and radiating.

spongodiscids: radiolaria that are discoidal and spongy in nature.

spongopylins: spongodiscid radiolaria with spongy lens without spines but with a pylome.

spongotrochins: spongodiscid radiolaria with a spongy lens and often with large, numerous spines.

spongurins: spongodiscids radiolaria with a spongy cylinder.

sporoblast: a cell formed by the division of a sporont in the life cycle of coccidians; in myxosporidians, the cell group within the pansporoblast envelope that will become the multicellular spore.

sporocyst: in the life cycle of most coccidians, the cyst formed within the oocyst.

sporogenesis: sporulation.

sporogony: formation of sporocysts and sporozoites.

sporonts: the zygote within the oocyst that will form the sporocyst.

sporoplasm: infective ameboid stage contained within the spore of micro- and myxosporidians.

sporozoites: the product of sporogony that is the motile infective stage in gregarine and coccidian life cycles.

sporulation: the process of spore or sporozoite production.

spumellarians: polycystine radiolaria with spherical shapes.

standing crop: the total amount of stored energy in living material existing at any given instant in a given area or volume.

stasis: periods of nonevolution.

stenohaline: denoting organisms that cannot tolerate wide variations in salinity.

stimulating metabolite: metabolite that is synthesized by the host but that if brought into the metabolic pool, advances cell cycle by reducing time and energy otherwise expended on their synthesis.

stylodictids (porodiscids, spongodiscids): spumellarian radiolaria with disks with rings of lattice, chambered or pored.

stylosphaerids: spumellarian radiolaria with latticed shells with polar spines.

subarctic convergence: arctic or polar convergence.

sulcal region: region containing the longitudinal furrow on the body of dinoflagellates and the longitudinal flagellum.

sulfuretum: a habitat strongly influenced by anaerobic conditions and by bacteria associated with the microbial sulfur cycle; this habitat develops where large amounts of organic material accumulate in the sediment or at its surface in relatively sheltered areas.

supralittoral: the area from the high-tide mark to the landward edge of the spray zone.

surface microlayer: the layer, at the air–sea interface that typically exhibits higher concentrations of organics, bacteria, and protozoa.

swarmer: dispersive form in the life cycle of numerous protozoa.

sympatric: referring to the condition of two populations occupying the same territory.

sympatric speciation: production of a new species within a single population as a result of polyploidy.

syngamy: fusion of two pronuclei to form a diploid zygotic nucleus.

system: time-stratification unit.

systematics: the portion of biology that deals with the identification of organisms and their evolutionary relationships.

syzygy: side-by-side or end-to-end association of gamonts prior to formation of gametocysts and gametes, as in gregarines.

taphonomy: study of the processes that affect organic remains after death and prior to final burial.

tectinous wall type: single-chambered membranous walls of tests formed in certain foraminifera (e.g., *Allogromia*). The principal material is tectin (glycoprotein of the acid mucopolysaccharide type) plus protein, lipid, and sometimes ferric iron.

tectonism: the phenomena that relates to earth plate motion and associated earthquakes and sea floor spreading.

telosporids: an outdated class consisting of gregarine parasites.

terminal water: mixed waters of a latitudinal current in between large gyres.

test: a general term for a protistan lorica or shell.

test polymorphism: variations in the thickness or direction of coiling of various foraminifera shells related to various geologic and environmental forces.

tethyan: an organism that lived in the Tethys Sea 135 million years ago.

Tethys Sea: a branch of the primordial ocean, Panthalassa, that separated Laurasia from Gondwana.

thallus: a solid structure of nonvascular plants.

thecate: the lorica of certain algal forms (e.g., armored dinoflagellates).

theoperids: nassellarian radiolaria with small spherical, poreless cephali with one or more postcephalic chambers and usually one small apical spine on top of the cephalis.

theory of punctuated equilibria: the theory that suggests that speciation occurs in jumps between periods of stasis.

thermocline: the layer of water in which there is a rapid change of temperature with increased depth.

thermohaline: density differences brought about by variations in temperature and salinity.

tholonids: hagiastrid radiolaria with elliptical outer shells containing bulblike extensions and some spines.

thorax: radiolarian second segment.

tintinnid: a ciliate with a body generally conical or bell-shaped and poorly developed somatic ciliature; oral cilia are prominent, making almost a complete circle of polykinetids around broad end, used for locomotion and feeding; generally sessile and attached by an aboral process to inner wall of a lorica that is taxonomically characteristic of the suborder Tintinnina; mostly marine planktonic forms.

tooth: an outgrowth partially filling the aperture of a foraminiferan shell.

tooth plate: a wall that runs within the chamber of a foraminiferan to the aperture.

toxicysts: extrusive organelle one end of which penetrates the prey and the other end of which remains attached to the ciliate. Used to pull prey into buccal area of *Didinium*.

trichocyst: an extrusome organelle underlying the pellicle of many ciliates and dinoflagellates capable of discharging a threadlike filament; function unknown.

trichome: a many-celled, frequently branched filament of algae.

triospyrins: subdivision of acanthodesmiids, nassellarian radiolaria with sagittal ring surrounded by two latticed lateral lobes, apical and basal poles obvious; commonly with basal feet and apical spines.

trochospiral: chambers of a foraminiferan shell that are added in a spiral with translation along the axis of coiling.

trophozoite: the growing, vegetative stage of a protozoan; usually used in reference to sporozoa.

tubes: entrances larger than pores in radiolaria.

tunicate: a urochordate.

umbilicus: axis of coiling in foraminifera.

unilocular: single-chambered foraminiferan test.

uniserial: the chambers of a foraminiferan shell arranged in a single row.

upwelling: rising of deep, relatively cold, nutrient-rich water to surface at speeds of 1 to 5 m/d to replace water deflected offshore under the influence of prevailing trade winds and the effect of the earth's rotation.

uroid: posterior-most portion of a "naked" ameba.

varve: an annual deposit usually consisting of two layers, one of fine materials and the other of coarse.

velum: the membranes surrounding the pseudostome of many ciliates.

vorticellids: peritrichous ciliates characteristically possessing a contractile stalk and a bell-shaped body.

western boundary currents: flow from equator to temperate areas.

wormy halibut: a disease of muscular tissue of halibut caused by a myxosporidan, *Unicapsula muscularis*.

wrack: ruinous.

xenoma: a structural–physiological complex formed by actively multiplying intracellular parasites (e.g., microsporidia) and their host cells. The host cells increase abnormally in size.

xenosome: cytobiontic motile gram-negative bacteria found in *Paramecium acutum*.

yellow cells: cryptomonad or dinoflagellate xanthellae cytobionts.

Zerfall: G. decay.

zoochlorellae: green coccoid photobionts.

zoogeographies: referring to the geographic distribution of animals throughout the planet.

zoomastigophorean: a heterotrophic flagellate bearing no plastids; eating either holozoically or saprozoically.

zooxanthellae: yellowish or brown photobionts in animal hosts.

ACKNOWLEDGMENTS

Many thanks to O. Roger Anderson and John Marra of Lamont-Doherty Geological Observatory of Columbia University, and Janet Valeski, Quinnipiac College, for their valued assistance in creating this glossary.

Index